CHROMOSOME BIOLOGY

CHROMOSOME BIOLOGY

Rudi Appels
CSIRO, Division of Plant Industry
Canberra, ACT, Australia

Rosalind Morris
Department of Agronomy
University of Nebraska, Lincoln, Nebraska

Bikram S. Gill
Department of Pathology
Kansas State University, Manhattan, Kansas

Cedric E. May
Agricultural Research Institute
Wagga Wagga, NSW, Australia

Illustrations
Rudi Appels and Jon Raupp

Kluwer Academic Publishers
Boston/Dordrecht/London
1998

Distributors for North, Central and South America:
Kluwer Academic Publishers
101 Philip Drive
Assinippi Park
Norwell, Massachusetts 02061 USA
Telephone (781) 871-6600
Fax (781) 871-6528
E-Mail ⟨kluwer@wkap.com⟩

Distributors for all other countries:
Kluwer Academic Publishers Group
Distribution Centre
Post Office Box 322
3300 AH Dordrecht, THE NETHERLANDS
Telephone 31 78 6392 392
Fax 31 78 6546 474
E-Mail ⟨services@wkap.nl⟩

Electronic Services <http://www.wkap.nl>

Library of Congress Cataloging-in-Publication Data

Chromosome biology / Rudi Appels . . . [et al.].
p. cm.
Includes bibliographical references and index.
ISBN 0-412-02601-5
1. Chromosomes. 2. Cytogenetics. 3. DNA. I. Appels, Rudi.
QH600.C494 1998
572.8′7–dc21 97-15812
CIP

Printed on acid-free paper.

Printed in the United States of America

Contents

PART I INTRODUCTION TO THE STUDY OF CHROMOSOMES

PART II MITOSIS, MEIOSIS, AND CHROMOSOMAL KARYOTYPES

PART III CHROMOSOME STRUCTURE AND REARRANGEMENTS

PART IV CHROMOSOME POLYPLOIDY, ANEUPLOIDY, AND HAPLOIDY

PART V GENE LOCATIONS, GENETIC MANIPULATIONS, AND REARRANGEMENTS

PART VI DNA ORGANIZATION, STRUCTURE, AND REPLICATION

Preface

The close of the twentieth century is an exciting and stimulating period of discovery in the area of chromosome structure and function. In ''Chromosome Biology'' we target the many different disciplines that are being applied in the analysis of chromosomes in order to provide an admixture of cytogenetics, genetics and molecular biology for application to courses taught singly or cooperatively by cytogeneticists and molecular biologists. The book provides background reading for courses involving any one discipline. Our intent is to convey to students an appreciation of plant and animal improvement programs, the importance of mapping the human genome, particularly in relation to diseases, and the advantages of modern technology to analyzing the development of different life-forms.

At the start of each Chapter the main concepts discussed are listed so that the subject matter has a clear focus and, where possible, figures illustrate the key concepts throughout the book. Specific references are cited in the figure legends, or in the text, to justify the information provided. General books and reviews are listed at the end of each Chapter.

We are grateful to numerous colleagues for providing helpful critisisms and information; we particularly want to thank Drs G. Hart, D. Smyth, C. Gillies, N. Darvey and D. Hayman; external reviewers provided extensive comments on all or part of the manuscript and made major contributions towards the development of the book. Colleagues and reviewers cannot, however, be held responsible for the final content of the book, since this ultimately reflects the individual interests and biases of the authors.

Rudi Appels
Chief Research Scientist, CSIRO, Division of Plant Industry, Canberra, ACT, Australia
Rosalind Morris
Emeritus Professor, Department of Agronomy, University of Nebraska, Lincoln, NE, USA
Bikram S. Gill
Professor, Department of Plant Pathology, Kansas State University, Manhattan, KS, USA
Cedric E. May
Principal Research Scientist, NSW Agriculture, Wagga Wagga, NSW Australia
W. John Raupp
Associate Scientist, Department of Plant Pathology, Kansas State University, Manhattan, KS, USA

I
Introduction to the Study of Chromosomes

1

Introduction

- Chromosome analyses are central to studies of the cell cycle, genome-mapping projects, and genetic-transformation experiments.
- A basic distinction between life-forms is whether or not the genomic DNA is enclosed within a nucleus.
- The Domain Eukarya includes eukaryotic organisms with nuclei and long, linear DNA molecules in chromosomes; the Domains Bacteria and Archaea include prokaryotes, which usually have circular DNA molecules forming protosomes.
- Many of the subcellular components of eukaryotic cells, in particular chloroplasts and mitochondria, are considered to be derived from prokaryotes.

The study of chromosome number, structure, function, and behavior in relation to gene inheritance, organization, and expression is an integral part of the science of cytogenetics. Chromosomes are recognized as the bearers of deoxyribonucleic acid (DNA) in eukaryotes and provide the basis for the orderly transmission of genetic information from generation to generation. In bacteria, the genetic material also consists of DNA, and in this book, we define the unit of inheritance in prokaryotes as the protosome. Chromosomes, in particular, have been actively studied since the mid-1800s. The rapid advances in plant and animal genetics and improvement, in the latter part of the twentieth century, have stimulated a renewed interest in cytogenetics and chromosome research.

1.1 THE VALUE OF CHROMOSOME RESEARCH

First, the central features of chromosome structure and function are defined at the molecular level. Familiar morphological features common to all chromosomes have been examined using high-resolution microscopy and molecular techniques. Cytogenetic, genetic, and molecular analyses have uncovered the way in which genes are organized within chromosomes, defined their structure and function, determined the effects of gene organization on function, and discovered the mechanisms that control their activity.

Second, many aspects of chromosome behavior in cell division have been clarified. The processes of cell division, whether for vegetative growth, as a result of mitosis, or for reproduction by forming haploid gametes as a result of meiosis, have been studied in organisms as diverse as humans, frogs, grasshoppers, flies, maize, wheat, and yeast. From these studies, it has been possible to build a detailed understanding of chromosomal behavior during cell division.

Third, whole-chromosome studies as a part of cytogenetics are having a major impact on genome-mapping projects. The microscopic identification of chromosome regions is providing a framework for molecular biologists to sequence the entire genome of different organisms, including humans. In agriculture, horticulture, forestry, animal improvement, and human health, cytogenetic studies of chromosomes are the major impetus for the molecular analysis of significant regions of chromosomes.

Fourth, detailed molecular-cytogenetic maps provide the basis for the design of DNA probes that can be used to analyze individuals, families, and populations. Extensive nucleotide-sequence databases of a variety of DNA fragments isolated from plant and animal genomes allow the analysis of molecular variation in the genes of individuals of the population. The completion of projects such as the Human Genome sequencing project will also provide a molecular basis to uncover variation in organisms that are significant economically but for which complete genome-sequence data are not available.

Fifth, understanding chromosome structure provides the means to analyze the secondary changes that occur during the processes of chromosome engineering and transformation. Many contemporary manipulations of cells and chromosomes to introduce new DNA may cause secondary changes in chromosome structure. Whole-chromosome studies are essential to understand these changes and to help in eliminating unwanted abnormalities.

Sixth, molecular biology can now provide the technology for the construction of simple chromosomes. The ability to synthesize chromosomes using molecular techniques is opening up a new era for analyzing chromosome behavior and for determining the function of large sections of DNA.

1.2 LIFE-FORMS AND THEIR GENETIC MATERIAL

A primary distinction among major life-forms can be made when the structure, function, and organization of their DNA and genetic material are studied. This distinction separates the so-called higher organisms, such as fungi, plants, and animals, from the so-called lower organisms, such as bacteria. The early definition of ''higher'' and ''lower'' organisms referred to perceived levels of complexity in cellular organization. This definition can, however, be ambiguous and the present-day use of ''eukaryotes'' and ''prokaryotes'' is clearer because it refers to the presence or absence of nuclei, respectively. The genetic material of eukaryotes is contained in distinct, subcellular, membrane-bound structures, of which the principal organelle is the nucleus. Lesser amounts of DNA are present in the mitochondria and, in the case of plants, in the chloroplasts (Fig. 1.1). Prokaryotes do not have distinct subcellular structures.

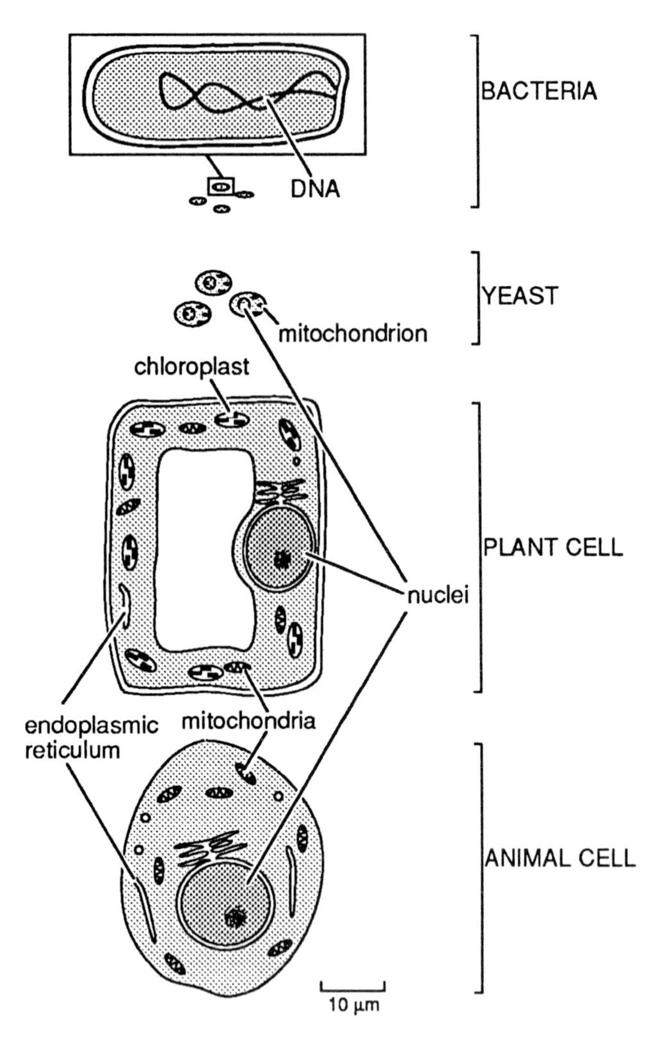

Fig. 1.1. The approximate, relative sizes of bacterial, yeast, plant, and animal cells. Bacterial cells do not have internal, membrane-bound compartments and generally have a much simpler structure than eukaryotic cells. Eukaryotes (yeast, plants, and animals) are characterized by the presence of a nucleus and organelles such as mitochondria, chloroplasts, and an endoplasmic reticulum.

The genetic material within the nuclei of eukaryotes consists of varying numbers of linear DNA molecules complexed with highly conserved proteins, in chromosomes that divide in either of two ways. These divisional processes are called mitosis for vegetative growth or meiosis for the production of gametes. In contrast, the DNA of prokaryotes is often in the form of a circular molecule, which can divide by replication in time with the fission of the prokaryotic cell. This clear distinction in the organization, arrangement, and function of the genetic material is the primary reason for the separation of eukaryotes, containing chromosomes, from prokaryotes, with their protosomes.

Although the genetic relationships between different life-forms are a much discussed topic, the main source of information concerning evolutionary relationships is the DNA sequence of genes coding for specific products common to all life-forms. Of such genes, ribosomal RNA genes are clearly identifiable in most organisms, and comparative analyses of the nucleotide sequences of these genes have been used to compare evolutionary relationships.

The phylogenetic tree in Fig. 1.2 was derived from nucleotide-sequence comparisons. This tree has three dominant branches, two of which, the Archaea and the Bacteria, are prokaryotes. The third branch, the Eukarya, considered to be almost as ancient, contains all organisms with nuclei and chromosomes, the eukaryotes.

It was originally suggested that present-day organisms of the Archaea are representative of the earliest life-forms because they grow in high-salt, high-sulfur, or high-temperature environments, conditions that are thought to have existed during the cooling phase of the earth. The true Bacteria, in contrast, are found in more temperate, or perhaps less hostile, conditions. Early on in the evolution of life on earth, the first organisms of the Eukarya were formed. DNA sequence evidence indicates that the Eukarya are more closely related to the Archaea than they are to Bacteria, and the sampling of prokaryotic organisms isolated from picoplankton has shown that over 30% of these microorganisms are also Archaeal in origin. Organisms of the Archaea therefore represent a major component of the ocean's biota. The Eukarya themselves are divided into five major branches, namely the single-celled organisms of the Archezoa and the Protista, and three Kingdoms of multicelled organisms, Plants, Fungi, and Animals.

1.3 CHROMOSOMES AND PROTOSOMES

The chromosomes of the Eukarya have been traditional subjects for cytogenetic research because their chromosomes are large enough to be examined by the light microscope. Since the early 1940s, however, progress in cytogenetics has included studies on bacterial gene expression and transmission. This led to the concept of bacterial DNA as a chromosome, despite the major differences between bacterial DNA and eukaryotic chromosomes. Attempts have been made to distinguish between prokaryotic and eukaryotic genetic material on a nomenclatural basis. Suggestions have included Chromosome with a capital C for eukaryotes, as compared to chromosome with a small c for prokaryotes. The term ''genophore'' has been applied to the DNA in prokaryotes, whereas other authors refer to bacterial chromosomes, the bacterial genome, or bacterial DNA. Throughout the present text, the term ''protosome'' is used to describe the hereditary unit of prokaryotic, mitochondrial, and chloroplast DNA. The term is distinctive, yet still shows its roots in that *proto* = first, and *some* = body, and thus relates to

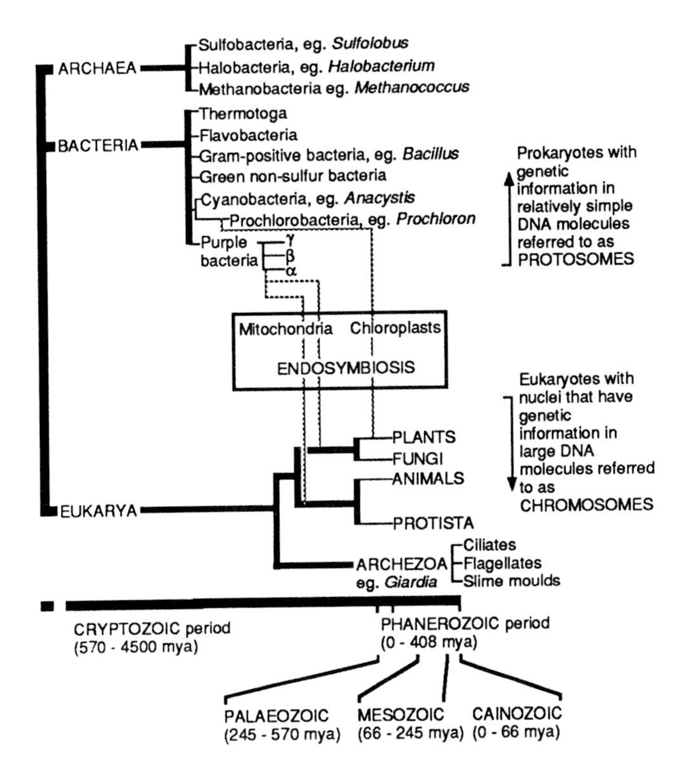

Fig. 1.2. The phylogenetic relationships among different life-forms as deduced from ribosomal DNA sequence analyses (Sogin et al., 1989; Woese et al., 1990). Examples of the γ, β and α purple bacteria are *Escherichia*, *Pseudomonas*, and *Agrobacterium*, respectively. The time scale is a very approximate guide for the timing of critical events. The endosymbiotic origin of mitochondria and chloroplasts in eukaryotes is a much discussed topic, and the existence of organisms such as chlorarachniophyte algae, which apparently retain the nucleus (nucleomorph) and chloroplast of the endosymbiont, provide a precedent for the endosymbiotic origin of eukaryotic organelles and valuable material for evolutionary analyses (McFadden et al., 1994). The fossil record indicates that only marine life was present in the early Paleozoic, with amphibians, insects, and terrestrial plants appearing as this period progressed. In the early Mesozoic, mammals, dinosaurs, and reptiles are found and in the later part of the period, the first records of angiosperms appear. In the Cainozoic (or Cenozoic), the first grasses are evident. Only in the latter part of this period do the first records of humans appear. Because viruses survive by using many products from their host, including ribosomal RNA, these particular life-forms are not included in the figure. (mya = millions of years ago)

the term ''chromosome'' used for eukaryotes where *chromo* = colored in response to a dye.

1.4 THE ORIGIN OF CELLULAR ORGANELLES

Although the evolutionary origins of the eukaryotic cell and chromosomes are not clear, the suggestion of an endosymbiotic origin for the eukaryotic cell is becoming widely accepted. According to this idea, all of the internal components of the eukaryotic cell were originally derived from the fusion or uptake of different types of bacterial organisms. Most of these components have by now been so extensively modified that their origins are completely obscured. Thus, although the nucleus might originally have been bacterial in nature, its principal present-day function is as a repository for the chromosomal material and gene action.

A bacterial origin for the mitochondria and chloroplasts of eukaryotic cells is more easily recognized. With the exception of the single-celled Archezoa, such as *Giardia lamblia* (Fig. 1.3), typical eukaryotic cells contain mitochondria to carry out the function of oxidative respiration. Plant cells also contain plastids, most commonly seen as chloroplasts, which function in photosynthesis. Both of these organelles contain their own DNA, arranged in small, circular molecules replicating more or less in time with the organelles themselves and with division of the mother cell. A single chloroplast or mitochondrion can have tens of copies of its DNA molecule and, therefore, tens of copies of the genes on these molecules. Some DNA sequences in these organelles are clearly related to the DNA of representatives of the true Bacteria (*see* Fig. 1.2). The genes of mitochondrial DNA coding for RNA and some protein molecules have sequences that are closely related to genes in the α-group of purple bacteria. Further, based on size and some other physical characteristics, it is likely that there have been at least two introductions of organisms undertaking the role of mitochon-

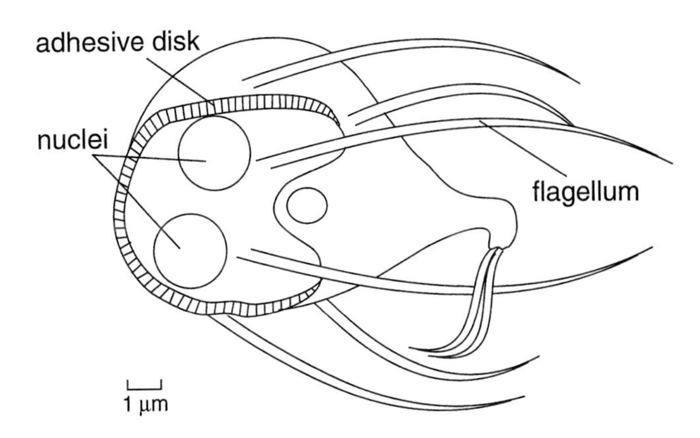

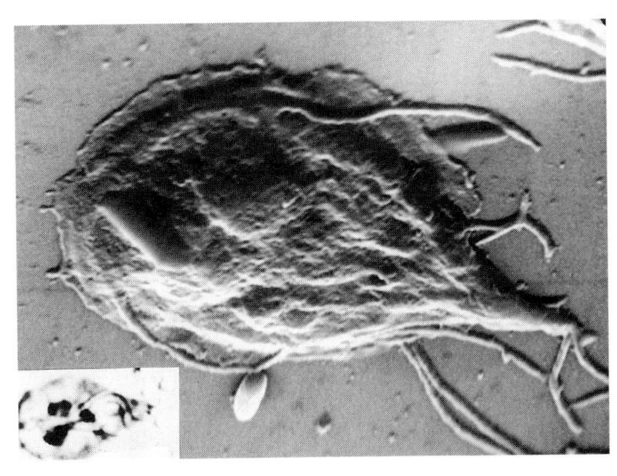

Fig. 1.3. A diagrammatic representation (top), an electron micrograph (bottom), and a light-microscopic image (insert) of the Archezoan organism *Giardia lamblia*. This organism is unusual in having two nuclei and appears to represent a very early form of eukaryote (Kabnick and Peattie, 1991). Analysis of ribosomal RNA sequences from *G. lamblia* indicates that the organism has more similarities to prokaryotes than eukaryotes, hence the idea that it represents a ''missing link'' in the early evolution of eukaryotes. In addition, the organism has no mitochondria and is an obligate anaerobe (Schofield and Edwards, 1991). Consequently, *G. lamblia* is a major parasite of the intestinal tract, including human beings, causing symptoms such as diarrhea and abdominal cramps (The electron micrograph of *G. lamblia* was kindly supplied by E.M. Edwards.)

dria. This argues for a polyphyletic origin for these organelles, as shown in Fig. 1.2. Chloroplasts are generally considered to be monophyletic in origin, based on both DNA sequences and size. The closest bacterial relatives of chloroplasts would appear to be marine organisms, originally designated Prochlorophytes but recently renamed Prochlorobacteria. These marine organisms are closely related to what were originally known as the blue-green algae, but are now called Cyanobacteria.

1.5 CONCLUSIONS

The developing understanding of the evolution of genomes from a wide range of organisms, as well as the origins of chloroplasts and mitochondria, provide an important adjunct to the structural analysis of protosomes and chromosomes. In addition, the continued structure/function analysis of genomes will contribute concepts that elaborate on how organisms evolved. The evolutionary tree shown in Fig. 1.2 is only one of many that may be drawn, and discussion will continue on whether or not mitochondria have a biphyletic origin or, for example, on whether *Giardia* may have lost the mitochondria it once had as a result of its parasitic lifestyle. The work described in the remainder of this book cannot be easily separated from problems related to the evolution of life, but general texts that consider evolutionary problems more directly are in the Bibliography.

BIBLIOGRAPHY

General

CAVALIER-SMITH, T. 1987. Molecular evolution—eukaryotes with no mitochondria. Nature 326: 332–333.

SCHOFIELD, P., EDWARDS, M. 1991. *Giardia*: animalcules a-moving very prettily. Today's Life Sci. (July 1991): 36–40.

Section 1.2

ALLSOPP, A. 1969. Phylogenetic relationships of the Procaryota and the origin of the eucaryotic cell. New Phytol. 68: 591–612.

CAVALIER-SMITH, T. 1989. Molecular phylogeny—Archaebacteria and Archezoa. Nature 339: 100–101.

DELONG, E.F., WU, K.Y., PRÉZELIN, B.B., JOVINE, R.F.M. 1994. High abundance of Archaea in Antarctic marine picoplankton. Nature 371: 695–697.

MCFADDEN, G.I., GILSON, P.R., HOFMAN, C.J.B., ADCOCK, G.J., MAIER, U.G. 1994. Evidence that an amoeba acquired a chloroplast by retaining part of an engulfed eukaryotic alga. Proc. Natl. Acad. Sci. USA 91: 3690–3694.

OLSEN, G.J. 1994. Archaea, Archaea, everywhere. Nature 371: 657–658.

SOGIN, M.L., GUNDERSON, J.M., ELWOOD, H.J., ALONSO, R.A., PEATTIE, D.A. 1989. Phylogenetic meaning of the kingdom concept: an unusual ribosomal RNA from *Giardia lamblia*. Science 243: 75–77.

WOESE, C.R. 1981. Archaebacteria. Sci. Am. 244: 98–122.

WOESE, C.R., KANDLER, O., WHEELIS, M.L. 1990. Towards a natural system of organisms: proposal for the domains Archaea, Bacteria and Eucarya. Proc. Natl. Acad. Sci. USA 87: 4576–4579.

Section 1.4

GRAY, J. 1986. Molecular botany—wonders of chloroplast DNA. Nature 322: 501–502.

GRAY, M.W. 1983. The bacterial ancestry of plastids and mitochondria. Bioscience 33: 693–699.

GRAY, M.W., CEDERGREN, R., ABEL, Y., SANKOFF, D. 1989. On the evolutionary origin of the plant mitochondrion and its genome. Proc. Natl. Acad. Sci. USA 86: 2267–2271.

KABNICK, K.S., PEATTIE, D.A. 1991. *Giardia*: a missing link between prokaryotes and eukaryotes. Am. Sci. 79: 34–43.

MORDEN, C.W., GOLDEN, S.S. 1989. *psbA* genes indicate common ancestry of prochlorophytes and chloroplasts. Nature 337: 382–384.

TURNER, S., BURGER-WIERSMA, T., GIOVANNONI, S.J., MUR, L.R., PACE, N.R. 1989. The relationship of a prochlorophyte *Prochlorothrix hollandica* to green chloroplasts. Nature 337: 380–382.

YANG, D., OYAIZU, Y., OYAIZU, H., OLSEN, G.J., WOESE, C.R. 1985. Mitochondrial origins. Proc. Natl. Acad. Sci. USA 82: 4443–4447.

2

A Historical Perspective on Chromosome Structure, Function, and Behavior

- The pre-1900 period established the cell theory, the concept of cell lineage, the microscopic structure of chromosomes, mitotic and meiotic cell divisions, and the existence of genes in nuclear chromatin.
- Studies during the 1900–1950 period recognized the linear order of genes in chromosomes, established the "one gene–one protein" concept, used karyotyping to investigate genetic and evolutionary relatedness, related changes in chromosome structure to mutations, provided evidence for transposable elements, and proved that DNA was the chemical component of genes.
- The 1950–1996 period established the physical structure of chromosomes, including the three-dimensional configuration of DNA and the histone proteins bound to it, the banding of whole chromosomes, and the distribution of specific DNA sequences.
- The universality of the genetic code was established in this period, as were techniques for transforming bacteria, plants, and animals, and ever-increasing uses for the polymerase chain reaction.
- Worldwide DNA sequence databases and associated computer technology for utilizing the information were devised, culminating, for the moment, in the sequencing of the entire yeast genome.

The excitement engendered by increased research activity on chromosomes is built upon a framework of knowledge erected over the past 300 years. The isolation of specific DNA fragments and gene sequences, their locations on chromosomes, and the development of massive computer data banks of nucleotide sequences are rapidly expanding research activities worldwide. The study of chromosomes provides a common thread linking all of these investigations, whether carried out in plants, humans, fungi, animals, or any of the other life-forms on earth. To introduce the rich history of cytogenetics, developments in chromosome research have been separated into three periods, pre-1900, 1900–1950, and 1950–1996. The reliance of modern cytogenetics on the observations, concepts, and principles developed in earlier insights is highlighted. In this way, the evolution of the science of cytogenetics can be described from the earliest microscopic observations to present-day investigations with the latest applications of biotechnology. The scientists named in the historical tables usually had one or more of the following qualities: a genius for invention or discovery, keen powers of observation, and/or the ability to synthesize concepts from preexisting information. However, it is not always possible to list all of the scientists involved in a particular discovery.

2.1 THE PRE-1900 PERIOD

The cellular structure of plants and animals was recognized almost as soon as the microscope was invented at the end of the sixteenth century (Table 2.1). It is of interest that one of the first organisms investigated with the microscope was *Giardia* (*see* Fig. 1.3), described as "animalcules a moving-very prettily." The recognition of separate organelles within cells required significant improvements in glass quality and lens manufacture, so it was another 150 years before the nucleus was observed. Shortly thereafter, the cell theory was proposed, stating that cells and their nuclei were the basic units of structure and function in living organisms. This knowledge, in turn, led to the associated theory, some 20 years later, that all cells are derived from preexisting cells, the cell lineage theory. The appreciation of these two theories established the importance of the individual cell in development, heredity, and evolution, in that present-day cells must trace their ancestry in an unbroken lineage to the first-ever cell. These ideas provided the impetus for the detailed study of cell division and embryology, as well as the starting point for the science of cytogenetics, with the technology becoming available with major improvements in microscopy (Fig. 2.1). Cells with large nuclei such as the amphibian egg cell (Fig. 2.2), or organisms with few chromosomes such as *Ascaris*, the horse threadworm, were favored for study.

By the end of the nineteenth century, both the mitotic and meiotic cell divisions had been described, and sexual reproduction had been associated with the fusion of egg and sperm in both animals and plants. Chromosomes had been observed at certain stages of the cell cycle, often with distinctive morphological traits such as relative sizes and the positions of constrictions. The constancy of

Table 2.1 Chronology of Chromosome Studies to the Year 1900

~1600	Janssen and Janssen, father and son, lay claim to the invention of the compound microscope.
1665	Hooke describes the cork cells of plants as "empty vessels."
1677	van Leeuwenhoek observes animal spermatozoa and in 1681 describes what has since been recognized as the Archezoan parasite *Giardia*, when examining his own stools.
1694	Camerarius publishes on sexual reproduction in plants and produces an artificial hybrid between hemp and hop.
1752	Maupertuis applies mathematical probabilities to genetic studies of polydactyly in humans, a century before Mendel.
1760s	Kölreuter makes reciprocal crosses between plants to show that each parent makes an equal contribution to the offspring.
1831	Brown gives an account of the nuclei in the ova of orchids, and their apparent disappearance when the pollen tube enters the ovum.
1835	von Mohl describes cell division.
1838/39	Schleiden and Schwann publish on the importance of cells in plants and animals, respectively, leading to the proposal of the cell theory.
1842	Nägeli shows that cells multiply by division.
1848	Hofmeister studies the meiotic chromosomes of pollen mother cells.
1858	The theory of cell lineage is proposed by Virchow.
1859	Darwin publishes the *Origin of Species.*
1866	Mendel's paper, "Experiments in plant hybridization," in which he applies mathematical logic to the inheritance of phenotypic traits in garden peas, is published.
1866	Haeckel suggests that the nucleus is the vehicle of inheritance.
1871	Miescher describes the chemical composition of "nuclein" isolated from the nuclei of pus cells.
1873	Abbé and Helmholtz independently demonstrate that the power of the light microscope to resolve two points depends on the wavelength of the light.
1875	Hertwig demonstrates that fertilization in the sea urchin involves the union of nuclei from both egg and sperm.
1875	Fertilization in plants in described by Strasburger.
1881	Balbiani describes puffing in the salivary-gland chromosomes of insects.
1882	By observing mitosis in detail, Flemming shows that it includes a lengthwise cleavage of chromatin.
1883/4	van Beneden shows that during meiosis, the number of chromosomes contributed to the egg and the sperm are half the total number present in the fertilized egg. The egg and sperm are defined as haploid and the fertilized egg as diploid.
1875–1887	The combined studies of Boveri, Hertwig, Roux, Strasburger, and Weismann result in the germplasm theory, and the general acceptance of nuclear chromatin as the physical basis for inheritance.
1888	Waldeyer coins the term "chromosome."
1888	Boveri describes meiotic chromosome pairs in both maternal and paternal germline tissues.
1889	Altmann provides biochemical evidence for the division of "nuclein" into nucleic acid and protein.
1892	Rückert discovers lampbrush chromosomes and suggests that pairing and exchange of genetic material occurs between paternal and maternal chromosomes during meiosis.
1896	Wilson lays the foundation for the chromosome theory of inheritance in his classic book *The Cell in Development and Inheritance.*
1898	Montgomery publishes an analysis of nucleoli, favoring their origin from cytoplasmic activity.

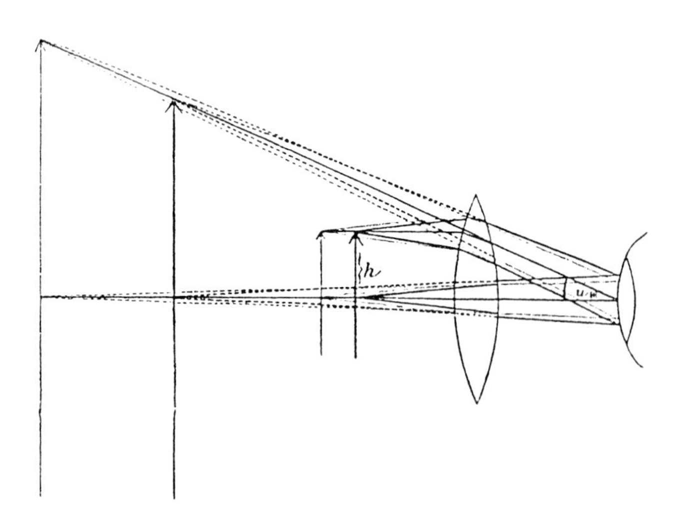

Fig. 2.1. The basic tool of cytogenetics is the light microscope. This diagram is taken from Abbé (1884) and illustrates the basis for defining the magnification of a lens system. Much of our present understanding of how we observe objects in the light microscope, and the limitations on resolution, stems from the work of E. Abbé.

their appearance and behavior suggested that they were structures of basic importance to cell biology. The nucleus, and later the chromosomes, were designated as the vehicles for the hereditary material, although the genetic significance of nucleic acid isolated from cells was not recognized until well into the twentieth century. Even so, the statement was made by Wilson in 1896, "There is no doubt that the morphological differentiation of parts within the cell is accompanied by corresponding chemical differentiation . . . . This fact is most conspicuous in the case of the nucleus and the chromatophore which contain, *and have the power of manufacturing, certain substances . . .* ."

2.2 THE 1900–1950 PERIOD

During the early part of the present century (Table 2.2), the full appreciation of Mendel's research on the inheritance of visible phenotypic traits led to new advances in understanding the genetics of a wide range of organisms. The central Mendelian concept of alleles as different forms of the same gene that segregate during gamete formation became established as the law of segregation. The independent partitioning of the alleles of different genes in the progeny of crosses between different individuals became known

Table 2.2. Chromosome Studies from 1900 to 1950

1900	de Vries, Tschermak, and Correns independently rediscover, recognize, and publicize the significance of Mendel's studies on pea genetics.
1902	Garrod and Bateson suggest that the human disease alkaptonuria is due to a single recessive gene and the loss of an enzyme.
1902/3	Sutton and Boveri independently show the relationship between Mendelian inheritance and chromosome behavior during meiosis.
1905	Stevens accurately describes the X and Y sex chromosomes of the beetle *Tenebrio*.
1906	Sex linkage of wing color is illustrated in the moth *Abraxas* by Doncaster and Raynor, and linkage in sweet peas is reported by Bateson and Punnett.
1907	The phenotypic effects of polyploidy in *Oenothera*, the evening primrose, are described by Lutz.
1907	Chromosome doubling is induced in mosses by Marchal and Marchal.
1908	Gates observes meiotic "ring" configurations in *Oenothera*, later attributed to breaks and exchanges between nonhomologous chromosomes.
1909	Maternal inheritance of plastids is demonstrated in the variegated plant *Mirabilis* by Correns.
1909	Janssens publishes the chiasmatype hypothesis to explain genetic exchanges between homologous chromosomes by chiasma formation.
1910	Morgan reports on the sex-linked inheritance of the white-eye mutation in *Drosophila* and shows that genetic recombination occurs between this and other sex-linked genes.
1913	Sturtevant produces the first chromosome map showing the linear arrangement of six sex-linked genes and their relative spacing.
1913/14	By combining the work of Bridges and Metz, it is realized that the two large and one small autosomes of *Drosophila* correspond to two large and one small genetic linkage groups.
1915	Morgan, Sturtevant, Muller, and Bridges publish the seminal book *The Mechanism of Mendelian Heredity*.
1917	Winge proposes that new species can arise by crosses between species, followed by chromsome doubling in the hybrids.
1917	By genetic analysis of *Drosophila* stocks, Bridges demonstrates the presence of duplications and deletions.
1920	Using the plant *Datura*, Blakeslee, Belling, and Farnham correlate specific phenotypic changes with the presence of an extra dose of single individual chromosomes.
1924	Polycentric chromosomes with many centromeres are observed in the roundworm *Ascaris megalocephala*, by Walton.
1924	Feulgen and Rossenbeck develop the Feulgen reaction to specifically detect DNA and chromosomes in tissues.
1926	Sturtevant shows that chromsome inversions cause a reduction in the number of crossovers.
1927	Independent research by Muller on *Drosophila* and, in 1928, by Stadler on barley and maize shows that X-rays increases mutation rates and chromosome aberrations.
1927	S. Navashin publishes drawings of chromosomes based on measurements during cell division to compare chromosome sets of different species.
1928	Based on studies of the liverwort *Pellia*, Heitz names the dark-staining regions of chromosomes heterochromatin, and lighter-staining regions euchromatin.
1928	Randolph names the supernumerary chromosomes of maize B chromosomes to distinguish them from the normal set of A chromosomes.
1929	McClintock publishes a physical map of maize pachytene chromosomes.
1931	Decisive proof that genetic recombination between homologous chromosomes is accompanied by the exchange of cytologically visible markers is obtained from *Drosophila* by Stern, and from maize by Creighton and McClintock.
1931	McClintock shows that specific genes are lost when parts of chromosomes are deleted.
1932	Knoll and Ruska publish the first description of the electron microscope.
1932	Darlington publishes the textbook *Recent Advances in Cytology*.
1933	Painter publishes a spreading technique for the analysis of the polytene chromosomes of *Drosophila* larval salivary-gland cells.
1934	In *Crepis* hybrids, M. Navashin shows that the satellited nucleolus-organizer regions of one species can be suppressed in the presence of the same region from another species.
1937	Blakeslee shows that colchicine has the ability to double the chromosome number of plants and of interspecies hybrids, helping to restore fertility.
1938	McClintock shows that variegation for marker genes in maize can be caused by the loss of ring chromosomes, formed by the fusion of broken ends, that contain the respective genes. She also finds that phenotypic variegation can result from the bridge–breakage–fusion cycle that results when chromosomes with two centromeres form chromatin bridges at anaphase and break under stress.
1941	Using X-ray-induced mutants of *Neurospora*, Beadle and Tatum show that specific genes direct the synthesis of specific proteins, and postulate the "one gene–one enzyme" theory.
1944	Avery, MacLeod, and McCarty induce transformation of bacteria using highly purified DNA preparations.
1947	McClintock publishes her first report on transposable elements.
1947	Auerbach and Robson describe the effects of chemical mutagens, confirming earlier studies by Oehkers in 1943.
1947	Caspersson uses microspectrophotometry to define nucleic acid:protein ratios in cells.
1949	Barr and Bertram discover a densely stained inclusion, the "Barr body," in the nuclei of female cats that is absent from the nuclei of male cells.

Fig. 2.2. An early study of chromosome division and mitotic chromosome structure using the light microscope. A photomicrograph of anaphase in the egg of the amphibian *Toxopneustes variegatus* (Wilson, 1896). Studies such as this were basic to establishing our present understanding of cell structure and, in particular, the behavior of chromosomes.

as the law of independent assortment. These laws were then related to the observed behavior of chromosomes in the meiotic cell divisions of reproductive tissues, and the correlation of gene transmission and chromosome behavior became known as the chromosome theory of heredity. The cytological basis for independent assortment, as well as apparent exceptions due to linkage of genes on the same chromosome, was firmly established by 1915. By the 1920s, the physical basis of inheritance was explained by the arrangement of genes in linear arrays on both sex chromosomes and autosomes. The ordering of these genes and the apparent distances between them could be determined by appropriate crosses.

In the late 1920s and 1930s, cytogenetic observations demonstrated that alterations in the number and structure of chromosomes could occur naturally or could be induced by exposing organisms or their parts to X-irradiation. As a result of parallel genetical and cytological studies on maize (*Zea mays*), Jimson weed (*Datura stramonium*), and the fruit fly (*Drosophila melanogaster*), changes in chromosome number or structure were correlated with changes in phenotypic characters.

Analyses of chromosome morphology reached their height in the study of maize meiotic–prophase chromosomes in pollen mother cells, and of the salivary-gland chromosomes of *Drosophila* (Fig. 2.3). These studies provided exquisite detail about the cytological structure of chromosomes, revealed the existence of heterochromatin, and showed that it was possible to arrange chromosomes into karyotypes. Studies in *Drosophila* demonstrated that heterochromatin was apparently genetically inert material that could influence the expression of genes located in nearby euchromatin.

The discovery that the drug colchicine, extracted from the plant *Colchicum autumnale*, could increase the chromosome number of cells by inhibiting spindle formation during cell division led to a flurry of activity in the experimental production of polyploid plants with more than the normal number of chromosomes. However, this approach had very limited success in animals.

Biochemical analysis of the nucleoprotein complex in the nuclei of eukaryotes showed that the main components were deoxyribonucleic acid, or DNA, and a basic protein named histone. With the development of microscope spectrophotometry (Fig. 2.4), the light-absorption data obtained directly from cellular components were shown to be compatible with the conclusions from biochemical studies.

The 1940s were marked by several landmark discoveries. Treatment of the bread mold *Neurospora crassa* with X-rays induced nutritional mutants that lacked the ability to survive on minimal medium alone. These mutant auxotrophs required the addition of specific nutrients for growth, whereas prototrophs did not. These

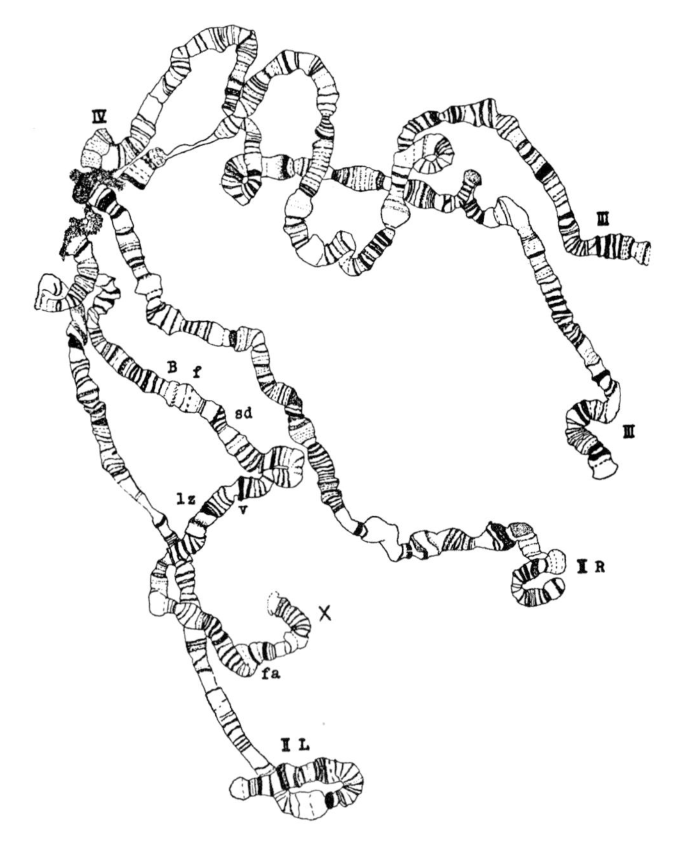

Fig. 2.3. Discovery of the potential of dipteran salivary-gland chromosomes for cytogenetic studies. A camera lucida drawing of the salivary-gland chromosomes of the fruit fly *Drosophila melanogaster* (Painter, 1934). This drawing shows the unusual features of the salivary-gland chromosomes, including somatic pairing of homologs, a greatly increased volume due to multiple chromatid replications, and a distinctive linear pattern of bands formed by the precise pairing of homologous chromomeres. Painter stated, ". . . we must realize that the utilization of these giant chromosomes for solving many vexing problems of cytology and genetics is comparable to the forging of a new tool and the important thing is to use this tool." This precision in band formation made it possible to localize the breakpoints that result in chromosome abnormalities such as duplications, deletions, translocations, and inversions.

Fig. 2.4. Microscope spectrophotometry has been used to provide direct measurements on the physical characteristics of regions within a single cell. One of the prototype microscopes built at the Karolinska Institute, Stockholm, under the direction of Prof. T. Caspersson is shown. (Photograph kindly supplied by Prof. N.R. Ringertz.)

results led to the "one gene–one enzyme" hypothesis, subsequently modified and confirmed as the "one gene–one polypeptide" concept. Transformation studies on the bacterium *Pneumococcus pneumoniae* demonstrated that highly purified DNA from one strain of bacterium could be transferred into another, giving the transformed strain a new function that was characteristic of the strain from which the DNA was isolated. Further, the transforming factor was destroyed if the DNA was degraded by enzymes, proving that DNA is the source of hereditary information and the chemical component of genes. In maize, certain events in the cell cycle were shown to destabilize the genetic material, as observed by color variegation in seeds and leaves, leading to the concept that units in the chromosome were capable of movement elsewhere in the genome. These units were suggested to be capable of creating mutations in genes, as well as chromosome abnormalities, and were named transposable, or mobile, elements (Fig. 2.5). The ability of certain chemicals to induce gene mutations, and in some cases chromosome breakage, was demonstrated by exposing *Drosophila* males to short, sublethal doses of mustard gas, which induced a high frequency (7.5%) of lethal mutations in the X chromosome. Such treatments were designed to provide a more convenient source of mutagens than irradiation. In later investigations, all of these discoveries were found to apply to a wide range of organisms.

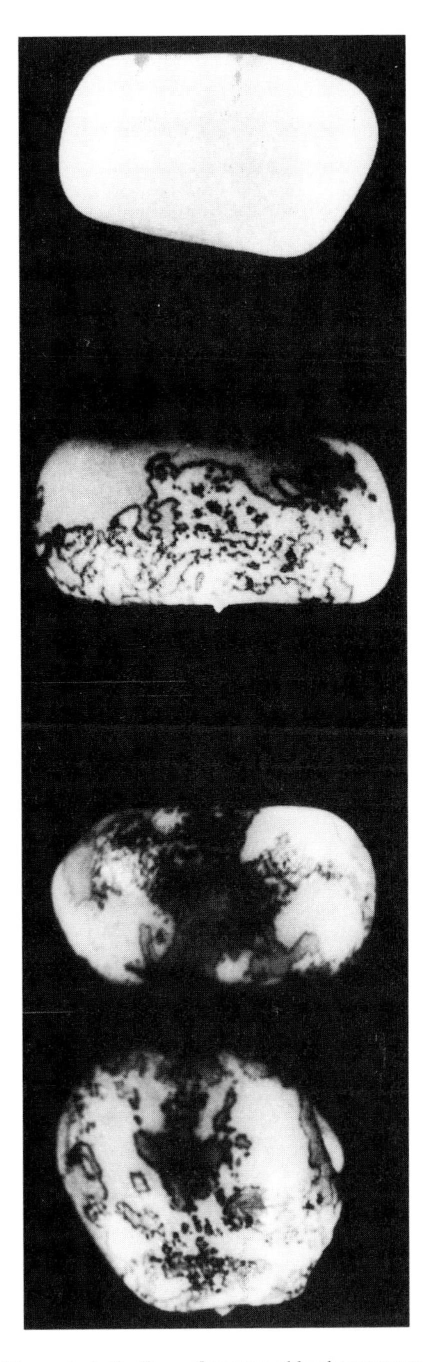

Fig. 2.5. The early indications of transposable chromosome elements came from investigations of unstable genetic conditions in maize. The photographs show mottled kernels (top photograph is a normal control), which were interpreted as resulting from the *Ac–Ds* system of transposable elements (McClintock, 1951). Transposable elements were discovered during investigations of the genetic instability associated with chromosome aberrations that create bridges at meiosis and cause chromosome breakage. McClintock presented her data and interpretations "... for whatever value they may have in giving focus to thoughts regarding the basic genetic problems concerned with nuclear organization and genic functioning." However, it was many years before the focus became clear enough for others to understand the far-reaching impact of her discovery.

2.3 THE 1950–1998 PERIOD

This period signalled a rapid advance in the application of biochemical and molecular developments to the study of cytogenetics, including chromosome structure and function (Table 2.3). It was

Table 2.3 Major Advances in Chromosome Structure and Function from 1950 to 1996

Year	Advance
1950	Chargaff establishes that DNA contains equimolar amounts of the bases adenine (A) and thymidine (T), and of the bases guanine (G) and cytosine (C).
1953	Hsu and Pomerat develop a hypotonic treatment for animal cells to simplify the production of chromosome spreads for karyotype analysis.
1953	Watson and Crick publish their interpretation of the DNA fiber X-ray diffraction data of Wilkins, Franklin, and colleagues.
1954	Sears classifies the chromosomes of common wheat into seven homoeologous groups based on compensating aneuploids.
1954	Gey, Berg, and Gey report on the establishment of the HeLa strain of epidermal carcinoma cells in tissue culture. Hsu characterizes the chromosomal complements of these cells within the range of 80–100 chromosomes per cell.
1956	Tjio and Levan obtain the correct chromosome number of $2n = 46$ for humans.
1957	Taylor, Woods, and Hughes demonstrate semiconservative DNA replication by labeling *Vicia faba* chromosomes with ^{3}H-thymidine.
1957	Minsky files U.S. patent #3013467 for the modern confocal microscope.
1958	Ford, Jacobs, and Lajtha show that, in humans, males are XY and females are XX.
1959	Trisomy of a human chromosome (later identified as #21) is linked to the genetic disability Down's syndrome, by Lejeune, Turpin, and Gautier.
1960	Nowell shows that phytohemagglutinin can stimulate cell division in human leukocytes.
1961	Lyon hypothesizes that the Barr bodies of mammalian female nuclei are, in fact, inactivated X chromosomes.
1961	Nirenberg and Matthaei demonstrate the dependence of ribosomes on messenger RNA and the coding of polyphenylalanine by poly-U, the first triplet of the genetic code to be deciphered.
1961	Jacob and Monod publish their model for the control of protein synthesis, based on the action of a repressor protein binding to a DNA control region preceding the 5′-end of the genes controlling the levels of messenger RNA.
1962/3	DNA is discovered in choloroplasts by Ris and Plaut, and in mitochondria by Nass and Nass.
1964	Littlefield specifically selects hybrid somatic cell lines in tissue culture using selective media and induced mutants.
1965	Harris and Watkins, Okada and Murayama, and Weiss and Ephrussi independently demonstrate the production of interspecific cell hybrids.
1967	Kornberg and colleagues demonstrate the relatively error-free synthesis of infectious ϕX174 DNA using purified DNA polymerase and polynucleotide-joining.
1968	Britten and Kohne demonstrate the existence of repetitive DNA sequences in mammalian DNA.
1968	Jordan, Saedler, and Starlinger show that some mutations in *E. coli* are caused by the insertion of transposable elements.
1969/70	Working independently, Pardue and Gall, Buongiorno-Nardelli and Amaldi, and Jones carry out *in situ* hybridization of radioactively labeled RNA probes to chromosome preparations.
1970	Temin and Mizutani, and Baltimore independently demonstrate that reverse transcriptase can yield a DNA molecule from an RNA template.
1970	Smith and Wilcox publish the discovery of the restriction endonuclease HindIII.
1971	Groups led by Ruddle, Bodmer, Miller, Siniscalco, and Bootsma begin the systematic analysis of hybrid cell lines for the purpose of mapping human chromosomes.
1973	Hewish and Burgoyne provide evidence for the nucleosome structure of chromatin.
1973	Peacock and colleagues establish the existence of distinct blocks of tandemly repeated DNA sequences in *Drosophila* heterochromatin.
1974	The first transformation of eukaryotic DNA into bacterial cells, by insertion into plasmids, is achieved by Morrow and colleagues.
1974	Moses and Counce refine the technique for spreading synaptonemal complexes for electron-microscopic analysis, initially investigated by Comings and Okada in 1970.
1975	Sanger and Coulson publish a DNA sequencing procedure based on synthesizing a DNA copy of a cloned, single-stranded DNA fragment.
1976	Hilliker establishes that *Drosophila* heterochromatin is not completely devoid of genes.
1977	Breathnach, Mandel, and Chambon publish their findings on the presence of intervening sequences in the gene coding for chicken ovalbumin.
1977/78	Sanger and colleagues, and Fiers and colleagues publish the complete nucleotide sequence of the DNA viruses ϕX174 and SV40, respectively.
1978	Blackburn and Gall publish the sequence of the termini of extrachromosomal ribosomal RNA genes from *Tetrahymena*.
1980	Botstein and colleagues point out the potential usefulness of DNA restriction fragment-length polymorphisms, or RFLPs, in genetic mapping.
1980	Chilton and Schell and their colleagues independently demonstrate that segments of the Ti plasmid of *Agrobacterium* are incorporated into plant genomic DNA.
1981	Cech, Zaug, and Grabowski report the self-splicing activity of *Tetrahymena* ribosomal RNA.
1982	Palmiter and colleagues transform mice by injecting DNA directly into egg-cell nuclei.
1982	Ward and colleagues publish the methodology for the *in situ*, nonradioactive detection of DNA sequences on chromosomes.
1983	Hall and colleagues announce the introduction of a gene into a plant (sunflower) using *Agrobacterium*.
1983	Federoff, Wessler, and Shure clone the maize transposable elements *Ac* and *Ds*.
1984	Fedoroff, Furtek, and Nelson clone a plant gene, the *Bronze* locus of maize, which was defined phenotypically by inactivation through the insertion of a transposable element.
1985	Saiki and colleagues report on the successful amplification of specific DNA segments using the DNA polymerase chain reaction or PCR.
1985	Greider and Blackburn publish the characterization of an enzymatic activity responsible for adding telomeres.
1987	Yeast artificial chromosomes (YACs) are constructed by Burke, Carle, and Olson.
1988	The Human Genome Organization, HUGO, is formed to coordinate the sequencing of the entire human genome.
1989	Report on the RNA editing activity in mitochondria of protozoa by the insertion or deletion of U residues, and three laboratories (Gray and colleagues, Weil and colleagues, and Brennicke and colleagues) report on the conversion of C's to U's in plant mitochondria.
1991	Michelmore and colleagues publish on the recovery of DNA markers closely linked to resistance genes, using the polymerase chain reaction primed by random primers to analyze the DNA bulked from a segregating population of individuals originating from a single cross.
1992	Oliver heads a consortium of laboratories to publish the complete DNA sequence of yeast chromosome 3.
1995	The complete sequences of two bacterial protosomes from *Haemophilus influenzae* and *Mycoplasma genitalia* are published by Fleischmann, Fraser, Venter, and their associates.
1996	Goffeau heads a consortium of over 100 laboratories to sequence the entire yeast genome.
1998	The EU Arabidopsis Genome group analyses 1.9 Mb of contiguous DNA from chromosome 4 of *Arabidopsis thaliana*.

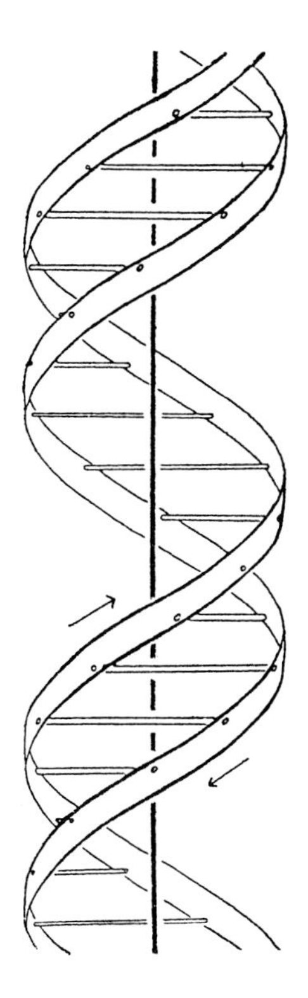

Fig. 2.6. The molecular structure of DNA. The DNA double helix as depicted in Watson and Crick (1953). This interpretation of DNA structure, with sugar-phosphate chains on the outside and nucleotide base pairs on the inside, provided the basis for molecular biology. Watson and Crick suggested that the structure may indicate "... a possible copying mechanism for the genetic material."

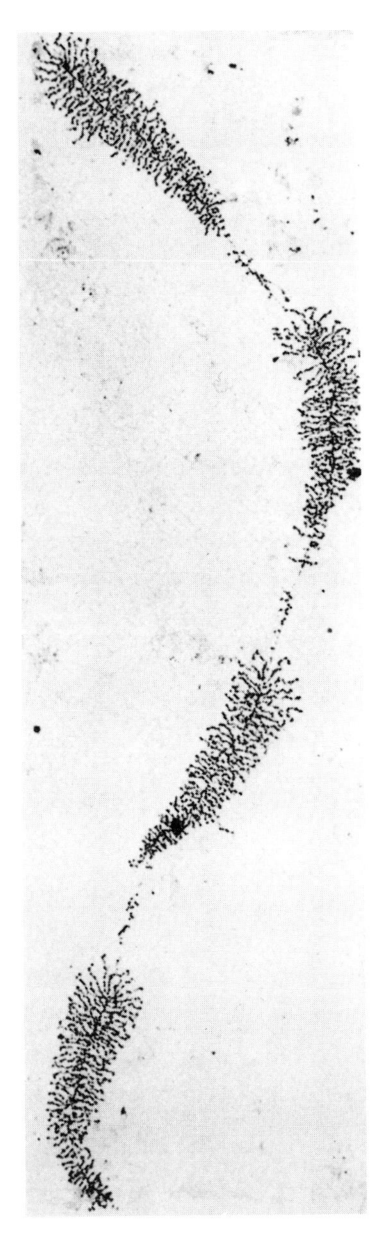

Fig. 2.8. Transcription of ribosomal RNA genes. Miller and Beatty (1969a and b) discovered that when the nuclei of actively growing cells were lysed or broken open onto an aqueous surface, the interphase material could be transferred to a grid and prepared for observation in the electron microscope. This photograph shows the "Christmas trees" that were considered to result from the process of transcription by the movement of RNA polymerase enzymes along a central DNA molecule. The steadily lengthening RNA molecules form the "branches" of the trees. Each RNA gene is separated from the next by a length of nontranscribed spacer DNA.

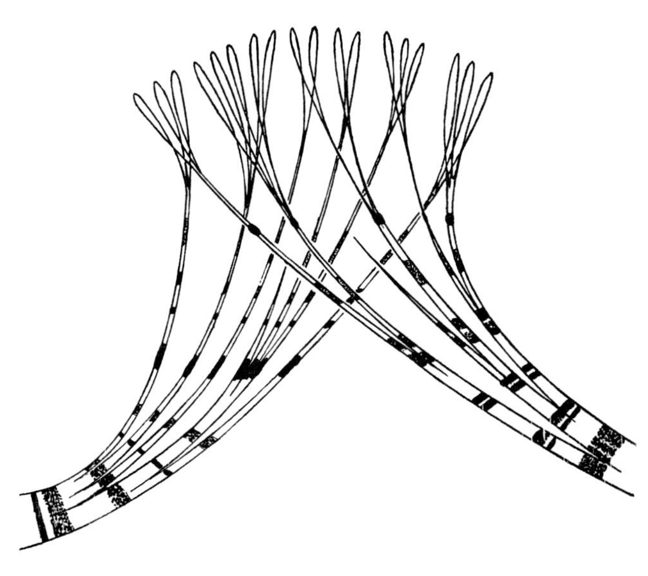

Fig. 2.7. Chromosome puffing in the midge *Chironomus*. The drawing is a diagrammatic representation of the unraveling of a polytene chromosome to give a Balbiani ring, showing the structural modification that led to the formation of the chromosome puff (Beermann, 1956). The puffs were considered to "... obviously indicate changes, most probably increases, in the activity of gene loci." Molecular biological techniques confirmed this idea and defined the products of some of the puffs in *Drosophila*, as well as *Chironomus*.

generally accepted that DNA was the source of genetic information. The determination of the chemical structure of the DNA molecule (Fig. 2.6) revealed the basis for semiconservative replication, which was subsequently demonstrated both cytogenetically and biochemically. The genetic code and how DNA codes for proteins using triplet nucleotide codons to determine the amino–acid composition of proteins were elucidated. In classical experiments using bacteria, the operon/repressor model for interpreting the control of gene expression was developed. This model explained how a single gene, coding for a repressor protein, could control the expression of a distant group of genes if these were arranged in a block, or operon.

In eukaryotes, chromosome puffing was correctly interpreted as a site of gene expression (Fig. 2.7). In addition, the identification of the nucleolar-organizer region, most often seen by early cytogeneticists as separating a satellite region from the remainder of the chromosome, was shown to be the physical site for ribosomal RNA genes. A combination of molecular and cytological techniques provided details of the structure of ribosomal RNA genes, and electron microscopy allowed the observation of the decoding of DNA into RNA by transcription (Fig. 2.8). The structure of the unusual lampbrush chromosomes in amphibian egg cells, first reported 70 years earlier (Fig. 2.9), also became more clearly understood in terms of gene transcription.

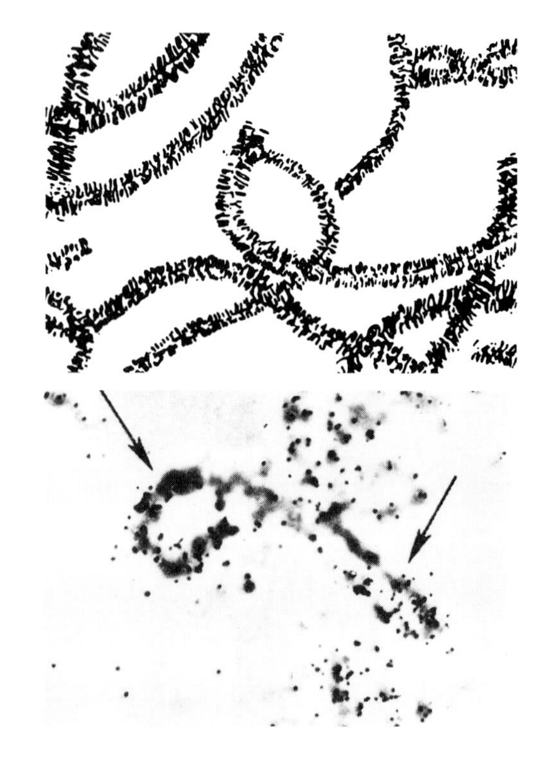

Fig. 2.9. The discovery of lampbrush chromosomes. **Top:** a drawing showing diplotene chromosome pairs in the primary oocyte of the shark, *Pristiurus* (Rückert, 1892). This meiotic stage can last for several years in the females of some vertebrates and invertebrates. The fuzzy appearance of the chromosomes, which Rückert compared to lamp-cleaning brushes, is caused by pairs of loops extending in all directions. **Bottom:** Many years later, Gall and Callan (1962) showed that the DNA in the loops was transcribing RNA. In this photograph of a loop pair of an oocyte chromosome of the newt *Triturus cristatus*, the incorporation of ^{3}H-uridine into RNA has proceeded approximately halfway around each loop to the points marked by the arrows. Rückert was therefore seeing "genes in action" in his observations of these fascinating chromosomes.

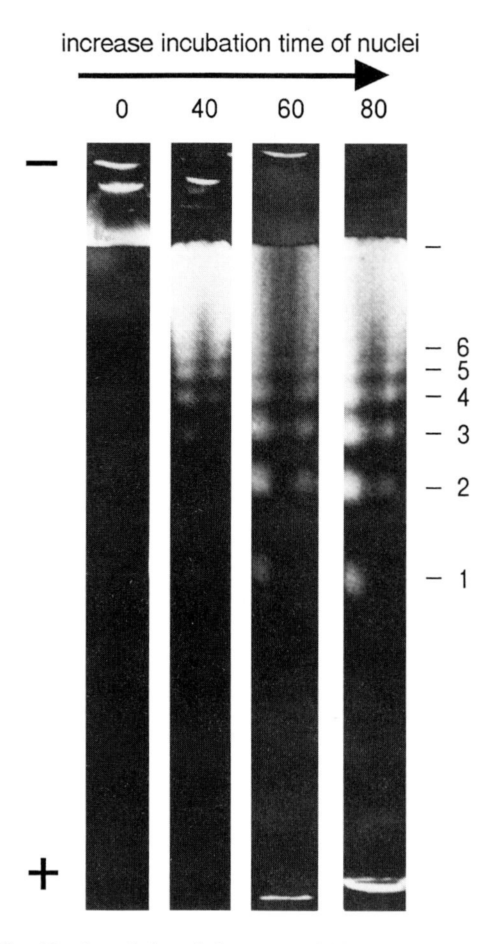

Fig. 2.10. The degradation of chromosomal DNA provided the first indication as to how DNA was actually arranged in chromosomes. The suggestion for the existence of a basic unit of chromosome structure came from interpreting the pattern of DNA degradation during the isolation of nuclei from rat liver (Hewish and Burgoyne, 1973). The photograph shows DNA fragments stained with the ultraviolet (UV)-fluorescent dye ethidium bromide, after electrophoresis in an agarose gel to separate fragments of different sizes. **Top:** the lanes are marked 0, 40, 60, and 80 to indicate the length of time in minutes that the nuclei were incubated at 37°C before carrying out a DNA preparation. The numbers 1–6, (right-hand side) indicate the relative size of the DNA band (e.g., monomer, dimer, trimer, etc.). It was found that the fragments of DNA occurred in multiples of approximately 200 base pairs, leading to the suggestion that this length of fragment is the basic unit of DNA folding. These basic units were named nucleosomes. Electron-microscopic observations on the DNA–protein complex present in nuclei and chromosomes (Olins and Olins, 1974) subsequently demonstrated the presence of small knobs along the length of DNA, consistent with the nucleosome concept.

The physical structure of chromosomes was analyzed in detail. Histone proteins were sequenced and shown to be highly conserved between different organisms, leading to the idea that DNA–histone complexes could form the core building blocks for chromosomes. Biochemical and electron-microscopic studies provided evidence for the existence of basic units of DNA folding called nucleosomes (Fig. 2.10), in which histone proteins form the center with DNA wrapping around the outside. The high content of repetitive-sequence DNA in the nuclei of eukaryotes was also uncovered (Fig. 2.11), raising many new questions about how genes were integrated into the chromosome as a whole. The manipulation of chromosomes of *Drosophila* to create chromosomes of different lengths, proved that they consist of a single strand of DNA, which changes in length concomitantly with changes in the cytological length of the chromosome.

Spreading techniques for human chromosomes, including phytohemagglutinin stimulation of cell division in lymphocytes, provided the technical advances that allowed an accurate chromosome number for humans to be determined (Fig. 2.12). Novel staining procedures for chromosomes led to the discovery of banding patterns that revealed details about their substructure (Fig. 2.13) and enabled homologous chromosomes to be identified. Heterochromatic regions of chromosomes were shown to contain specific classes of repetitive DNA, and it was recognized that genes controlling the expression of phenotypic traits could also be found in the heterochromatin. Chromosome engineering in wheat (Fig. 2.14) provided new sources of disease resistance, which could be transferred into commercial wheat cultivars, thus demonstrating one of the practical applications of cytogenetics.

The cloning of the first eukaryotic DNA segment into bacteria was reported in 1974. The DNA was comprised of fragments of the ribosomal RNA genes from *Xenopus*, inserted into a special class of bacterial DNA called a plasmid (Fig. 2.15). The technology provided a major breakthrough in the molecular analysis of the genomes of all life-forms. Based on cloning technology, rapid advances were made in the analysis of the sequence structure of DNA, and the molecular nature of transposable elements as well as how

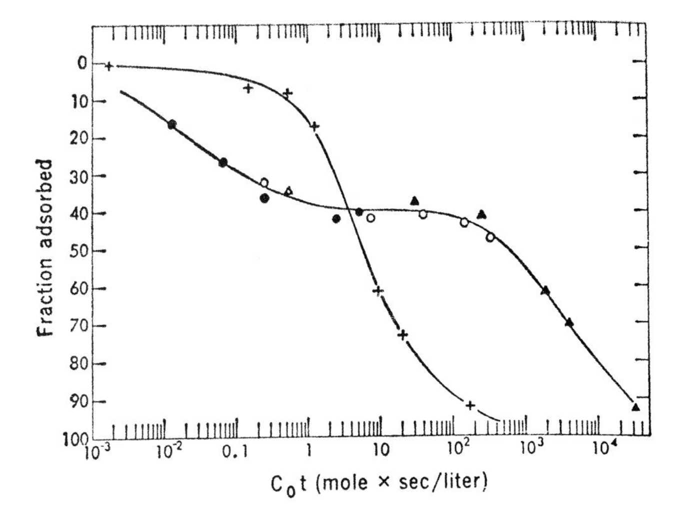

Fig. 2.11. The reassociation of single strands of DNA. After double-stranded molecules of DNA have been separated into their complementary single strands by a process of denaturation, the double-stranded molecule can spontaneously re-form following normal physical laws. This diagram (from Britten and Kohne, 1968) demonstrates that whereas *Escherichia coli* DNA (+) follows a renaturation curve typical of a bimolecular reaction, calf-thymus DNA (●, ○, ▲) shows a much more complex renaturation curve. A portion of the DNA renatured very rapidly, from which it was deduced that this particular DNA was composed of repetitive arrays of DNA sequences.

10μ

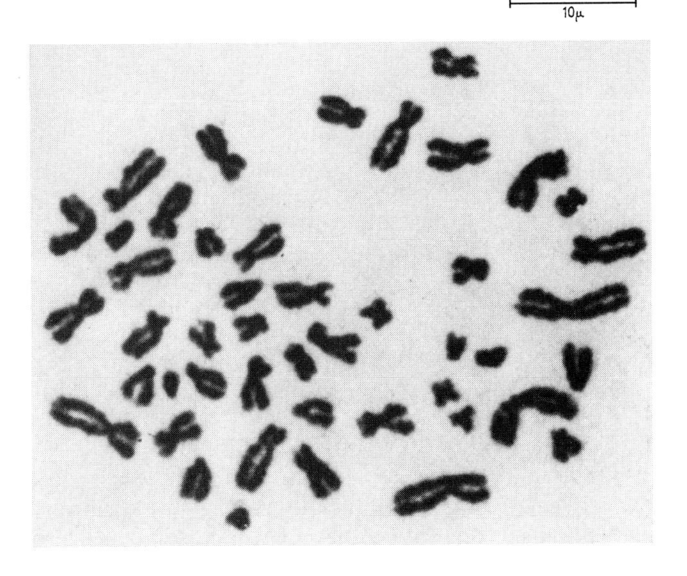

Fig. 2.12. The discovery of the correct number of chromosomes in human cell nuclei. Tjio and Levan (1956) discovered that the nuclei from dividing embryonic lung fibroblast cells contained 46 chromosomes. Numerous studies have since confirmed their finding that humans normally have 46 chromosomes in somatic cells. Tumor cells have varying numbers of chromosomes and can contain double, or even four times, the stemline number (Levan, 1956).

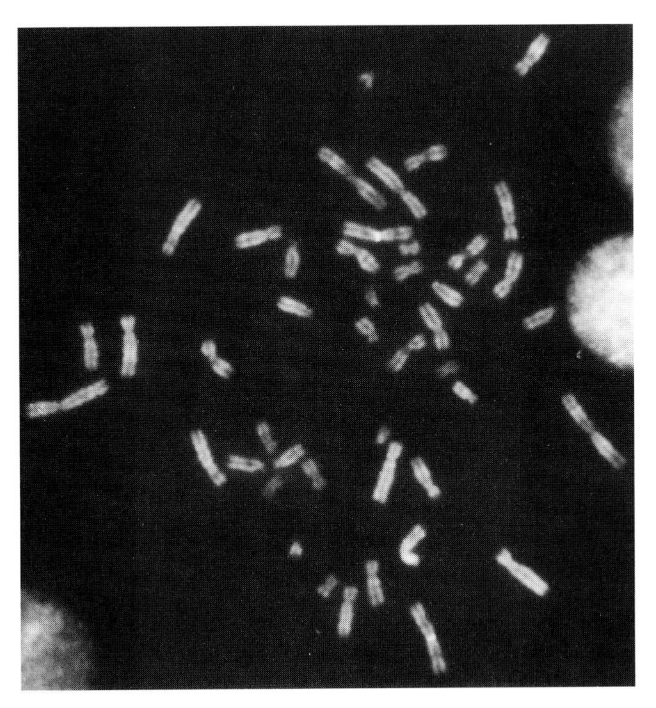

Fig. 2.13. Chromosome banding using fluorescent staining to aid in the identification of individual chromosomes. Caspersson and co-workers (1970a and b) showed that human chromosomes could be individually identified after staining chromosome preparations with DNA-binding dyes such as quinacrine hydrochloride. This photograph (kindly supplied by Dr. L. Zech) shows a routine preparation of lymphocyte chromosomes from a human male.

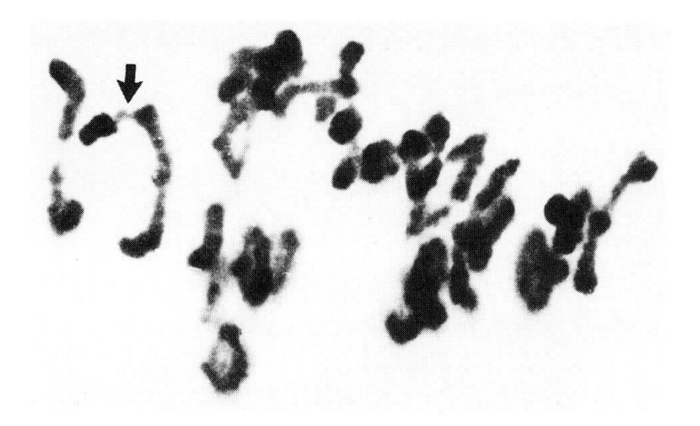

Fig. 2.14. A photograph of a wheat meiotic cell nucleus close to metaphase I showing a wheat–rye translocation chromosome (Sears, 1972). Using materials such as this, methods have been developed to transfer chromosome segments from one species to another. The bivalent (arrow) has a normal wheat chromosome 6B (lower part of bivalent) paired with a translocated chromosome consisting of the long arm of 6B and the short arm of the rye chromosome 5R (to left of arrow). The translocation resulted from manipulations that introduced chromosomes 6B and 5R in single doses into an individual plant. In this singular, unpaired state, centromeres can misdivide transversely during meiosis. The divided centromeres can then reunite to give translocated chromosomes, consisting of one wheat arm and one rye arm.

they caused mutations. In addition, the structure of specific sequences that control the expression of genes advanced our understanding of how the information held in DNA is decoded. The fragmented nature of many genes was demonstrated (Fig. 2.16).

In the 1980s, a rapid expansion of molecular/genetic maps was made possible, as variations in DNA sequences between individuals were used as markers for the analysis of genetic crosses. This information was combined with earlier work that utilized phenotypic traits and specific proteins, and enzymes, for genetic analysis. An exciting advance was the transformation of eukaryotic life-forms by the introduction and expression of new, foreign DNA gene sequences. These techniques further stimulated the study of cytogenetics in that they provided a new means of modifying chromosomes. The creation of yeast artificial chromosomes (YACs) was particularly relevant, because it offered an alternative approach to understanding the complexities of chromosome behavior in cells. The polymerase chain reaction (PCR), combined with specific or random primers, was also a technological breakthrough that provided many new opportunities for the analysis of specific regions of the genome.

The rapid advances in computer technology in the 1950–1996 period have not only made the extensive analysis of DNA sequence databases possible but have also completely changed the basic tool of the cytogeneticist, namely the light microscope. Image-enhancement technology has begun to replace the human eye as the primary means of collecting information, with the result that the microscope has become a much more versatile tool.

The achievements of the 1950–1996 period emphasize the wide range of scientific fields that have had an impact on chromosome studies in cytogenetics. It is clear that this interdisciplinary approach will continue in the future. Computer technology, molecular biology, genetics, cytology, physics, and chemistry converge, for the cytogeneticist, in unraveling secrets of chromosomes. Many of these techniques were used in the determination of the entire DNA nucleotide sequences of the 16 chromosomes of the budding yeast *Saccharomyces cereviseae.*

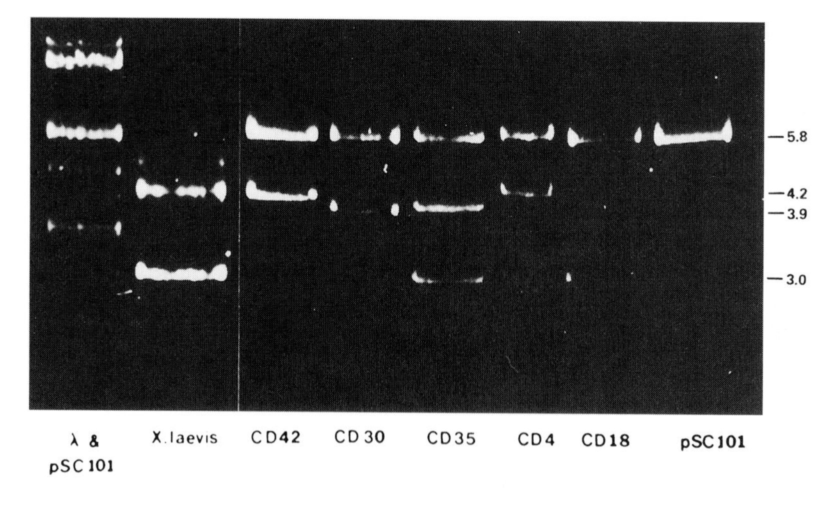

Fig. 2.15. The cloning and reduplication of eukaryotic DNA in bacteria. Small, circular DNA molecules, called plasmids, provide the vectors for DNA cloning. In this instance, the plasmid vector pSC101 was used to clone EcoR1 digested DNA fragments from *Xenopus* ribosomal RNA genes. Bacteria containing the plasmids with the cloned DNA sequences were then identified and cultured. These cultures were again digested with the EcoR1 restriction endonuclease and the DNA fragments were observed on an agarose gel after electrophoresis and staining with ethidium bromide. The photograph (Morrow et al., 1974) shows the vector pSC101 band of 5.8 kb (right lane), DNA length markers (left lane), uncloned *Xenopus* rDNA cut with EcoR1 (*X. laevis*), and the various newly made plasmids (lanes marked with CD numbers) containing the cloned *Xenopus* rDNA EcoR1 fragments corresponding to the major band in the uncloned *Xenopus* rDNA.

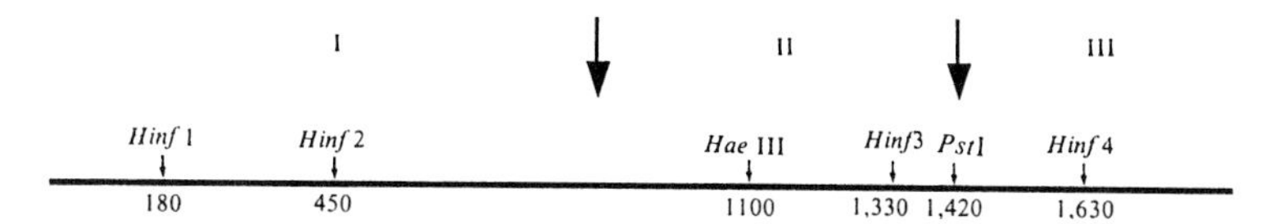

Fig. 2.16. The molecular structure of an ovalbumin gene. Early studies of gene structure revealed a number of unusual features, one of the more unexpected being the interruption of the gene sequence by unrelated DNA sequences, later named introns. One of the first genes to be characterized in this way was that coding for ovalbumin (Breathnach et al., 1977). This diagram summarizes the structure of the ovalbumin gene as it appears in a messenger RNA ready for translation into protein. The structure was determined using the technique of restriction–endonuclease site mapping. The large arrows in the figure indicate the positions of two introns, which are normally removed during the processing of the transcribed RNA into messenger RNA.

BIBLIOGRAPHY

General

BURNHAM, C.R. 1962. Discussions in Cytogenetics. Burgess Publishing Co., Minneapolis, MN.

LEWIS, K.R., JOHN, B. 1963. Chromosome Marker. J & A Churchill Ltd., London.

MORGAN, T.H., STURTEVANT, A.H., MULLER, H.J., BRIDGES, C.B. 1915. The Mechanism of Mendelian Heredity. Henry Holt and Co., New York, NY.

SCHULZ-SCHAEFFER, J. 1980. Cytogenetics: Plants, Animals, Humans. Springer-Verlag, New York.

STURTEVANT, A.H. 1966. A History of Genetics. Harper and Row, New York, NY.

SWANSON, C.P. 1957. Cytology and Cytogenetics. MacMillan and Co. Ltd., London.

SWANSON, C.P., MERZ, T., YOUNG, W.J. 1967. Cytogenetics. Prentice-Hall Inc., Englewood Cliffs, NJ.

WILSON, E.B. 1896. The Cell in Development and Heredity. MacMillan Co., New York.

Section 2.1

ABBÉ, E. 1884. Note on the proper definition of the amplifying power of a lens or a lens-system. J. Roy. Microscopy Soc. 4: 348.

ALTMANN, R. 1889. Über Nuklein Säuren. Arch. Anat. Physiol. Lpz., Physiol. Abt. 524.

BALBIANI, E.G. 1881. Sur la structure du noyau des cellules salivaires chez les larves de *Chironomus*. Zool. Anz. 4: 637–641, 662–666.

BENEDEN, E. VAN. 1883. L'appareil sexuel femelle de l'Ascaride mégalocéphale. Arch. Biol. 4: 95.

BOVERI, T. 1887. Über Differenzierung der Zellkerne während der Furchung des Eies von *Ascaris megalocephala*. Anat. Anz. 2: 688–693.

BROWN, R. 1828. A brief account of microscopical observations on the particles contained in pollen of plants. Arnold Arboretum, Harvard University Coll. Works B81 II.

BROWN, R. 1831. Observations on the organs and mode of fecundation in Orchideae and Asclepideae. Trans. Linn. Soc. London 16: 685–745.

CAMERARIUS, R.J. 1694. De sexu plantarum epistola. In: Ostwald's Klassiker der Exacten Wissenschaften. Verlag & Engelmann, Leipzig. 105: 1899.

DARWIN, C. 1859. The Origin of Species by Means of Natural Selection, or the Preservation of Favoured Races in the Struggle for Life. John Murray, London, UK.

FLEMMING, W. 1882. Zellsubstanz, Kern-und Zellteilung. Vogel, Leipzig.

HAECKEL, E. 1866. Entstehung der ersten Organismen. Generelle Morphologie der Organismen 1: 167–190.

HERTWIG, O. 1875. Beiträge zur Kenntnis der Bildung, Befruchtung und Theilung des Tierischen Eies. Abh. Morph. Jb. 1: 347.

HERTWIG, O. 1884. Das Problem der Befruchtung und der Isotropie des Eies, eine Theorie der Vererbung. Zeit. Med. Naturwiss. 18: 21–23.

HOFMEISTER, W. 1848. Über die Entwickelung des Pollens. Bot. Ztg. 6: 425–434.

HOOKE, R. 1665. Micrographia or some physiological descriptions on minute bodies by magnifying glasses. London. (Facsimilis edition published by Gunther, R.T., 1938. In: Early Science in Oxford. XIII. The life and works of R. Hooke. Oxford, UK.)

KÖLREUTER, J.G. 1761–1766. Vorläufige Nachricht von einigen das Geschlecht der Pflanzen betreffenden Versuchen, und Beobachtungen, nebst Fortsetzungen 1, 2 und 3. In: Ostwald's Klassiker der Exacten Wissenschaften No 41. Verlag & Engelmann, Leipzig.

KÖLREUTER, J.G. 1761. Geschlecht der Pflanzen. Arnold Arboretum Harvard University G K81.3.

LEEUWENHOEK, A. van. 1667. In: Letter to Viscount Brouncker, Nov 1677. Phil. Trans. 12: 1040.

MAUPERTUIS, P.L.M. de. 1752. In: Oeuvres de Maupertuis: lettres sur divers sujets. Nouvelle Edition 1768. J.M. Bruyset, Lyon.

MENDEL, G.J. 1866. Versuche über Pflanzenhybriden. Verhandl. Naturforsch. Ver., Brünn., Bd. 4: 3–47.

MIESCHER, F. 1871. Über die chemische Zusammensetzung der Eiterzellen. F. Hoppe-Seyler's Med. Chem. Untersuch. Berlin. 4: 441–460.

MOHL, H. von. 1835. Über die Vermehrung der Pflanzenzelle durch Theilung. Dissertation Tübingen Universität.

Nägeli, C. von. 1842. Zur Entwicklungsgeschichte des Pollens bei den Phanerogamen. Orell Füssli u Comp: Zurich, Switzerland.

Roux, W. 1883. Über die Bedeutung der Kernteilungsfiguren. Eine hypothetische Erterung. In: Ostwald's Klassiker der Exacten Wissenschaften. Verlag & Engelmann, Leipzig.

Rückert, J. 1892. Zur Entwicklungsgeschichte des Ovarialeies bei Seelachiern. Anat. Anz. 7: 107–158.

Schleiden, M.J. 1838. Beiträge zur Phytogenesis. Joh. Müllers Arch. Anat. Physiol. Wiss. Med. pp. 137–176.

Schwann, T. 1839. Mikroskopische Untersuchungen über die Übereinstimmung in der Struktur und dem Wachstum der Tiere und Pflanzen. In: Ostwald's Klassiker der Exacten Wissenschaften No 176. Verlag & Engelmann, Leipzig.

Strasburger, E. 1876. Über Zellbilding und Zellteilung. 2nd Ed. Nebst Untersuchungen über Befruchtung. H. Dabis, Jena.

Strasburger, E. 1884. Neue Untersuchungen über den Befruchtungsvorgang bei den Phanerogamen als Grundlage für eine Theorie der Zeugung. Gustav Fischer, Jena.

Virchow, R. 1858. Die Cellularpathologie in ihrer Begründung auf physiologische und pathologische Gewebelehre. Verlag A. Hirschwald, Berlin.

Waldeyer, W. 1888. Über Karyokinese und ihre Beziehung zu den Befruchtungsvorgängen. Arch. Mikrosk. Anat. 32: 1.

Weismann, A. 1883. Über die Vererbung. Gustav Fischer, Jena.

Wilson, E.B. 1896. The Cell in Development and Inheritance. The Macmillan Co., New York.

Section 2.2

Auerbach, C., Robson, J.M. 1946. Chemical production of mutations. Nature 157: 302.

Auerbach, C., Robson, J.M. 1947. Tests of chemical substances for mutagenic action. Proc. Roy. Soc. Edinburgh Ser. B 62: 284–291.

Avery, O.T., MacLeod, C.M., McCarty, M. 1944. Studies on the chemical nature of the substance inducing transformation of pneumococcal types. Induction of transformation by a desoxyribonucleic acid fraction isolated from *Pneumococcus* Type III. J. Exptl. Med. 79: 137–158.

Barr, M.L., Bertram, E.G. 1949. A morphological distinction between neurons of the male and female, and the behaviour of the nucleolar satellite during accelerated nucleoprotein synthesis. Nature 163: 676–677.

Bateson, W., Saunders, E.R. 1902. Experimental studies in the physiology of heredity. Report to the Evolution Committee of the Royal Society of London. 1: 1–160.

Beadle, G.W., Tatum, E.L. 1941. Genetic control of biochemical reactions in *Neurospora*. Proc. Natl. Acad. Sci. USA 27: 499–506.

Blakeslee, A.F. 1937. Redoublement du nombre de chromosomes chez les plantes par traitement chimique. Compt. Rend. Acad. Sci. Paris 205: 476–479.

Blakeslee, A.F., Belling, J., Farnham, M.E. 1920. Chromosomal duplication and Mendelian phenomena in *Datura* mutants. Science 52: 388–390.

Boveri, T. 1902. Über mehrpolige Mitosen als Mittel zur Analyse des Zellkerns. Verhandl. Deut. Physiol. Med. Gesellsch. zur Würzburg. 35: 67–90.

Bridges, C.B. 1913. Nondisjunction of the sex chromosomes of *Drosophila*. J. Exptl. Zool. 15: 587–606.

Bridges, C.B. 1917. Deficiency. Genetics 2: 445–465.

Caspersson, T. 1947. The relations between nucleic acid and protein synthesis. Symp. Soc. Exptl. Biol. 1: 127–151.

Correns, C. 1900. G. Mendels Regel über das Verhalten der Nachkommenschaft der Rassenbastarde. Berlin Deutsch. Bot. Gesselsch. 18: 158–167.

Correns, C. 1909. Vererbungs versuche mit blass (gelb) grünen und buntblättrigen Sippen bei *Mirabilis jalapa*, *Urtica pilulifera*, und *Lunaria annua*. Zeits. ind. Abst. Vererbs. 1: 291–329.

Creighton, H.B., McClintock, B. 1931. A correlation of cytological and genetical crossing-over in *Zea mays*. Proc. Natl. Acad. Sci. USA 17: 492–497.

Darlington, C.D. 1932. Recent Advances in Cytology. Churchill, London.

Doncaster, L., Raynor, G.H. 1906. Breeding experiments with Lepidoptera. Proc. Zool. Soc. London 1: 125–133.

Feulgen, R., Rossenbeck, H. 1924. Mikroskopisch-chemischer Nachweis einer Nucleinsäure vom Typus der Thymonucleinsäure und die darauf beruhende elektive Färbung von Zellkernen in mikroskopischen Präparaten. Hoppe-Seyler Z. Physiol. Chem. 135: 203–248.

Garrod, A.E. 1902. The incidence of alkaptonuria: a study in chemical individuality. Lancet ii: 1616–1620.

Gates, R.R. 1908. A study of reduction in *Oenothera rubrinervis*. Bot. Gaz. 46: 1–34.

Heitz, E. 1928. Das Heterochromatin der Moose. I. Jahrb. Wiss. Bot. 69: 762–818.

Janssens, F.A. 1909. Spermatogénèse dans les Batraciens. V. La théorie de la chiasmatypie. Nouvelle interprétation des cinèses de maturation. La Cellule 25: 387–411.

Knoll, M., Ruska, E. 1932. Das Elektronenmikroskop. Zeit. Physik 78: 318–339.

Lutz, A.M. 1907. A preliminary note on the chromosomes of *Oenothera lamarckiana* and one of its mutants, *O. gigas*. Science 26: 151–152.

McClintock, B. 1929. Chromosome morphology of *Zea mays*. Science 69: 629–630.

McClintock, B. 1931. Cytological observations of deficiencies involving known genes, translocations and an inversion in *Zea mays*. Missouri Agr. Exp. Sta. Res. Bull. 163: 1–30.

McClintock, B. 1938a. The production of homozygous deficient tissues with mutant characteristics by means of the aberrant behavior of ring-shaped chromosomes. Genetics 23: 315–376.

McClintock, B. 1938b. The fusion of broken ends of sister half-chromatids following chromatid breakage at meiotic anaphases. Missouri Agr. Exp. Sta. Res. Bull. 290: 1–48.

McClintock, B. 1947. Cytogenetic studies of maize and *Neurospora*. Carnegie Inst. Wash. Yearb. 46: 146–152.

Marchal, Él, Marchal, Ém. 1907. Aposporie et sexualité chez les mousses I. Bull. Acad. Roy. Belgique 1907: 765–789.

Metz, C.W. 1914. Chromosome studies in the Diptera. J. Exptl. Zool. 17: 45–58.

Morgan, T.H. 1910. The method of inheritance of two sex-limited characters in the same animal. Proc. Soc. Exptl. Biol. Med. 8: 17–19.

MORGAN, T.H., STURTEVANT, A.H., MULLER, H.J., BRIDGES, C.B. 1915. The Mechanism of Mendelian Heredity. Henry Holt and Co., New York.

MULLER, H.J. 1927. Artificial transmutation of the gene. Science 66: 84–87.

NAVASHIN, M. 1934. Chromosome alterations caused by hybridization and their bearing upon certain general genetic problems. Cytologia 5: 169–203.

NAVASHIN, S. 1927. Zellkernidimorphisms bei *Galtonia candicans* Des. und einigen verwandten Monokotylen. Deutsch Botan. Gesell. Berichte. 45: 415–428.

OEHLKERS, F. 1943. Die Auslösung von Chromosomenmutationen in der Meiosis durch Einwirkung von Chemikalien. Zeit. Ind. Abstamm. Vererbungsl. 81: 313–341.

PAINTER, T.S. 1933. A new method for the study of chromosome rearrangements and the plotting of chromosome maps. Science 78: 585–586.

PAINTER, T.S. 1934. A new method for the study of chromosome aberrations and the plotting of chromosome maps in *Drosophila melanogaster*. Genetics 19: 175–188.

RANDOLPH, L.F. 1928. Chromosome numbers in *Zea mays* L. Cornell Univ. Exp. Sta. Mem. 117

STADLER, L.J. 1928a. Mutations in barley induced by X-rays and radium. Science 68: 186–187.

STADLER, L.J. 1928b. Genetic effects of x-rays in maize. Proc. Natl. Acad. Sci. USA. 14: 69–75.

STERN, C. 1931. Zytologisch-genetische Untersuchungen als Beweise für die Morgansche Theorie des Factorenaustausches. Biol. Zentralbl. 51: 547–551.

STEVENS, N.M. 1905. Studies in spermatogenesis with especial reference to the "accessory chromosome." Carnegie Inst. Wash. Pub. 36: 1–33.

STURTEVANT, A.H. 1913. The linear arrangement of six sex-linked factors in *Drosophila*, as shown by their mode of association. J. Exptl. Zool. 14: 43–59.

STURTEVANT, A.H. 1926. A crossover reducer in *Drosophila melanogaster* due to inversion of a section of the third chromosome. Biol. Zentralbl. 46: 697–702.

SUTTON, W.S. 1903. The chromosomes in heredity. Biol. Bull. 4: 231–251.

TSCHERMAK, E. 1900. Über künstliche Kreuzung bei *Pisum sativum*. Berlin Deutsch. Botan. Gesellsch. 18: 232-239.

VRIES, H. DE. 1900. Das Spaltungsgesetz der Bastarde. Berlin Deutsch. Botan. Gesellsch. 18: 83–90.

WALTON, A.C. 1924. Studies on nematode gametogenesis. Zeit. f Zellen- und Gewebelehre 1: 167–239.

WINGE, Ö. 1917. The chromosomes. Their number and general importance. Compt. Rend. Trav. Lab. Carlsberg 13: 131–275.

Section 2.3

APPELS, R. 1988. A historical perspective on the study of chromosome structure and function. In: Chromosome Structure and Function, Impact of new concepts (Gustafson, J.P., Appels, R., eds.). 18th Stadler Genetics Symp., Plenum Press, New York. pp. vii-xii.

ASHBURNER, M. 1973. Temporal control of puffing activity in polytene chromosomes. Cold Spring Harb. Symp. Quant. Biol. NY. 38: 655–662.

BALTIMORE, D. 1970. RNA-dependent DNA polymerase in virions of RNA tumour viruses. Nature 226:1209–1211.

BEERMANN, W. 1956. Nuclear differentiation and functional morphology of chromosomes. Cold Spring Harb. Symp. Quant. Biol. NY. 21: 217–232.

BEERMANN, W., CLEVER, O. 1964. Chromosome puffs. Sci. Am. 210: 50–58.

BOTSTEIN, D., WHITE, R.L., SKOLNICK, M., DAVIS, R.W. 1980. Construction of a genetic linkage map in man using restriction fragment length polymorphisms. Am. J. Hum. Genet. 32: 314–331.

BREATHNACH, R., MANDEL, J.L., CHAMBON, P. 1977. Ovalbumin gene is split in chicken DNA. Nature 270: 314–319.

BRITTEN, R.J., KOHNE, D.E. 1968. Repeated sequences in DNA. Science 161: 529–540.

BURKE, D.T., CARLE, G.F., OLSON, M.V. 1987. Cloning of large segments of exogenous DNA into yeast by means of artificial chromosome vectors. Science 236: 806–812.

CASPERSSON, T., ZECH, L., JOHANSSON, C. 1970a. Analysis of the human metaphase chromosome set by aid of DNA-binding fluorescent agents. Exp. Cell. Res. 62: 490–492.

CASPERSSON, T., ZECH, L., JOHANSSON, C., MODEST, E.J. 1970b. Identification of human chromosomes by DNA-binding fluorescent agents. Chromosoma 30: 215–227.

CECH, T.R. 1983. RNA splicing: three themes with variations. Cell 34: 713–716.

CECH, T.R. 1986. RNA as an enzyme. Sci. Am. 255: 64–75.

CECH, T.R., ZAUG, A.J., GRABOWSKI, P.J. 1981. In vitro splicing of ribosomal RNA precursor of *Tetrahymena*: involvement of a guanosine nucleotide in the excision of the intervening sequence. Cell 27:487–496.

CHARGAFF, E. 1950. Chemical specificity of the nucleic acids and mechanisms of their enzymatic degradation. Experimentia 6: 201–209.

CHILTON, M.D. 1983. A vector for introducing new genes into plants. Sci. Am. 248: 50–59.

CHILTON, M.D., SAIKI, R.K., YADAV, N., GORDON, M.P., QUETIER, F. 1980. T-DNA from *Agrobacterium* Ti plasmid is the nuclear DNA fraction of crown gall tumor cells. Proc. Natl. Acad. Sci. USA. 77: 4060–4064.

CHILTON, M.D., TEPFER, D.A., PETIT, A., DAVID, C., CASSE-DELBART, F., TEMPE, J. 1982. *A. rhizogenes* inserts T-DNA into the genomes of host plant root cells. Nature 295: 432–434.

COMINGS, D.E., OKADA, T.A. 1970. Whole mount electron microscopy of meiotic chromosomes and the synaptonemal complex. Chromosoma 30: 269–286.

CORVELLO, P.S., GRAY, M.W. 1989. RNA editing in plant mitochondria. Nature 341: 662–666.

EPHRUSSI, B., WEISS, M.C. 1965. Interspecific hybridization of somatic cells. Proc. Natl. Acad. Sci. USA 53: 1040–1042.

EU ARABIDOPSIS GENOME PROJECT. 1998. Analysis of 1.9 Mb of contiguous sequence from chromosome 4 of *Arabidopsis thaliana.* Nature 391: 485–488.

FEDEROFF, N.V. 1984. Transposable genetic elements in maize. Sci. Am. 250: 84–98.

FEDOROFF, N.V., FURTEK, D.B., NELSON, O.E. JR. 1984. Cloning of the *bronze* locus in maize by a simple and generalizable procedure using the transposable controlling element *Ac*. Proc. Natl. Acad. Sci. USA 81: 3825-3829.

FEDOROFF, N., WESSLER, S., SHURE, M. 1983. Isolation of the transposable maize controlling elements *Ac* and *Ds*. Cell 35: 235–242.

FIERS, W., CONTRERAS, R., HAEGEMAN, G., ROGIERS, R., VAN DE VOORDE, A., VAN HEUVERSWYN, J., VAN HERREWGHE, J., VOLCKAERT, G., YSEBAERT, M. 1978. Complete nucleotide sequence of SV40 DNA. Nature 273: 113–120.

FLEISCHMANN, R.D., et al. (39 others). 1995. Whole-genome random sequencing and assembly of *Haemophilus influenzae* Rd. Science 269: 496–512.

FORD, C.E., JACOBS, P.A., LAJTHA, L.G. 1958. Human somatic chromosomes. Nature 181: 1565–1568.

FRALEY, R.T., ROGERS, S.G., HORSCH, R.B., SANDERS, P.R., FLICK, J.S., ADAMS, S.P., BITTNER, M.L., BRAND, L.A., FINK, C.L., FRY, J.S., GALLUPPI, G.R., GOLDBERG, S.B., HOFFMANN, N.C., WOO, S.C. 1983. Expression of bacterial genes in plant cells. Proc. Natl. Acad. Sci. USA 80: 4803–4807.

FRANKLIN, R.E., GOSLING, R.G. 1953. Molecular configuration in sodium thymonucleate. Nature 171: 740–741.

FRASER, C.M., GOCAYNE, J.D., WHITE, D., ADAMS, M.D., CLAYTON, R.A., FLEISHMANN, R. D., BULT, C.J., KERLAVAGE, A. R., SUTTON, G., KELLEY, J.M., FRITCHMAN, J.L., WEIDMAN, J.F., SMALL, K.V., SANDUSKY, M., FUHRMANN, J., NGUYEN, D., UTTERBACK, T.R., SAUDER, D.M., PHILLIPS, C.A., MERRICK, J.M., TOMB, J.F., DOUGHERTY, B.A., BOTT, K.F., HU, P-C., LUCIER, T.S., PETERSON, S.N., SMITH H. D., HUTCHINSON III, C.A., VENTNER, J.C. 1995. The minimal gene complement of *Mycoplasma genitalia*. Science 270: 397–403.

GALL, J.G, CALLAN, H.G. 1962. H^3 uridine incorporation in lampbrush chromosomes. Proc. Natl. Acad. Sci. USA 48: 562–570.

GOULIAN, M., KORNBERG, A., SINSHEIMER, R.L. 1967. Enzymatic synthesis of DWA. xxiv. Synthesis of infectious phage ϕX174 DNA. Proc. Natl. Acad. Sci. USA. 58: 2321–2328.

GREIDER, C.W., BLACKBURN, E.H. 1985. Identification of a specific telomere terminal transferase activity in *Tetrahymena* extracts. Cell 43: 405–413.

HARRIS, H., WATKINS, J. 1965. Hybrid cells derived from mouse and man: artificial heterokaryons of mammalian cells from different species. Nature 205: 640–646.

HEWISH, D.R., BURGOYNE, L.A. 1973. Chromatin sub-structure. The digestion of chromatin DNA at regularly spaced sites by a nuclear deoxyribonuclease. Biochem. Biophys. Res. Commun. 52: 504–510.

HILLIKER, A.J. 1976. Genetic analysis of the centromeric heterochromatin of chromosome 2 of *Drosophila melanogaster*: deficiency mapping of ems-induced lethal complementation groups. Genetics 83: 765–782.

HIESEL, R., WISSINGER, B., SCHUSTER, W., BRENNICKE, A. 1989. RNA editing in plant mitochondria. Science 246: 1632–1634.

HSU, T.C. 1952. Mammalian chromosomes in vitro. I. The karyotype of man. J. Hered. 43: 167–172.

HSU, T.C. 1954. Cytological studies on HeLa, a strain of human cervical carcinoma. I. Observations on mitosis and chromosomes. Texas Rept Biol. Med. 12: 833–846.

JACOB, F., MONOD, J. 1961. Genetic regulatory mechanisms in the synthesis of proteins. J. Mol. Biol. 3: 318–356.

JONES, G.H., CRAIG-CAMERON, T. 1969. Analysis of meiotic exchange by tritium autoradiography. Nature 223: 946–947.

JONES, K.W. 1970. Chromosomal and nuclear location of mouse satellite DNA in individual cells. Nature 225: 912–915.

JORDAN, E., SAEDLER, H., STARLINGER, P. 1968. O° and strong polar mutations in the gal operon are insertions. Mol. Gen. Genet. 102:353–363.

KAVENOFF, R., ZIMM, B.H. 1973. Chromosome-sized DNA molecules from *Drosophila*. Chromosoma 41: 1–27.

KOURILSKY, P., CHAMBON, P. 1978. The ovalbumin gene: an amazing gene in eight pieces. Trends Biochem. Sci. 3: 244–247.

LANDRY, B.S., KESSELI, R.V., FARRARA, B., MICHELMORE, R.W. 1987. A genetic map of lettuce (*Lactuca sativa* L.) with restriction fragment length polymorphism, isozyme, disease resistance and morphological markers. Genetics 116: 331–337.

LEJEUNE, J., TURPIN, R., GAUTIER, M. 1959. Le mongolisme, premier exemple d'aberration autosomique humaine. Ann. Génét. Hum. 1: 41–49.

LEVAN, A. 1956. Chromosome studies on some human tumors and tissues of normal origin, grown in vivo and in vitro at the Sloan-Kettering Insitute. Cancer 9: 648–663.

LITTLEFIELD, J.W. 1964. Selection of hybrids from matings of fibroblasts in vitro and their presumed recombinants. Science 145: 709–710.

LYON, M.F. 1961. Gene action in the X-chromosome of the mouse (*Mus musculus* L.). Nature 190: 372–373.

MCCLINTOCK, B. 1951. Chromosome organization and genic expression. Cold Spring Harb. Symp. Quant. Biol. 16: 13–47.

MAXAM, A.M., GILBERT, W. 1977. A new method for sequencing DNA. Proc. Natl. Acad. Sci. USA 74: 560–564.

MESELSON, M.S., STAHL, F.W. 1958. The replication of DNA in *Escherichia coli*. Proc. Natl. Acad. Sci. USA 44: 671–682.

MESELSON, M., YUAN, R. 1968. DNA restriction enzyme from *E. coli*. Nature 217: 1110–1114.

MILLER, O.J., ALLDERDICE, P.W., MILLER, D.A., BREG, W.R., MIGEON, B.R. 1971a. Human thymidine kinase gene locus: assignment to chromosome 17 in a hybrid of man and mouse cells. Science 173: 244–245.

MILLER, O.J., COOK, P.R., KHAN, P.M., SHIN, S., SINISCALCO, M. 1971b. Mitotic separation of two human X-linked genes in man–mouse somatic cell hybrids. Proc. Natl. Acad. Sci. USA 68: 116–120.

MILLER, O.L. JR., BEATTY, B.R. 1969a. Visualization of nucleolar genes. Science 164: 955–957.

MILLER, O.L. JR., BEATTY, B.R. 1969b. Portrait of a gene. J. Cell Physiol. 74: 225–957.

MORROW, J.F., COHEN, S.N., CHANG, A.C.Y., BOYER, H.W., GOODMAN, H.M., HELLING, R.B. 1974. Replication and transcription of eukaryotic DNA in *Escherichia coli*. Proc. Natl. Acad. Sci. USA 71: 1743–1747.

NASS, M.M.K., NASS, S. 1963. Intramitochondrial fibers with DNA characteristics. I. Fixation and electron staining reactions. J. Cell Biol. 19: 593–611.

NIRENBERG, M.W., MATTHAEI, J.H. 1961. The dependence of cell-free protein synthesis in *E. coli* upon naturally occurring or synthetic polyribonucleotides. Proc. Natl. Acad. Sci. USA 47: 1588–1602.

Nowell, P.C. 1960. Phytohemagglutinin: an initiator of mitosis in cultures of normal human leukocytes. Cancer Res. 20: 462–466.

Okada, Y., Murayama, F. 1965. Multinucleated giant cell formation by fusion between cells of two different strains. Exp. Cell Res. 40: 154–158.

Olins, A.L., Olins, D.E. 1974. Spheroid chromatin units (υ bodies). Science 183: 330–334.

Oliver, S.G., et al. (147 others). 1992. The complete sequence of yeast chromosome III. Nature 357: 38–46.

Palmiter, R.D., Brinster, R.L., Hammer, R.E., Trumbauer, M.E., Rosenfeld, M.G., Birnberg, N.C., Evans, R.M. 1982. Dramatic growth of mice that develop from eggs microinjected with metallothionein-growth hormone fusion genes. Nature 300: 611–615.

Paran, I., Kessell, R., Michelmore, R. 1991. Identification of restriction fragment length polymorphisms and random amplified polymorphic DNA markers linked to downy mildew resistance genes in lettuce, using near-isogenic lines. Genome 24: 1021–1027.

Pardue, M.L., Gall, J.G. 1969. Molecular hybridization of radioactive DNA to the DNA of cytological preparations. Proc. Natl. Acad. Sci. USA 64: 600–604.

Pardue, M.L., Gall, J.G. 1970. Chromosomal localization of mouse satellite DNA. Science 168: 1356–1358.

Peacock, W.J., Brutlag, G., Goldring, R., Appels, R., Hinton, C.W., Lindsley, D.L. 1973. The organization of highly repeated DNA sequences in *Drosophila melanogaster* chromosomes. Cold Spring Harb. Symp. Quant. Biol. 38: 405–416.

Ris, H., Plaut, W. 1962. Ultrastructure of DNA-containing areas in the chloroplast of *Chlamydomonas*. J. Cell Biol. 13: 383–391.

Ruddle, F.H. 1973. Linkage analysis in man by somatic cell genetics. Nature 242: 165–169.

Ruddle, F.H., Chapman, V.M., Ricciutti, F., Murnana, M., Klebe, R., Meera Khan, P. 1971. Linkage relationships of seventeen human gene loci as determined by man-mouse somatic cell hybrids. Nature New Biol. 232: 69–73.

Saiki, R.K., Scharf, S., Faloona, F., Mullis, K.B., Horn, G.T., Erlich, H.A., Arnheim, N. 1985. Enzymatic amplification of β-globin genomic sequences and restriction site analysis for diagnosis of sickle cell anaemia. Science 230: 1350–1354.

Saiki, R.K., Gelfand, D.H., Stoffel, S., Scharf, S.J., Higuichi, R., Horn, G.T., Mullis, K.B., Erlich, H.A. 1987. Primer-directed enzymatic amplification of DNA with a thermostable DNA polymerase. Science 239:487–491.

Sanger, F., Coulson, A.R. 1975. A rapid method for determining sequences in DNA by primed synthesis with DNA polymerase. J. Mol. Biol. 94: 441–448.

Sanger, F., Air, G.M., Barrell, B.G., Brown, N.L., Coulson, A.R., Fiddes, J.C., Hutchison, C.A., Slocombe, P.M., Smith, M. 1977. Nucleotide sequence of bacteriophage ΦX174 DNA. Nature 265: 687–695.

Sears, E.R. 1954. The aneuploids of common wheat. Missouri Agr. Exptl. Sta. Res. Bull. 572: 1–59.

Sears, E.R. 1972. Chromosome engineering in wheat. In: Stadler Genet. Symposia, Vol. 4 (Kimber, G., Redei, G.P., eds.). University of Missouri Press, Columbia. pp. 23–38.

Siniscalco, M., Klinger, H.P., Eagle, H., Koprowski, H., Fujimoto, W.F., Seegmiller, J.E. 1969. Evidence for intergenic complementation in hybrid cells derived from two human diploid strains each carrying an X-linked mutation. Proc. Natl. Acad. Sci. USA 62: 793–799.

Smith, H.O. 1970. Nucleotide sequence specificity of restriction endonucleases. Science 205: 455–462.

Smith, H.O., Nathans, D. 1973. A suggested nomenclature for bacterial host modification and restriction systems and their enzymes. J. Mol. Biol. 81: 419–423.

Smith, H.O., Wilcox, K.W. 1970. A restriction enzyme from *Haemophilus influenzae*. I. Purification and general properties. J. Mol. Biol. 51: 379–391.

Southern, E.M. 1975. Detection of specific sequences among DNA fragments separated by gel electrophoresis. J. Mol. Biol. 98: 503–517.

Taylor, J.H., Woods, P.S., Hughes, W.L. 1957. The organization and duplication of chromosomes as revealed by autoradiographic studies using tritium-labeled thymidine. Proc. Natl. Acad. Sci. USA 43: 122–128.

Temin, H.M., Mizutani, S. 1971. RNA-dependent DNA polymerase in virions of Rous sarcoma virus. Nature 226:1211–1213.

Tjio, J.H., Levan, A. 1956. The chromosome number in man. Hereditas 42: 1–6.

Watson, J.D., Crick, F.H.C. 1953a. Molecular structure of nucleic acids. A structure for deoxyribose nucleic acid. Nature 171: 737–738.

Watson, J.D., Crick, F.H.C. 1953b. Genetic implications of the structure of deoxyribonucleic acid. Nature 171: 964–967.

Weiss, M.C., Ephrussi, B. 1966. Studies of interspecific (rat × mouse) somatic hybrids. I. Isolation, growth and evolution of the karyotype. Genetics 54: 1095–1109.

Wilkins, M.H.F., Stokes, A.R., Wilson, H.R. 1953. Molecular structure of deoxypentose nucleic acids. Nature 171: 738–740.

Yeast Genome. 1996. *See* http://www.mips.biochem.mpg.de or http://genome__www.stanford.edu/ yeast/genome__seq.

Zaug, A.J., Cech, T.R. 1986. The intervening sequence RNA of *Tetrahymena* is an enzyme. Science 231: 470–475.

3

Microscopes: Basic Tools for Cytogenetics

- The standard light microscope has many variations based on manipulating the path of light through the optical system.
- The wavelength of light is a major determinant of resolution in the microscope, and the shorter ''wavelength'' of electrons provides greater magnification with the electron microscope.
- Fluorescence provides a versatile basis for tracing specific features of chromosomes.
- Computer capture of microscopic images has enhanced the analytical power of microscopes.

Although the foundations for the design of the compound light microscope were established in the latter half of the nineteenth century, this instrument still provides the primary means for the routine observation of chromosomes. In recent years, the analytical power of the light microscope has been enriched by improved detection techniques developed in the physical sciences. The application of the confocal principle, for example, allows the light microscope to be used for optically sectioning nuclei, and this technique has become particularly powerful as the tools derived from molecular biology provide the means for observing specific segments of chromosome structure. The development of the electron microscope and its derivative, the scanning electron microscope, provide much greater resolution than the light microscope in observing biological structures. The application of electron microscopy was significant, for example, in establishing the existence of chromatin folding, and the presence of specialized structures in meiosis, as well as in observing gene transcription. In this chapter, the present range of microscopic techniques are discussed. The process of preparing cells for microscopic observation needs to be carefully understood, as improper preparation can lead to errors in the interpretation of the image; this problem is discussed in Chapter 19.

3.1 LIGHT MICROSCOPY

3.1.1 The Standard Light Microscope

The optical system of a basic light microscope is shown in Fig. 3.1. This figure emphasizes that lenses are used to focus light onto the specimen by means of the condenser and to observe the specimen using the objectives and eyepieces. Traditionally, the human eye is an integral part of the optics. The primary magnification of a microscope is derived from the objective lens, with a secondary magnification coming from the eyepieces. Although the magnification offered by a microscope is important, the amount of detail that can be distinguished is ultimately dependent on the resolution of the lens system. When light passes through a specimen, it interacts with it in one of two ways. It can either undergo scattering, or diffraction, as it passes opaque edges, or it can pass through freely as so-called zero-order light. The diffraction phenomenon means that, at the primary image focal plane (*see* Fig. 3.1), the image of a pinhole is not a single spot of light, but a central point of light surrounded by a series of rings of decreasing intensity. This means that when two pinholes or two points in a specimen are very close together, they appear to merge into each other. The diffraction phenomenon therefore places a limit on how close two points can be in a specimen and still be resolved as two separate points. The amount of diffraction is also dependent on the wavelength of light, λ, illuminating the specimen, an important variable in determining the resolving power of a microscope. The main variable determining resolution, however, is the numerical aperture (NA) of the objective lens (and the matching condenser lens). The numerical aperture is defined by

$$\mathrm{NA} = n \sin u,$$

where n is the refractive index of the medium between the specimen and objective lens and u is one-half of the angle made by the effective cone of light rays entering the objective (Fig. 3.2). An increase in u means that more of the diffracted light passing through the specimen is collected by the objective lens, thus improving the numerical aperture of the objective lens. If n is increased using an immersion liquid, so that it is closer to the refractive index of the glass coverslip covering the specimen, more of the light passing through the specimen enters the objective lens because the immersion liquid reduces the refraction of light that normally occurs at a boundary with air (*see* Fig. 3.2).

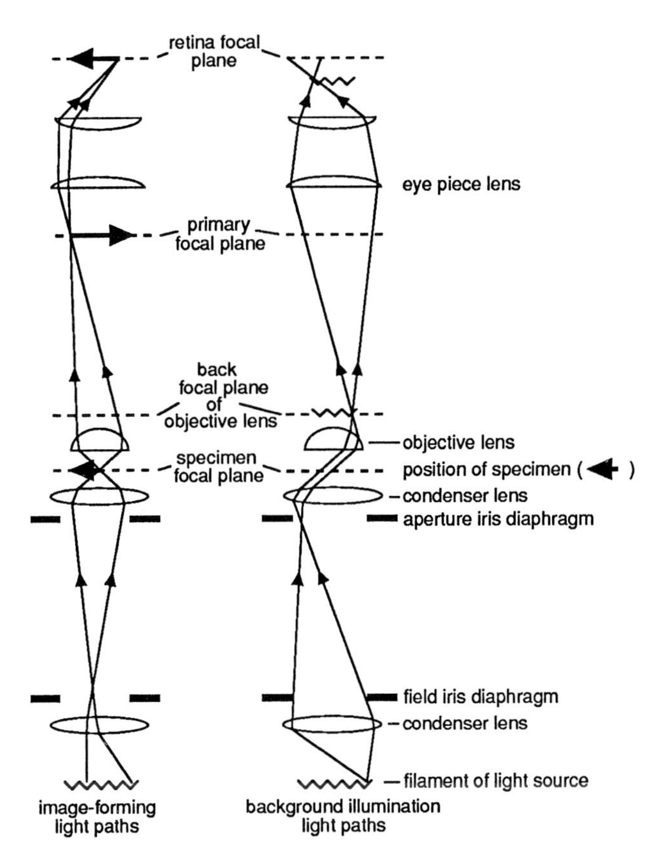

Fig. 3.1. The optical system of a standard compound microscope. Two light paths are shown. The diagram on the left side follows the rays of light that focus the image of the specimen on the retina of the eye. The diagram on the right side follows the rays of light that provide the background illumination. This background or Kohler illumination focuses the image of the light-source filament in front of the focal plane of the retina (top right) and, thus, is not actually "seen" in any detail.

The resolution of an optical system can now be defined by the minimum distance (d) that can exist between two points while still resolving them as two points, as summarized by the equation:

$$d = \frac{\lambda}{\mathrm{NA}_{\mathrm{objective}} + \mathrm{NA}_{\mathrm{condenser}}}$$

From this relationship, it is evident that the resolution of the microscope increases as d decreases. The equation further illustrates that decreasing the wavelength of light (λ) used to observe the specimen improves the resolving power, although in the traditional light microscope, this is not a major variable. Because the refractive index (n) of the medium between the objective lens and the specimen affects the numerical aperture, the medium can be modified to improve the resolution of a microscope. Immersion oil, for example, with $n = 1.52$, can significantly improve resolution when, using suitably designed objective lenses, it is used to fill the space between the specimen and the objective lens (*see* Fig. 3.2, lower portion).

3.1.2 Microscopy Using Ultraviolet Light

Microscopes using short-wavelength ultraviolet light (UV), with a range of approximately 300 nm (1 nm = 0.001 μm or 10 Å), rather than the usual ~500 nm, can lead to an almost twofold increase in resolution. These microscopes have quartz rather than glass lenses, because glass blocks the passage of UV light. UV irradiation is especially useful when the specimens have been stained with fluorochrome compounds, which absorb the UV radiation and reemit the energy at a wavelength in the visible region.

Fluorescence microscopy is widely exploited in studies on chromosomes (e.g., *see* Fig. 2.13). With the use of fluorochromes that bind to defined structures, the ability to detect extremely small features in a specimen is greatly enhanced because the features become a source of radiation. Consequently, features smaller than 100 nm (shown by electron-microscopic analysis) have been observed using fluorescent microscopy. However, the intensity of emitted light is very low and so the development of highly sensitive electronic-detection systems has advanced the utilization of this technique.

Dark-field illumination and incident-light excitation are two

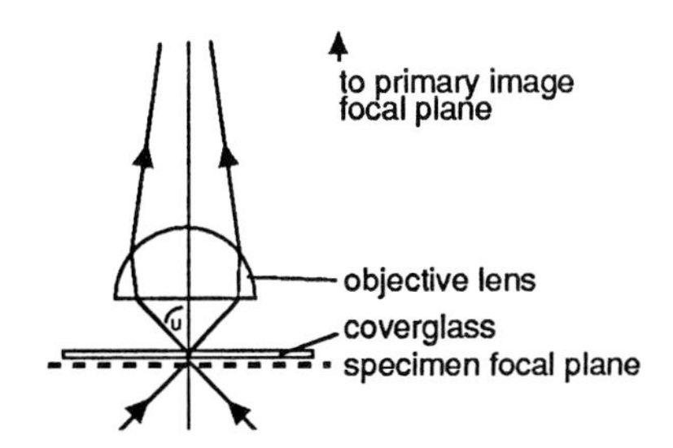

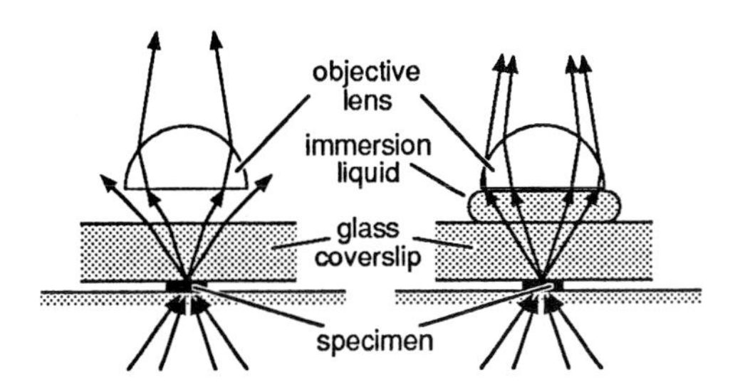

Fig. 3.2. The importance of the objective lens. **Top:** The numerical aperture of the objective lens (NA) is proportional to the refractive index (n) of the medium between the lens and the specimen. Objective lenses **(bottom)** designed for use with water ($n = 1.33$) or oil ($n = 1.52$) (right side) have a greater resolving power than those designed for use in air ($n = 1.00$) (left side). The designation of an objective lens as an immersion type (water or oil) is marked on the outside of the lens, together with a number of other markings (e.g., 40×/0.65 indicates the primary magnification/numerical aperture, 160/0.17 indicates the mechanical tube length/thickness of the coverslip), whereas abbreviations indicate the optical quality of the objective lens: Plan = flat, Apo = apochromatic (all corrections available), EF = flattened, Ach. = achromatic, Fl. = high color correction.

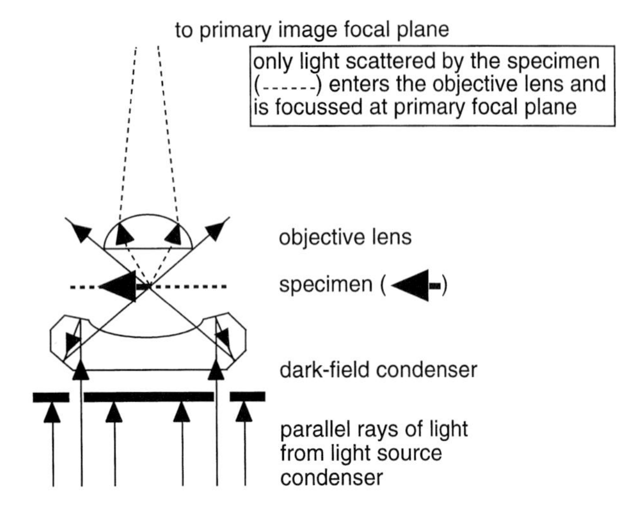

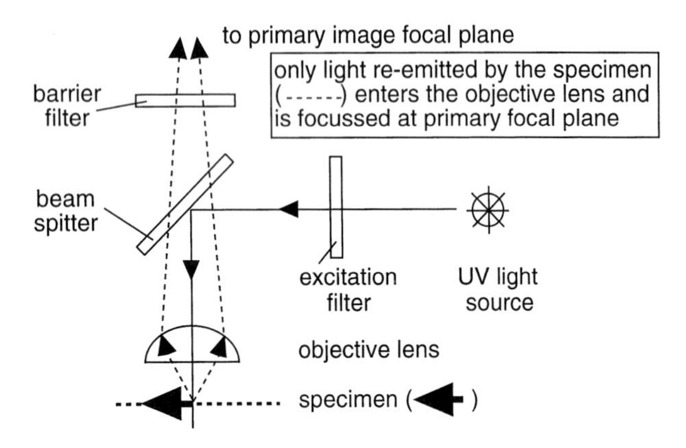

Fig. 3.3. Light paths for dark-field or epi-illumination microscopy. **Top:** The light paths for dark-field illumination of a microscopic specimen. The arrangement mechanically excludes light from the central region of the condenser so that zero-order light does not even enter the objective. The presence of a physical feature in the specimen that diffracts light or, in the case of fluorescence, emits light results in rays entering the objective lens for viewing. All the features of the specimen that cause light to enter the objective lens are then seen against a dark background. **Bottom:** A diagram illustrating epi-illumination, the most common form of illumination used for fluorescence microscopy. A beam splitter directs UV light of a wavelength determined by the excitation filter through the objective lens onto the specimen. The incident light excites fluorescence in those parts of the specimen that have been selectively labeled by fluorochromes and the emitted light is collected by the objective lens. Standard lenses are then used to focus the image on the retina of an observer, or on the surface of an electronic sensing device.

forms of specimen illumination that are used in fluorescence microscopy (Fig. 3.3). The excitation and barrier filters are important components in fluorescent microscopy in that the excitation filters are used to select the wavelength of UV light that can specifically induce fluorescence in the specimen, whereas the barrier filters select the wavelength that is emitted light by the sample and characterizes the fluorochrome.

3.1.3 The Phase-Contrast Microscope

The differential absorption of visible light as it passes through a specimen leads to changes in the amplitudes of the light waves, which can be used to provide contrast in microscopic observations. Because the contents of many living cells are virtually transparent, contrast is usually obtained by fixing the cells and treating them with various dyes. An alternative way of obtaining contrast is to exploit the fact that the different paths taken by light as it passes through different parts of a specimen result in phase changes of the light waves. These changes in phase of light are exploited in the phase-contrast microscope by observing the interference, and resulting changes in amplitude of light, between the zero-order light and the light waves after they have passed through the unstained specimen. Figure 3.4 summarizes the light paths in a phase-contrast microscope.

3.1.4 The Interference-Contrast Microscope

The interference-contrast microscope also allows unstained specimens to be studied and can be used with either transmitted

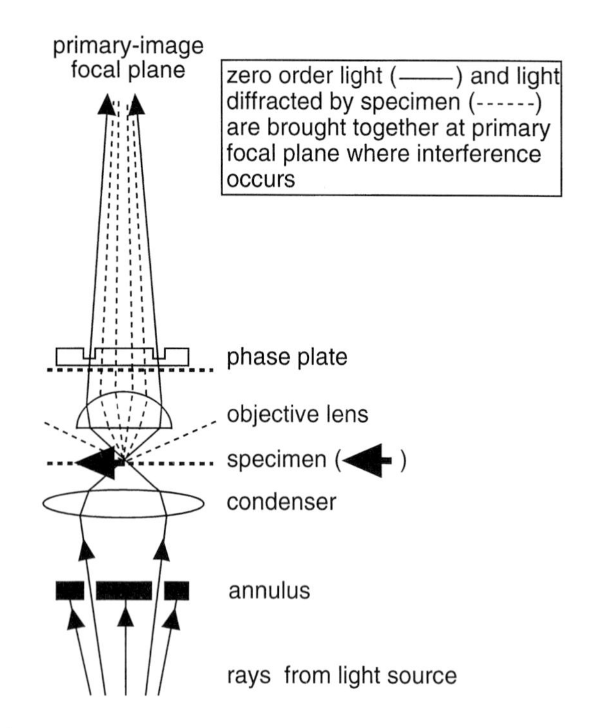

Fig. 3.4. Light paths for the phase-contrast microscope. With phase contrast, a mechanical annulus is used to limit the light reaching the condenser, permitting only a ring of light to pass through. Zero-order light, unaffected by the specimen, enters the objective and passes through an analogous ring in the glass-phase plate that is not as thick as the rest of the glass. This brings about a change of phase that differentiates this light from the light passing through the rest of the plate. At the same time, the light that enters the objective as a result of diffraction from features in the specimen (dotted lines) passes through the phase plate inside the ring through which the zero-order light passes. Upon emergence from this plate, this light, therefore, has a phase different from the zero-order light, in addition to small phase changes resulting from passage through the specimen. When these light paths mix at the primary-image focal plane, interference occurs between the different phases, allowing the edges of objects in the specimen to be more clearly distinguished.

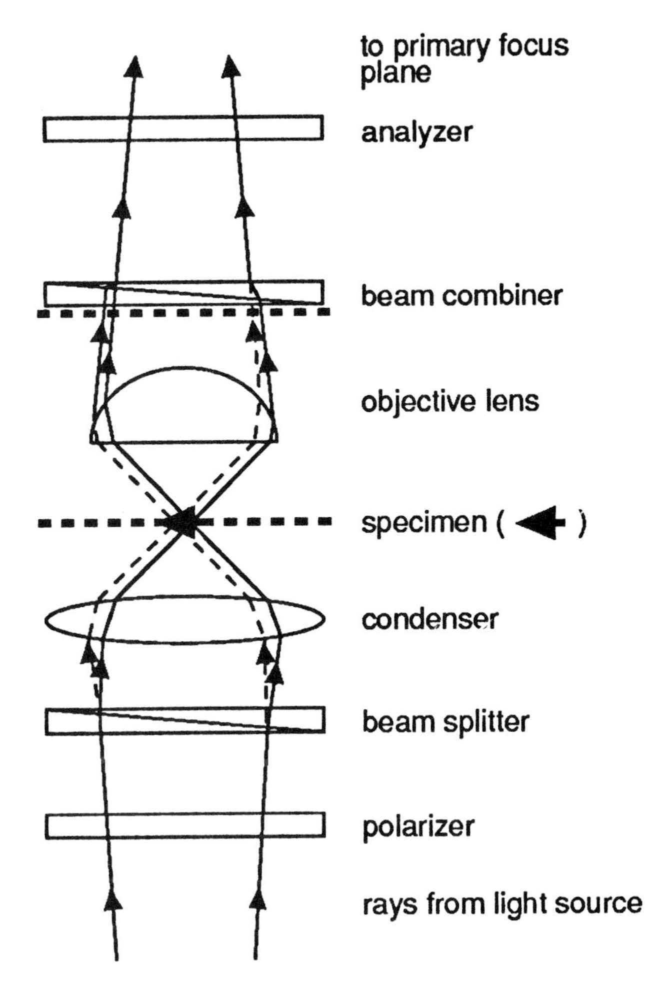

Fig. 3.5. The interference-contrast microscope. With interference contrast, rays of polarized light are split into two components by a beam-splitter or Wollaston prism. When the two components pass through a uniform part of the specimen, they are only slightly separated from one another and undergo the same phase change as shown in the figure. Consequently, when the components are later recombined, no interference occurs. If, however, one of the components passes through part of the specimen and the other does not, one component will undergo a greater phase change than its partner and will show interference when they are recombined. This interference allows the specimen to be observed. The use of polarized light, through the incorporation of polarizer and analyzer components, gives a shadow effect that helps to highlight components of the specimen.

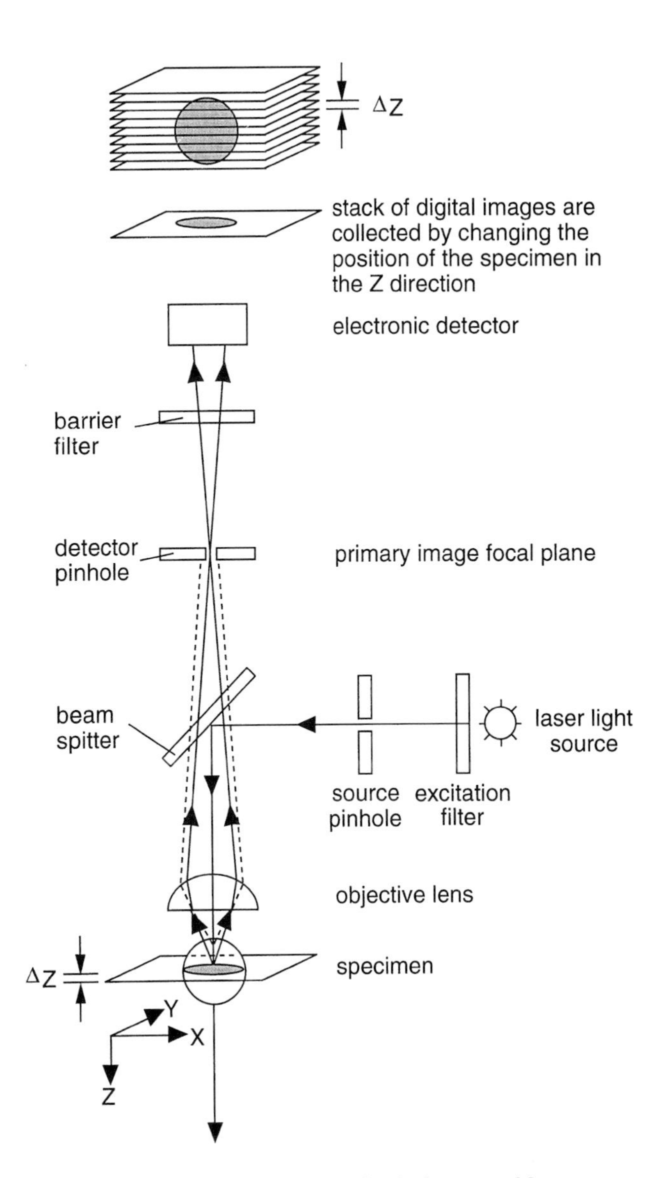

Fig. 3.6. The basis for the modern confocal microscope. Many confocal microscopes use a laser (Light Amplification by Stimulated Emission of Radiation) as the source of light (Minsky, 1957). Light beams of appropriate excitation wavelength reach the specimen by epi-illumination and react with fluorochrome-labeled portions of the specimen to give fluorescence. The rays of light indicated by dashed lines demonstrate the effect of the confocal pinhole in excluding light from those parts of the specimen that are not in the precise plane of focus. The rays of light indicated by solid lines are emitted by those parts of the specimen that are in the precise focal plane to pass through the pinhole and are recorded by the detection device. For simplicity, the additional lenses that focus the image on the detection device are not shown. The specimen is scanned by a beam of light in the *X-Y* directions of a given focal plane to record an image. Images from different focal planes in the *Z* direction are then stored in a computer to form a stack of digital images, which can be used to reconstruct a three-dimensional image of the specimen. Although the intense laser beam can cause photobleaching, this problem can be minimized by efficient data collection, as well as the use of antioxidants in the sample preparation. (Prof. N.R. Ringertz is thanked for his suggestions concerning this figure.)

or reflected light (Fig. 3.5). Even though this microscope is more complex than a phase-contrast microscope, it has little if any advantage if thinly sectioned materials are being analyzed. With thick sections, however, it produces an image that is free of the halo effect associated with the phase-contrast microscope. The interference-contrast microscope can be used to measure the relative dry mass of different parts of the specimen that are under study.

3.1.5 The Confocal Microscope

The effective resolution of a conventional light microscope is diminished by blurring of the image, due to light entering the viewing field from planes above and below the plane of focus. This effect is a particularly serious problem in fluorescence microscopy. One way of eliminating out-of-focus light is by the use of confocal pinholes near the light source as well as the light detector (Fig. 3.6). The exploitation of the confocal pinhole concept, along with the development of suitable light sources and image-detection systems, has led to the commercial availability of confocal microscopes. Because the confocal principle results in the collection of light from the plane of focus only, confocal microscopes can be used to make a series of optical serial sections through the specimen under examination (*see* Fig. 3.6). Sophisticated computer-program development has been an integral part of the establishment of the confocal microscope and is used for optical three-dimensional (3D) reconstructions of the specimen. In practice, a beam of laser light is focused on the specimen and used to scan it in an *X-Y* plane. The image is collected by a video camera for storage in a computer. Scanning in the *Z* direction builds up the 3D image in the computer database so that the computer image can be manipulated further. The confocal microscope is extensively used to examine specimens treated with fluorescent probes.

3.2. ELECTRON MICROSCOPY

Table 3.1 compares some of the properties of the electron and light microscopes. Instead of light, the electron microscope makes use of the wavelike properties of electrons. The source of electrons is an electron gun (Fig. 3.7) and the wavelengths of the electrons are determined by the accelerating voltage (up to 400,000 V). At a voltage of 100,000 V, the approximate wavelength of the electron beam is 0.004 nm. Because the resolution of a microscope is directly proportional to the wavelength of the ''light'' used to study the specimen, it is clear that the much reduced wavelength of the electron beam leads to greater resolution. Although the full theoretical increase in resolution is not achieved due to aberrations in the electromagnetic lenses, the electron microscope provides much greater detail about the structure of fixed and stained biological materials than the light microscope.

3.2.1 The Transmitting Electron Microscope

Thinly sectioned material, stained with heavy metal-containing compounds, is used to determine internal structure using the transmitting electron microscope (TEM), as illustrated in Fig. 3.7.

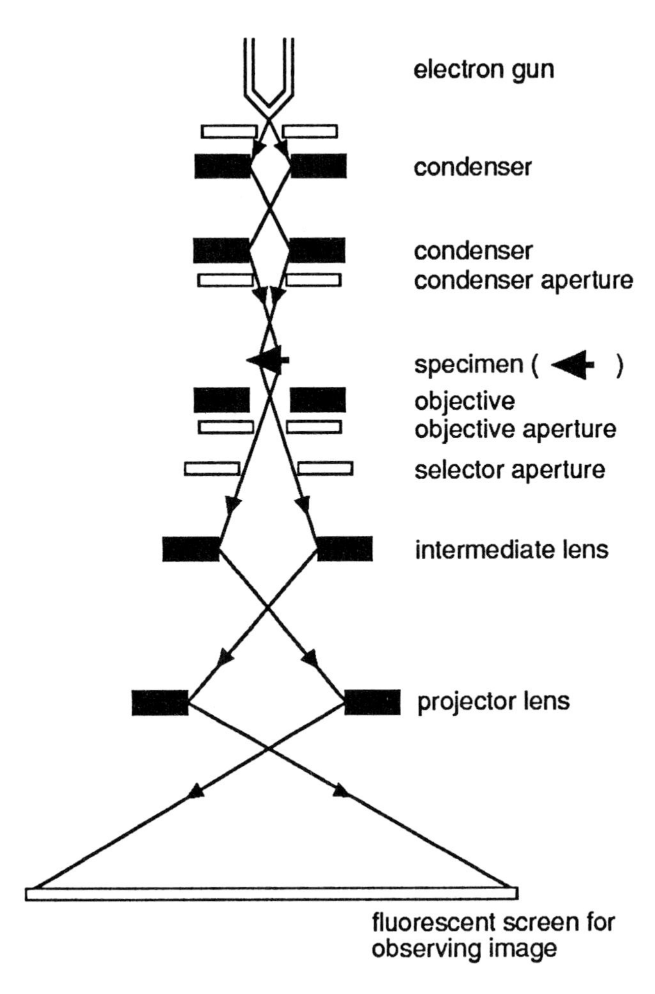

Fig. 3.7. The transmitting electron microscope (TEM). This diagram summarizes the paths of the electrons as they are focused by the various electromagnetic lenses. The interior of the microscope is under a high vacuum to prevent the loss of electrons caused by collisions with the various molecules present in air.

Table 3.1 Comparison of the Electron and Light Microscopes

Variable	Electron Microscope	Light Microscope
Illuminating beam	Electron beam: λ = approximately 0.004 nm.	Light beam: λ = 200–750 nm.
Specimen environment	Vacuum, to increase the distance electrons travel.	Atmosphere.
Specimen preparation	Fixed and thinly sectioned to allow electrons to pass through TEM. Fixed and gold-coated to produce secondary electrons (SEM).	Fixed and sectioned or squashed. Mounted with minimal distortion (3D studies).
Resolution	0.1 nm	Visible light: 200 nm; UV light: 100 nm.
Magnification	$10–10^6$ × (continuous), large molecules can be seen.	10–2000 × (by changing objectives).
Focusing	Electrical.	Mechanical.
Types of information	Transmitted image (TEM). Secondary electron image (SEM). Secondary X-rays characteristic of elements in specimen. Backscatter electrons reflecting atomic mass at surface.	Transmitted and reflected images. Absorption properties of specimen. Dry mass (interference contrast). Fluorescence images.
Contrast	Scattering absorption, diffraction, phase.	Absorption, reflection, diffraction, phase.
Monitor	Projection screen, photographic film, solid-state detection device.	Human eye, photographic film, solid-state detection device.

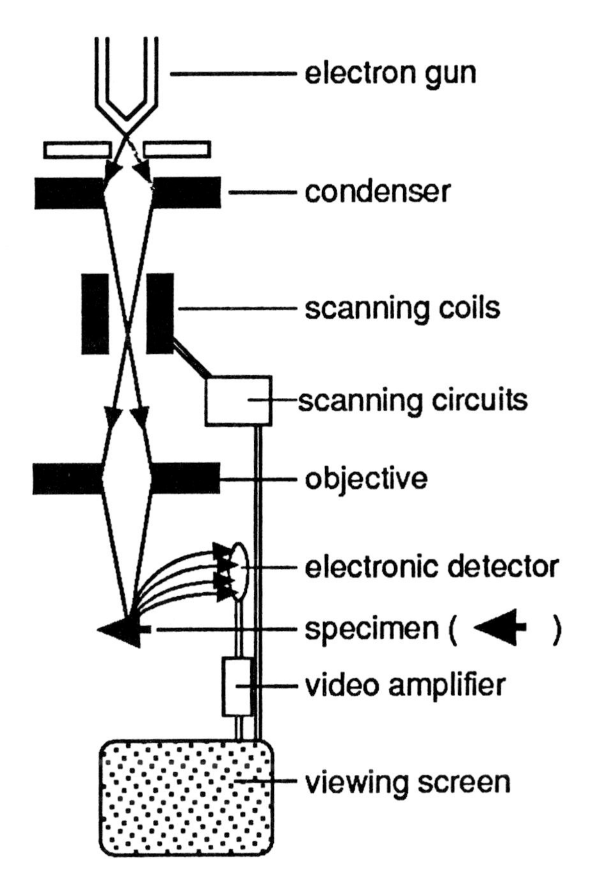

Fig. 3.8. The scanning electron microscope (SEM). This diagram summarizes the paths of the electrons as they are focused by various electromagnetic lenses. The scanning coils move the focused beam across the specimen, which has been coated with a metal such as gold, in order to release secondary electrons that are recorded by the detector device. The three-dimensional surface topography of the specimen is displayed on a video screen.

Three-dimensional information can be obtained from the specimen sections used for the TEM by tilting the sample (this is also true for light microscopy). Examination of the specimen at different angles and perspectives allows the 3D image to be developed. Information about the atomic composition of the specimen can also be obtained by analyzing secondary emissions resulting from the electron bombardment. Different elements release X-rays of characteristic wavelengths, which can be used as fingerprints to identify the amounts of different elements in the samples.

3.2.2 The Scanning Electron Microscope

The scanning electron microscope (SEM) (Fig. 3.8) physically scans the surface of a specimen with a finely focused electron beam, so that the primary resolution is determined by the actual diameter of the beam. The object is observed by the detection of secondary electrons released from the surface of the specimen, which has been coated with a thin layer of a metal such as gold. The resolution of the microscope is also limited by the energy of electrons from the electron gun, the number of scanning lines, and the detection system. A major advantage of the scanning electron microscope is that the depth-of-field is much greater than that of either the transmitting electron microscope or the light microscope. Consequently, the final images are similar to those that would be seen by the unaided eye, and the three-dimensional image greatly simplifies their interpretation.

3.3. SCANNING PROBE MICROSCOPY

The scanning tunneling microscope (STM, Fig. 3.9) is an example of a new class of microscope based on analyzing the surface of a specimen by scanning it with an imaging tip. The analysis of the surface is at ultrahigh, atomic levels of resolution. The principle on which the STM operates is based on the observation that when a metal tip is brought very close to a conducting surface, electrons tunnel through the gap between the tip and the surface when there is an appropriate voltage difference between them. The movement of electrons produces a tunneling current and is extremely sensitive to gap distance—it changes by a factor of 10 when the distance changes by 0.1 nm. This provides the basis for the extremely high vertical resolution of the scanning tunneling microscope. Other factors affecting resolution are the number of scanning lines and possibly the geometry of the metal-tip probe.

The atomic force microscope (AFM) scans the surface of specimens using a metal tip (curvature radius 10 nm) mounted on a cantilever, which is deflected according to the surface topography of the sample. The movements of the cantilever are recorded by a laser beam reflected from the lever into a photodetector. To build an image of the specimen, the intensity of the signal recorded by the photodetector is computed to indicate the Z-axis features of the sample, which are then combined with the X and Y axis coordinates, which define the position of the metal tip as it scans the specimen.

The heart of the scanning probe microscope is the nature of the probe and the accurate control over its scanning of the specimen surface, so that signals from the probe can be related to its position at any point in time. The probe can be as described above or a

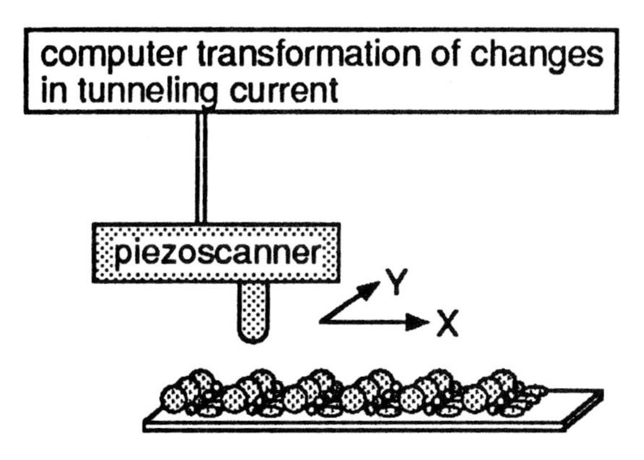

Fig. 3.9. An extremely simplified representation of scanning tunneling microscopy (STM). The basic recording mechanism for the STM is a piezoscanner device to which a fine metal probe has been attached. The molecular specimen is adsorbed on a graphite surface and scanned in the *X-Y* direction. Changes in the tunneling current between the probe and the specimen define the surface topography. Computer interpretation of the data, analogous to that used in analyzing data from X-ray crystallography, is required to build the 3D image of the specimen (Driscoll et al., 1990).

small ferromagnet so that an image of the magnetic field can be collected. Alternatively, the probe can be coated with polymers or macromolecules that provide molecular recognition of specific features of the surface and can, therefore, be utilized to recognize parts of chromosomes spread on a surface for analysis. The analysis of soft material, such as biological specimens, has been aided by the introduction of a tapping mode of probe movement, to replace a continuous, dragging motion across the surface.

3.4 FLOW CYTOPHOTOMETRY

Although flow cytophotometry was developed primarily for use in medical analytical laboratories to characterize different components in cell populations, it is also a valuable tool in chromosome and cytogenetic research. The principle on which the flow cytophotometer works is that biological particles in a liquid-flow stream scatter the light, through which they pass in single file, in a way that relates to their structure. If, in addition, the biological particles have been treated with a fluorochrome, they will radiate light of characteristic wavelength. The fluorescent and light-scattering properties of the particles can be measured, as shown in Fig. 3.10. The computing software associated with flow cytophotometers produces distribution patterns of the particles being analyzed, and, based on an electrical charge, can activate the sorting mechanism to collect specific subclasses of particles. Flow cytophotometry has been used to sort individual human metaphase chromosomes in order to provide the numbers needed for DNA isolations and to allow DNA sequences from specific chromosomes to be cloned. Flow cytophotometry has also been used to detect mutant cell populations, which are arrested in certain stages of the cell cycle (*see* Chapter 4). Other applications of the instrument include investigations of the variable polyploidy in plant cells, and measurement of the total amount of DNA in cells of different organisms. For these purposes, DNA-binding fluorochromes have been used, commonly including propidium iodide and 4′,6-diamidino-2-phenylindole (DAPI). Advances in technology are reducing the size of the laser light source and expanding the number of different fluorochromes that can be attached to the biological particles.

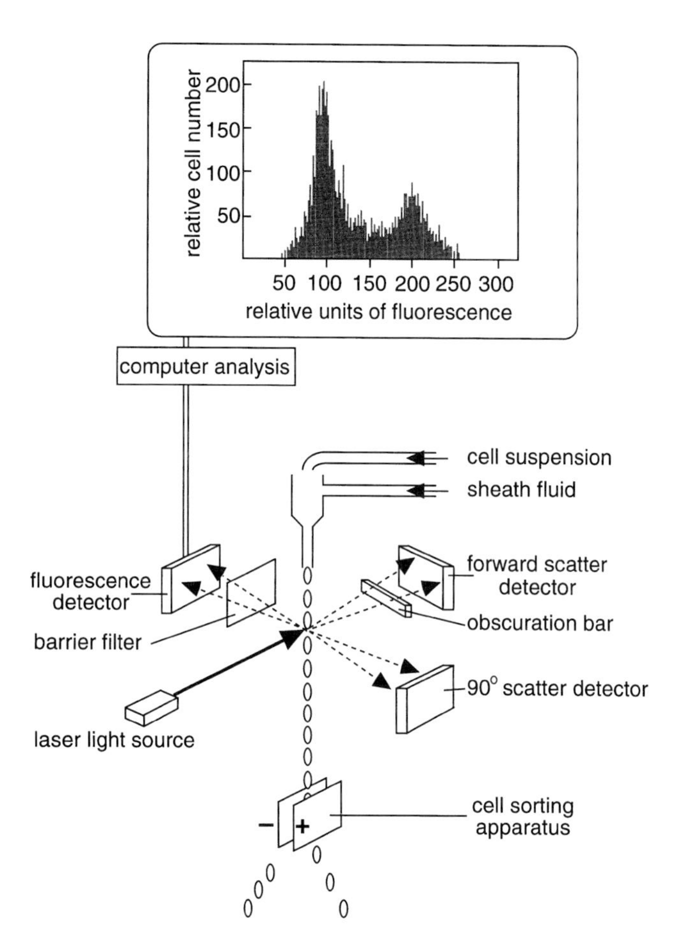

Fig. 3.10. A simplified diagram of a flow cytophotometer. The computer output from the machine (top) shows the distribution of particles in the liquid-flow stream according to the amount of fluorescent dye bound to the particles. In the example shown, the histogram on the computer screen summarizes the number of cells that fall into different categories as determined by the amount of fluorescence they exhibit (relative to a standard). If a DNA-specific fluorochrome is used to stain a population of cells and these are then passaged through the flow cytophotometer, the output provides a distribution of cells in the G1 through G2 phases of the cell cycle (as discussed in Chapter 4). Cells in G2 have undergone a complete cycle of DNA replication and their resulting DNA content—which is proportional to the fluorescence intensity resulting from the presence of the DNA-binding dye—is double that of cells in the G1 phase (Willman and Stewart, 1989.)

3.5 COMPUTER-ASSISTED IMAGE ANALYSIS

The electronic processing of an image can remove distracting and unwanted information, and assist in the interpretation, measurement, and analysis of the information present in a video image. The computer processes used to modify the image never add new information, and the original information is stored as an electronic image if alternate analytical procedures need to be applied to the image. The various steps in image analysis are summarized in Fig. 3.11.

Processing of electronic images includes point operations, averaging to reduce noise, smoothing, removing periodic noise, adding artificial colors, correcting shading, adjusting the background, aligning and superimposing images, sharpening the image, and computer focusing in collecting the original data. These procedures are complex, so only some are discussed.

Point operations refer to the processes that change the value of pixels, the individual components of a recorded image, by a mathematical transfer function that uses the original value of the respective pixel. The overall effect on the image is analogous to printing an image, captured on negative film, on photographic paper of varying hardness, in order to increase or decrease the amount of contrast in the image. This process can reveal details of darkly stained areas that may have been masked in the initial version of the image. In addition, false colors may be introduced to emphasize gradients in brightness.

Averaging to reduce noise created during the collection of an image is a common application of computer-assisted image analysis. In this procedure, random noise is removed by making many image acquisitions and averaging them, so that *only* the pixels present in every image are retained in the final image. Because the pixels resulting from random noise are not in the same position in every image, they are deleted from the final image.

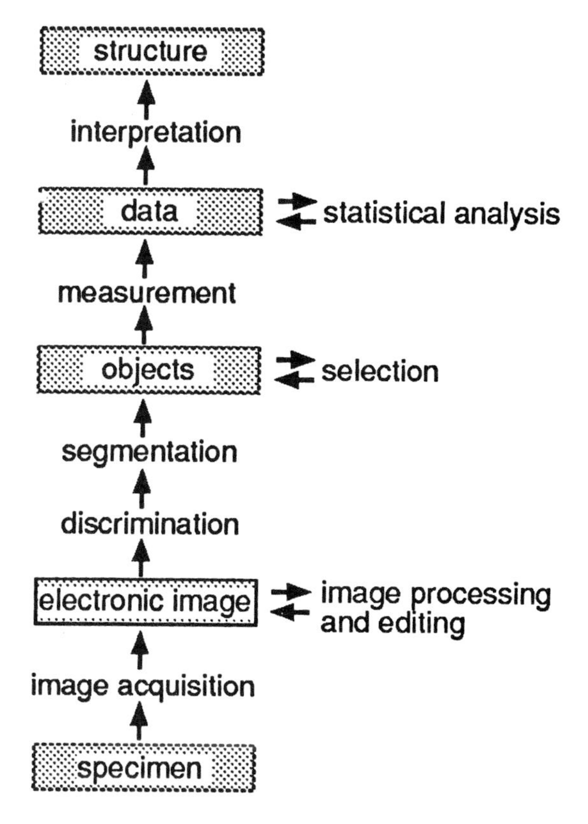

Fig. 3.11. A summary of the stages involved in the computer analysis of an electronic image to enhance structural detail and remove distracting information. Image enhancement is used extensively to analyze the very faint fluorescent signals emitted by the bound complexes of fluorochrome-labeled genes in chromosome preparations (Viegas-Pequignot et al., 1989) (*see also* Chapter 19).

Smoothing involves groups of pixels. Each pixel in a high-resolution electronic image is usually surrounded by eight neighbors. Because these groups of pixels are usually small relative to the size of the objects recorded in the image, each pixel and its eight neighbors often originate from the same object. The smoothing process replaces the value of the central pixel with the average of its neighbor simultaneously for all pixels in the image. The new image is stored in addition to the original image. Median filtering can also be used to achieve smoothing and tends to avoid the loss of sharp edges. In this process, the nine pixels of a cluster are ranked in order and the median value (fifth brightest pixel) is used to replace the pixel that was originally the central one in the cluster.

In the two-dimensional analysis of specimen preparations, labeled with multiple fluorochromes, a very useful procedure is to color code the images, so that they can be superimposed on one another. This process not only allows an overview of the distribution of fluorochromes but also shows the exact location of the different fluorochromes relative to each other. If necessary, a difference analysis can be carried out to determine whether or not two fluorochromes are bound to exactly the same region of a chromosome. If two colors do coincide exactly, a third color (or white) can be encoded to aid in efficiently locating all of the overlap regions. Chromosome image-analyzing systems have been developed to automate the scanning of metaphase chromosome spreads and record the karyotype.

An alternative to the confocal microscope for "sectioning" the specimen is based on using computer software to remove out-of-focus information at various planes of focus, and thus develop a three-dimensional analysis using a standard light microscope. The development of this alternative uses a highly sensitive solid-state camera such as a charge-couple device, or CCD, camera to collect information. The advantage of this alternative system is that it minimizes the photobleaching of fluorochromes associated with high-intensity laser light sources, which are used for confocal microscopy.

BIBLIOGRAPHY

Section 3.1

Agard, D.A., Hiraoka, Y., Shaw, P., Sedat, J. 1989. Fluorescence microscopy in three dimensions. In: Methods in Cell Biology Vol. 30 (Taylor, D.L., Wang, Y.L., eds.). Academic Press, San Diego, CA. pp. 353–377.

Bradbury, S. 1984. An Introduction to the Optical Microscope. Oxford University Press, Oxford.

Burk, G. 1984. Instrumentation and Techniques for Fluorescence Microscopy. H. Clark Printing, Sydney, NSW.

Inoue S. 1989. Foundations of confocal scanned imaging in light microscopy. In: Handbook of Biological Confocal Microscopy (Pawley, J.B., ed.). Plenum Press, New York. pp. 1–13.

Minsky. 1957. U.S. patent #3013467.

Pawley, J.B. (ed.) 1990. Handbook of Biological Confocal Microscopy. Plenum Press, New York.

Spencer, M. 1989. Fundamentals of Light Microscopy. Cambridge University Press, Cambridge.

Taylor, D.L., Wang, Y-L. (eds.). 1989. Fluorescent microscopy of living cells in culture. Part B—Quantitative fluorescence microscopy. In: Methods in Cell Biology Vol. 30. Academic Press, San Diego, CA.

Section 3.2

Grimstone, A.V. 1968. The Electron Microscope in Biology. Inst. Biol. Studies in Biol. Ser. No. 9. Edward Arnold (Publishers) Ltd., London.

Section 3.3

Arscott, P.G., Bloomfield, V.A.. 1990. Scanning tunnelling microscopy in biotechnology. Trends Biotech. 8: 151–156.

Arscott, P.G., Lee, G., Bloomfield, V.A., Evans, D.F. 1989. Scanning tunnelling microscopy of Z-DNA. Nature 339: 484–486.

Butt, H.J., Wolff, E.K., Gould, S.A.C., Dixon-Northern, B., Peterson, C.M., Hansma, P.K. 1991. Imaging cells with the atomic force microscope. J. Struct. Biol. 105: 54–61.

Driscoll, R.J., Youngquist, M.G., Baldeschwieier, J.D. 1990. Atomic-scale imaging of DNA using scanning tunnelling microscopy. Nature 346: 294–296.

Feng, L., Andrade, J.D., Hu, C.Z. 1989. Scanning tunneling microscopy of proteins on graphite surfaces. Scan. Microsc. 3: 399–410.

MUSIO, A., MARIANI, T., FREDIANI, C., SBRANA, I., ASCOLI, C. 1994. Longitudinal patterns similar to G-banding in untreated human chromosomes: evidence from atomic force microscopy. Chromosoma 103: 225–229.

OVERNEY, R.M. 1995. Nanobiological studies on polymers. Trends Polym. Sci. 3: 359–364.

Section 3.4

GALBRAITH, D.W. 1989. Analysis of higher plants by flow cytometry and cell sorting. In: International Review of Cytology Vol. 116. (Bourne, G.H., ed.) Academic Press, San Diego, CA. pp. 165–228.

WILLMAN, C.L., STEWART, C.C. 1989. General principles of multiparameter flow cytometric analysis: applications of flow cytometry in the diagnostic pathology laboratory. Semin. Diag. Pathol. 6: 3–12.

Section 3.5

FUKUI, K. 1986. Standardization of karyotyping plant chromosomes by a newly developed chromosome image analyzing system (CHIAS). Theoret. Appl. Genet. 72: 27–32.

RUSS, J.C. 1990. Computer-Assisted Microscopy. Plenum Press, New York.

VIEGAS-PEQUIGNOT, E., DUTRILLAUZ, B., MAGDELENAT, H., COPPEY-MOISAN, M. 1989. Mapping of single-copy DNA sequences on human chromosomes by *in situ* hybridization with biotinylated probes: Enhancement of detection sensitivity by intensified-fluorescence digital-imaging microscopy. Proc. Natl. Acad. Sci. USA 86: 582–586.

II
Mitosis, Meiosis, and Chromosomal Karyotypes

4

Chromosomes in the Mitotic Cell Cycle

- The cell cycle is comprised of a chromosome-replication cycle occurring in concert with a cytoplasm-replication cycle.
- The centrosomes determine the polarity of cell division, and centromeres (either well-defined chromosome regions or diffuse) play an active role in the movement of chromosomes to the poles.
- The cell cycle is regulated by a series of protein phosphorylation and dephosphorylation reactions controlled by cell-division-cycle (*cdc*) genes.
- Mutations in the *cdc* genes lead to uncontrolled cell growth and are the primary cause of many types of cancer.

The genetic material of organisms was originally studied by analyzing and comparing the structure and behavior of their chromosomes. Classical cytogenetic studies relied on these observations to relate genetical changes to physical changes in the genome and to establish the relationship between changes in chromosome structure and stages in the cell cycle. In this chapter, the mitotic cell cycle is examined to become familiar with the processes that are essential for cell division and growth of an organism. By observing chromatin throughout the cell cycle, "windows" are provided through which the structural organization of the genome can be investigated. In addition, the differentiated states of some cells in various organisms present unusual chromosome numbers and structures, which contribute to the detailed analysis of the genome.

4.1 THE CELL CYCLE CONSISTS OF TWO CYCLES OCCURRING IN CONCERT

The processes involved in the cell cycle (Fig. 4.1) were deduced using a combination of structural, physiological, biochemical, and genetical studies. The creation of two cells from one preexisting cell involves a chromosome cycle and a cytoplasm cycle. The different stages of the cell cycle occur in an interdependent sequence, with specific start and termination signals to determine the progression into the various stages of the nuclear cycle. The familiar cycle of cell division and chromosome replication alternates between a DNA synthesis or S phase and a mitotic M phase, separated by gaps designated G1 and G2 (Fig. 4.2). During the S phase, DNA synthesis provides the basis for new copies of each chromosome. During the M phase, the process of mitosis ensures that one copy of each of the duplicated chromosomes is faithfully distributed to two daughter cells. Cells that have exited from the mitotic cycle to initiate differentiation are in the so-called G0 phase because the nucleus is no longer involved in division.

4.1.1 The Chromosomes Replicate and Divide During Mitosis

Most multicellular organisms undergo open mitosis in which the nuclear membrane breakdown occurs. In contrast, yeast and other unicellular organisms have closed mitosis in the sense that the nuclear membrane persists. The typical stages of open mitosis (as shown in Figs. 4.3 and 4.4) are as follows:

Interphase. At the start of interphase, during G1, the DNA content of the nucleus is defined as 2C. This is standard for the somatic cells in diploid organisms with a diploid or $2n = 2x$ chromosome number. During the S phase, the DNA content of the interphase nucleus increases as DNA synthesis occurs, and by G2 the DNA content has doubled to a 4C content, although nuclear division has not yet occurred. In flow cytophotometry experiments carried out to determine the DNA content of nuclei, a continuum of DNA contents is observed between the peaks corresponding to the 2C and 4C contents (Fig. 4.5).

Although the M phase is the most readily visible part of the chromosome replication cycle, it takes up only approximately 10% of the time required to complete the cycle. Interphase, incorporating the G1, S, and G2 phases, occupies up to 90% of the cell cycle. Although interphase was traditionally considered a resting nuclear phase, at the biochemical and molecular levels it is actually the most active phase of the cell cycle. A typical human cell requires 24 h to divide and a yeast cell, 90 min.

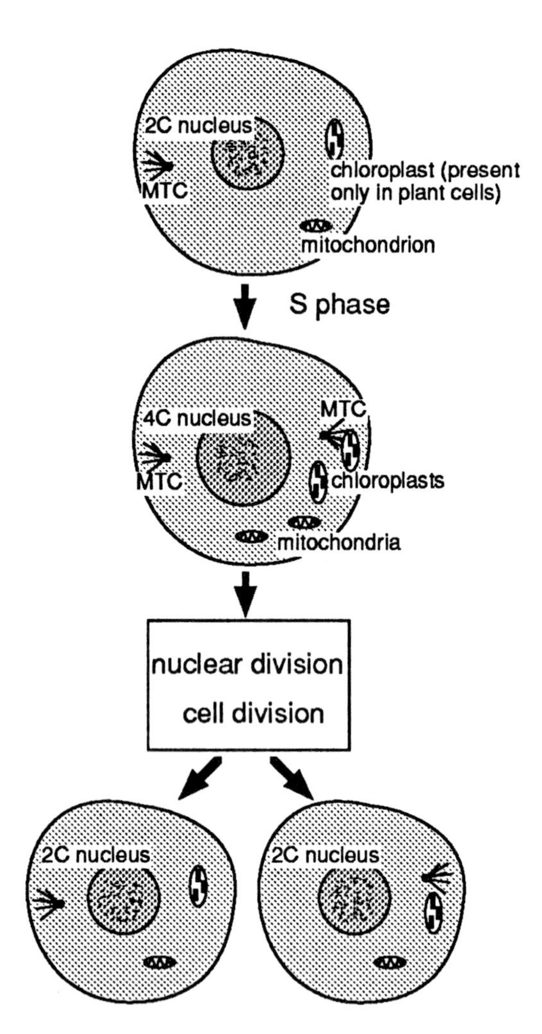

Fig. 4.1. Nuclear (mitotic) and cytoplasmic divisions. The mitotic cycle starts in G1 (top) when the nuclei have a 2C content of DNA. The cycle progresses through the S phase to give a nucleus with all the DNA replicated to the 4C level and a cell with duplicated cytoplasmic organelles and microtubule-organizing centers (MTC) (middle). Division of the nucleus and cytoplasm creates new G1 cells (bottom).

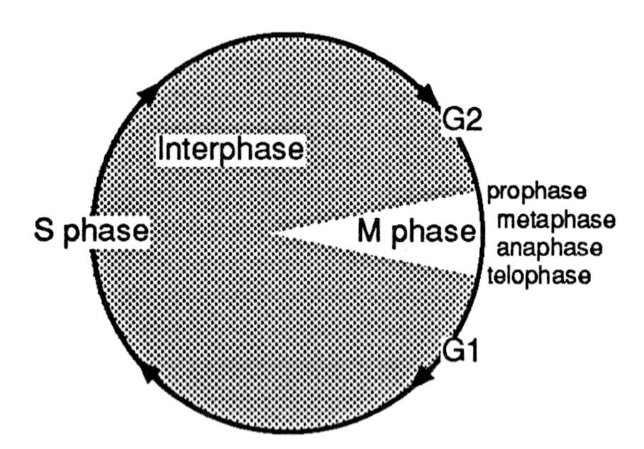

Fig. 4.2. Successive stages of growth in dividing cells. The S phase occupies 90% of the time required for cell division. Interphase (dark area of circle) includes G1, S, and G2.

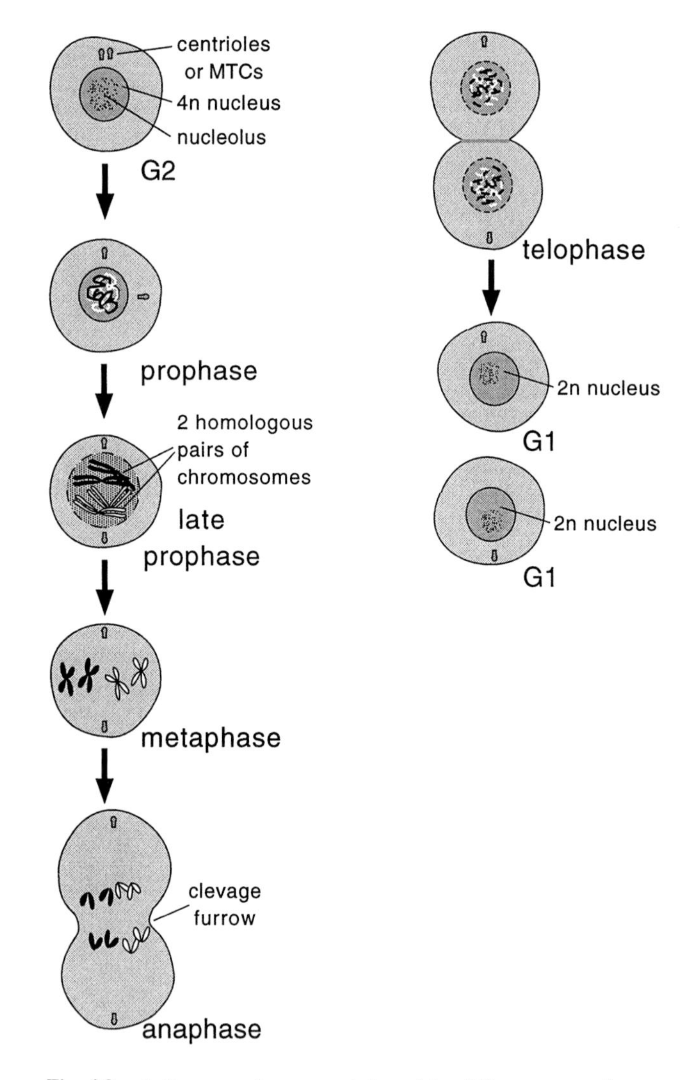

Fig. 4.3. A diagrammatic representation of the different stages of mitosis. A normal diploid nucleus has two copies or homologs of each chromosome, which usually do not pair during mitosis although they tend to lie near each other. The microtubule-organizing centers (MTC) are indicated. The nucleolus, the site of ribosomal RNA synthesis, disappears during the actual division.

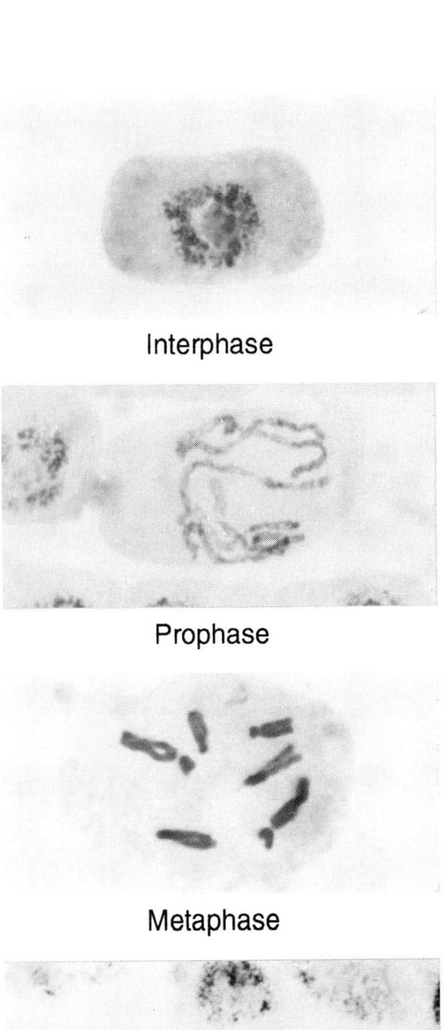

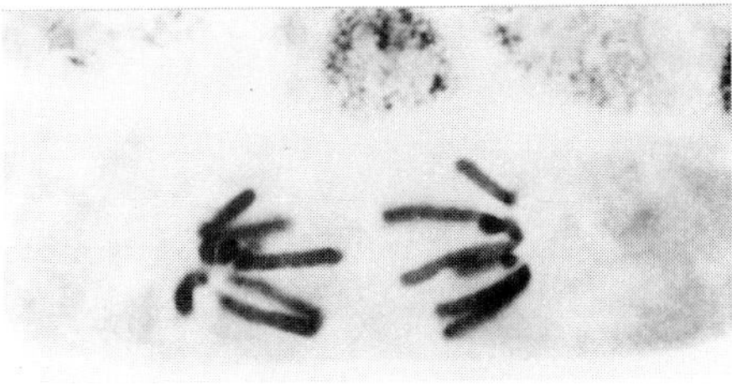

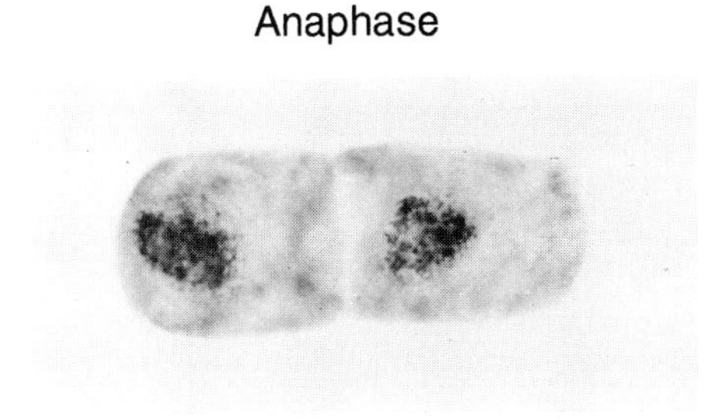

Fig. 4.4. Mitosis in root-tip cells of the plant *Crepis capillaris*. A polar view of metaphase is shown. (Photographs courtesy of Dr. B. Friebe.)

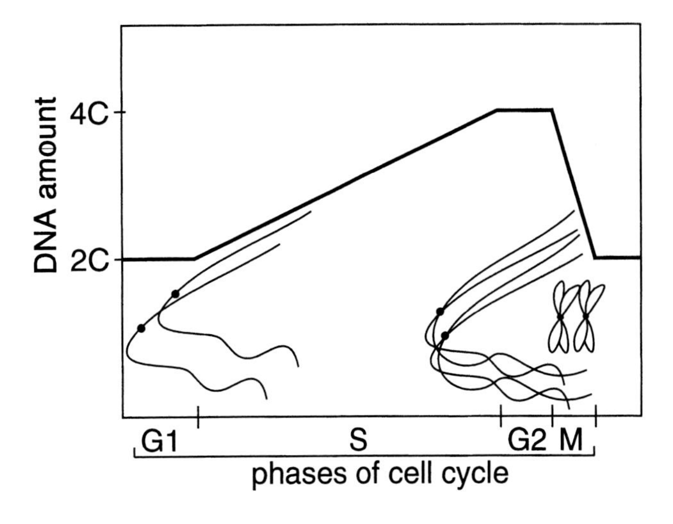

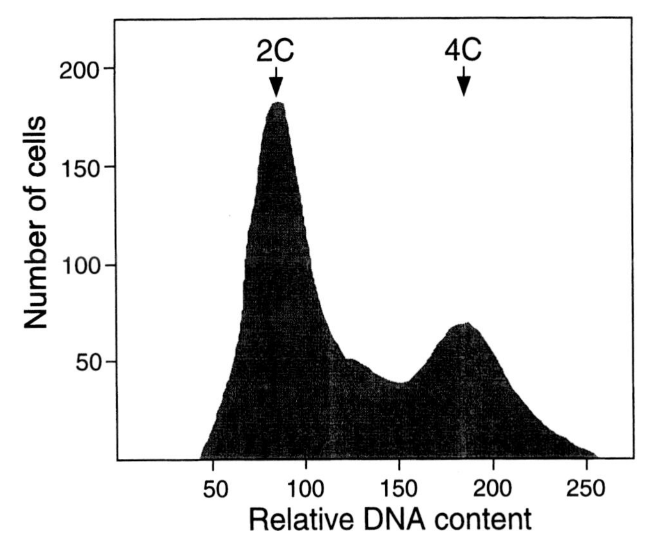

Fig. 4.5. The continuum of DNA contents in dividing cells. **Top:** The doubling of nuclear DNA content during the S phase as homologous chromosomes replicate. **Bottom:** The relative proportions of cells in various phases of the cell cycle in a population of dividing cells, as measured by flow cytophotometry. A fluorescent DNA-binding dye is generally used to provide an estimate of the relative DNA contents of the nuclei (*see also* Fig. 3.10).

Classical "landmarks" used in chromosome studies such as nucleoli and heterochromatin are observable in interphase nuclei. Furthermore, special chromosomes such as the inactive X chromosome of humans can also be observed in interphase nuclei as darkly staining, heteropycnotic bodies. With the advent of electron microscopy and molecular analyses, it has become possible to study the organization and behavior of the chromosome material within the interphase nucleus. For example, antibodies against proteins present in specific parts of the chromosome have been produced and are useful for studying the behavior of chromosomes at mitosis in detail (*see also* Chapter 20); antibodies located in the kinetochore, for example, have been crucial in the characterization of the centromere region. In addition, the availability of specific, cloned DNA sequences has made it possible to follow well-defined chromosome segments using *in situ* hybridization (a technique discussed in

Chapter 19). The synthetic activity during interphase includes RNA and protein syntheses, as well as DNA synthesis, and it is this intense biochemical activity that makes interphase of central importance in the nuclear division cycle.

Prophase. The chromosomes in the nucleus begin to contract in prophase and form long, threadlike structures (*see* Figs. 4.3 and 4.4). This represents the first sign, visible through the light microscope, that mitosis is in progress. The replicated strands of each chromosome, called sister chromatids, become visible, whereas the sites for high rates of ribosomal RNA synthesis, called nucleoli (*see* Chapter 20), begin to disappear. The sister chromatids are held together at the centromere, and at this stage, different structural features of chromosomes (discussed below) become evident. The phosphorylation of chromosomal proteins is closely correlated with the contraction process, and the enzyme topoisomerase II, also a chromosomal protein, is required to remove torsional stresses induced by the process (*see* Chapter 17). Inhibitors of topoisomerase II block chromosome condensation. The breakdown of the nuclear envelope signals the end of prophase.

Metaphase. At the beginning of the stage, also called prometaphase, contraction continues; each chromosome gains a distinct overall morphology, and individual chromosomes can often be identified. Further differentiation of the structure of chromosomes into dark-staining regions, called heterochromatin, and light-staining regions, called euchromatin (Fig. 4.6), can also be visible at this stage. The nucleoli may or may not be present. The kinetochore, located in the centromeric region of each chromosome, establishes contact with the spindle microtubules (Fig. 4.7).

At metaphase, the chromosomes reach their maximum degree of contraction (Fig. 4.8), and they align in a region referred to as the equatorial plate (*see* Fig. 4.3), which is midway between the spindle poles. Although most procedures stain chromosomes uniformly, specialized procedures have been developed to reveal differential staining along the length of chromosome arms. The analysis of mitotic chromosomes using differential staining techniques has become a powerful tool in genomic analyses of eukaryotes (*see* Chapter 6). Studies of this type require the spindle microtubules to be broken down, using a chemical such as colchicine (Fig. 4.9), followed by "spreading" of the chromosomes on glass slides.

A number of organisms such as mites and some plants (e.g. *Luzula*), do not have localized centromeres and are considered to have diffuse centromeres, or holokinetic chromosomes. Stud-

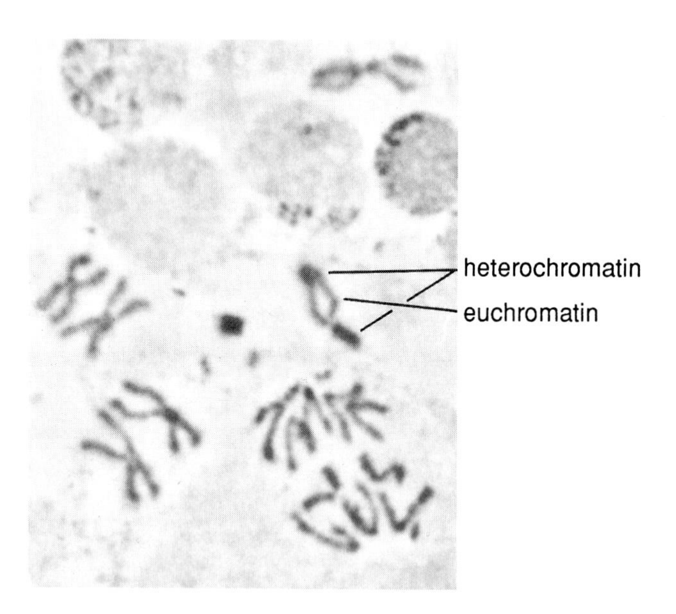

Fig. 4.6. *Drosophila* mitotic chromosomes from brain-cell nuclei showing euchromatin and heterochromatin. Large blocks of heterochromatin are present in *Drosophila* mitotic chromosomes, and homologous chromosome pairs often show the somatic side-by-side pairing exhibited by the chromosomes on the left-hand side of the photograph. (Photograph courtesy of Dr. A.J. Hilliker.)

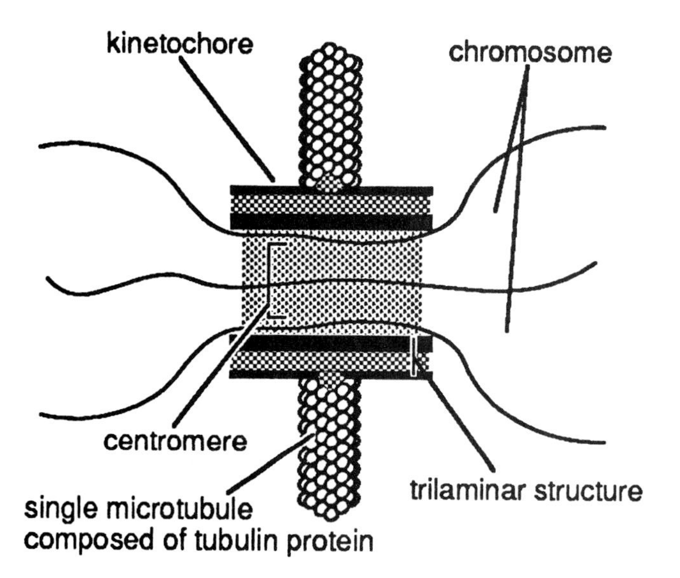

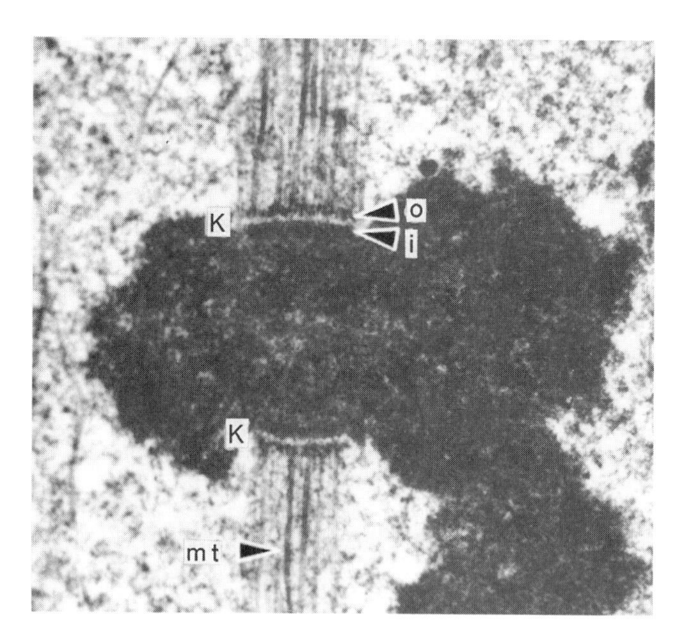

Fig. 4.7. The attachment of microtubules to the centromeric region of a chromosome. **Top:** A diagrammatic representation of the attachment of a single microtubule to the centromeric region of a chromosome via the kinetochore. **Bottom:** An electron micrograph of the attachment of microtubules to the centromere region of a metaphase chromosome in the alga *Oedogonium* (K = kinetochore, o = outer, i = inner, mt = microtubules). [Photograph from Schibler and Pickett-Heaps (1987).]

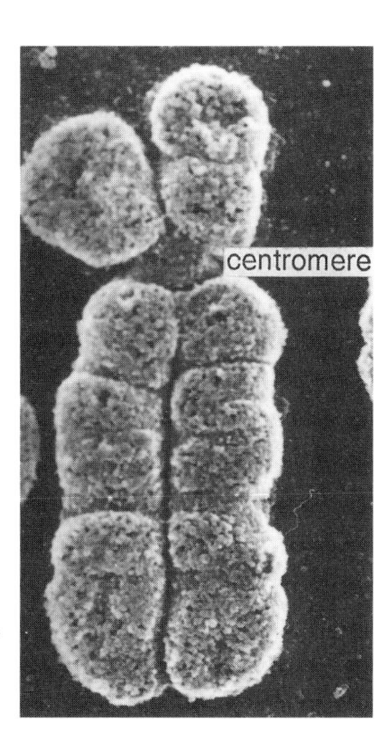

Fig. 4.8. Scanning electron micrograph of a human metaphase chromosome. [From Sumner (1991).]

ies at the electron-microscopic level have shown that a kinetochore-type structure is present in organisms with diffuse centromeres (Fig. 4.10). In Fig. 4.11, the mitotic process in organisms with holokinetic chromosomes is compared to the process with the more usual monocentromeric chromosomes.

Anaphase. The kinetochores separate and the sister chromatids move in opposite directions from the equatorial plate to the poles of the mitotic spindle; topoisomerase II is required to catalyze the disentanglement of chromatid arms. The movement of chromatids to opposite poles is brought about mainly by the shortening of the tubules attached to the kinetochore. The time for anaphase to be completed can vary from 2.5 min in chick fibroblasts to 25 min in a plant such as *Tradescantia* and suggests that the chromosomes move at velocities that vary from 0.7 to 2.6 μm/min. These rates are equivalent to 1 km in 730–2700 years. In preparation for cell division, the cell plate or phragmoplast forms at the equatorial plate in plant cells, whereas in animal cells, myosin and actin filaments begin to form a contractile ring inside the cell wall, in alignment with the equatorial plate.

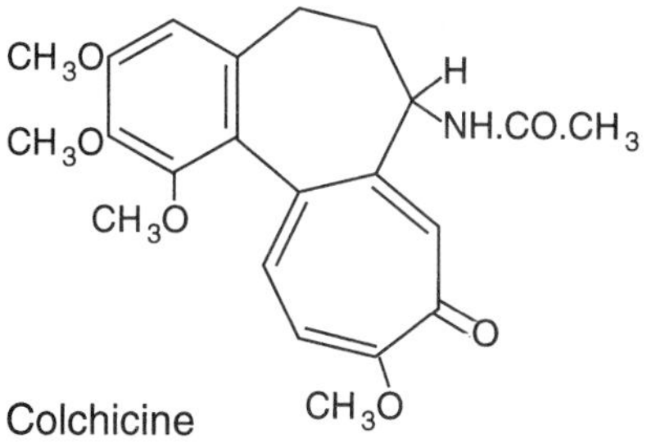

Telophase. After reaching the poles, the chromosomes decondense, the microtubules disperse, and the nuclear membrane reappears. The interphase of the next cell cycle now begins in these nuclei.

Cytokinesis. During mitosis, the doubling of all the cytoplasmic organelles also occurs and they are distributed to daughter cells in a process known as cytokinesis (*see* Fig. 4.1). The synthesis of protein is continuous throughout the cell cycle, except for a gap during the M phase of the chromosome-replicating cycle, when RNA synthesis is reduced. Cytokinesis is completed in animal cells when the central ring of myosin and actin fibers contracts to separate the daughter cells, and in plant cells when a cell plate is laid down in preparation for the separation of the daughter cells by a rigid cell wall (Fig. 4.12).

Colchicine

Fig. 4.9. The chemical structure of colchicine. This compound is extracted from seeds and corms of the autumn crocus (*Colchicum autumnale*). When applied in a paste or solution to seeds, or seedling meristems, it disorganizes the cell spindles by binding to the tubulin protein of the microtubules. Colchicine does not prevent chromosome or DNA replication but, because of its effect on the spindle, cell division is suppressed. As long as colchicine remains in contact with the cell, the chromosomes continue to replicate without cell division, resulting in increasing levels of polyploidy. When colchicine is removed, the spindle function is generally restored and cell division proceeds normally.

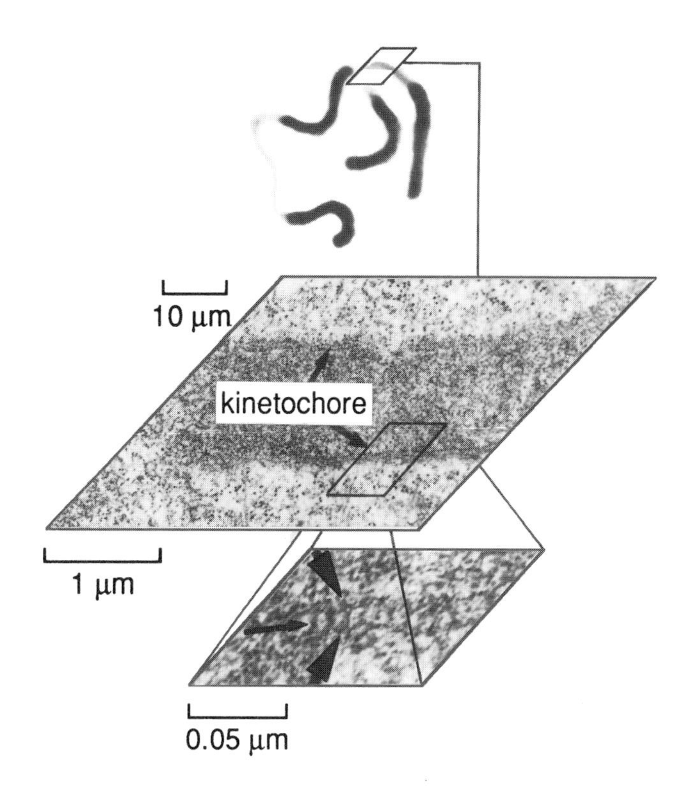

Fig. 4.10. Increasing magnifications of the kinetochore of a holokinetic chromosome. **Top:** A light micrograph of the single pair of chromosomes present in *Parascaris univalens*, showing the large blocks of heterochromatin that characterize the chromosomes of this species. **Middle:** An electron micrograph of a section of the mitotic chromosome to illustrate the kinetochore running along the length of the chromosome. **Bottom:** A higher magnification of the kinetochore, with the arrows indicating its boundaries. [Based on Goday et al. (1985).]

4.1.2 The Assembly of the Spindle Apparatus

The formation of the spindle apparatus in plant and animal cells is initiated by the microtubule organizing center, MTC (also called

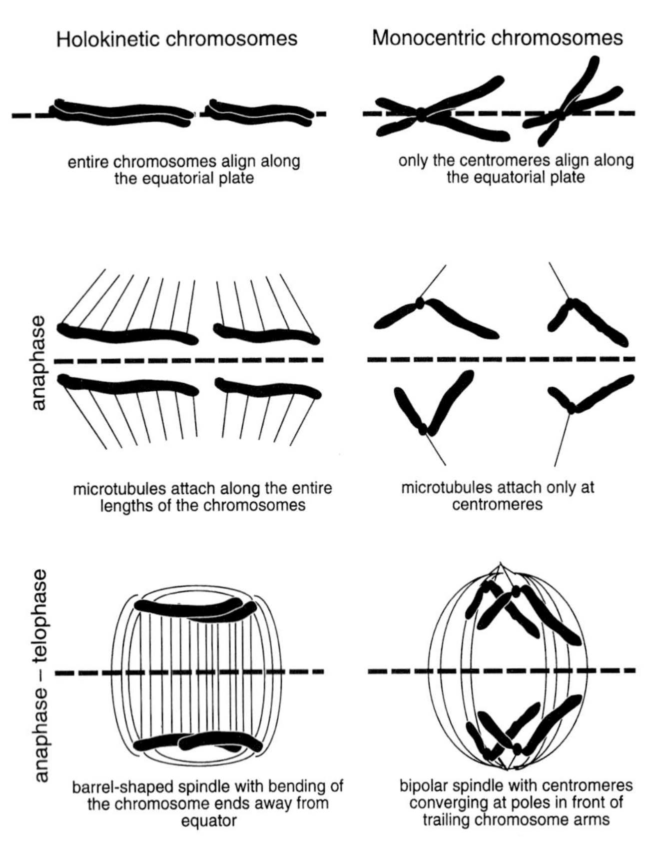

Fig. 4.11. A comparative representation of monocentric (right) and holokinetic (left) cell division. [Based on Wrensch et al. (1994).] The characteristic feature of holokinetic cell division is that a distinct centromere does not exist, a feature that is also associated with a distinct meiotic process (*see* Chapter 5).

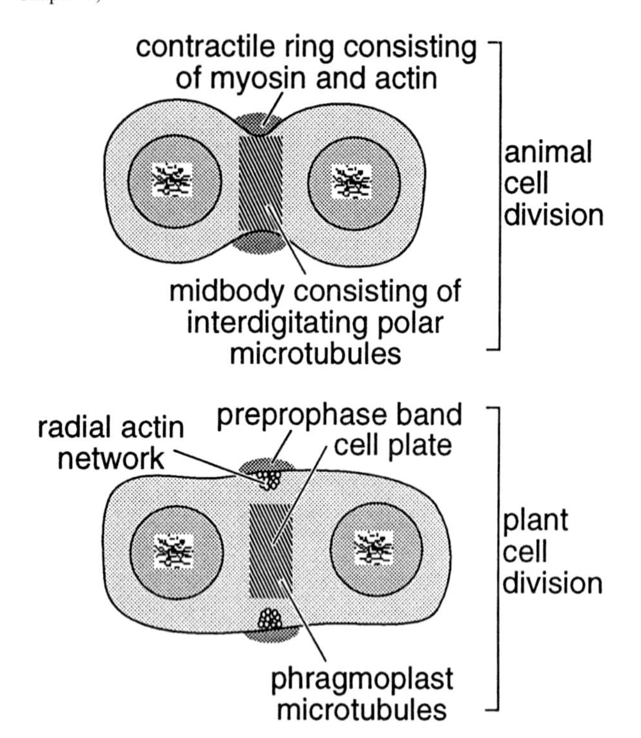

Fig. 4.12. A diagrammatic representation of cytokinesis in animals (top) and plants (bottom).

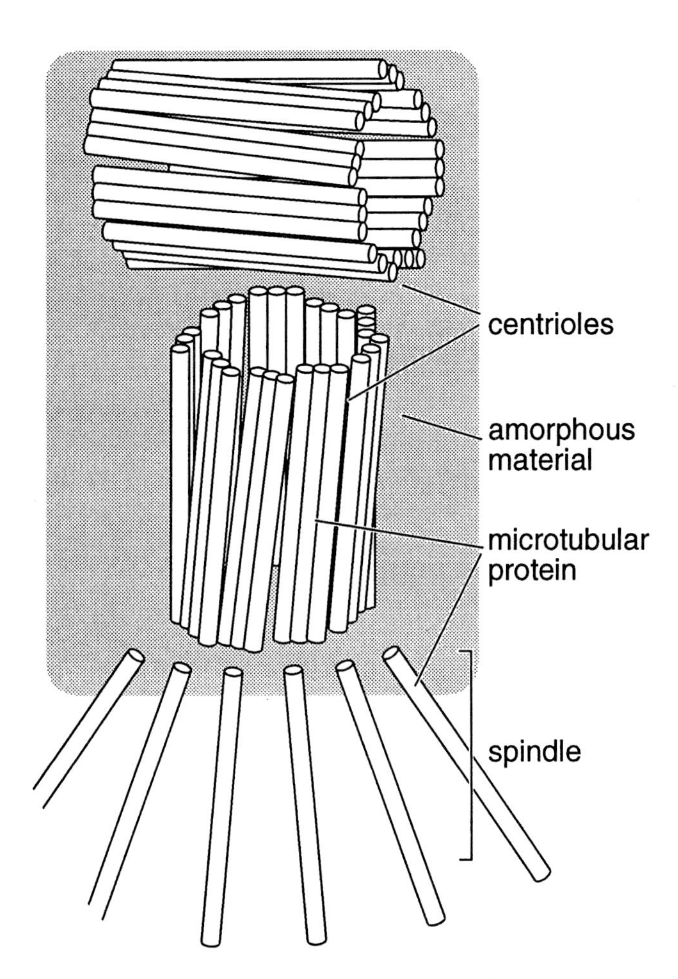

Fig. 4.13. A drawing of a highly magnified microtubule-organizing center to show the detailed structure of the centrioles, each of which has a ring of nine triplet microtubular structures.

centrosome). In animal cells, the MTC consists of two centrioles surrounded by amorphous material from which the microtubules of the spindle radiate (Fig. 4.13). Plant MTCs lack centrioles and have only amorphous material. MTCs duplicate during the S phase, and, at prophase, migrate to opposite poles to organize the spindle.

The microtubules are an assemblage of dimeric tubulin proteins, each dimer containing an α-tubulin and a β-tubulin subunit of molecular weight 50 kDa. Using antitubulin antibodies and immunofluorescent staining techniques, microtubules can be seen throughout the cell cycle. Chromosomes introduced into an environment of microtubules actively capture them for attachment to the kinetochore (Fig. 4.14). Motor-protein dyneins and kinesins are associated with kinetochores and help in the capture of the microtubules. Studies have also shown that as the nuclear membrane breaks down, an active interaction between microtubules and chromosomes is initiated, with the kinetochore as the focal point; the trilaminar, proteinaceous structure is attached to the centromeric region of the chromosome (*see* Fig. 4.7). Each kinetochore also duplicates during the S phase, as was demonstrated using immunofluorescent techniques.

The capture of microtubules by the kinetochores results in the orientation of the chromosomes on the equatorial plate. This process then allows the dual functions of the kinetochore to come into play, namely to maintain an attachment to the microtubules and to catalyze the polymerization/depolymerization reactions that lead to chromosome movement (Fig. 4.15). The ends of the microtubules attached to the kinetochore are defined as plus (+) ends because they grow faster than the minus (−) ends located at the poles. Another set, called polar microtubules, overlap at the equatorial plate and are held together by a protein matrix. A third set, the aster microtubules, radiate into the cytoplasm.

At anaphase, the (+) ends of the microtubules at the kinetochore begin to depolymerize, resulting in the poleward movement of sister chromatids; the (−) ends may possibly undergo some depolymerization as well. At telophase, laminar proteins, which were originally derived from the nuclear membrane, reassemble around individual chromosomes. As the chromosomes decondense, the laminar proteins coalesce to form the interphase nuclear membrane.

Recent work indicates that a class of tightly bound, passenger or inner-centromere proteins (INCENPs) are present in prometaphase. The INCENPs can be detected at the metaphase plate/plasma membrane furrow, using immunofluorescent techniques. It has been speculated that INCENPs play a role in spindle structure, chromosome segregation, and/or the process of cytokinesis, because they depart from the centromere at anaphase.

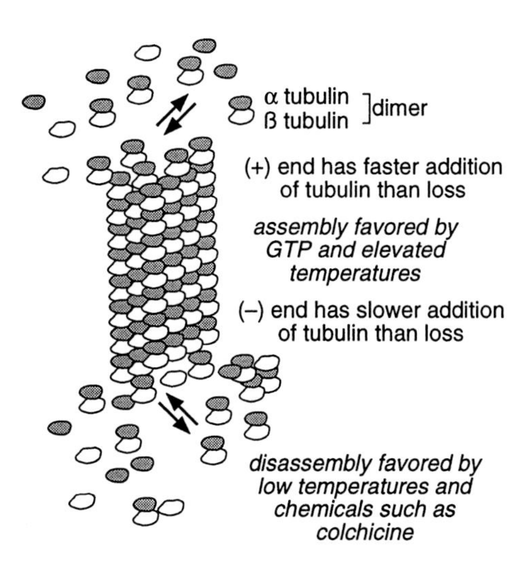

Fig. 4.15. A diagrammatic representation of the assembly and disassembly of microtubules. A number of the factors that affect the formation of microtubules are summarized to show that the process of association requires energy, in the form of guanosine triphosphate (GTP) or elevated temperatures, and is readily reversible. The X-ray structure of the α/β tubulin dimer has been determined and shows some structural similarities to an equivalent protein (FtsZ) in bacteria at the nucleotide-binding face (Burns, 1998)

4.1.3 The Plane of Cell Division Controls the Fate of Cells

The morphology of cells and their fate in development is determined by the cytoplasm- and centrosome-replication cycles. In animal cells, the plane of cell division is related to the orientation of the centrosomes, which organize the bipolar spindle. The contractile ring of myosin and actin fibers forms a furrow perpendicular to the spindle (*see* Fig. 4.12). Although the source for orienting the bipolar spindle is unclear, it has been suggested that the astral microtubules may provide the physical means for the orientation of centrosomes. Depending on the location of the cell-division plane, some cells will continue to divide while others will differentiate into specific cell types.

In plant cells, a preprophase band of actin protein is formed inside the cell wall and defines where the new cell wall will be formed. The spindle is organized perpendicular to the preprophase band. At anaphase, the band disappears, but remaining actin filaments guide the developing cell plate to the area defined by the preprophase band (*see* Fig. 4.12).

During cell differentiation, the chromosome- and cytoplasm-replication cycles may not be regulated coordinately. For example, the nucleus may enter a quiescent state called G0, whereas the cytoplasm will continue to increase in volume. Conversely, the chromosome-replication cycle may continue, in a modified form, so that extra rounds of DNA replication occur without cell division and lead to polytenization of the chromosomes. Cell fusion may result in many nuclei inside a single cell such as in the formation of muscle cells from myoblasts (Fig. 4.16), and should not be confused with events resulting from failure of cytokinesis.

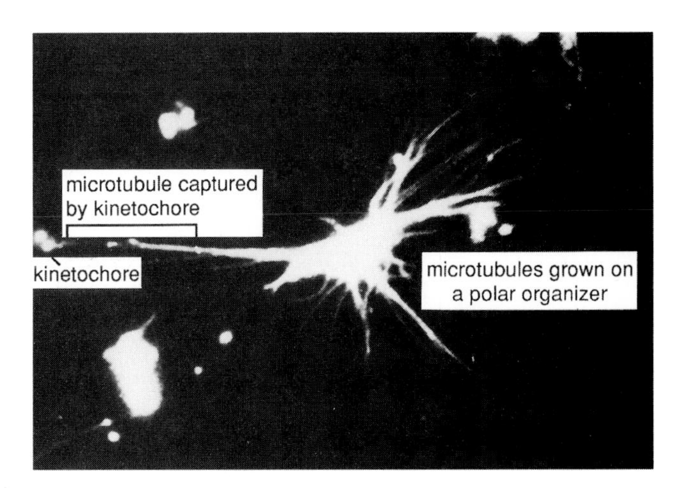

Fig. 4.14. The capture of microtubules by kinetochores *in vitro*. The microtubules were made visible using fluorescent antibody staining (Mitchison and Kirschner, 1985). Although D. Mazia is often cited for his statement of the role of chromosomes in mitosis "... the corpse at a funeral, ... provides a reason for the proceedings but does not take an active part in them," more recent studies indicate that chromosomes do play an active role in their own movement through their interactions with the spindle apparatus.

4.2 UNIVERSAL MECHANISMS REGULATE THE CELL CYCLE

Many types of experiment suggest that each step in the cell cycle is under the control of many genes that form part of a molecular clock controlling the cycle (Fig. 4.17). Regulatory molecules in the plasma membrane sense signals from adjacent cells that include

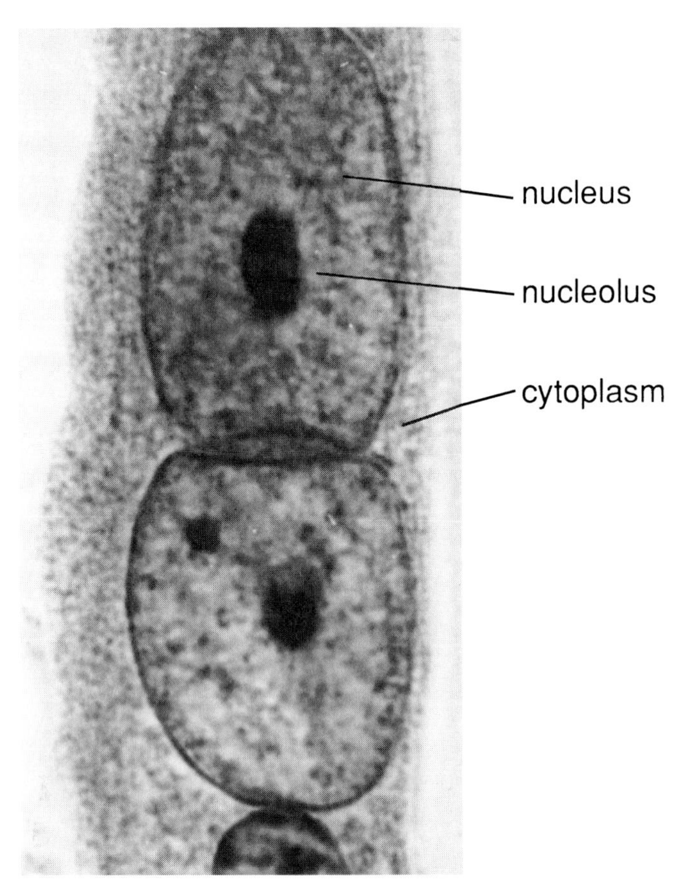

Fig. 4.16. A UV microphotograph of the multiple nuclei in a muscle cell formed from myoblasts in tissue culture. Because these cells form as a result of cell fusion, the presence of several nuclei is not due to nuclear division without cell division. (Photograph courtesy of Prof. N. R. Ringertz.)

growth regulators, and physiological and/or nutritional factors. Once stimulated, the plasma-membrane regulators initiate a cascade of events inside the cell, leading to a decision, or switch, called "start." The start decision, also referred to as "restriction point," initiates the transition from G1 to S and commits the cell to begin the chromosome-replication cycle. Another major switch point is the transition from G2 to mitosis. The regulatory molecules involved in both of these major switches have been identified and are the same in yeast, sea urchins, frogs, humans, and plants. The studies of these molecules, as discussed below, provide strong evidence for a universal set of mechanisms controlling eukaryotic cell-cycle regulation.

4.2.1 Kinase and Cyclin Proteins Regulate the Eukaryotic Cell Cycle

A maturation-promoting factor, now called mitosis-promoting factor or MPF, was originally identified in 1971 as a requirement for the start of meiosis in frog oocytes. An initially unrelated protein named cyclin was discovered in sea-urchin embryos, which were analyzed at the completion of mitotic divisions. In addition, the protein cdc2, discovered in budding yeast, was found to be required for the initiation of the cell-division cycle. Related studies in fission yeast and human cells showed that the cdc2 protein has a molecular weight of 34 kDa and is capable of catalyzing the phosphorylation of serine and threonine residues in proteins, which is characteristic for a kinase. This protein is now known as $p34^{cdc2}$ or cdc2 kinase. The relationship among MPF, cyclin, and $p34^{cdc2}$ in organisms that are widely divergent, became clear when the MPF factor was purified and demonstrated to be a complex of the $p34^{cdc2}$ and cyclin proteins (Fig. 4.18).

An *in vitro* system for the study of the chromosome-replication cycle, using cytoplasm from frog eggs, demonstrated that the addition of MPF protein to nuclei at G2 immediately promoted mitotic

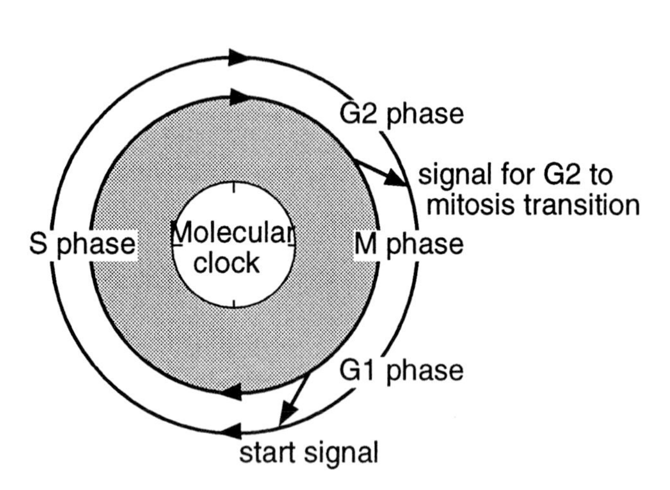

Fig. 4.17. A diagrammatic representation of the cell cycle to emphasize the clocklike nature of events and the timing of two critical events: the first to initiate the transition from G2 to metaphase, and the second, the transition from G1 to S phases.

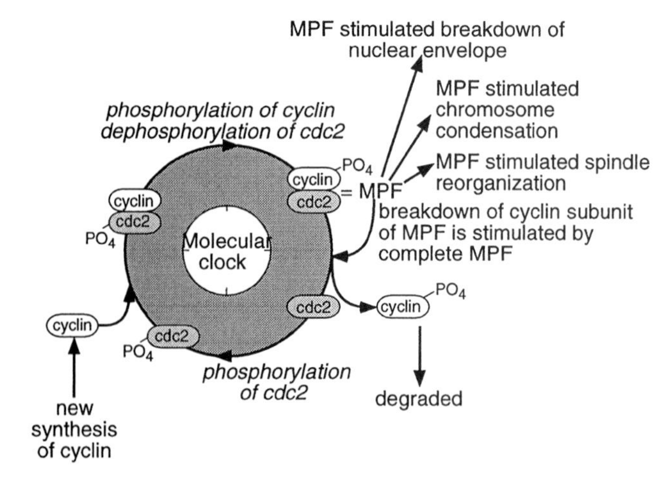

Fig. 4.18. Summary of the molecular events that determine progress through the cell cycle. MPF = maturation-promoting factor; cyclin and cdc2 are proteins involved in the control of the cell cycle; PO_4 = phosphorylation of protein. The cyclin is a B-type and its destruction is necessary for exit from mitosis and for the initiation of a new cycle. Several cdc proteins are required for the destruction of B-cyclin, which also involves conjugation with ubiquitin protein (Irniger et al., 1995; King et al., 1995). A number of other cyclin/cdc complexes, formed during G1 and S phases, are involved in cell cycle control (Jacobs, 1995), but these have not been included.

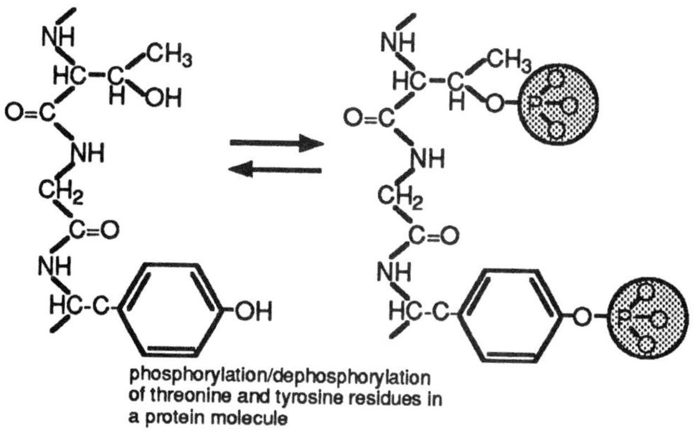

phosphorylation/dephosphorylation of threonine and tyrosine residues in a protein molecule

Fig. 4.19. The chemical structure of some amino acid residues that are phosphorylated or dephosphorylated during the course of the cell cycle. The amino acid residues most commonly phosphorylated are serine, threonine, and tyrosine. The reaction is carried out by protein kinases using ATP as the source of phosphate. The addition of a negative charge to a portion of the protein molecule that is normally uncharged can have major effects on the conformation of the protein at the site of phosphorylation. If, for example, phosphorylation of protein kinase occurs at the active site responsible for the phosphorylation reaction of other proteins, loss of activity could result. In addition, phosphorylated proteins may become more susceptible to the action of proteinases. This complex interplay of proteins appears to be modified in a cyclic manner through phosphorylation and dephosphorylation.

division. Furthermore, a drop in the level of cyclin signaled the end of mitosis. Another protein, cdc25, was needed to activate MPF. The interaction of these various proteins has been intensively studied, and it is now understood that the process of phosphorylation is the common bond between them.

The action of MPF is to phosphorylate proteins (Fig. 4.19) that, in turn, are involved in the initiation of mitosis. When cyclin is destroyed, in a reaction stimulated by the conjugation of a protein called ubiquitin, the MPF loses its ability to phosphorylate, and dephosphorylation of proteins begins under the action of phosphatases, thus leading to G1. The retinoblastoma susceptibility gene product (RB) is differentially phosphorylated during the cell cycle and has been suggested to control progression through the G1 phase because the loss of RB function can lead to uncontrolled cell division. It is, therefore, evident that the chromosome-replication cycle is very closely tied to a phosphorylation/dephosphorylation cycle involving p34^{cdc2} and cyclin. The regulation of the cyclin-dependent kinases (CDKs) is currently an area of intense research in order to establish the nature of checkpoint controls that coordinate the cell cycle. Several cyclins (A, B, D1, D2, D3, E, G) and inhibitors of CDKs (p15, p16, p21, p27) have been identified and represent a complex set of interactions controlling the cell cycle.

4.2.2 Cell-Fusion Experiments Indicate That Diffusible Cellular Components Control the Cell Cycle

Although the biochemical data for the phosphorylation/dephosphorylation cycle is largely based on studies of frog eggs and sea-urchin embryos, evidence from other animal cells and plant cells supports the model. Mutation analyses in yeast and animal cell-fusion experiments support the existence of diffusible, or trans-acting, cellular components that determine the different stages of the chromosome-replication cycle. The following cell-fusion experiments can now be understood in terms of the phosphorylation/dephosphorylation cycle model depicted in Fig. 4.18:

G1 cells fused with S cells. This fusion immediately induces the S phase in the G1 cells and is consistent with the role of p34^{cdc2} protein kinase as an activator of the S phase.

G1 cells fused with G2 cells. This fusion does not induce the S phase in the G1 cells. Thus, the S phase inducing factor is absent from G2 cells at the completion of the S phase.

S cells fused with G2 cells. No DNA synthesis occurs in these

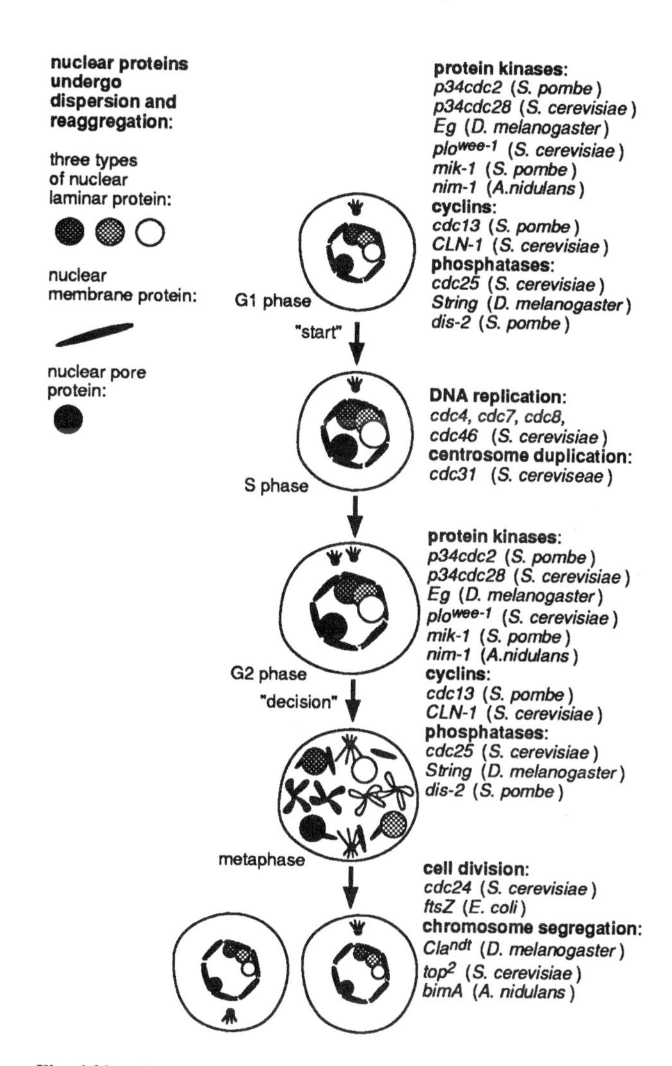

Fig. 4.20. A summary of the mitotic processes caused by the aggregation and disaggregation of nuclear proteins during the cell cycle (left and center columns), and by mutations that affect cell division (right column). The broad categories of protein activities that are involved in the cell cycle are indicated, as well as a selection of mutations from different organisms that have been used to characterize the respective stages in the cell cycle. The three types of laminar proteins are called lamin A, B, and C, which, in addition to forming a protein mesh underlying the nuclear membrane, also contribute to the internal nucleoskeleton (Hozák et al., 1995). The nuclear-pore protein is also composed of a complex group of proteins—as indicated by molecular genetic studies in yeast (Doye and Hurt, 1995). The terms "start" and "decision" are also called checkpoint controls (Nurse, 1997).

fusion products, indicating that G2 cells must accumulate a factor that blocks DNA replication.

M cells fused with G1, S, or G2 cells. In these combinations, the chromatin in the G1, S, or G2 nuclei undergoes premature chromosome condensation (PCC), and the nuclear membrane disappears. This implies the existence of a mitosis-promoting factor in animal M cells.

4.2.3 Overview of the Mutations and Proteins Important in Mitosis

A summary of the mutations that affect mitosis is given in Fig. 4.20. Although most of the mutations come from yeast, sufficient work in other organisms has been carried out to suggest that the genetic framework developed in yeast is widely applicable. The movement of some of the major nuclear proteins during the mitotic process is also indicated to emphasize the dynamic nature of the process.

4.2.4 Oncogenes Code for Proteins That Influence the Phosphorylation/Dephosphorylation Cycle

Cancers are caused by uncontrolled cell division resulting from the influence of factors that interfere with the normal cell cycle. The introduction of retroviruses into animal cells, for example, can lead to the integration of tumor-causing genes, oncogenes, into the DNA of the host. Detailed studies on these tumors show that even before infection, genes related in sequence to oncogenes (i.e., proto-oncogenes) can be assayed in the host DNA. The existence of proto-oncogenes has proven to be a general occurrence, and they have been shown to represent genes coding plasma-membrane receptor proteins or growth factors. An example of a cellular growth-factor gene is *c-sis*, coding for the platelet-derived growth factor, PDGF. The related oncogene is *v-sis*, which leads to uncontrolled cell growth. The biochemical activity associated with proto-oncogenes is the ability to phosphorylate and dephosphorylate other proteins, or bind chemical components such as guanosine triphosphate (GTP), that are required for the functioning of enzymes involved in the normal phosphorylation/dephosphorylation cycle. Developing concepts suggest that it is these activities that initiate uncontrolled cell division by interfering with the normal phosphorylation/dephosphorylation cycle.

BIBLIOGRAPHY

General

Cross, F., Roberts, J., Weintraub, H. 1989. Simple and complex cell cycles. Ann. Rev. Cell Biol. 5: 341–395.

He, D., Zeng, C., Brinkley, B.R. 1995. Nuclear matrix proteins as structural and functional components of the mitotic apparatus. Annu. Rev. Cytol. 162B:1–74.

Hyams, J.-S., Brinkley, B.R. (eds). 1989. Mitosis: Molecules and Mechanisms. Academic Press, New York.

John, P.C.L. 1996. The plant cell cycle: conserved and unique features in mitotic control. In: Progress in Cell Cycle Research (Meijer, L., Guidet, S., and Vogel, L., eds). Plenum Press, New York. 2:59–72.

Lu, B.C.K. 1996. Chromosomes, mitosis, and meiosis. In: Fungal Genetics, Principles and Practice (Bos, C.J., ed.). Marcel Dekker, Inc., New York. pp. 119–175.

Mitchison, T.J. 1989. Mitosis—from molecules to machine. Am. Zool. 29: 523–535.

Murray, A.W., Hunt, T. 1993. The Cell Cycle. W. H. Freeman & Co., New York.

Murray, A.W., Kirschner, M.W. 1991. What controls the cell cycle? Sci. Am. 246: 56–63.

Section 4.1

Barber, H.N. 1939. The rate of movement of chromosomes on the spindle. Chromosoma 1: 33–50.

Buendia, B., Karsenti, E. 1995. Regulation of centrosome function during mitosis. In: Advances in Molecular and Cell Biology. JAI Press Inc. 13: 43–67.

Burns, R. 1998. Synchronized division proteins. Nature 391: 121–122.

Dasso, M. 1993. RCC1 in the cell cycle: the regulator of chromosome condensation takes on new roles. Trends Biochem. Sci. 18: 96–101.

Galbraith, D.W., Hoskins, K.R., Maddox, J.M., Ayres, N.M., Sharma, D.P., Firoozabady, E. 1983. Rapid flow cytometric analysis of the cell cycle in intact plant tissues. Science 220: 1049–1051.

Goday, C., Ciofi-Luzzatto, A., and Pimpinelli, S. 1985. Centromere ultrastructure in germ-line chromosomes of *Parascaris*. Chromosoma 91: 121–125.

Lloyd, C. 1991. Probing the plant cytoskeleton. Nature 350: 189–190.

Mineyuki, Y., Aioi, H., Yamashita, M., Nagahama, Y. 1996. A comparative study on stainability of preprophase bands by the PSTAIR antibody. J. Plant Res. 109: 185–192.

Mitchison, T.J., Kirschner, M.W. 1985. Properties of the kinetochore in vitro. II. Microtubule capture and ATP-dependent translocation. J. Cell Biol. 101: 766–777.

Nicklas, R.B. 1988. Chance encounters and precision in mitosis. J. Cell. Sci. 89: 283–285.

Orr-Weaver, T.L. 1994. Developmental modification of the *Drosophila* cell cyle. Trends Genet. 10: 321–327.

Pfarr, C.M., Cone, M., Grissom, P.M., Hays, T.S., Porter, M.E., McIntosh, J.R. 1990. Cytoplasmic dynein is localized to kinetochores during mitosis. Nature 345: 263–265.

Pluta, A.F., Mackay, A.M., Ainsztein, A.M., Goldberg, I.G., Earnshaw, W.C. 1995. The centromere: hub of chromosomal activities. Science 270:1591–1594.

Roos, U-P., Guhl, B. 1996. A novel type of unorthodox mitosis in Amoebae of the cellular slime mold (Mycetozoan) *Acrasis rosea*. Europ. J. Protistol. 32: 171–189.

Schibler, M.J., Pickett-Heaps, J.D. 1987. The kinetochore fiber structure in the acentric spindles of the green alga *Oedogonium*. Protoplasma 137: 29–44.

Schubert, I., Oud, J.L. 1997. There is an upper limit of chromosome size for normal development of an organism. Cell 88: 515–520.

Steur, E.R., Wordewan, L., Schroer, T.A., Sheetz, M.P. 1990. Localization of cytoplasmic dynein to mitotic spindles and kinetochores. Nature 345: 266–268.

SUMNER, A.T. 1991. Scanning electron microscopy of mammalian chromosomes from prophase to telophase. Chromosoma 100: 410–418.

WRENSCH, D.L., KETHLEY, J.B., NORTON, R.A. 1994. Cytogenetics of holokinetic chromosomes and inverted meiosis: keys to the evolutionary success of mites, with generalizations on eukaryotes. In: Mites: Ecological and Evolutionary Analyses of Life History Patterns (Houck, M.A., ed.). Chapman and Hall, New York. pp. 282–343.

Section 4.2

ALBERTS, B., BRAY, D., LEWIS, J., RAFF, M., ROBERTS, K., WATSON, J.D. 1989. Molecular Biology of the Cell, 2nd ed. Garland Publishing Inc., New York.

BHAT, M.A., PHILP, A.V., GLOVER, D.M., BELLEN, H.J. 1996. Chromatid segregation at anaphase requires the *barren* product, a novel chromosome-associated protein that interacts with topoisomerase II. Cell 87: 1103–1114.

BROOKS, R., FANTES, P., HUNT, T., WHEATLEY, D. (EDS). 1989. The Cell Cycle. J. Cell Sci. Supp. 12. The Company of Biologists Ltd., Cambridge.

CHANG, F., NURSE, P. 1993. Finishing the cell cycle: control of mitosis and cytokinesis in fission yeast. Trends Genet. 9: 333–335.

CLEMENS, M.J., TRAYNER, I., MENAYA, J. 1992. The role of protein kinase C isoenzymes in the regulation of cell proliferation and differentiation. J. Cell Sci. 103: 881–887.

DATTA, N.S., WILLIAMS, J.L., CALDWELL, J., CURRY, A.M., ASHCRAFT, E.K., LONG, M.W. 1996. Novel alterations in CDK1/cyclin B1 kinase complex formation occur during the acquisition of a polyploid DNA content. Mol. Biol. Cell. 7: 209–223.

D'URSO, G., MARRACHINO, R.L., MARSHAK, D.R., ROBERTS, J.M. 1990. Cell cycle control of DNA replication by a homologue from human cells of the $p34^{cdc2}$ protein kinase. Science 250: 786–791.

DASSO, M. 1993. RCC1 in the cell cycle: the regulator of chromosome condensation takes on new roles. Trends Biochem. Sci. 18: 96–101.

DOYE, V., HURT, E.C. 1995. Genetic approaches to nuclear pore structure and function. Trends Genet. 11: 235–241.

FERREIRA, P., HEMERLY, A. 1994. Three discrete classes of *Arabidopsis* cyclins are expressed during different intervals in the cell cycle. Proc. Natl. Acad. Sci. USA 91: 11313–11317.

GAUTIER, J., SOLOMON, M.J., BOOHER, R.N., BAZAN, J.F., KIRSCHNER, M.W. 1991. cdc25 is a specific tyrosine phosphatase that directly activates $p34^{cdc2}$. Cell 67: 197–211.

GLOTZER, M., MURRAY, A.W., KIRSCHNER, M.W. 1991. Cyclin is degraded by the ubiquitin pathway. Nature 349: 132–138.

GOODRICH, D.W., WANG, N-P., QIAN, Y-W., LEE, E.Y-H.P., LEE, W-H. 1991. The retinoblastoma gene product regulates progression through the G1 phase of the cell cycle. Cell 67: 293–302.

GRAFI, G., LARKINS, B.A. 1995. Endoreduplication in maize endosperm: involvement of M phase-promoting factor inhibition and induction of S phase-related kinases. Science 269:1262–1264.

HARTWELL, L.-H., WEINERT, T-A. 1989. Checkpoints—controls that ensure the order of cell cycle events. Science 246: 629–634.

HIRT, H., HEBERLE-BOS, E. 1994. Cell cycle regulation in higher plants. Devel. Biol. 5: 147–154.

HOZÁK, P., SASSERVILLE, A.M-J., RAYMOND, Y., COOK, P.R. 1995. Lamin proteins form an internal nucleoskeleton as well as a peripheral lamina in human cells. J. Cell Sci. 108: 635–644.

IRNIGER, S., PIATTI, S., MICHAELIS, C., NASMYTH, K. 1995. Genes involved in sister chromatid separation are needed for B-type cyclin proteolysis in budding yeast. Cell 81: 269–277.

JACOBS, T.W. 1995. Cell cycle control. Ann. Rev. Plant Physiol. Mol. Biol. 46: 317–339.

KAMB, A. 1995. Cell cycle regulators and cancer. Trends Genet. 4: 136–140.

KING, R.W., PETERS, J-M., TUGENDREICH, S., ROLFE, M., HIETER, P., KIRSCHNER, M.W. 1995. A 20s complex containing cdc27 and cdc16 catalyzes the mitosis-specific conjugation of ubiquitin to cyclin B. Cell 81: 279-288.

MACNEILL, S.A., CREANOR, J., NURSE, P. 1991. Isolation, characterization and molecular cloning of new mutant alleles of the fission yeast $p34^{cdc2+}$ protein kinase gene: identification of temperature-sensitive G2-arresting alleles. Mol. Gen. Genet. 229: 109–118.

MANANDHAR, G., APOSTOLAKOS, P., GALATIS, B. 1996. Nuclear and microtubular cycles in heterophasic multinuclear *Triticum* root-tip cells induced by caffeine. Protoplasma 194: 164–176.

NURSE, P.M. 1990. Universal control mechanisms regulating an onset of M-phase. Nature 344: 503–508.

NURSE, P. 1997. Checkpoint pathways come of age cell 91, 865–867.

PFARR, C.M., CONE, M., GRISSOM, P.M., HAYS, T.S., PORTER, M.E., MCINTOSH, J.R. 1990. Cytoplasmic dynein is localized to kinetochores during mitosis. Nature 345: 263–265.

ROUSSEL, P., ANDRÉ, C., COMAI, L., HERNANDEZ-VERDUN, D. 1996. The rDNA transcription machinery is assembled during mitosis in active NORs and absent in inactive NORs. J. Cell Biol. 133(2)L: 235–246.

STEUER, E.R., WORDEMAN, L., SCHROER, T.A., SHEETZ, M.P. 1990. Localization of cytoplasmic dynein to mitotic spindles and kinetochores. Nature 345: 266–268.

TERZOUDI, G.I., PANTELIAS, G.E. 1997. Conversion of DNA damage into chromosome damage in response to cell cycle regulation of chromatin condensation after irradiation. Mutagenesis 12: 271–276.

WAGNER, S., GREEN, M.R. 1991. A transcriptional tryst. Nature 352: 189–190.

ZHANG, K., LETHAM, D.S., JOHN, P.C.L. 1996. Cytokinin controls the cell cycle at mitosis by stimulating the tyrosine dephosphorylation and activation of $p34^{cdc2}$-like H1 histone kinase. Planta 200: 2–12.

5

Meiosis and Gamete Formation: A View of Chromosomes in the Germline

- Eukaryotic somatic cells carry two copies of each chromosome, and meiosis must intervene to produce gametes with only a single copy of each chromosome.
- During meiosis, homologous chromosomes pair and exchange segments (crossovers) between chromatids before separating and eventually entering different gametes.
- The synaptonemal complex is a structure usually required for the intimate pairing that leads to crossing over between homologous pairs of chromosomes.
- Chiasmata are the cytological manifestations of crossing over and are essential for regular chromosome behavior during meiosis.
- The enzyme topoisomerase II is responsible for unraveling interlocked chromosome segments.
- Many genes are involved in the successful completion of meiosis, and mutations in these genes disrupt specific stages of the process.

The meiotic cell division is much more specialized than the mitotic cell division. Whereas the effect of mitosis is to increase the number of cells and allow growth of an organism, the function of meiosis is to generate gametes. Through the reductional process of meiosis, gametes are haploid in chromosome number, containing only one copy of each chromosome, in contrast to the somatic cells of diploid organisms. During subsequent sexual reproduction, when a female gamete is fertilized by a male gamete, homologous chromosomes are regrouped into a common nucleus, and the $2n$ chromosome number is reconstituted. The development of haploid germ cells includes both meiotic and postmeiotic events. In animals, this sequence is called spermatogenesis in males and oogenesis in females; the comparable terms in higher plants are microsporogenesis and megasporogenesis, respectively.

In this chapter, the basic features of the meiotic divisions and postmeiotic events are presented, including abnormal types of meiosis, followed by detailed discussions of the meiotic chromosome-pairing process, crossing over between chromosomes, and genetic controls of meiosis.

5.1 THE MEIOTIC AND POSTMEIOTIC CYCLES

The $2n$ germline cells divide mitotically until committed to meiosis, after which they are called meiocytes. During the division leading into meiosis, meiocytes enter an S phase that is much longer than the normal mitotic S phase. Meiosis consists of two consecutive nuclear divisions called meiosis I and meiosis II (Fig. 5.1). In meiosis I, the centromeres holding the sister chromatids together function as units, and the homologs of each chromosome pair move to opposite poles. This division is reductional because the chromosome complement of the $2n$ nucleus is halved, but some exceptions are discussed later. Meiosis II is similar to a mitotic division in that the split centromeres of each chromosome separate and sister chromatids move to opposite poles. This is called an equational division because the daughter cells receive the same number of chromosomes as the parental prophase II cell. Meiotic stages can be observed through the light and electron microscopes. The sequences of meiotic events in an insect, a plant, and a fungus are illustrated in Figs. 5.2–5.4, respectively.

5.1.1 Meiosis I

Premeiotic interphase. In the S phase preceding meiosis I, more than 99% of the DNA is replicated as each chromosome forms a pair of sister chromatids. During this stage, the cell and nuclear volumes increase to such an extent that the meiocytes are easily distinguishable from the surrounding somatic cells.

Prophase I. The chromosomes become visible for the first time since the previous mitotic division. Several cytologically distinct stages have been named, based on light-microscope observations, to aid in understanding the complexities of the early phases of meiosis I:

Leptotene stage or Leptonema (Gk. *leptos* = thin, *tainia* = band). The chromosomes appear as very long, single strands, but the sister chromatids, although present, are not yet detectable. The chromosomes give the appearance of being highly jumbled and are commonly differentiated into dark- and light-staining regions (*see* Figs 5.2 and 5.3). Electron micro-

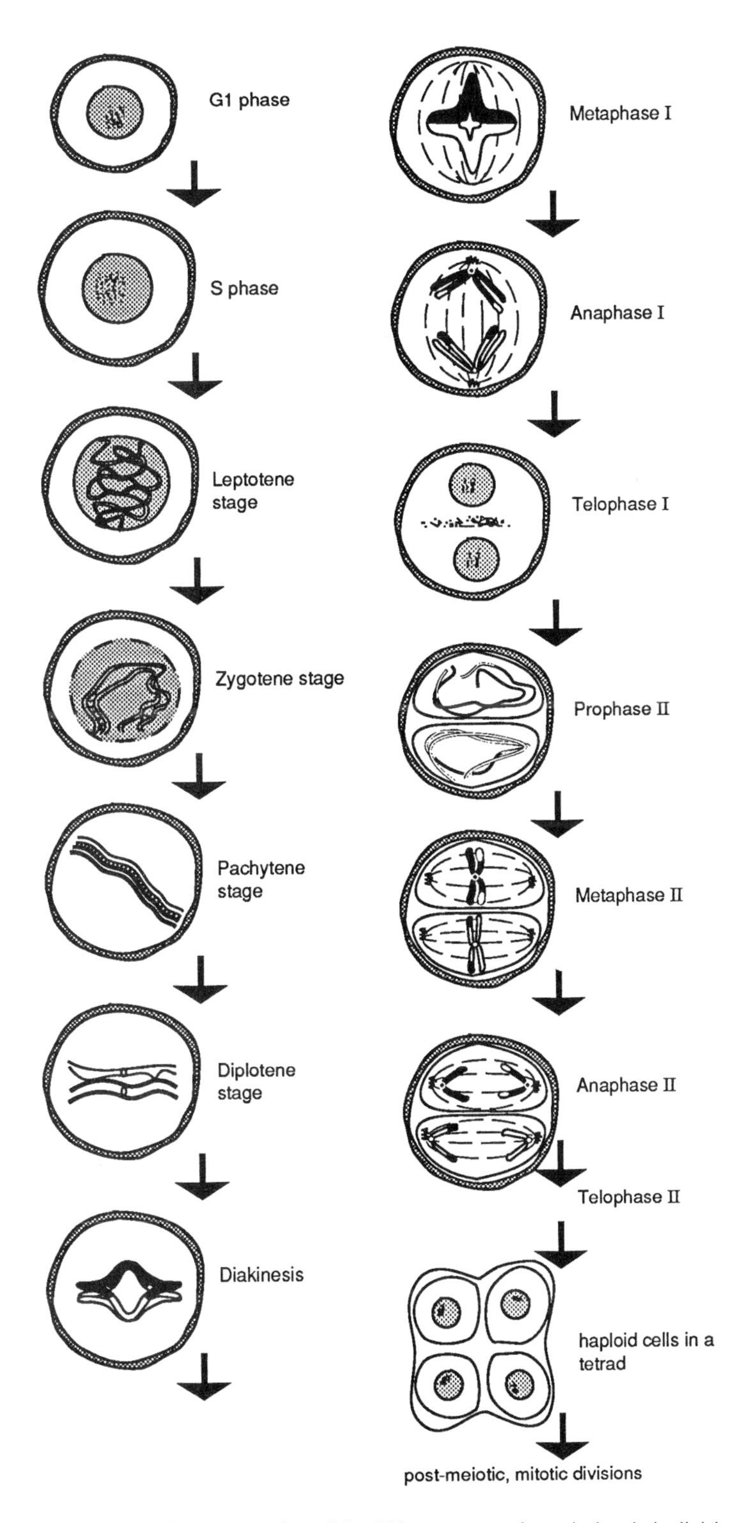

Fig. 5.1. A diagrammatic representation of the different stages of a typical meiotic division in plants, using one chromosome pair, as deduced from observations using the light microscope. Meiosis I includes the leptotene stage through telophase I, and meiosis II includes prophase II through telophase II.

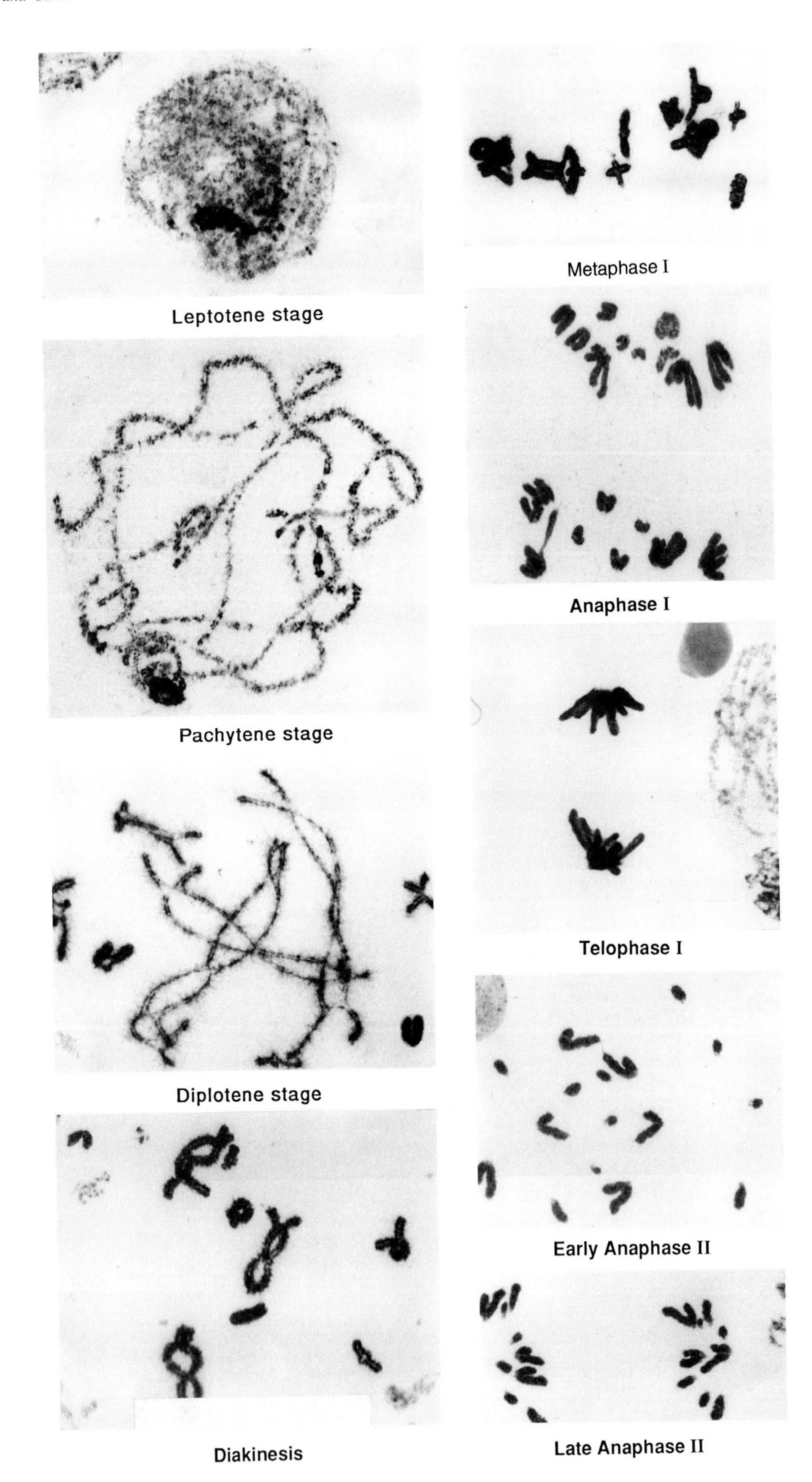

Leptotene stage
Pachytene stage
Diplotene stage
Diakinesis
Metaphase I
Anaphase I
Telophase I
Early Anaphase II
Late Anaphase II

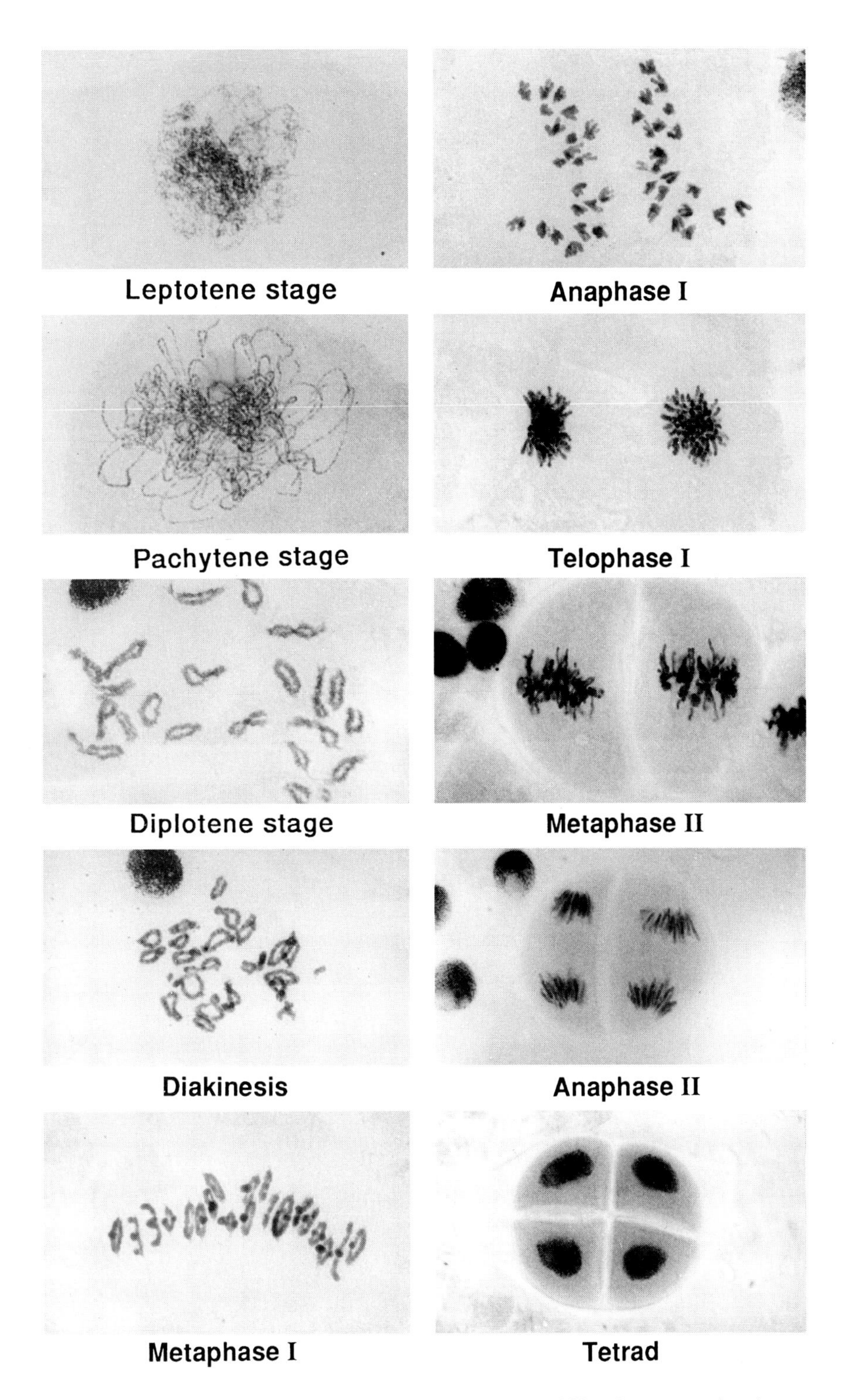

Fig. 5.3. Different meiotic stages in wheat (*Triticum aestivum*). Although common wheat is a hexaploid, diploid like pairing is usually observed in which all 21 pairs of homologous chromosomes form bivalents (bottom left) rather than hexavalents. (Photographs courtesy of Dr. B. Friebe.)

Fig. 5.2. The different stages of meiotic divisions in a male grasshopper (*Chorthippus parallelus*). Early observations on diplotene grasshopper chromosomes provided direct evidence that crossing over occurs at the four-strand (chromatid) stage and that the chiasmata arise as a result of the breakage and reunion of non-sister chromatids. (Photographs courtesy of Dr. David Shaw.)

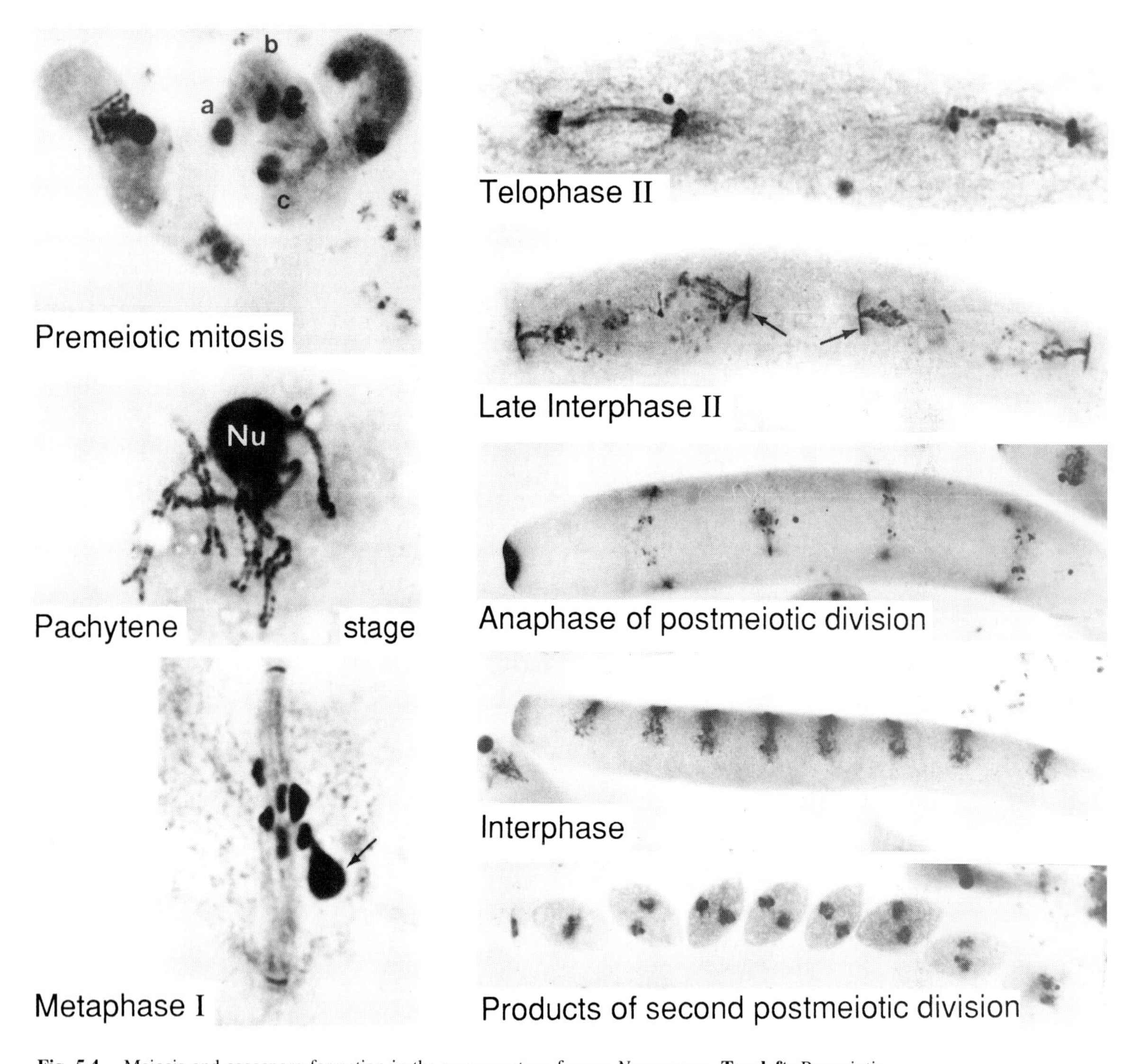

Fig. 5.4. Meiosis and ascospore formation in the ascomycetous fungus *Neurospora*. **Top left:** Premeiotic mitosis—the last premeiotic conjugate mitosis showing a three-celled "crozier" in which a = apical cell, b = penultimate cell, and c = stalk cell. The binucleate penultimate cell gives rise to an ascus. **Center left:** Pachytene—the seven pairs of homologous chromosomes are completely synapsed and the nucleolus (Nu) is at its maximum size. **Bottom left:** Metaphase I–the condensed chromosomes orient on the spindle halfway between the spindle–pole bodies. The nucleolus is still present (arrow). **Top right:** Telophase II—the two second-division spindles are aligned in tandem, parallel to the ascus wall. **Upper right:** Postmeiotic late interphase II—the enlarged double spindle–pole body plaques (arrows) are formed at this stage. **Center right:** Postmeiotic anaphase—all four spindles are oriented across the ascus. **Lower right:** Postmeiotic division interphase—the sister nuclei reorient in a linear manner with spindle–pole body plaques facing the same side. **Bottom right:** Products of second postmeiotic division—binucleate immature ascospores. The mature ascospores undergo additional mitoses so that each ascospore may eventually contain 30 or more nuclei. (Photographs courtesy of Dr. N. B. Raju.)

scopy indicates that additional structures called axial elements, comprised of 30 and 90KD proteins and RNA, begin to be deposited between the sister chromatids of each chromosome at this stage.

Zygotene stage or Zygonema (Gk. *zygos* = yoked). Homologous chromosomes begin to align at many points along their lengths in a process called synapsis, thus forming bivalents. Synapsis appears to be facilitated by the axial elements of homologous chromosomes, which, after being pulled closer together by transverse filaments, move to the outside of the synapsed regions, where they are modified to form lateral elements of the synaptonemal complex (Fig. 5.5). In both leptonema and zygonema, the ends of the synaptonemal complex, usually corresponding to the ends of the chromosomes or telomeres, are attached to the nuclear membrane. Movement of the telomeres leads to their clustering, the so-called "bouquet stage" in plants and the "synizetic knot" in animals (Fig. 5.6). Some interlocking of bivalents is observed at the zygotene stage in both diploid and polyploid organisms. A small amount of DNA synthesis, Zyg-DNA,

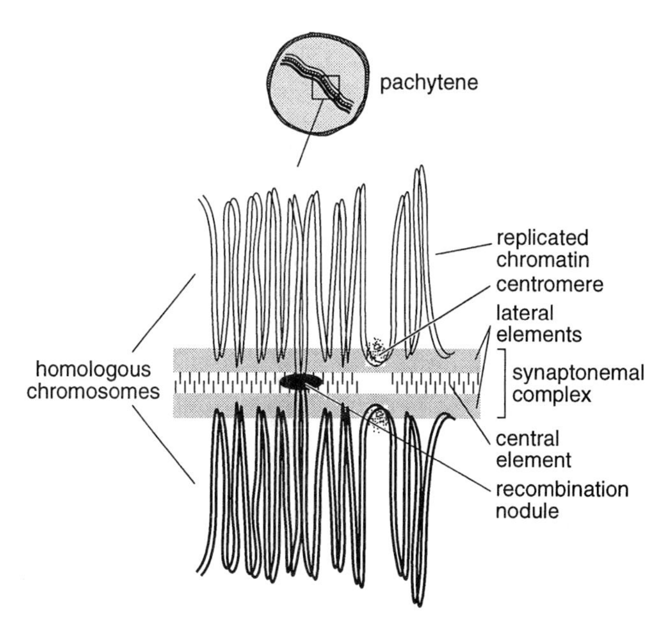

Fig. 5.5. A diagrammatic representation of an electron micrograph to show the intimate pairing between homologous chromosomes at the pachytene stage via the synaptonemal complex. Note the dense recombination nodule within the synaptonemal complex and the more diffuse centromeres. (Adapted from Pearlman et al., 1992.)

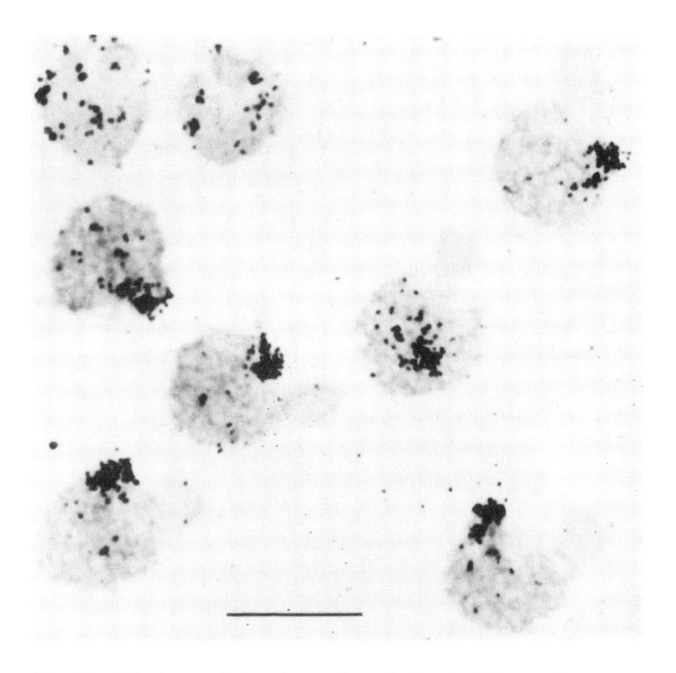

Fig. 5.6. The bouquet stage in prophase I of meiosis in cereal rye showing the clustering of telomeres within a small area of the nuclear envelope. *In situ* hybridization of pollen mother cells was carried out with a ^{3}H-labeled repetitive-sequence probe that detects the telomeric heterochromatin on the ends of the rye chromosomes (Appels et al., 1981).

has been measured biochemically in both plants and animals at this stage. Zygotene nodules are also present and have been speculated to be the sites of DNA homology searches.

Pachytene stage or Pachynema (Gr. *pachys* = thick). Homologs have completed their synapsis by this stage. Each bivalent consists of four chromatids, which are shorter and less tangled than in earlier stages. Individual sister chromatids are not usually distinguishable, and the chromosomes show differentiation into heterochromatic (darkly staining) and euchromatic (lightly staining) regions. In some organisms, these regions are suitable for the identification of individual chromosomes. The clustering of telomeres is no longer evident, and a structure called the nucleolus is clearly visible. Electron microscopy shows that the synapsed chromosomes are held together by a tripartite synaptonemal complex consisting of one central element between two lateral elements. Located within the synaptonemal complex are electron-dense recombination nodules, which are believed to be the physical sites at which successful crossing over has occurred (*see* Fig. 5.5). A crossover at this stage is the physical exchange of genetic material between paternal and maternal chromatids, resulting in a chiasma and leading to recombination between paternal and maternal genes.

Diffuse stage. In some animals, fungi, and plants, pachynema is followed by a stage where the chromatin has a diffuse appearance.

Diplotene stage or Diplonema (Gr. *diplos* = double). By this stage, the kinetochores associated with the chromosomal centromeres have duplicated. As the synaptonemal complex breaks down, homologous chromosomes appear to repulse each other and are eventually held together only by chiasmata (singular = chiasma) at the sites of crossing over between nonsister chromatids. In many animals, including humans, the female meiocytes may be arrested at the diplotene stage for prolonged periods, during which chromosomes may decondense and assume a lampbrush appearance while genes are actively transcribed.

Diakinesis (Gr. *dia* = through, *kinesis* = movement). This is the last stage of prophase I, during which the bivalents reach their maximum stage of condensation (*see* Fig. 5.3). Some chiasmata located away from the ends of the chromosomes appear to move closer to the telomeres. This stage has been valuable in cytogenetic work for estimating crossover frequencies from chiasma counts per nucleus or chromosome.

Metaphase I. The bipolar spindle appears and the kinetochores establish contact with the spindle fibers. The bivalents are cooriented at the metaphase plate, with only the chiasmata holding the homologs together (Fig. 5.7). The centromeres of bivalent homologues are positioned toward opposite poles. This stage of meiosis is extensively used to assess multivalent formations, the mode of disjunction of chromosomes, and chiasma frequencies. The orientation of each homologue in a bivalent is at random with respect to the poles of the cell.

Anaphase I. The homologous chromosomes move to opposite poles (Fig. 5.8). Each homolog consists of two chromatids held together by the kinetochores, which are located in the centromere region.

Telophase I. The movement of the chromosomes to the respective poles is completed and is accompanied by the uncoiling

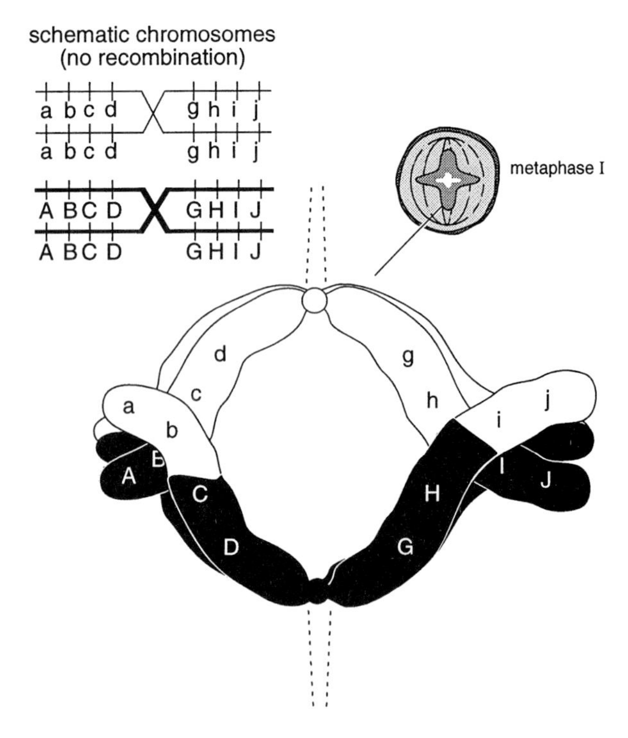

Fig. 5.7. A diagrammatic representation of a bivalent at metaphase I to show the result of two crossover events on the opposite arms of a homologous pair of chromosomes. The right end of the white chromatid without a crossover is hidden. **Top left:** The parental-type, noncrossover chromatids summarizing the genes and their loci (A, B, etc.) that were used to detect the crossover events.

of the chromosomes and re-formation of nucleoli and nuclear membranes. Division of the cytoplasm (i.e., cytokinesis) may occur at this stage or may be delayed until meiosis II is completed.

Interkinesis. No DNA replication occurs prior to meiosis II.

5.1.2 Meiosis II

The second phase of meiosis is generally very short. It involves the separation of the kinetochores and the chromatids that comprise each chromosome at this stage of meiosis.

Prophase II. From their uncoiled state in the preceding interstage nucleus, the chromatids condense into easily recognized structures.

Metaphase II. The spindle appears and the chromosomes prepare for separation of sister chromatids. The orientation of single chromosomes in the spindle, with split centromeres facing opposite poles, is called auto-orientation to distinguish the process from the co-orientation of bivalents at metaphase I.

Anaphase II. The kinetochores separate longitudinally, and the movement of the chromatids of each chromosome to opposite poles completes anaphase II (*see* Fig. 5.8). Genetically, the separation of alleles may be reductional or equational, depending on whether or not crossing over has occurred between the centromere and a gene locus. In Fig. 5.8, the separation of AA from aa (anaphase I) or B from b (anaphase II) is defined as reductional, whereas the separation of Bb from Bb (anaphase I) and A from A, or a from a (anaphase II) is equational.

Telophase II. The chromatids reach their respective poles and uncoil, and the nuclear membrane re-forms. Cytokinesis follows, and the fate of the resulting four haploid cells depends on whether the organism is an animal or plant, and male or female (*see* below).

Prophase I

A B

A B

a b

a b

Anaphase I

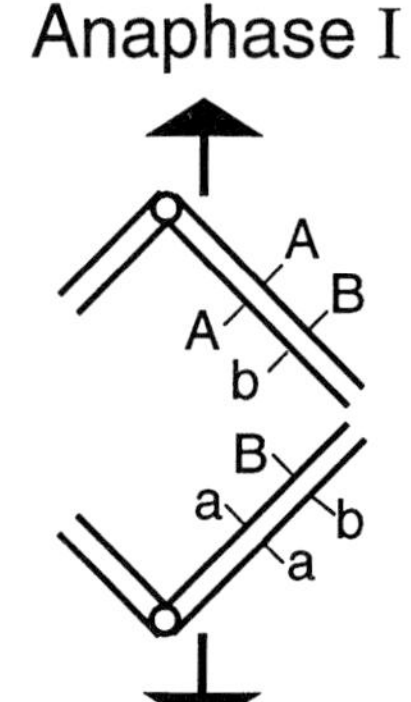

Anaphase II

A B A b

B a b a

Fig. 5.8. The cytological (chromosomes or chromatids) and genetic (alleles AA, aa; BB, bb) consequences of anaphase I and anaphase II segregations. Genetically, as there is no crossover between the centromere and the A locus, the resulting AA ↔ aa disjunction at anaphase I (center) is variously described as reductional, chromosomal, or first-division segregation. The A-A ↔ a-a disjunction at anaphase II (bottom) is variously described as equational, chromatid, or second-division segregation. For the B locus, because there is a crossover between the centromere and the B-b alleles, it undergoes equational segregation at AI (Bb ↔ Bb) and reductional segregation at AII (B-b ↔ B-b). Cytologically, AI is reductional and AII is equational. Such distinctions are important in tetrad analysis and genetic-linkage calculations.

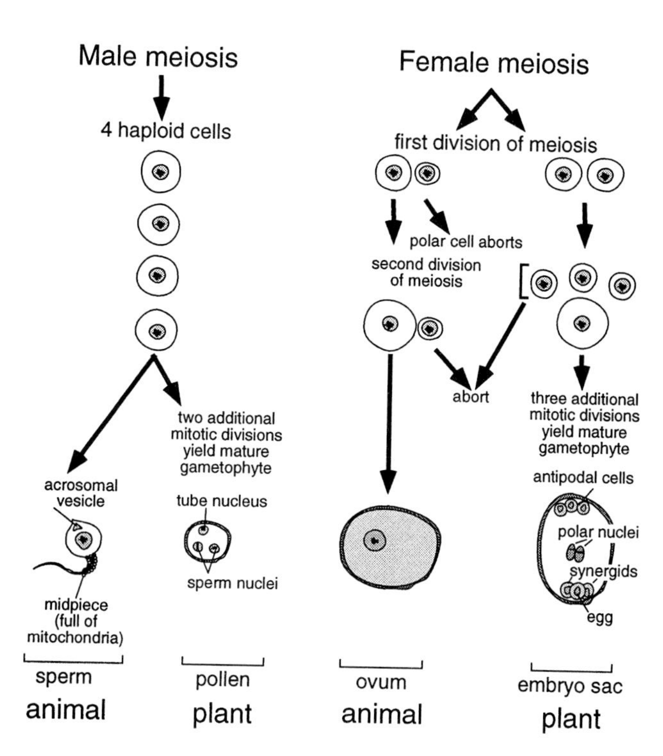

Fig. 5.9. A comparison of male and female gametogenesis in plants and animals. Note that in plants, the haploid cells undergo additional mitotic divisions to form mature male (pollen) and female (embryo sac) gametophytes.

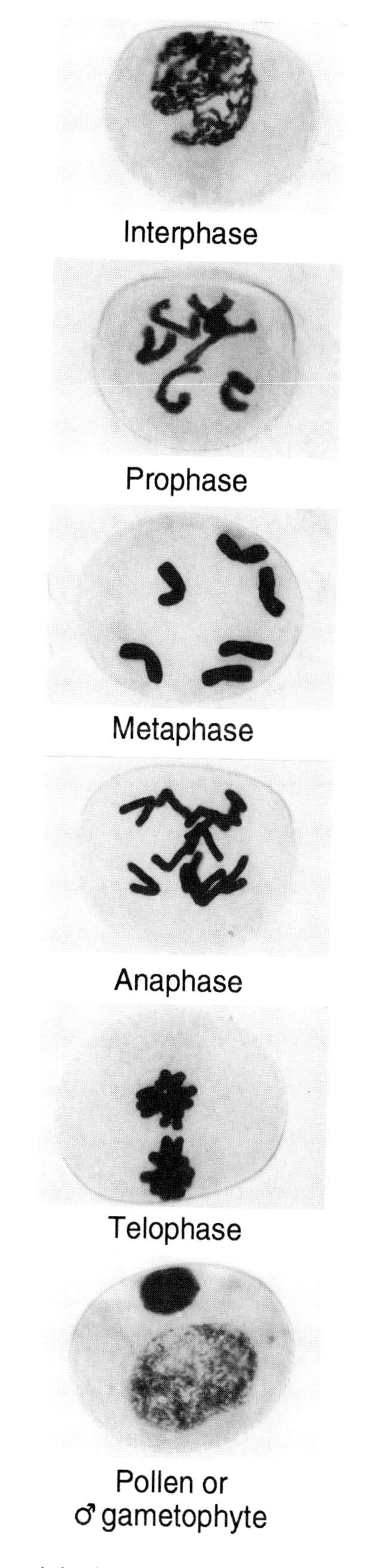

Fig. 5.10. Postmeiotic mitosis in pollen cells of *Crepis capillaris*. The metaphase stage is a polar view. A further division involving one of the nuclei in the bottom photo produces two sperm nuclei (*see* Fig. 5.9). (Photographs kindly supplied by Dr. B. Friebe.)

5.1.3 Postmeiotic Events

The major difference between animals and plants is the lack of mitotic divisions in animals (Fig. 5.9). In the male organs of animals, the products of meiosis (spermatids) differentiate into sperm; in female organs, the meiotic divisions result in a single large egg cell, which matures into an ovum, and two or three polar bodies, which abort. In microsporogenesis of plants, each haploid nucleus undergoes a mitotic division to form a tube nucleus and a generative nucleus (Fig. 5.10). The generative nucleus divides to form two male gametes, and the final structure with three nuclei is the male gametophyte or pollen grain. On the female side, three megaspores abort, and the nucleus in the remaining megaspore undergoes three successive divisions to form the embryo sac or female gametophyte.

In lower eukaryotes, the products of meiosis also undergo mitoses (*see* Fig. 5.4), and the haploid end products can form a dominant part of the life cycle of the organism. In some Ascomycetes (e.g., *Neurospora crassa*, shown in Fig. 5.4), the immediate haploid end products of this process are maintained in a linear array, which reflects the orientation of meiotic divisions (Fig. 5.11).

5.1.4 Achiasmate Meiosis

Achiasmate meiosis has been observed in worms, insects, and plants. In a vast majority of cases, it is meiosis in the male that is achiasmatic, and well-known examples include *Drosophila* and *Fritillaria* species (Fig. 5.12). The meiotic process is similar to a normal meiosis except that prophase I is shorter, and the diffuse, diplotene, and diakinesis stages are not distinguishable. As the name suggests, chiasmata do not occur at meiosis I even though the chromosomes are paired and lying parallel to each other. The pachytene-stage bivalents can have abnormal synaptonemal complexes, but often none are observed. In organisms where synaptonemal complexes do occur, the structures have sometimes persisted until metaphase I and may assist in the process of chromosome segregation. In the cases where no synaptonemal complexes are

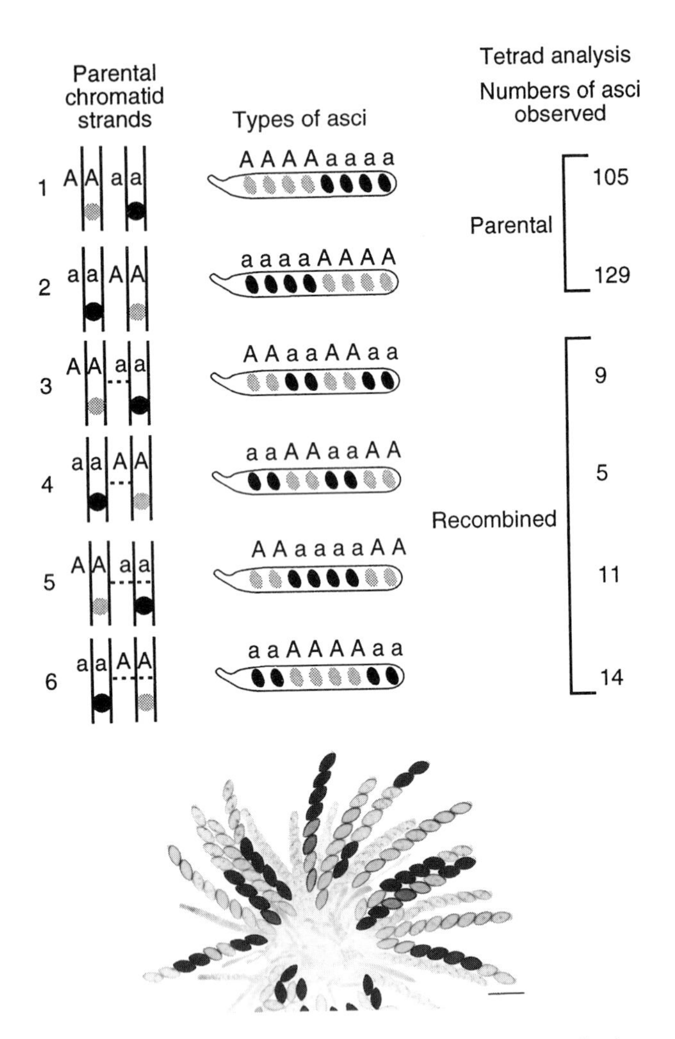

Fig. 5.11. Top: An example of tetrad analysis in *Neurospora* for the mating-type alleles [data from Lindegren (1932) and redrawn after Rédei (1982)]. Asci types 1 and 2 result from random chromosome segregation (no crossing over) and are parental type. Asci types 3–6 result from crossovers between the gene locus and the centromere. From these data, recombination between the A locus and the centromere is estimated to be 14.3%. It is calculated by dividing asci with crossovers (9 + 5 + 11 + 14 = 39) by all asci examined (273) and multiplying by 100.

Bottom: Photo showing a rosette of maturing asci of *Neurospora crassa* from a cross heterozygous for the gene *cys-3*, which has pleiotropic effects on cysteine biosynthesis and on ascospore maturation, resulting in the pigmentation observed in the ascospores. Mature asci show four black and four white spores, the white spores receiving the *cys-3* mutation. Asci with all spores unpigmented are immature. The ascus at the top center and the two asci at the upper left show first-division segregation for *Cys-3* versus *cys-3*, with four adjacent black spores and four adjacent white spores. The remaining mature asci show second-division segregation patterns resulting from crossing over between *cys-3* and the centromere, with 2:4:2 or 2:2:2:2 alignments of black versus white spores. Scale bar = 50 μm. (Photograph courtesy of Dr. N. B. Raju.)

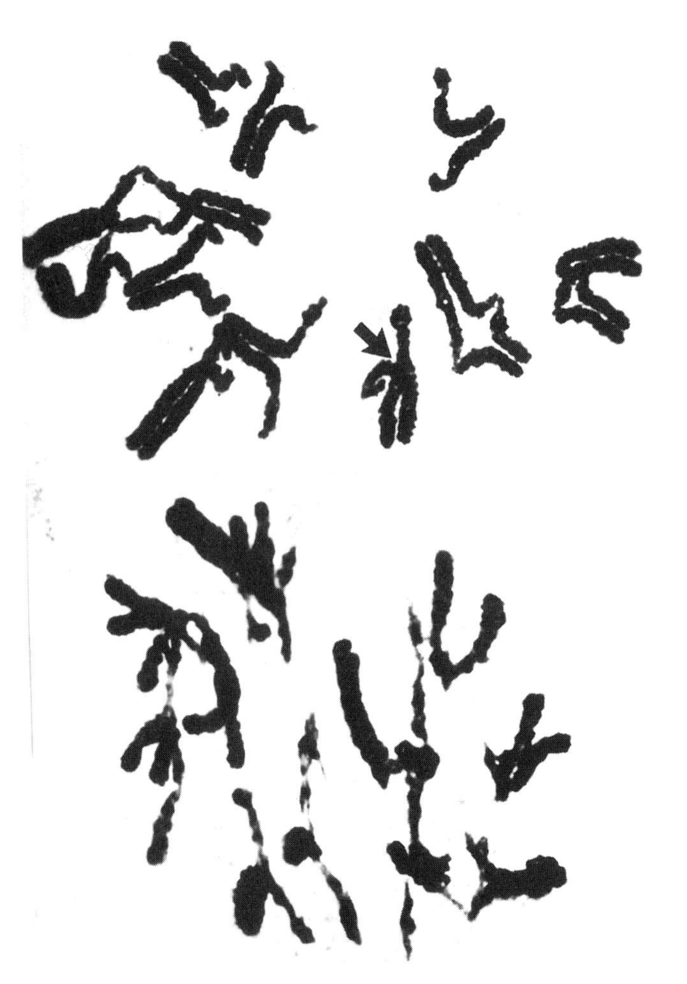

Fig. 5.12. Meiosis is achiasmate in the anthers (male germ cells) and chiasmate in the ovules (female germ cells) of *Fritillaria*. **Top:** A pollen mother cell showing 11 achiasmate bivalents at metaphase I. Arrow denotes rare concealed chiasma made visible by separation of homologous chromosomes. **Bottom:** An egg mother cell in early anaphase I showing 11 bivalents with many chiasmata. (Photographs courtesy of Dr. S. Noda.)

observed, alternative mechanisms must exist for holding homologous chromosomes together. Careful studies in male *Drosophila* have provided evidence for a low level of genetic recombination (1%), but synaptonemal complexes are not observed.

In contrast to the achiasmate nature of male meiosis in the above species, female meiosis shows normal synaptonemal complexes, crossing over, and chiasma formation (*see* Fig. 5.12). In evolutionary terms, achiasmate meiosis appears to be a secondarily derived character because, with rare exceptions, only one sex within a species shows the modification. It is also observed in a diverse group of organisms, indicating its multiple and independent origins from chiasmate meiosis.

5.1.5 Meiosis in Organisms With Multiple Or Diffuse Centromeres

In some insects and plants, the microtubules are attached to chromosomes at multiple, discrete sites, and such chromosomes are called polycentric. Alternatively, the microtubules may appear to have a continuous distribution (diffuse) along the length of the chromosome, in which case they are referred to as holocentric or holokinetic. Some, but not all, organisms with diffuse centromeres undergo inverted meiosis (i.e., meiosis I is equational and meiosis II is reductional). Whether meiosis I is equational or reductional is determined by the orientation of the bivalents with respect to the equatorial plate; meiosis I is equational if the orientation of the bivalents is parallel to the plate (called equatorial orientation) and reductional if bivalents lie at right angles to the plate (called axial orientation). The normal and inverted meiotic processes are compared in Fig. 5.13.

A typical example of inverted meiosis is seen in the plant genus *Luzula* (Fig. 5.14). At diakinesis, the chiasmata are clearly observed at interstitial positions in the three bivalents, but they are terminalized by metaphase I. The bivalents assume an equatorial orientation on the metaphase I plate and undergo equational division. Each bivalent separates into mirror-image "half-bivalents," which appear to have relics of "half-chiasmata" still attached at anaphase I. Nonsister chromatids stay joined and resynapse at prophase II, co-orient at metaphase II, and undergo reductional division at anaphase II. Genetically, the haploid gametes produced from an inverted meiosis are indistinguishable from those produced by a normal meiosis.

In those organisms with holokinetic chromosomes with clearly

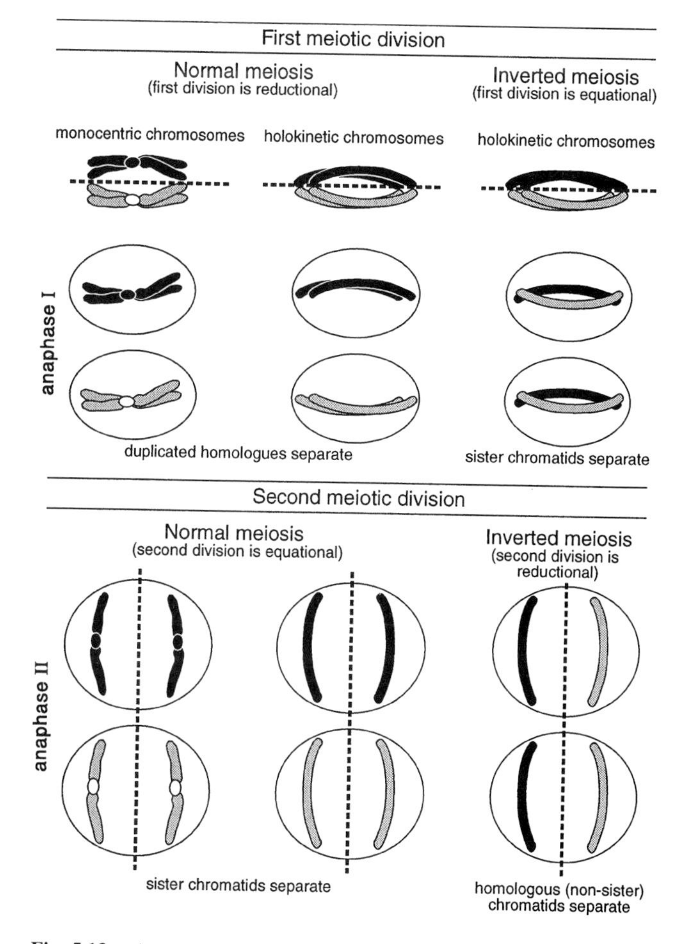

Fig. 5.13. A comparison of normal meiosis in organisms with monocentric chromosomes (left-hand side) with normal (center) or inverted (right-hand side) meioses in organisms with holokinetic (diffuse) chromosomes. Only one bivalent is depicted. (Redrawn from Wrench et al., 1994.)

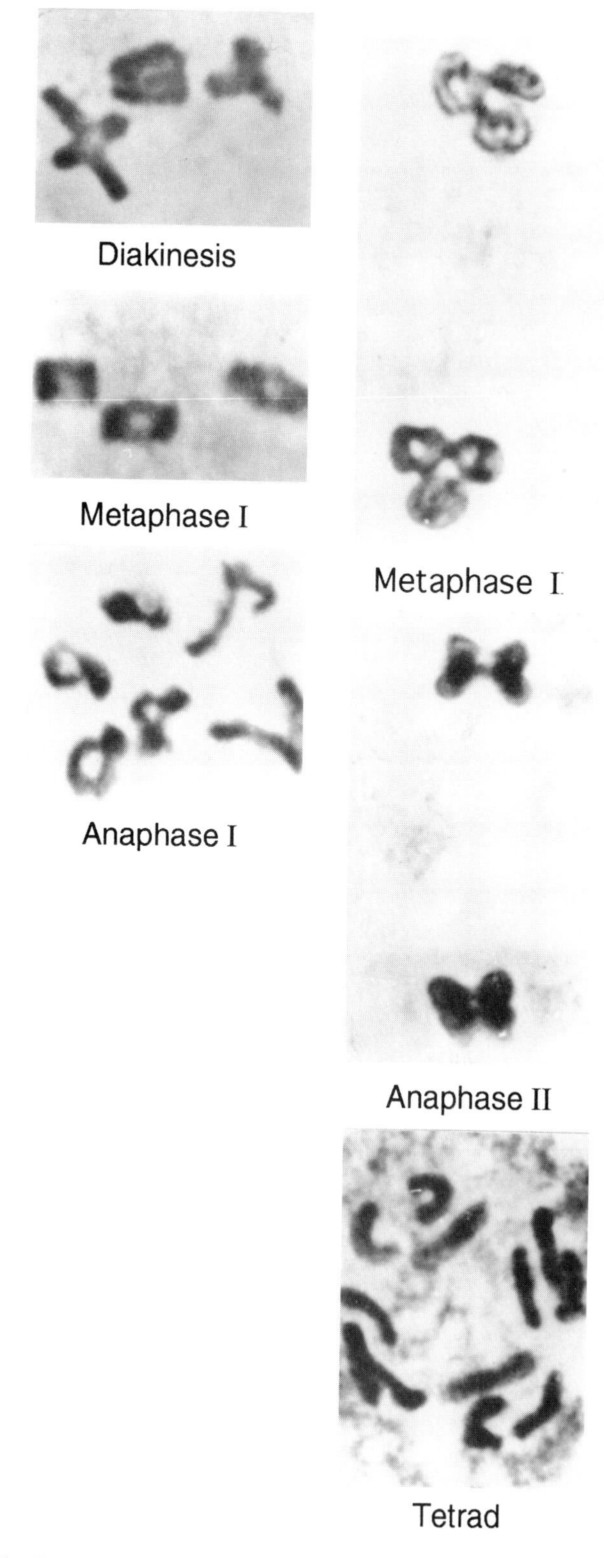

Fig. 5.14. Meiosis in *Luzula purpurea* ($2n = 6$) with holokinetic chromosomes. Key features include chiasma terminalization between the diakinesis stage and metaphase I; mirror-image separation of sister chromatids held by relic chiasmata at anaphase I; pairing of homologous but nonsister chromatids at metaphase II; reduction division at anaphase II (note spindles are in horizontal direction and the third chromosome overlaps the other two at anaphase II); and a tetrad of microspores, each with three chromosomes. (Adapted from Nordenskiöld, 1962.)

distinguishable sex chromosomes, inverted meiotic behavior applies only to autosomal bivalents, whereas the sex chromosomes have a normal meiotic behavior.

In organisms with diffuse centromeres, the trilaminar kinetochore plate can be seen along the length of the chromosome in mitotic divisions using the electron microscope (*see* Fig. 4.10). It is possible that such kinetochore structures, and the associated kinetic activity, could interfere with the normal meiotic behavior of chromosomes, especially chiasma formation and terminalization, and chromosome segregation during meiosis I. As an adaptation to meiotic divisions, holokinetic chromosomes do not generally show the characteristic kinetochore plate along the length of the chromosomes during meiosis, as shown by electron microscopy. Instead, the microtubular and centromere activities tend to be limited to the terminal regions of chromosomes (Fig. 5.15); in the case of *Parascaris*, this leads to orientation of the bivalent parallel to the equatorial plate, resulting in inverted meiosis. In the nematode *Caenorhabditis elegans,* kinetic activity is randomly restricted to only one telomeric end of a bivalent, and the resulting axial orientation relative to the equatorial plate results in a normal reductional meiosis I division.

5.1.6 Other Variations in the Meiotic Process

In the scale insect *Icerya purchasi*, the males are haploid; therefore, meiosis I is bypassed completely and meiosis II is equational. In mealy-bug (*Pseudococcus obscurus*) males, the paternal haploid set of chromosomes is genetically inactive and has a heterochromatic appearance. No pairing occurs between the maternal and paternal sets of chromosomes, and meiosis I is equational. Meiosis II is reductional and unipolar, with the maternal chromosome set forming sperm while the paternally derived, heterochromatinized chromosome set disintegrates. These particular variations have profound genetic consequences; in the mealy bug, for example, any new mutation that arises in the male chromosomal DNA is not transmitted to following generations.

Some of the complexities of insect meiosis are illustrated in the Hessian fly (*Mayetiola destructor*), a most destructive insect pest of wheat (Figs. 5.16 and 5.17). The somatic cells in males have six chromosomes (two pairs of autosomes + X1 + X2) and in females, eight chromosomes (two pairs of autosomes + X1 pair + X2 pair). The germline cells in both sexes have 8 S (somatic) chromosomes and a variable number (~30) of E (eliminated) chromosomes. In males, meiosis is achiasmate, and all E chromosomes, along with the paternally derived set of S chromosomes, are eliminated in one of the cells resulting from the meiosis I division. The other cell, with maternally derived S chromosomes, undergoes meiosis II to form two sperm cells. In females, meiosis is chiasmate. Each ovum contains a haploid set of S chromosomes and a variable number of E chromosomes. On fertilization, each zygote contains eight S chromosomes and a variable number of E chromosomes. During embryogenesis, the germline in both sexes has the same chromosome constitution, but somatic cells in males have fewer chromosomes than those in females.

5.2 CHROMOSOME SYNAPSIS AND THE SYNAPTONEMAL COMPLEX

5.2.1 The Synaptonemal Complex Forms Between Chromosomes Located Close to Each Other

Sister chromatids of each chromosome at leptonema organize axial cores, which, following chromosome alignment, become the lateral elements associating homologous chromosome pairs. Transverse elements begin to connect the lateral elements of chromo-

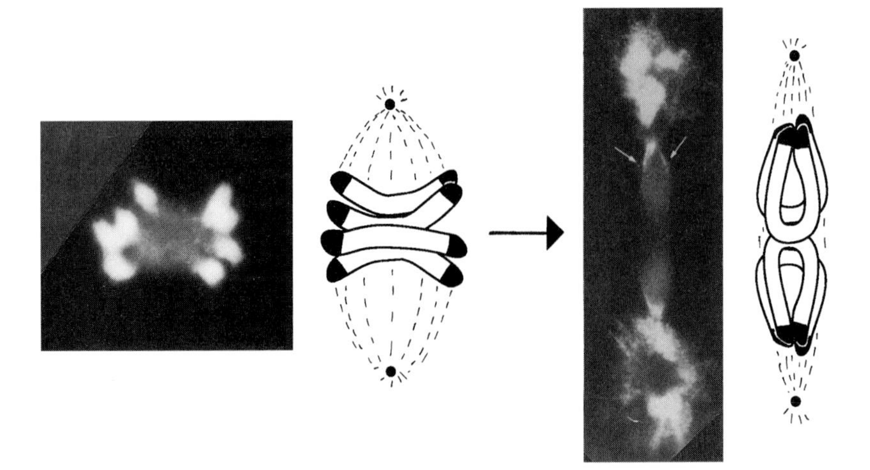

Fig. 5.15. Meiosis in holokinetic chromosomes of *Parascaris univalens* ($2n = 2$). Although kinetochore structures are observed along the entire length of the mitotic chromosomes (see Fig. 4.10), no such structure is observed in meiotic chromosomes and kinetic activity is restricted to the telomeric ends of both chromosomes. **Left side:** At metaphase I, the parallel orientation with the equatorial plate, as shown in the accompanying diagrammatic interpretation, results in an equational meiosis I division. **Right side:** Telophase I with accompanying diagrammatic interpretation. Indirect immunofluorescence staining of tubulin strands demonstrates their attachment to the telomeres of the chromosomes. (Photographs courtesy of Dr. C. Goday.)

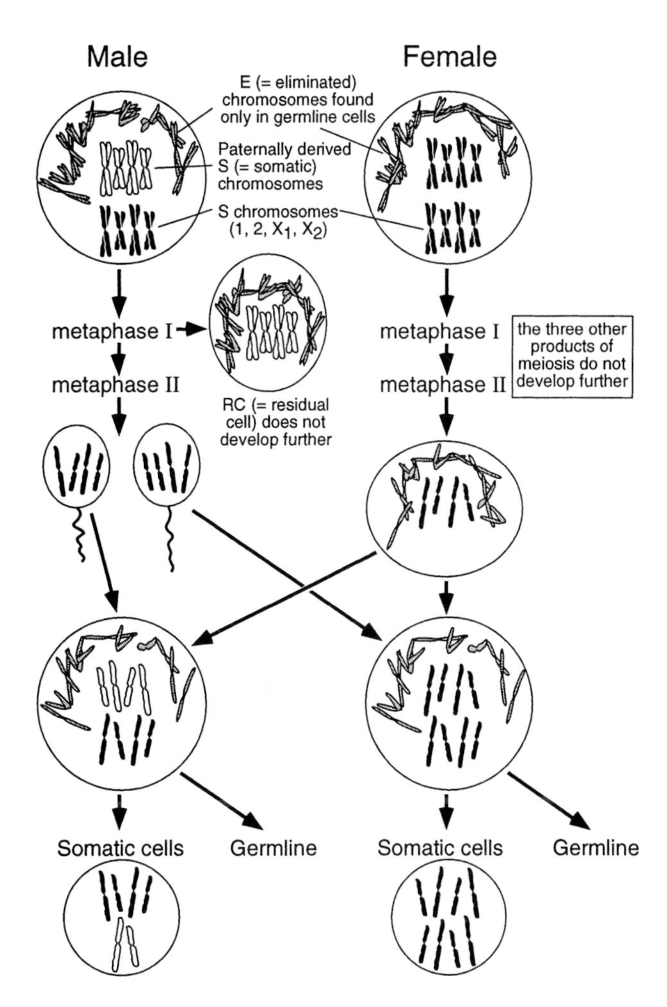

Fig. 5.17. A diagrammatic representation of the chromosome cycle in Hessian fly. In contrast to males, the female meiosis is chiasmate, and E chromosomes are retained in the mature egg nucleus. *Note:* In the male zygote, the paternal chromosome complement is shown in gray, which will be eliminated during male meiosis. In male somatic cells, paternally derived X1 and X2 are eliminated. (Adapted from Stuart and Hatchett 1988.)

somes progressing into zygonema, when homologous pairs are less than 100 nm apart. The lateral elements and transverse elements together form the synaptonemal complex, which at pachynema has a tripartite, ribbonlike structure, 0.2 μm or 200 nm wide (Fig. 5.18). The uniform structure is occasionally interrupted by the presence of spherical bodies, about 100 nm in diameter, called recombination

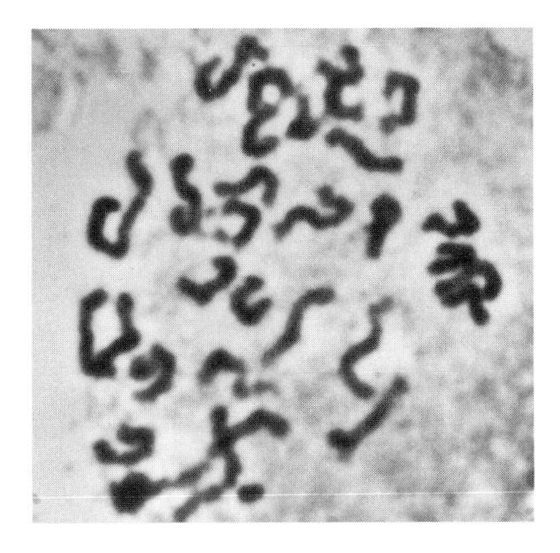

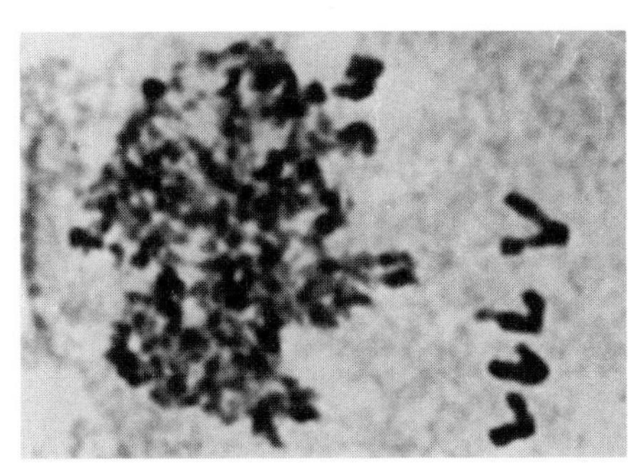

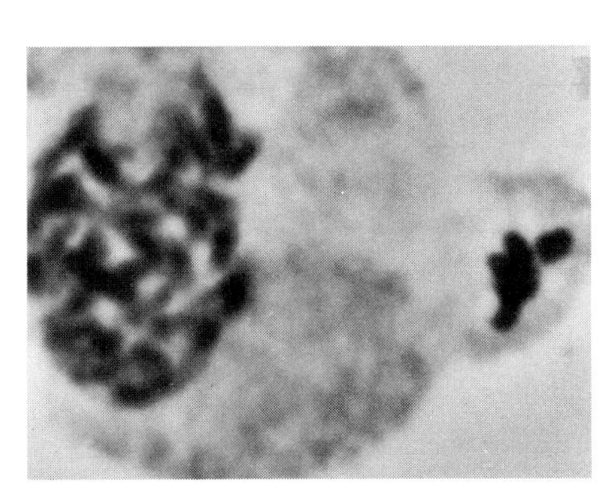

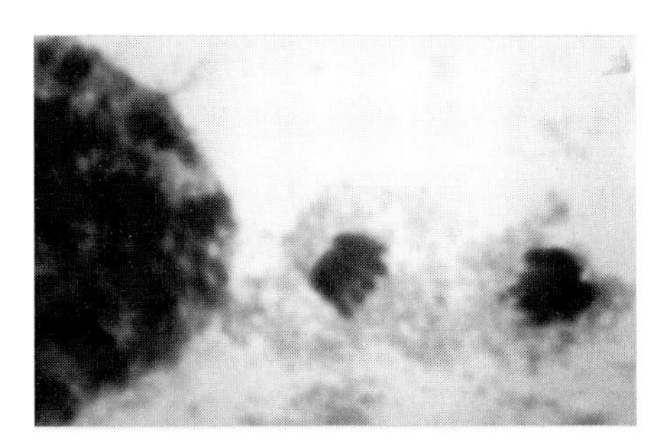

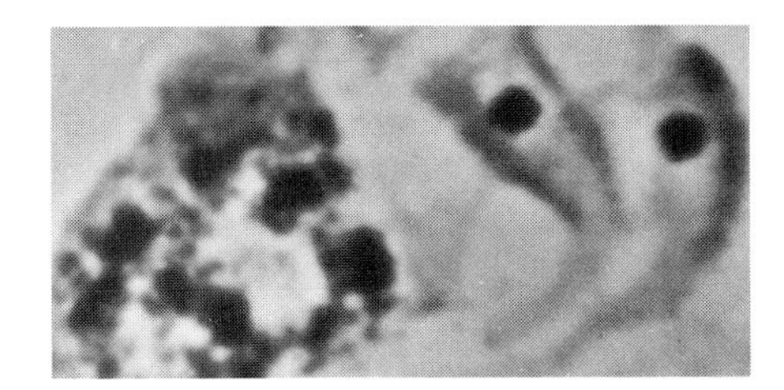

Fig. 5.16. Meiosis in Hessian fly males. Meiosis in males is achiasmate, and spermatogenesis is accompanied by elimination of the haploid paternal set of chromosomes and E chromosomes. **Top:** Metaphase I with four S chromosomes (right side) separating from the E chromosomes. **Upper:** Anaphase I with the four maternally derived S chromosomes, which will eventually bud off and undergo meiosis II, after which the four paternally derived chromosomes along with the E chromosomes will perish. **Center:** Telophase I showing unequal cytokinesis. **Lower:** Telophase II. **Bottom:** Spermatogenesis showing the small nuclei of the spermatids and the degenerating chromatin of the large cell. (Photographs courtesy of Dr. J. Stuart and Dr. J. H. Hatchett.)

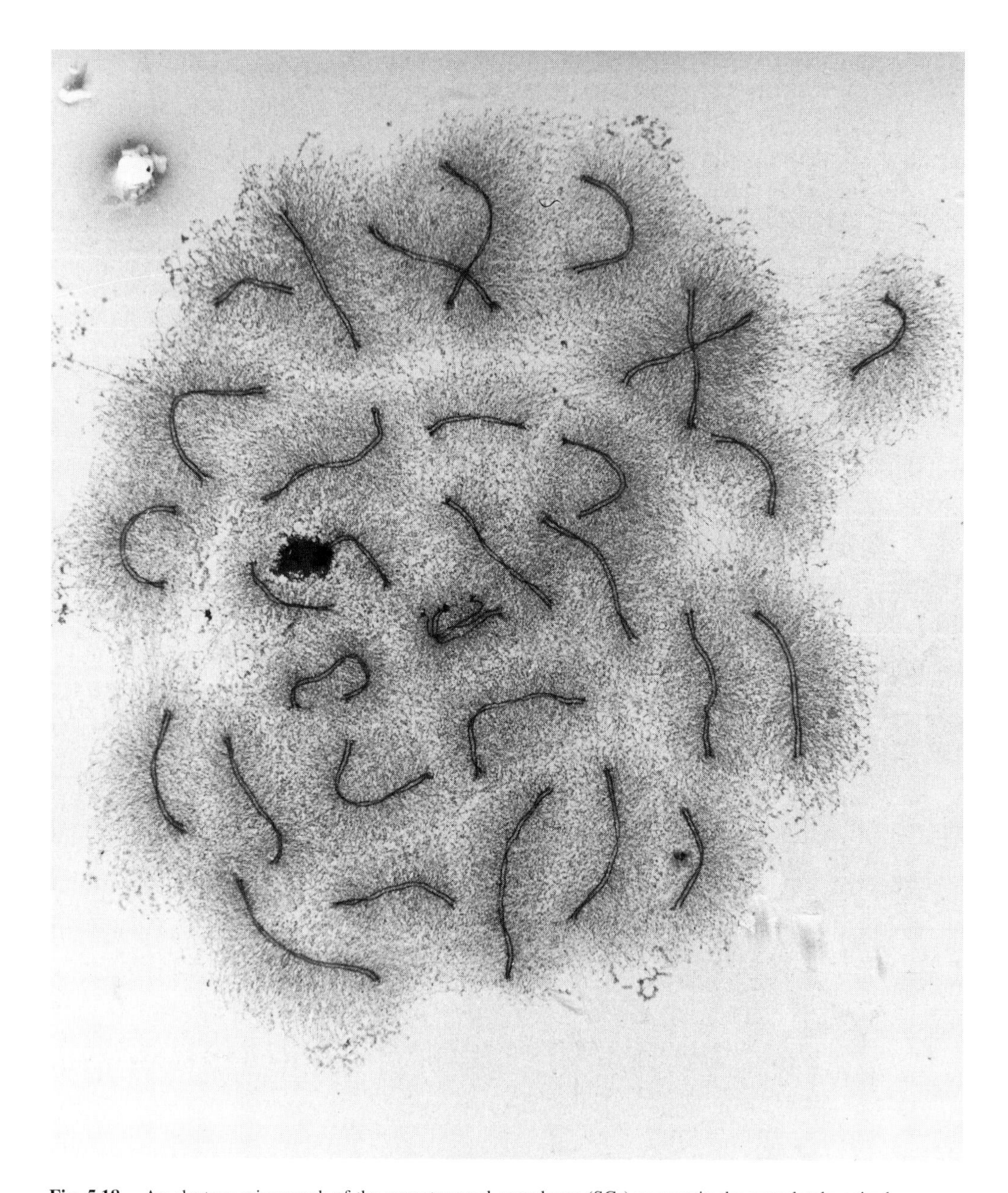

Fig. 5.18. An electron micrograph of the synaptonemal complexes (SCs) present in the completely paired pachytene-stage chromosomes of the moth *Hyalophora columbia*. The SCs are organized by the parallel alignment of the axial cores of homologous chromosomes. Note that the bulk of the chromosomal DNA loops out from the SC core and is thus not intimately associated with the SCs. This particular moth was trisomic for one chromosome (center of the plate). (Photograph courtesy of Dr. Weith and Dr. Traut.)

nodules (Fig. 5.19). They first appear at zygonema, where they are more numerous than in pachynema, and persist until early diplonema.

The synaptonemal complex is mainly composed of protein. The ability to isolate large quantities of synaptonemal complexes in favorable organisms has provided the opportunity to generate monoclonal antibodies to proteins that are specific for these complexes. Several of these proteins have been identified by this procedure.

In the diplotene stage, the synaptonemal complex between paired chromosomes breaks down, but the homologous chromosomes are prevented from taking random positions because of the presence of chiasmata. Remnants of the synaptonemal complex remain at the chiasma sites and may play a role in chiasma maintenance.

As is clear from Fig 5.18, more than 99% of chromosomal DNA lies in looped domains outside the synaptonemal complex. In experiments where the synaptonemal complex from rats was isolated and DNase-treated, a small amount of DNA (less than 1%) was embedded in it (Fig. 5.20). Following proteinase digestion and phenol extraction of the synaptonemal complex, DNA with a length distribution of 50–500 base pairs has been cloned. It consists of short and long interspersed repeat sequences and tandemly repeated GT/CA sequences. It is speculated that these DNA sequences may be involved in homology search and recombination functions. The sequence classes of DNA that are enriched in this DNase-resistant

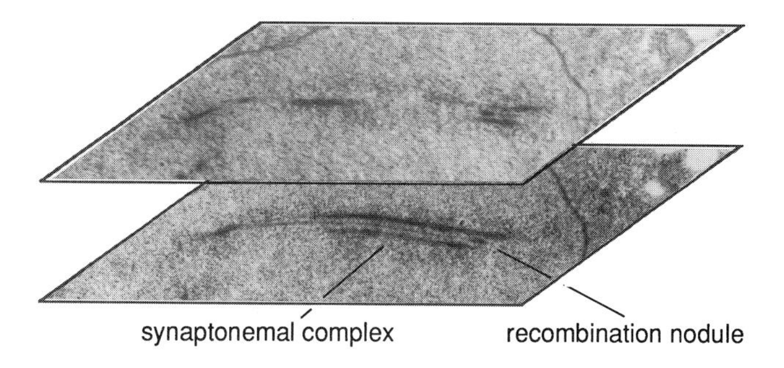

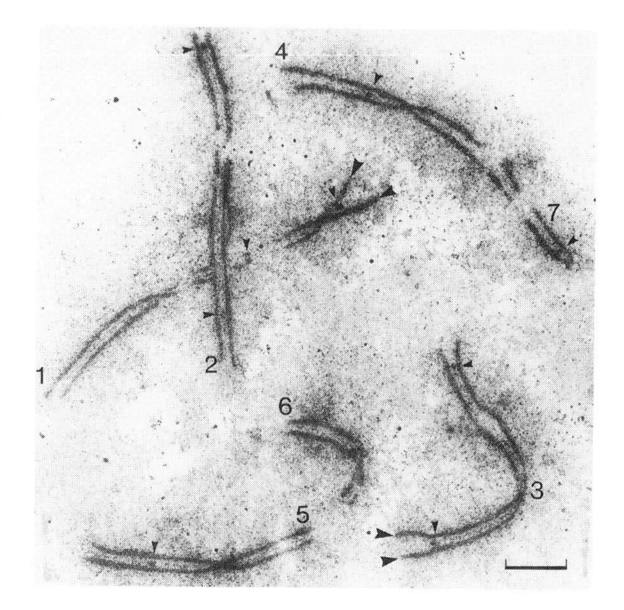

Fig. 5.19. Electron micrographs of recombination nodules in the synaptonemal complexes of pachytene-stage chromosomes of *Neurospora crassa*. The recombination nodules (some indicated by arrows) can be seen in both serial sections (top) and the spread chromosomes (bottom). The large arrowheads (bottom photo) show telomeres that have begun to dissociate, and the numbers indicate the designations of chromosomes in the *Neurospora* karyotype (Lu, 1993). Bar = 1 μm.

material include $d(GT)_n$, as well as long and short interspersed repetitive sequences. The $d(GT)_n$ class is highly transcribed in *Drosophila hydei* testes.

5.2.2 The Synaptonemal Complex Is a Flexible Association Between Chromosome Pairs

The synaptonemal complex forms between chromosome pairs that are juxtapositioned. It does not appear to be the primary basis for homologous chromosome recognition, because synaptonemal complexes also form between nonhomologous chromosomes, and these instances are often referred to as heterologous pairing. A certain level of heterologous pairing and synaptic error is observed in individuals with normal chromosome complements (Table 5.1; Fig. 5.21). Interestingly, in a study of human meiosis, the synaptic errors were much lower in normal males than in subfertile males.

Heterologous pairing is much more common in situations where one or more chromosomes have no homologous partner. Thus, univalent chromosomes in haploid individuals show extensive heterologous pairing and may also show fold-back pairing on themselves. The X–Y sex chromosomes often show heterologous pairing beyond the region of strict homology.

A different form of heterologous pairing is observed in heteromorphic bivalents that differ by an inversion or are of unequal length. In these cases, perfectly homologous synapsis gives way to heterologous synapsis following a process called synaptic adjustment, which refers to the movement of chromosomes relative to each other, after an initial alignment of chromosome segments. Paired chromosome sections that are structurally heterozygous are resolved into perfectly matched bivalents by late pachynema as a result of the process of synaptic adjustment. The synaptonemal complex is present during synaptic adjustment, so it does not represent a rigid binding between paired chromosomes.

5.2.3 DNA Topoisomerase II Unravels Interlocked Bivalents

The necessity for avoiding nonhomologous chromosome entanglements was, for many years, the basis for many models of homol-

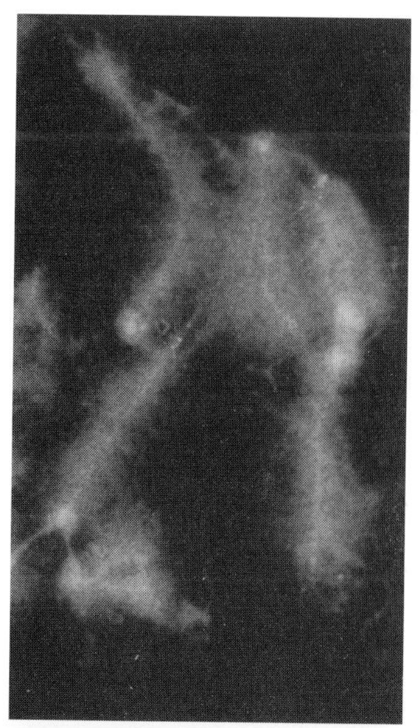

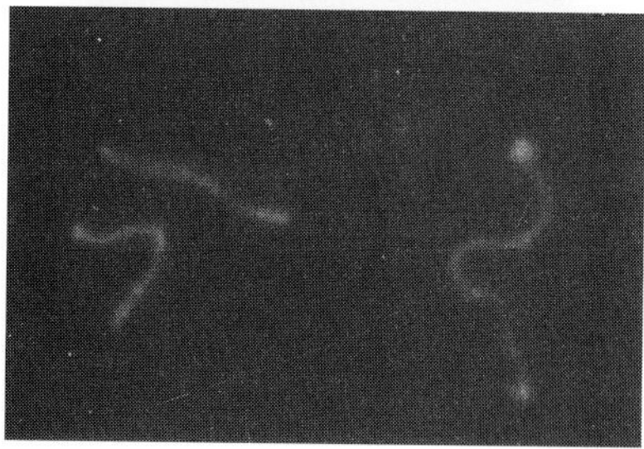

Fig. 5.20. Experimental proof that a certain fraction of the DNA is embedded in the synaptonemal complexes of rat pachytene-stage chromosomes where it is DNase-resistant. The chromosome spreads were stained with the fluorescent DNA-binding dye, DAPI. The chromatin loops and the axial elements (top) fluoresce brightly against the dark background. After extensive DNase treatment and staining with an anti-DNA antibody (bottom), it can be seen that some residual DNA is still intimately associated with the SC. (Reproduced from Pearlman et al., 1992.)

Table 5.1. Reports of Nonhomologous Synapsis and Other Synaptic Irregularities Based on Electron Microscopy of Synaptonemal Complexes

Synaptic Configuration	Species
Foldback synapsis	1× barley, wheat, maize, tomato, lily, *Tradescantia*, *Allium*, 3× *Bombyx*, trisomic and XO mouse.
Mismatched kinetochores	Wheat, 4× *Allium, Festuca* hybrid, potato, 3× *A. sphaerocephalon*, *Choealtis conspersa*, *Macropus rufus*
Straight synapsis due to duplications, deletions, and inversion heterozygotes	Maize, yeast, *Ephestia kuhniella*, chicken, mouse, *Keracris scurra*
Triradial configurations, self-synapsis in rings (spiral synapsis), and synapsis between nonhomologs	Tomato, petunia, snapdragon, maize and barley haploids, tomato trisomics, maize, 6× wheat, *Lolium* hybrids, lily, *Tradescantia*, humans, 3× *Bombyx*, 1× *Physarum polycephalum*, mouse, XO mouse, 3× *A. sphaerocephalon*
Asynapsis	*Coprinus cinereus*, *Tradescantia*, chicken, *Ellobius talpinus*, 3× *A. sphaerocephalon*, aneuploid and euploid mice
Multiple synapsis and partner exchanges	3× lily, 4× potato, 6× wheat, 4× *Allium*, 3× *C. cinereus*, 3× *Lolium*, 4× yeast, aneuploid bulls, 3× and euploid chickens, human trisomy 21, 3× *Bombyx*, 3× *A. sphaerocephalon*

Source: Sherman et al. (1989).

ogous chromosome alignment in preparation for pairing. However, electron-microscopic studies on the nuclei at the zygotene stage from plants and animals have now shown that bivalent interlocking is widespread (Fig. 5.22). Progression to pachynema resolves all interlocks. The properties of the enzyme DNA topoisomerase II provide a molecular mechanism for the separation and resolution of interlocked DNA helices, by catalyzing the reversible breakage and rejoining of double-stranded DNA (*see* Fig. 17.28) in both mitosis and meiosis. The enzyme is present throughout meiosis, with a peak of activity at the pachytene stage, followed by a decrease in activity at diplonema. Immunolocalization studies, using fluorochrome-tagged antibodies to topoisomerase II, have indicated that this enzyme is primarily located in the synaptonemal complex and associated chromatin. Topoisomerase II appears to play an additional role in chromatin condensation during prophase, as well as in the segregation of crossover chromatids.

Studies in yeast have identified a protein, designated Sgs1, that interacts with a region of the topoisomerase II protein located near the C-terminus. This region is not required for activity of the topoisomerase *per se*. The Sgs1 is a helicase type of protein, which most likely helps to remove steric constraints that prevent topoisomerase II from functioning *in vivo*. The Sgs1/topoisomerase complex therefore brings together two enzymes that act in concert to untangle DNA molecules.

5.2.4 Presynaptic Chromosome Alignment at Zygotene Begins at Multiple Sites

Knowledge of the basic mechanisms involved in homologous chromosome recognition and pairing is incomplete. The homologous chromosome alignment at the zygotene stage could conceivably begin at one point (e.g., near the centromeres or telomeres) and proceed zipperlike to completion, or it could begin simultaneously at many points. Experiments with well-characterized cytogenetic systems, which allow reshuffling of chromosome segments, indicate that at zygonema, homologous chromosomes must align at many sites along their lengths. In organisms where one member of each of two pairs of chromosomes is modified by exchanging sections with another chromosome, a cross-shaped configuration is observed in paired chromosomes at the pachytene stage. This

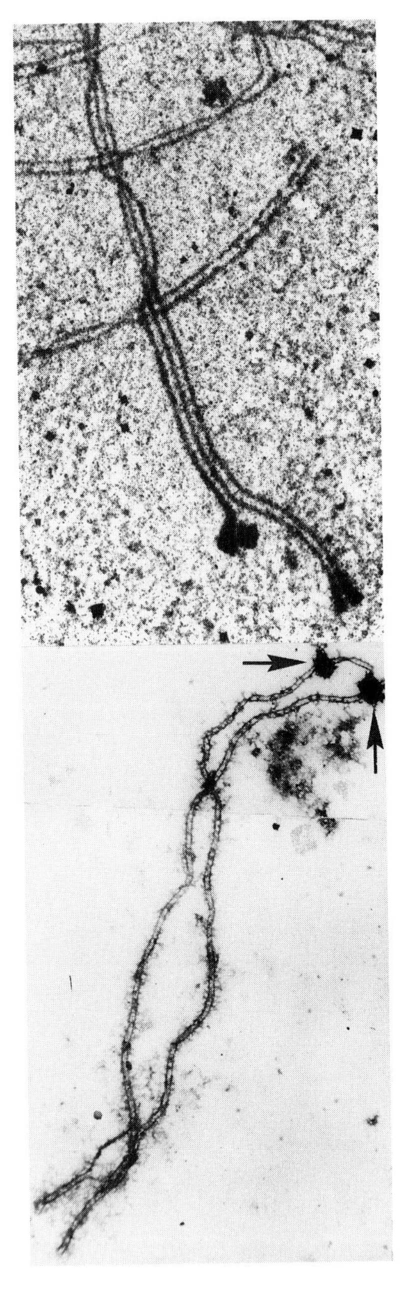

Fig. 5.21. Heterologous pairing and other synaptic irregularities observed from electron microscopic analyses of synaptonemal complexes. **Top:** Triple synapsis in human oocytes. **Bottom:** Mismatched kinetochores (arrows), pairing partner switches, and triple synapsis (lower left) in tetraploid potato. (Photographs kindly supplied by Dr. R. M. Speed and Dr. S. M. Stack.)

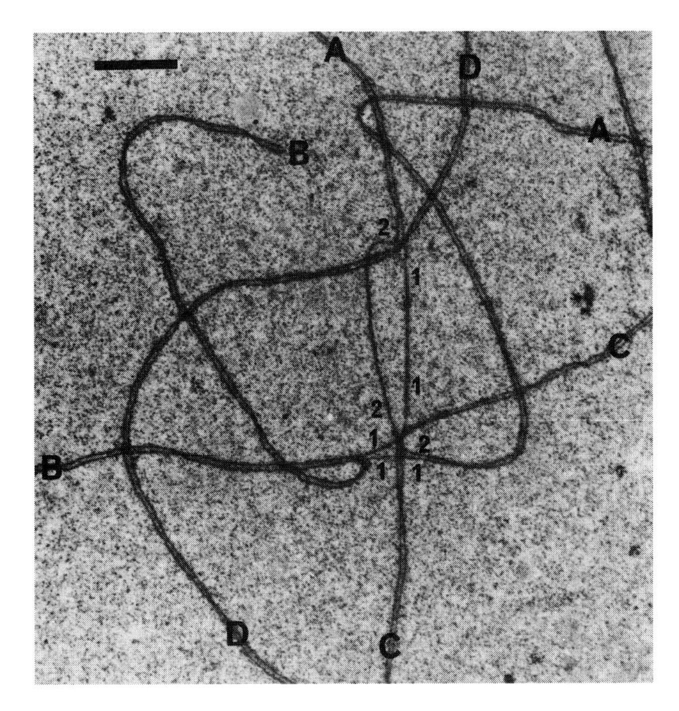

Fig. 5.22. Bivalent interlocking of the synaptonemal complexes of rye pachytene chromosomes. A, B, C, and D indicate four different chromosomes. The 1's and 2's trace different strands of the A-chromosome complex, showing its interlocking with the synaptonemal complexes of B, C, and D. (Photograph courtesy of Dr. C.B. Gillies.)

configuration is a direct result of chromosome pairing along the entire lengths of chromosome arms.

Electron-microscopic analyses, both in three-dimensional serial sections and surface-spread preparations, on a wide range of organisms consistently indicate that synaptonemal-complex formation is initiated at multiple sites within homologous pairs of chromosomes (Fig. 5.23). Further synapsis proceeds by extension from existing paired sites. However, within the multiple-initiation sites, there are primary and secondary centers of synapsis. The primary synapsis begins at the distal ends of the chromosomes, as is evident from the observed congregation of telomeres in one area of the zygonema nucleus (bouquet stage) in a large number of organisms. Thus, at mid-zygonema, with 60% synapsis, most of the buckles (unpaired segments) are observed at interstitial sites of chromosomes.

Chromosome structure may further affect initiation of synapsis. Electron-microscopic analysis of tomato chromosomes indicated that synapsis is essentially complete in distal euchromatic regions before proximal heterochromatic regions initiate synapsis.

5.2.5 Mechanisms of Homologous Recognition

It is known that synaptonemal complexes cannot form until chromosomes are aligned within 0.2 μm of each other. Thus, there must exist mechanisms that bring about initial alignment of homologous chromosomes prior to synaptonemal-complex formation. The problem of homologous-chromosome recognition is further complicated by the fact that almost all organisms share many low- to high-copy-number repetitive DNA-sequence families, located at multiple sites on all chromosomes. The process of pairing at meiosis therefore requires the strict selection of partner chromosomes. The possible mechanisms responsible for this selection process fall into four classes:

Permanent physical association. According to this hypothesis, homologous chromosomes maintain physical associations throughout mitotic and meiotic cell divisions. Thus, the processes in the mitotic division immediately preceding meiosis result in alignment of homologous chromosomes by somatic association. There is experimental evidence both for and against somatic association. In one set of experiments, the application of high temperature or colchicine treatment to disrupt physical associations at premeiotic interphase was correlated with a disruption of chromosome pairing in meiosis, when the respective agents were no longer present. The chromosome movements associated with the formation of the ''bouquet stage'' or synizetic knot at the zygotene stage involve the attachment of chromosome ends to nuclear membranes, and may be involved in chromosome alignment in preparation for synapsis.

Autonomous recognition. In this process, gross physical features of the chromosome are envisaged to provide sites for the recognition of homology. The recognition process may involve distinct oscillatory movements and, thus, lead to the alignment of chromosome segments with similar levels of differential coiling, identical blocks of heterochromatin, matching pairing centers, and/or similar mass. Direct experimental evidence for the existence of autonomous pairing is not available.

Recognition via pairing proteins. The idea that proteins are involved in chromosome pairing was stimulated by the finding that the synaptonemal complex is composed of RNA and protein

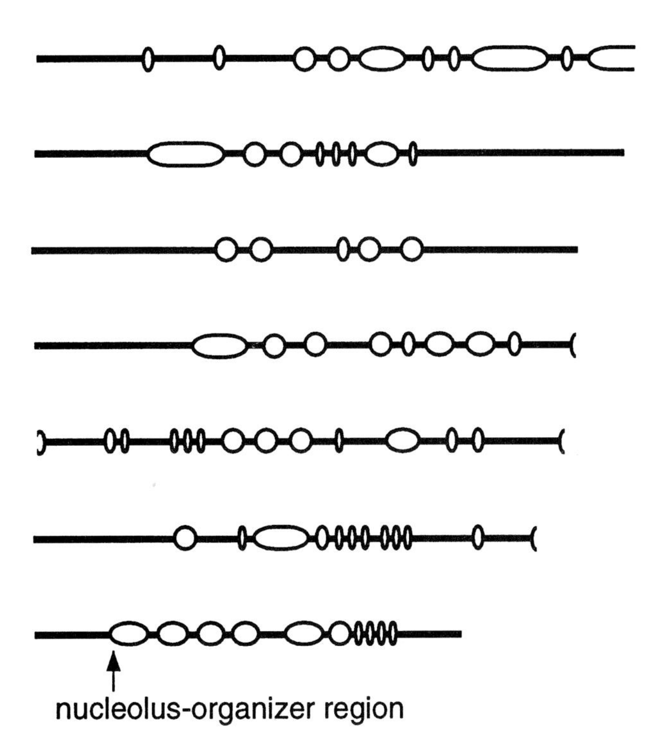

Fig. 5.23. Idiograms of the zygotene-stage chromosomes of rye (*Secale cereale*) showing multiple sites of initiation and formation of the synaptonemal complex in the different chromosomes. The open gaps indicate unpaired regions of the complexes; the closed lines indicate fully paired regions. (Redrawn from Gillies, 1985.)

complexes. The possibility exists that specific-pairing proteins bind to particular DNA sequences and undergo a change in conformation, or allosteric change, thus forming the basis of recognition.

Direct DNA–DNA recognition. The possibility of direct DNA–DNA interaction to establish homologous pairing was first raised by the discovery that a small fraction (0.2%) of the meiotic–prophase nuclear DNA remained unreplicated until the zygotene–pachytene stages. It was postulated that the unreplicated DNA, distributed along the length of the chromosome, contained single-strand breaks, which served as recognition points to form duplexes with the respective DNA regions in the other member of the homologous chromosome pair. These interacting sections of DNA could then provide a focus for the formation of a synaptonemal complex, after which they would be replicated. *In vitro* and *in vivo* biochemical studies indicate that single-stranded DNA "feelers" could be involved in the search for homology between chromosomes (Fig. 5.24). Genetic studies utilizing translocation chromosomes of *Caenorhabditis elegans* have revealed that chromosomes contain localized homology-recognition regions (HRRs) that are involved in the early meiotic events leading to chromosome pairing. In *Ustilago maydis*, an enzyme similar to the *E. coli* recA protein, called rec 1 protein, has been found to catalyze "synapsis" between a double-stranded DNA helix and an invading single strand of DNA that is complementary to one of the strands in the double-stranded DNA. Similar activities have been identified in yeast and human cells. The developing concept is that the interactions between DNA helices involved in chromosome pairing, gene conversion, and genetic recombination are different facets of a single molecular phenomenon, which begins with a DNA homology search for presynaptic chromosome alignment. The nodules observed in the synaptonemal complexes of the zygotene and pachytene stages may be the physical locations of these DNA–DNA interactions. The enzymes needed to catalyze these reactions are currently the subject of intense research.

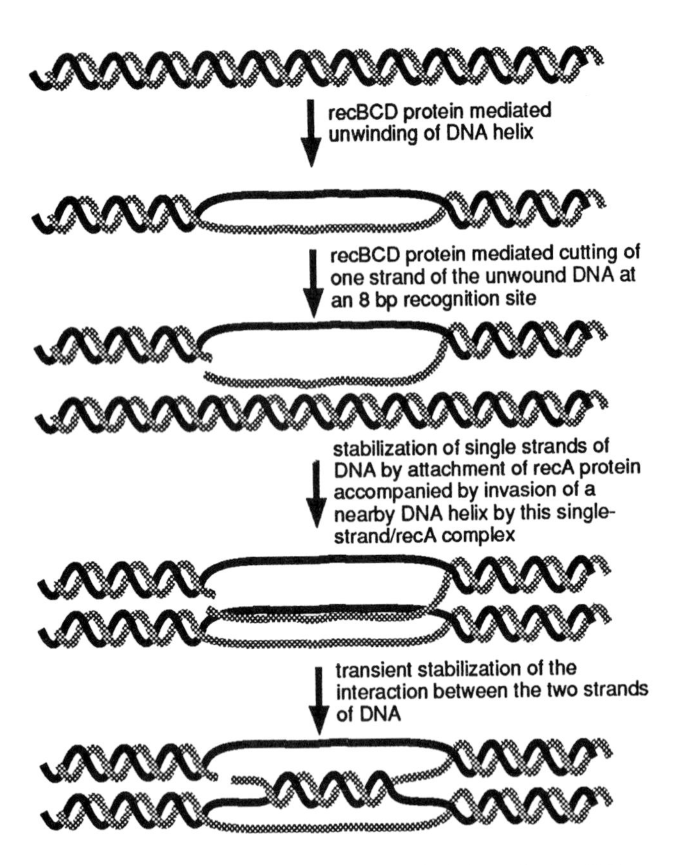

Fig. 5.24. A mechanism of homology search via heteroduplex DNA formation. The accessibility of DNA to the homology search/recombination processes may be related to the level of transcriptional activity in the respective region of the chromosome, because transcription disrupts chromatin structure. The recBCD enzyme and recA protein were discovered in *E. coli* and are among many enzymes required for homologous recombination.

5.3 THE MECHANISM OF CROSSING OVER AND GENETIC RECOMBINATION

The genetic and physical evidence for recombination between homologous chromosomes during meiosis was established in the early 1900s. Cytologically, chiasmata were identified as points where nonsister chromatids had recombined through crossing over. The molecular mechanisms underlying recombination take place during pachynema, when the synaptonemal complex is fully formed and synaptic adjustment is complete.

5.3.1 Crossing Over Occurs at the Four-Strand Stage During Pachynema

Crossing over that leads to genetic recombination can be deduced to occur between nonsister chromatids at the so-called four-strand stage of meiosis, because a single crossover between a pair of chromosomes produces two gametes of the recombinant type and two of the parental type. Furthermore, the chiasmata resulting from crossovers involving two nonsister chromatids can be directly observed at the diplotene stage in favorable cases such as male-grasshopper meiosis (*see* Fig. 5.2), thus indicating that crossing over must have occurred before diplonema. The genetic consequences of environmental shocks such as temperature changes and the application of chemicals, as well as the effects of specific mutations, favor pachynema as the stage where crossing over occurs. This conclusion is consistent with the observation that at this stage the pairing between chromosomes is fully resolved and free of nonhomologous associations. Finally, the recombination nodules, the postulated physical sites of crossing over, are clearly observed at the pachytene stage.

5.3.2 The Recombination Process

Recombination may be part of the homology search by interacting DNA helices beginning at zygonema, possibly at the sites where zygotene-stage nodules are found. Following a break, or nick, in the sugar-phosphate backbone of a DNA helix, a single-stranded segment of DNA is free to invade the helix of a neighboring, nonsister DNA molecule in search of homology (*see* Fig. 5.24). In bacteria, the invasion process, or heteroduplex formation, is catalyzed by the product of the *recA* locus. The recA protein is composed of four subunits, each of which has a molecular weight of 38,000 Da. The protein hydrolyzes ATP to provide the energy for catalyzing the heteroduplex formation, and movement of the invading single DNA strand along the receptor DNA helix by branch migration. These activities are basic prerequisites for any model of re-

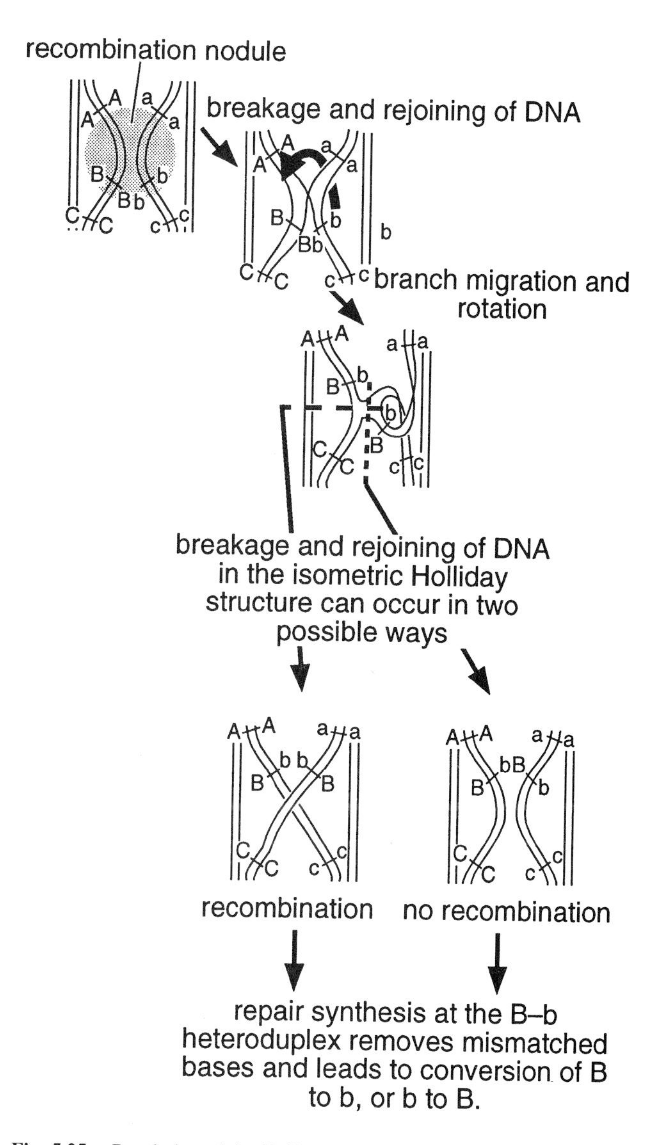

Fig. 5.25. Resolution of the Holliday junction in a recombination nodule. The diagrams show a possible mechanism of recombination between the DNA strands using three gene loci. The classical Holliday model, as well as the Meselson–Radding and the double-chain break models, have been compared in detail by Stahl (1994).

combination. It is also possible that recA catalyses the interaction of two intact DNA helices after partial unwinding.

The net result of the activities leading to genetic recombination is a cross-strand configuration interconnecting the two DNA helices that is often called a Holliday junction (Fig. 5.25) even though several different mechanisms have been postulated to give rise to this final structure. The visible result of the exchange process is a chiasma in the diplotene stage of meiosis. Branch migration before completion of the recombination process means that a heteroduplex at the B/b locus can form whether or not recombination finally occurs between the A/a and C/c loci, and thus accounts for the observation that gene conversion can occur as an integral part of the recombination process. In the final structures shown in the Fig. 5.25 diagrams, the B/b heteroduplex can be converted by normal repair mechanisms to either a BB or bb homoduplex. The isometric Holliday structure can undergo two possible breakage and rejoining events, with only one leading to recombination.

More recent models for recombination have elaborated on the above model and included a double-chain break in one of the DNA helices, followed by 5′–3′ exonuclease action to produce single-strand tails. Both single-strand tails are then postulated to invade an adjoining helix, followed by DNA repair, to form a pair of Holliday junctions that subsequently can be resolved.

The recombination nodules may define the sites of DNA–protein complexes involved in recombination. The number and position of recombination nodules at the pachytene stage closely correlate with the number of recombination events that are expected to have occurred. As a result of these correlations, recombination nodules have been postulated to represent the DNA–protein complex at the site of crossing over.

5.3.3 Relationship Between Genetic Recombination and Cytological Chiasmata

Several types of experimental evidence clearly show that genetic recombination involves physical exchange (crossing over) between nonsister chromatids. Early classical experiments used both genetically and physically marked chromosomes (Fig. 5.26). The differential labeling of chromatids with BrdU has provided the

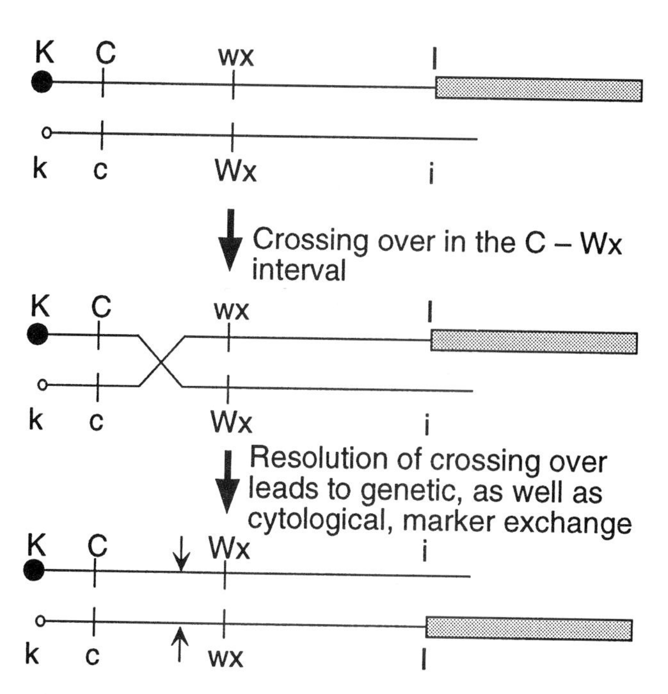

Fig. 5.26. A clear demonstration of the correspondence between cytological chiasmata and genetic recombination. In this diagrammatic representation of a pair of maize chromosomes, K/k = heterochromatic knobs of differing sizes, C /c = colored versus colorless seeds, Wx /wx = nonwaxy versus waxy seeds, I/i = interchanged (translocated) versus normal chromosomes, and the shaded area represents a heterochromatic block involved in the translocation. The presence or absence of all these characters can easily be determined. This experiment provided the first proof that the genetic exchange of markers C and Wx involved an actual physical interchange between nonsister chromatids marked with the cytological markers K and I (Creighton and McClintock, 1931). Only the crossover chromatids are shown, although there are also two parental, noncrossover chromatids.

basis for directly correlating the presence of a chiasma and the breakage–rejoining of DNA strands. Figure 5.27 shows a pair of chromosomes at diakinesis and the exchange of nonsister chromatids at the center of the chiasma.

There also exists a large body of earlier literature on the one-to-one correspondence between a chiasma and a crossover event. Thus, chiasma counts (assuming chiasma terminalization does not occur) have been used to predict total crossovers and genetic recombination. However, recent data on genetic mapping with molecular markers show that genetic maps far exceed the length predicted from chiasma counts (Table 5.2). Although there is little doubt that all chiasmata result from crossovers, the conclusion from recent data is that chiasmata counts underestimate levels of recombination. This may be due to the fact that only a limited number of cytological and genetical markers were used in earlier studies (as compared to recent studies), and the genetic recombination potential was not fully assayed.

Recent work comparing genetic and physical maps in wheat shows that recombination is nonrandomly distributed along the chromosome length. The proximal 50% of each arm surrounding a centromere has a low frequency of recombination, which is extremely high in the distal ends of chromosomes. The distal parts of the chromosomes contain a higher density of genes and, thus, recombination is apparently more frequent in gene-rich regions. In addition, one or two tiny chromosome regions, which are hot spots for genetic recombination, occur in each arm. Double crossovers are common in these regions, and the juxtaposed chiasmata, if they occur, are not resolved under the microscope. In addition, crossovers also occur in chromosome regions where no chiasmata are observed. The emerging view appears to be that not all crossovers mature into chiasmata and that other novel recombination mechanisms may operate that undermine the postulated 1 : 1 correspondence between chiasmata and crossovers.

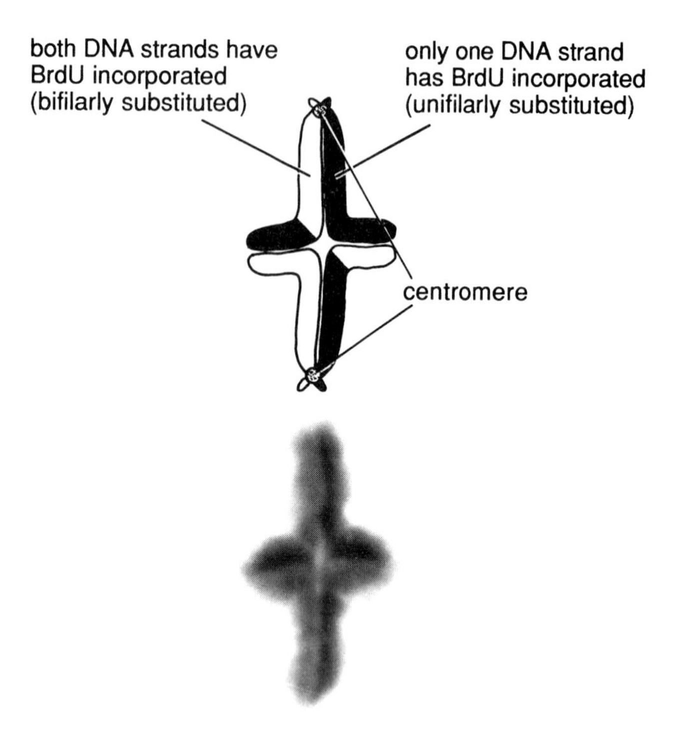

Fig. 5.27. The differential labeling of chromatids with BrdU to indicate the actual physical site of crossing over and recombination between chromatids (Jones, 1987). The diagram and photograph are visual proof that chiasmata arise from the breakage and fusion of nonsister chromatids and that chiasmata do not always terminalize. (Photograph courtesy of G. H. Jones.)

Table 5.2. Estimated Number of Crossovers in Various Plant Species, Based on Genetic Maps, Far Exceed Those Based on Chiasma Counts, Indicating a Lack of Correspondence Between Cytological Chiasmata and Genetic Recombination

	Estimated Number of Crossovers	
Organism	Chiasma Counts	Genetic Maps
Brassica campestris	10.0–18.5	37.0
Brassica oleracea	12.8–14.8	22.2
Hordeum vulgare	13.5–15.6	22.7
Lactuca sativa	14.6–20.7	28.1
Lycopersicon esculentum	16.2–17.0	25.5[a]
Oryza sativa	18.9–27.6	36.7
Pisum sativum	10.3–18.1	29.3
Solanum tuberosum	13.2–14.1	20.7
Zea mays	17.4–25.0	35.8

[a] RFLP map based on an interspecific cross.

Source: Data from Nilsson et al. (1993) and Gill et al. (1997).

5.3.4 The Distribution of Chiasmata

The number, position, and movement of chiasmata are under genetic control. Chiasmata may have a random or nonrandom distribution in a chromosome pair and can vary between closely related species. For example, in *Allium fistulosum*, the chiasmata are located proximally, near the centromere, whereas in *A. cepa*, they are located distally, near the ends of chromosomes. The actual distribution may also be related to the sex of an organism. In *Triturus helveticus*, the chiasmata are located distally in females but proximally in males.

The number of chiasmata observed at diakinesis is generally greater than the number observed at the more condensed stage, metaphase I. This has been argued to result from the terminalization of the chiasmata. In some organisms, such as wheat and rye, however, the terminal positions of chiasmata at diakinesis reflect regions of high genetic recombination.

The occurrence of a chiasma in a given position within a chromosome arm has been shown to reduce the likelihood of another chiasma occurring near it. This is called positive chiasma interference. In certain situations, the opposite effect, negative chiasma interference, can occur. The involvement of two chromatids in a chiasma event usually reduces the likelihood that these same chromatids will be involved in another chiasma, and this phenomenon has been named chromatid interference. Interference is limited to a particular chromosome arm and does not extend across the centromere.

5.3.5 Mechanisms for Chiasma Maintenance at Metaphase I

Besides involvement in the recombination process, other functions of chiasmata are to co-orient bivalents at metaphase I and to ensure proper chromosome segregation during meiosis I. The importance of chiasma maintenance until the end of metaphase I for normal chromosome behavior is emphasized by the fact that it is under specific genetic control. Meiotic mutants in corn and yeast that undergo normal crossing over, but have impaired chiasma maintenance, display nondisjunction at anaphase I, where both members of a bivalent pass to the same pole, causing irregular distributions. It has been suggested that sister-chromatid cohesiveness, involving associations between heterochromatic regions of chromosomes, is important in preserving chiasmata. However, it is difficult to explain the effects of single-gene mutations on chiasma maintenance by such mechanisms.

In yeast, it is possible to construct artificial chromosomes, which have crossing over but undergo nondisjunction because they are presumably unable to maintain chiasmata. This could be due to their small size, or the absence of specific sequences or genes that are involved in the process.

5.4 DISTRIBUTIVE SYSTEM OF SEGREGATION

The phenomenon of distributive pairing appears to provide an extra mechanism for ensuring normal chromosome segregation, particularly of achiasmate or unpaired chromosomes. It has been suggested that there are two pairing phases in meiosis:

1. "Exchange pairing" between homologous chromosomes, which leads to crossing over between specific chromosomes.
2. "Distributive pairing" between homologs or nonhomologs, which ensures segregation of those chromosomes that fail to form chiasmata.

In *Drosophila* females, the smallest chromosome 4, does not form chiasmata, yet at anaphase I, the chromosome 4 pair undergoes a normal reductional division. Similarly, the X-chromosome pair fails to form chiasmata in 5% of meiocytes, yet always segregates reductionally.

Two mutants in *Drosophila*, namely *nod* (no distributive disjunction) and *ncd* (nonclaret disjunction) have been shown to interfere with distributive segregations. The *ncd* locus affects eye color as well as chromosome disjunction. Both the *nod* and *ncd* loci have been cloned and have been shown to code for proteins that are part of the kinesis protein, or protein heavy-chain superfamily involved in microtubule formation. The network of microtubules in cells provides a basic skeleton for maintaining the orderly positioning of organelles and chromosomes. In the case of distributive pairing, microtubules have been observed to radiate between an autosomal pair and the single X chromosome in males, and may ensure distributive segregation of the X chromosome. The hypothesis that arises from these studies is that a specific protein may be involved in the formation of interchromosomal microtubules, which form the basis for distributive segregation at anaphase I.

5.5 MANY GENES CONTROL THE MEIOTIC PROCESS

There are hundreds of genes in plants and animals that affect meiotic divisions and/or chromosome behavior during meiosis. Mutants of these loci arose spontaneously or, more often, were recovered from mutagen treatments. Large inventories of mutants exist in yeast (e.g., *Saccharomyces cerevisiae* and *Schizosaccharomyces pombe*), fungi (e.g., *Neurospora crassa*), animals (e.g., *Drosophila melanogaster*), and plants (e.g., *Zea mays, Vicia faba, Pisum sativum*). The vast majority of mutants behave as single genes (monogenic) that have lost their function (recessive) and cause both male and female sterility.

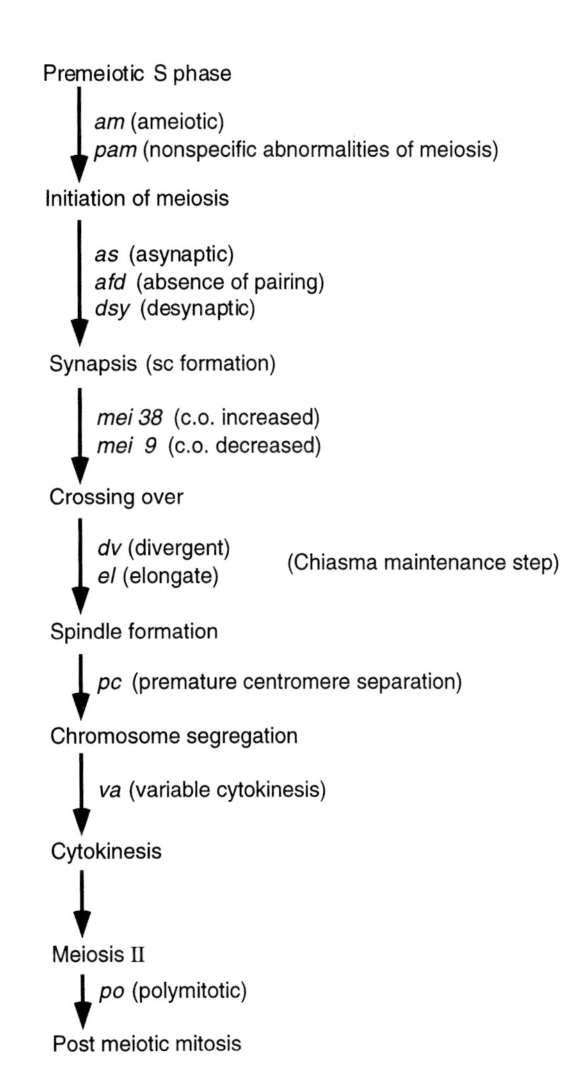

Fig. 5.28. A summary of key meiotic steps showing some of the mutants that affect them, from studies of meiosis and mutants of corn (Golubovskaya, 1979, 1989; Curtis and Doyle, 1991).

The mutants have been characterized with respect to the nature of the block they cause in the sequence of meiotic events, using the light and electron microscopes. Genetic studies, in which individuals carrying meiotic mutants are intercrossed, have been used to determine the effects of these mutants on genetic recombination, and also have suggested that many pathways affect meiosis. Several genes have been characterized biochemically. A small number have been isolated in bacteria, and the gene products have been identified, using the techniques of molecular biology.

The stepwise progression of meiosis is summarized in Fig. 5.28, together with gene mutations that control different steps. These mutant genes can be grouped into the following six broad categories:

1. Genes controlling entry into meiosis, resulting in the elimination of meiosis I and meiosis II.

2. Genes that eliminate meiosis I *or* meiosis II. This group includes genes that can lead to the substitution of meiosis I for meiosis II or vice versa.
3. Genes affecting synapsis by causing failure or malformation of the synaptonemal complex (asynaptic mutants), or precocious dissolution of the synaptonemal complex and/or chiasmata (desynaptic mutants). These mutations affect crossing over as well as chromosome disjunction at anaphase I. In polyploids, there are several *pairing homoeologous* (*Ph*) genes, which suppress synapsis between chromosomes that are not perfectly homologous (*see* Fig. 10.15). These genes are important in the formation of polyploids, but their mechanism of action is not clear at present.
4. Genes affecting chromosome disjunction through malfunction of the spindle apparatus or premature separation of bivalents.
5. Genes causing abnormal behavior of chromosomes such as adhesiveness and fragmentation.
6. Genes affecting cytokinesis. Some genes affect cell division in one sex only. Other genes mimic natural variations of the meiotic process found in diverse groups of organisms.

Many of these mutant genes will be useful in unraveling the biochemistry of meiosis and will also shed light on the evolutionary significance of the specialized meiotic process.

BIBLIOGRAPHY

General

ALBERTS, B., BRAY, D., LEWIS, J., RAFF, M., ROBERTS, K., WATSON, J.D. 1989. Molecular Biology of the Cell, 2nd ed. Garland Publishing Inc., New York.

GILLIES, C.B. (ed.). 1989. Fertility and Chromosome Pairing: Recent Studies in Plants and Animals. CRC Press, Boca Raton, FL.

JOHN, B. 1990. Meiosis. Cambridge University Press, Cambridge.

MOENS, P.B. (ed.) 1987. Meiosis. Academic Press, New York.

Section 5.1

ALBERTSON, D.G., THOMSON, J.N. 1993. Segregation of holokinetic chromosomes at meiosis in the nematode *Caenorhabditis elegans*. Chromosome Res. 1: 15–26.

APPELS, R., DENNIS, E.S., SMYTH, D.R., PEACOCK, W.J. 1981. Two repeated DNA sequences from the heterochromatic regions of rye (*Secale cereale*) chromosomes. Chromosoma 84: 265–277.

CHAPMAN, M., MULCAHY, D.L. 1997. Confocal optical sectioning for meiotic analysis in *Oenothera* species and hybrids. Biotech. Histochem. 72: 105–110.

GODAY, C., PIMPINELLI, S. 1989. Centromere organization in meiotic chromosomes of *Parascaris univalens*. Chromosoma 98: 160–166.

LINDEGREN, C.C. 1932. The genetics of *Neurospora*—II. Segregation of the sex factors in asci of *N. crassa*, *N. sitophila*, and *N. tetrasperma*. Bull. Torrey Bot. Club 59:119–138.

MCCORMICK, S. 1991. Molecular analysis of male gametogenesis in plants. Trends Genet. 7: 298–303.

NODA, S. 1975. Achiasmate meiosis in the *Fritillaria* Japonica group. I. Different modes of bivalent formation in the two sex mother cells. Heredity 34: 373–380.

NOKKALA, S., NOKKALA, C. 1997. The absence of chiasma terminalization and inverted meiosis in males and females of *Myrmus miriformis* Fn. (Corizidae, Heteroptera). Heredity 78: 561–566.

NORDENSKIÖLD, H. 1962. Studies of meiosis in *Luzula purpurea*. Hereditas 48: 503–519.

ORR-WEAVER, T.L. 1995. Meiosis in *Drosophila*: seeing is believing. Proc. Natl. Acad. Sci. USA 92: 10443-10449.

RAJU, N.B. 1980. Meiosis and ascospore genesis in *Neurospora*. Eur. J. Cell Biol. 23: 208–223.

RÉDEI, G.P. 1982. Genetics. Macmillan Publishing Co., Inc., New York.

ROEDER, G.S. 1995. Sex and the single cell: meiosis in yeast. Proc. Natl. Acad. Sci. USA. 92: 10450–10456.

ROSS, K.J., FRANSZ, P., JONES, G.H. 1996. A light microscopic atlas of meiosis in *Arabidopsis thaliana*. Chromosome Res. 4: 507–516.

STUART, J., HATCHETT, J.H. 1988. Cytogenetics of the Hessian fly: II. Inheritance and behavior of somatic and germ-line-limited chromosomes. J. Hered. 79: 190–199.

WRENSCH, D.L., KETHLEY, J.B., NORTON, R.A. 1994. Cytogenetics of holokinetic chromosomes and inverted meiosis, keys to the evolutionary success of mites, with generalizations on eukaryotes. In: Mites: Ecological and Evolutionary Analyses of Life History Patterns (Houck, M.A., ed.). Chapman & Hall, New York. pp. 282–343.

Section 5.2

BASS, H.W., MARSHALL W.F., SEDAT, J.W., AGARD, D.A., CANDE, W.Z. 1997. Telomeres cluster de novo before the initiation of synapsis: a three-dimensional spatial analysis of telomere positions before and during meiotic prophase. J. Cell Biol. 137: 5–18.

BURNHAM, C.R., STOUT, J.T., WEINHEIMER, W.H., KOWLES, R.V., PHILLIPS, R.L. 1972. Chromosome pairing in maize. Genetics 71: 111–126.

CONRAD, M.N., DOMINGUEZ, A.M., DRESSER, M.E. 1997. Ndj1p, a meiotic telomere protein required for normal chromosome synapsis and segregation in yeast. Science 276: 1252–1255.

DOBSON, M.J., PEARLMAN, R.E., KARAISKAKIS, A., SPYROPOULOS, B., MOENS, P.B. 1994. Synaptonemal complex proteins: occurrence, epitope mapping and chromosome disjunction. J. Cell Sci. 107: 2749–2760.

ENGELS, W.R., PRESTON, C.R., JOHNSON-SCHLITZ, D.M. 1994. Long-range cis preference in DNA homology search over the length of a *Drosophila* chromosome. Science 263: 1623–1625.

GILLIES, C.B. 1985. An electron microscopic study of synaptonemal complex formation at zygotene in rye. Chromosoma 92: 165–175.

GILLIES, C.B. 1989. Chromosome pairing and fertility in polyploids. In: Fertility and Chromosome Pairing: Recent Studies in Plants and Animals (Gillies, C.B. ed.). CRC Press Inc., Boca Raton, FL. pp. 109–176.

HABER, J.E., LEUNG, W-Y., BORTS, R.H., LICHTEN, M. 1991. The frequency of meiotic recombination in yeast is independent of the number and position of homologous donor sequences: implications for chromosome pairing. Proc. Natl. Acad. Sci. USA 88: 1120–1124.

HASENKAMPF, C.A. 1996. The synaptonemal complex—the chaperone of crossing over. Chromosome Res. 4: 133-140.

HEYTING, C., DIETRICH, A.J.J., MOER, P.B., DETTMERS, R.J., OFFENBERG, H.H., REDEKER, F.J.W., VINK, A.C.G. 1991. Synaptonemal complex proteins. Genome 31: 81–87.

HUIJSER, P., BECKERS, L., TOP, B., HERMANS, M., SINKE, R., HENNING, W. 1990. Poly(dC-A) poly (dG-T) is highly transcribed in testes of *Drosophila hydei*. Chromosoma 100: 48–55.

KEEGAN, K.S., HOLTZMAN, D.A., PLUG, A.W., CHRISTENSON, E.R., BRAINERD, E.E., FLAGS, G., BENTLEY, N.J., TAYLOR, E.M., MEYN, M.S., MOSS, S.B., CARR, A.M., ASHLEY, T., HOEKSTRA, M.F. 1996. The Atr and Atm protein kinases associate with different sites along meiotically pairing chromosomes. Genes & Development 10: 2423–2437.

LOIDL, J. 1990. The initiation of meiotic chromosome pairing: the cytological view. Genome 33: 759–778.

LU, B.C. 1993. Spreading the synaptonemal complex of *Neurospora crassa*. Chromosoma 102: 464–472.

MAGUIRE, M.P. 1988. Interactive meiotic systems: chromosome structure and function. In: Chromosome Structure and Function, Impact of New Concepts (Gustafson, J.P., Appels, R., eds.). 18th Stadler Genetics Symp., Plenum Press, New York. pp. 117–144.

MAGUIRE, M.P., REISS, R.W. 1994. The relationship of homologous synapsis and crossing over in a maize inversion. Genetics 137: 281–288.

MCKEE, B., HAFERA, L., VRANA, J.A. 1992. Evidence that intergenic spacer repeats of *Drosophila melanogaster* rRNA genes function as X–Y pairing sites in male meiosis, and a general model for achiasmate pairing. Genetics 132: 529–544.

MCKIM, K.S., GREEN-MARROQUIN, B.L., SEKELSKY, J.J., CHIN, G., STEINBERG, C., KHODOSH, R., HAWLEY, R.S. 1998. Meiotic synapsis in the absence of recombination. Science 279: 876–878.

MOENS, P.B. 1990. Unravelling meiotic chromosomes: topoisomerase II and other proteins. J. Cell Sci. 7: 1–3.

PADMORE, R., CAO, L., KLECKNER, N. 1991. Temporal comparison of recombination and synaptonemal complex formation during meiosis in *S. cerevisiae*. Cell 66: 1239–1256.

PEARLMAN, R.E., TSAO, N., MOENS, P.B. 1992. Synaptonemal complexes from Dnase-treated rat pachytene chromosomes contain (GT)n and LINE/SINE sequences. Genetics 130: 865–872.

PLUG, A.W., CLAIRMONT C.A., SAPI E., ASHLEY, T., SWEASY, J.B. 1997. Evidence for a role for DNA polymerase β in mammalian meiosis. Proc. Natl. Acad. Sci. USA 94: 1327–1331.

ROEDER, G.S. 1990. Chromosome synapsis and genetic recombination. Trends in Genetics 6: 385–389.

ROSE, D., THOMAS, W., HOLM, C. 1990. Segregation of recombined chromosomes during meiosis I requires topoisomerase II. Cell 60: 1009–1017.

SHERMAN, J.D., STACK, S.M., ANDERSON, L.K. 1989. Two-dimensional spreads of synaptonemal complexes from Solanaceous plants. IV. Synaptic irregularities. Genome 32: 743–753.

SCHERTHAN, H., WEICH, S., SCHWEGLER, H., HEYTING, C., HÄRLE, M., CREMER, T. 1996. Centromere and telomere movements during early meiotic prophase of mouse and man are associated with the onset of chromosome pairing. J. Cell Biol. 134: 1109–1125.

SMITH, A.V., ROEDER, G.S. 1997. The yeast Red1 protein localizes to the cores of meiotic chromosomes. J. Cell Biol. 136: 957–967.

SPEED, R.M. 1989. Heterologous pairing and fertility in humans. In: Fertility and Chromosome Pairing: Recent Studies in Plants and Animals (Gillies, C.B., ed.). CRC Press Inc., Boca Raton, FL. pp. 1–35.

VON WETTSTEIN, D., RASMUSSEN, S.W., HOLM, P.B. 1984. The synaptonemal complex in genetic segregation. Annu. Rev. Genet. 18: 331–413.

WATT, P.M., LOVIS, E.J., BORTS, R.H., HICKSON, I.D. 1995. Sgs1: a eukaryotic homolog of *E. coli* RocQ that interacts with topoisomerase II in vivo and is required for faithful chromosome segregation. Cell 81: 253–260.

WEITH, A., TRAUT, W. 1980. Synaptonemal complexes with associated chromatin in a moth. Chromosoma 78: 275-291.

ZETKA, M., ROSE, A. 1995. The genetics of meiosis in *Caenorhabditis elegans*. Trends Genet. 11: 27–31.

Section 5.3

CREIGHTON, H.B., MCCLINTOCK, B. 1931. A correlation of cytological and genetical crossing-over in *Zea mays*. Proc. Natl. Acad. Sci. USA 17: 492–497.

GILL, B.S., GILL, K.S., FRIEBE, B., ENDO, T.R. 1997. Expanding genetic maps: reevaluation of the relationship between chiasmata and crossovers. In: Chromosomes Today, Vol. 12 (Henriques-Gil, N., Parker, J.S., Puertas, M.J., eds.). Chapman & Hall, New York. pp. 283–298.

GILL, K.S., GILL, B.S., ENDO, T.R., BOYKO, E.V. 1996. Identification and high-density mapping of gene-rich regions in chromosome group 5 of wheat. Genetics. 143: 1001–1012.

HASTINGS, P.J. 1992. Mechanism and control of recombination in fungi. Mut. Res. 284: 97–110.

JANG, J.K., MESSINA, L., ERDMAN, M.B., ARBEL, T., HAWLEY, R.S. 1995. Induction of metaphase arrest in *Drosophila* oocytes by chiasma-based kinetochore tension. Science 268: 1917–1919.

JONES, G.H. 1987. Chiasmata. In: Meiosis (Moens, P.B., ed.). Academic Press, New York. pp. 213–244.

KABUCK, D.B., GUACCI, V., BARBER, D., MAHON, J.W. 1992. Chromosome size-dependent control of meiotic recombination. Science 256: 228–232.

LICHTEN, M., GOLDMAN, A.S.H. 1995. Meiotic recombination hotspots. Ann. Rev. Genet. 29:423–444.

LINDAHL, K.F. 1991. His and hers recombinatorial hotspots. Trends Genet. 7: 273–276.

MOORE, S.P., RICH, A., RISHEL, R. 1989. The human recombination strand exchange process. Genome 31: 45–52.

NILSSON, N-O., SÄLL, T., BENGTSSON, B. 1993. Chiasma recombination data in plants: are they compatible? Trends Genet. 9: 344–347.

OTT, J. 1997. Testing for interference in human genetic maps. J. Mol. Med. 75: 414–419.

QUEVEDO, C., DEL CERRO, A.L., SANTOS, J.L., JONES, G.H. 1997.

Correlated variation of chiasma frequency and synaptonemal complex length in *Locusta migratoria*. Heredity 78: 515–519.

STACK, S.M., ROELOFS, D. 1996. Localized chiasmata and meiotic nodules in the tetraploid onion *Allium porrum*. Genome 39: 770–783.

STAHL, F.W. 1994. The Holliday junction on its thirtieth anniversary. Genetics 138: 241–246.

STAHL, F.W. 1996. Meiotic recombination in yeast: coronation of the double-strand-break repair model. Cell 87: 965–968.

SYBENGA, J. 1996. Recombination and chiasmata: few but intriguing discrepancies. Genome 39: 473–484.

WU, T-C., LICHTEN, M. 1994. Meiosis-induced double-strand break sites determined by yeast chromatin structure. Science 236: 515–518.

Section 5.4

BICKEL, S.E., ORR-WEAVER, T.L. 1996. Holding chromatids together to ensure they go their separate ways. BioEssays 18: 293–300.

CARPENTER, A.T.C. 1991. Distributive segregation: motors in the polar wind. Cell 64: 885–890.

ENDOW, S.A. 1992. Meiotic chromosome distribution in *Drosophila* oocytes: Roles of two kinesin-related proteins. Chromosoma 102: 1–8.

FUGE, H. 1997. Nonrandom chromosome segregation in male meiosis of a Sciarid fly: elimination of paternal chromosomes in first division is mediated by non-kinetochore microtubules. Cell Motility and the Cytoskeleton 36: 84–94.

GILLIES, C.B. 1975. Chromosome pairing and recombination. In: The Eukaryote Chromosome (Peacock, W.J., Brock, R.D., eds.). ANU Press, Canberra, Australia. pp. 313–325.

HAWLEY, S., THEARKAUF, W.E. 1993. Requiem for distributive segregation: achiasmate segregation in *Drosophila* females. Trends Genet. 9: 310–316.

KARPEN, G.H., LE, M-H., LE, H. 1996. Centric heterochromatin and the efficiency of achiasmate disjunction in *Drosophila* female meiosis. Science 273: 118–122.

MURPHY, T.D., KARPEN, G.H. 1995. Interactions between the *nod+* kinesin-like gene and extracentromeric sequences are required for transmission of a *Drosophila* minichromosome. Cell 81: 139–148.

WILLIAMS, B.C., GATTI, M., GOLDBERG, M.L. 1996. Bipolar spindle attachments affect redistributions of ZA10, a *Drosophila* centromere/kinetochore component required for accurate chromosome segregation. J. Cell Biol. 134: 1127–1140.

ZHANG, P., KNOWLES, B.A., GOLDSTEIN, L.S.B., HAWLEY, R.S. 1990. A kinesin-like protein required for distributive chromosome segregation in *Drosophila*. Cell 62: 1053–1062.

Section 5.5

CURTIS, C.A., DOYLE, G.G. 1991. Double meiotic mutants of maize: implications for the genetic regulation of meiosis. J. Hered. 82: 156–163.

ENGEBRECHT, J., VOELKEL-MEIMAN, K., ROEDER, G.S. 1991. Meiosis-specific RNA splicing in yeast. Cell 66: 1257–1268.

LI, Y.F., NUMATA, M., WAHLS, W.P., SMITH, G.R. 1997. Region-specific meiotic recombination in *Schizosaccharomyces pombe*: the *rec11* gene. Molec. Micro. 23: 869–878.

GOLUBOVSKAYA, I.N. 1979. Genetic control of meiosis. Int. Rev. Cytol. 58: 247–290.

GOLUBOVSKAYA, I.N. 1989. Meiosis in maize: *mei* genes and conception of genetic control of meiosis. Adv. Genet. 26: 149–192.

HOEKSTRA, M.F., DEMAGGIO, A.J., DHILLON, N. 1991. Genetically identified protein kinases in yeast. II. DNA metabolism and meiosis. Trends Genet. 7: 293–297.

HOBOLTH, P. 1983. Chromosome pairing in allohexaploid wheat var. Chinese Spring: transformation of multivalents into bivalents, a mechanism for exclusive bivalent formation. Carlsberg Res. Commun. 46: 129–173.

PAGE, A.W., ORR-WEAVER, T.L. 1996. The *Drosophila* genes *grauzone* and *cortex* are necessary for proper female meiosis. J. Cell Sci. 109: 1707–1715.

RAJU, N.B. 1992. Genetic control of the sexual cycle in *Neurospora*. Mycol. Res. 96: 241–262.

SEITZ, L.C., TANG, K., CUMMINGS, W.J., ZOLAN, M.E. 1996. The *rad9* gene of *Coprinus cinereus* encodes a proline-rich protein required for meiotic chromosome condensation and synapsis. Genetics 142: 1105–1117.

SHERIDAN, W.F., AVALKINA, N.A., SHAMROV, I.I., BATYGINA, T.B., GOLUBOVSKAYA, I.N. 1996. The *mac1* gene: controlling the commitment to the meiotic pathway in maize. Genetics 142: 1009–1020.

SPIELMAN, M., PREUSS, D., LI, F-L., BROWNE, W.E., SCOTT, R.J., DICKINSON, H.G. 1997. TETRASPORE is required for male meiotic cytokinesis in *Arabidopsis thaliana*. Development 124: 2645–2657.

WALLIS, J.W., CHREBET, G., BRODSKY, G., ROLFE, M., ROTHSTEIN, R. 1989. A hyper-recombination mutation in *S. cerevisiae* identifies a novel eukaryotic topoisomerase. Cell 58: 409–419.

6

Chromosome Morphology and Number

- A karyotype describes the chromosome complement of an individual or species in terms of number, size, and morphology of its chromosomes.
- Karyotypes can be based on mitotic or meiotic chromosomes and are enhanced by chromosome-banding techniques.
- Giant chromosomes occur in specialized tissues due to extra cycles of DNA replication without cell division.
- Sex chromosomes carry the genes that determine the sex of an organism, and they show unusual sex-specific variation in chromatin structure, behavior, and gene expression.
- Extra chromosomes, referred to as B chromosomes, occur in some organisms and although not essential for survival, they contain genes that enhance their retention and transmission to offspring.

The term karyotype is used to describe the chromosome number of an organism, as well as the size, arm ratio, and other landmark features of individual chromosomes. A diagrammatic representation of the chromosomes, based on observations and measurements in a number of cells, is an idiogram, which provides a starting point for detailed analyses of DNA sequence organization, gene order, and chromosome structure and function, especially with respect to cytological landmarks. Karyotypes are also useful for interpreting the arrangement of chromosomes within interphase nuclei. Chromosomes at the prometaphase or metaphase stage of the mitotic cell cycle are the most useful in somatic or mitotic karyotype analysis; certain stages of the meiotic cell cycle are also suitable for constructing meiotic karyotypes.

6.1 MITOTIC KARYOTYPE ANALYSIS

Any tissue containing rapidly dividing cells is suitable for mitotic karyotype analysis after treatment with drugs such as colchicine or other spindle-inhibiting agents, which arrest cells in metaphase. In plants, the root or shoot meristems are the main source of dividing cells. In animals, the most common source of dividing tissue is cultured lymphocytes stimulated to divide by phytohemagglutinin (PHA). Chemicals that stimulate or synchronize cell divisions, break down the spindle, and stain the chromosomes provide the basis for studying mitotic chromosomes for karyotype analysis (Table 6.1). To prepare chromosomes for microscopic examination after appropriate pretreatment, arrested mitotic cells can be either squashed or gently dropped onto a glass slide. This procedure lyses the mitotic cells and disperses the chromosomes within a limited area.

6.1.1 Number and Morphology of Chromosomes Are Usually Constant for an Organism

The somatic chromosome number ($2n$) for a species is usually constant, although some exceptions do occur, as discussed later. However, the chromosome number may vary widely among diverse groups of eukaryotic organisms, from the minimum of $2n = 2$ in an ant species, *Myrmecia pilosula* (Fig. 6.1), and $2n = 4$ in the Australian daisy *Brachycome dichromosomatica* (Fig. 6.1), *Haplopappus gracilis*, and *Zingeria biebersteiniana* (Poaceae), to $2n > 1000$ in some ferns and palms (Fig. 6.2). The individual chromosomes that constitute the karyotype of a species have distinctive morphological features. In constructing a karyotype, the largest pair of chromosomes is usually designated as number 1 and the rest of the chromosome pairs are numbered in order of their relative diminishing size, to the smallest pair.

The position of the centromere or primary constriction relative to the ends of the chromosome is an important parameter because it determines the arm ratio of a chromosome. It may be terminal (telocentric), almost terminal (acrocentric), subterminal, median to submedian, or metacentric to submetacentric with chromosome arms of more or less equal length (Fig. 6.3).

A prominent landmark that is often distinguishable on one or more chromosome pairs in a karyotype is called a satellite (*see* Figs. 6.1 and 6.4). Satellites are created by secondary constrictions, which define the locations of ribosomal RNA genes (*see* Chapter 20). In some species, tertiary constrictions can also occur, and although these are not as prominent as the primary and secondary constrictions, they often correspond to the sites of cold-sensitive heterochromatic regions within the genome (Fig. 6.4).

Table 6.1. Chemicals Used to Observe Chromosomes

Category	Chemical
Stimulation of cell division	Phytohemagglutinin stimulates lymphocytes to divide
Cell synchronization	Hydroxyurea; amethopterin (folic acid analog); cold treatment
Spindle dispersion and spreading of chromosomes	Colchicine; α-bromonaphthalene; 8-hydroxyquinoline; hypotonic treatment (20–30% isotonic).
Staining chromosomes	Feulgen (DNA-specific staining); Giemsa; carmine

Source: Barch et al. (1991) and Sharma and Sharma (1994).

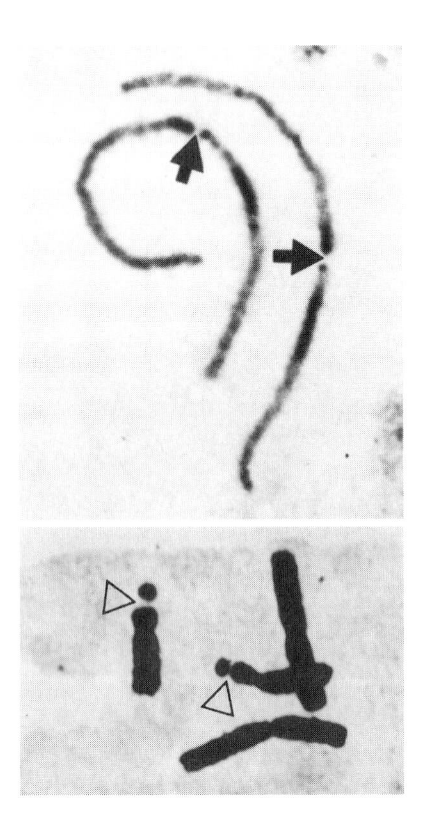

Fig. 6.1. This and Fig. 6.2 demonstrate variations in chromosome number in eukaryotic organisms. Among chromosome complements, the lowest number among animals (top) is found in an ant, *Myrmecia pilosula* ($2n = 2$; Crosland and Crozier, 1986), and one of the lowest among plants (bottom) is the daisy *Brachycome dichromosomatica* ($2n = 4$; Watanabe et al., 1975). Note the primary constriction (solid arrows) at the site where the centromere divides each chromosome into two arms, which may be described as short or long based on size, or more simply, as left and right. In the daisy, in addition to the primary constriction, a secondary constriction (open arrows) shows the position of the nucleolus-organizer region, which separates the distal satellite from the short arm of each chromosome.

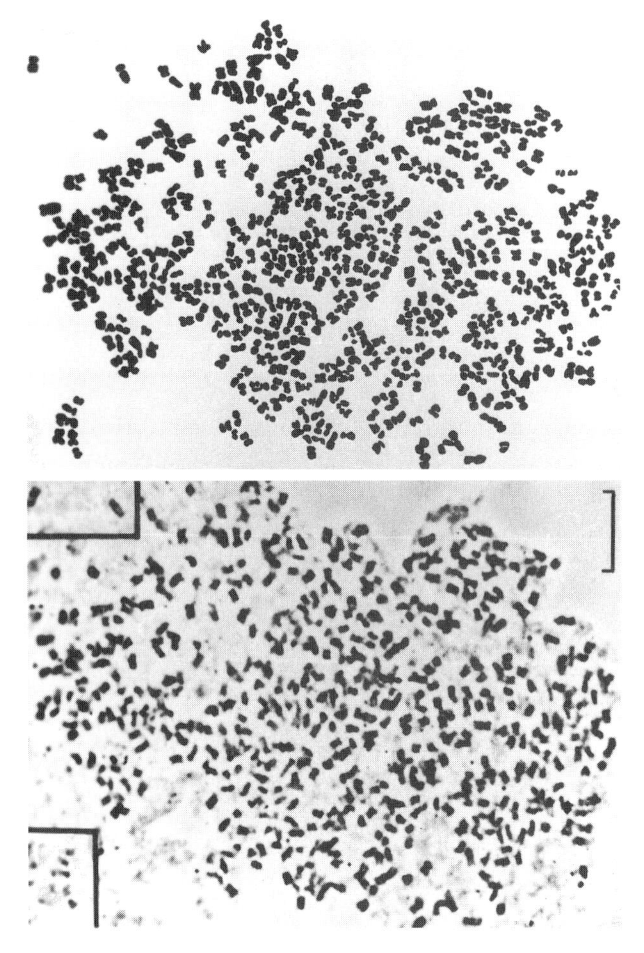

Fig. 6.2. Among the highest-numbered chromosomal complements, the palm *Voanioala gerardii* has a $2n$ complement of approximately 596 chromosomes. A photograph (bottom) and a diagrammatic representation (top) of the chromosomes (Johnson et al., 1989). Bar = 10 μm.

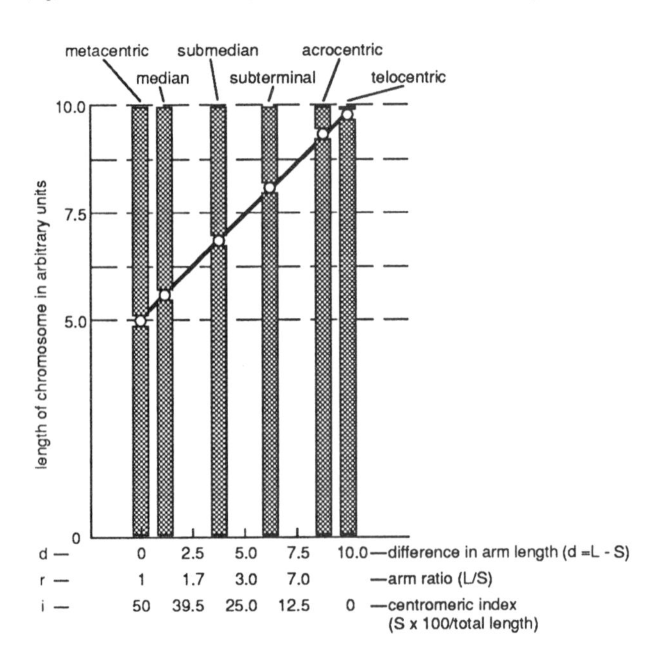

Fig. 6.3. Chromosome arm-length ratios are a useful parameter for the classification of chromosomes into different types. (Adapted from Levan et al., 1964.)

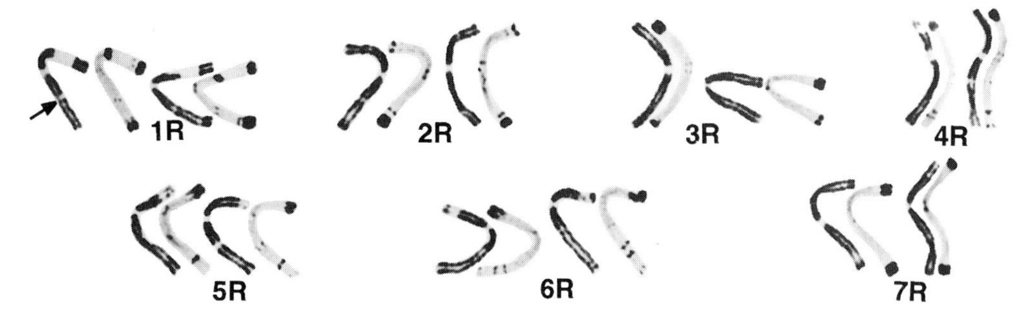

Fig. 6.4. The karyotype of cereal rye (*Secale cereale*), $2n = 14$. The chromosome preparations were made from root tips pretreated with ice water (0–4°C) for 24 h and stained by a sequential acetocarmine and C-banding procedure (left and right, respectively, for each chromosome). Note that the cold treatment, clearly illustrated for chromosome 1R, reveals tertiary constrictions in the long arm (arrow) in addition to the primary (centromere) constriction and a secondary constriction in the short arm. The chromosomes 2R to 7R show primary and tertiary constrictions. The C-banding reveals the location of constitutive heterochromatin bands (C bands). Some C bands correspond to the location of tertiary constrictions and represent cold-sensitive heterochromatin, first described in *Trillium* by Darlington and La Cour (1940). Note that most rye chromosomes are of similar size. However, on analyzing rye chromosome-addition lines to wheat, Gill and Kimber (1974) were able to identify individual rye chromosomes using C-banding patterns and cytological parameters such as arm ratios. [Figure adapted from Gill and Sears (1988) from an original photograph prepared by T. R. Endo.]

Chromosomes may also vary in relative and absolute sizes. Among diverse groups of organisms, the size may range from submicroscopic, as in many fungi, to mitotic–metaphase chromosomes of 20 μm or greater in size. Chromosome size can also vary between closely related organisms, and in many such cases, the size variation is caused by amplification or deletion of the repetitive fraction of DNA. For a particular organism, the chromosome size differences may be small (*see* Fig. 6.4), or chromosomes can be clearly separated into two size classes, in which case they form a bimodal karyotype (Fig. 6.5). The small chromosomes of yeast and many other fungi are best resolved by pulsed-field gel electrophoresis (Fig. 6.6).

6.1.2 Prophase Chromosomes Reveal Additional Cytological Landmarks

The earliest, and most biologically meaningful, differentiation of chromosome structure came from observations on the behavior of chromatin in early prophase chromosomes. With standard dyes such as carmine, certain parts of chromosomes were found to maintain a darker-staining appearance in interphase, forming heteropycnotic bodies. The darkly staining chromatin was called heterochromatin to distinguish it from the lighter-staining euchromatin. The differentiation of chromosomes into heterochromatic and euchromatic regions is most clearly observed in mitotic prometaphase chromosomes, as well as in meiotic pachytene chromosomes.

The euchromatic regions of prophase chromosomes, particularly pachytene chromosomes in meiotic cells and polytene chromosomes in insect salivary glands, show darkly staining structures called chromomeres (*see* Fig. 6.16). The size and distribution of chromomeres are also reproducible features of chromosomes and possibly arise from the differential coiling of the chromatin.

Prophase chromosomes also show characteristic ends, which were initially called telochromomeres, but were later called telomeres. Telomeres have special properties that are essential to the stability of chromosomes, as demonstrated by the fact that X-ray-induced terminal deletions lead to chromosome instability. Furthermore, broken chromosomes that have healed contain telomeric DNA sequences at the break sites.

6.2 SPECIAL STAINING TECHNIQUES REVEAL CYTOLOGICAL LANDMARKS IN METAPHASE CHROMOSOMES

Although mitotic–metaphase chromosomes are used extensively in karyotype analysis, they are highly condensed structures and do not always reveal clear distinctions between regions such as chromomeres, euchromatin, and heterochromatin. Although prophase chromosomes can provide some differentiation of the chromosome structure, they can be difficult to analyze due to their extended and overlapping nature. Banding procedures on mitotic chromosomes, introduced in the early 1970s, provide a procedure for differentially staining metaphase chromosomes and are widely applied to animal and plant material (Fig. 6.7, Table 6.2).

A fluorochrome, quinacrine mustard, was the first compound to reveal a reproducible banding structure, or Q-banding, within metaphase chromosomes (*see* Fig. 6.7; also Fig. 2.13). Q-banding results in bright-fluorescent bands against a dull-fluorescent background and was quickly applied to human prenatal diagnosis. In 1971, a C-banding procedure (C = constitutive) clearly differentiated between euchromatin and heterochromatin in metaphase chromosomes. At about the same time, the G-banding (G = Giemsa) procedure revealed a chromomere-like differentiation along the length of the chromosomes. Later, an R-banding (R = reverse) technique preferentially stained euchromatin in a way that was the reverse of G-banding. These banding techniques had a profound impact and heralded the birth of molecular cytogenetics by providing clear landmarks that could be used to relate the physical structure of chromosomes to genetic linkage groups. The ease of collecting mitotic–metaphase chromosomes and revealing their

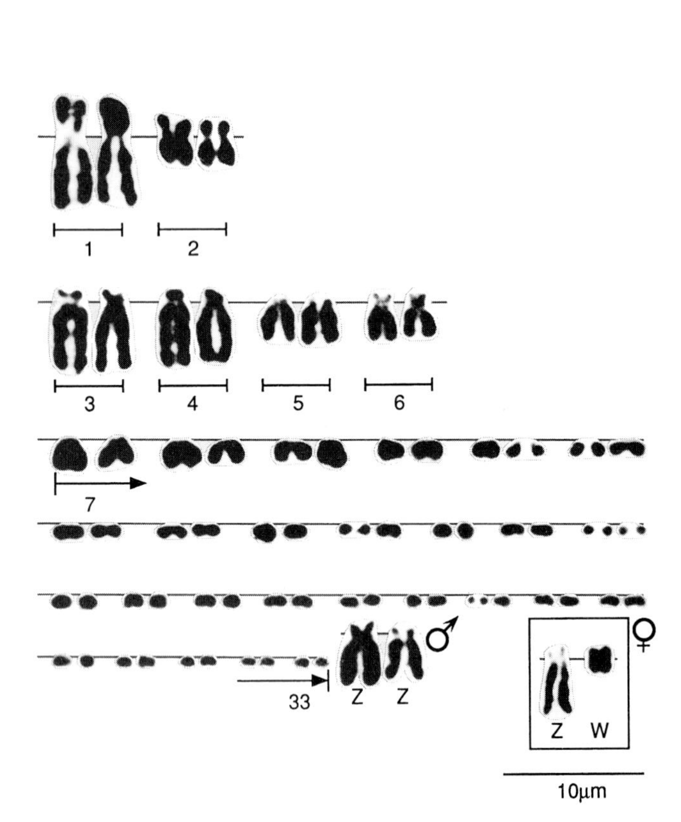

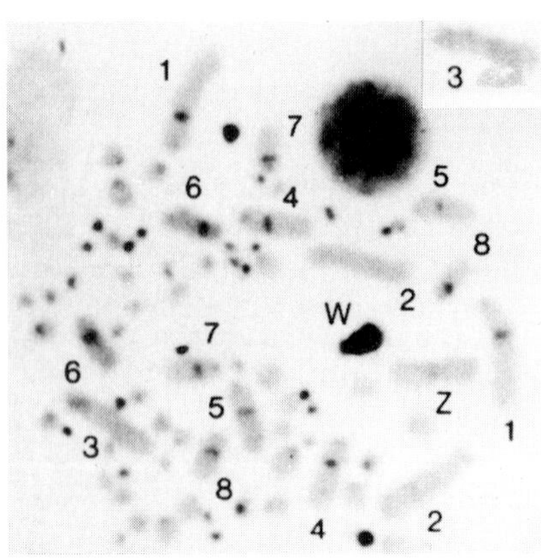

Fig. 6.5. The chromosomes of birds. In contrast to organisms with similar-sized chromosomes, birds have chromosomes that fall into two distinct size classes. The upper panel illustrates a karyotype from *Turdoides striatus striatus* (Ray-Chaudhuri et al., 1969) where 34 pairs of chromosomes are classified into a bimodal karyotype of 7 large chromosome pairs (including the sex chromosomes) and 27 pairs of small chromosomes. The galah parrot *Cacatua roseicapillus* (lower panel) has nine pairs of relatively large, easily detectable chromosomes (including the Z and W sex chromosomes), and a large number of virtually indistinguishable, very small chromosomes. (Christidis et al., 1991). The chromosomes were C-banded (*see* Fig. 6.8) to help in differentiating between chromosomes.

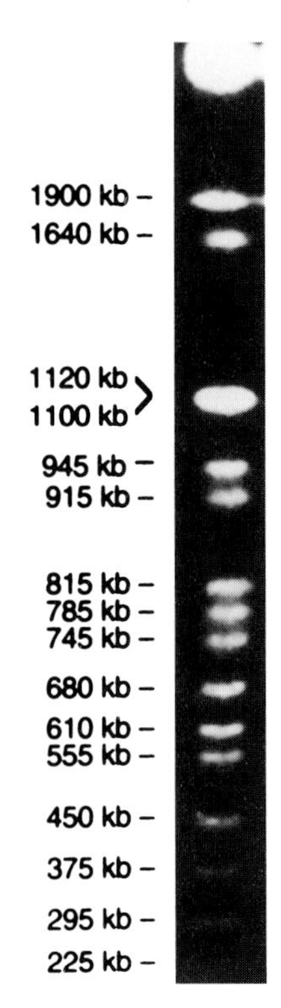

Fig. 6.6. The use of pulsed-field gel electrophoresis to separate the 16 pairs of yeast (*Saccharomyces cerevisiae*) chromosomes. These are generally too large to be resolved by standard gel electrophoresis and too small for microscopic analysis (Carle and Olson, 1985). The chromosomes vary in size from 225 to 1900 kb and two of the pairs (VII and XV) are almost identical in size. Such long lengths of DNA are separated by regularly changing the direction of the electric current within the separating gel. [Photo from Pharmacia Biotech Catalog (1996).] The nucleotide sequence of all these chromosomes have been determined (*see* Section 24.3, Chapter 24).

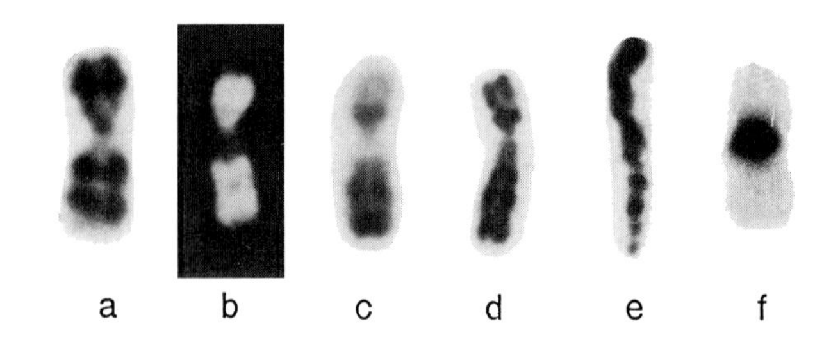

Fig. 6.7. Various chromosome-banding methods reveal the substructure of human chromosome No. 9 at mitotic metaphase except (e), which shows meiotic pachytene morphology (Bickmore and Sumner, 1989). The chromosome has been labeled by (a) G-banding, (b) Q-banding, (c) R-banding, (d) replication banding where early-replicating regions are dark, (e) pachytene chromomeres, and (f) C-banding. Note that the G-banding (a) and Q-banding (b) patterns are very similar and somewhat resemble the chromosome pattern found in pachytene (e) chromosomes. Note also that R-banding (c) and replication banding (d) resemble one another. The heterochromatic centromeric region stained by C-banding (f) is not labeled by the other procedures. The dimorphism of chromatin, with heterochromatic (dark staining) and euchromatic (light staining) regions, was first described by Heitz (1928).

Table 6.2. Chromosome-Banding Techniques

Name of Technique	Basis of Technique	Organism Studied
Q-banding	Binding of AT-specific fluorochromes such as quinacrine and 4′,6-diamidino-2-phenylindole (DAPI), to DNA	Reptiles, birds, mammals
G-banding	Giemsa staining after incubation in warm SSC[a] or trypsin	Fish, amphibia, reptiles, birds, mammals
R-banding	Giemsa staining after incubation in hot buffer	Mammals
Replication banding	Incorporation of BrdU during either early or late S phase followed by Giemsa staining	Plants, vertebrates
C-banding	Giemsa staining after alkali treatment	Most plants and animals

[a] 0.15 *M* sodium chloride, 0.015 *M* sodium citrate.

Source: After Bickmore and Sumner (1989).

banding patterns means that "difficult" organisms with many chromosomes, such as human beings and wheat, are now model systems for cytogenetic analysis.

6.2.1 C Bands Reveal the Location of Constitutive Heterochromatin

The technique of C-banding can be applied universally to reveal the location of heterochromatin in mitotic–metaphase chromosomes. In a typical procedure, chromosomes that have been spread on glass slides are sequentially treated with an acid or alkali, followed by incubation in a salt buffer and staining with Giemsa. Studies in diverse organisms such as humans, *Drosophila*, and wheat, have demonstrated that this staining technique differentiates substructure within the chromosome. The C bands correspond to the locations of constitutive heterochromatin in pachytene chromosomes. In interphase nuclei, C bands appear as heteropycnotic bodies. This aspect of the chromosomal substructure shows variation between different populations of a species, as summarized for wheat in Fig. 6.8.

The mechanism of C-banding is not clear, but it is known that the procedure preferentially stains regions of chromosomes enriched in highly repetitive DNA sequences (Fig. 6.9). It is possible that the procedure may differentially extract DNA from euchromatin as a result of a greater susceptibility of euchromatin to degradation by the acid (or alkali) pretreatments. A somewhat greater concentration of DNA remaining in the heterochromatic regions could then lead to their increased staining by Giemsa. It is also possible that proteins associated with DNA influence the staining properties of the chromatin remaining after the pretreatments.

6.2.2 G, R, and Q Bands Reveal Chromomere Structure

In animals, G- and Q-banding procedures for mitotic–metaphase chromosomes appear to reveal the chromomere structure that was seen previously only in prophase chromosomes. The G bands of chromosomes are most commonly observed by treating chromosome preparations in either saline sodium citrate solution (Fig. 6.10) or the enzyme trypsin, followed by staining with Giemsa dye. G-Banding generally provides a better resolution of chromosome substructure than other procedures and, in prophase chromosomes, may reveal as many as a thousand bands per genome. The locations of G bands in prophase chromosomes coincide with the locations of chromomeres in conventional preparations.

The mechanism of G-banding is unclear, but the G bands appear to be the sites of AT-rich DNA and DNA that tends to replicate later in the cell cycle. G bands are thought to be relatively scarce in DNA sequences coding for genes, and they may contain proteins, specific for AT-rich DNA, that may be responsible for the differential staining reaction (Table 6.3). In human chromosomes, the regions between G bands (G-negative regions) are the locations for Alu sequences, which are representative of a short, interspersed type of repetitive DNA sequence. Representatives of the L1 class of long, interspersed DNA sequences tend to be excluded from the regions where Alu sequences are located. This supports the conclusion that G-banding reflects the sequence composition of nuclear genomic DNA; regions of DNA with a characteristic base-pair composition are also referred to as isochores.

Chromosomes treated at high temperatures (such as 86°C) in a salt solution, followed by staining with Giemsa dye, develop R bands, which seem to correspond to regions between chromomeres and thus are the reverse of G-banding (Fig. 6.11). Although the procedure stains regions that are rich in actively transcribed genes, and also regions where the corresponding DNA is often enriched in GC content, the actual basis for the R-banding effect is not understood.

Many chemicals interact with DNA in a sequence-specific manner, providing stains that often reflect the characteristic DNA sequence composition of specific chromosome regions. Compounds that are particularly useful in this area of research include a number of antibiotics, which exert their action by binding to DNA and fluorescing when illuminated by ultraviolet (UV) light. The banding revealed by staining chromosomes with the antimalarial agent quinacrine hydrochloride or quinacrine mustard is a result of the DNA sequence-binding specificity of the quinacrine moiety. When applied to human chromosomes, the procedure generates a Q-banding pattern that is valuable in the identification of chromosomes (*see* Fig. 2.13). Quinacrine has a preference for AT-rich segments of DNA, and varying amounts of the dye, bound to different parts of the chromosomes, are considered to reflect the base composition of the DNA located in these regions. The differential fluorescence along the chromosome after quinacrine staining can also be modified by the differential quenching effects of other agents, so that the final pattern of bands observed is probably the result of a combination of differential binding as well as differential quenching, or suppression, of fluorescence. Rules of nomenclature cover the description of banding patterns (*see* Chapter 24 for standard karyotypes of a range of organisms).

6.2.3 Combinations of Stains Can Enhance the Resolution of Bands

In general, chromomycin and actinomycin-D have a preference for GC-rich DNA, whereas compounds such as quinacrine, DAPI

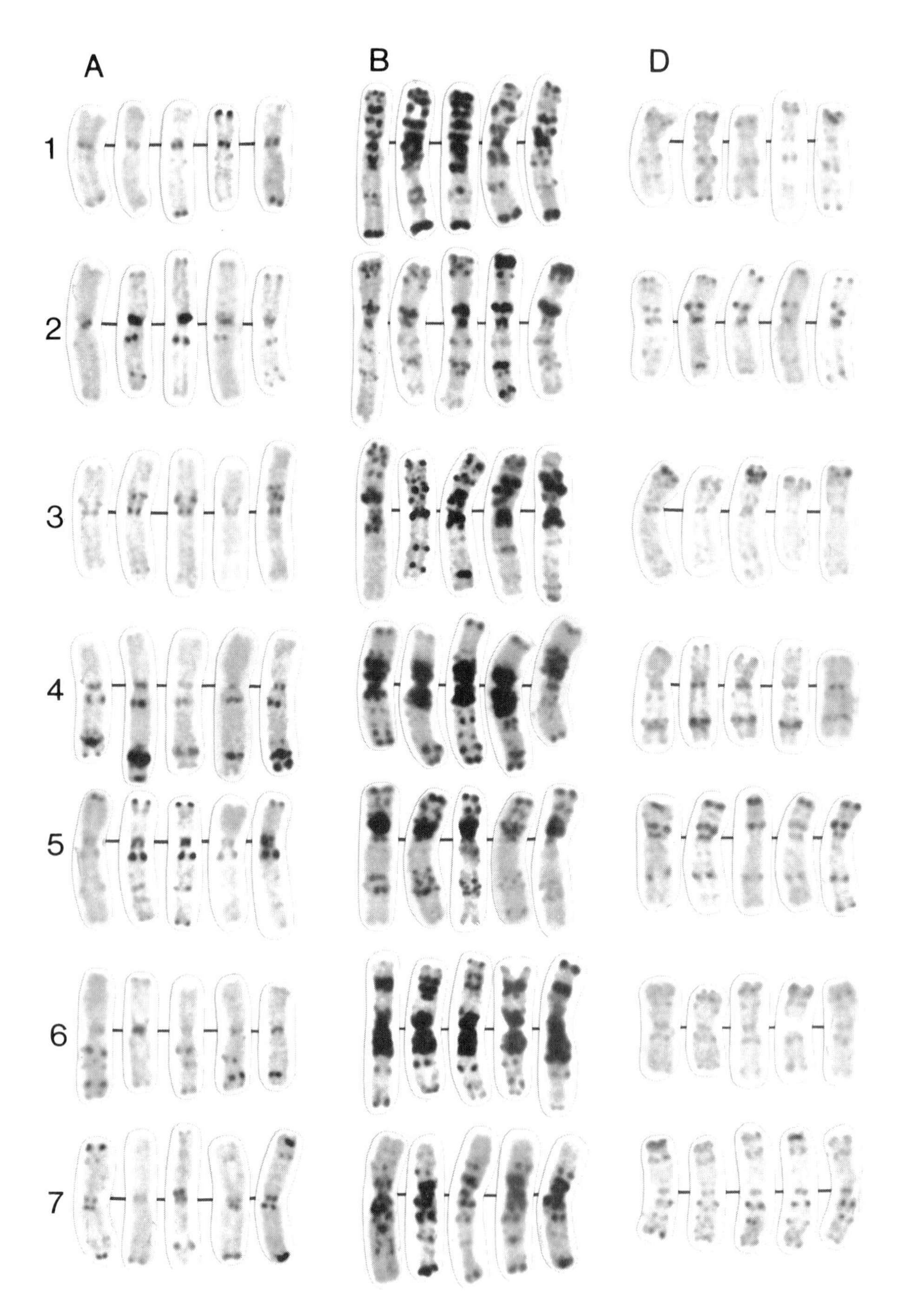

Fig. 6.8. The 21 chromosomes of 5 different cultivars of common wheat (*Triticum aestivum*; $2n = 42$) labeled by C-banding (Friebe and Gill, 1994). In animals as well as plants, C-banding is the most used method of chromosome identification because in contrast to G- or Q-banding patterns, which are usually quite conservative and relatively unchanging, C bands are often highly polymorphic, as demonstrated in this figure. The seven chromosomes in each of the A, B, and D genomes are derived from different ancestors and, in general, the B genome chromosomes are the most heavily labeled. A nomenclature system to describe the C-banded wheat chromosomes was proposed by Gill et al. (1991).

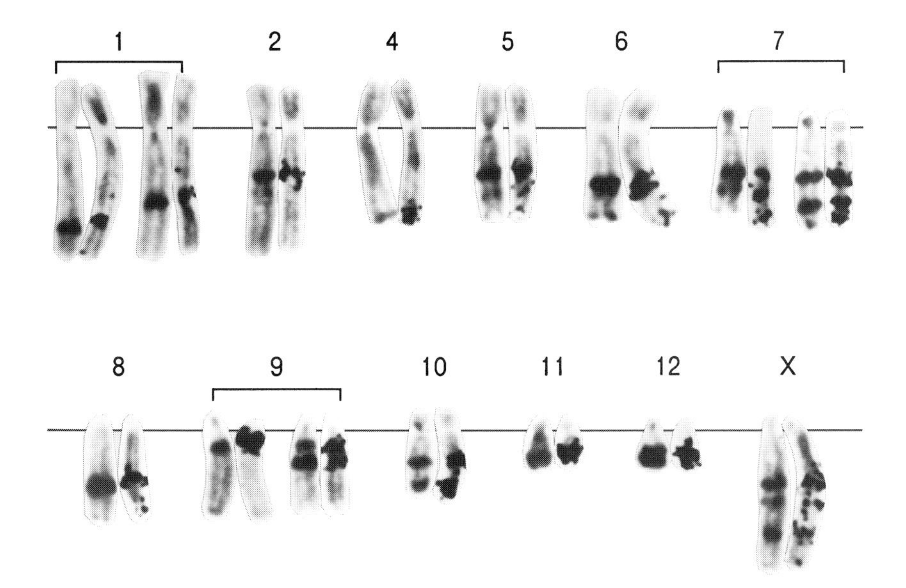

Fig. 6.9 Chromosomes from the grasshopper *Caledia captiva,* showing correspondence between the location of heterochromatin as detected by C-banding (left-hand chromosome of each pair), and by a family of highly repetitive DNA sequences (with units 168 bp long), detected by *in situ* hybridization with a ^{3}H-labeled RNA probe (right-hand chromosome of each pair). Where polymorphisms occur—as shown for chromosomes 1, 7, and 9—the correspondence between heterochromatin and localization of repetitive sequence is maintained (Arnold, 1985). Similar results from many other organisms demonstrate that C bands are the sites of late-replicating, highly repetitive DNA sequences (John et al., 1986). The chromosomes are arranged in order of decreasing size except for the X chromosome, which is the third longest and shown at that position in the published literature, but labeled as X.

(4′-6-diamidino-2-phenylindole), and Hoechst 33258 have a preference for AT-rich DNA. However, the preferential staining of certain parts of the chromosome by DNA-binding compounds reflects not only the AT or GC content of the DNA but also the nature of the actual sequence. For example, the secondary constrictions of human chromosomes 1, 9, and 16 contain a simple sequence called satellite II that is AT rich, yet these constrictions do not show strong fluorescence after staining with quinacrine. The major component of mouse centromeric heterochromatin also contains long tracts of AT base pairs, but stains poorly with quinacrine. The terminal heterochromatin of *Secale cereale* (rye) is stained by

Table 6.3. Comparative Properties of G-Positive (Heterochromatic) and G-Negative (Euchromatic) Bands

G-Positive Bands	G-Negative Bands
Contain AT-rich DNA and fluoresce with AT-specific fluorochromes	Contain GC-rich DNA and fluoresce with GC-specific fluorochromes
Negative staining in R-banding pachytene chromomeres	Positive staining in R-banding interchromomeric regions
Condense early in cell cycle	Condense late in cell cycle
Replicate late in cell cycle	Replicate relatively early in cell cycle
Low density of genes	Contain a wide range of genes
Contain long tandem arrays of repetitive sequences, and long interspersed repetitive DNA sequences (LINES)	Contain short interspersed repetitive sequences (SINES)

Source: After Bickmore and Sumner (1989).

Hoechst 33258 but not by any of the other DNA-binding compounds available.

The application of another molecule, such as actinomycin-D, to quinacrine-stained chromosomes has been found to enhance the relative fluorescence of certain regions. The basis of this enhancement is the contrasting DNA-binding specificities of quinacrine (AT preference) and actinomycin-D (GC preference), combined with the fact that actinomycin-D can accept light energy from quinacrine. Where actinomycin-D binds, it therefore quenches the background fluorescence of quinacrine resulting from low-level, ''nonspecific'' sets of interactions, and the characteristic yellow-green fluorescence persists only in the normally quinacrine-bright regions. The remainder of the karyotype has the characteristic red fluorescence of actinomycin-D. An example of the modification of the relatively uniform fluorescence staining of human chromosomes by the fluorochrome DAPI using actinomycin-D is shown in Fig. 6.12. Similarly, the staining of chromosomes by chromomycin, which has a preference for GC-rich DNA, can be modified by the application (counterstaining) of molecules with AT specificity for binding to DNA.

Thus, it is evident that a number of different pairs of DNA-binding compounds can be utilized to differentially stain chromosome regions, either to gain more information about the DNA composition of specific regions of the chromosomes or to develop novel banding patterns for chromosome identification purposes. When both compounds used in the staining are fluorescent, the final appearance of bands depends on the wavelength of the light used to illuminate the chromosome preparations. The heterochromatic Y chromosome of *Drosophila melanogaster* has been investigated in detail by staining with quinacrine and the dye Hoechst 33258. The

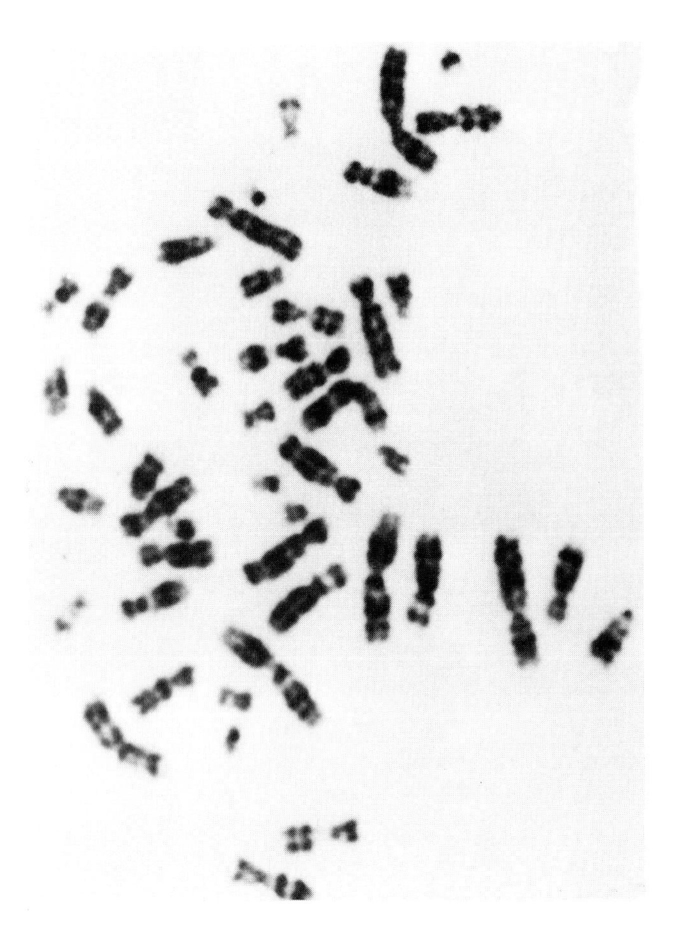

Fig. 6.10. Human mitotic–metaphase chromosomes labeled by G-banding (Bickmore and Sumner, 1989).

bands differentiate regions within heterochromatin and have provided cytological markers for the mapping of fertility genes located on the Y chromosome (*see* Chapter 11).

6.2.4 Many Specialized Techniques Reveal Functional Domains in Metaphase Chromosomes

A variety of procedures have been developed to visualize functional domains in metaphase chromosomes. Staining procedures based on the use of silver nitrate ($AgNO_3$) can result in the deposition of silver atoms onto RNA-binding proteins (RNPs). High concentrations of RNPs in a particular chromosomal region often reflect intense transcriptional activity, and in the case of ribosomal genes, the concentration of RNPs is sufficiently great for the resulting silver deposits to be clearly visible with the light microscope. The technique has been applied to both plants and animals (Fig. 6.13) to observe the nucleolus-organizer region.

Chromosome preparations treated with restriction endonucleases can result in distinct banding patterns after staining with Giemsa dye. Depending on the restriction endonuclease used, the final banding pattern can be similar to either G- or C-banding (Fig. 6.14). The enzyme micrococcal nuclease induces G bands when it is allowed to act on chromosome preparations for a short period of time, but C-type bands when it acts for a longer period of time. Under these conditions, it would appear that the DNA in euchroma-

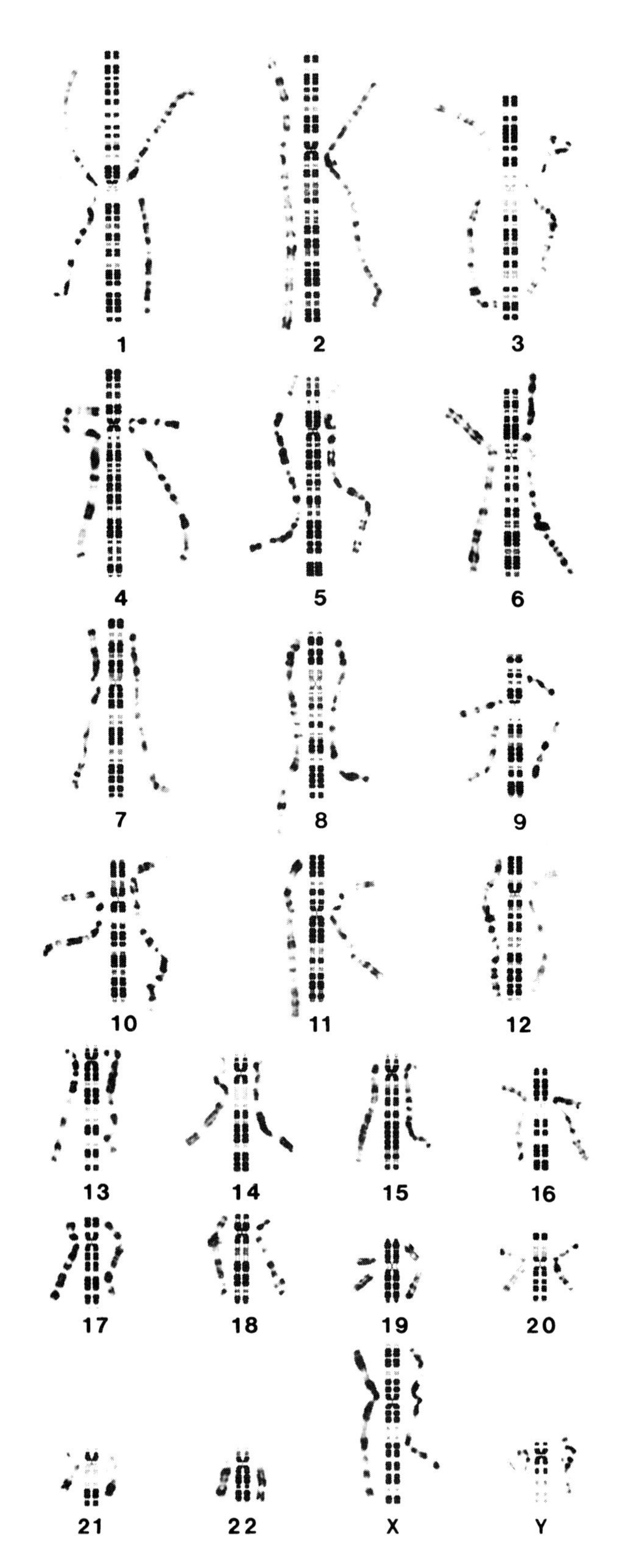

Fig. 6.11. A highly detailed map of R bands on human chromosomes (Drouin and Richer, 1989). Note that these details are more easily observed in chromosome preparations that are less contracted. R-banding patterns are the converse of G-banding patterns and generally correspond to gene-rich regions of chromosomes. The diagrammatic representation of the banding for each pair is shown in the middle.

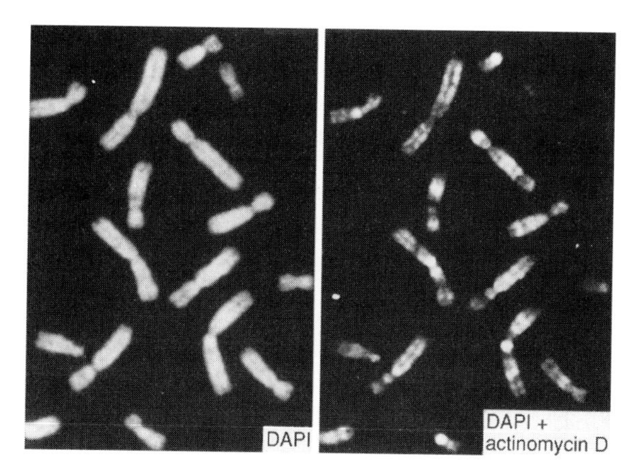

Fig. 6.12. A partial spread of human chromosomes stained either with DAPI, or with DAPI counterstained with a second DNA-binding compound, actinomycin-D (Schweizer, 1981). The actinomycin-D molecule binds to DNA in a sequence-specific manner and quenches the fluorescence associated with the binding of the DAPI molecule. The result is a banding pattern on the chromosomes that reflects the DNA-binding specificities of both molecules. The same cell is shown for both treatments.

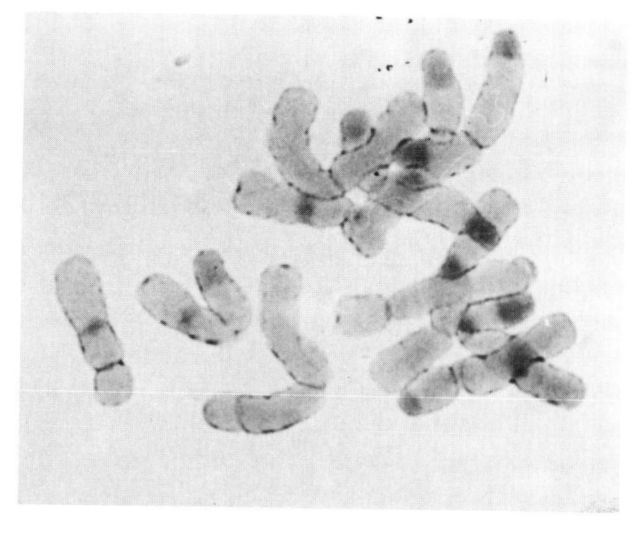

Fig. 6.14. The chromosomes of the liliaceous plant *Scilla siberica* ($2n = 12$) after restriction endonuclease treatment and Giemsa staining (Lozano et al., 1991).

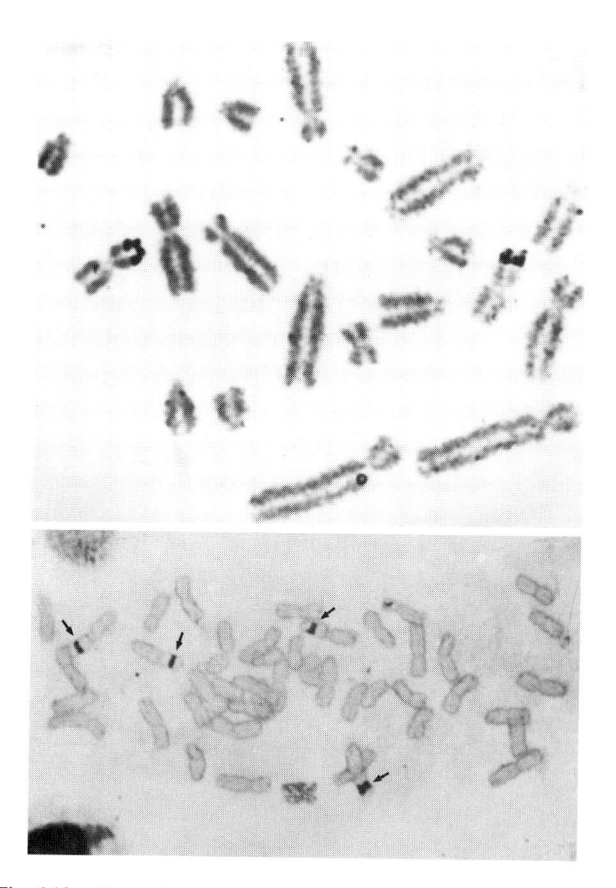

Fig. 6.13. The nucleolus-organizer regions (NORs) of (top) a female red kangaroo (*Macropus rufus*, $2n = 18 + XX$) and (bottom) of wheat (*Triticum aestivum*, $2n = 42$), detected by silver staining (*see* Adolph et al., 1990). The NORs of the kangaroo are located toward the ends of the metacentric X chromosomes. In wheat, the major NORs are located on chromosomes 1B and 6B (arrows). The 1B location is homoeologous to the location of the NOR of rye on chromosome 1R, seen in Fig. 6.4. [Photographs kindly supplied by Dr. B. Friebe (wheat) and Dr. D. Hayman (kangaroo).]

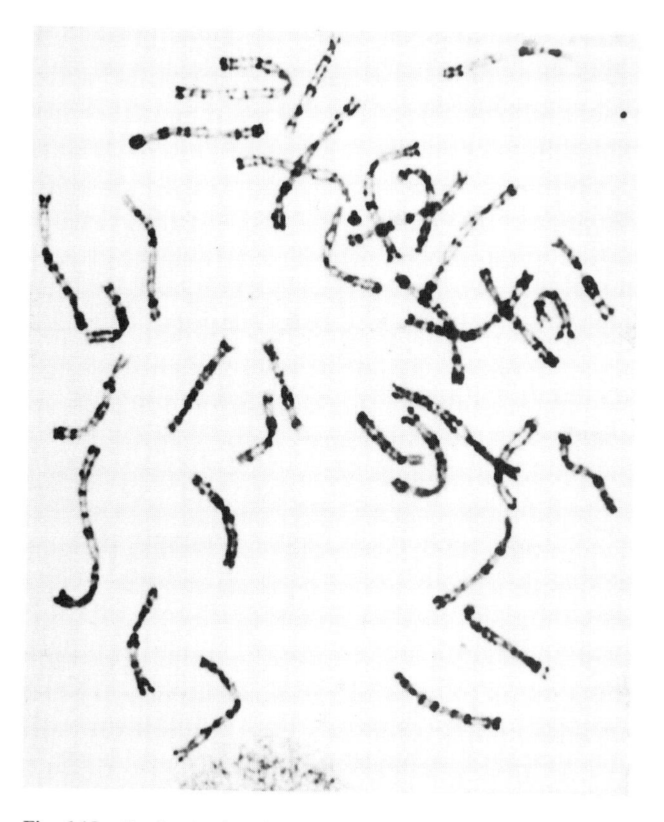

Fig. 6.15. Replication banding of the chromosomes of the toad *Xenopus laevis* using BrdU labeling and Giemsa staining (Schmid and Steinlein, 1991). The uridine analog is only incorporated into the replicating chromosomes for a limited period of time, and the differential incorporation of BrdU is subsequently observed by the staining procedure.

tin is more accessible to degradation than the DNA in heterochromatin.

Sister-chromatid labeling by ^{3}H-thymidine or 5-bromo-2′-deoxyuridine (BrdU), during DNA replication has been widely used to study sister-chromatid exchanges. An extension of this technology is to restrict BrdU labeling to late in the S phase, and then stain the chromosomes to detect the BrdU incorporation. An alternative is to limit labeling to early in the S phase because heterochromatic regions (C bands) are typically late-replicating, and the replication bands often correspond to C band patterns (Fig. 6.15). This replication banding can reveal many additional late-replicating regions, which can be used to characterize mitotic chromosomes.

6.3 MEIOTIC KARYOTYPE ANALYSIS

The chromosomes at the pachytene and diplotene stages of meiotic prophase are highly differentiated and offer the opportunity for high-resolution studies of chromosome structure and behavior. At pachynema, when homologous chromosomes are still synapsed tightly, even the smallest structural differences between the homologs are visible as characteristic perturbations in synapsis under the light or electron microscope. In this respect, meiotic–chromosome analysis is a much more powerful cytogenetic tool than mitotic–chromosome analysis. The major disadvantage is that not many organisms have reproductive tissues favorable for the study of meiotic prophase.

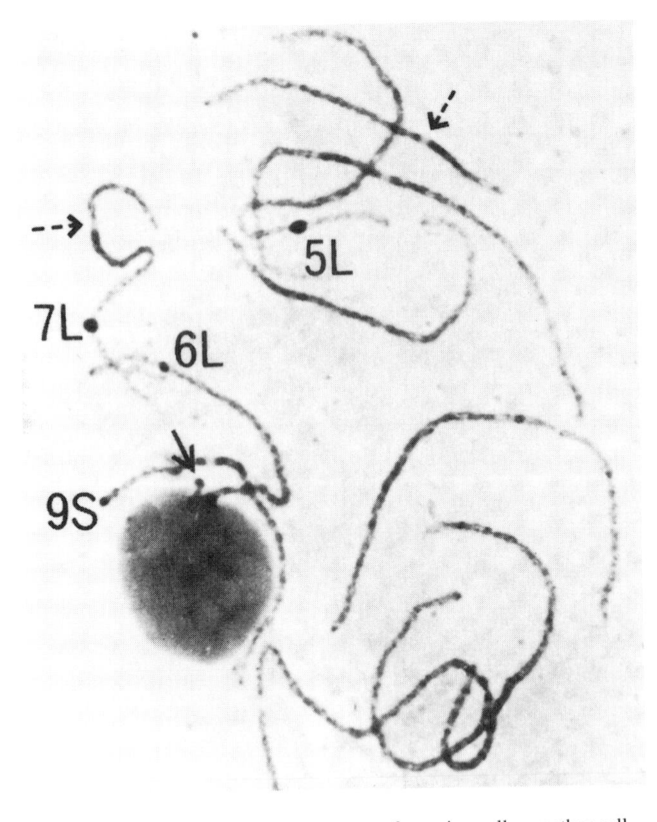

Fig. 6.16. The pachytene chromosomes of a maize pollen mother cell showing a range of morphological landmarks (Pryor et al., 1980). These include the centromere (arrow with broken line); differences in the total lengths and arm ratios; the position, size, and shape of heterochromatic knobs on the long arms of chromosomes 5, 6, and 7, and the short arms of 6 and 9; the clear attachment of the nucleolus-organizer region on the short arm of chromosome 6 to the nucleolus (arrow with solid line); and the presence of prominent chromomeres at specific sites on the chromosomes.

6.3.1 Origins of Cytogenetics Are Based on Pachytene and Metaphase I Chromosome Analyses

The combination of cytological observations and analyses of gene segregations relied on the discovery of pachytene chromosomes in maize pollen mother cells (Fig. 6.16). Squashed preparations of these chromosomes showed the positions of all the classical landmarks such as primary and secondary constrictions, heterochromatin, euchromatin, chromomeres, and telomeres, with a clarity and detail not possible in mitotic chromosomes. As a result, early studies on maize pachytene chromosomes allowed the genetic dissection of specific regions such as the heterochromatin near the secondary constrictions of chromosomes, the correlation of genetical and cytological maps, and the analysis of instability caused by transposable elements. Tomato, another favorable organism for pachytene analysis, shows a clear distinction between proximal (near the centromere) heterochromatin and euchromatin (e.g., *see* Fig. 11.16).

A major breakthrough in the electron-microscopic analysis of pachytene karyotypes came from the development of a surface-spreading technique (*see* Figs. 5.18 and 5.19). The structure that is actually observed in these karyotypes is the synaptonemal complex, which joins the homologous chromosomes at meiotic prophase. The positions of the centromeres can be determined so that arm ratios can be measured. Using this technique, structural rearrangements can be analyzed at a high level of resolution in most organisms. The spreading technique has been modified so that the chromatin of the paired chromosomes (rather than just the synaptonemal complex) remains attached to the spread material. This allows the technique of *in situ* hybridization to be used to localize specific DNA sequences in the chromatin.

Although pachytene karyotypes provide the most detail in the linear resolution of chromomeres, the analysis can be tedious and is not possible in many organisms, due to poor spreading and/or the presence of large numbers of chromosomes. Metaphase I, on the other hand, can generally be readily analyzed, with individual chromosomes identifiable by banding techniques (Fig. 6.17). Whereas mitotic karyotypes can be used to compare the chromosomes of two given species, the metaphase I karyotype can be used to analyze the F_1 hybrids resulting from crossing the two species, to determine degrees of pairing similarity between various chromosomes.

6.3.2 Giant Lampbrush Chromosomes and Diplotene Karyotypes

In oocytes of certain vertebrates, and sometimes in spermatocytes of some invertebrates, the meiotic cycle is arrested at the diplotene stage for periods up to several months or even years. In the case of sharks, some birds, and especially amphibians (*Rana temporaria*, a frog; *Triturus viridescens*, a newt), the diplotene chromosomes expand greatly in size and develop symmetrical loops that give the "lampbrush" appearance to the chromosomes (Fig. 6.18). The loops result from high levels of transcriptional activity. Newly synthesized RNA, which is present on the loops,

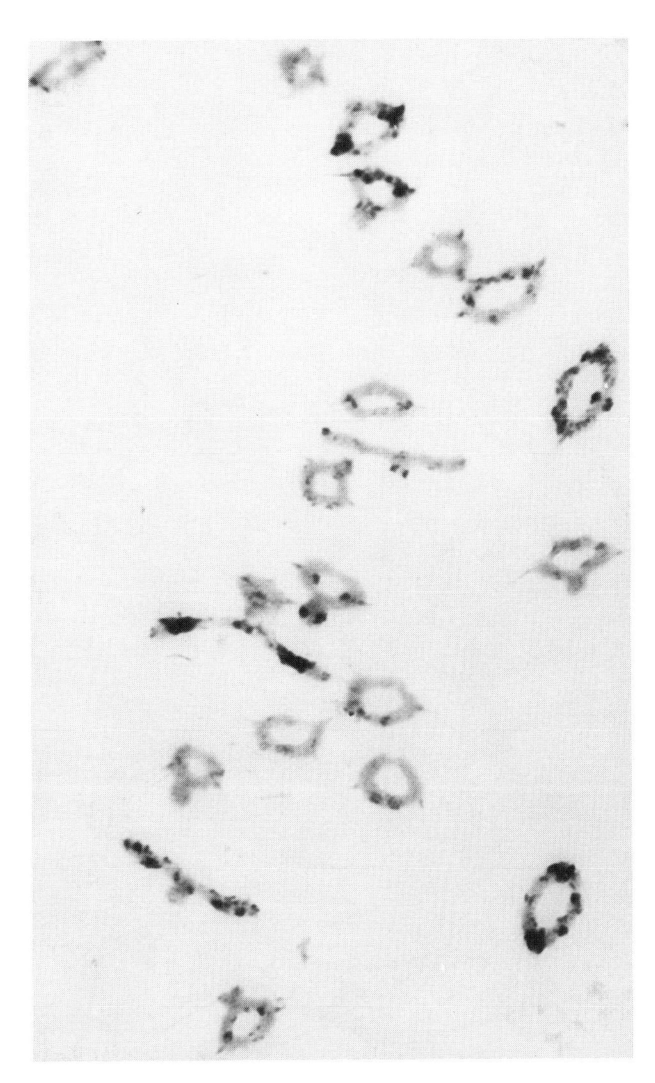

Fig. 6.17. C-banding of the 21 bivalents of wheat at metaphase I of meiosis. Note the seven more heavily labeled bivalents of the B genome (compare with Fig. 6.8) and the diagnostic labeling shown by many of the A and D genome bivalents. (Photograph kindly supplied by Dr. B. Friebe.)

represents the transcription of both genes and neighboring repetitive DNA sequence families. Each chromomere of a chromosome can give rise to one to nine loops. Because each loop has a reproducible size, a characteristic pattern is observed for each chromosome, thus allowing chromosome identification. Each diplotene bivalent also has a fixed number of chiasmata. In newts and frogs, the largest chromosomes may reach a length of 1000 μm. At the end of diplotene, the lampbrush appearance disappears as the chromosomes enter a diffuse stage before progressing into metaphase I.

In other plants and animals, diplotene chromosomes may not attain the lampbrush appearance, yet are still suitable for karyotypic analysis. The diplotene stages in grasshopper spermatocytes provide particularly fine details of sister-chromatid structure and their involvement in chiasmata associations (*see* Fig. 5.2). Among plants, *Arabidopsis thaliana*, with one of the smallest known DNA contents of any plant, can be karyotyped using diplotene chromosomes (Fig. 6.19).

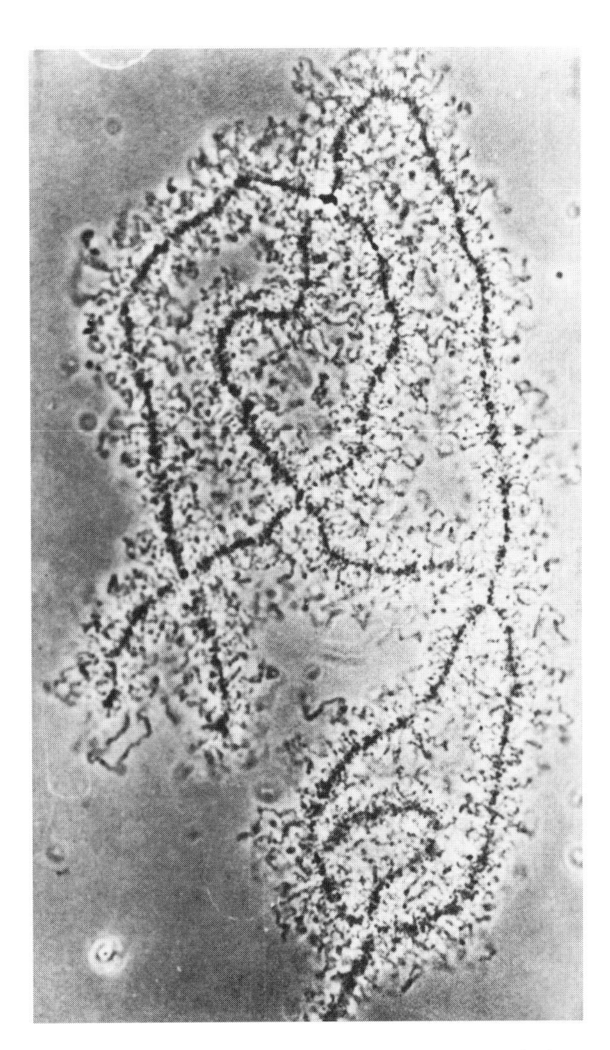

Fig. 6.18. Meiotic bivalents of the newt, *Triturus cristatus*, in the diffuse diplotene or lampbrush stage (Gall, 1966). The loops of exposed chromatin are actively synthesizing RNA. (*See also* Figs. 2.9 and 17.29).

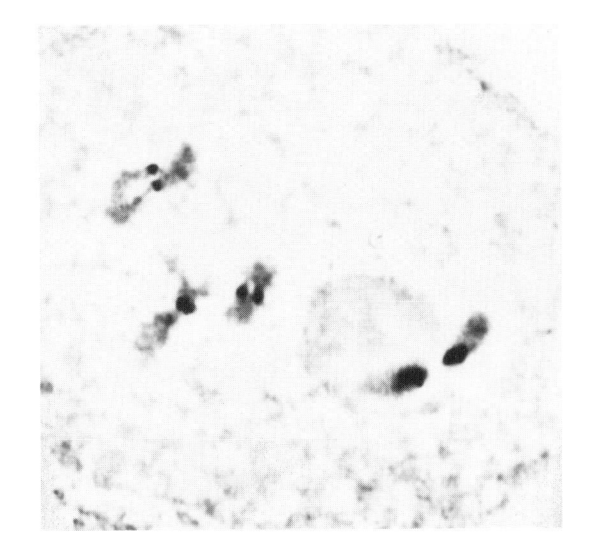

Fig. 6.19. The five bivalents of *Arabidopsis thaliana* at the diplotene stage of meiosis. The diplotene chromosome arrangements can be individually identified and have been used to identify different trisomic types (Sears and Lee-Chen, 1970). (Photograph kindly supplied by Dr. B. R. Tyagi.)

6.4 POLYTENY PRODUCES GIANT CHROMOSOMES IN SOME DIFFERENTIATED TISSUES

In the 1880s, it was found that the salivary glands of dipteran insects contained giant chromosomes. It was not until the 1920s, however, that carmine staining of squash preparations was developed, and the detailed analyses of the structure of the salivary-gland chromosomes of *Drosophila melanogaster* could be carried out (Fig. 6.20; *also* Fig. 2.3). The formation of polytene chromosomes is characterized by DNA replications taking place during the interphase stage, but because mitosis is arrested at the G2 stage, the replicated chromatids of each chromosome remain attached to one centromere, so the chromosomes become multistranded. This process is called endoreduplication and can lead to final levels of 1024 DNA strands in the polytene chromosomes of *D. melanogaster*; in other insects, the level of polyteny can be from 16 to 32 times greater than this. The polytene chromosomes usually are found in differentiated tissues with special functions and have provided a wealth of cytogenetic information, correlating changes in chromosome structure, such as duplications, with genetic changes. Extremely small structural changes in the chromosomes can be readily analyzed, and with molecular techniques, the DNA structure of several of the bands (chromomeres) and interband regions have been studied in detail. In *D. melanogaster*, the centromeric heterochromatin fails to undergo endoreduplication, and the respective regions from each of the four chromosome pairs remain fused together to form a chromocenter (*see* Fig. 6.20).

In plants, polytene-type chromosomes have been reported in the ovary tissues that nourish the developing embryo, and examples include the suspensor cells in *Phaseolus*, the antipodal cells in poppy (Fig. 6.21) and wheat, and the endosperm cells in maize. The polytene-type chromosomes in plants are the result of up to eight rounds of endoreplication in the absence of cell division. The replicated chromosomes do not show the degree of synapsis of homologous chromatin that is characteristic of salivary-gland polytene chromosomes in insects, so they are difficult to analyze cytologically.

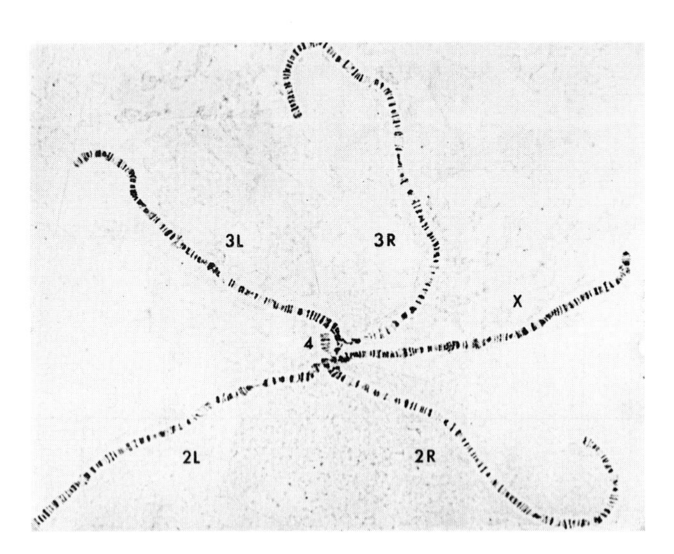

Fig. 6.20. Polytene chromosomes from the larval salivary glands of *Drosophila melanogaster*. Note the banding patterns, consisting of chromomeres, of the three autosomes and the X chromosome, and the heterochromatin of the chromocenter to which all of the chromosomes are attached (Lefevre, 1976).

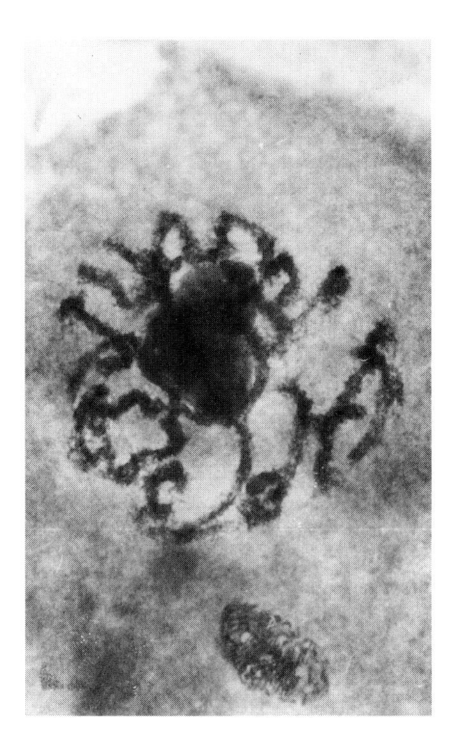

Fig. 6.21. An antipodal cell from an embryo of the poppy, *Papaver somniferum* ($2n = 22$), showing polytene-type chromosomes. These chromosomes do not show the precise alignment of endoreduplicated DNA observed in the polytene chromosomes of insects (*see* Fig. 6.20). (Photograph kindly supplied by Dr. B. Friebe.) Endoreduplication has also been reported in animal tissue-culture cells in response to stress such as exposure to radioactivity and BrdU labeling (Wolff and Perry, 1974; *see also* Fig. 18.3).

6.5 SEX CHROMOSOMES

In most animals and some plants, sex is determined by specialized sex chromosomes (called X, Y, Z, or W, depending on the organism), which show unique variations in number, structure, function, and behavior. In these chromosomes, the regions that are normally euchromatic may behave as heterochromatic bodies in interphase. This type of heterochromatin is called facultative to distinguish it from the constitutive or permanently heterochromatic parts of the chromosomes. Usually, females are the homogametic sex and carry a pair of sex chromosomes (XX in most animals as well as plants). In these organisms, the males (XY in most animals and plants) are the heterogametic sex and may show variation in sex-chromosome number such as having only one X (XO). The male-determining Y chromosome is usually much smaller than the X and carries fewer genes. There are many variations of the basic XX/XO or XY scheme of sex determination, including cases in *Lepidoptera* (moths and butterflies) and some birds, where females are the heterogametic sex (*see* Fig. 6.5). In addition, the number of X chromosomes in XO organisms may vary from one to five. XY organisms may also have up to eight X chromosomes along with one Y.

Diversity in sex-chromosome number and morphology can be illustrated in the genus *Drosophila*. Among the various *Drosophila* species, the X and Y are of equal size in 60% of the species, Y is somewhat smaller than X in 30%, and Y is tiny in 9%; in less than 1% of the species, there is no Y. The XO condition is considered to be derived from XY by loss of the Y chromosome. The *Drosophila miranda* male is X(1–2)Y(1–2), and the increase in numbers of X and Y chromosomes occurs by translocations with autosomes.

6.5.1 Sex Determination in Different Organisms

In *Drosophila*, the ratio between X and autosomal, or A, chromosomes determines the sex, irrespective of XX/XO or XX/XY constitution. Individuals with a 1 : 1 ratio between X and autosomes (XX/2A or XXY/2A) are females and those with a 1 : 2 X/A ratio (XO/2A or XY/2A) are males. Individuals with other ratios, for example, 2X/3A or 3X/4A, are intersexes. However, the Y chromosome in *Drosophila* does contain fertility factors and rRNA genes, so it is essential for normal sperm development.

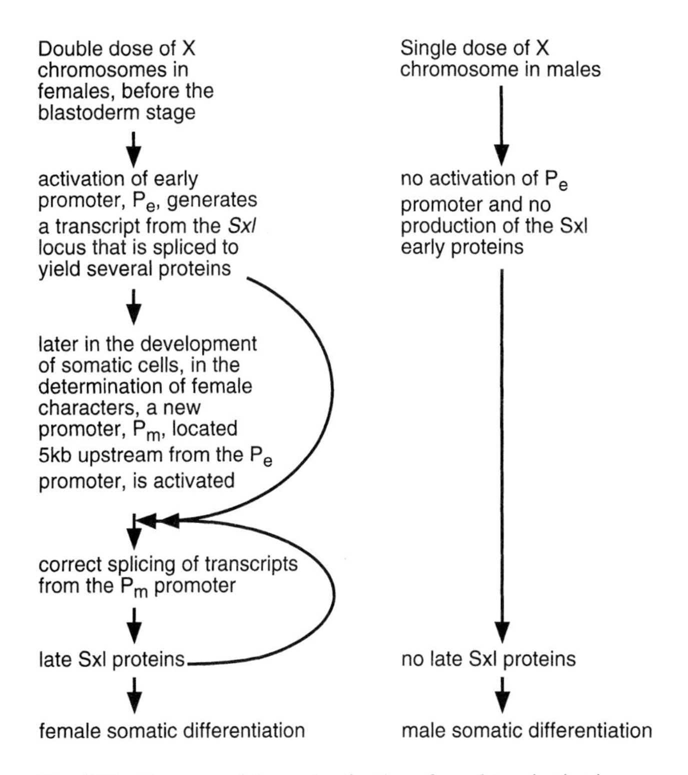

Fig. 6.22. Summary of the mode of action of sex determination in *Drosophila*. The *Drosophila* Y chromosome is not male-determining, as it is in mammals, although it does carry fertility factors (*see* Fig. 20.6). The organisms *Drosophila* and *Caenorhabditis* are widely used as models for molecular studies in animals (*see* Chapter 24) and both share the feature that the primary sex-determining signal is the ratio of X chromosomes to autosomes (X/A). When the organism has one pair of X chromosomes per diploid complement (X/A = 1), the phenotype is female for *Drosophila* and hermaphroditic for *Caenorhabditis*. When there is a single X chromosome per diploid complement (X/A = 0.5), the phenotype is male in both organisms. There is a narrow threshold for the X/A value, and in *Drosophila,* the master gene that is affected by the ratio is at the *Sxl* locus. In *Caenorhabditis*, a master gene may be located at the *sdc* locus. Although the organisms share similarities in sex determination, the actions of the master genes in the regulatory cascade are probably quite different. The structure of promoters and the nature of splicing are discussed in Chapter 16. [Adapted from Hodgkin (1990) and Cline (1993).]

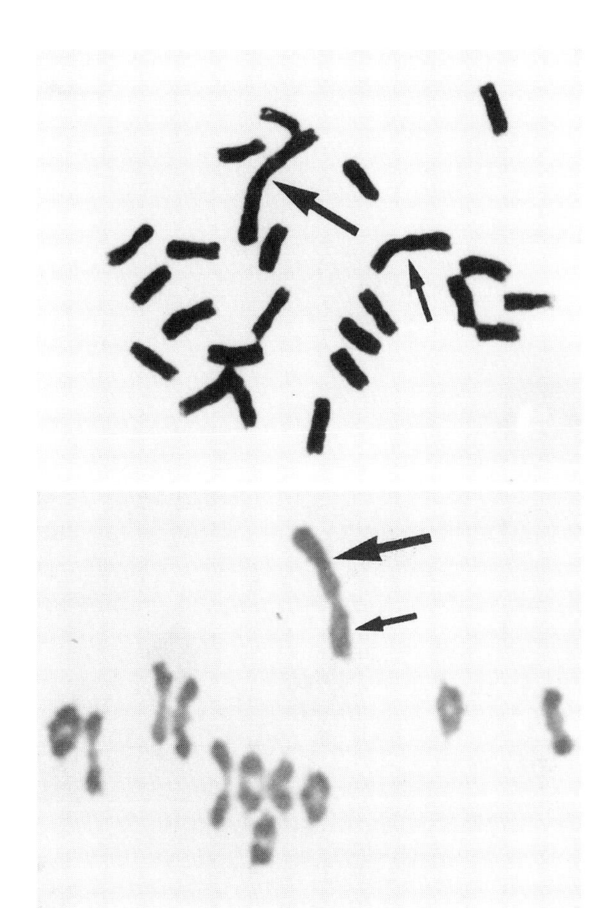

Fig. 6.23. Mitotic metaphase (top) and meiotic metaphase I (bottom) in a male plant of evening campion, *Melandrium album* ($2n = 22 + XY$). The bottom photo shows 11 pairs of autosomal bivalents and an XY heteromorphic bivalent with the 2 chromosomes associated by their terminal regions. The unequal-sized X and Y chromosomes (large and small arrows, respectively) in both photos are telocentrics. [The photograph of mitosis is from Cirpercescu et al. (1990), and the photograph of meiosis was kindly supplied by Dr. B. Friebe.]

The specific genes that trigger *Drosophila*-type and human-type sex determination have been identified. In *Drosophila*, the gene *Sxl* (sex-lethal) controls somatic sexual differentiation, and also the dosage compensation of the X chromosome, by its effect on many subordinate regulatory genes with more specialized functions. *Sxl* is turned on in diplo-X individuals and turned off in haplo-X individuals. The postulated molecular mechanism of its mode of action is illustrated in Fig. 6.22.

In organisms such as humans, mice, and the plant *Melandrium* (Fig. 6.23), the Y chromosome determines maleness, irrespective of the number of autosomal sets or X chromosomes. *Sry* (sex-determining region) genes, located on the Y chromosome, have been cloned both in humans (designated *SRY*) and mice (designated *Sry*). The *Sry* gene is identical to a previously described gene, *Tdy* (testis-determining gene) and is expressed in testis-specific gonadal somatic cells, but is absent in *Tdy*-defective mouse mutants. Mutations in the human *SRY* gene, leading to loss of function, account for sex reversal to females in two XY humans, whereas experiments in which an active *Sry* gene is introduced into XX female mice have shown that sex reversal to males is induced. Microinjection of a 14.6-kb segment of DNA from the mouse *Sry* locus into XX embryos results in the development of phenotypically male transgenic mice. It is evident, therefore, that the *Sry* gene is similar to *Sxl* of *Drosophila*, in providing a primary trigger, which controls a complex cascade of many other downstream sex-determining genes (*see* Fig. 6.22).

6.5.2 Structure and Behavior of Sex Chromosomes

The evolution of the heterogametic sex has given rise to several features of sex-chromosome structure and behavior that are unusual. The cytogenetic structure of the human X chromosome is shown in Fig. 6.24 and, in particular, emphasizes a region on the X chromosome that contains genes also present on the Y chromosome.

In XO or XY individuals, many genes that are present in a single dose are in a double dose in XX individuals. Thus, to maintain the same levels of gene product in XY (or XO) and XX individuals, either the genes in XO or XY individuals need to be expressed at twice the level or mechanisms must exist to reduce the expression of genes by one-half in XX individuals. In *Drosophila*, cis-acting elements provide the basis for a compensation mechanism that results in a twofold increase of X-linked gene products in males; the *Sxl* gene is a major controlling element. In *Drosophila*, as well as other organisms, there must also exist special mechanisms for pairing and disjunction of XY chromosomes and the normal movement of unpaired X chromosomes during meiosis; determining the basis of these mechanisms is the subject of intense current research. In *Drosophila*, for example, the ribosomal-gene region is a key region of homology between the X and Y chromosomes in males and is the site of pairing between these chromosomes.

In mammals, the gene-dosage problem in XY versus XX individuals is overcome by inactivation of one of the X chromosomes in the somatic cells of females, so that only a single X chromosome remains active in any somatic cell, or oogonium, regardless of how many X chromosomes are present. The process results in dosage compensation for X-linked genes in chromosomally XX females and XY males. In humans, the Barr body is an X chromosome that is condensed, and inactive, in the somatic nuclei of XX cells. The inactivation of the X chromosomes occurs in early embryos (4–6 days of gestation in mice) and is thought to be initiated at an X-inactivation center by the trans-action of a product encoded by an autosome. The choice of maternal or paternal X for inactivation is random. Only the X chromosome that expresses the X-inactivation center, carrying the *Xist* locus, is inactivated and it is therefore a cis-acting function. The *Xist* locus on the active X is silent and fully methylated at the CpG dinucleotides in the promoter region, whereas the *Xist* locus on the inactive X is functioning as a result of complete lack of methylation in this region; the role of methylation in control of gene expression is covered in Chapter 17. Once the X chromosome to remain active has been chosen, it is maintained in subsequent cell divisions by a process of imprinting, which may involve methylation of the DNA of the X chromosome that remains inactive. In many marsupials, the inactivation is usually directed more specifically to the paternally derived X chromosome.

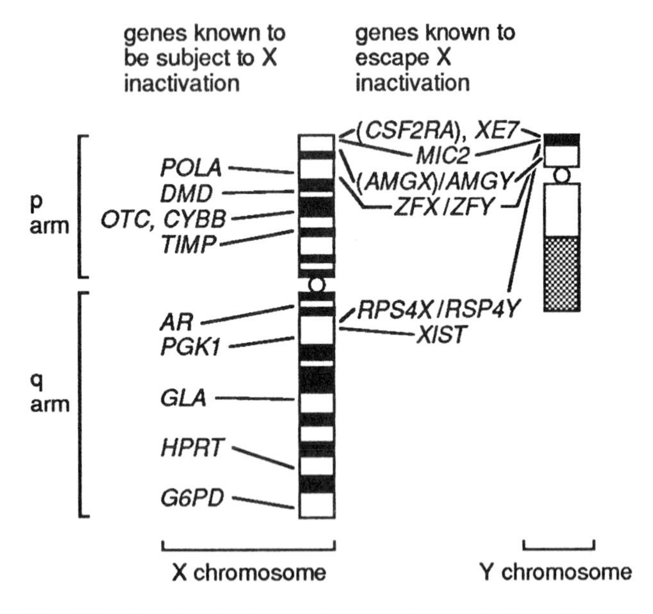

Fig. 6.24. Shared and unique molecular markers on the X and Y chromosomes of humans (Zinn et al., 1993; Disteche, 1995). The black bands on the diagrammatic representations of the chromosomes are standard chromosome-banding landmarks (*see* Chapter 24).

As indicated in Fig. 6.24, the entire X is not inactivated, and this incomplete inactivation is now considered to be the basis for explaining the phenotype of human females that contain only a single X chromosome. Monosomy for the X chromosome is associated with Turner's syndrome, and the symptoms of the defect most likely result from the reduced dosage of the genes that are duplicated on the Y chromosome and normally escape inactivation on the X chromosome. Certain loci that escape X inactivation, such as *RSP4*, have functional equivalents on the Y chromosome and code for proteins that are basic in cell biology. A reduced amount of the S4 ribosomal protein encoded by the *RSP4* locus could, in principle, result in the marked phenotypic changes associated with monosomy for the X chromosome in Turner's syndrome. These symptoms include anatomical abnormalities as well as metabolic defects. Although a number of loci that escape X inactivation have functionally equivalent genes on the Y chromosome, exceptions do occur and gene dosage is therefore not critical for all loci.

In the meiocytes of XX female, the inactive X condition is reversed, so that both of the X chromosomes are transcriptionally active and able to undergo pairing and crossing over. In the meiocytes of XY male, the single X chromosome becomes condensed for the first time and is transcriptionally inactive during prophase. The process of meiotic sex-chromosome inactivation (MSCI) in heterogametic (XY) males is considered to be quite distinct from the somatic X inactivation in XX females discussed above. Whereas somatic X inactivation is related to dosage compensation, the function of MSCI is related to the different status of the two sex chromosomes with respect to pairing and recombination. The homologous X chromosomes must pair and recombine not only to ensure proper segregation at anaphase I but also to prevent the accumulation of deleterious mutations.

6.6 B CHROMOSOMES

Many organisms contain a variable number of accessory or supernumerary chromosomes (B chromosomes) in addition to the normal (A) complement of chromosomes (Fig. 6.25). The B chromosomes are usually small in size and variable in number. In some organisms, they have a deeply staining, heterochromatic appearance after treat-

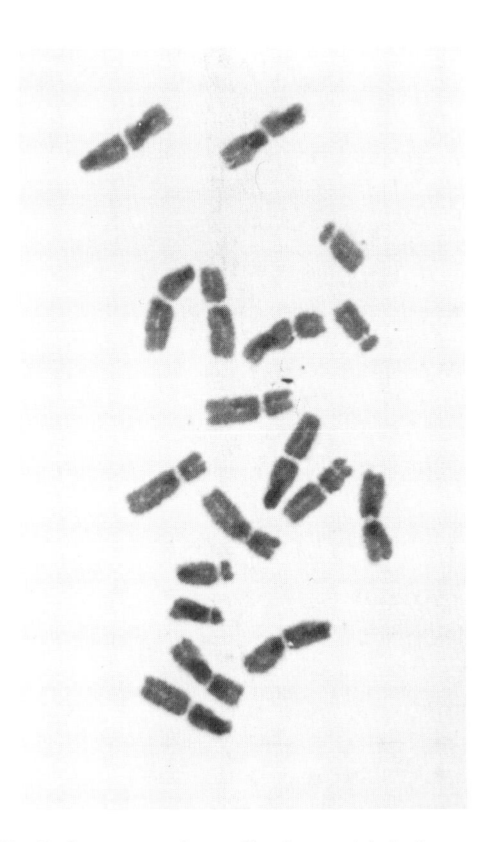

Fig. 6.25. B chromosomes in rye (*Secale cereale*). A chromosome squash of a root-tip cell in metaphase of mitosis showing $2n = 14 + 4$ smaller, acrocentric B chromosomes. Rye plants can contain up to eight B chromosomes, although fertility is severely reduced with this many extra chromosomes. (Photograph kindly supplied by Dr. B. Friebe.)

ment with standard stains, whereas in other organisms, they can be lightly staining, with telomeric and centromeric regions of heterochromatin. B chromosomes range in appearance from acrocentric in rye (*see* Fig. 6.25) and maize, to submetacentric in the parasitic wasp (*Nasonia vitripennis*) and metacentric in *Allium cernuum* and *Crepis capillaris*. They usually have little phenotypic effect on the organisms in which they occur, although they are not genetically inert, and the reason for their continued persistence in organisms, as well as their origins, are the subjects of current research.

6.6.1 Unusual Behavior of B Chromosomes

In an organism with B chromosomes, not all cells carry them and they are most likely to be present in mitotically active somatic cells and in germ cells. In the wheat-related plant *Triticum tripsacoides*, for example, the B chromosomes cannot be detected in root-tip meristematic cells, but are present in pollen mother cells. In *Crepis capillaris*, there are two B chromosomes in root-tip and aerial vegetative cells, but four in germ cells.

During meiosis and/or postmeiotic mitotic divisions in plants, the segregation behavior of B chromosomes enhances their retention and transmission to offspring. The most dramatic behavior is demonstrated during the development of the male gametophyte in plants. In rye, for example, the B chromosomes undergo nondisjunction of sister chromatids in the first mitotic division of microspores and are preferentially included in the generative nucleus, which divides to form two sperm nuclei. In maize, nondisjunction of a B chromosome occurs in the second mitotic division to give one sperm nucleus with two B's and one with none. The sperm nucleus with the B chromosomes preferentially fertilizes the egg cell. Nondisjunction of B chromosomes also occurs on the female side in rye (but not in maize), during the first mitotic division of the basal megaspore. In some plant species, the B chromosomes are preferentially recovered in the basal megaspore at the end of meiosis. All of these mechanisms increase the chances for transmission of B chromosomes.

6.6.2 Genetic Functions Associated with B Chromosomes

Although B chromosomes are transcriptionally inert relative to high levels of transcription from A chromosomes, in stages of the cell cycle such as meiosis, several important genetic functions have been assigned to them. The B chromosomes often carry functional ribosomal RNA genes (*Nor* locus), and genes that affect A chromosome pairing, chiasma formation, and genetic recombination. The level of transcriptional activity from B chromosomes generally correlates with the degree of DNA methylation. For example, in grasshoppers, where the B-chromosome *Nor* locus is inactive, the DNA in this region is methylated, whereas a translocation that moves the *Nor* locus to an A chromosome reduces the level of methylation and increases the activity at the *Nor* locus. Another example of a biological activity that is correlated with the presence of a B chromosome is found in *Festuca*, where an increase in B chromosome number is associated with increased frequencies of chiasmata in autosomal bivalents. In maize, the regions near A–B translocation breakpoints show enhanced recombination.

When *Triticum tripsacoides* lines with or without B chromosomes are crossed to other *Triticum* species, reduced pairing is observed in hybrids with B chromosomes relative to those without B chromosomes. This effect on pairing suggests that B chromosomes carry genes that determine the degree of pairing between homologous or homoeologous chromosomes in a way that is analogous to the *Ph* gene in wheat (*see* Chapter 10). In maize, deletions of sections of B chromosomes have been used to identify regions of these chromosomes carrying genes that influence poleward movement during anaphase and mitotic nondisjunction. Identification of small chromosome regions in this way opens up the possibility of using transposable elements to create mutations in the respective genes (*see* Chapter 7). The insertion of a transposable element into a putative gene that influences, for example, mitotic nondisjunction, provides a molecular tag for isolating the gene and further characterizing its functions in transformation experiments.

6.6.3 Origin of B Chromosomes May Be Related to Cytoplasmic-Specific Introgression During Interspecific Hybridization

Most organisms show no pairing between the A and B chromosomes in meiosis, indicating that B chromosomes have an origin independent of the A chromosomes. Numerous hypotheses have been discussed to account for the origin of B chromosomes. One interesting, more recent, suggestion is that they arise from A-chromosome fragmentation during interspecific hybridization and that some of these fragments are retained because they restore compatibility between the nucleus and the cytoplasm. Nucleo-cytoplasmic interactions are important in the evolution of plants by interspecific hybridization, which is discussed in detail in Chapter 22. Arguments in favor of this concept of B-chromosome origin for plants include the fact that B chromosomes are usually found in cross-pollinated species, contain repetitive-sequence families that are not present in the A chromosomes, and carry essential genes such as those coding for ribosomal RNA.

BIBLIOGRAPHY

General

BARCH, M.J. (ed.) 1991. The ACT Cytogenetics Laboratory Manual, 2nd ed. Raven Press, New York.

CALLAN, H.G. 1986. Lampbrush Chromosomes. Springer-Verlag, Berlin.

FUKUI, K., NAKAYAMA, S. (eds.). 1996. Plant chromosomes: Laboratory methods. CRC Press, Boca Raton, FL.

GUSTAFSON, J.P., APPELS, R. (eds.) 1988. Chromosome Structure and Function: The Impact of New Concepts. 18th Stadler Genetics Symposium. Plenum Press, New York.

JAUHAR, P. (ed.). 1996. Methods of genome analysis in plants. CRC Press, Boca Raton, FL.

JONES, K. 1978. Aspects of chromosome evolution in higher plants. In: Advances in Botanical Research, Vol. 6. Academic Press, New York.

JONES, R.N., REES, H. 1982. B Chromosomes. Academic Press, New York.

MILLER, O.J. 1995. The fifties and the renaissance in human and mammalian cytogenetics. Genetics 139: 489-494.

SHARMA, A.K., SHARMA, A. 1994. Chromosome Techniques—A Manual. Hardwood Academic Publisher, Langhorne, PA.

SUMNER, A.T. 1990. Chromosome Banding. Unwin and Hyman, London.

TSUNEWAKI, T. (ed.). 1996. Plant genome and plastome: their structure and evolution. Kodansha Scientific Ltd., Tokyo, Japan.

Section 6.1

BENNETT, M.D., SMITH, J.B., SEAL, A.G. 1986. The karyotype of the grass *Zingeria biebersteiniana* ($2n = 4$) by light and electron microscopy. Can. J. Genet. Cytol. 28: 554–562.

BROWN, S.W. 1966. Heterochromatin (cytological observations). Science 151: 417–425.

BURNS, J.A. 1966. The heterochromatin of two species of *Nicotiana*. J. Hered. 57: 43–47.

CARLE, G.F., OLSON, M.V. 1985. An electrophoretic karyotype for yeast. Proc. Natl. Acad. Sci. USA 82: 3756-3760.

CHRISTIDIS, L., SHAW, D.D., SCHODDE, R. 1991. Chromosomal evolution in parrots, lorikeets and cockatoos (Aves: Psittaciformes). Hereditas 114: 47–56.

CROSLAND, M.W.J., CROZIER, R.H. 1986. *Myrmecia pilosula*, an ant with only one pair of chromosomes. Science 231: 1278.

DARLINGTON, C.D., LACOUR, L.F. 1940. Nucleic acid starvation of chromosomes in *Trillium*. J. Genet. 40: 185-213.

GILL, B.S., SEARS, R.G. 1988. The current status of chromosome analysis in wheat. In: Chromosome Structure and Function, Impact of New Concepts (Gustafson, J.P., Appels, R., eds.). 18th Stadler Genetics Symp., Plenum Press, New York. pp. 299–321.

JOHNSON, M.A.T., KENTON, A.Y., BENNETT, M.D., BRANDHAM, P.E. 1989. *Voanioala gerardii* has the highest known chromosome number in the monocotyledons. Genome 32: 328–333.

KAMISUGI, Y., FURUYA, N., IIJIMA, K., FUKUI, K. 1993. Computer-aided automatic identification of rice chromosomes by image parameters. Chromosome Res. 1: 189–196.

LEVAN, A., FREDGA, K., SANDBERG, A.A. 1964. Nomenclature for centromeric position on chromosomes. Hereditas 52: 201–220.

MILLS, D., MCCLUSKEY, K. 1990. Electrophoretic karyotypes of fungi: the new cytology. Molec. Plant Microbe Int. 3: 351–357.

RAY-CHAUDHURI, R., SHARMA, T., RAY-CHAUDHURI, S.P. 1969. A comparative study of the chromosomes of birds. Chromosoma 26: 148–168.

VAN DER PLOEG, L.H.T., SMITS, M., PONNUDURAI, T., VERMOULEN, A., MEUWISSEN, J.H.B.TH., LANGLEY, G. 1985. Chromosome-sized DNA molecules of *Plasmodium falciparum*. Science 229: 658–661.

WATANABE, K., CARTER, C.R., SMITH-WHITE, S. 1975. The cytology of *Brachycome lineariloba*. 5. Chromosome relationships and phylogeny of the race A cytodemes ($n = 2$). Chromosoma 52: 383–397.

WERNER, J., KOTA, R.S., GILL, B.S. 1992. Distribution of telomeric repeats and their role in the healing of broken chromosome ends in wheat. Genome 35: 844–848.

Section 6.2

ARNOLD, M. L. 1985. Patterns of evolution of highly repeated DNA from *Caledia captiva*. Ph.D. thesis, Australian National University, Canberra.

BERNARDI, G. 1985. The mosaic genome of warm-blooded vertebrates. Science 228: 953–957.

BIANCHI, M.S., BIANCHI, N.O., PANTELIAS, G.E., WOLFF, S. 1985. The mechanism and pattern of banding induced by restriction endonucleases in human chromosomes. Chromosoma 91: 131–136.

BICKMORE, W.A., SUMNER, A.T. 1989. Mammalian chromosome banding—an expression of genome organization. Trends in Genetics 5: 144–148.

BURKEHOLDER, G. 1988. The analysis of chromosome organization by experimental manipulation. In: Chromosome Structure and Function, Impact of New Concepts (Gustafson, J.P., Appels, R., eds.). 18th Stadler Genetics Symp., Plenum Press, New York. pp. 1–52.

DROUIN, R., RICHER, C-L. 1989. High-resolution R-banding at the 1250-band level. II. Schematic representation and nomenclature of human RBG-banded chromosomes. Genome 32: 425–439.

FRIEBE, B., GILL, B.S. 1994. C band polymorphism and structural rearrangements detected in common wheat (*Triticum aestivum*). Euphytica 78: 1–5.

GILL, B.S., FRIEBE, B., ENDO, T.R. 1991. Standard karyotype and nomenclature system for description of chromosome bands and structural aberrations in wheat. Genome 34: 830–839.

GILL, B.S., KIMBER, G. 1974. Giemsa C-banding and the evolution of wheat. Proc. Natl. Acad. Sci. USA 71: 4086–4090.

HEITZ, E. 1928. Des heterochromatin der moose. I. Jahrb. Wiss. Bot. 69: 762–818.

HSU, T.C. 1973. Longitudinal differentiation of chromosomes. Annu. Rev. Genet. 7: 153–176.

JOHN, B., APPELS, R., CONTRERAS, N. 1986. Population cytogenetics of *Atractomorpha similis*. II. Molecular characterisation of the distal C-band polymorphisms. Chromosoma 94: 45–58.

KOELFLING, M., MILLER, D.A., MILLER, O.J. 1984. Restriction enzyme banding of mouse metaphase chromosomes. Chromosoma 90: 128–132.

LACADENA, J.R., CERMENT, M.C., ORELLANA, J., SANTOS, J.L. 1984. Evidence for wheat-rye nucleolar competition (amphiplasty) in triticale by silver-staining procedure. Theoret. Appl. Genet. 67: 207–213.

LOZANO, R., SENTIS, C., FERRANDEZ-PIQUERAS, J., JEJON, M.R. 1991. *In situ* digestion of satellite DNA of *Scilla siberica*. Chromosoma 100: 439–442.

ROWLEY, J.D. 1974. Identification of human chromosomes. In: Human Chromosome Methodology (Yunis, J.J., ed.). Academic Press, New York. pp. 17–46.

SCHMID, M., STEINLEIN, C. 1991. Chromosome banding in Amphibia. XVI. High resolution replication banding patterns in *Xenopus laevis*. Chromosoma 101: 123–132.

SCHWEIZER, D. 1981. Counterstain-enhanced chromosome banding. Human Genet. 57: 1–14.

TORRE, J., DE LE, MITCHELL, A.R., SUMNER, A.T. 1991. Restriction endonuclease/nick translation of fixed mouse chromosomes: A study of factors affecting digestion of chromosomal DNA *in situ*. Chromosoma 100: 203–211.

WARBURTON, D., HENDERSON, A.S. 1979. Sequential silver staining and *in situ* hybridization on nucleolus organizing regions in human cells. Cytogenet. Cell. Genet. 24: 168–175.

WOLFF, S., PERRY, P. 1974. Differential Giemsa staining of sister chromatids and the study of sister chromatid exchanges without autoradiography. Chromosoma 48: 341–353.

Section 6.3

GALL, J.G. 1966. Lampbrush chromosomes. In: Methods in Cell Physiology, Vol. II (Prescott, D.M., ed.). Academic Press, New York. pp. 37–60.

GALL, J., CALLAN, H., WU, Z., MURPHY, C. 1991. Lampbrush chromosomes. Meth. Cell. Biol. 36: 150–166.

PRYOR, A., FAULKNER, K., RHOADES, M.M., PEACOCK, W.J. 1980. Asynchronous replication of heterochromatin in maize. Proc. Natl. Acad. Sci. USA 77: 6705–6709.

SEARS, L.M.S., LEE-CHEN, S. 1970. Cytogenetic studies in *Arabidopsis thaliana*. Can. J. Genet. Cytol. 12: 217–223.

SOLOVEI, I.V., JOFFE, B.I., GAGINSKAYA, E.R., MACGREGOR, H.C. 1996. Transcription on lampbrush chromosomes of a centromerically localized highly repeated DNA in pigeon (*Columba*) relates to sequence arrangement. Chromosome Res. 4: 588–603.

THOMAS, J.B., KALTSIKES, P.J. 1974. A possible effect of heterochromatin on chromosome pairing. Proc. Natl. Acad. Sci. USA 71: 2787–2790.

Section 6.4

BERGHELLA, L., DIMITRI, P. 1996. The heterochromatic *rolled* gene of *Drosophila melanogaster* is extensively polytenized and transcriptionally active in the salivary gland chromocenter. Genetics 144: 117–125.

LEFEVRE, G. 1976. A photographic representation and interpretation of the polytene chromosomes of *Drosophila melanogaster* salivary glands. In: Genetics and Biology of *Drosophila* Vol. 1a (Ashburner, M., Novitski, E., eds.). Academic Press, New York. pp. 32–66.

NAGL, W. 1981. Polytene chromosomes of plants, giant chromosomes. Int. Rev. Cytol. 73: 21–53.

STUART, J., HATCHETT, J.H. 1988. Cytogenetics of the Hessian fly. I. Mitotic karyotype analysis and polytene chromosome correlations. J. Hered. 79: 184–189.

ZHANG, P., SPRADLING, A.C. 1995. The *Drosophila* salivary gland chromocenter contains highly polytenized subdomains of mitotic heterochromatin. Genetics 139: 659–670.

Section 6.5

BROWN, C.J., BALLABIO, A., RUPERT, J.L., LAFRENIEVE, R.G., GROMPE, M., TONLORENZI, R., WILLARD, H.F. 1991. A gene from the region of the human X inactivation centre is expressed exclusively from the inactive X chromosome. Nature 349: 38–44.

CAPEL, B., SWAIN, A., NICOLLS, S., HACKER, A., WALTER, M., KOOPMAN, P., GOODFELLOW, P., LOVELL-BADGE, R. 1993. Circular transcripts of the testis-determining gene *Sry* in adult mouse testis. Cell 73: 1019–1030.

CHARLESWORTH, B. 1991. The evolution of sex chromosomes. Science 251: 1030–1033.

CIRPERCASCU, D.D., VEUSKENS, J, MOURAS, A., YE, D., BRIQUET, M., NEGRUTIU, I. 1990. Karyotyping *Melandrium album*, a dioecious plant with heteromorphic sex chromosomes. Genome 33: 556–562.

CLINE, T.W. 1993. The *Drosophila* sex determination signal. How do flies count to two? Trends Genet. 9: 385–390.

DISTECHE, C.M. 1995. Escape from X inactivation in human and mouse. Trends Genet. 11: 17–22.

GOODFELLOW, P.N., LOVELL-BADGE, R. 1993. *SRY* and sex determination in mammals. Annu. Rev. Genet. 27: 71–92.

GORMAN, M., BAKER, B.S. 1995. How flies make one equal two: dosage compensation in *Drosophila*. Trends Genet. 10: 376–380.

KOOPMAN, P., GUBBAY, J., VIVIAN, N., GOODFELLOW, P., LOVELL-BADGE, R. 1991. Male development of chromosomally female mice transgenic for *Sry*. Nature 351: 117–121.

LYON, M.F. 1991. The quest for the X-inactivation center. Trends Genet. 7: 69 – 70.

LYON, M.F. 1992. Some milestones in the history of X-chromosome inactivation. Annu. Rev. Genet. 26: 17–28.

MCKEE, B.D., HANDEL, M.A. 1993. Sex chromosomes, recombination and chromatin conformation. Chromosoma 102: 71–80.

MIGEON, B.R. 1994. X-chromosome inactivation: molecular and genetic consequences. Trends Genet. 10: 219–259.

OGAWA, A., SOLOVEI, I., HUTCHISON, N., SAITOH, Y., IKEDA, J-E., MACGREGOR, H., MIZUNO, S. 1997. Molecular characterization and cytological mapping of a non-repetitive DNA sequence region from the W chromosome of chicken and its use as a universal probe for sexing Carinatae birds. Chromosome Res. 5: 93–101.

Oostra, B.A., Venkerk, A.J.M.H. 1992. The fragile X syndrome: isolation of the FMR-1 gene and characterization of the fragile X mutation. Chromosoma 101: 381–387.

Riggs, A.D., Pfeifer, G.P. 1992. X-chromosome inactivation and cell memory. Trends Genet. 8: 169–174.

Schwartz, S., Depinet, T.W., Leana-Cox, J., Isada, N.B., Karson, E.M., Park, V.M., Pasztor, L.M., Sheppard, L.C., Stallard, R., Wolff, D.J., Zinn, A.B., Zurcher, V.L., Zackowski, J.L. 1997. Sex chromosome markers: characterization using fluorescence *in situ* hybridization and review of the literature. Am. J. Med. Genet. 71: 1–7.

Solari, A.J. 1989. Sex chromosome pairing and fertility in the heterogametic sex of mammals and birds. In: Fertility and Chromosome Pairing: Recent studies in plants and animals (Gillies, C.B., ed.). CRC Press, Boca Raton, FL. pp. 77–107.

Zinn, A.W., Page, D.C., Fisher, E.M.C. 1993. Turner syndrome: the case of the missing sex chromosome. Trends Genet. 9: 90–93.

Section 6.6

Alfenito, M.R., Birchler, J.A. 1993. Molecular characterization of a maize B chromosome centric sequence. Genetics 135: 589–597.

Carlson, W., Roseman, R.R. 1992. A new property of the B chromosome of maize. Genetics 131: 211–223.

Friebe, B. 1989. Nucleolar activity of B chromosomes in *Allium cernuum* (Alliaceae). Plant Syst. Evol. 163: 87–92.

Ishak, B., Jaafar, H., Maetzad, J.L., Rumpler, Y. 1991. Absence of transcriptional activity of the B-chromosomes of *Apodemus peninsulae* during pachytene. Chromosoma 100: 278–281.

Jiménez, M.M., Romera, R., González-Sánchez, M., Puertas, M.J. 1997. Genetic control of the rate of transmission of rye B chromosomes. III. Male meiosis and gametogenesis. Heredity 78: 636–644.

Jones, R.N. 1995. B chromosomes in plants, Tansley Review No. 85. New Phytol. 131: 411–434.

López-Leon, M.D., Cabraso, J., Camacho, J.P.M. 1991. A nucleolus organizer region in a B chromosome inactivated by DNA methylation. Chromosoma 100: 134–138.

Nur, U., Warren, J.H., Eickbush, D.G., Burke, W.D., Eickbush, T.H. 1988. A "selfish" B chromosome that enhances its transmission by eliminating the paternal genome. Science 240: 512–514.

Ohta, S. 1991. Phylogenetic relationship of *Aegilops mutica* Boiss. with the diploid species of congeneric *Aegilops–Triticum* complex, based on the new method of genome analysis using its B-chromosomes. Mem. Coll. of Agricult.; Kyoto Univer. 137: 1–116.

III
Chromosome Structure and Rearrangements

7

Structural Stability of Chromosomes

- Chromosome breakage can be caused by external factors such as radiation or certain chemicals, and by internal factors such as aging and transposable elements.
- The efficient repair of damaged chromosomes is crucial for survival and utilizes enzymatic activities that recognize and remove lesions in DNA, carry out repair synthesis, or transfer homologous DNA from an undamaged chromosome to a damaged one.
- Sister-chromatid exchanges involving unrepaired lesions are associated with aging and some disease syndromes in humans.
- Certain regions of chromosomes, the fragile sites, are more susceptible to breakage than others.
- Transposable elements are segments of DNA that are capable of moving from one place to another in the genome and, in so doing, they can cause extensive disruption to chromosome structure.
- Gametocidal chromosomes induce breaks in other chromosomes during the mitotic divisions that occur in the postmeiotic maturation of gametes.

It is a large step from understanding DNA structure and gene function at the molecular level to defining the nature of genetic change in whole chromosomes. In Chapter 20, landmarks that are well-known to cytogeneticists are discussed in molecular terms, but it is evident that these represent only a small fraction of the genetic material in the chromosomes. Although gene structure and function are also considered in that chapter, it should again be emphasized that genes usually comprise only a small proportion of the genome. In this chapter, we shall pursue the behavior of whole chromosomes in order to analyze the range of factors that cause changes in the structure and number of chromosomes. We will also analyze the major factors that contribute to the stability of chromosomes and in particular, the ability of chromosomes to repair damaged segments, particularly DNA molecules, through the action of enzymes. This capability applies equally to the complex chromosomes of eukaryotic organisms and to the simple protosomes of prokaryotes. These repair activities complement mechanisms that filter out certain agents before they can cause damage to the DNA. For example, in human skin, the ability to absorb ultraviolet (UV) light before it causes damage to DNA is controlled by the production of melanin pigment; in plant epidermal cells, harmful UV is absorbed by flavonoids; and in blue-green algae, it is absorbed by derivatives of amino acids.

7.1 CHROMOSOME STRUCTURAL STABILITY

The constancy of inherited characteristics that distinguish the major groups of living organisms depends on the integrity of chromosomes and genes. Chromosomes and their respective DNA molecules are replicated with remarkable accuracy through infinite cell and organism generations. Thus, for a specific type of organism, we expect to find the same chromosome number and structure in different cells of the same organ or tissue, whether in one individual or different individuals of the same species. Genes are expected to replay their program of information with enough fidelity that offspring resemble parents in the gross features of development and differentiation.

In prokaryotes, the circular form of the protosome may of itself contribute to stability in that there are no exposed ends. Furthermore, because prokaryotes lack a spindle mechanism, they are not involved in the elaborate maneuvers of mitosis and meiosis. On the other hand, the linear chromosomes of eukaryotes have evolved a number of protective features to maintain their integrity during cell division. Their telomeres prevent random end-to-end joining, and the close association of histone and nonhistone proteins with DNA provides general structural stability to the chromosome framework. Entanglement of chromosomes is prevented by the action of DNA topoisomerases, which break and rejoin DNA helices to resolve interlocked strands of chromosomes. There is also the stabilizing effect of having two homologous copies of each chromosome present in most cells for most of the life cycle. A loss of function in a part of one homolog can be rescued by the normal functioning of the partner chromosome.

The evolution of karyotypes encompasses numbers ranging from a single protosome in bacteria to over 1000 chromosomes in some plant species. Although the differences among organisms in number, size, and other characteristics of chromosomes are presently difficult to explain, it can be argued that they are directly relevant to the attainment of cytological, developmental, and genetic stability.

7.2 CHROMOSOME INSTABILITY AND REPAIR

Although the inherent structural properties of chromosomes give them a high degree of constancy, transverse breaks or changes in chromosome number can occur under a variety of stressful conditions, which can result in the formation of different types of structural or quantitative chromosomal changes. The harmful effects of many of these changes, such as losses of chromosome segments, are commonly eliminated by death in the gametic or early embryonic stages. Changes that do not have a drastic effect on viability are valuable experimentally in determining the modification of gene function or the physical locations of genes, and in moving genes within or between genomes.

It is estimated that about 6 in every 1000 live human births and up to 60% of human spontaneously aborted fetuses have inherited some kind of chromosome abnormality. The causal agents in most cases are difficult or impossible to trace, although, in some instances, environmental factors such as increased use of chemicals in agriculture, industry, and medicine may contribute. Some plant and insect species use novel reproductive strategies to maintain and transmit particular changes in chromosome structure and number. Persistence of these chromosomal aberrations in natural populations indicates that they confer some adaptive advantages.

7.2.1 Timing of Chromosome Breakage

The kinds of chromosome structural aberrations that occur depend on whether or not the chromosomes have duplicated at the time a break occurs. When cells in the G1 phase are exposed to X-rays, the aberrations observed in the first-division cycle after the treatment are mainly of the chromosome type, where both of the sister chromatids have a break in the same place as a result of replication after the break was induced (Fig. 7.1). If the radiation is applied to cells in the S or G2 phase, the chromatid type of aberration results because the affected chromosome region has already replicated at the time of treatment (Fig. 7.2). Ultraviolet radiation and most chemicals produce only the chromatid type of aberration because the lesions they produce must interact with replicating DNA during the S phase.

7.2.2 Major Groups of Chromosome Abnormalities

Two major groups of chromosome aberrations are generally distinguished. Spontaneous aberrations are usually attributed to unknown causes because the point in time or the generation when the chromosome breaks occurred is unknown. Possible causes of spontaneous breaks are natural radiations such as cosmic rays, or stresses on chromosome threads during meiosis due to interlocking of bivalents or association of nonhomologous heterochromatic regions. On the other hand, induced aberrations are the result of an experimental procedure or some other known condition. For example, if a sample of seeds or insects is treated with a clastogenic (chromosome-breaking) agent such as ionizing radiation, any statistically significant increase in the frequency of chromosome abnormalities over an untreated sample from the same population can be attributed to the radiation.

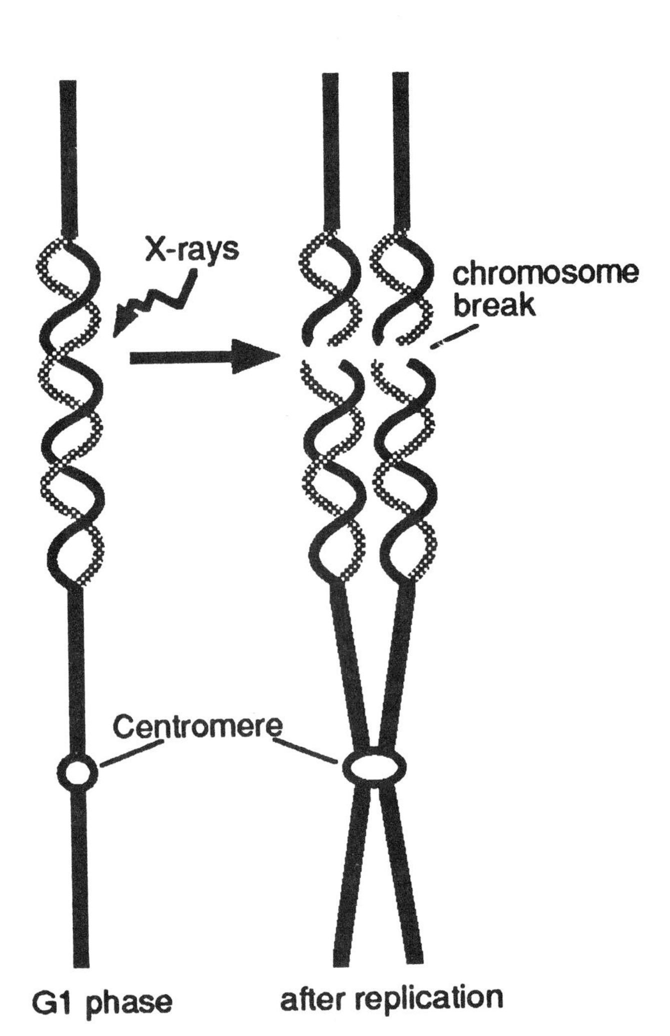

Fig. 7.1. Representation of one type of chromosome break as a result of applying X-rays in the G1 phase of the cell cycle. After replication, both the old and newly replicated chromatids and the DNA double helices will all have a break.

7.2.3 Repair of Damaged Chromosomes

Induced aberrations can be used to study the repair of damaged chromosomes because they are available in much greater numbers than spontaneous aberrations. Quantitative data can be obtained to compare a treated group of cells or individuals with an untreated or control group, or to make comparisons among different treatments.

The initial target is usually the DNA molecule, where the damage may be in the form of single-strand breaks (SSBs), double-strand breaks (DSBs), or chemical changes in nucleotide base pairs such as the formation of thymidine dimers. DSBs are lethal in some organisms because the DNA molecule is severed into two pieces, which lack structural support. However, in other organisms, repair systems within cells can correct the damage to affected strands. There are four known repair mechanisms:

1. In the process of excision–repair, the defective segment of DNA is removed and the resulting gap repaired by synthesizing new DNA (Fig. 7.3). The failure to fill a gap could result in a base injury becoming a SSB, or a SSB becoming a DSB after replication. To achieve excision–repair, the concerted action of four enzymes is required (Fig. 7.4). Enzymatic activities specifically target damaged bases (G*

in Fig. 7.4) or base mismatches and produce gaps that are then processed as shown in Fig. 7.4. In the case of mammalian cells, an enzyme exists to target G·T mismatches and to remove the T residue; thus, it could also correct changes resulting from deamination of C residues to give a moiety closely related to T residues (*see* Fig. 23.11). A long-range repair mechanism has been described in *Escherichia coli*, where excision of the DNA strand carrying the mismatched base is initiated from the nearest GATC sites that are methylated. Following this degradation, the strand is resynthesized by DNA polymerase.

2. Recombination repair or postreplication repair can repair a lesion that has occurred in one strand of DNA. During replication, the damaged strand produces a new strand with a gap opposite the lesion where replication was not able to take place (Fig. 7.5). At the same time, the unaffected strand

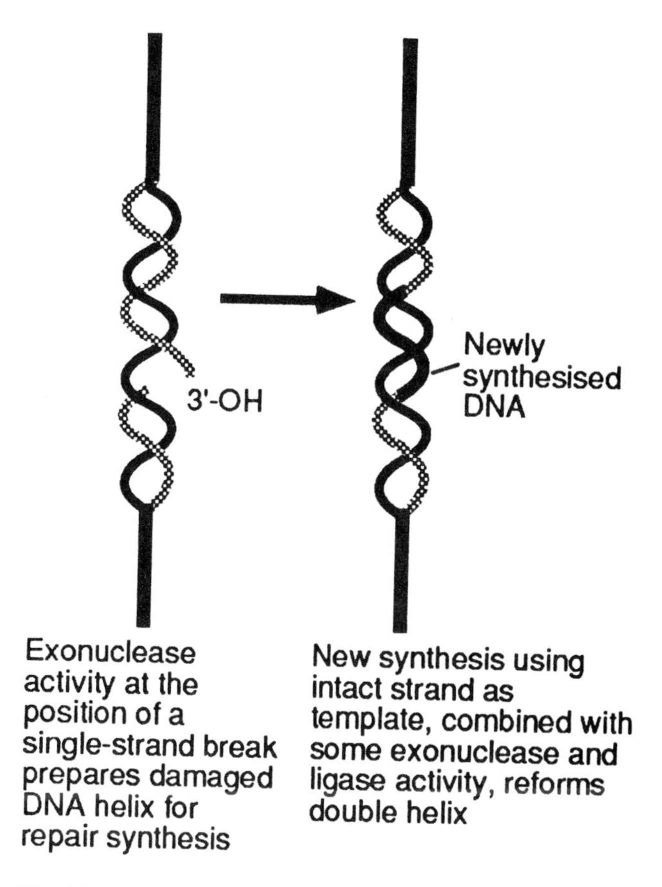

Fig. 7.3. A simplified diagram illustrating excision–repair of a damaged DNA molecule. (Modified from Friedberg, 1985.)

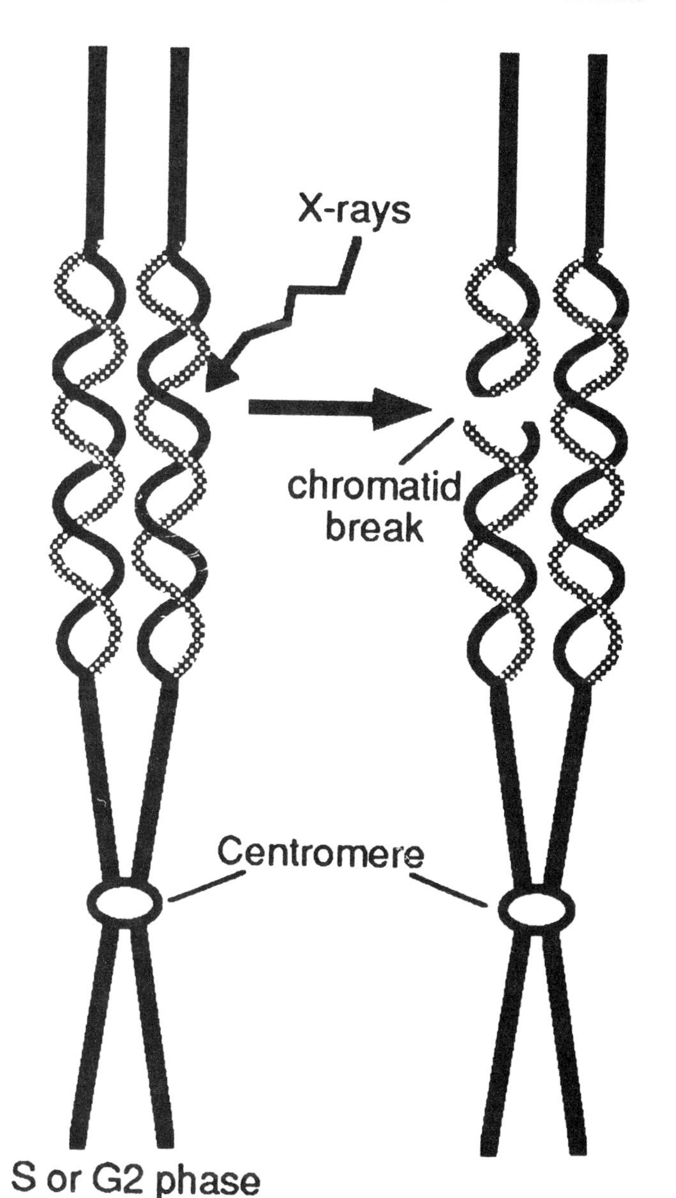

Fig. 7.2. A second type of chromatid break resulting from the application of X-rays after replication. Only one chromatid and its DNA double helix will have a break.

produces a normal, new strand, and provides a source of DNA that can be transferred to the deficient strand, followed by gap filling. The gap left in the DNA that donated the undamaged DNA segment is repaired by repair synthesis. A strand with a lesion still in the system could at some point lead to a chromosome break, but it is often removed by repeating the postreplication repair process. The amount of repair varies according to the time available, and cells that have a longer division cycle, such as mouse spermatogonial stem cells, have a higher amount of repair activity. A similar mode of repair has been found in bacteria, in which a protein (recA) acts to catalyze the invasion of a single DNA strand into an intact double helix and thereby initiate an exchange with one of the double strands. If the invading strand has a break, the exchange could involve an aberration in a process similar to those suggested earlier (*see* Fig. 7.5).

3. External factors such as energy from radiation can also initiate photoreactivation repair of the damage produced by some clastogenic agents. The effect of photoreactivation on DNA after UV irradiation is discussed in Section 7.4.2.
4. The SOS response comprises a series of activities that are activated when DNA replication is blocked by the presence of a damaged base, and results in the insertion of a random base in the new DNA strand, at the position of damage to the parental DNA strand. The LexA and RecA proteins are central in the response of *E. coli* to damage to its DNA. The LexA protein acts to repress the transcription of a number of different genes, whereas the RecA protein acts as a protease

to cleave the LexA protein (and reduce its activity). In addition, the RecA protein binds single-stranded DNA, an activity that is an essential part of the DNA repair process.

Excision–repair and postreplication repair occur in mammals, including humans, and in most forms of life, but there are differences among species in repair frequencies. Inhibitors of the DNA polymerases involved in the repair process also change the frequencies of different types of aberrations. For example, when a marsupial cell line is exposed to X-rays, then treated in G2 with an inhibitor of DNA polymerase repair activity, there are fewer exchange aberrations than in cells treated with X-rays alone. Inhibitors that prevent SSBs from undergoing repair cause them to be changed to DSBs after replication. Some inhibitors also slow down the repair process and provide an opportunity for more aberrations to occur. Imbalances in deoxyribonucleotide-precursor pools can also lead to misincorporation and inhibition of repair synthesis.

Recognition of lesions in DNA occurs at several levels. The stalling of a RNA polymerase complex, during the transcription of a normally actively transcribed gene, can occur as a result of a thymidine dimer formed during UV irradiation of the template strand (*see* Section 7.4.2). In human transcription complexes, one of the polypeptides involved in the complex is the same as one of the excision–repair cross-complement (ERCC) genes identified in the analyses of DNA repair in cells of patients suffering from xeroderma pigmentosum (*see* Section 7.3). It is therefore possible that a stalled RNA polymerase complex forms a focus for additional repair activities to aggregate on the DNA and form a "repairosome," which contains all the activities to remove the lesion in the DNA. Experimental evidence for the coupling of transcription and DNA repair has come not only from characterization of the genes involved in DNA repair but also from the observation that thymidine dimers located in an efficiently transcribed gene are removed five times faster than in the rest of the genome. Another level of recognition of DNA lesions involves a highly conserved mismatch-repair protein (MutS), identified from studies in *E. coli* and *Salmonella typhymurium*. The MutS protein was discovered during the characterization of mutations that affected DNA repair; sequence comparisons indicate that analogous proteins exist in yeast and higher eukaryotes. A human cancer, hereditary nonpolyposis colorectal carcinoma (HNPCC), is associated with a deficiency in a protein that is closely related in sequence to the MutS

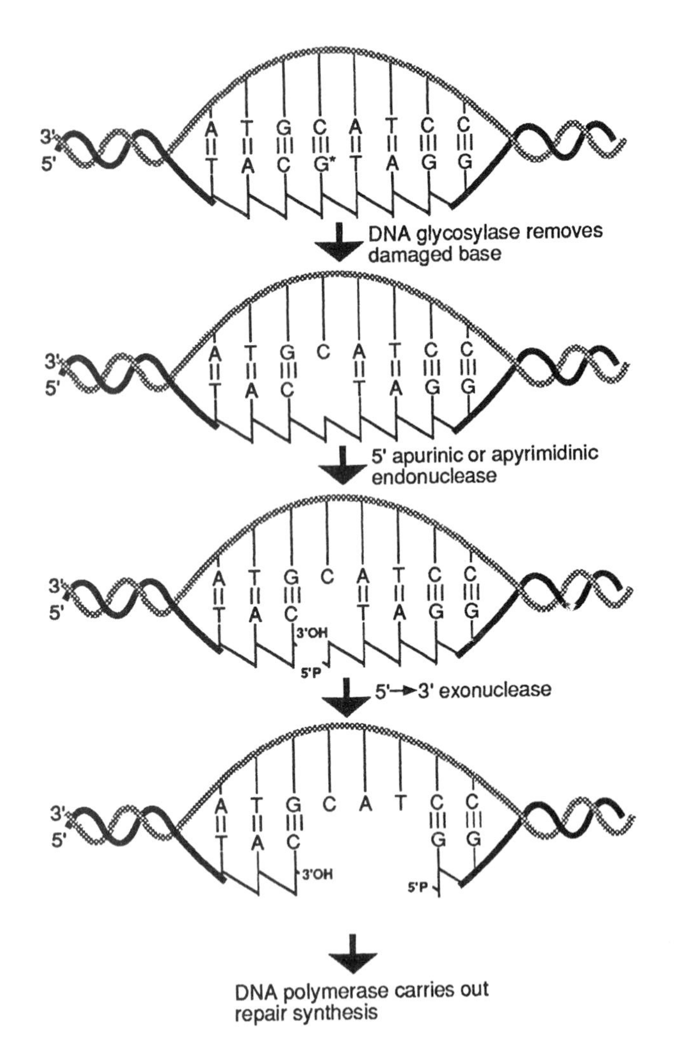

Fig. 7.4. A diagrammatic explanation of the four steps in excision–repair, showing the consecutive action of four enzymes. (Modified from Friedberg, 1985.)

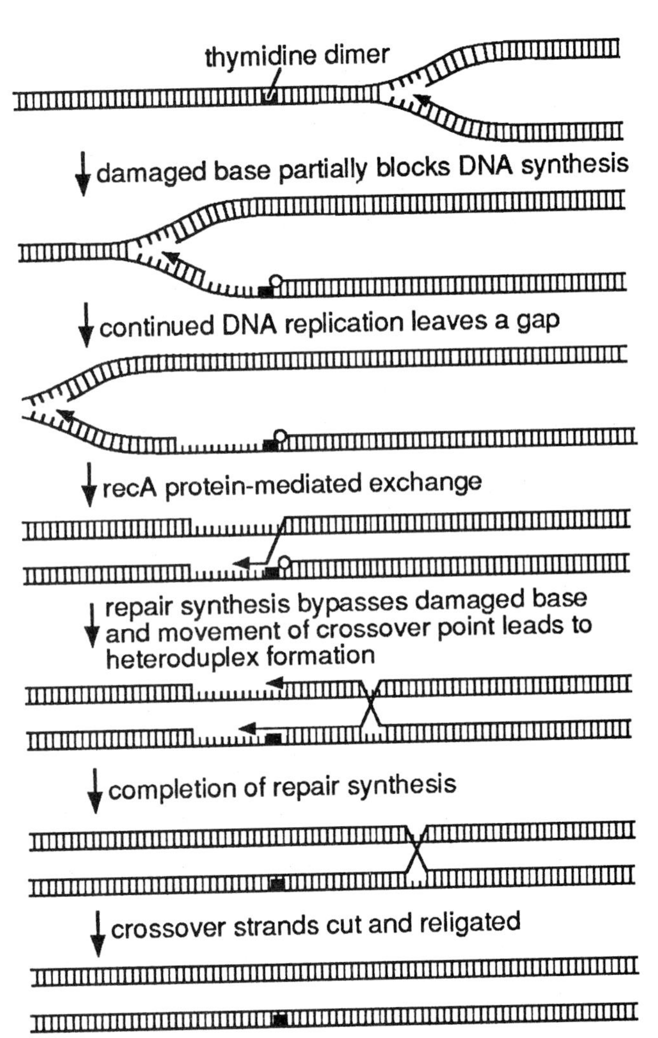

Fig. 7.5. The sequence of steps involved in the repair of DNA by recA protein (Friedberg 1985). The DNA strand that invades the adjoining double helix in a recA-mediated exchange is shown by the arrow in the fourth panel.

protein. The first step in the repair of a mismatch lesion in the DNA is considered to be the binding of the MutS protein, with subsequent binding of other proteins that provide the enzymatic activities to carry out the repair.

7.3 SISTER-CHROMATID EXCHANGE

The detection of sister-chromatid exchange (SCE) is facilitated by growing cells in the presence of 5-bromo-deoxyuridine (BrdU), a thymidine analog, which causes sister chromatids to become differentially stained (Fig. 7.6). This allows exchanges between the chromatids to be clearly detected. Although the BrdU can itself induce SCEs in low frequency, they are significantly increased by treatment of the cells with clastogenic agents such as chemicals or UV irradiation, but not by ionizing radiations. In repair-deficient cells of the Chinese hamster, unrepaired lesions induced by chemical treatment persist to the end of the third-division cycle and result in an increased frequency of SCEs, whereas repair-proficient human cells given the same treatment have fewer SCEs.

The way in which exchanges between sister chromatids occur is not well understood, and although SCE frequency is affected by environmental conditions such as temperature, an intriguing question is whether the lesions are permanently repaired when the exchanges occur. In both plant and animal cells, SCEs occur more frequently in euchromatin than in heterochromatin, often at the borderline between the two types of chromatin. Further, their occurrence tends to be proportional to chromosome length, although larger chromosomes seem to have more SCEs than expected.

Human twins can be used to differentiate between the effects of genetic background and age on the frequency of sister-chromatid exchanges. In older groups of twins, SCEs are more frequent than in a younger group, and within age groups, no significant difference in SCE frequency occurs between monozygotic and dizygotic twins.

Sister-chromatid exchanges have been associated with several human disease syndromes. Individuals with xeroderma pigmentosum, who are extremely sensitive to sunlight due to a defect in excision–repair of UV-induced pyrimidine dimers and postreplication DNA, have what is considered to be a normal level of spontaneous SCEs. In epidermal cells cultured from such individuals, however, significant increases in SCEs occur after UV irradiation or treatment with monofunctional alkylating agents. In another human disorder, Bloom's syndrome (BS), which also involves sunlight sensitivity, the amount of DNA ligase is significantly reduced and is correlated with the observation that cultured BS cells have a spontaneous frequency of sister-chromatid exchanges 10 to 15 times that of cells from normal controls (Fig. 7.7). The exposure of blood lymphocytes from BS patients to ethylmethane sulfonate (EMS) also gives a greater increase in SCEs than exposure of cells from normal humans.

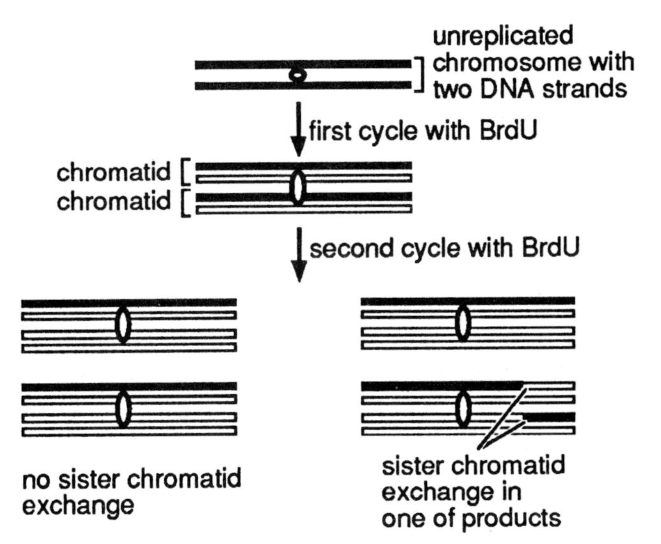

Fig. 7.6. The microscopic detection of sister-chromatid exchanges in mitotic cells grown in the presence of BrdU. Details of the basis for the differential staining of BrdU-containing chromatids are described in Chapter 18.

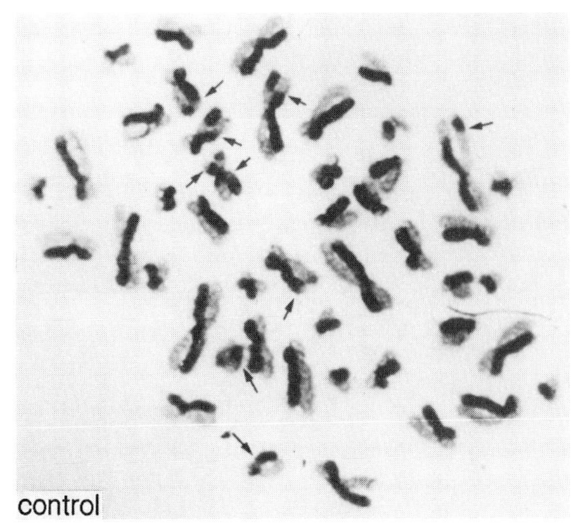

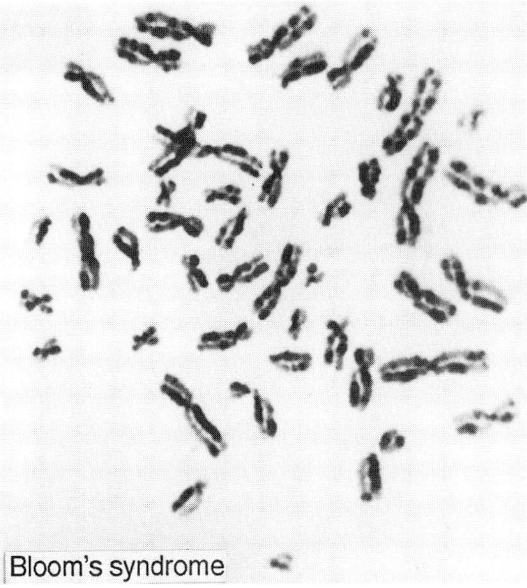

Fig. 7.7. Normal human chromosomes (upper) compared to the chromosomes of a Bloom's syndrome lymphocyte (lower) that show a marked increase in SCEs. The cells were grown in the presence of bromodeoxyuridine (BrdU), then stained with Hoechst 33258, followed by Giemsa, to detect the SCEs (Chaganti et al., 1974). Arrows in the upper photo mark points of exchange between sister chromatids.

7.4 AGENTS AND CONDITIONS INDUCING CHROMOSOME BREAKAGE

In the latter 1920s, the discovery was made that X-rays could induce transverse breaks in the chromosomes of *Drosophila melanogaster* and maize. Since that era, the number of clastogenic agents has proliferated to include all types of ionizing radiation, nonionizing radiation such as ultraviolet light, and many diverse chemicals. In addition, various inherent or external conditions can produce chromosome instability.

7.4.1 Ionizing Radiations

Ionizing radiations can induce chemical damage to the purine and pyrimidine bases of the DNA molecules, as well as to the deoxyribose sugar molecules (Fig. 7.8). Another type of damage involves breaks in the phosphodiester backbone of one of the DNA strands to give single-strand breaks (SSBs), or of both strands at the same place to give double-strand breaks (DSBs). The proportion of SSBs to DSBs depends on the linear energy transfer (LET) of the radiation. With low LET, SSBs are 10 to 20 times as frequent as DSBs, whereas with high LET, DSBs predominate. Base damage is more frequent than strand breaks, and some SSBs could result from damaged bases that are not completely repaired. Most of the SSBs are repaired quite rapidly, but some of those not repaired can be changed to DSBs by single-strand endonucleases. DSBs are needed to produce a chromosome or chromatid aberration. Ionizing radiations differ from most of the other clastogenic agents in producing DSBs directly (i.e., without a delayed effect, which would allow some of the initial damage to be repaired), and so they are an efficient source of aberrations.

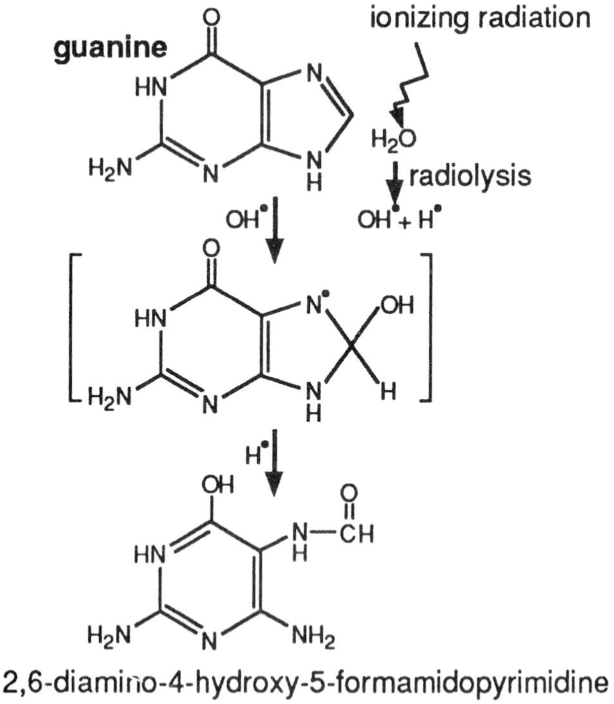

guanine

ionizing radiation

H_2O radiolysis $OH^{\bullet} + H^{\bullet}$

2,6-diamino-4-hydroxy-5-formamidopyrimidine

Fig. 7.8. An example of the effect of ionizing radiation on a single DNA base, guanine, and its biochemical transformation into a novel pyrimidine (Chetsanga et al., 1981).

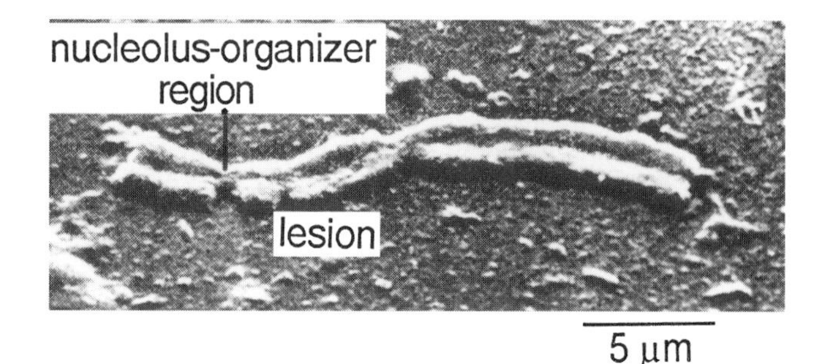

Fig. 7.9. One of the nucleolus-organizer chromosomes of *Vicia faba* showing a radiation-induced lesion in one chromatid near the nucleolus-organizer region (Scheid and Traut, 1971).

Two main concepts have been developed to explain what happens when an ionizing radiation contacts chromosomes: breakage–reunion and exchange. The breakage–reunion theory, advanced in the early years of X-ray studies, argues that breaks occur at the moment of contact. The broken parts then behave in one of the following ways:

1. Do not rejoin, leading to the loss of the piece of chromosome without a centromere
2. Rejoin in the original sequence by restitution
3. Rejoin with adjacent breaks in the same or different chromosomes to give new arrangements

The exchange theory argued that the first effect of the radiation is not a break but a lesion (Fig. 7.9), which can either be repaired or can interact with other nearby lesions to produce exchanges. Although the exchanges can occur within a chromosome that is looped on itself, or between chromosomes if they are close together, the exact mechanism of strand breakage, with respect to whether 3′-OH or 3′-P is exposed for exchange, has not been determined. At a molecular level, an abundant nuclear protein (Ku) has been implicated in the repair of DSBs caused by ionizing radiation. This protein, a heterodimer with subunits 70 and 86kDa, binds to the broken ends of the DNA in a sequence-independent manner and, together with protein kinases, forms a complex that initiates the repair of the DSB.

7.4.2 Ultraviolet Radiation

Ultraviolet radiation, at shorter wavelengths of about 260 nm, close to the maximum absorbed by DNA, causes changes in the structure of purines and pyrimidines. One of the more important effects in the production of chromosome aberrations is the formation of pyrimidine dimers, most often involving adjacent thymines (Fig. 7.10). If the dimers persist, the DNA is prevented from replicating, and aberrations such as deletion of chromatin result. Thymine dimers are removed by excision–repair if the cells are in the dark. If the repair process is incomplete, and a thymine dimer is removed from a strand leaving a gap, this gap could persist as a

SSB. A DSB can also result if an endonuclease nicks the intact strand opposite the gap, leading to a chromatid type of aberration.

When UV irradiation of cells or tissues is followed by exposure to short-wavelength visible light or to long-wavelength UV light, the enzyme DNA photolyase breaks the bonds between the thymine dimers by monomerization. This light-induced enzyme activity is present in the cells of many organisms (plants and animals) and is known as photoreactivation–repair. In plants, photoreactivation reduces the number of UV-induced DNA lesions and chromosome aberrations. However, it causes a greater decrease in the number of pyrimidine dimers than in the number of chromosome aberrations, indicating that not all the induced dimers result in aberrations. The DNA photolyase protein from *E. coli* has been crystallized and used to determine the relative positions of the key cofactors, the light-harvesting molecule 5,10-methenyl-tetrahydrofolylpolyglutamate (MTHF) and the electron transfer molecule flavin-adenine dinucleotide (FADH), attached to the protein. The amino-acid sequence of DNA photolyase has homology to the blue-light photoreceptor in plants, and thus the detailed structural studies on photolyase have broader implications.

The four different ways for removing thymine dimers are summarized in Fig. 7.11. The pathways are determined by combining mutation analysis with biochemical and chemical studies, in a wide range of organisms, to determine the basis for the various possible repair mechanisms. A first step in mutation studies is to determine that a DNA repair activity is involved rather than a change in the protective pigments that reside in cells.

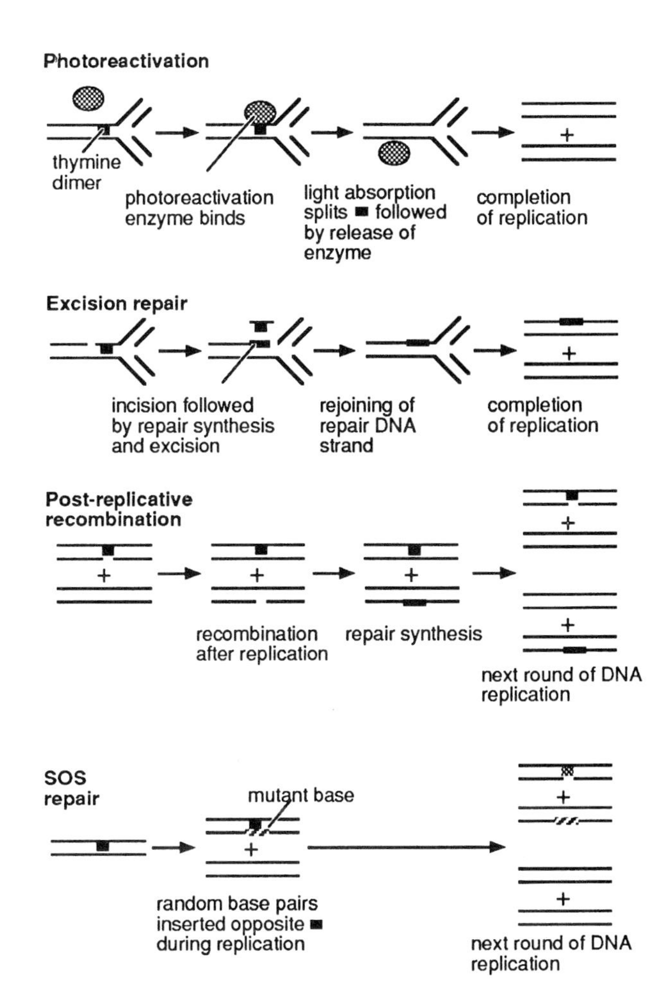

Fig. 7.11. Four possible mechanisms for the removal and repair of thymine dimers from double-stranded DNA (Friedberg, 1985).

7.4.3 Chemically Induced Damage to Chromosomes

The first evidence that chemicals can induce chromosome aberrations was obtained in the 1940s, using nitrogen and sulfur mustards. Since that discovery, many chemicals have been tested and found to be clastogenic. Some of the chemicals used in chemotherapy treatments for cancer, as well as others, are carcinogenic, so there is great interest in finding out how they interact with DNA and other cell components.

A few chemicals are similar to ionizing radiation in inducing the chromosome type of aberration when applied to cells in the G1 phase, but the chromatid type when applied in the S or G2 phases. Examples of such chemicals are the antibiotic streptonigrin and the purine derivative 8-ethoxycaffeine. Most chemicals differ from ionizing radiation and resemble ultraviolet radiation in producing only chromatid aberrations regardless of whether they are applied before or during DNA and chromosome replication. If chemicals of this type are applied to cells during the G1 phase, the conversion of lesions to breaks is delayed until the S phase. This delay gives an opportunity for some of the lesions to be repaired before replication occurs, but when lesion repair is slowed down by a DNA repair inhibitor such as cytosine arabinoside, there is an increase in aberrations. This suggests that the presence of previously induced lesions may cause interruptions in the replication of DNA during the S phase, leading to simultaneous breaks, some of which can produce chromatid aberrations. Sister-chromatid exchanges (SCEs) are induced by many chemicals, and they occur with higher frequencies than structural chromosome aberrations. For this reason, an increase in SCE frequency in human cells can give an initial indication of the effects of exposure to a chemical agent, but additional studies should be made on possible mutagenic or clastogenic effects.

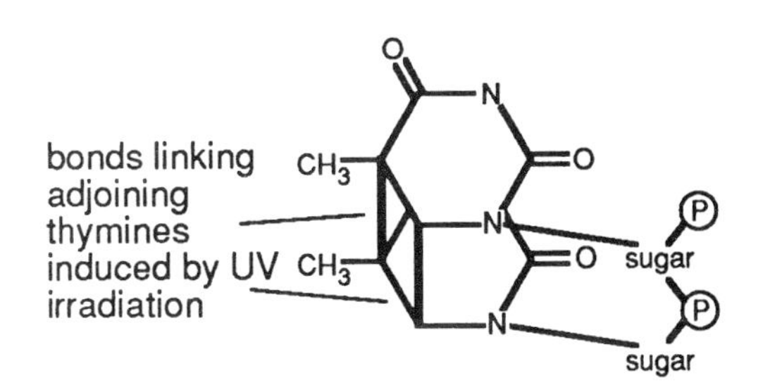

Fig. 7.10. The structure of a thymine dimer induced by the action of UV light on adjacent thymidine residues. (Modified from Lehninger, 1982.)

The different types of lesions produced in DNA by various chemical agents are summarized in Fig. 7.12, along with the effects of ionizing radiations and ultraviolet radiation. Alkylating agents are a prominent group of chemicals with chromosome-breaking ability. They are classified as monofunctional, bifunctional, or

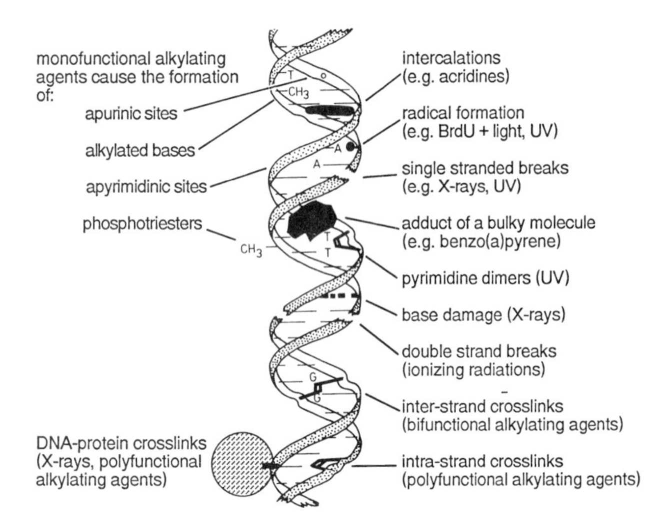

Fig. 7.12. A diagram of the DNA double helix showing the primary lesions produced by different types of chemicals and radiations. The single letters inside the spiral refer to the different bases (Natarajan, 1984).

polyfunctional based on the number of alkyl groups (ethyl or methyl, with the general formula C_nH_{2n+1}) that react with the DNA molecule. Monofunctional alkylating agents such as ethyl methanesulfonate (EMS), methyl methanesulfonate (MMS), and dimethyl sulfate (DMS) have one reactive site on one strand of the DNA double helix. For example, EMS causes alkylation of guanine by adding the ethyl group CH_3CH_2 to the N-7 position of this base (*see* Fig. 17.1), causing it to pair with thymine instead of cytosine and leading to a base substitution. In an attempt to correct the mispairing, the alkylated guanine may be excised from the DNA molecule by enzymic action in a process called depurination, leaving an apurinic site. If the gap is not repaired, a single-strand break may occur in one of the DNA strands and result in a double-strand break after replication. The same sequence can happen with loss of a pyrimidine base through depyrimidation, giving an apyrimidinic site; however, the pyrimidine bases are lost less often than the purine bases. Bifunctional and polyfunctional alkylating agents are more effective clastogens than monofunctional alkylating agents because they react with two or more sites either within one DNA strand (intrastrand cross-links), between the two DNA strands (interstrand cross-links), or between DNA and protein. Cross-links, especially those of the interstrand type, can interfere with the replication of DNA and cause strand breaks. In addition, bifunctional alkylating agents such as nitrogen and sulfur mustards may act like monofunctional agents in reacting with DNA at only one site.

The N-7 atom of guanine (*see* Fig. 17.1) is more likely to be alkylated than any other site on any of the bases. All the nitrogen mustards induce monofunctional alkylation at this site, as well as interstrand cross-links between the N-7 reactive sites of two guanines. When different monofunctional alkylating agents are compared for reactivity with DNA base sites, alkylation of N-7 guanine predominates in the production of chromosomal aberrations and SCEs, whereas O-6 guanine alkylation lesions give rise to point mutations. Alkylating agents may also affect the sugar–phosphate backbone of DNA or RNA by causing disruption of the phosphodiester bonds through the formation of phosphotriesters, and resulting in breaks in the strands.

Two other types of chemically induced lesions that can result in chromosomal aberrations are included in Fig. 7.12. Organic compounds containing planar-ring systems such as the acridine dyes and the antibiotic actinomycin-D (*see* Fig. 17.15) can be inserted between DNA base pairs. The intercalation causes the DNA molecule to elongate and become more rigid, so that during replication, a base pair may be added or deleted in the new strand, causing a reading-frame shift for the region downstream from the point of addition or deletion. This creates an opportunity for misrepair of the lesion and strand breakage. Polycyclic aromatic hydrocarbons, represented by benzo(a)pyrene in Fig. 7.12, have indirect effects on DNA by being metabolized to diol epoxides. Some of these derivatives react mostly with the DNA purines to form bulky adducts and distort the structure of the double helix in the process. Faulty repair of the lesions can cause strand breakage.

7.4.4 Chromosomal Changes Resulting from Tissue Culture

Chromosome structural and numerical changes can occur when plant tissues or cells are cultured *in vitro*. Somaclonal variation is the term applied to changes arising in cultured somatic tissues such as embryos or protoplasts. Gametoclonal variation describes changes occurring in cultured cells from the gametophytic pathway such as microspores. In such studies, it is essential to know the chromosome constitution of the plants used as a control group, so that any aberrations already present are not added to those induced by the tissue-culture procedures. Two factors that affect the frequency of aberrations are the genotypes of the donor plants from which the tissues were obtained, and the length of time that the cells are cultured—in general, the longer the time, the higher the frequency of aberrations.

7.4.5 Genetic Predisposition to Chromosomal Breakage

Chromosome instability can be caused by various states within an organism. A number of viable but severely defective inherited conditions in humans are associated with high frequencies of spontaneous chromosome abnormalities. In some cases, these frequencies can be increased even more by exposing the cells to a clastogenic agent. One example is ataxia telangiectasia (Louis–Bar syndrome), which is due to an autosomal recessive gene on chromosome 11 and occurs in a frequency of 1 in every 40,000 live births. The syndrome affects many body functions, particularly the nervous and immune systems, and increases the risk for cancer. The frequency of spontaneous chromosome abnormalities is high but is increased markedly by X-rays and by chemicals that break DNA strands. Defects in DNA synthesis or repair have been implicated as causes of chromosomal abnormalities in at least some human disease syndromes.

Chromosomes also have localized segments that are vulnerable to breakage because of structural or chemical composition. There are heritable fragile sites in human chromosomes, which are transmitted like codominant genes and appear as nonstaining gaps in chromosome preparations, involving one or both sister chromatids (Fig. 7.13). Although different types of fragile sites have been defined based on their frequencies in human populations and their reactions to pH and certain chemicals, they all correlate with the occurrence of a low frequency of acentric fragments and deficient chromosomes as well as rearrangements. Rare fragile sites occur in only a small fraction of people, whereas common fragile sites

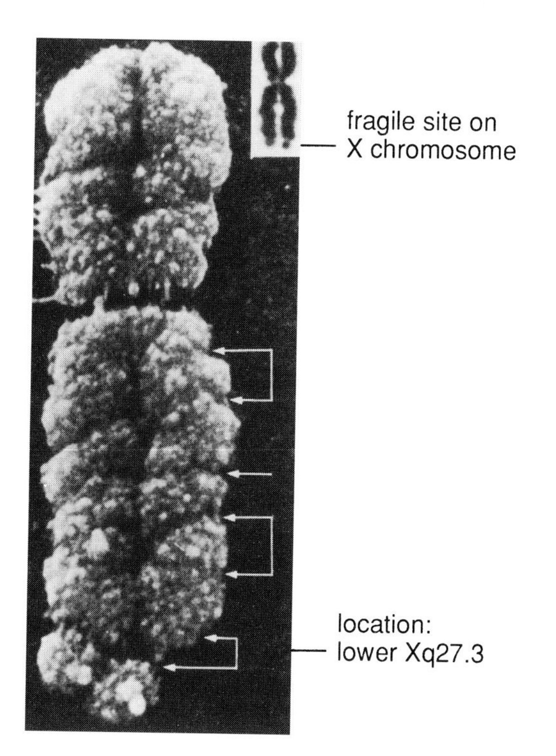

Fig. 7.13. The fragile X chromosome of humans as seen with a scanning electron microscope. A light-microscopic image is shown in the inset. The chromosome has a complete gap in the fragile site of the chromatid on the right where the terminal fragment is completely detached. The arrows indicate the locations of G bands (Harrison et al., 1983). The fragile site consistently occurs at the lower Xq27.3 band, where molecular studies have shown that a short array of CGG nucleotides is amplified in varying amounts. The tandem array of CGG trinucleotides is an example of a microsatellite DNA sequence, and instabilities in the size of such regions are associated with a range of other human disorders (Richards and Sutherland, 1992; Jiricny, 1994).

are found in all individuals. Cells must be cultured in suitable media for fragile sites to become visible, and breakage occurs more frequently at many fragile sites when the cells are cultured in a low-folate medium. The folate sensitivity is associated with the incorporation of deoxyuridine (dU) into DNA, because the enzyme thymidylate synthetase, which converts dU monophosphate to thymidine monophosphate, is inhibited in a low-folate medium. The accumulation of dU and a shortage of thymidine for DNA synthesis result in DNA misrepair during G2 and an increase in aberrations.

The fragile X (Martin–Bell) syndrome is a mental retardation defect in humans that is associated with the expression of a rare fragile site at Xq27.3 (*see* Fig. 7.13). In molecular studies, sequence-target sites (STSs; *see* Chapter 21, Section 21.7.1) for this region have been isolated by microdissecting chromosomes, then using the isolated DNA as probes for the isolation of large DNA clones covering this entire region. The cloning technology utilized the construction of yeast artificial chromosomes (YACs) and *in situ* hybridization to confirm the correct chromosomal location of the isolated DNA sequences. The cytological abnormality (*see* Fig. 7.13) was in this way shown to be related to the enlargement, by amplification, of a CGG trinucleotide repetitive region including from 2–60 repeats to over 200 repeats. The amplification occurs in the 5′-upstream region of a gene named *FMR1*, coding for a RNA-binding protein, and correlates with the inactivation of this gene, leading to the mental retardation syndrome associated with the fragile X site. DNA methylation (*see* Chapter 17, Section 17.2) of the amplified CGG trinucleotide has been suggested as a mechanism by which expression of the *FMR1* gene could be suppressed.

7.4.6 Chromosome Breakage and the Aging Process

The effect of age in producing chromosome aberrations can be tested effectively using seeds. The loss of seed viability with age in barley (*Hordeum vulgare*) and peas (*Pisum sativum*) is associated with an increase in chromosome aberrations, mainly at the chromatid level. Older barley seeds also have less DNA polymerase activity, and this could contribute to misrepair of lesions during chromosome replication. High temperatures and high humidity during seed storage can interact with the aging process in causing chromosomal breaks.

Many studies have been made on cultured cells from humans and other mammals to find out if there is an association between chromosome aberrations and aging. The types of cells commonly used are lymphocytes from humans, and liver or bone-marrow cells from animals. In some human studies, where the cells were not exposed to a clastogenic agent in culture, the frequencies of chromosome breaks, chromatid-like lesions, and chromosome rearrangements show increases in older individuals. Losses of the inactive X chromosome in females and of the Y chromosome in males are also related to the age of the subjects. It has also been suggested that fragmentation of regions of the genome, such as the nucleolus leads to increased amounts of extra-chromosomal circles of ribosomal DNA (rDNA). The unscheduled replication and accumulation of these rDNA circles leads to aging in organisms such as yeast.

7.5 TRANSPOSABLE ELEMENTS AND CHROMOSOME INSTABILITY

Transposable elements are now known to be widespread in different organisms, and where they have been well characterized, they share the common features of target duplication upon insertion, the presence of open reading frames (ORFs), and usually terminal repeated sequences. The elements are specialized DNA sequences that reside in certain chromosomes but can move from the initial donor site to a second recipient site, often close to the first site. The movement of the elements causes chromosome aberrations at the sites of insertion in diverse organisms such as maize, *Drosophila melanogaster*, yeast (*Saccharomyces cerevisiae*), and bacteriophages. Other terms used to describe these fascinating genetic entities, depending on the organism and the investigator, are mobile elements, controlling elements, transposons, and insertion sequences.

7.5.1 The Genetics of Transposable Elements

Transposable elements were first detected by virtue of their chromosome-breaking ability in maize (Fig. 7.14), and the genetic characterization of this phenomenon demonstrated the existence of a two-element system. *Activator* (*Ac*) turns on a switch controlling the activities of a second element, *Dissociation* (*Ds*), named for

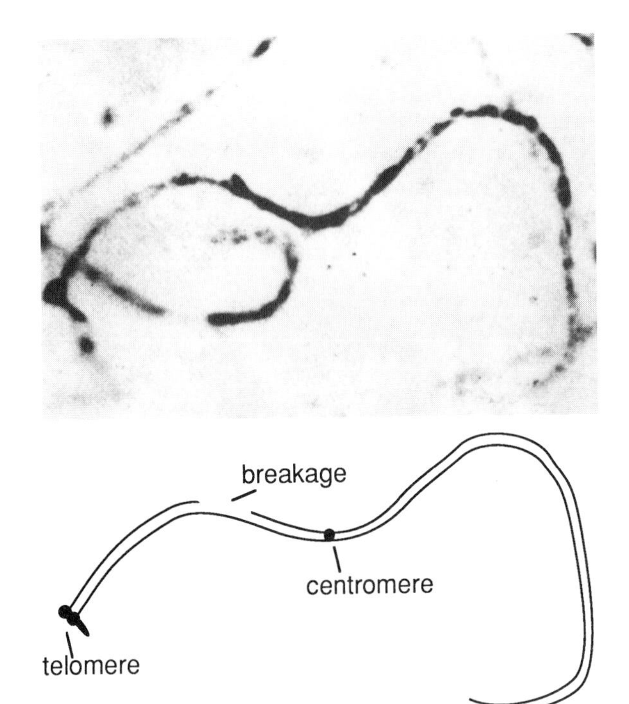

Fig. 7.14. Chromosome breakage induced by transposable elements in maize. A light-microscope photograph (top) and diagram (bottom) of a chromosome 9 bivalent at pachytene (McClintock, 1951). The gap in the upper homolog was caused by the relocation of a *Ds* element, which induces a breakage at the point of insertion if a second element, *Ac*, is present. The lower homolog does not have *Ds*.

its ability to break chromosomes at its insertion site. With at least one dose of *Ac* in the cell, *Ds* can move to another site and cause the chromosome to break where it is inserted. The result is loss of the acentric piece beyond *Ds*, leaving a deficient, centric chromosome. *Ds* exists in two states, one of which gives more breaks and allows fusions of sister chromatids, whereas the other gives fewer breaks and no fusions of sister chromatids. The difference in the two states is due to the number of *Ds* elements.

Some 30 years after the genetic studies in maize, molecular research has provided a framework for interpreting the properties of the transposable elements. The DNA structure of the *Ac* and *Ds* elements, together with the proposed mechanism of insertion into a target site in the genomic DNA, is summarized in Fig. 7.15.

The mechanism for the movement of *Ds* elements, particularly where this leads to chromosome breakage, has been deduced from structural analyses and is summarized in Fig. 7.16. The special feature of chromosome-breaking *Ds* elements is that they are "double *Ds*," containing two identical copies of a 2-kb (kilobase) *Ds* element originally derived from *Ac* by deletion (*see* Fig. 7.16b). The critical feature of the double-*Ds* element is that it has two extra copies of the terminal sequences, because it has one *Ds* element inserted into another *Ds* element. Chromosome breaks result from pairing between the terminal repeats at one end of the double-*Ds* element (*see* Fig. 7.16b), as the transposing enzyme transposase attempts to excise this complex element.

Transposable elements that accompany aberrant chromosome events have phases of inactivity and activity for excision and insertion. They seem to be activated by conditions causing stress to the chromosomes such as chromosome aberrations that produce dicentric bridges and stretch the chromatin, extreme temperatures, virus infections, and cell culture on synthetic media. A detailed proposal for the mode of excision of transposable elements has been suggested to account for the modified target site that remains after the element has been excised. The footprint of the transposable, or mobile, element suggests that several steps occur in the excision process, as summarized in Fig. 7.17.

Some transposable-element systems in *D. melanogaster* induce hybrid dysgenesis, a syndrome of adverse traits in hybrids between interacting strains. Natural populations of *D. melanogaster* strains distributed globally, as well as other *Drosophila* species, show hybrid dysgenesis behavior. There is good evidence that the transposable elements producing hybrid dysgenesis are repetitive DNA sequences, 5000–6000 nucleotides in length, which are distributed throughout the *Drosophila* genome. The increased levels of mutations and chromosome rearrangements, due to activation of the transposable elements, is the cause of traits such as sterility, which is associated with the syndrome.

The two best-known hybrid dysgenesis systems are *P–M* (paternal–maternal), with males contributing the transposable elements called *P* factors (Fig. 7.18), and *I–R* (inducer–reactive), with males contributing the transposable elements called *I*. The mechanism by

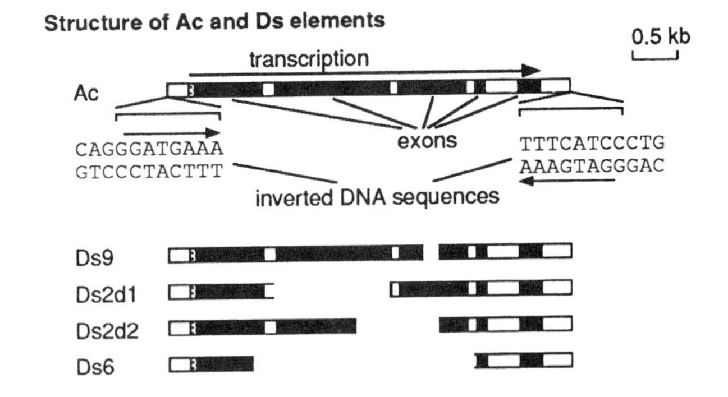

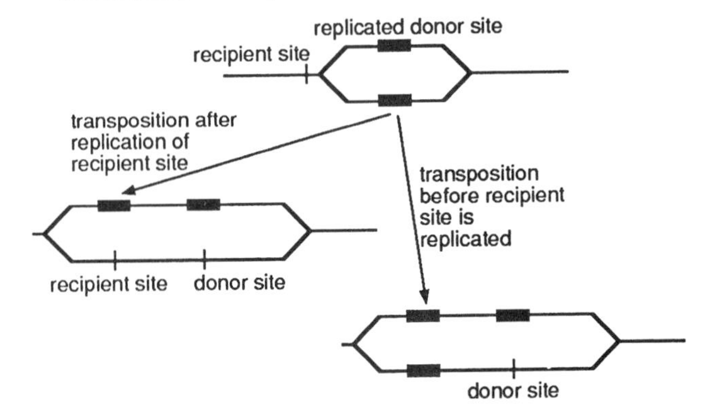

Fig. 7.15. The molecular structure of the *Ac* and *Ds* elements from maize. The *Ac* element (top) contains five exons encoding for a single transposase, as well as the characteristic inverted-repeat sequences at the ends of the element. The *Ds* elements beneath are derived from *Ac* elements by various deletions. The consequences of the timing of transposition of an element relative to the time of replication are indicated in the lower part of the diagram. (Modified from Federoff, 1989.)

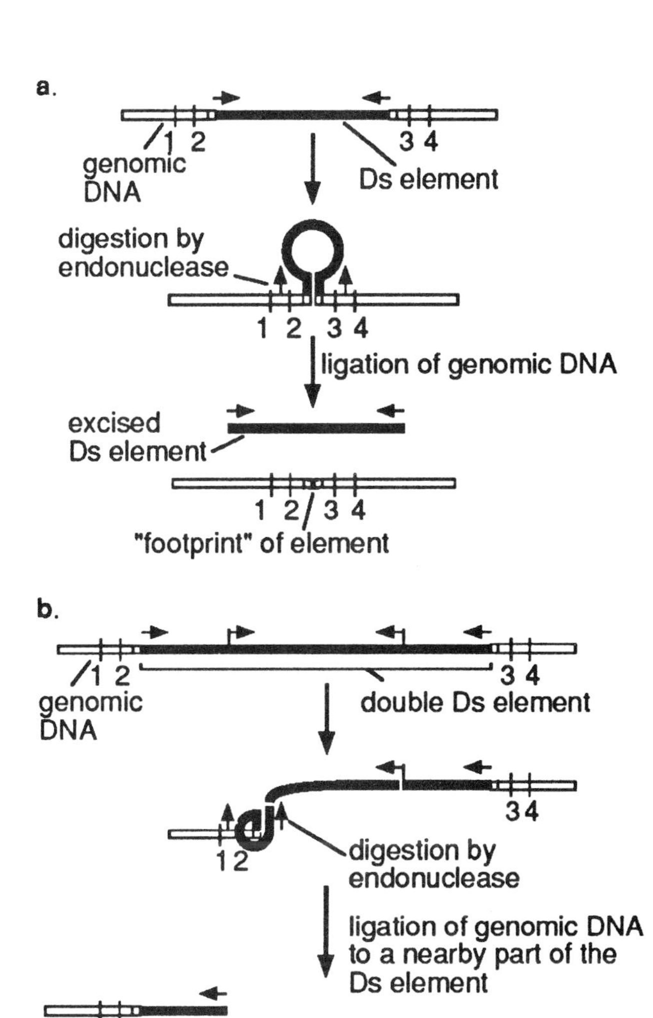

Fig. 7.16. A diagrammatic summary of (a) the mechanism of excision of a normal *Ds* element, and (b) a ''double'' *Ds* element that contains one *Ds* element inserted into another. (Based on Döring and Starlinger, 1986: Federoff, 1989).

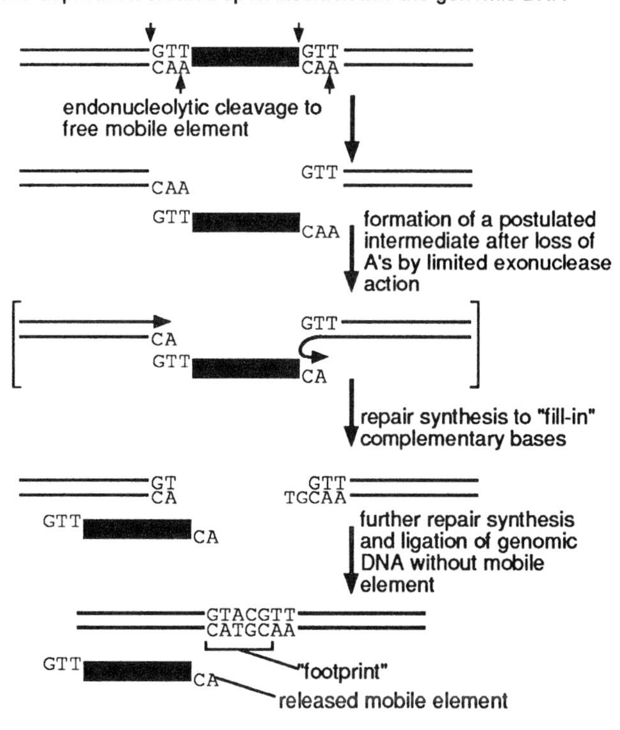

Fig. 7.17. A diagrammatic representation of the postulated mechanism for excision of a transposable element, marked by small arrows, from a genomic site. The excision leaves a seven-base-pair ''footprint'' in the genomic DNA that marks the previous position of insertion of the element which created the target-site duplication.

which an *I* factor transposes is postulated to be a combination of the DNA repair enzymes and the transcription product from the *I* factor unit (Fig. 7.19). There are several similarities in the two systems. Dysgenic traits appear in hybrids from crossing *P* males with *M* females or *I* males with *R* females. The reciprocal cross in each system does not cause hybrid dysgenesis. This indicates that cytoplasm of *M* or *R* strains is needed, because cytoplasmic factors are transmitted mainly with the egg cells through the females. The phenomenon was discovered when males derived from wild populations were crossed to laboratory-strain females. Together with the influence of the age of the flies and external factors such as temperature, the phenomenon of hybrid dysgenesis is consistent with the activation of transposable elements under stress. In both systems, the dysgenic traits are confined to transmission through the germline, because there is no evidence for transposition in somatic tissues. The major difference between the two systems is

Phenotypic consequences of P element insertion

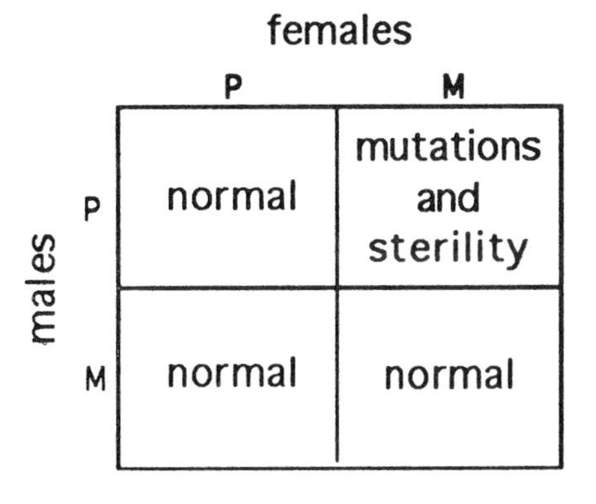

Fig. 7.18. Diagram showing the results of crosses in *Drosophila melanogaster* between paternal (*P*) strains carrying transposable elements and maternal (*M*) strains that are susceptible to the *P* elements. When *P* elements are combined with *M* cytoplasm, the *P* elements are activated and cause a range of genetic abnormalities known as hybrid dysgenesis (upper-right quadrant). Crosses within the *P* or *M* strains, or between *P* females and *M* males, produce normal progeny (Kidwell, 1986). (*See* Chapter 22, section 22.3.1.)

that hybrid-dysgenesis traits occur in both sexes in the *P–M* system, but only in females in the *I–R* system.

Both *P* and *I* elements have been located on each of the four chromosomes of *D. melanogaster*, and are associated with localized sites or hot spots of mutations and chromosome breaks. In the *P–M* system, the hot spots on a *P* chromosome disappear when their segment is replaced by a homologous segment from the *M* chromosome. Many visible and lethal mutations are associated with chromosome rearrangements in both systems. A study of about 1000 *P*-induced rearrangements showed that two-break inversions and reciprocal translocations were common, but three- to five-break aberrations also occurred, indicating a flurry of transposition events. Most of the rearrangements showed net losses of *P* sites at the breakpoints, due to excision or the coming together of two separated *P* elements. Other rearrangements had net gains of *P* sites, resulting from either the splitting of a *P* element into two sites or duplication during a rearrangement event.

7.5.2 Transposable Elements in Prokaryotes

Transposable elements in prokaryotes are characterized by a great diversity in structure, and their movement occurs during DNA replication or may involve essentially no DNA synthesis. Transposable elements occur in both gram-positive and gram-negative bacteria, and usually confer resistance to antibiotics. In common with eukaryotic transposable elements, most of the prokaryotic elements, including bacteriophages, feature an inverted-repeat sequence at their ends, and the creation of a duplication of the 5-bp

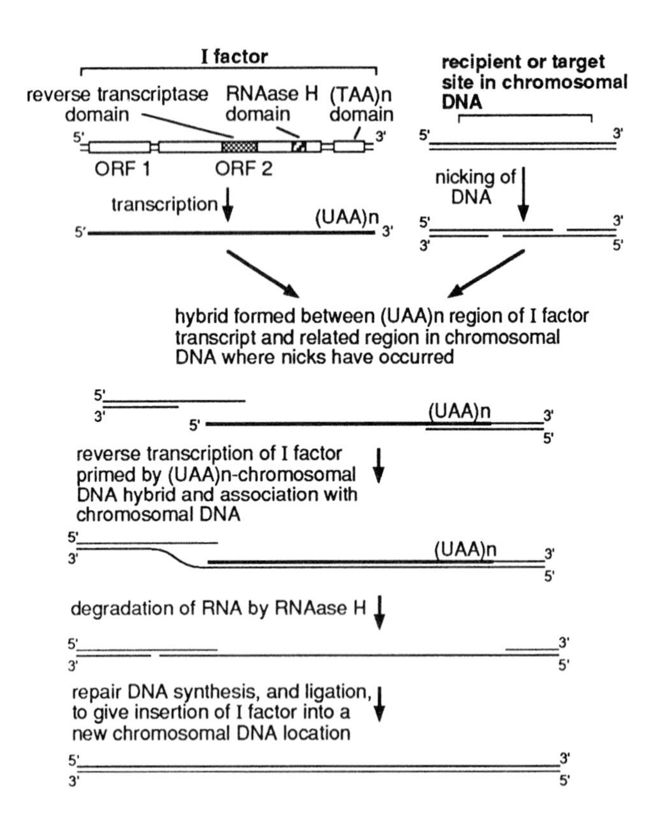

Fig. 7.19. A diagrammatic explanation of the transposition of a *Drosophila* I factor via a RNA intermediate (Finnegan, 1989).

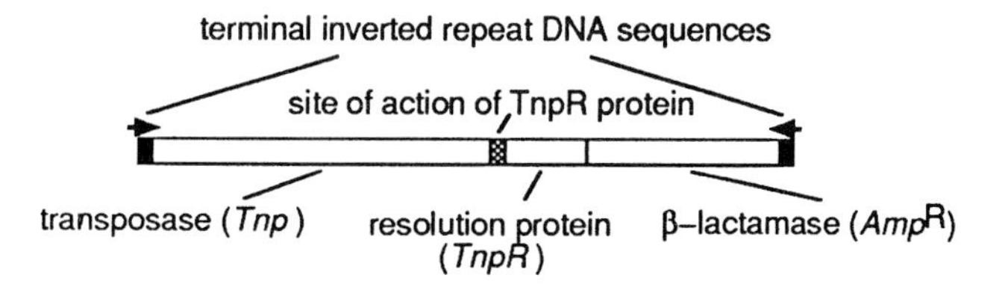

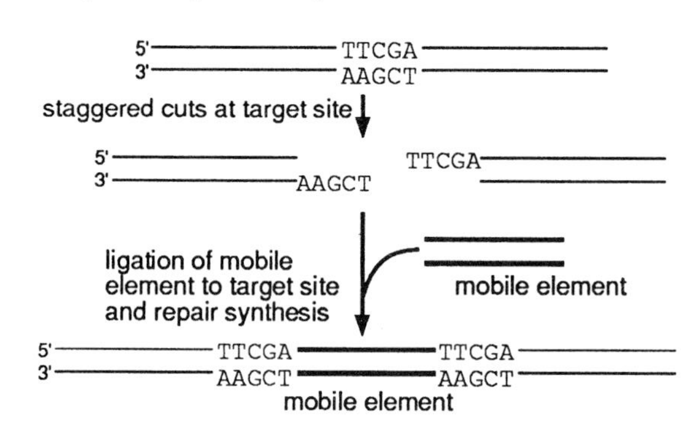

Fig. 7.20. The structure of a bacterial transposable element (top) and its transposition into a genomic DNA site, resulting in a duplication of the genomic target site (lower diagrams).

target sequence upon insertion into the host DNA (Fig. 7.20). The target sequence can vary from being highly specific for some transposable elements to essentially random for other elements, and movement of the elements can be accompanied by deletions, inversions, and duplications. During replication of host DNA carrying the transposable element, cointegrate structures can be formed, and deletions and inversions can take place (Fig. 7.21).

7.6 DISPERSED REPEATED DNA SEQUENCES AND CHROMOSOME INSTABILITY

Repeated DNA sequences are commonly found in eukaryotic chromosomes. If homologous repeated sequences are at the same site on sister chromatids or homologous chromosomes, crossing over between them gives no loss, gain, or structural rearrangement of chromosome segments. As has been found in *Drosophila,* yeasts, and humans, however, if the sequences are at different sites, they are sometimes aligned so that an unequal type of crossing over occurs between them, resulting in various types of chromosome aberrations (Fig. 7.22, *see also* Fig. 20.23). The relationship between the asymmetrical pairing of repeated sequences and the alignment of other homologous parts of the chromosome is not known. Hence, the terms ectopic pairing, ectopic crossing over, exchange, and recombination are used to describe the behavior of repeated sequences (see Fig. 7.22).

At least some of the repeated DNA sequences that promote ectopic defects are transposable elements, which can occur in multiple, dispersed copies due to their ability to move between chromosomal sites. In yeast, for example, two *Ty* (transposon yeast) elements had the same orientation on a homologous pair of chromosomes but were separated by about 21 kb of DNA. Crossing over between the *Ty* elements gave a deleted region in one chromosome and a duplication for the same region in the other chromosome (*see* Fig. 7.22c). Ectopic recombination occurred in about 1% of the diploid yeast cells during meiosis, but it was rare in the haploid cells during mitosis.

Transposable elements have been involved in both intrachromosomal and interchromosomal rearrangements in *D. melanogaster* as a result of ectopic recombination. One study involved a pair of X chromosomes in which there were 26 copies of the transposon *roo* in one chromosome and 21 copies in the other. The *white-eye* locus and flanking genes were used as markers to screen over 400,000 X chromosomes. Ectopic exchanges between repeated DNA sequences produced rearrangements that accounted for 25 phenotypic changes involving the marker genes, and two-thirds of these involved exchanges between two copies of *roo*. Deletions were easy to detect in this research, whereas most duplications, translocations, and inversions could not be detected. All except one of the detectable rearrangements resulting from exchanges between copies of *roo* were deletions of varying sizes similar to those shown in Figs. 7.22a and 7.22c. The exception was an inversion, which was thought to result from pairing and exchange between two copies of *roo* in reverse orientation on the same X chromosome, similar to that shown in Fig. 7.22b.

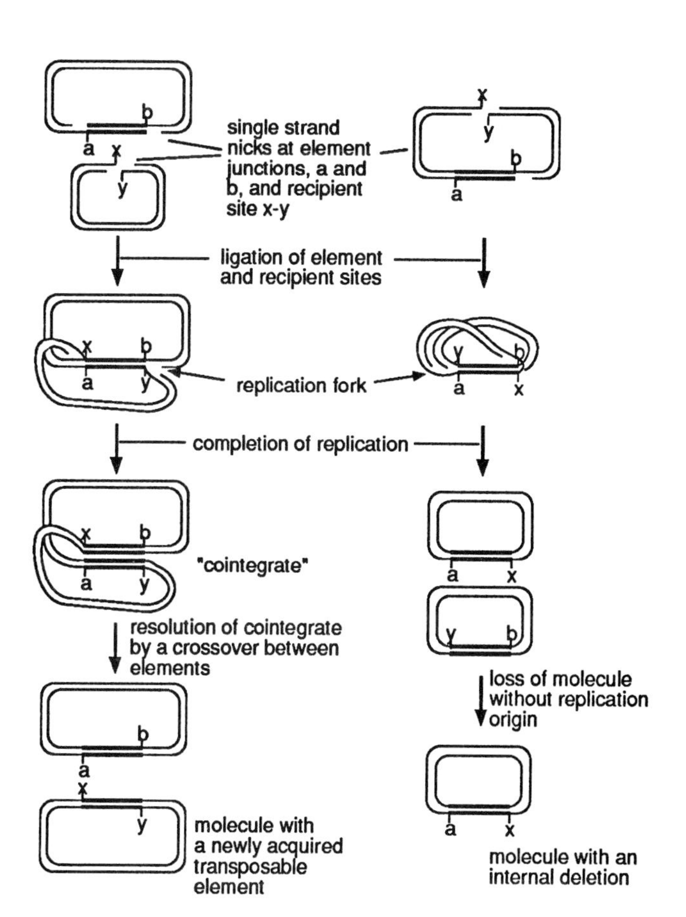

Fig. 7.21. A summary diagram showing the consequences of transposition between two circular DNA molecules or within a circular DNA molecule (the molecules are depicted as rectangles in the diagrams).

a. Intrachromosomal crossover between direct repeat DNA sequences to create deletions

b. Intrachromosomal crossover between inverted repeat DNA sequences to create inversions

c. Interchromosomal, or sister chromatid, crossover between direct repeated DNA sequences to give a deletion and duplication

d. Crossover between DNA repeated sequences with the same orientation relative to the centromeres, on non-homologous chromosomes, to give a reciprocal translocation

e. Crossover between DNA repeated sequences with opposite orientation relative to the centromeres, on non-homologous chromosomes, to give a dicentric chromosome and an acentric fragment

Fig. 7.22. The effects of pairing and crossing over between repeated sequences (black arrows) and the orientation of intervening sequences (gray arrows). Nonhomologous chromosomes are represented by lines of differing widths and filled or clear centromeres. (a) Intrachromosomal crossing over between direct repeats in the same orientation, leading to a ring and a rod chromosome with complementary deletions (right). (b) Intrachromosomal crossing over between a direct and an inverted repeat, leading to an inversion of the intervening segment. (c) Unequal crossing over between direct repeats in homologous chromosomes or sister chromatids, leading to a deletion and a duplication. (d) Crossing over between repeats—with the same orientation relative to their centromeres—on nonhomologous chromosomes, leading to a reciprocal translocation. (e) Crossing over between repeats—with opposite orientations relative to their centromeres—on nonhomologous chromosomes, leading to a dicentric chromosome and an acentric fragment. (Adapted from Petes and Hill, 1988).

7.7 GAMETOCIDAL CHROMOSOMES

Another chromosome-breaking condition occurs when wheat (*Triticum aestivum*) is crossed with wild members of the genus *Triticum*, followed by backcrosses of the hybrids to wheat, in order to eliminate all but one of the alien chromosomes. Some of these alien chromosomes carry a gene(s) or some other factor that induces chromosome breaks during the gametophytic cycle, or around the time of fertilization. The kinds of structural aberrations observed in the progenies of the backcrosses include deletions (Fig. 7.23), translocations, ring chromosomes, and dicentric chromosomes. It is only by observing the behavior of the alien chromosome during meiosis that its effects can be interpreted. If a wheat plant has a complete complement of wheat chromosomes plus one dose of the alien chromosome, the latter is an unpaired univalent during meiosis. As a result, it will be included in one-half or less of the gametes, depending on how often it lags behind during both anaphase stages. However, during meiosis, some kind of interaction occurs between the alien chromosome and the wheat chromosomes, so that gametes without the alien chromosome have aberrant wheat chromosomes, whereas gametes with the alien chromosome have normal wheat chromosomes. A clear example is seen when a chromosome from the wild species *Triticum longissimum* is added to the wheat complement. No chromosomal abnormalities are observed in male or female meiotic cells. During the first postmeiotic division, however, megaspores and microspores without the alien chromosome have many acentric fragments (Fig. 7.24). At least some of these wheat fragments are lost during later mitotic divisions in the development of pollen grains or embryo sacs. Thus, many of the resulting gametes without the alien chromosome abort due to the deletions, and the transmission of the alien chromosome through the gametes that have intact wheat chromosomes is assured.

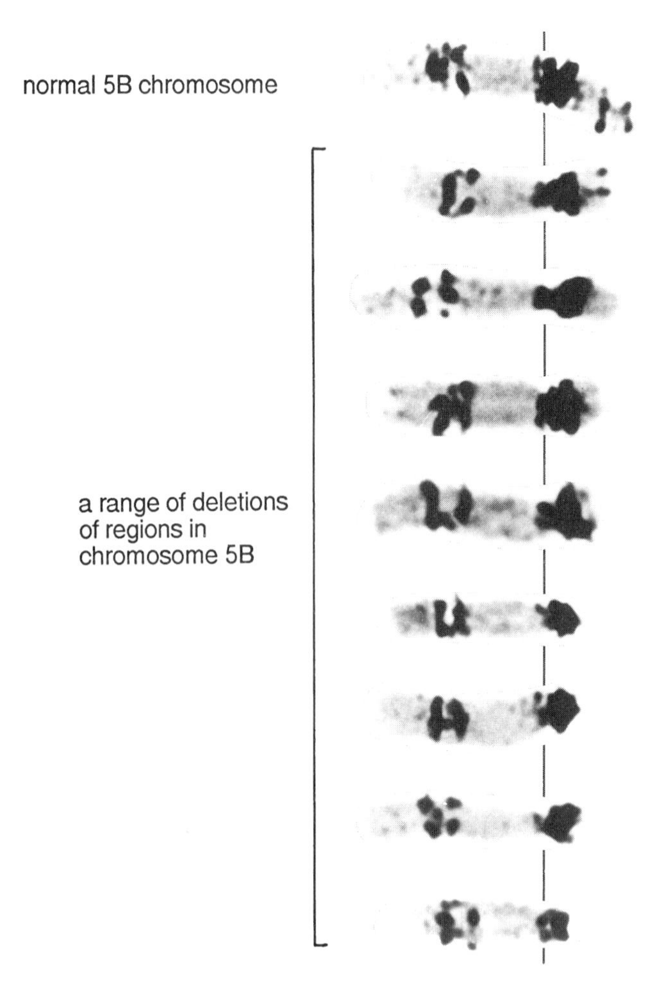

Fig. 7.23. A comparison of structural changes in wheat mitotic–metaphase chromosomes induced by the presence of a gametocidal chromosome. A normal chromosome 5B (top) is followed by eight photos of chromosome 5B carrying deletions of varying sizes in the short arm (right-side). Centromeres are marked by vertical lines. C-banding was used to locate the deletions (Endo, 1988; Endo and Gill, 1996). Molecular analyses have shown that the ends of the deleted chromosomes are "healed" by the addition of telomeric sequences (Tsujimoto et al., 1997; *see also* Chapter 20, Section 20.4).

Fig. 7.24. The cytological effect of adding a gametocidal chromosome from *Triticum longissimum* to wheat. When the gametocidal chromosome is present as a univalent during meiosis, many of the postmeiotic cells that lack this chromosome show lagging wheat chromosomes or wheat-chromatid fragments that completely lack centromeres, as shown in this anaphase cell at the first postmeiotic division in the embryo sac. Cells in which the gametocidal chromosome has been retained have normal wheat chromosomes. This difference in the two types of cells seems to be programmed during meiosis (Finch et al., 1984).

The alien chromosomes responsible for the above effects are called gametocidal chromosomes because of the gametic abortion, and resulting sterility, that they cause. At least one wheat chromosome carries a gene that suppresses the gametocidal effect of an alien chromosome but enhances its chromosome-breaking ability. Some wild *Triticum* chromosomes are effective only when transmitted through the male parent, similar to the *P* and *I* elements of *Drosophila*, whereas others are effective through both parents.

BIBLIOGRAPHY

General

Evans, H.J. 1988. Mutation cytogenetics: past, present and future. Mut. Res. 204: 355–363.

Federoff, N.V. 1989. Maize transposable elements. In: Mobile DNA (Berg, D.E., Howe, M.M., eds.). American Society for Microbiology, New York. pp. 377–411.

Sankaranarayanan, K. 1982. Genetic Effects of Ionizing Radiation in Multicellular Eukaryotes and the Assessment of Genetic Radiation Hazards in Man. Elsevier Biomedical Press, Amsterdam.

Section 7.2

Baker, S.M., Bronner, E., Zhang, L., Plug, A.W., Robatzek, M., Warren, G., Elliott, E.A., Yu, J., Ashley, T., Arnheim, N., Flavell, R.A., Liskay, R.M. 1995. Male mice defective in the DNA mismatch repair gene PMS2 exhibit abnormal chromosome synapsis in meiosis. Cell 82: 309–319.

CHASAN, R. 1994. A ray of light on DNA repair. Plant Cell 6: 159–161.

DE WIND, N., DEKKER, M., BERNS, A., RADMAN, M., RIELLE, H. 1995. Inactivation of the mouse Msh2 gene results in mismatch repair deficiency, methylation tolerance, hyperrecombination and predisposition to cancer. Cell 82: 321–330.

DRAPKIN, R., SANCAR, A., REINBERG, D. 1994. Where transcription meets repair. Cell 77: 9–12.

DROUIN, R., GAO, S., HOLMQUIST, G.P. 1996. Technologies for detection of DNA damage and mutations (Pfeifer, G.P., ed.). Plenum Press, New York. p. 37.

JIRICNY, J. 1994. Colon cancer and DNA repair: have mismatches met their match? Trends Genet. 10: 164–168.

KIMBALL, R.F. 1987. The development of ideas about the effect of DNA repair on the induction of gene mutations and chromosomal aberrations by radiation and by chemicals. Mut. Res. 186: 1–34.

MOORE, R.C., RANDELL, C., BENDER, M. A. 1988. An investigation using inhibition of G2 repair of the molecular basis of lesions which result in chromosomal aberrations. Mut. Res. 199: 229–233.

RADMAN, M., WAGNER, R. 1993. Mismatch recognition in chromosomal interactions and speciation. Chromosoma 102: 369–373.

Section 7.3

CHAGANTI, R.S.K., SCHONBERG, S., GERMAN, J. 1974. A manyfold increase in sister chromatid exchanges in Bloom's syndrome lymphocytes. Proc. Natl. Acad. Sci. USA 71: 4508–4512.

HEIM, S., JOHANSSON, B., MERTENS, F. 1989. Constitutional chromosome instability and cancer risk. Mut. Res. 221: 39–51.

SCHVARTZMAN, J.B. 1987. Sister-chromatid exchanges in higher plant cells: past and perspectives. Mut. Res. 181: 127–145.

WAKSVIK, H., MAGNUS, P., BERG, K. 1981. Effects of age, sex and genes on sister chromatid exchange. Clin. Genet. 20: 449–454.

WOLFF, S. 1981. Induced chromosome variation. In: Chromosomes Today, Vol. 7 (Bennett, M.D., Bobrow, M., Hewitt, G., eds.). George Allen & Unwin, London. pp. 226–241.

Section 7.4

BENDER, M.A., GRIGGS, H.G., BEDFORD, J.S. 1974. Mechanisms of chromosomal aberration production. III. Chemicals and ionizing radiation. Mut. Res. 23: 197–212.

CHETSANGA, C.J., LOZON, M., MAKAROFF, C.M., SAVAGE, L. 1981. Purification and characterization of *Escherichia coli* formamidopyrimidine–DNA glycolase that excises damaged 7-methylguanine from deoxyribonucleic acid. Biochemistry 20: 5201-5207.

DOURADO, A.M., ROBERTS, E.H. 1984. Chromosome aberrations induced during storage in barley and pea seeds. Ann. Bot. 54: 767–779.

EVANS, D.A. 1989. Somaclonal variation—genetic basis and breeding applications. Trends Genet. 5: 46–50.

HARRISON, C.J., JACK, E.M., ALLEN, T.D., HARRIS, R. 1983. The fragile X: a scanning electron microscope study. J. Med. Genet. 20: 280–285.

HEARTLEIN, M.W., PRESTON, R.J. 1985. An explanation of interspecific differences in sensitivity to X-ray-induced chromosome aberrations and a consideration of dose-response curves. Mut. Res. 150: 299–305.

LARKIN, P.J. 1987. Somaclonal variation: history, method, and meaning. Iowa State J. Res. 61: 393–434.

LEHNINGER, A.L. 1982. The Principles of Biochemistry. Worth Publishers, New York.

MANDEL, J.L., HEITZ, D. 1992. Molecular genetics of the fragile-X syndrome: A novel type of unstable mutation. Curr. Opinions Genet. Devel. 2: 422–430.

MAYER, P.J., LANGE, C.S., BRADLEY, M.O., NICHOLS, W.W. 1989. Age-dependent decline in rejoining of X-ray-induced DNA double-strand breaks in normal human lymphocytes. Mut. Res. 219: 95–100.

MULLER, H.J. 1927. Artificial transmutation of the gene. Science 66: 84–87.

NATARAJAN, A.T. 1984. Origin and significance of chromosome aberrations. In: Mutations in Man (Obe, G., ed.). Springer-Verlag, Berlin. pp. 156–176.

PARK, H-W, KIM, S-T., SANCAR, A., DIESENHOFER, J. 1995. Crystal structure of DNA photolyase from *Escherichia coli*. Science 268: 1866–1872.

REIDY, J. A. 1988. Role of deoxyuridine incorporation and DNA repair in the expression of human chromosomal fragile sites. Mut. Res. 200: 215–220.

RICHARDS, R.I., SUTHERLAND, G.R. 1992. Fragile X syndrome: the molecular picture comes into focus. Trends Genet. 8: 249–253.

SCHEID, W., TRAUT, H. 1971. Visualization by scanning electron microscopy of achromatic lesions (''gaps'') induced by X-rays in chromosomes of *Vicia faba*. Mut. Res. 11: 253–255.

SINCLAIR, D.A., GUARENTE, L. 1997. Extrachromosomal rDNA circles—a cause of aging in yeast. Cell 91: 1033–1042.

SINGH, N. P., DANNER, D.B., TICE, R.R., BRANT, L., SCHNEIDER, E.L. 1990. DNA damage and repair with age in individual human lymphocytes. Mut. Res. 237: 123–130.

SIOMI, H., CHOI, M., SIOMI, M.C., NUSSBAUM, R.L., DREYFUSS, G. 1994. Essential role for KH domains in RNA binding: Impaired RNA binding by a mutation in the KH domain of FMR1 that causes fragile X syndrome. Cell 77: 33–40.

SLAGBOOM, P.E., VIJG, J. 1989. Genetic instability and aging: theories, facts, and future perspectives. Genome 31: 373–385.

STADLER, L.J. 1931. The experimental modification of heredity in crop plants 1. Induced chromosome irregularities. Sci. Agr. 11: 557–572.

SUTHERLAND, G.R. 1991. Chromosomal fragile sites. Genet. Anal. Tech. Analy. 8: 161–166.

WEAVER, D.T. 1995. What to do at an end: DNA double-strand-break repair. Trends Genet. 11: 388—392.

Section 7.5

BREGLIANO, J-C., KIDWELL, M.G. 1983. Hybrid dysgenesis determinants. In: Mobile Genetic Elements (Shapiro, J.A., ed.). Academic Press, Orlando, FL. pp. 363–410.

BUCHETON, A. 1990. *I* transposable elements and *I–R* hybrid dysgenesis in *Drosophila*. Trends Genet. 6: 16–21.

DÖRING, H.-P., STARLINGER, P. 1986. Molecular genetics of transposable elements in plants. Annu. Rev. Genet. 20: 175–200.

ENGELS, W.R. 1983. The P family of transposable elements in *Drosophila*. Annu. Rev. Genet. 17: 315–344.

FINNEGAN, D.J. 1985. Transposable elements in eukaryotes. Int. Rev. Cytol. 93: 281–326.

FINNEGAN, D.J. 1989. The *I* factor and *I–R* hybrid dysgenesis in *Drosophila melanogaster*. In: Mobile DNA (Berg, D.E., Howe, M.M., eds.). American Society for Microbiology, New York. pp. 503–529.

GIERL, A., SAEDLER, H. 1989. Maize transposable elements. Annu. Rev. Genet. 23: 71–85.

KIDWELL, M.G. 1986. P–M mutagenesis. In: *Drosophila*: A Practical Approach (Roberts, D.B., ed.). IRL Press, Oxford. pp. 59–81.

MCCLINTOCK, B. 1951. Chromosome organization and genic expression. Cold Spring Harb. Symp. Quant. Biol. 16:13–47.

PETERSON, P.A. 1987. Mobile elements in plants. CRC Crit. Rev. Plant Sci. 6: 105–208.

PRUDHOMMEAU, C., PROUST, J. 1990. *I–R* hybrid dysgenesis in *Drosophila melanogaster*; nature and site specificity of induced recessive lethals. Mut. Res. 230: 135–157.

WEIL, C.F., WESSLER, S.R. 1990. The effects of plant transposable element insertion on transcription initiation and RNA processing. Annu. Rev. Plant Physiol. Mol. Biol. 41: 527–552.

WOODRUFF, R. C., SLATKO, B.E., THOMPSON, J.N. JR. 1983. Factors affecting mutation rates in natural populations. In: The Genetics and Biology of *Drosophila*, Vol 3c (Ashburner, M., Carson, H. L., Thompson Jr., J. N., eds.). Academic Press, New York. pp. 37–124.

Section 7.6

MONTGOMERY, E.A., HUANG, S-M., LANGLEY, C.H., JUDD, B.H. 1991. Chromosome rearrangement by ectopic recombination in *Drosophila melanogaster*: genome structure and evolution. Genetics 129: 1085-1098.

PETES, T.D., HILL, C.W. 1988. Recombination between repeated genes in microorganisms. Annu. Rev. Genet. 22: 147–168.

ROEDER, G.S. 1983. Unequal crossing over between yeast transposable elements. Mol. Gen. Genet. 190: 117–121.

YEN, P.H., LI, X.-M., TSAI, S.-P., JOHNSON, C., MOHANDAS, T., SHAPIRO, J. 1990. Frequent deletions of the human X chromosome distal short arm result from recombination between low copy repetitive elements. Cell 61: 603–610.

Section 7.7

ENDO, T.R. 1988. Chromosome mutations induced by gametocidal chromosomes in common wheat. In: Proc. 7th Int. Wheat Genet. Symp. (Miller T.E., Koebner R.M.D., eds.). IPSR, Cambridge. pp. 259–265.

ENDO, T.R. 1990. Gametocidal chromosomes and their induction of chromosome mutations in wheat. Jpn. J. Genet. 65: 135–152.

ENDO, T.R., GILL, B.S. 1996. The deletion stocks of common wheat. J. Hered. 87: 295–307.

FINCH, R.A., MILLER, T.E., BENNETT, M.D. 1984. ''Cuckoo'' *Aegilops* addition chromosome in wheat ensures its transmission by causing chromosome breaks in meiospores lacking it. Chromosoma 90: 84–88.

Tsujimoto, H., Yamada, T., Sasakuma, T. 1997. Molecular structure of a wheat chromosome end healed after gametocidal gene-induced breakage. Proc. Natl. Acad. Sci. USA 94: 3140–3144.

8

Losses and Gains of Chromosome Segments

- Chromosome deletions or duplications that do not affect vital functions can be maintained in an organism, where they can be characterized by chromosome banding or *in situ* hybridization techniques, as well as by pairing behavior at meiosis.
- Some deletions or duplications have phenotypes that are typical of dominant mutations. Deletions, in particular, are involved in numerous human abnormalities.
- Chromosome deletions can be terminal—at least when first produced—or interstitial, and duplications may be adjacent to the original segment or displaced on the same or a different chromosome.
- Losses or gains of chromosome segments may arise from unequal crossingover, from other chromosome aberrations such as reciprocal translocations or as the direct effect of a chromosome-breaking agent.
- Deletions usually have more severe effects on viability and fertility than duplications.

Karyotypes evolve through numerical and structural changes, and what may appear to be the standard chromosome complement in present-day organisms could have been structurally abnormal in ancestral individuals. The reverse situation occurs when a newly arisen aberration creates an abnormal chromosome in comparison with the ancestral chromosome. The choice of a standard karyotype is, therefore, based on comparative observations of chromosomes of related species and any putative ancestors still in existence, as well as chromosome behavior in the hybrids of crosses between related species. Although the evidence may not be decisive, it helps to choose a representative karyotype to be the standard of comparison within a species, genus, or family.

In this chapter and Chapter 9, structural chromosome aberrations are discussed with respect to their behavior during cell divisions and their effects on organisms. The breaks that produce these aberrations may involve either chromosomes or chromatids, depending on the stage in the cell cycle when they occur, but we shall usually apply the term chromosome in referring to aberrations. The restitution of breaks by rejoining of the original sequences will not be considered, as this does not lead to aberrant chromosomes.

8.1 THE RANGE OF STRUCTURAL CHANGES

Chromosome aberrations involving structural changes have a continuum in size from those that are visible within the magnification range of conventional light microscopes, to those that are detectable only at the molecular level and are defined by new combinations of particular DNA sequences. In both classical- and molecular-cytogenetic studies, the behavior of aberrant chromosomes, and the loss or gain of DNA sequences, can often be correlated with visible phenotypic effects and gene segregations.

If both members of a pair of homologous chromosomes have the same aberration, it is homozygous, whereas if one homolog is aberrant and the other is normal, the aberration is heterozygous. Aberrations can be maintained in the heterozygous state by matings between unrelated individuals, which are unlikely to carry the same aberration. Thus, one individual contributes a gamete with an aberrant chromosome, and the other individual, a gamete with the normal version of the same chromosome. Homozygous aberrations result from self-pollination of plants with a heterozygous aberration, or from intercrosses between closely related plants or animals that are heterozygous for the same aberration, so that some offspring can receive an aberrant chromosome through both male and female gametes.

8.2 DELETIONS AND DUPLICATIONS OF CHROMOSOME SEGMENTS

A deletion is any loss of material from a structurally complete chromosome. The size of deletions can range from the loss of base pairs in a DNA molecule to the loss of most of a chromosome.

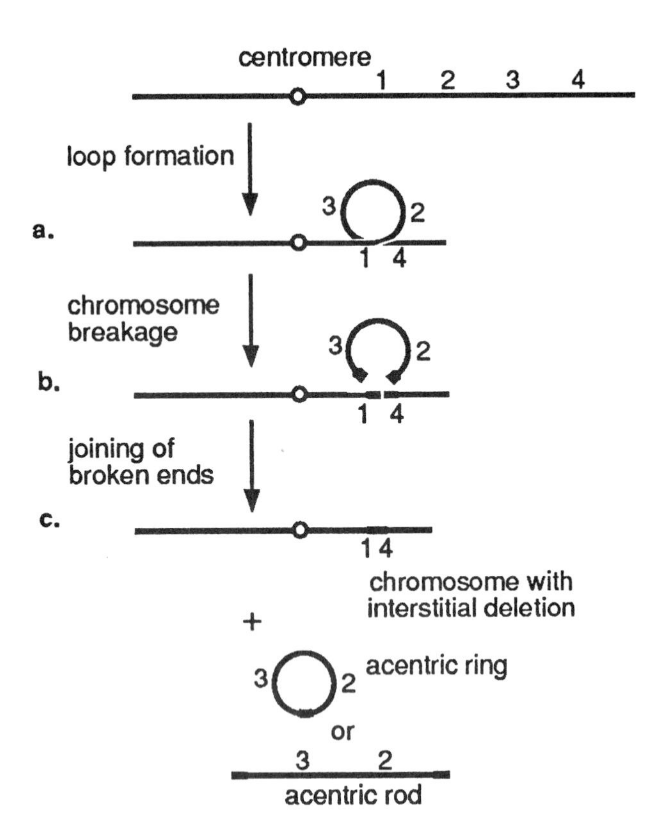

Fig. 8.1. The formation of an interstitial deletion. From top to bottom: A chromosome loops on itself with the centromere outside the loop. Two breaks then occur at the base of the loop to give four broken ends (1–4) that can behave in different ways. Ends 2 and 3 either remain unjoined, forming an acentric rod, or join to form an acentric ring chromosome. Ends 1 and 4 may join to produce a centric chromosome with an interstitial deletion.

Terminal deletions or deficiencies occur when a chromosome is broken into two pieces (*see* Fig. 7.1). Barring restitution or union with another broken chromosome, the end fragment without a centromere (acentric) is lost during cell division because it cannot become attached to spindle fibers for poleward movement. It may be included fortuitously in the nucleus because of its location in the cell, but eventually it will be left in the cytoplasm during a future cell division. The unstable end of the centric chromosome—that is, the remainder of the chromosome that retains the centromere—must undergo a repair process to restore stability (*see* Chapter 20, Section 20.4).

Interstitial or intercalary deletions require two breaks per chromosome to produce four unstable ends (Fig. 8.1). If the breaks occur in the same arm, as shown, the intervening segment with two unstable ends is lost because it has no centromere. Fusion then has to occur between the other two unstable ends, otherwise the chromosome would have a terminal deletion. When the two breaks occur in opposite arms (Fig. 8.2), the centric piece has two unstable ends, which sometimes fuse to form a centric ring chromosome. As usual, the end pieces are lost regardless of whether they fuse or remain separate. A human centric ring chromosome is shown in Fig. 8.3. Centric ring chromosomes are retained through cell divisions because of their centromeres, but they lack the distal parts of both arms; therefore, cell viability is jeopardized unless the loss is compensated by the presence of an extra dose of the pertinent chromosome.

Duplications are extra segments of varying sizes that are integrated into the structure of chromosomes and increase their lengths. Duplications are classified in the following ways according to their locations in the chromosomes:

> **Tandem duplications** have the duplicated segment directly adjacent to the original segment.
>
> **Displaced duplications** are separated from their homologous segments by varying distances and can be located within the same chromosome or on a different chromosome, which may be homologous or nonhomologous with the one that carries the original segment.

Deletions and duplications often arise from the meiotic behavior of chromosomes involved in heterozygous inversions or heterozygous reciprocal translocations (*see* Chapter 9). They also may occur as the complementary products of a single event such as unequal

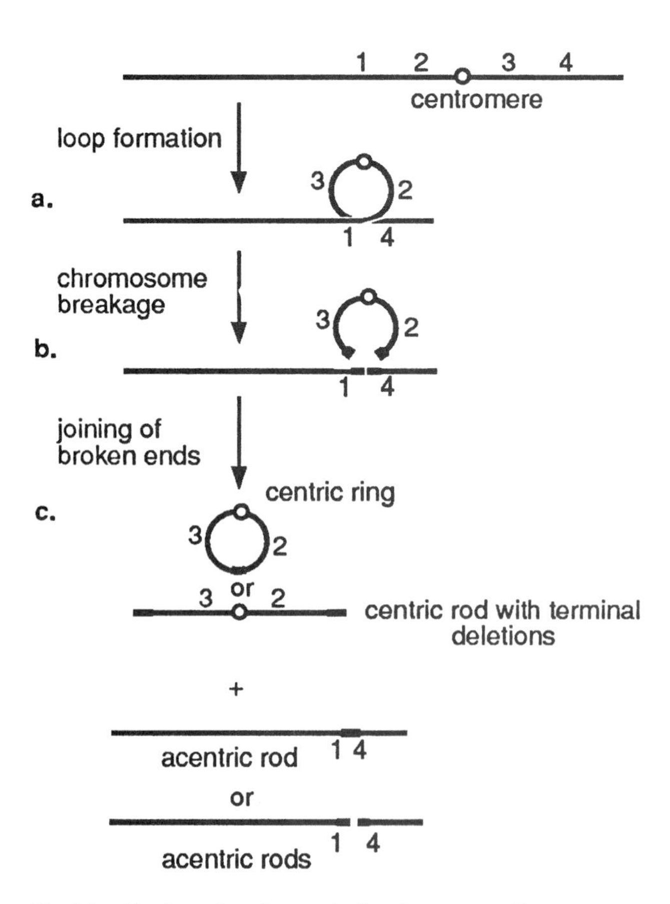

Fig. 8.2. The formation of a centric ring chromosome. From top to bottom: A chromosome loops on itself with the centromere inside the loop. Two breaks occur at the base of the loop to give four broken ends (1–4) that can behave in different ways. Ends 2 and 3 either join to give a centric ring chromosome with terminal deletions, or they remain unjoined to give a centric rod chromosome with terminal deletions. Ends 1 and 4 are on acentric fragments, whether they join or not.

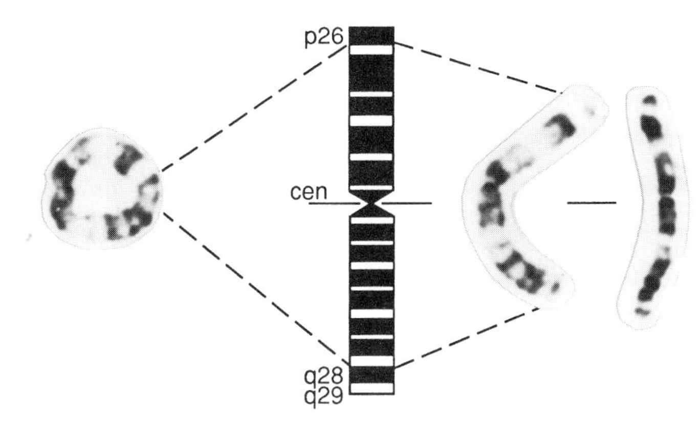

Fig. 8.3. A human centric ring chromosome 3 (left), compared with two normal chromosomes 3 (right), and an idiogram of chromosome 3 (center). The approximate locations of the breakpoints that subsequently fuse to give the ring are indicated by dotted lines (McKinley et al., 1991).

or ectopic crossing over, in which segments of unequal size are exchanged between chromosomes. This situation occurs if homologous chromosomes are not precisely aligned during synapsis due to an attraction between repeated DNA sequences (Fig. 8.4; *see also* Fig. 7.22c).

Deletions and duplications can be induced directly by clastogenic agents, which increase the chances of simultaneous breaks, and thereby of interstitial deletions or displaced duplications, both of which require more than one break. For example, a displaced duplication induced by ionizing radiation needs a minimum of three breaks: two in one chromosome to excise the segment, and a third at the insertion site in the same chromosome or a different chromosome. Three fusions of the broken ends are also needed to repair the breaks and prevent loss of chromatin.

8.3 DETECTION OF DELETIONS AND DUPLICATIONS

The visibility of deletions and duplications depends on their size, their heterozygosity or homozygosity, and the chromosome segments involved. The development of chromosome-banding (*see* Chapter 6, Section 6.2) and *in situ* hybridization (*see* Chapter 19) techniques has facilitated the detection of these aberrations in mitotic chromosomes. Otherwise, they are difficult or impossible to observe with light microscopy unless they are large and markedly alter the length of a chromosome or its arm ratio.

Although mitotic chromosomes usually do not pair, a significant exception is found in the salivary-gland polytene chromosomes of dipterous insects such as *Drosophila* and *Sciara*, where homologous chromosomes do pair during mitosis. The polytene chromosomes have a built-in banding pattern resulting from multiple replications of each chromosome, with the linear subdivisions staying together and chromomeres forming the bands (*see* Fig. 6.20). Thus, if one homolog has an interstitial deletion and the other is complete, the latter forms a lateral loop so that homologous parts can come together in the rest of the chromosome pair. With a heterozygous duplication, the homolog with the extra segment forms the loop. The locations of all the bands in the salivary-gland chromosomes of *Drosophila melanogaster* have been precisely determined, and deleted or duplicated segments can be accurately located. Although polytene chromosomes occur in certain tissues of other organisms, including some angiosperms, their usefulness is limited due to lack of accessibility to the tissues in which they occur. Moreover, the clarity of their banding patterns is poor in these instances. Consequently, in most organisms apart from dipterous species, other methods must be used to locate aberration breakpoints such as the labeling techniques mentioned at the beginning of this section.

The types of meiotic configurations involving chromosomes heterozygous for deletions and duplications also depend on the lengths deleted and inserted. If these segments are very short, they may be difficult to detect cytologically. The most reliable indicators of small deletions are the loss of physical markers such as chromomeres, the loss of chromosome bands or DNA-hybridization regions, or changes in phenotype due to the loss of gene markers. With longer terminal deletions and close synapsis in the regions present, the normal chromosome projects beyond the deleted homo-

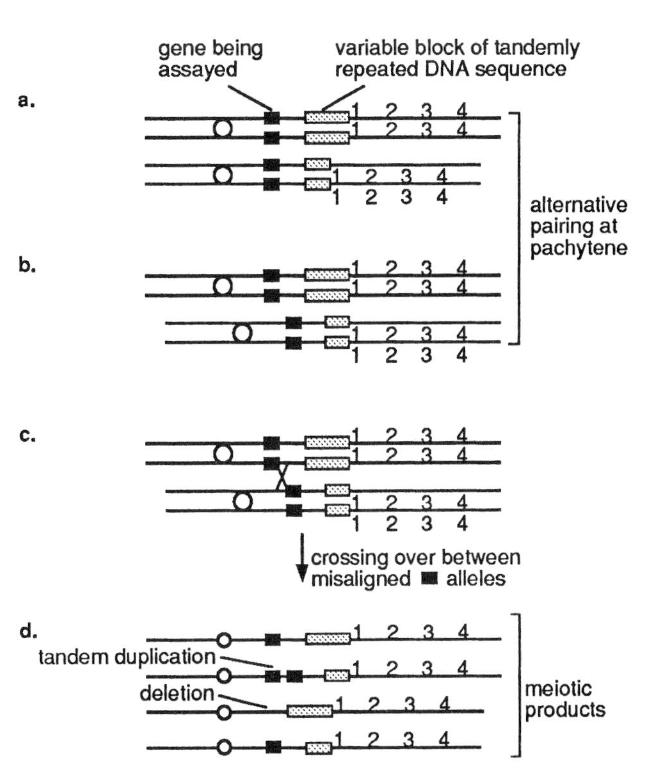

Fig. 8.4. The formation of a tandem duplication and a deletion caused by the pairing of homologous chromosomes that differ in size because of the presence of variable numbers of repeated DNA sequences. The alleles of the gene under study are aligned in (a) but misaligned in (b). In (c), unequal crossing over occurs between the alleles of interest caused by the misalignment shown in (b). Finally, (d) shows the products of meiosis with two parental-type chromosomes and two recombinant chromosomes (one with a tandem duplication, the other with a deletion), both resulting from the unequal crossover shown in (c).

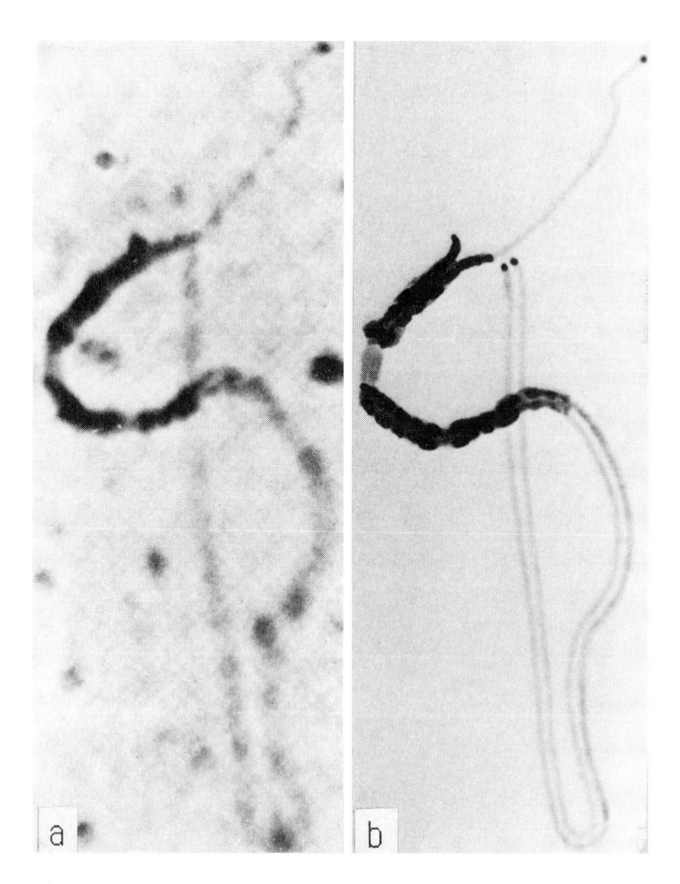

Fig. 8.5. A terminal deletion of the short arm of one of the homologs of a tomato chromosome 1 bivalent at pachytene (photomicrograph, left; interpretive drawing, right). The centromere is the gray body between two large heterochromatic segments (Khush and Rick, 1968).

log as a thinner strand when compared with the paired region (Fig. 8.5).

If heterozygous interstitial deletions and heterozygous tandem duplications are long enough, the loops that they can form during meiotic pachytene are similar to those in the mitotic polytene chromosomes of the Diptera. The loop is formed by the normal homolog in the case of a deletion (Fig. 8.6), and by the chromosome with the added segment in the case of a duplication. The loops vary in size, depending on the lengths of the deleted or duplicated segments. If the segments are very short, the loops may not form, and there is either no pairing or nonhomologous pairing in the aberrant region. The presence of a loop cannot be used to distinguish between a deletion and a duplication unless there is a physical or genetic marker that is absent or duplicated. Evidence from mitotic observations and the effects on fertility and phenotype need to be combined with the meiotic configurations in deciding on the type of aberration.

8.4 BEHAVIOR OF MODIFIED CHROMOSOMES DURING CELL DIVISION

With the exception of centric rings, chromosomes with deletions or duplications do not cause any problems during mitotic divisions unless they disrupt genes or gene complexes that govern chromosome behavior. In such cases, the loss of genes usually has a more severe effect than the extra doses. Cell functions can be affected if a deletion includes genes concerned with some aspect of cell metabolism or with the transfer of genetic information such as in the nucleolus-organizer region. The effect is more severe with a homozygous deletion where both homologs lack essential genes. Added genes in the form of duplications may prolong cell division by creating an unbalanced condition in the gene content of the nucleus.

During meiosis, the normal and aberrant homologs of a heterozygous deletion or duplication usually synapse at least in parts of their lengths, although some regions may pair nonhomologously or not at all (Fig. 8.7). If the chromosomes behave normally during the meiotic divisions, half of the gametes receive a normal chromosome, and the other half, a chromosome with the deletion or duplication. Tandem duplications in the heterozygous or homozygous state are prone to misalignment of the duplicated segments. Unequal crossing over between the misaligned segments can increase the amount of chromatin in one direction and decrease it in the other, thereby changing the composition of the crossover chromatids (*see* Fig. 8.4, and Section 8.7 for the evolutionary significance of these changes).

A centric ring chromosome pairs with the homologous part of its normal (rod) partner during meiosis (Fig. 8.8). If there is no crossing over between the ring and the rod, they separate at anaphase I and half the gametes receive the ring chromosome. A crossover between a ring and a rod chromatid joins the two chromatids and their centromeres in a dicentric structure, which forms a bridge at anaphase I as the two centromeres move toward opposite poles. The stretched bridge breaks, generally at a random position, so that the broken pieces may have more or less material added or deleted (*see* Fig. 8.8). Certain combinations of double crossovers between

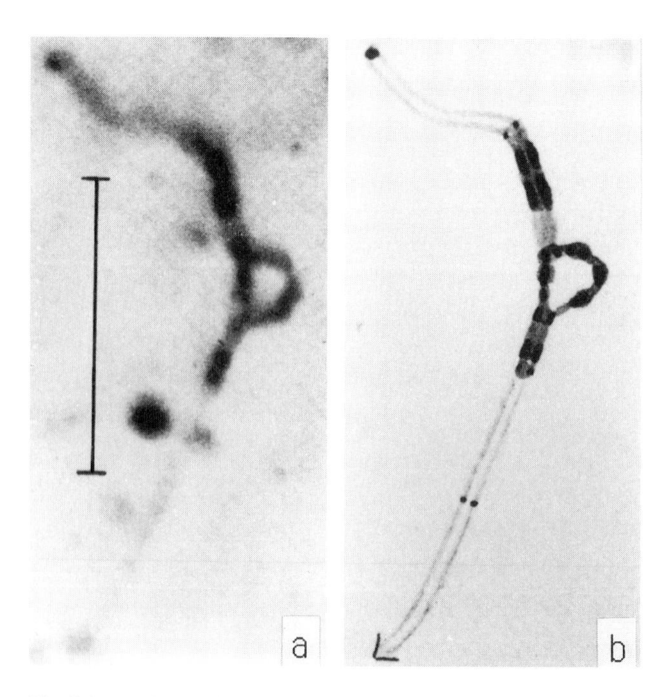

Fig. 8.6. An interstitial deletion in the long arm of one homolog of a tomato chromosome 9 pair at pachytene (photomicrograph, left scale = 10μ, interpretive drawing, right). The loop (just below the centromere) is in the normal homolog and all homologous parts are paired (Khush and Rick, 1968).

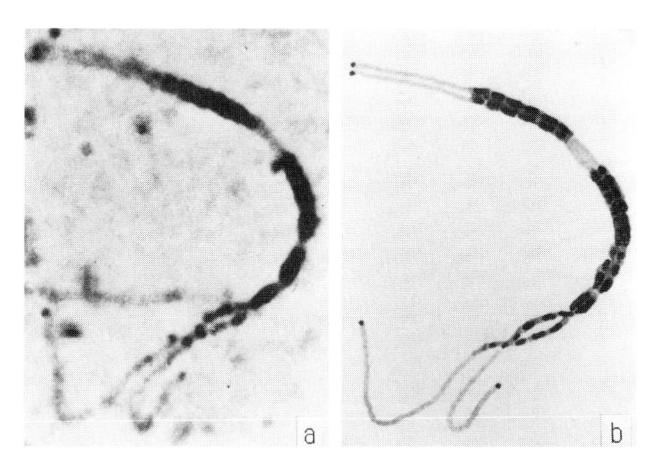

Fig. 8.7. An interstitial deletion in the long arm of one homolog of a tomato chromosome 10 pair at pachytene (photomicrograph, left; interpretive drawing, right). The deletion has resulted in an extensive lack of pairing in the distal region of the long arm. The centromere is the gray body between two large heterochromatic segments (Khush and Rick, 1968).

ring and rod chromatids, or single crossovers between ring-sister chromatids, lead to bridge formation at anaphase II.

In mitotic divisions, some centric ring chromosomes behave normally, and the sister chromatids pass to opposite poles without changing in size. Others are unstable and may be lost, or they may increase or decrease in size. Changes in size are usually attributed to somatic crossing over between the sister chromatids of a ring chromosome, and the formation of a dicentric, double bridge at anaphase (Fig. 8.9). Unless the bridges break exactly halfway between the centromeres, the two rings formed after the fusion of broken ends are unequal in size, and these inequalities in size can be a major source of instability. This deviant behavior has been observed in such diverse organisms as humans, maize, and *Drosophila*. In maize, small ring chromosomes are lost more frequently than larger ones involving the same chromosome, because they are more likely to lack genes required for survival. The aging of *Drosophila* females results in a considerable somatic loss of otherwise stable ring chromosomes. In humans, a number of abnormal syndromes are caused by centric-ring-chromosome deletions (*see* Section 8.6.1).

8.5 EFFECTS OF DELETIONS AND DUPLICATIONS ON VIABILITY AND FERTILITY

8.5.1 Effects on Viability

Deletions are more likely to have lethal effects on an organism than duplications, because the absence of genes usually has a more severe effect than the presence of additional genes. However, the severity of effect for both types of aberrations depends on their size, gene content, heterozygosity or homozygosity, and the organisms in which they occur. A heterozygous deletion produces a hemizygous (one-dose) state for genes in the corresponding segment on the normal homolog. If the deletion includes a number of genes that are unable to function in a single dose, lethality or major defects can result. Deletions may be tolerated if they are small and/or involve genes with minor functions, or if the missing DNA sequences are repeated elsewhere in the chromosome complement. Most genes are present in more than two doses in most polyploid-plant species, and this situation enables them to tolerate deletions more readily (Fig. 8.10). However, humans and most higher ani-

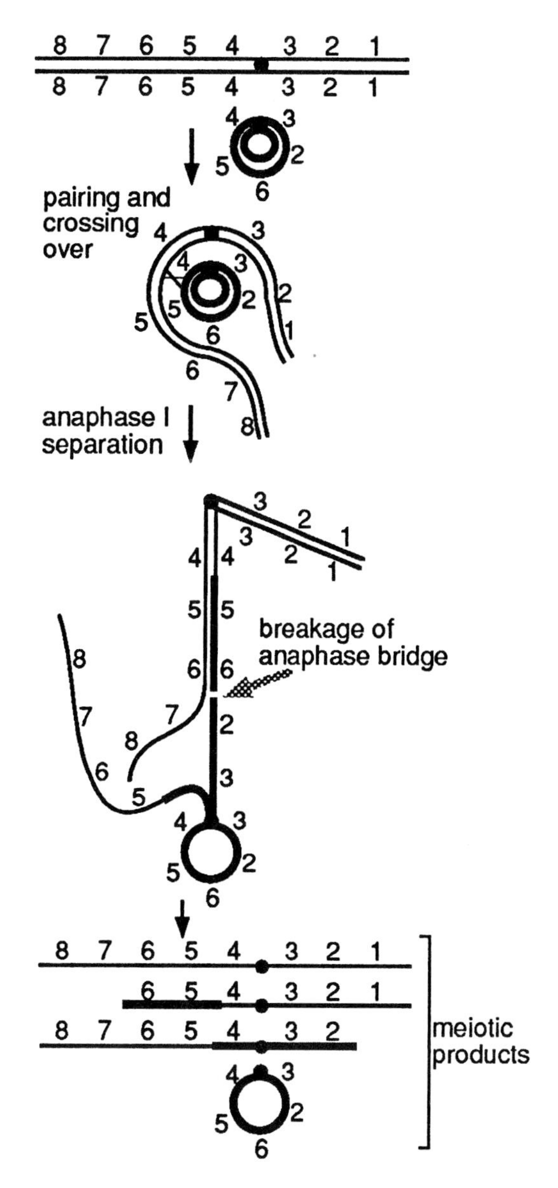

Fig. 8.8. A diagrammatic representation of the results of crossing over between (a) rod and ring chromosomes; (b) pachytene configuration of pairing between a complete rod chromosome and a ring chromosome that lacks terminal segments 1 and 7–8, with a crossover taking place between a rod and a ring chromatid; (c) at anaphase I, the two crossover chromatids form a dicentric bridge with a stress-induced break occurring between positions 6 and 2; and (d) the meiotic products again include one of each of the parental-type chromosomes, but the crossover has converted one ring chromatid into parts of two rod chromosomes, both with deletions. It is assumed that the broken ends "heal" by forming new telomeres.

mals are diploids and thus more vulnerable to gene loss, because they carry only two doses of most of the functioning, structural genes. Deletions that are large enough to be detected with a light microscope are usually heterozygous, because the normal homolog gives complete or partial genetic compensation for the deficient homolog. Most homozygous deletions are lethal in the early stages of the life cycle, because the same genes are missing in both homologs.

Duplications are most likely to have effects on viability if they are homozygous and involve several genes with essential functions. These effects can be due to an unbalance created by the extra doses of these genes in relation to the rest of the gene complement. Displaced duplications not only introduce changes in the locations of genes but may also disrupt the gene sequences on either side of the duplicated segments. Thus, the new arrangement may cause phenotypic changes by affecting the functioning of the displaced genes, as well as the functioning of the sequences into which it has been introduced (*see* Section 8.6.2).

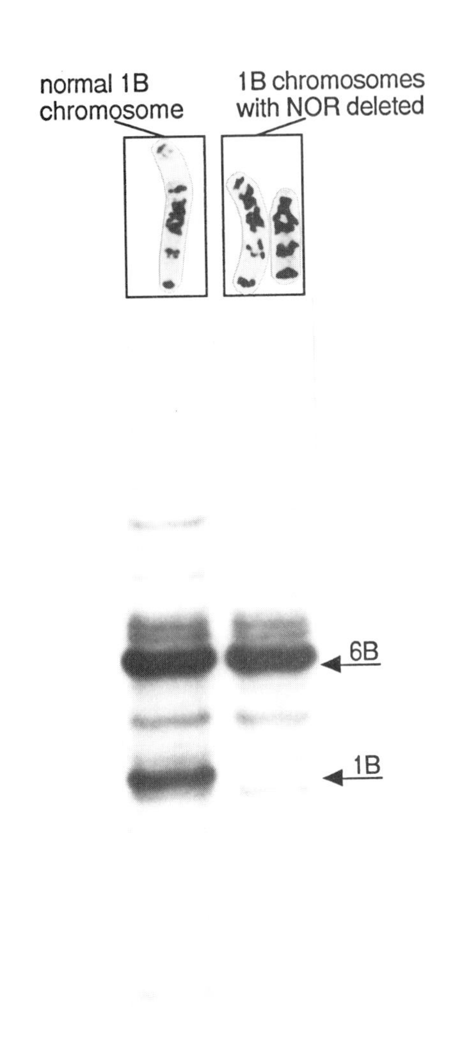

Fig. 8.10. The detection of a terminal deletion from the short arm of chromosome 1B of wheat. The deletion was originally detected through the absence from the autoradiograph of chromosome 1B-specific ribosomal-DNA sequences (labeled 1B; bottom right). The remaining rDNA sequences are specific to the *Nor* locus on chromosome 6B (labeled 6B). The loss of the 1B *Nor* locus had little effect owing to the presence of a *Nor* locus on the short arm of chromosome 6B. The photographs (top) of G-banded chromosomes (Kota et al., 1993) show the normal (left) and 1B modified chromosomes (right). The loss of all the genes distal to the *Nor* locus was compensated by the other Group 1 homoeologs, and plants with the terminal deletion showed no deleterious phenotypic effects. (C.E. May, personal communication.)

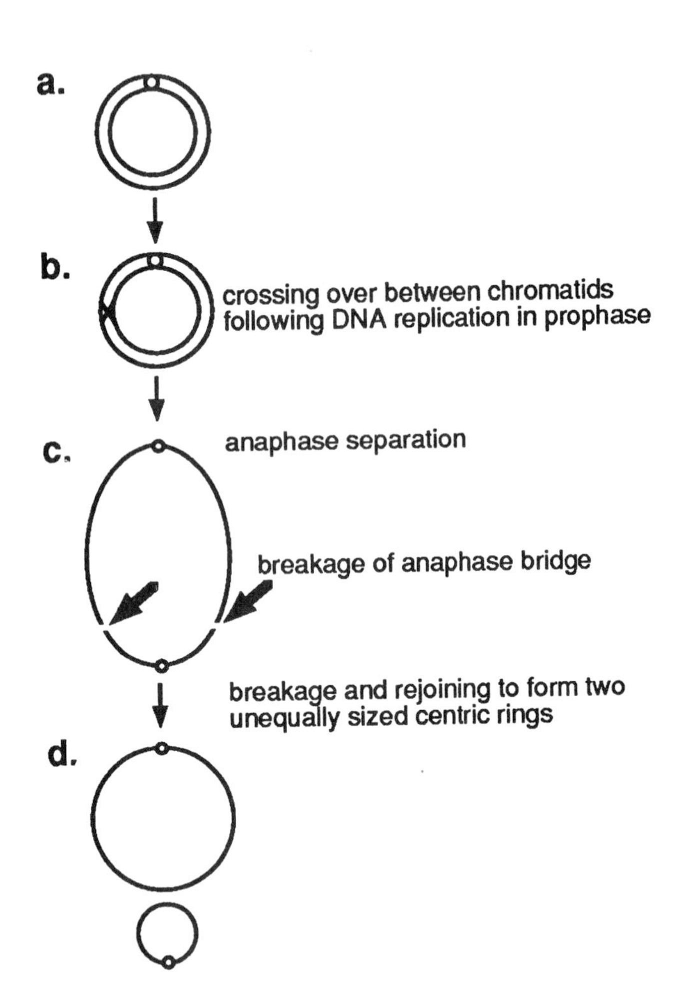

Fig. 8.9. The formation of unequal-sized ring chromosomes during mitosis. In a replicating chromosome (a), a mitotic crossover occurs between sister chromatids (b); (c) a double-sized ring is formed at anaphase, and two random breaks (arrows) result from the movement of the two centromeres in opposite directions. The broken ends rejoin to form two rings that differ in size from each other and from the original ring chromosome (d).

8.5.2 Effects on Fertility

Deletions are more apt to cause reduced fertility than duplications. In higher plants, the gametophytic phase of the life cycle is particularly sensitive to the loss of genes because of the series of postmeiotic divisions that occur in the development of eggs and pollen. In diploids, most of the chromosome complement is needed for the development of normal gametes; therefore, a plant that is heterozygous for a deletion is expected to have about 50% aborted pollen and an equal frequency of aborted eggs, detected as reduced seed set. However, some deletions can be transmitted through the eggs even though they are lethal to pollen. A possible reason for this difference is that there are hundreds of developing pollen grains in an anther, and they compete for available nutritive substances. Those with an incomplete chromosome complement may lack some vital genes governing nutrient uptake or may otherwise be unable to compete with normal pollen. By contrast, there is only one egg cell per embryo sac and because it is nourished by its surrounding cells, the synergids, it has a better chance of functioning with a deletion.

In higher animals and humans, the meiotic products develop into sperm or egg cells without further divisions, so that abnormalities occurring during meiosis or during the development of sperm or egg cells are tolerated. In these organisms, deletions and some duplications have lethal effects on the zygote or developing embryo.

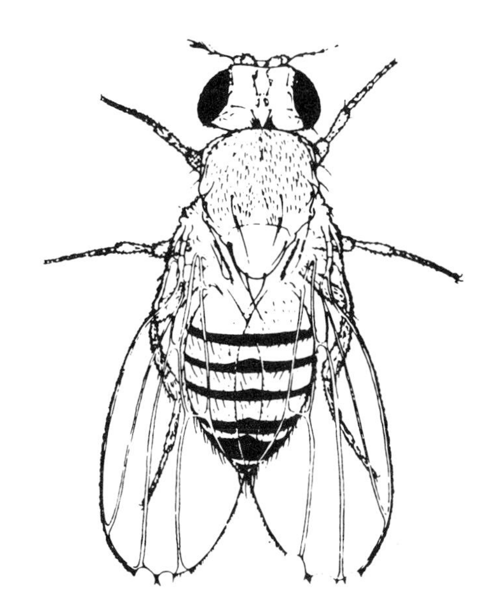

Fig. 8.11. A drawing of a *Drosophila melanogaster* female showing the Notch phenotype at the tip of the wings (King, 1965).

8.6 PHENOTYPIC EFFECTS OF DELETIONS AND DUPLICATIONS

Losses or gains of chromosome segments can cause deviations from the standard phenotypes depending on the genes that are located in the missing or added segments, and their roles in the development and functioning of an individual. Deletions and duplications that are large enough to be detected with a light microscope may involve many genes, but even those that are submicroscopic in size can have pronounced effects if they change the dosage of genes with essential functions. In general, missing genes have more pronounced and deleterious effects than duplicated genes.

Transmissible deletions and duplications usually occur in the heterozygous state because they are transmitted through one parent, except for matings between close relatives or self-fertilization in organisms where it is possible such as many plant species. When heterozygous, they behave like dominant mutations because the chromosome carrying a deletion or duplication dominates over the standard chromosome to produce a mutant phenotype.

8.6.1 Deletions

In *D. melanogaster*, several dominant phenotypes have been associated with heterozygous deletions, which can be detected by missing bands in the salivary-gland polytene chromosomes (*see* Section 8.3). The first to be discovered was Notch (*N*), so-called because the edges of the wings are nicked (Fig. 8.11). Many spontaneous and induced occurrences of the Notch phenotype are caused by deletions of varying lengths in the X chromosome (*see* Figs. 20.2 and 20.3 for a detailed description of the *N* locus). These deletions are lethal in males, which have only one X chromosome, and in homozygous females (*N/N*), so only heterozygous females (*N/+*) survive to express the Notch phenotype. Other dominant phenotypes of *Drosophila* that are due to heterozygous deletions include many Minute mutants, a general term referring to mutations that are dominant in their bristle effect and lethal when homozygous. At a molecular level, the mutations cause reductions in transfer RNA activity with severe effects on protein synthesis. The heterozygote has short, thin bristles with a slower developmental period than the wild type. Deletions on each of the four chromosomes produce similar Minute phenotypes, which may be accompanied by a range of phenotypic defects, including small body size, rough eyes, missing aristae, thin wings, abnormal venation, missing bristles, and sterility. The widespread chromosomal distribution of the Minute deletions reflects the distribution of transfer RNA genes.

In plants, the gametophytic phase eliminates deletions unless they are quite small or involve nonessential segments. Even if transmitted, it has been difficult to associate deletions with dominant mutations by traditional cytological procedures. Although polytene chromosomes occur in certain plant tissues, the synapsis of homologs is not as regular as in dipterous polytene chromosomes, so the banding pattern is less reliable. Banding techniques for mitotic chromosomes also do not give the resolution of the dipterous polytene chromosomes, so molecular procedures are the best approach for detecting small deletions.

Generally, homozygous deletions have not been recovered in higher animals, humans, or cross-fertilizing plants for the reason given at the beginning of this section. However, a limited number of small, homozygous deletions were obtained in maize, a cross-fertilizing species, by controlled self-pollinations. The effects of these deletions on phenotype simulate the recessive mutations of genes that are normally present in those segments and not essential for the viability of the organism. In self-fertilizing plant species, the same deletion can be present in both male and female gametes but is rarely transmitted through male gametes, and not always through female gametes.

In humans, heterozygous deletions are responsible for many abnormal phenotypes or syndromes, and some of them are observable with a light microscope (Fig. 8.12). Centric ring chromosomes, which lack the distal parts of both arms, are a clearly visible cytological defect. They have been found for each chromosome of the human genome and are sometimes associated with named disease syndromes. The phenotypes of persons with the same ring chromosome vary all the way from normal to severe mental and physical disabilities. This wide range in effects can be caused by the size of the rings—the larger the rings, the smaller the deleted portions, and vice versa—and how much crossing over occurs between the sister chromatids.

Many deletions in humans are submicroscopic in size and require molecular techniques for detection. *In situ* hybridization using nonradioactive probes is considered indispensable in clinical cytogenetics to search for minute chromosome aberrations that are either missed or not clearly defined with conventional cytological methods. Particularly relevant procedures are the use of fluorescence *in situ* hybridization (FISH), whereby several probes can be used simultaneously, and the polymerase chain reaction (PCR), which can amplify DNA sequences in interphase or metaphase cells spread on glass slides (*see* Chapter 19).

The following examples illustrate the use of molecular procedures to associate human-disease syndromes with deletions:

Miller–Dieker syndrome. A microdeletion in the short arm of chromosome 17 (17p33) was detected in persons with this syndrome, a disease characterized by brain and facial abnormalities, by *in situ* hybridization. In one study, normal relatives showed hybridization signals on the short arms of both No. 17 chromosomes, whereas three persons with the disease had only one 17 labeled.

DiGeorge syndrome (DGS). This syndrome, which includes absence or hypoplasia of the thymus and the parathyroids, cardiac malformations, and facial abnormalities, is caused by several kinds of chromosomal abnormalities but is most often associated with deletions in the long (q) arm of chromosome 22 at band 11. Molecular tests, including FISH, have been used to establish a molecular-cytogenetic map of the critical region within 22q11 that produces DGS when the genes are present in one dose because of a deletion in that region in one homolog. These tests also identified several low-copy repeat families in the same chromosome region, and it is possible that they provide the milieu for the occurrence of deletions by causing instability.

X-linked ichthyosis. The enzyme steroid sulfatase has an essential role in the desulfonation of cholesterol sulfate, a critical step in skin metabolism. A deficiency of the steroid sulfatase (*STS*) gene causes abnormal skin development and is caused mainly by deletions that include part or all of the 146-kb *STS* locus in the distal part of the p arm of the X chromosome. A human genomic probe, CRI-S232, indicated that S232-like sequences were closely linked on either side of the *STS* locus. Using molecular probes, a study of 27 patients with ichthyosis, and known to have deletions of the entire *STS* gene, identified three groups according to the lengths of the deletions (Fig. 8.13). The single person in Group A had the longest deletion and was afflicted with mental retardation in addition to ichthyosis. In both the Group A patient and the two patients in Group B, the breakpoints for the deletions were not within the S232 sequences. All the deletions in the 24 Group C patients had both breakpoints in the S232 sequences flanking the *STS* locus. This suggests that many of the deletions of this locus may be due to unequal crossing over between the repeated S232 sequences.

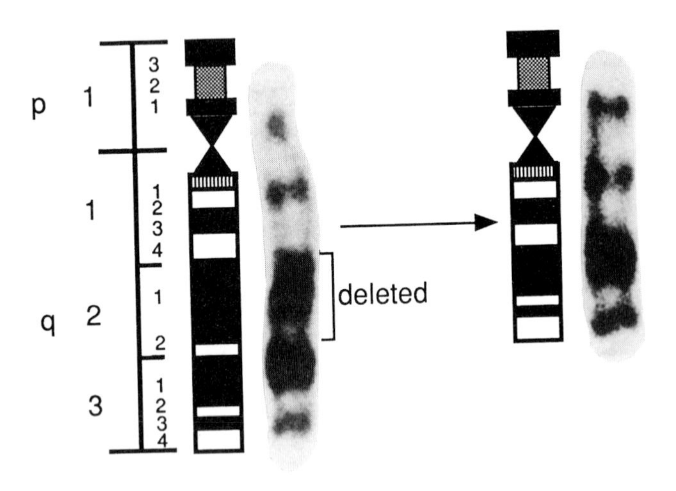

Fig. 8.12. Deletions of human chromosomes. The idiogram and photomicrograph on the left side show a G-banded chromosome 13 from lymphocyte nuclei isolated from cell cultures of a normal human male. Those on the right are from a male with severe mental retardation and show the deletion of bands 21 and 22 from chromosome 13. The bands deleted from the normal chromosome are indicated by a bracket (Dean et al., 1991).

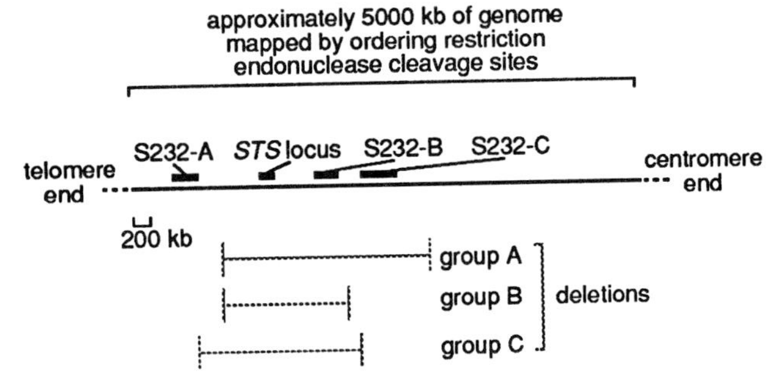

Fig. 8.13. Deletion breakpoints in the X chromosomes of persons with steroid sulfatase deficiency. A restriction map is shown at the top with the locations of the steroid sulfatase (*STS*) locus and S232 sequences around the locus. The bars beneath the map indicate the extent of deletions in three groups of patients with steroid sulfatase deficiency (Yen et al., 1990).

Thalassemia. Deletions resulting from unequal crossing over after mispairing of homologous segments also account for many cases of thalassemia, a blood disorder that causes anemia and other destructive effects. Thalassemias are characterized by reduced or no synthesis of one or more globin chains (*see* Section 8.7). There are two major subdivisions, α-thalassemias and β-thalassemias, depending on the genes that are affected. Deletions account for more α-thalassemias than β-thalassemias, and the severity of the effects increases as the number of deleted α-globin genes increases. In two types of β-thalassemia, called Hb Lepore (Hb = hemoglobin) and Hb Kenya, a deletion is associated with fusion between two globin genes. In Hb Lepore, the two genes, δ and β, are closely linked and also closely related. An unequal crossover gives one chromatid with a fused δ-β gene and deletes the two separate genes. The fused gene produces a single β-like chain instead of both a δ and a β chain. There are several variants of Hb Lepore depending on the site of the unequal crossover, which determines where the switch occurs from the δ to the β amino-acid sequence in the globin chain. Hb Kenya involves a fusion between the β gene and one of the two γ genes, which are farther apart on the chromosome and more distantly related, so a greater misalignment had to occur to cause their fusion. Molecular studies are continuing in efforts to find the underlying causes of thalassemia in all its forms (*see* Chapter 16, Section 16.8.2).

8.6.2 Duplications

In humans, molecular techniques have been useful in aligning duplicated regions with disease syndromes. For example, the Charcot–Marie–Tooth syndrome (CMT) is a disorder of the peripheral nervous system, with an incidence of 1:2500 and different modes of inheritance. One form of the disease, CMT1A, shows dominant inheritance and is usually associated with a tandem, submicroscopic duplication of a 1.5-Mb region on the short arm of chromosome 17. Another autosomal dominant but less frequent disorder, hereditary neuropathy with liability to pressure palsy (HNPP), has been associated with a deletion of the same size as the duplication in the same region of chromosome 17. This finding supports the theory that unequal crossing over during meiosis (*see* Fig. 8.4) produced both the CMT1A duplication and the HNPP deletion.

A number of duplications in *D. melanogaster* produce new and usually dominant phenotypes. A classical example is Bar eye (symbolized by *B*), which is narrower than the wild-type eye because of a reduction in the number of facets. The Bar phenotype is caused by a tandem duplication in the X chromosome, and observations of this chromosome in the salivary glands of Bar flies show that the duplication involves region 16A, which includes several bands. Wild-type (standard) flies, with round eyes, have one dose of 16A on each of the X chromosomes in females and one dose on the single X in males. Females heterozygous for Bar (*B*/+) have two doses of 16A in tandem on one homolog and one dose on the other homolog, whereas homozygous females (*B*/*B*) have two doses of 16A on each homolog. The various doses produce different phenotypes, with the homozygotes having narrower eyes than the heterozygotes. In a homozygous Bar female (*B*/*B*), there is sometimes misalignment of the duplicated regions at the time of synapsis, as follows:

16A 16A	instead of	16A 16A
16A 16A		16A 16A

In the alignment on the left, an unequal crossover between the first 16A regions in each line (chromatid) puts three doses of 16A on one chromatid and one 16A on the other. In a mating between this female and a wild-type male, union between a female gamete with the triple dose of 16A and a X-bearing male gamete with one dose of 16A gives a female offspring with four doses of 16A and a new phenotype (Ultrabar or double-Bar), which has smaller eyes than homozygous Bar. A similar gametic union involving the other crossover product with one 16A results in a reversion to a wild-type female.

Ultrabar provided the first report of a stable type of position effect. Ultrabar and homozygous Bar have different phenotypes but the same number of 16A regions. However, Ultrabar has the distribution 16A 16A 16A / 16A between the two homologs, whereas homozygous Bar has the following distribution: 16A 16A / 16A 16A. The effect of the two extra 16A regions is greater when three rather than two of them are adjacent to each other on the same chromosome.

Displaced duplications are found in both diploid and polyploid plant species, and one way of detecting them is by modified segregations for the genes in the duplicated regions (Fig. 8.14). Some mutant phenotypes in plants are caused by the presence of tandem duplications. For example, since the early part of the twentieth century, two anthocyanin loci of maize, *A* and *R*, have been known to give some unusual mutation patterns. Genetic studies indicated that each locus consists of tandem, duplicated segments and that some of the mutant phenotypes can be attributed to unequal crossing over between the segments and the resultant changes in their numbers.

Molecular techniques are being used increasingly to disclose the presence of duplications in plants. For example, Knotted (Fig. 8.15), a dominant leaf abnormality of maize that controls the fate of cells in leaf development, has several variant forms conditioned by different alleles at the *Kn1* locus on chromosome 1. One of these mutant alleles, *Kn1–0*, is attributed to a tandem duplication that includes 17 kb of DNA and the entire coding region for this gene, based on DNA sequencing and restriction analyses of genomic clones (*see* Fig. 8.15). The recessive allele *kn1* gives normal leaf development and does not have the duplication. Mutations of the *Kn1*/*kn1* locus are analogous to the homoeotic mutants of *Drosophila* that affect development in that

(a) all mutations of the locus are dominant;
(b) the locus codes for a transcriptional control factor;
(c) the mutant phenotype results from the new position of the 5′-upstream control regions next to the duplicated gene.

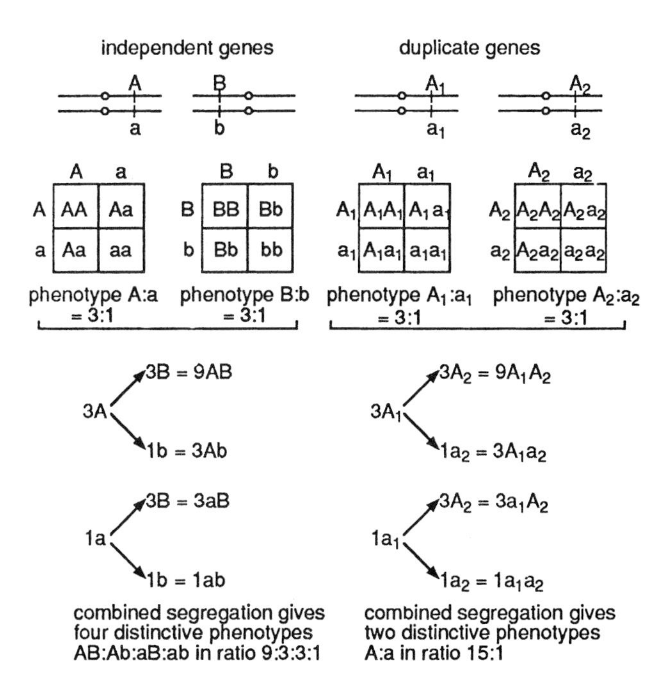

Fig 8.14 Two-locus F_2 segregations for independent genes assorting independently to give 9:3:3:1 segregations, and duplicate genes assorting independently to give 15:1 ratios. The *A*/*a* and *B*/*b* loci (left side) have different functions, whereas the A_1/a_1 and A_2/a_2 loci (right side) duplicate each other's functions. Duplicated genes often give additional phenotypes with smaller differences that are dependent on the number of dominant alleles present. The small circles represent centromeres.

8.7 EVOLUTIONARY SIGNIFICANCE OF DUPLICATIONS

The Bar-eye mutation in *D. melanogaster*, described in the previous section, provided the first evidence that new genes in the form of tandem duplications could arise from unequal crossing over be-

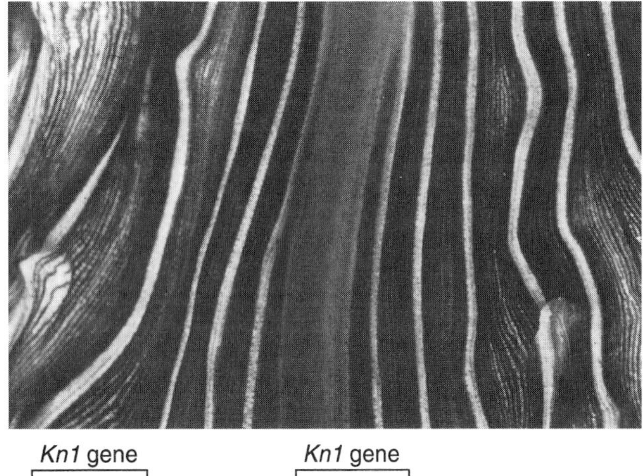

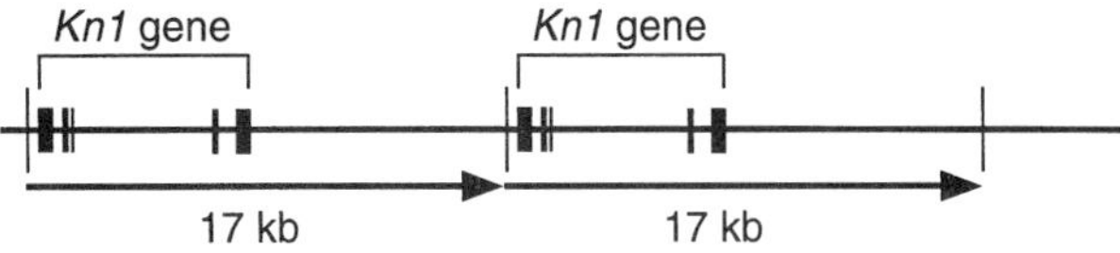

Fig. 8.15. The Knotted mutation in maize. **Top:** Part of a leaf blade showing the "knots" produced by abnormal vein development. **Bottom:** Genetic diagram of the 17-kb duplication (marked by arrows) that is associated with the *Kn1–0* mutant allele. The dominant phenotype does not result from the actual duplication of sequences but rather from the new *position* of the 5′-upstream control regions next to the duplicated genes. The altered expression of the KN1 protein causes changes in the differentiation of cells along the lateral veins of the leaf blade. Knotted mutations can also be induced by insertion of the transposable element (*Mu*) into one of the introns, for this also leads to altered KN1 protein expression during leaf development (Green et al., 1994).

tween misaligned chromosome segments. In addition, the discovery of the Ultrabar phenotype demonstrated that the cycle of unequal crossing over and production of a new phenotype could recur. These milestones laid the groundwork for the concept that duplications have an important role in the evolution of new genes. The following sequential steps are postulated:

1. A single ancestral locus is duplicated either through fortuitous mispairing or a breakage–union event such as the exchange of unequal terminal segments involving the same arm of two homologs.
2. The resulting tandem duplication, in either a heterozygous or homozygous state, increases the likelihood that mispairing and unequal crossing over will occur, thereby producing additional copies of the original locus.
3. Duplication of the locus provides the opportunity for functional diversity, because as long as one copy retains the original function, other copies can mutate to acquire modified functions without having the disruptive effects that a mutated, single locus might have.

Thus, cycles of duplication of the original locus or its copies lead over evolutionary time to clusters of closely linked genes with related functions, but each gene may be activated at a different stage in the development of an end product. Some genes that seem to have a common origin because of their functions are located on different chromosomes. They could have been moved from the site of the original locus by chromosome rearrangements or by a transposable-element system (*see* Chapter 7, Section 7.5 for a discussion of transposable elements).

The hemoglobin genes clearly demonstrate the significance of duplications in the evolution of essential genes. Hemoglobin consists of four molecules of globin–protein chains and four molecules of the nonprotein heme group. There are six kinds of globin chains—alpha (α), beta (β), gamma (γ), delta (δ), epsilon (ϵ), and zeta (ζ)—and the types and proportions of the different chains vary according to the stage in the life cycle. In humans, about 97% of the hemoglobin of normal adults is $\alpha_2\beta_2$, and the rest is divided between $\alpha_2\delta_2$ and $\alpha_2\gamma_2$.

The amino-acid sequences of the globin chains are controlled by two clusters of globin genes, known as the α cluster and the β cluster. In humans, the α cluster is located on the short arm of chromosome 16 over a 28-kb region and includes three functional genes, ζ and two α genes ($\alpha 1$ and $\alpha 2$). The β cluster is located on the short arm of chromosome 11 over a 50-kb region and includes five functional genes, ϵ, two γ genes ($G\gamma$ and $A\gamma$), δ, and β (*see* Fig. 16.25). The amino-acid sequence of each type of chain is determined by a different gene, but in the following three cases, there is a functional relationship between each closely linked pair of genes: $\alpha 1$ and $\alpha 2$ code for the same protein; Gγ and Aγ differ only in one amino acid (glycine versus alanine) at position 136 of their coding sequence; and the β and δ genes differ in only about 7% of their sequences. The most likely explanation for these paired-gene relationships is that tandem duplications, resulting from unequal crossing over between mispaired genes on homologous chromosomes, have occurred comparatively recently from an evolutionary standpoint. Therefore, the duplicates have not had time for extensive divergence either spatially or functionally.

Further evidence for duplications in the α and β gene clusters is the presence of nonfunctional genes called pseudogenes, which are symbolized by the Greek letter ψ. They resemble functional genes in their structure but have lost the ability for transcription and/or translation because of the accumulation of mutations. The β cluster includes one β-like pseudogene (*see* $\psi\beta$ in Fig. 16.25) not far from the functional β gene. The most likely origin for the pseudogene was by a duplication of the β gene, which maintained its function, allowing the duplicate to evolve into a pseudogene because deleterious mutations could be retained. Similar origins can explain one ζ pseudogene and two α pseudogenes in the α cluster.

All the globin genes have three exons and two introns (*see* Fig. 16.25 for the structure of the β gene). Two other proteins that are related to the globins and to each other are myoglobin, found in animals, and leghemoglobin, initially found in leguminous plants. Both are monomeric, oxygen-binding proteins. The single myoglobin gene in humans has essentially the same three-dimensional structure as the globin genes, and the leghemoglobin genes have the three-exon structure, but with an extra intron (Fig. 8.16). These similarities among the globins, myoglobin, and leghemoglobin indicate that they evolved from a common ancestral gene.

The relationship of the α and β gene clusters in different groups of organisms is helpful in tracing the evolution of the globin genes from a single ancestral gene. The single-globin chains of the protein myoglobin and of the cytostomes (lampreys and hagfish) preceded the duplication of the ancestral gene, which was followed by further evolutionary changes to give the α and β clusters. These two clusters were most likely linked at first, as is still the case in the frog (*Xenopus*), but with further duplication, divergence, and transposition, the two clusters were separated on different chromosomes, which is the present situation in birds and mammals.

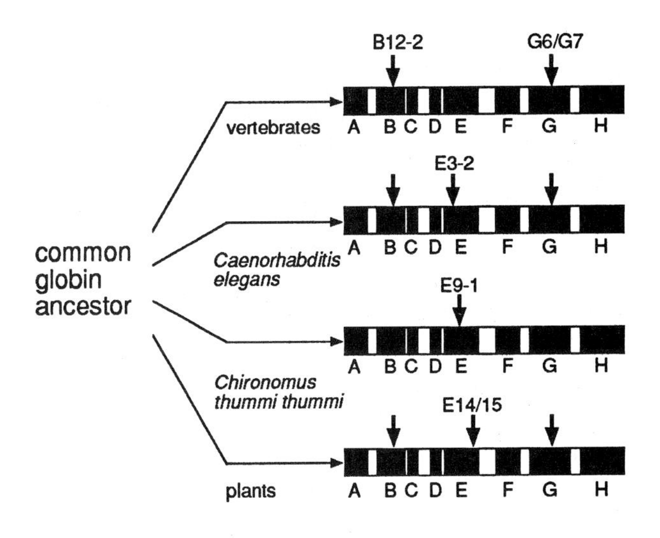

Fig. 8.16. Diagram showing the evolution of the globin genes from one ancestral gene, based on current knowledge concerning the organization and interrelationships of the globin genes in different groups of organisms (Stoltzfus and Doolittle 1993; Lewin 1994). The structure and evolution of the globin genes has been interpreted on the basis of conserved three-dimensional features of the globin protein. In particular, conserved α-helices have been defined and are indicated in the figure as A, B, C, D, E, F, G, and H. The positions of introns (vertical arrows) in the genes are indicated in relation to the conserved α-helices. Thus, the introns occurring in all vertebrate, plant, and many invertebrate globin genes are located at positions B12-2 (i.e., after the second base in the codon specifying amino acid 12 in helix B) and G6/G7 (i.e., between the codons specifying amino acids 6 and 7 of helix G). (*see* Chapter 24, Section 24.10 for a description of *Caenorhabditis*.)

BIBLIOGRAPHY

General

Adinolfi, M., Crolla, J. 1994. Nonisotopic *in situ* hybridization: clinical cytogenetics and gene mapping applications. Adv. Human Genet. 22: 187–255.

King, R.C. 1965. Genetics, 2nd ed. Oxford University Press, New York.

Levitan, M. 1988. Textbook of Human Genetics, 3rd ed. Oxford University Press, New York.

Lewin, B. 1994. Genes V. Oxford University Press, New York.

Section 8.2

Day, J.P., Marder, B.A., Morgan, W.F. 1993. Telomeres and their possible role in chromosome stabilization. Environ. Mol. Mutagenesis 22: 245–249.

McKinley, M., Colley, A., Sinclair, P., Donnai, D., Andrews, T. 1991. *De novo* ring chromosome 3: a new case with a mild phenotype. J. Med. Genet. 28: 536–538.

Strathdee, G., Harrison, W., Riethman, H.C., Goodart, S.A., Overhauser, J. 1994. Interstitial deletions are not the main mechanism leading to 18q deletions. Am. J. Human Genet. 54: 1085–1091.

Section 8.3

Khush, G.S., Rick, C.M. 1968. Cytogenetic analysis of the tomato genome by means of induced deficiencies. Chromosoma 23: 452–484.

Roberts, P.A. 1976. The genetics of chromosome aberration. In: The Genetics and Biology of *Drosophila*, Vol. 1a (Ashburner, M., Novitski, E. eds.). Academic Press, New York. pp. 67–184.

Section 8.4

Khush, G.S., Rick, C.M. 1968. Cytogenetic analysis of the tomato genome by means of induced deficiencies. Chromosoma 23: 452–484.

Section 8.5

Kota, R.S., Gill, K.S., Gill, B.S., Endo, T.R. 1993. A cytogenetically based physical map of chromosome 1B in common wheat. Genome 36: 548–554.

Section 8.6

Carlson, W.R. 1988. The cytogenetics of corn. In: Corn and Corn Improvement, 3rd ed. (Sprague, G.F., Dudley, J.M., eds.). American Society Agronomy, Madison, WI. pp. 259–343.

Dean, J.C.S., Simpson, S., Couzin, D.A., Stephen G.S. 1991. Interstitial deletion of chromosome 13: prognosis and adult phenotype. J. Med. Genet. 28: 533–535.

Green, B., Walko, R., Hake, S. 1994. *Mutator* insertions in an intron of the maize *knotted1* gene result in dominant suppressible mutations. Genetics 138: 1275–1285.

Hake, S. 1992. Unravelling the knots in plant development. Trends Genet. 8: 109–114.

Kidd, S., Lockett, T.J., Young, M.W. 1983. The Notch locus of *Drosophila melanogaster*. Cell 34: 421–433.

Metzenberg, A.B., Wurzer, G., Huisman, T.H.J., Smithies, O. 1991. Homology requirements for unequal crossing over in humans. Genetics. 128: 143–161.

Roa, B.B., Lupski, J.R. 1994. Molecular genetics of Charcot–Marie–Tooth neuropathy. Adv. Human Genet. 22: 117–152.

Scambler, P.J. 1993. Deletions of human chromosome 22 and associated birth defects. Curr. Opinions in Genet. Devel. 3: 432–437.

Veit, B., Vollbrecht, E., Mathers, J., Hake, E. 1990. A tandem duplication causes the *Kn1–0* allele of Knotted, a dominant morphological mutant of maize. Genetics 125: 623–631.

Yen, P.H., Li, X.-M., Tsai, S.-P., Johnson, C., Mohandas, T., Shapiro, L.J. 1990. Frequent deletions of the human X chromosome distal short arm result from recombination between low copy repetitive elements. Cell 61: 603–610.

Section 8.7

Stoltzfus, A., Doolittle, W.F. 1993. Slippery introns and globin gene evolution. Curr. Biol. 3: 215–217.

Vogel, F., Motulsky, A.G. 1986. Human Genetics: Problems and Approaches. Springer-Verlag, Berlin.

9

Rearrangements of Chromosome Segments

- Inversions rearrange segments within or between arms of single chromosomes, whereas reciprocal translocations exchange segments between chromosomes.
- Inversions and translocations can be detected by changes in mitotic chromosome structure and by distinctive meiotic configurations.
- The frequency of aborted gametes or zygotes resulting from heterozygous inversions or translocations depends on the number and locations ofcrossovers, as well as the orientation of the translocation rings at metaphase I.
- Inversions and translocations rearrange genes and create new genes, depending on the positions of the breakpoints, and can lead to cancers.
- Chromosome rearrangements have significant roles in some natural populations.

Reciprocal translocations, also called interchanges, and inversions result from linear repatterning of the chromosomes after a minimum of two breaks. Breakpoints for inversions are confined to one chromosome, whereas those for a reciprocal translocation are in different, usually nonhomologous chromosomes, resulting in an exchange of segments. Single breaks in one chromosome cannot give translocations or inversions because they do not create enough unstable ends.

9.1 HOW CHROMOSOME REARRANGEMENTS ARE DETECTED

Heterozygous rearrangements, which have one normal and one rearranged chromosome for each pair involved, may be detected initially by a reduction in male and female gametic fertility in plants, and by spontaneous abortions in humans and higher animals. Phenotypes are usually not affected by rearrangements if there is no loss or gain of chromatin, unless the movement of genes to new locations changes their functions and sometimes produces variegation, a type of position effect. In humans, viable offspring with abnormal phenotypes may be caused by extra doses of chromosome segments, which are derived from a parent with a heterozygous translocation through irregular meiotic behavior.

Chromosome aberrations other than rearrangements also cause reduced fertility; therefore, the next step is cytological observation of dividing mitotic or meiotic cells to determine if a rearrangement is present. If a mitotic karyotype has distinctive bands, occurring naturally as in dipteran salivary-gland chromosomes or revealed by staining methods (*see* Chapter 6, Section 6.2), inverted segments or translocation transfers can cause changes in banding sequences. Some rearrangements cause noticeable changes in arm ratios (i.e., the relative lengths of the two arms of a chromosome), but the normal arm ratios must be well established to recognize the deviations. An example of a translocation involving human chromosomes illustrates changes in both banding patterns and arm ratios (Fig. 9.1). Meiotic configurations are distinctive for heterozygous inversions and translocations, and are discussed in Sections 9.2.1 and 9.3.1, respectively.

Molecular techniques, notably *in situ* hybridization (*see* Chapter 19), detect submicroscopic rearrangements and locate breakpoints at the DNA-sequence level. In the case of a translocation, for example, appropriate DNA probes can be tested for hybridization with DNA sequences on the chromosomes involved in the translocation. The breakpoint on a chromosome can be localized between two sequences by determining which sequence has been transferred to the translocated chromosome and which one has been left behind. This is a valuable procedure in identifying chromosome segments transferred between plant species or genera (*see* Chapter 15, Section 15.6).

9.2 INVERSIONS

Inversions are either confined to one arm of a chromosome (paracentric type) or they involve both arms (pericentric type) and include the centromere (Fig. 9.2). Arm ratios are not changed by paracentric inversions, but they can be changed by pericentric inversions unless the breaks in the two arms are equidistant from the

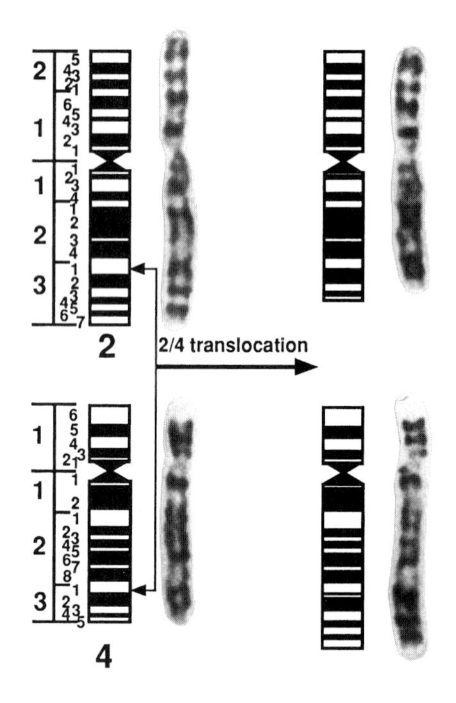

Fig. 9.1. Human mitotic chromosomes, showing a heterozygous reciprocal translocation between chromosomes 2 and 4 (Walpole and Mulcahy, 1991). The normal chromosomes are shown on the left and the translocated chromosomes on the right. Each quadrant has a chromosome diagram on the left and a photomicrograph on the right. The breakpoints (arrows) are in band 31 in the long (q) arm of chromosome 2, and in band 31 in the q arm of 4. Note the physical difference in length between the normal and the translocated chromosomes.

centromere. Inversions have a wide range in size, from those easily observable by light microscopy to minute inversions requiring characterization by molecular techniques. They can be heterozygous or homozygous (Fig. 9.3), but when they are induced by a clastogenic agent such as radiation, the homozygous state is sometimes not

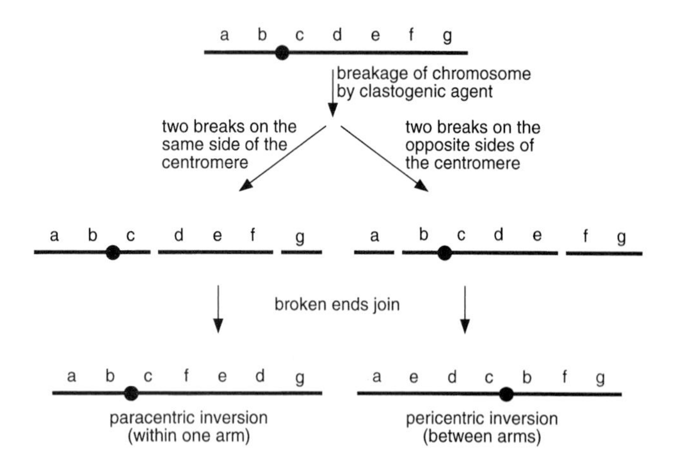

Fig. 9.2. The formation of two types of inversion from two breaks within one chromosome. The centromere is outside the inversion in the paracentric type, but inside the inversion in the pericentric type. Inversions were first detected genetically by their effect on genetic recombination (Muller, 1916) and by inverted linear arrays of the same genes in different species of *Drosophila* (Sturtevant, 1921).

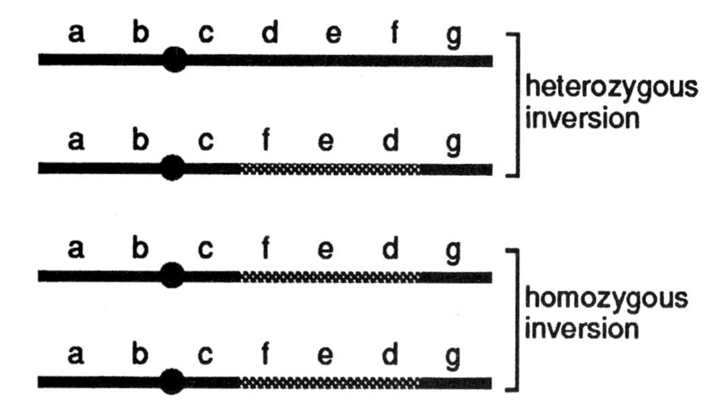

Fig. 9.3. A diagrammatic representation of homologous chromosome pairs from different individuals showing the distinction between heterozygous and homozygous inversions. The inverted segment is the sequence f–e–d.

viable because of accompanying lethal, recessive mutations or extra breaks that cause deletions. These changes become homozygous along with the inversion, whereas they are covered by a normal allele or segment when an inversion is heterozygous.

9.2.1 Behavior of Inversions During Mitosis and Meiosis

Inverted chromosomes function normally during mitotic divisions unless rearrangement of genes controlling chromosome movement disturbs their expression. During meiosis, there are no synaptic problems with homozygous inversions unless repositioning genes that control pairing has a disruptive effect. If heterozygous inversions are long enough, they form distinctive loop configurations so that synapsis can occur throughout the chromosome lengths (Fig. 9.4). In nonpaired regions (Figs. 9.4 and 9.5), which often occur around the breakpoints because of the changes in sequence, there is no crossing over, and genetic linkages between breakpoints and genes in these regions are tighter than usual for the physical distances involved. For paired regions, the number and locations of crossovers determine whether or not the resulting gametes have a complete chromosome complement. In comparing the diagrams in Figs. 9.6 and 9.8 through 9.10, the following guidelines make it easier to follow each chromatid through meiosis:

1. For a pericentric inversion, start with the centromere and follow the arms of each of the four chromatids to the ends. This suggestion applies to both crossover and non crossover chromatids.
2. For a paracentric inversion, do the same for chromatids without crossovers. For chromatids with one or more crossovers, start at a crossover point and follow the reciprocal exchange to each end of each chromatid.

If no crossing over occurs between the inversion breakpoints because of the short distance between them, the normal chromosome separates from the inverted chromosome at anaphase I. After sister-chromatid separation at anaphase II, half the meiotic products have a normal chromosome and the other half have an inverted chromosome. There are no deleted or duplicated segments.

Examples of the effects of single crossovers within heterozygous inversions on meiotic behavior of the chromosomes involved are given in Figs. 9.6 and 9. 8. When a single crossover occurs anywhere within the pachytene loop of a paracentric or pericentric inversion, the crossover chromatids have both a deletion and a duplication, and are usually not transmitted because of gametic or zygotic abortion. A single crossover within a paracentric inversion produces an acentric fragment and a dicentric chromatid, which forms a chromatin bridge at anaphase I (Fig. 9.7, *see also* Fig. 9.6). The fragment lacks orientation with respect to the poles, and may be separated from the bridge or attached by a chiasma to the homologous end of one of the complete chromatids. These distinctive configurations do not occur after a single crossover within a pericentric inversion.

With longer inversions, two or more crossovers can occur simultaneously within or near the inversion loop. Because more that two crossovers are rare, we shall confine this discussion to two crossover points. Using the crossover point labeled X1 in Fig. 9.6, a second crossover may involve the same two chromatids (X2), only one of them (X3 or X5), or the other two chromatids (X4). The respective terms for these double crossovers are two-strand (chromatid), three-strand, and four-strand doubles. None of these double crossovers involving pericentric inversions give anaphase bridges, whereas three-strand and four-strand doubles involving paracentric inversions do result in bridges.

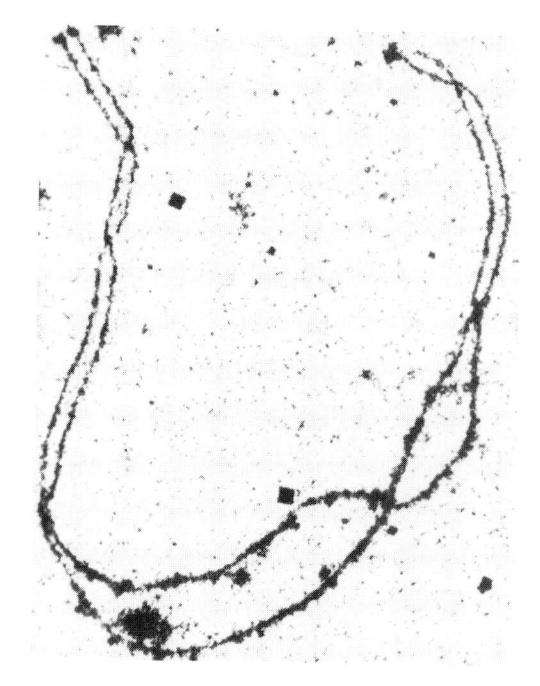

Fig. 9.5. An electron micrograph of a pachytene spread of human chromosome 1 showing an extensive lack of pairing within a heterozygous inversion (Chandley et al., 1987).

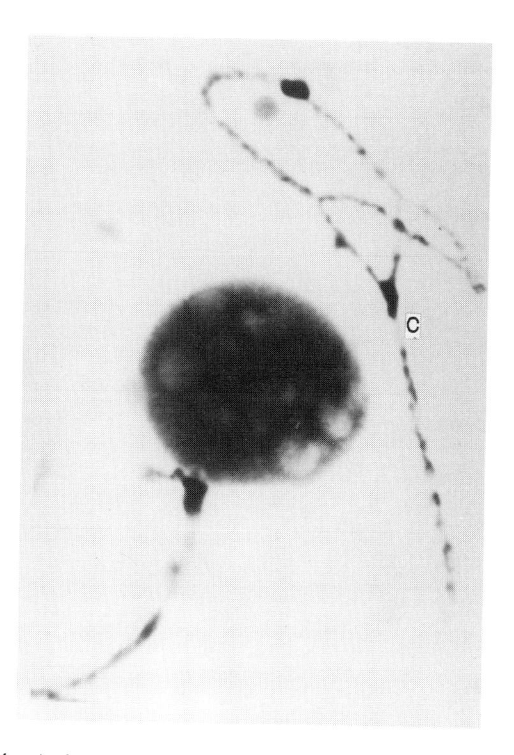

Fig. 9.4. A photomicrograph of a heterozygous paracentric inversion in the long arm of maize chromosome 7 showing the pachytene–loop configuration above the nucleus (Morris, unpublished). The homologues are tightly paired in the short arm (below the loop) with the inversion in the long arm beyond the heterochromatic region adjacent to the centromere (c). The inverted sequence in one homolog has paired with the normal sequence in the other except around the breakpoints, and the homologs are paired in the terminal region of the long arm (to the right of the loop). Note the heterochromatic knob inside the loop. The nucleolus organizer of chromosome 6 is in contact with the nucleolus. The first observations on the meiotic behavior of heterozygous inversions were made in maize by McClintock (1931).

An example of a two-strand double crossover within a paracentric inversion is marked by X1 and X2 (*see* Fig. 9.6). The gametes derived from this meiotic cell each have a complete chromosome complement, and those containing the crossover chromatids provide transmissible genetic recombination for the *C–c* locus in relation to the other loci. If the second crossover also occurs between *B–b* and *C–c*, the original gene arrangement is maintained. The same results are obtained for a pericentric inversion with the same two crossovers. A three-strand double can have both crossovers within the inversion loop such as X1 and X3 (*see* Fig. 9.6), or one crossover inside the loop and one outside but near a breakpoint such as X1 and X5 (Fig. 9.9 pachytene; also, compare the anaphase I diagram in Fig. 9.9 with the top photo in 9.11). In either case and for both types of inversions, the meiotic products will include a complete, noncrossover chromatid, a complete crossover chromatid, and two incomplete chromatids. The crossover chromatid marked 2–3* in Fig. 9.9 anaphases I and II has a transmissible recombination for *A* in relation to the other genes, however, the genes that are recombined will vary depending on the locations of the crossover points. A four-strand double within the loop (Fig. 9.10 pachytene; also compare the anaphase I diagram in Fig. 9.10 with the bottom photo in Fig. 9.11) gives complete abortion with both paracentric and pericentric inversions.

Studies of some paracentric inversions indicate that after a single crossover or a three-strand double crossover within the inverted region, the anaphase I bridge does not break in female meiotic cells. Instead, it is oriented so that it moves into a polar body in the dipteran species *Drosophila melanogaster* and *Sciara impatiens* and avoids the basal megaspore in maize. Thus, the egg nucleus receives only a complete, noncrossover chromatid, which has the normal or inverted sequence. This can be confirmed by the lower amount of fertilized egg abortion in *Drosophila* and of ovule abortion in maize. There is no crossing over in *Drosophila* males, hence no bridge formation. On the male side in maize, all meiotic products have the potential to develop into pollen grains, so the cells with bridges are not eliminated from the germline, and they contribute to a higher amount of sterility than on the female side.

A summary of the effects of heterozygous inversions on transmitted genetic recombination is given in Table 9.1. These effects apply equally to pericentric and paracentric inversions. Genetic recombinants are not transmitted after single crossovers (the most frequent type) and four-strand doubles, and have reduced transmission rates after three-strand doubles. Thus, for genes that are within or close to heterozygous inversions, the genetic variability arising from crossing over is greatly reduced.

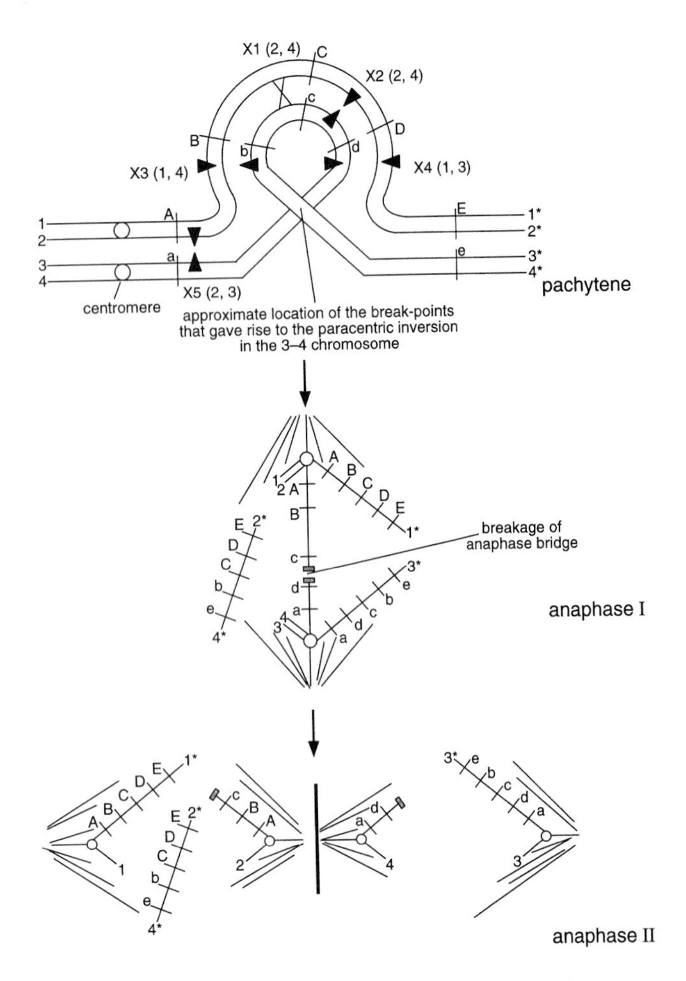

Fig. 9.6. The meiotic behavior resulting from the presence of a heterozygous paracentric inversion in which a single crossover (e.g., X1) has taken place within the pachytene-loop configuration (top). The chromatid ends are numbered, with an asterisk beside the numbers at one end for clarity, and the letters represent five gene pairs. A dicentric chromatin bridge is formed at anaphase I, with the two centromeres moving to opposite poles (middle). A rupture in the bridge gives two incomplete strands, so one-half of the meiotic products lack some segments of this chromosome (bottom). The nonoriented, 2*–4* acentric fragment commonly lags during the second meiotic division (left cell) and is left in the cytoplasm. The fate of the broken strands and fragments was explored in depth by McClintock (1938). Examples using other crossover points (X2 through X5) are marked by arrowheads. These are discussed in the text and shown in Figs. 9.9 and 9.10. (*see* Burnham, 1962; Schulz-Schaeffer, 1980; Sybenga, 1975, 1992.)

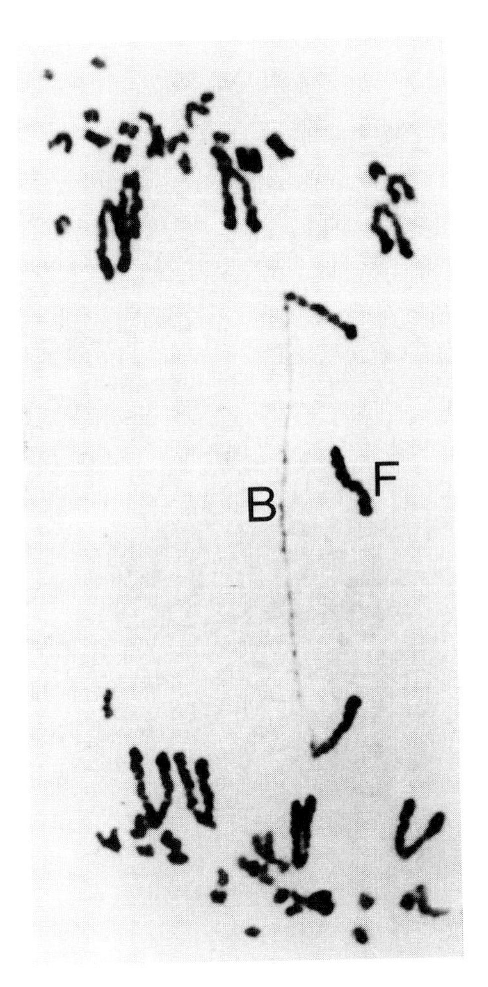

Fig. 9.7. A dividing cell of *Agave stricta* showing an anaphase I chromosome spread with a dicentric chromatid bridge (B) and an acentric fragment (F) (Brandham, 1969; Sybenga, 1975). This configuration is the result of a crossover within a heterozygous paracentric inversion (*see* Fig. 9.6).

9.2.2 Effects of Heterozygous Inversions on Fertility

The amount of gametic or zygotic abortion resulting from heterozygous inversions depends on the length of inverted segments, the number and kinds of crossovers inside or near the breakpoints, and the number of meiotic products with deleted chromosome segments (*see* Table 9.1). Quantitative measurements of abortion are most readily obtained in plants because they usually produce thousands of pollen grains, which can be used as units in measuring reduced fertility. In maize, pollen abortion for different inversions ranges from around 10%, which is similar to the level due to environmental or developmental conditions, to close to 50%. The amount of ovule abortion is expected to be the same as the amount of pollen abortion for pericentric inversions and can be used if pollen development is affected by adverse environmental conditions. The frequency of ovule abortion for some paracentric inversions may be considerably lower than the frequency of pollen abortion, depending on the behavior of the anaphase I bridges (*see* previous section).

9.3 RECIPROCAL TRANSLOCATIONS

Reciprocal translocations rearrange segments of different chromosomes in many ways depending on the distribution of breakpoints along the lengths of the chromosomes and the degree of randomness in the union of broken pieces. Persistent translocations must involve the attachment of an acentric segment of one chromosome to the centric segment of another chromosome (Fig. 9.12). If two centric pieces unite, the resulting dicentric chromosome has deletions for part of one arm of both chromosomes (*see* Fig. 9.12). Persistent translocations can be heterozygous or homozygous (Fig. 9.13), but some induced translocations cannot be maintained in the homozygous state due to lethal mutations, deletions, or other adverse effects.

Three other types of translocations are insertions, Robertsonian translocations, and B–A translocations. The last type involves essential (designated A) chromosomes and accessory (B) chromosomes (for a discussion of B chromosomes, *see* Chapter 6, Section 6.6).

Insertions require three breaks: two within an arm in the chromosome donating a segment, and one in the recipient chromosome (Fig. 9.14). Insertions are sought in order to transfer genes for valuable traits from one species or genus to another (*see* Chapter

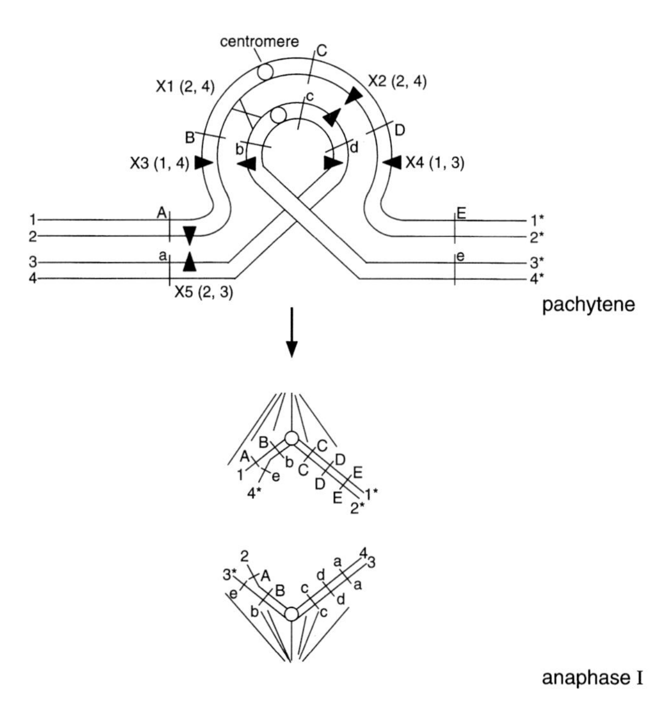

Fig. 9.8. The meiotic behavior resulting from a heterozygous pericentric inversion with a single crossover (e.g. X1) within the pachytene-loop configuration. (*see* Fig. 9.6 for explanations of numbers and letters.) At anaphase I, each chromatid is monocentric, so there is no bridge formation. However, the crossover chromatids 2–4 and 2*–4* each have a duplication for one end and a deletion for the opposite end, and one-half of the meiotic products are unbalanced.

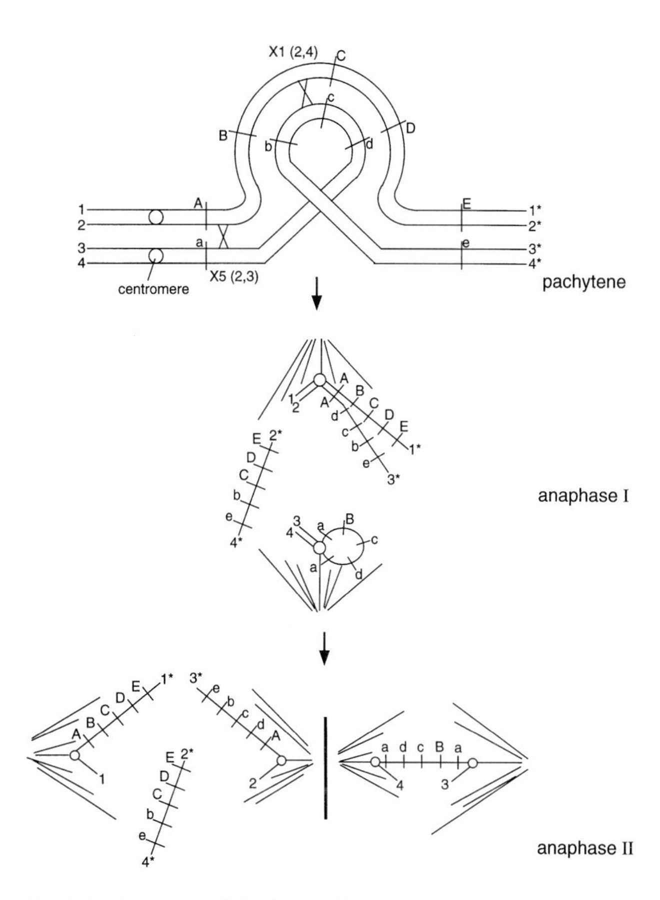

Fig. 9.9. The meiotic behavior resulting from a heterozygous paracentric inversion with a three-strand double crossover (e.g., X1 inside the pachytene loop between strands 2 and 4, and X5 outside the loop between strands 2 and 3). (*see* Fig. 9.6 for explanations of numbers and letters.) The anaphase I chromatin loop with one centromere forms a dicentric bridge at anaphase II (bottom right) when the two halves of the centromere move toward opposite poles. Assuming that the acentric fragment 2*–4* is lost as usual, breakage of the bridge gives incomplete complements for two of the four meiotic products.

15), but they are rare because of the specific requirements for their origin.

Robertsonian translocations, named after the scientist who discovered them, result from a break near or within the centromere region of two nonhomologous, acrocentric chromosomes. Fusions that bring together the long arms constitute the transmissible Robertsonian translocation, and the short-arm fragments are often lost from nuclei by lagging. The long-arm translocated chromosome is dicentric if both breaks occur in the short arms, followed by the union of the centric parts (Fig. 9.15a), but it is monocentric if the breaks occur in opposite arms of the two chromosomes, followed by the union of the long-arm segments. In both modes of origin, the genomic chromosome number is reduced by one. A balanced, heterozygous Robertsonian translocation includes one translocated chromosome with two long arms and the two acrocentric chromosomes from which it was derived.

Robertsonian translocations occur frequently in insects and mammalian species with acrocentric chromosomes in their complements, and they are the predominant chromosomal aberration in cattle and sheep. The most studied Robertsonian translocation in cattle combines parts of chromosomes 1 and 29 and occurs in many breeds throughout the world.

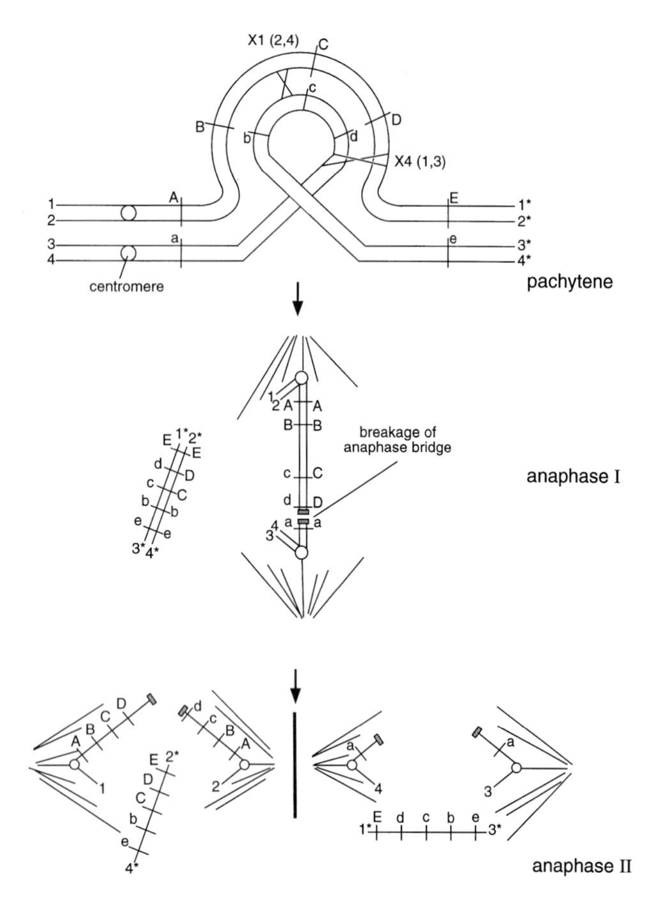

Fig. 9.10. The meiotic behavior resulting from a heterozygous paracentric inversion with a four-strand double crossover within the pachytene loop (e.g., X1 and X4). (*see* Fig. 9.6 for explanations of numbers and letters.) Breaks in both of the dicentric bridges at anaphase I result in deficient strands at anaphase II, and all the meiotic products lack part of the chromosome, unless an acentric fragment is fortuitously included in a nucleus and supplies the missing segments.

In the human karyotype, the six chromosomes of the D and G groups (including the Y chromosome) are acrocentric, but three of these chromosomes, 13, 14, and 21, are involved in Robertsonian translocations much more frequently than the others, and the most common interchange is between the long (q) arms of 13 and 14 (Fig. 9.15b). Most of the Robertsonian translocations studied in humans have a dicentric chromosome, indicating that they arose by breaks in the short arms of both acrocentrics, which each have a nucleolus organizer region (NOR) with a satellite distal to it (*see* Fig. 9.15b). When nucleoli are being formed, the satellites of the acrocentrics come together and could facilitate exchanges.

Although acrocentric chromosomes are much less frequent in plants than in animals, karyotypic comparisons in different plant species of the same genus suggest that some metacentric chromo-

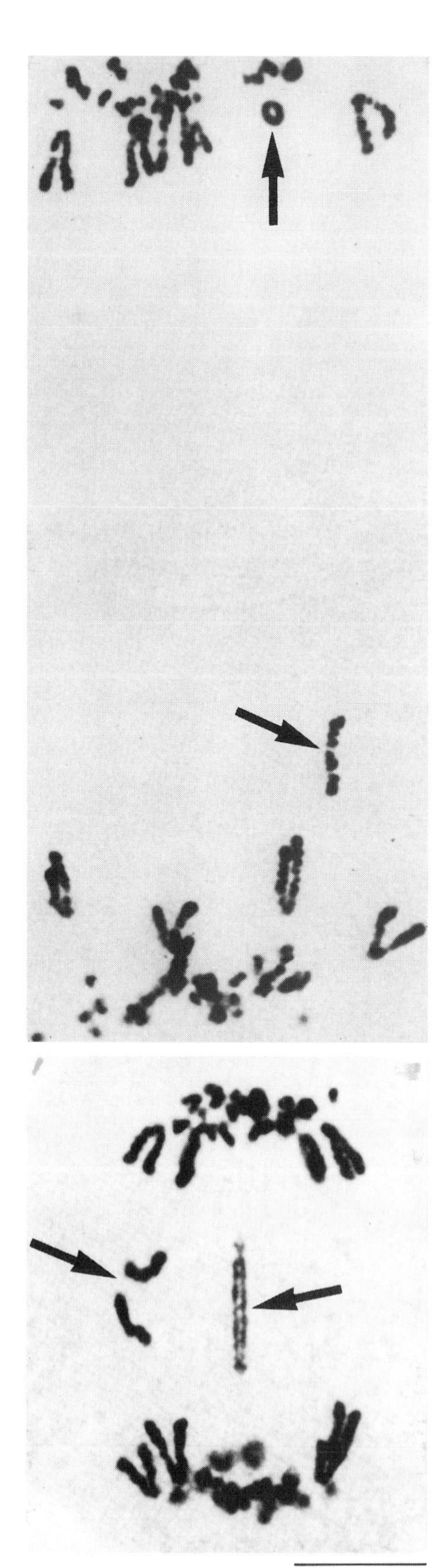

Fig. 9.11. Meiotic anaphase I configurations produced by two crossovers involving a heterozygous paracentric inversion. **Top:** A loop centric chromosome (upper arrow) and acentric fragment (lower arrow) resulting from a three-strand double crossover at pachytene (Brandham, 1969; *see* Fig. 9.9). The ends of the loop chromosome are not visible in this photo. **Bottom:** Two dicentric bridges (right arrow) and two

Fig. 9.11. *(continued)* acentric fragments (left arrow), resulting from a four-strand double crossover within the inversion (Sybenga, 1975; *see* Fig. 9.10).

Table 9.1. Effects of Crossing Over, Within or Near Heterozygous Paracentric or Pericentric Inversions, on Viability and Transmissible Genetic Recombination

Type of Crossover	Gametes or Zygotes Viable (%)	Gametes or Zygotes Nonviable (%)	Viable Gametes/ Zygotes with Genetic Recombination (%)
Single	50	50	0
Two-strand double	100		50
Three-strand double			
Both inside loop	50	50	50
One outside loop	50	50	50
Four-strand double		100	0

somes are formed by Robertsonian translocations. For example, in the genus *Gibasis* the base chromosome numbers are $x = 5$ with three acrocentric and two metacentric chromosomes, and $x = 4$ with one acrocentric and three metacentric chromosomes. It is probable that the extra metacentric chromosome in the group with $x = 4$ came from a Robertsonian translocation that joined two of the acrocentric long arms present in the group with $x = 5$.

B–A translocations occur in species that have B chromosomes in

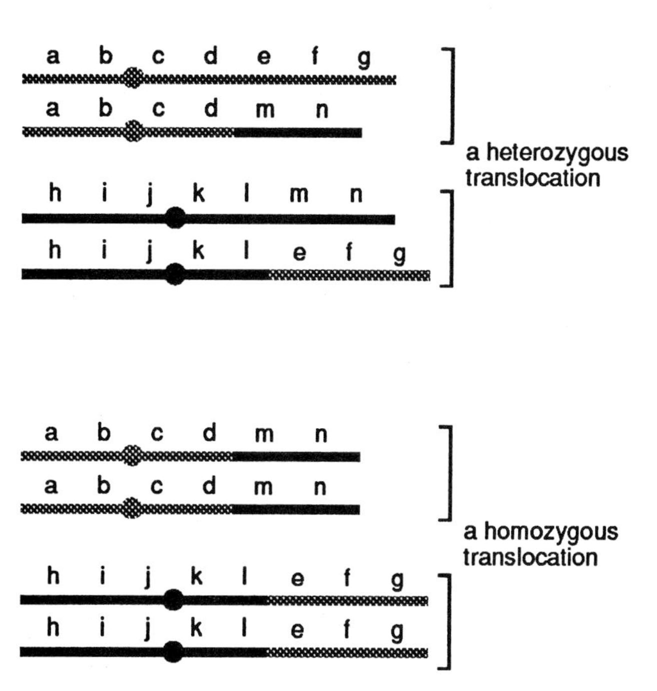

Fig. 9.13. Diagrams of two pairs of chromosomes to show the difference between heterozygous and homozygous balanced reciprocal translocations.

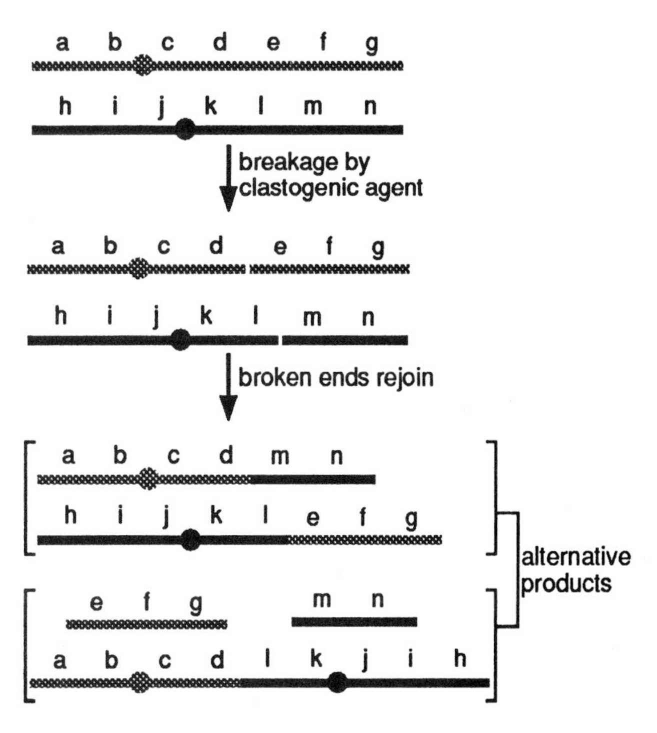

Fig. 9.12. The formation of translocations between two nonhomologous chromosomes (top). One breakpoint in each chromosome is indicated (center). The exchange of terminal segments gives rise to a balanced reciprocal translocation without loss of chromatin (upper alternative). Alternatively, the union of the two centric pieces can give an unbalanced translocation, consisting of a dicentric chromosome, and two acentric fragments, which are lost from the nucleus during subsequent division (lower alternative). If the two centromeres move to opposite poles at anaphase, the dicentric chromosome forms a bridge that usually breaks under stress, leading to a further loss of genes. Cells with deleted segments may die or be outcompeted in division rates by cells with a complete karyotype.

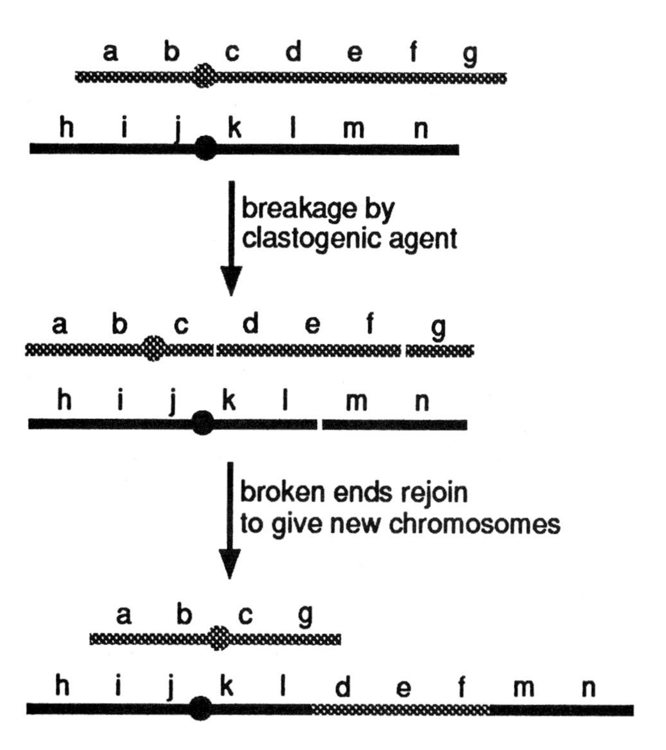

Fig. 9.14. The steps needed for insertion of an interstitial segment (*def*) from one chromosome into another. The acentric (g) and centric (abc) parts of the donor chromosome (white centromere) may unite as shown or remain separate, but in either case part of the chromosome is missing. A chromosome-breaking or clastogenic agent such as ionizing radiation is one way to increase the chances of obtaining useful gene transfers (Sears, 1969).

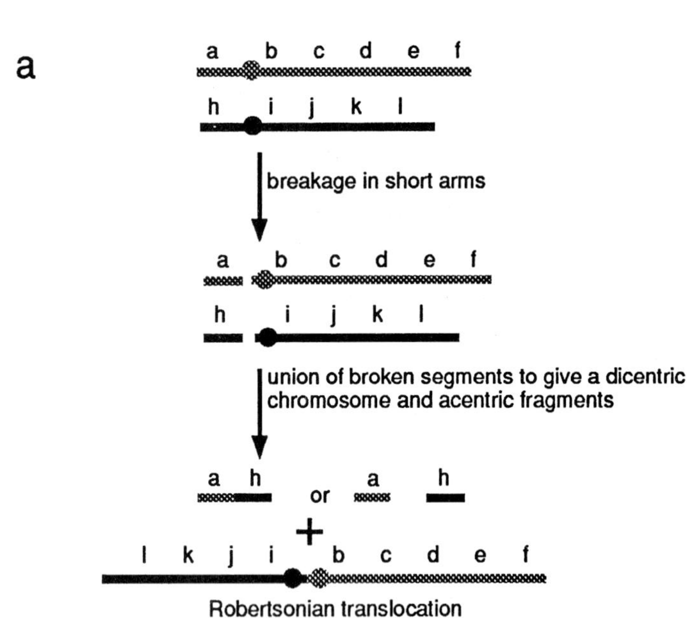

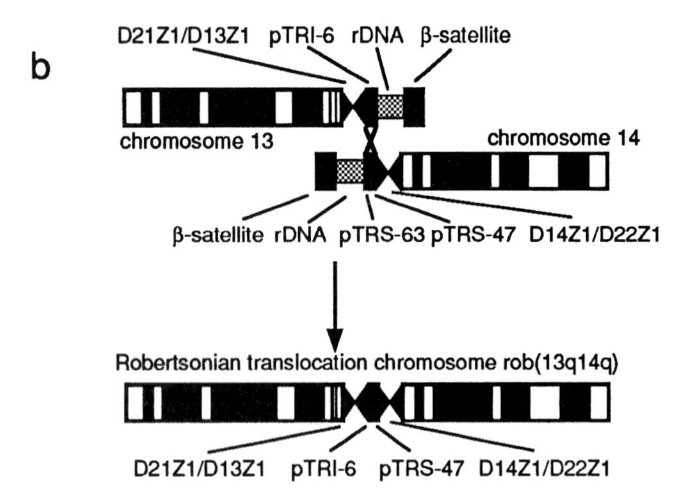

Fig. 9.15. (a) A sequence of events that produces the predominant type of Robertsonian translocation in human acrocentric chromosomes. The chromosome with two centromeres close together seems to move normally through cell division, indicating that the two centromeres act as a single unit. The two short-arm, acentric fragments, which may unite, are dispensable because they are heterochromatic and are usually lost during some cell division. Robertson (1916) first suggested that fusions could occur between two chromosomes to form a compound chromosome and thereby decrease chromosome number. (b) Diagram of a Robertsonian translocation that joins the long (q) arms of human chromosomes 13 and 14, each of which has a nucleolus-organizer region marked by rDNA. (Redrawn from Han et al. 1994). The centromeres are marked by α-satellite DNA probes D14Z1/D22Z1 and D21Z1/D13Z1, which show that the translocated chromosome, rob(13q14q), is dicentric. The use of four other DNA probes indicates that the breakpoints in the short (p) arms of both chromosomes are between the pTRS-47 and pTRS-63 sequences in chromosome 14, and between the pTRI-6 and

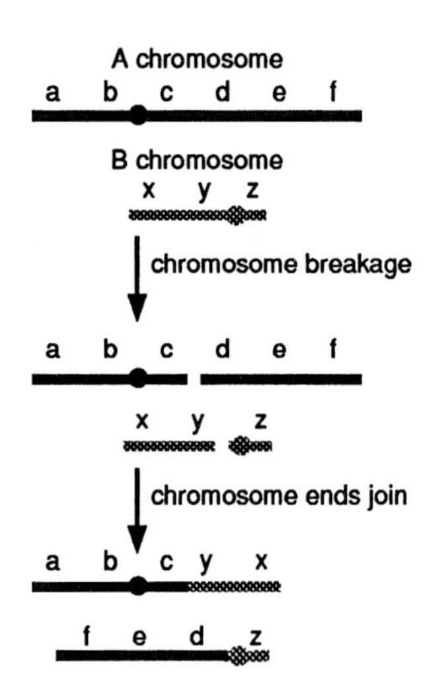

Fig. 9.16. Breaks and exchanges between A and B chromosomes to give reciprocal A–B translocations. B chromosomes are usually smaller and more heterochromatic than A chromosomes. Roman (1947) used X-rays to produce translocations between A and B chromosomes in maize, and, by using gene markers on the A translocated segments, he found that nondisjunction of the B centromeres occurred during division of the generative cell in the male gametophyte (*see* Chapter 13, Fig. 13.7).

at least some individuals. B chromosomes do not pair with A chromosomes, but breaks and exchanges can occur between them (Fig. 9.16). Detailed studies on the behavior and cytogenetic applications of B-A translocations have been confined to maize, where the first ones were induced by X-rays (*see* Chapter 13, Section 13.6.2).

9.3.1 Behavior of Translocations During Mitosis and Meiosis

During mitosis, the chromosomes of heterozygous or homozygous balanced translocations behave independently and segregate normally in anaphase. During meiosis, homozygous translocations form bivalents, which progress through the stages to produce functional gametes and zygotes. Heterozygous translocations achieve homologous pairing by forming a four-branched cross configuration at pachytene, with the lengths of the branches being determined by the sizes of the translocated segments (Figs. 9.17 and 9.18). Nonpaired areas often occur around the breakpoints because of the changes in sequence, resulting in tighter linkages for genes in these regions. If a branch is too short for crossing over to occur, the ends separate when the homologs repel each other during diplotene, resulting in a chain rather than a ring (Fig. 9.19).

The later behavior of each chromosome in the pachytene configuration can be determined by starting with the centromere in each crossover or noncrossover chromatid and following each arm to the end (Fig. 9.20). Crossovers in the distal regions of chromosome arms do not change chromosome structure, whereas a crossover in an interstitial segment (i.e., the region between a centromere and a breakpoint) rearranges segments distal to the crossover point (*see* Fig. 9.23).

Fig. 9.15. *(continued)* rDNA sequences in chromosome 13. The sequences with question marks have not been identified. The advantage of molecular techniques to pinpoint the precise locations of breakpoints is evident in this study.

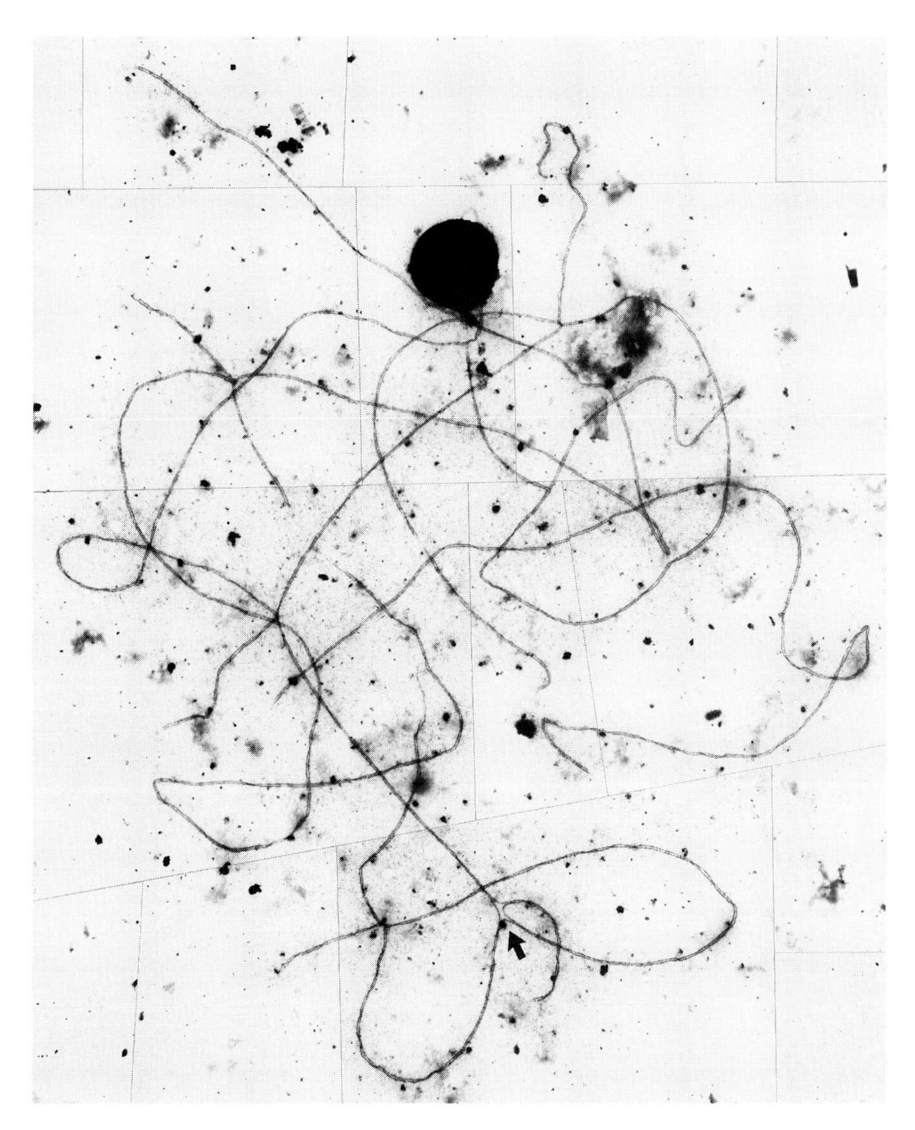

Fig. 9.17. An electron micrograph of pachytene chromosomes showing a four-branched configuration (bottom, marked by arrow) resulting from a heterozygous reciprocal translocation between two rye chromosomes. Close homologous pairing in the four branches occurs near the exchange of segments (electron micrograph kindly supplied by R. Giraldez, unpublished).

The power of the centromeres in guiding the translocation complex to the metaphase I plate is indicated by the usually regular alignment, with two centromeres facing one pole and two facing the other pole (Figs. 9.21a and 9.21b). There are two types of adjacent orientations, in which centromeres of adjacent chromosomes face the same pole (*see* Fig. 9.21a). With adjacent 1, nonhomologous centromeres face the same pole, whereas with adjacent 2, homologous centromeres face the same pole. The ring must make a twist if alternate chromosomes are to go to the same pole, and two metaphase I orientations are shown in Fig. 9.21b. There is a difference in the diagonal alignment of the homologous versus nonhomologous centromeres in the two diagrams, but both orientations have the same chromosomes poised for movement to opposite poles, with no deletions or duplications. One viewpoint is that these two orientations stem from one three-dimensional configuration, which can be rotated during the squash procedure, when preparing material for microscopic examination, to form either of the two alternate orientations in the two-dimensional plane. A differing viewpoint in support of two types of alternate orientation is based on chromosome morphological markers (Fig. 9.22) and numerical differences in the frequencies of what are called alternate 1 and alternate 2 in this study. Because of these different viewpoints, the alternate orientations in Fig. 9.21b have not been numbered. A rarer orientation shows only two of the four centromeres aligned with respect to the poles (Fig. 9.21c) and has been a source of trisomics in barley (*see* Chapter 11, Section 11.4.5).

A number of factors affect the types and frequencies of ring orientations on the metaphase I plate. Chromosome traits favoring alternate arrangements are uniform lengths, median or submedian centromeres, and movement of chiasmata to the ends of the chromosomes during diplotene. These conditions give a ring greater flexibility to make the twist for alternate orientation. If the chias-

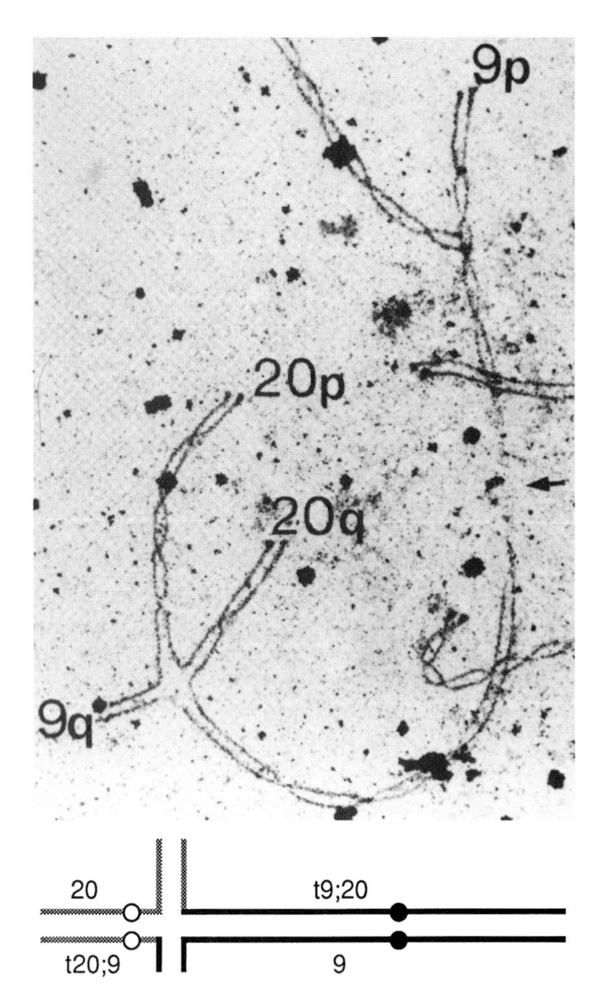

Fig. 9.18. An electron micrograph of a heterozygous reciprocal translocation between human chromosomes 9 and 20 showing close pairing throughout the configuration. The position of the centromere of chromosome 9 is indicated by the arrow (Chandley et al., 1986). A diagrammatic representation of this configuration is shown beneath the photograph (not to scale).

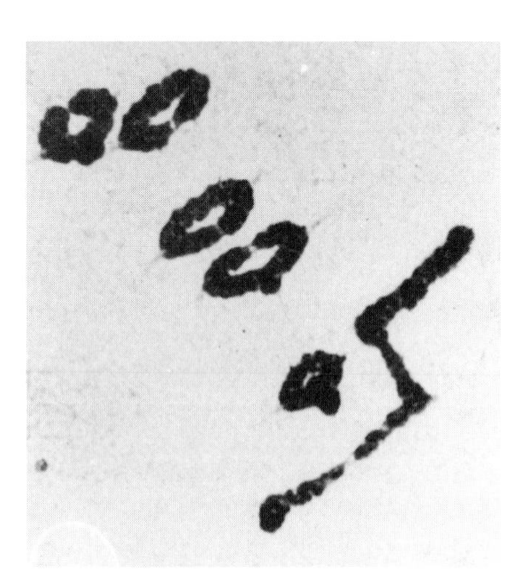

Fig. 9.19. Metaphase I of meiosis in barley showing five bivalents and a chain-of-four chromosomes induced by X-radiation (Caldecott and Smith, 1952).

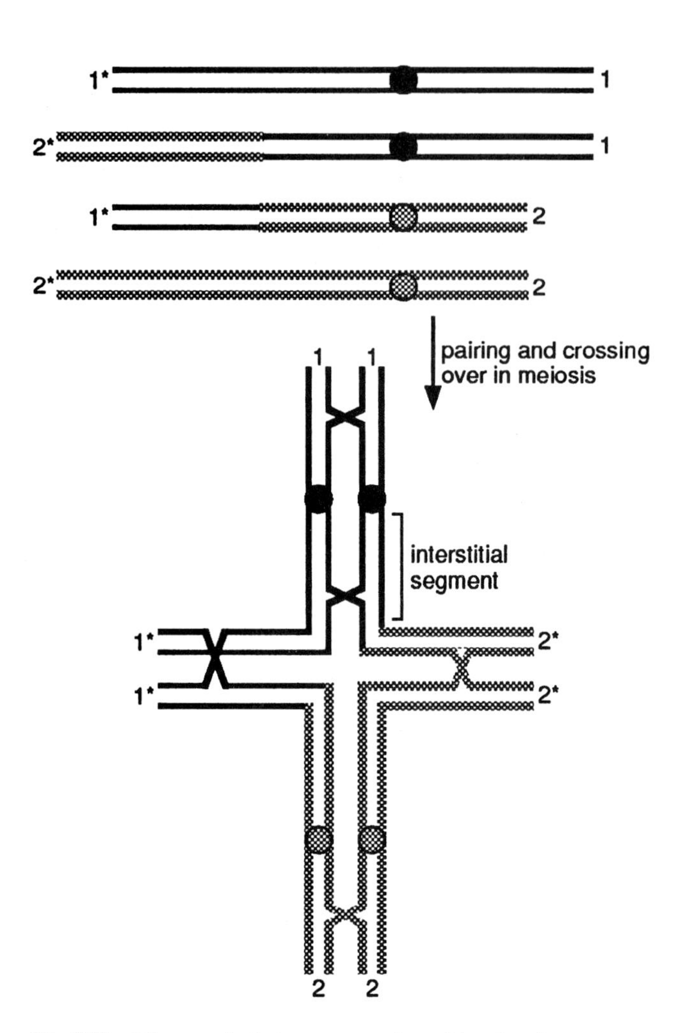

Fig. 9.20. Diagram of a heterozygous reciprocal translocation involving two pairs of nonhomologous chromosomes (top), and the pachytene configuration (bottom) with the breakpoints at the center of the "cross." The chiasmata toward the ends of the chromosome arms are needed to retain a ring formation at later stages of meiosis. A crossover is shown in one of the two interstitial segments. Measurements of pachytene configurations were used to determine the breakpoints of spontaneous and induced translocations in maize (Longley, 1961), whereas in *Drosophila,* the salivary-gland chromosomes are more suitable for locating breakpoints (Roberts, 1976). Biotin-labeled probes can also be used to identify the chromosomes involved in translocations and the sites of exchanges (Simpson et al., 1990).

mata remain closer to the centromeres, the ring is more rigid, and adjacent orientations are more likely to occur. Alternate orientations are predominant in several plant species, notably in the genus *Oenothera*, where translocations involving most or all of the chromosome set have evolved naturally (*see* Section 9.6).

The outcome in the gametes from the different anaphase I segregations can now be summarized. With no crossing over in interstitial segments, alternate segregations result in gametes with a complete complement of chromosomes and genes, and both types of adjacent segregation result in deficient gametes. A crossover in an interstitial segment, followed by alternate segregation, precludes the transmission of the crossover chromatids because the gametes containing them have an incomplete complement (Fig. 9.23), and this is also the situation for adjacent 2 segregation if it occurs. However, with adjacent 1 segregation, gametes with the crossover chromatids have a complete complement, and the genetic recombination is transmissible (*see* Fig. 9.23).

Robertsonian translocations. A heterozygous Robertsonian translocation has a pachytene configuration of three chromosomes, consisting of the translocated chromosome and the two acrocentric chromosomes, from which it was derived. If chiasmata retain the chain-of-three, it can be aligned on the metaphase I plate for alternate or adjacent segregation. The anaphase I segregations in Fig. 9.24 are for a chain derived from the type of translocation shown in Fig. 9.15a. The adjacent segregations give unbalanced gametes, with an increase or decrease in the dosage of some genes depending on their locations.

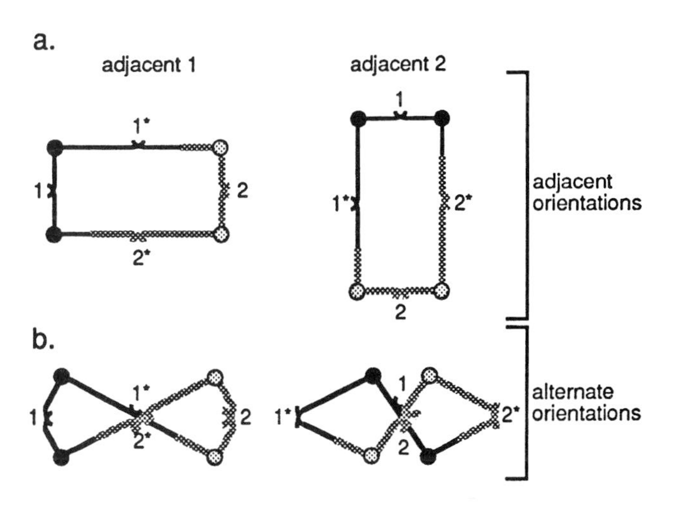

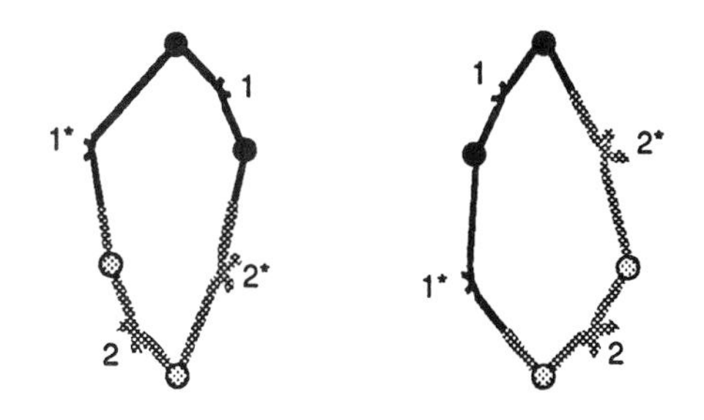

Fig. 9.21. The orientation of translocation rings at metaphase I given no crossing over in the interstitial segments (*see* Fig. 9.20). The poles are in the vertical plane. Only one chromatid of each chromosome is shown because sister chromatids are identical. Chromosome ends are marked by numbers and chiasmata by crossed ends. The adjacent orientations (a) have segmental deletions and duplications, whereas the alternate orientations (b) have complete chromosome complements. The orientations in (c) can give 3:1 distributions to the poles at anaphase I (*see* Schulz-Schaeffer, 1980; Sybenga, 1992).

Fig. 9.22. Drawings of rings-of-four at metaphase I in cotton (*Gossypium hirsutum*) (modified from Endrizzi, 1974). 1 = smaller, nontranslocated (standard) chromosome; 2 = standard larger chromosome; 1′ and 2′ = translocated chromosomes; homologous centromeres are 1 and 1′, and 2 and 2′. The four orientations are distinguished by a size difference in the two standard chromosomes and by the presence of a large chiasma knob at the ends of chromosomes 1 and 1′. From left to right, the orientations as used in this study are adjacent 1, alternate 1, adjacent 2, and alternate 2. With the adjacent 1 orientation and both types of alternate orientation, homologous centromeres face opposite poles, whereas with the adjacent 2 orientation, they face the same pole. Note that nonhomologous centromeres are diagonally opposite in the alternate 1 orientation, whereas homologous centromeres are in that position with the alternate 2 orientation. The frequencies of the four types of orientation and segregation depend on a number of variables involving the behavior of the ring in relation to the nucleus and the cell (Sybenga and Rickards, 1987).

B–A translocations. When a plant homozygous for a reciprocal translocation between a B and an A chromosome, such as the one in Fig. 9.16, is crossed with a plant lacking both the translocation and B chromosomes, the progeny have a three-chromosome meiotic configuration (Fig. 9.25) consisting of the two translocated chromosomes and the normal A chromosome, which is homologous with the one in the translocation. Because B chromatin is not needed for normal cell functions, we need to consider only the segregation of A chromatin into the gametes. Studies of B–A translocations in maize indicate that the A and A^B (*see* Fig. 9.25 legend for explanation of symbols) chromosomes usually pass to opposite poles at anaphase I. The B^A chromosome pairs with A part of the time, but passes at random to either pole regardless of its pairing behavior. Thus, the following four types of spores are expected at the end of meiosis: A, A^B B^A, A B^A, and A^B. The first two types have a complete complement of A chromosomes and are transmissible. The A B^A combination has a duplication for the translocated segment from A and usually functions through the female, but less often through the male. A^B spores have the corresponding deletion for the translocated segment from A, and they abort unless the deleted segment is very short and nonessential.

9.3.2 Effects of Heterozygous Translocations on Fertility

Fertility is influenced by the relative proportions of alternate and adjacent chromosome segregations from the rings or chains, together with the amount of crossing over in the interstitial seg-

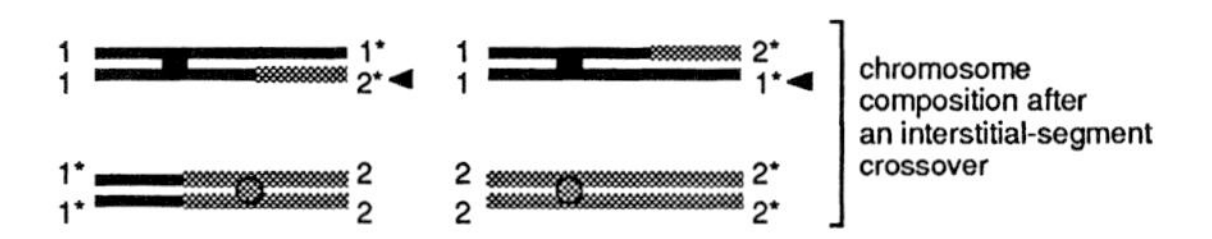

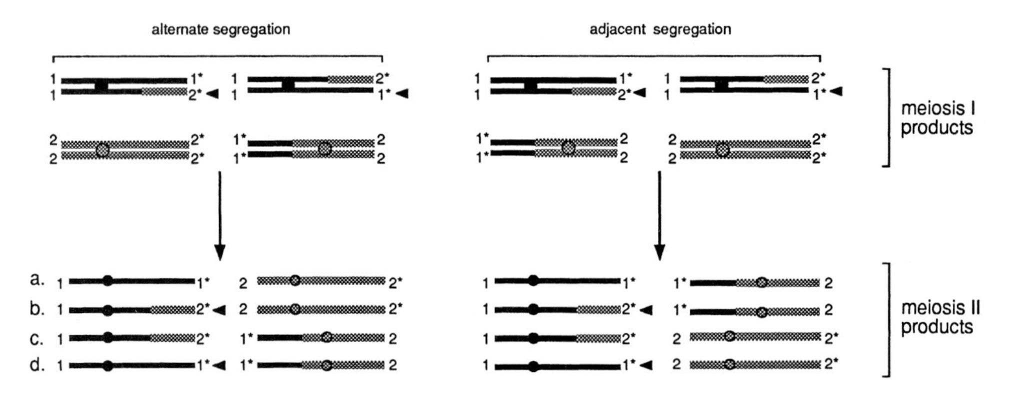

Fig. 9.23. The consequences of a single crossover in one of the interstitial segments of a heterozygous translocation. The four chromosomes at the top are in the same quadrants as the pachytene configuration shown in Fig. 9.20. The chromatids resulting from the crossover in the interstitial segment shown in Fig. 9.20 are marked by arrowheads. The meiosis I products (center) result from chromosome distributions to the poles after alternate and adjacent 1 segregations. The adjacent 2 segregation is not shown because there is evidence from both maize and watermelon that it rarely occurs after crossing over in an interstitial segment. Meiosis II products, b and d (bottom left) from alternate segregation, include the crossover chromatids but have segmental deletions and duplications. Products b and d from adjacent I segregation (bottom right) have complete complements. Pollen abortion in *Datura stramonium,* resulting from four differing translocations that generally separate through alternate segregation, was attributed to crossing over in the interstitial segments (Burnham and Stout, 1983).

ments. In animals and humans, gametes with deleted and duplicated segments can function in fertilization and are usually combined with gametes carrying the full chromosome complement to give trisomy for one segment and monosomy for another segment. This chromosome condition often is lethal in the early developmental stages of the offspring.

In plants, gametes receiving deleted and duplicated segments usually abort. In maize, sorghum, and pea, the frequency of alternate : adjacent segregation for four-chromosome translocations is about 1 : 1 and agrees with the frequency of normal : aborted pollen. In other plant species, with four or more chromosomes in the translocation complexes (barley, rye, *Datura*, and *Oenothera*), the segregation is mostly alternate and the amount of pollen abortion is much lower. At least in *Datura* and *Oenothera*, low sterility is associated with short interstitial segments, which have no crossing over and do not contribute to an increase in gamete abortion. Chain-forming translocations (*see* Fig. 9.19) usually cause lower sterility than ring-forming translocations, because half of the adjacent segregations have a deletion for the short terminal part of one chromosome that was translocated to the other chromosome, and gametes with this small deletion are often functional in plants. B–A translocations in maize give around 25% sterility because the A B^A combination mentioned in the previous section is viable in female gametes.

The sterility from Robertsonian translocations depends on the frequency of adjacent segregations from the chain-of-three (*see* Fig. 9.24), resulting in unbalanced gametes. Whereas the trisomic state for most of one chromosome would probably be viable in diploid plants, the monosomic state for one chromosome causes gametic abortion. The monosomic state and usually the trisomic state cause spontaneous abortions in humans and higher animals.

9.4 PHENOTYPIC EFFECTS OF CHROMOSOME REARRANGEMENTS

Both inversions and reciprocal translocations have the potential to change phenotypes by creating position effects, whereby genes moved to new locations are altered in expression, but care must be taken to separate position effects from other conditions that can affect phenotypes. For example, if one of the breakpoints involved in a rearrangement occurs within a gene, its components are separated and it may lose its function, thus allowing an alternate allele to be expressed. Another cause for a change in phenotype is a small deletion or duplication near a rearrangement breakpoint.

Phenotypic changes can occur as indirect effects of rearrangements if gametes or zygotes with duplicated and deleted chromosome segments are viable, because there are changes in gene dosage in those segments. The deleted portions must be small and dispensable for survival of the individual. Robertsonian translocations fit this situation because losses of the short arms of the participating acrocentric chromosomes are tolerated. When this type of translocation involves the human chromosome 21, it accounts for 2–5% (in different studies) of individuals with Down's syndrome, which is attributed to a trisomic dose of the terminal part of band 21q22. The children of a parent, usually the mother, who is heterozygous for a Robertsonian translocation (*see* Fig. 9.24) involving chromosome 21 will develop Down's syndrome if two long arms of 21 enter the same parental gamete as a result of adjacent segregation,

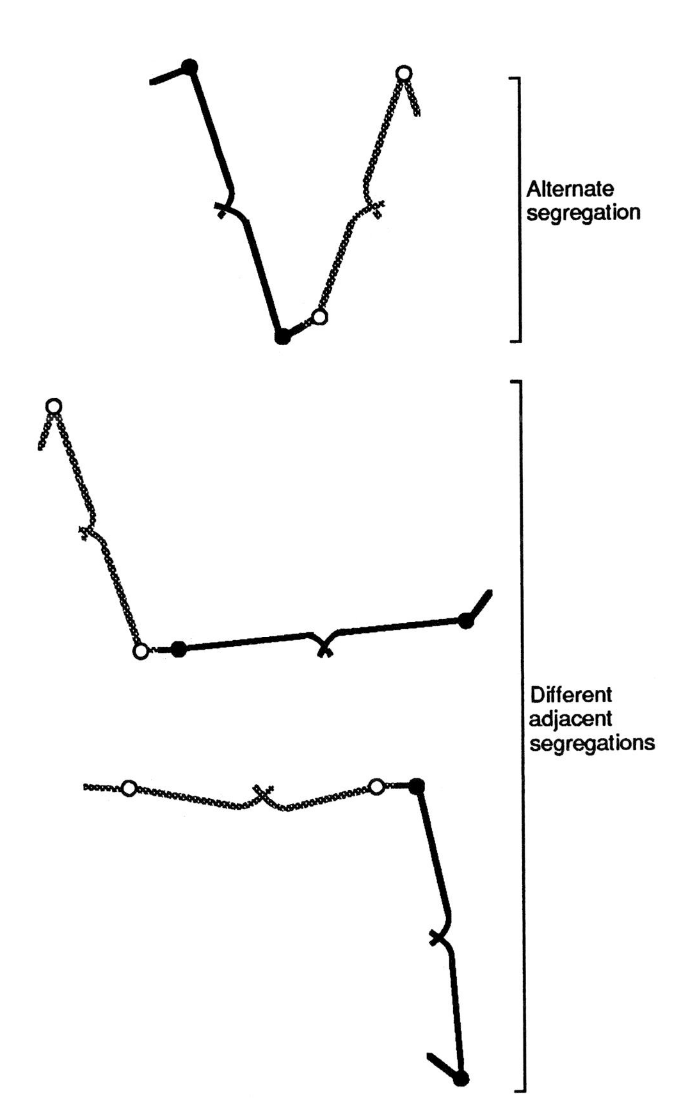

Fig. 9.24. Types of segregation for a heterozygous Robertsonian translocation at anaphase I. Only one chromatid of each chromosome is shown. Alternate segregation (upper) sends the translocated chromosome to one pole, and the two acrocentric chromosomes to the other pole. All the meiotic products should be functional because the short arms are dispensable (*see* Fig. 9.15a). With both types of adjacent segregation (lower), an entire chromosome is missing in half the gametes and its long arm is duplicated in the other half. The deletion of a chromosome causes gametic or embryonic abortion in diploid organisms. In humans, when a gamete with the duplicated arm meets a euploid gamete, the trisomic state for this arm causes abortion or an abnormal phenotype depending on the chromosome (Jacobs et al., 1987; Levitan, 1988).

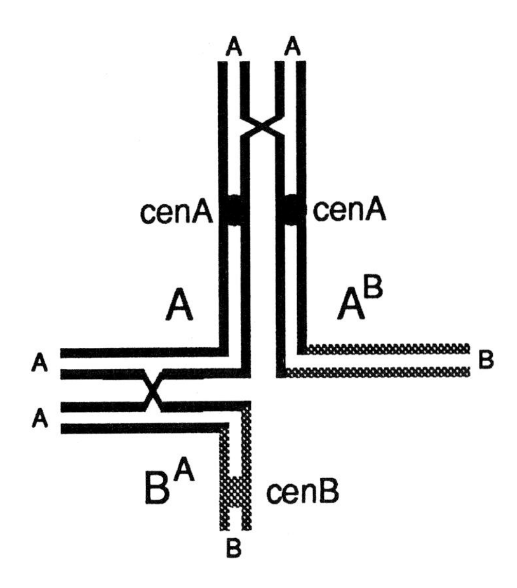

Fig. 9.25. Diagrammatic representation of a heterozygous translocation between A and B chromosomes at pachytene. The translocated chromosome with a B acentric fragment attached to the centric part of the A chromosome is designated A^B (upper right), and the other translocated chromosome with the A acentric fragment attached to the centric B segment is B^A (lower left). A different A-chromosome segment can be added to a B–A translocation by crossing it with an A–A translocation that has one breakpoint in the same A arm but at a different location. A crossover in the differential segment between the two breakpoints joins the two translocations and attaches a new A segment to a B centromere (Rakha and Robertson, 1970; Birchler, 1991).

and this gamete unites with a gamete carrying one dose of the long arm. The usual cause of Down's syndrome is trisomy for a complete chromosome 21 (*see* Fig. 11.9).

9.4.1 Chromosomal Rearrangements and Position–Effect Variegation

Associations of rearrangements with unstable phenotypic changes, called position–effect variegation, are often difficult to document. One approach is to use a heterozygous translocation with appropriate breakpoints so that a gene, normally located in euchromatin, is relocated next to heterochromatin on one of the translocated chromosomes, whereas a differentiating allele remains on a standard chromosome and is stable. If the relocated gene has a variegated effect, it can be returned to its normal location in the standard chromosome through crossing over between the locus and the breakpoint. Assuming that position–effect variegation is involved, the stability of this gene should be restored, whereas its allele is transferred to the translocated chromosome and should become unstable, giving variegation. Tests of this kind have demonstrated position–effect variegation in *Drosophila* and mice, but the only clear case in plants is in *Oenothera blandina* for the P^s and P^r alleles at a locus affecting sepal pattern.

Another way of separating a gene from its association with heterochromatin is by an induced secondary rearrangement. However, reversions to normal phenotype may be due to mutations induced by the clastogenic agent, whereas in the translocation method, the reciprocal products of crossing over can be distinguished from mutation.

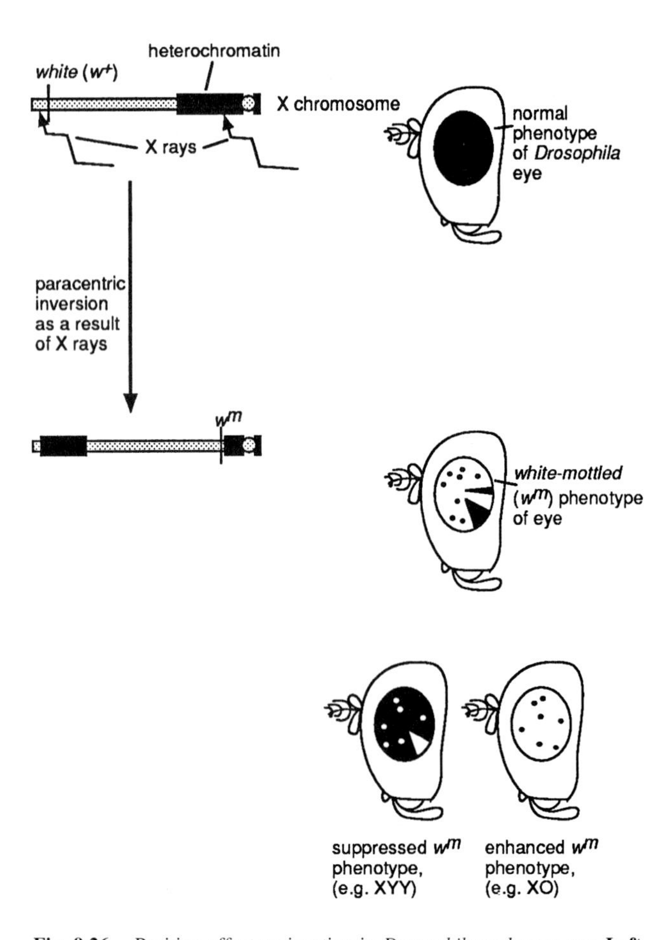

Fig. 9.26. Position-effect variegation in *Drosophila melanogaster*. **Left top:** X-ray induction of a paracentric inversion with one breakpoint in euchromatin near the normal allele (w^+) of the white-eye locus and the other breakpoint in the heterochromatin near the centromere (*see* bent arrows). Red eye color, due to the normal functioning of w^+, is depicted by a uniformly dark eye (top right). **Middle row:** The *white-mottled* (w^m) mutant (right) is caused by the inactivation of w^+ in its new position next to the heterochromatin (left). This allele was inactivated in the initial stages of eye development, giving a white background, but its function was restored at early (dark wedges) to late (dark dots) stages, depending on conditions in individual cells. **Bottom:** The addition of an extra Y chromosome (left side) restores the w^+ function to give dark eyes except for a few white dots and the occasional white wedge where w^+ is suppressed. The absence of the Y chromosome (right side) results in an enhancement of w^+ inactivity, so that it functions much later in eye development to give only a few dark dots on a white background (Henikoff, 1990).

Position–effect variegation is affected by various environmental and genetic conditions. A notable environmental effect is temperature, with higher temperatures preventing variegation and lower temperatures enhancing it. Most cases of position–effect variegation involving euchromatic genes in *Drosophila* are suppressed by extra Y chromosomes, which are largely heterochromatic, and enhanced in the absence of Y chromosomes (Fig. 9.26). But, heterochromatic genes that show variegation when moved to euchromatin, respond in a reverse manner to dosage changes in the Y chromosome. Variegation is modified by numerous loci in euchromatin, including histone genes, which indicate that at least some of the modifiers have other functions.

The distance between heterochromatin and the affected genes varies with different rearrangements, and in some cases involves more than 50 bands on a salivary-gland chromosome in *Drosophila*. The variegated effect also can extend to genes farther removed from the heterochromatin than the main gene being tested. This spreading effect is called heterochromatinization. One hypothesis to explain this effect assumes that heterochromatic proteins, encoded by the modifying genes mentioned above, spread into euchromatin, which becomes partially heterochromatic, thereby inactivating the normal (+) alleles at certain stages of development. An alternative concept is that transposable elements, located within heterochromatin, may induce position–effect variegation by moving into nearby euchromatin, then away from it, during somatic cell divisions.

The association of variegation with rearrangements and heterochromatin was first uncovered for some X-ray-induced mutant alleles, called *white-mottled* (w^m), at the white-eye (w) locus on the sex (X) chromosome of *D. melanogaster* (*see* Fig. 9.26). Variegation at this locus also was produced by transformation experiments, in which the normal allele (w^+) was inserted near centromeric or telomeric heterochromatin. The w^m alleles, like many variegating genes, are recessive (i.e., they must be combined with w to be expressed). The genotype w^+w^m gives stable, red eyes. These recessive, *white*-mottled mutants are examples of the effects of bringing heterochromatin close to genes located in euchromatin in the same chromosome (cis-inactivation).

Whereas position–effect variegation can be induced in most euchromatic *Drosophila* genes, some genes remain stable when placed near heterochromatin. On the other hand, the heterochromatic gene *light*(lt) shows cis-inactivation when moved to euchromatin, suggesting that cis-inactivation, in general, relates to a change in the local environment that gives normal expression to a gene. Dominant mutations at the brown-eye locus (bw) in *Drosophila* indicate that trans-inactivation between chromosomes also can occur (Fig. 9.27). The dominant mutant (bw^D) allele arises from the relocation of heterochromatin to the vicinity of the bw^+ allele by a chromosome rearrangement, resulting in cis-inactivation of the locus, analogous to the position–effect variegation at the white locus. The unusual feature of the bw^D allele is that if there is a bw^+ allele on the homologous chromosome, there seems to be a spreading effect of the heterochromatin across homologues (trans-inactivation), and the bw^+ allele is prevented from producing the wild-type red eyes. The dominance of bw^D over the bw^+ allele suggests that the relative positioning of the locus is important.

The dominant position–effect variegation at the brown-eye locus in *Drosophila* has relevance for a human neurological disorder, Huntington's disease, which is caused by a dominant mutation (H) on chromosome 4. Indications of variegation are that onset of this disease in adults covers a wide range in age, from 20 to 70 years, and that the severity of the neurological degeneration varies as the disease progresses. Because of striking parallels between H and bw^D, it has been suggested that Huntington's disease is due to a dominant position–effect variegation, which results from trans-inactivation of the normal allele (H^+) of a structural gene (or genes) governing neurological functions. The effect is analogous to that illustrated in Fig. 9.27 for bw^D effects on bw^+.

Molecular studies have sought to find the underlying causes of position–effect variegation. Late replication and reduced DNA content within heterochromatic regions seem to be involved, at least in some cases. Gene cloning indicates that both suppression of gene activity and reduction in gene-copy number can result from placement of genes next to heterochromatin. Of particular interest are the loci that enhance or suppress variegation, and the kinds of

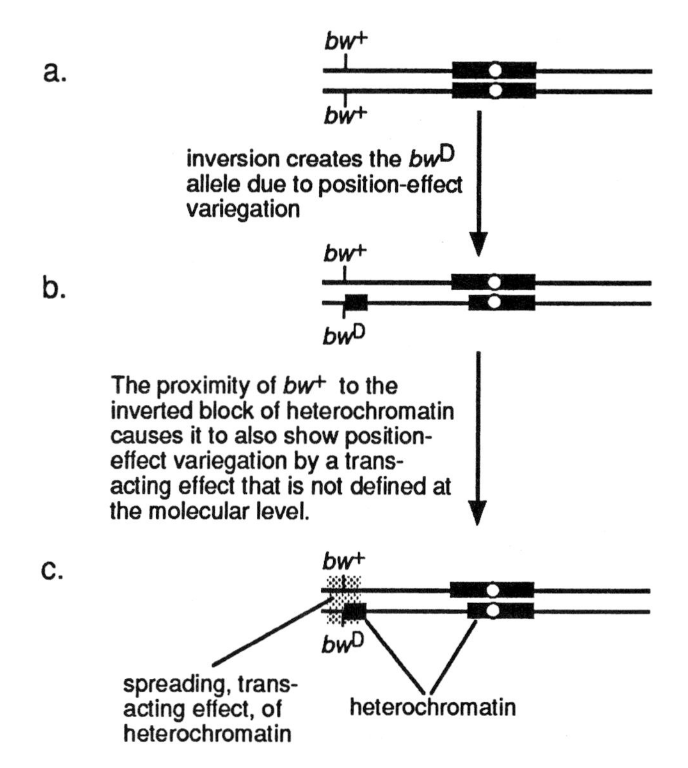

Fig. 9.27. A model for position–effect variegation due to cis- and trans-inactivation at the brown-eye locus in *Drosophila melanogaster*. (a) Chromosome 2 mitotic homologues with the normal allele (bw^+) on each chromosome. (b) Cis-inactivation of bw^+ on the lower chromosome caused by an inversion that positions heterochromatic DNA adjacent to the gene. The spreading effect of the heterochromatin produces a dominant mutant, bw^D. (c) Subsequent trans-inactivation caused by somatic pairing and a spreading of the heterochromatic effect from the lower to the upper chromosome. The normal bw^+ allele is inactivated and dominated by bw^D (Henikoff, 1990; Henikoff et al., 1995).

proteins that they encode (*see* Chapter 20, Section 20.1.4). Additional molecular evidence is needed to substantiate trans-inactivation as an explanation for dominant position–effect variegation in the two cases discussed above, one in *Drosophila* and the other in humans.

9.5 RELATIONSHIPS BETWEEN CHROMOSOMAL ABERRATIONS AND CANCER IN HUMANS

The suggestion that chromosomal aberrations might be involved in the development of abnormal growths such as tumors was made at the end of the nineteenth century, but it was not experimentally confirmed until some 70 years later. With the development of chromosome-banding techniques, it could be determined if an aberration involving one or more identified chromosomes was consistently associated with a particular kind of cancer. Some chromosome changes occur sporadically and their role, if any, in the development of cancer cells from normal cells is difficult to determine. In other cases, a specific aberration is present consistently in the abnormal cells of individuals with the same type of cancer. Studies of these consistent relationships demonstrate the value of the interplay between cytogenetic and molecular methods.

The first evidence of the role of a chromosome aberration in promoting cancerous cells was the consistent presence of an abnormal G-group chromosome in individuals with chronic myelogenous leukemia (CML), a malignant disease involving the leukocytes (white blood cells) in the bone marrow. The aberrant chromosome is called the Philadelphia (Ph^1) chromosome because of the city where the discovery was made, and initially the only abnormality detected was a deletion of about half of the long arm. When chromosome-banding techniques became available, the Ph^1 chromosome, identified as 22, was found to have a small segment from chromosome 9 attached to its shortened long arm, whereas its missing segment had been transferred to the long arm of 9 (Fig. 9.28). Over 90% of individuals afflicted with CML carry this reciprocal translocation, and some of the remaining cases have three-way translocations involving 9, 22, and a third chromosome.

The molecular nature of the Philadelphia chromosome became apparent when the *c-abl* oncogene (*c* designates a cellular form and *v* a viral form), which is involved in leukemias, was investigated in its proto-oncogenic (inactive) form in the DNA from CML patients. The *c-abl* gene was known, from gene-mapping studies, to be located on chromosome 9, so when a DNA sequence was cloned carrying both the *c-abl* gene and a sequence known to be specific to chromosome 22, interest in the Philadelphia chromosome was heightened. At the level of RNA transcripts, a new 8-kb *c-abl* product could be assayed in addition to the normal 6- and 7-kb products. The new 8-kb RNA transcript was composed of a 5′ end from the chromosome 22 DNA segment joined to the *c-abl* gene from exon 2 onward, and was shown to code for a CML-specific protein called p210. Because the chromosome 22 DNA segment did not derive from a known gene region, it was named the *breakpoint cluster region (bcr)* gene. The significant aspect of this translocation is that the segment of chromosome 9 with the *c-abl* gene had been transferred to chromosome 22 (Fig. 9.29). The indication that variation in the position of the translocation breakpoints in the *c-abl* and/or *bcr* genes may correlate with variation in the clinical features of CML has stimulated continuing research in this area.

The Philadelphia (9 : 22) chromosome has been found in acute forms of leukemia, but in lower frequencies than in CML cases, and in some occurrences its frequency is related to the age of the afflicted persons. For instance, this chromosome is found in 2–6% of childhood cases of acute lymphocytic leukemias, and in 17–25% of adult cases. The *c-abl* gene appeared to be transferred to chromosome 22 in all these cases, but the *bcr* gene was not always rearranged.

Some reciprocal translocations involving specific chromosomes are consistently associated with malignancies of the two kinds of lymphocytes: B cells derived from bone marrow in mammals and T cells derived from the thymus. Both types of cells play essential roles in immunological (antigen–antibody) reactions. Before discussing a human B-cell tumor called Burkitt's lymphoma, some relevant information on the human immunological system will be reviewed.

There are five major classes of immunoglobulin (Ig) molecules in the gamma-globulin fraction of serum proteins. Each Ig molecule contains four polypeptide chains; the two heavy chains have about twice as many amino acids as the two light chains (*see* Chapter 19, Fig. 19.3). A locus for the heavy chains has been mapped on the q arm of chromosome 14 at band 32 (14q32), and loci for the light chains on 2p11 and 22q11. In stem cells from which mature B cells develop, the Ig heavy-chain locus (*IgH*) on chromosome 14 consists of a constant (C) region, which has essentially the same

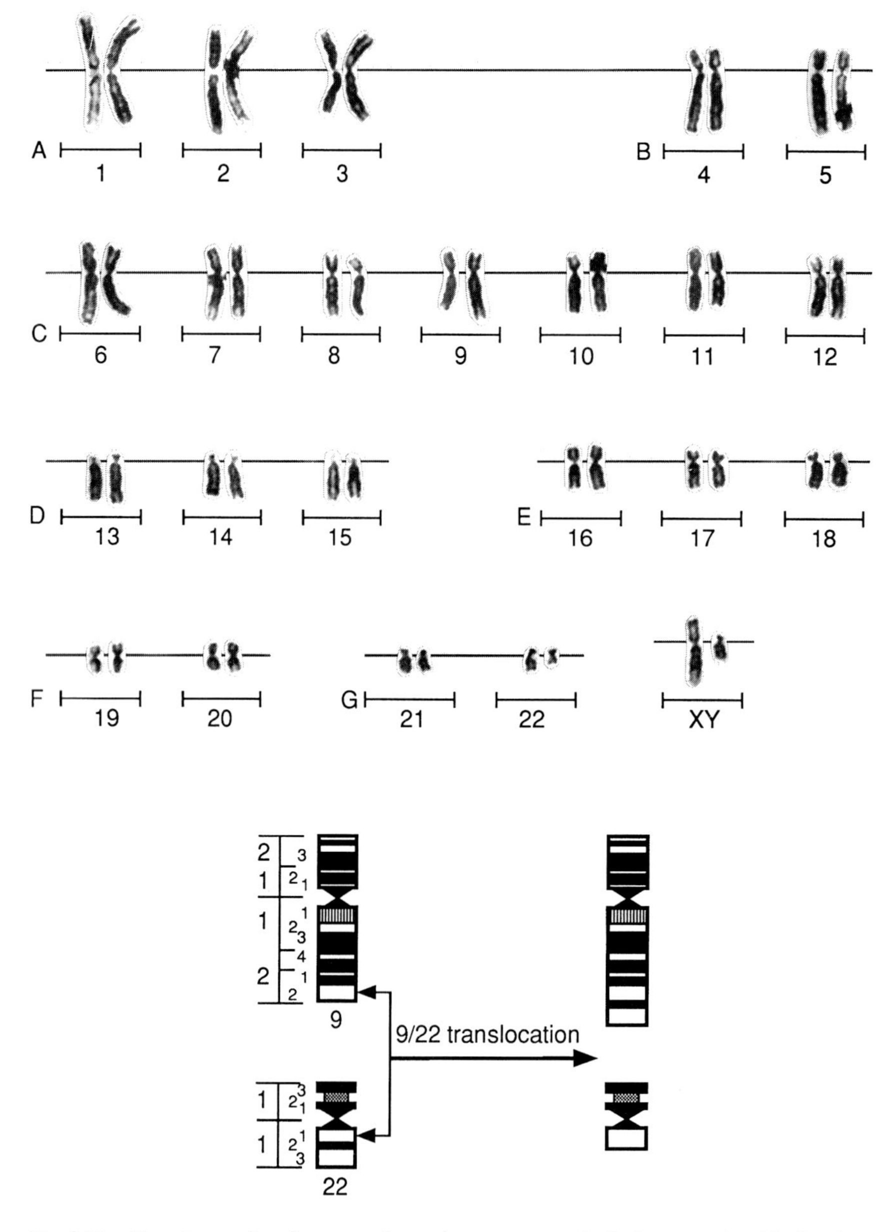

Fig. 9.28. Top: A metaphase karyotype from a bone-marrow cell of a human male with chronic myelogenous leukemia (CML) (Rowley, 1980). The shortened Philadelphia chromosome (right side, pair No. 22) has a shorter long arm than its homolog because of a reciprocal translocation with chromosome 9 (right side, pair No. 9), which has gained chromatin in its long arm. **Lower:** A diagram of the translocation between chromosomes 9 and 22 (Rosson and Premkumar Reddy, 1988). The normal chromosomes are on the left and the translocated chromosomes on the right. The arrows indicate where the breaks and exchanges took place. As a result of this translocation, a gene (*c-abl*) is transferred from section 22 of the long arm of chromosome 9 to chromosome 22 (bottom right), a movement that involves this gene with CML and the Philadelphia chromosome.

amino-acid sequence in all Ig molecules of the same class, and a variable region, which has different amino-acid sequences depending on the antigenic specificities of the Ig molecules (Fig. 9.30). The recombination events between different segments of the variable region contribute to the tremendous amount of antibody diversity in the immune system.

Ig heavy chains are divided into several classes and subclasses based on the structure of the constant region, which includes nine C segments that give heavy chains their antigenic characteristics (isotypes). Generally, only the two C segments next to the V/D/J exon can be expressed (*see* Fig. 9.30). However, after a more mature B cell has been stimulated by the binding of antigen to its membrane immunoglobulin, rearrangements can take place between a V/D/J exon and a nonadjacent C sequence by deletion of intervening C sequences (Fig. 9.31). This process is called isotype switching and is another source of antibody diversity.

Burkitt's lymphoma has two forms based on geographic distribution, both affecting the B cells: endemic Burkitt's lymphoma in

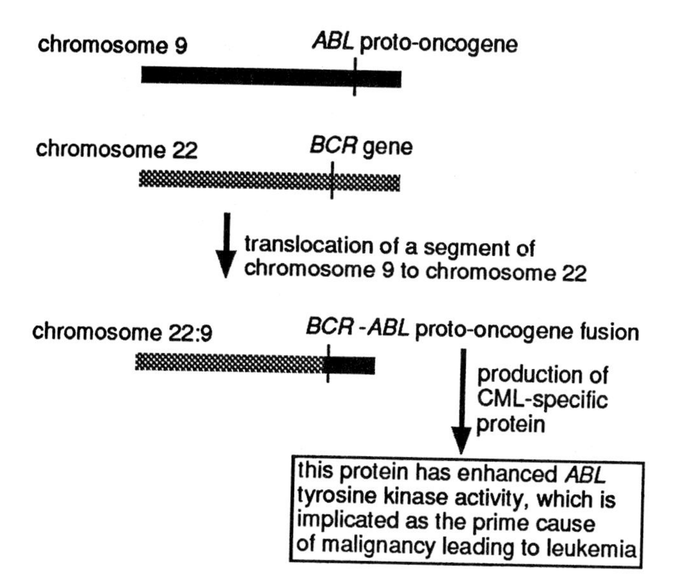

Fig. 9.29. A molecular interpretation of the translocation giving rise to the Philadelphia chromosome and chronic myelogenous leukemia, CML (Dobrovic et al., 1991). *See* text for explanation of gene symbols.

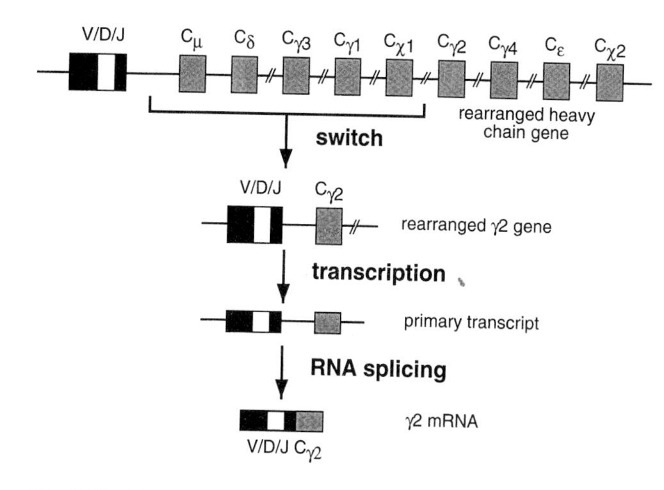

Fig. 9.31. Further rearrangement of the immunoglobulin heavy-chain locus (*IgH*) by isotype switching. This switch places a different C sequence (C_γ) adjacent to the V/D/J exon (Stites and Terr, 1991) by deleting the intervening C sequences.

equatorial Africa and sporadic Burkitt's lymphoma, a less common form, in Europe and North America. The tumor cells of both the endemic and the sporadic forms have a translocation between chromosome 8 and one of three other chromosomes, 2, 14, or 22, with breakpoints at the sites of the *Ig* loci. A cellular proto-oncogene, *c-myc*, which plays a role in growth control of cells, is located on chromosome 8 (band q24), which is the breakage point for each of the three translocations. The *c-myc* locus includes three exons with intervening introns. In the 8–14 translocations, all or part of *c-myc* is transferred to chromosome 14 along with the distal part of the 8q arm (Fig. 9.32). In the other two translocations, the distal segment of 2p or 22q, including an *Ig* locus, is transferred to chromosome 8. Thus, each of the three translocations brings the *c-myc* gene and the *Ig* loci in close proximity on the same chromosome.

Molecular studies have helped to clarify the origin of the translocations, and their effect on *c-myc* in relation to the human immune system and the development of Burkitt's lymphoma. The 8–14 translocation has been analyzed in greater depth than the other two because of its higher frequency. By cloning and sequencing the breakpoints of the 8–14 translocation from many cases of Burkitt's

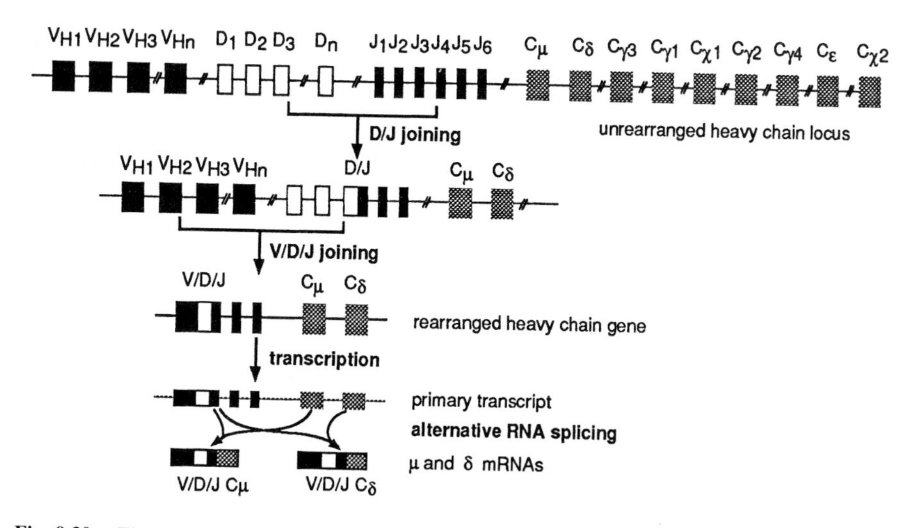

Fig. 9.30. The molecular structure and rearrangements of the immunoglobulin heavy-chain locus (*IgH*) on human chromosome 14 (Stites and Terr, 1991). This abbreviated model (top) shows 4 of several hundred variable (V) segments, 4 of about 50 diversity (D) segments, the 6 joining (J) segments, and 9 constant (C) sequences, each consisting of several exons. The D/J joining, via the enzyme recombinase, involves D3 and J4 (top two rows), which, in turn, join with V_{H2} to give a V/D/J exon (third row). The presence of the combined V/D/J exon enables transcription to take place, and the intervening segments are then deleted by enzymic action. The primary transcript can now be spliced to generate mRNAs that encode (in this example) μ or δ heavy chains with identical V domains.

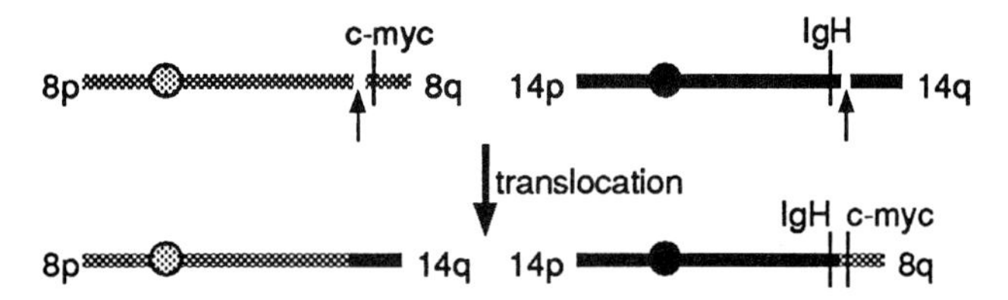

Fig. 9.32. A diagram of a reciprocal translocation event between chromosomes 8 and 14 that occurs in the tumor cells of many patients with Burkitt's lymphoma (Stites and Terr, 1991). **Upper:** Chromosomes 8 and 14 showing the locations of the breakpoints (arrows pointing up) and the *c-myc* and *IgH* genes. **Lower:** The two translocated chromosomes showing *c-myc* next to *IgH* on 14–8. In this location, *c-myc* becomes deregulated and oncogenic (Haluska et al., 1987).

lymphoma, the breakpoints on chromosome 14 have been found to occur at several sites within the *Ig* locus. This suggests that the breaks are induced as a result of some type of error at the time of rearrangements within the *Ig* locus. Most of the translocations in the endemic form of Burkitt's lymphoma are thought to arise during the rearrangement of the V, D, and J regions in the early stages of B-cell development, perhaps by malfunctioning of the recombinase system. In the sporadic form, the translocation breakpoints seem to result from mistakes during isotype switching. These conclusions are based on molecular analyses of the breakpoints in the two forms, and on the Ig characteristics of the tumor surfaces, which relate to the timing of rearrangement errors.

Another interesting difference between the endemic and sporadic forms of Burkitt's lymphoma relates to the positions of the breakpoints within the *c-myc* gene. Analyses of DNAs from tumors with the 8–14 translocation by Southern blotting showed that the translocation caused rearrangement of the *c-myc* gene in some tumors, but not in others. Generally, the endemic tumors had breakpoints outside the *c-myc* locus, thus, all three exons in the c-myc gene were transferred to chromosome 14. Most of the sporadic tumors had breakpoints within the intron located between the first two exons, and exons 2 and 3 were transferred to 14. Exon 1 is thought to be noncoding, but exons 2 and 3 encode for the *c-myc* protein. Because the gene protein can be produced by both types of transfers, the oncogenic activity of *c-myc* is thought to occur because its control over cell growth is deregulated when it is transferred to chromosome 14 next to the *Ig* locus. The result is a proliferation of cells to form a tumor.

9.6 CHROMOSOME REARRANGEMENTS IN NATURAL POPULATIONS

Inversions and reciprocal translocations arise in natural populations as a result of external or internal chromosome-breaking agents. Although the frequency of these aberrations is generally low, populations of some species have noticeably higher frequencies, which may be due to some localized condition. For instance, when transposable elements (*see* Chapter 7, Section 7.5), which seem to be almost universal in their occurrence, are activated under stress, some of them cause chromosome breakage and, as a result, a burst of aberrations.

Changes in chromosome-arm ratios by pericentric inversions or reciprocal translocations can be used, along with other attributes, to distinguish different populations. Although paracentric inversions do not change arm ratios, they can change the morphology of a chromosome arm with a nucleolus organizer. If breakpoints in the satellite region and in the other part of the same arm are not equidistant from the secondary constriction, changes in length of these two segments are detectable in mitotic chromosomes if a standard karyotype is available for comparison.

The permanency of inversions or translocations depends on whether they improve the fitness of natural populations in relation to their environment. A major disadvantage of heterozygous rearrangements in sexually reproducing populations is the partial sterility, caused by unbalanced gametes, which makes individuals with a heterozygous aberration less fit than those with either a homozygous aberration or standard chromosomes. This is an unusual situation, called underdominance, as opposed to overdominance, where the heterozygote is more fit than either homozygote. Underdominance can be averted if the chromosome aberrations behave in specific ways during meiosis. For example, heterozygous translocations with no crossing over in interstitial segments and alternate segregation of the chromosomes from the ring or chain at anaphase I will not cause any reduction in fertility (*see* Section 9.3). The sterility produced by heterozygous inversions results from crossing over within or near the inverted regions; therefore, conditions that prevent crossing over in these regions such as lack of pairing or nonhomologous, straight (as opposed to loop) pairing will prevent sterility (*see* Section 9.2). This behavior is expected with short inversions because they may not be able to form the pachytene loop, which brings homologous parts together. For longer inversions, which usually have homologous pairing, one of the following three conditions that prevent sterility must exist if the inversions are to persist in populations:

1. The inversions occur in chromosome regions that have no chiasma formations.
2. The heterozygous inversions cause a redistribution of chiasmata to regions outside the inverted segments.
3. Nonhomologous, straight pairing within the heterozygous inversions occurs with a high frequency.

There is evidence that all of these conditions exist in grasshopper (order Orthoptera) populations, where the type of pairing can be observed exceptionally clearly at pachytene.

In *D. melanogaster* populations, pericentric inversions are rare compared to paracentric inversions because of the difference in their meiotic behavior in females—there is no crossing over in males. Heterozygous paracentric inversions are associated with high fertility and lack of underdominance, because the meiotic anaphase bridges, which consist of the crossover chromatids with deletions and duplications, are excluded from the egg cells. On the other hand, heterozygous pericentric inversions are usually underdominant because crossover chromatids do not form bridges, so they can be included in egg cells and cause sterility. However, a large pericentric inversion, including almost half of chromosome 2, lacked underdominance for fertility, and tests for recombination of gene markers showed that almost no crossing over occurred in the inverted region. Nonhomologous pairing of the inverted and standard segments was one possible explanation for the underdominance, but it could not be confirmed because of the poor quality of female meiotic preparations.

From a genetic standpoint, heterozygous inversions of both types can retain allelic arrangements of the genes between or adjacent to their breakpoints because of the absence or elimination of recombinations. In specific environments, these inversion-enforced linkages may provide adaptive advantages over standard chromosomes, in which the allelic group can be recombined by crossing over. New inversions that include a desirable linkage group have an advantage in becoming established as permanent heterozygotes in individuals and eventually in populations.

If heterozygous translocations are to preserve blocks of genes in populations, they must have certain characteristics, which are exemplified in the plant genus *Oenothera* ($2n = 14$). Translocations have occurred spontaneously in the evolution of this genus, but the number of chromosomes involved varies among species from none or a low number in the center of origin, Mexico or Central America, to the whole complement in the eastern two-thirds of North America. The larger rings are believed to have developed from intercrosses between species with smaller rings. Large rings or chains are usually oriented at metaphase I to give alternate segregation at anaphase I, which is facilitated because *Oenothera* chromosomes are quite uniform in size with median centromeres, and are therefore flexible in movement. The breaks that produced the translocations tended to occur in heterochromatin near the centromeres, making the exchanged pieces equal in length and thereby retaining uniformity in chromosome size. Crossing over does not occur in the interstitial segments because they are short and located in heterochromatin; therefore, there are no unbalanced gametes from crossing over in this region when followed by alternate segregation (*see* Section 9.3.1).

Oenothera species with large rings are self-pollinating and largely true-breeding for their characteristics, but when crosses are made between some of the species, the progenies segregate into two phenotypic groups, called Renner complexes after the German scientist who first described them. When the genetic and cytological aspects of the genus were compared, it was realized that the two Renner complexes of a species are due to the segregation of two chromosome groups from the large translocation rings during meiosis. The allelic arrangement in each complex is usually transmitted intact through the egg or sperm, which implies that there is very little recombination in the regions containing the genes that differentiate the two complexes. The ring or chain configurations are maintained by chiasmata in the distal regions of the chromosomes.

The anomaly that *Oenothera* species with two Renner complexes can be true-breeding is due to zygotic or gametophytic lethal conditions, which prevent the transmission of specific complexes or homozygous combinations depending on the species. *Oenothera lamarckiana*, a European species that apparently arose from hybridization between two introduced North American species, has 12 of its 14 chromosomes in a ring, and its two complexes are designated *gaudens* and *verlans*. Self-pollination gives 50 % nonviable seeds, but fertility is normal when it is crossed with some other species. The nonviable seeds from selfing contain two like complexes, either *gaudens.gaudens* or *velans.velans*, whereas the viable seeds have one dose of each complex. This difference is caused by at least one recessive, lethal gene, or possibly a deletion, at a different locus in each complex. If two doses of the same complex come together at fertilization, the lethal gene for that complex is homozygous and is expressed as a zygotic lethal. The *gaudens.velans* combination is viable because each complex has the dominant allele of the lethal gene in the other complex.

Mechanisms in other *Oenothera* species act during the gametophytic phase to prevent gametes with a particular Renner complex from participating in fertilization. On the male side, there are genes that kill or inactivate pollen containing one of the two complexes in a species. There may be incompatibility between the style and pollen tubes containing a specific complex, or competition between pollen tubes containing different complexes, with one type participating in fertilization. On the female side, the two Renner complexes are distributed in a 2:2 ratio to the linear tetrad of cells at the end of meiosis. Competition between two complexes results in the development of the embryo sac by the megaspore with the stronger complex, regardless of its position in the tetrad, thus preventing the transmission of the other complex. If one complex is not transmitted via pollen for one of the reasons mentioned above and the other complex is not transmitted through the egg, the perpetuation of the two-complex genotype is assured. These lethal effects on pollen, or on seeds in the case of zygotic lethals, do not reduce population size because *Oenothera* plants produce an abundance of pollen and seeds.

The survival of multichromosome translocations in *Oenothera* natural populations indicates that the genes preserved in the Renner complexes provide adaptive advantages through heterotic effects. As the large rings evolved, the incorporation of balanced lethal systems maintained heterozygosity and heterosis, and self-pollination replaced cross-pollination, thus ensuring that the genetic system would be transmitted intact. The disadvantage is that genetic variability through recombination or the expression of recessive mutations is minimized and may affect the future evolutionary development of the genus.

Permanent, heterozygous translocations including all or most of the chromosomes occur in other genera of the Onagraceae family, to which *Oenothera* belongs, and in genera of several other plant families. In animals, they have been reported in two scorpion genera, *Tityus* and *Isometrus*, and in isolated populations of the American cockroach, *Periplaneta americana*. Characteristics of the *Oenothera* translocation system are established or still evolving in these other cases, but with some variations. For example, in *Rhoeo spathacea* (syn. *discolor*), $2n = 12$, with all the chromosomes in the translocation complex (Fig. 9.33), the large rings or chains separate into smaller chains more frequently and have more nondisjunction due to adjacent segregation of two or more chromosomes. *Rhoeo* chromosomes are not as uniform in length or centromere placement as the *Oenothera* chromosomes, probably due to unequal exchanges of segments. Nonetheless, *Rhoeo* apparently has enough of the requirements to maintain permanent heterozygosity.

The association of translocations with sex determination has been found in dioecious species of the genus *Viscum* (East African mistletoes). In *V. fischeri*, females ($2n = 22$) have only bivalent formations at meiosis and $n = 11$ in the gametes, but males ($2n = 23$) have 7 bivalents and a chain-of-9 which segregates 5:4 to give gametes that are $n = 12$ or $n = 11$. Union of each kind of male gamete with a female gamete ensures that dioecism will be perpetuated.

9.7 SOME SPECIFIC USES OF RECIPROCAL TRANSLOCATIONS

9.7.1 Source of Stable Duplications

The rationale in using translocations to produce stable duplications is to increase the dosage of small chromosome segments containing desirable genes. The method has been applied to *Drosophila*, maize, and barley. The initial requirement is to have available a number of translocations involving the same two nonhomologous chromosomes, with the physical positions of the breakpoints accurately determined. It is an advantage if one or both of the translocated chromosomes carry genes for desirable traits. The translocation lines selected for crosses must have the breakpoints in the same arms of both chromosomes (Fig. 9.34). If the breakpoints of the two translocations are at about the same point in one arm, and at two points on opposite sides of a pertinent locus (e.g., locus 7 in Fig. 9.34a) in the other arm, one type of resulting F_1 gamete has a duplication for the locus. If fertilization brings together two gametes with the same chromosome constitution, the progeny has four doses of the locus. It is also possible to have breakpoints on opposite sides of desirable loci in both arms so that some gametes can have a duplication for both loci.

The effects of this localized increase in gene dosage can vary, with some duplications producing effects on viability and fertility. However, barley lines with duplications for segments of two chromosomes gave higher grain yields than their translocation parents. In species with a large supply of appropriate translocations, it is worthwhile to test many duplicated segments for their effects on traits of importance in plant breeding. When genes are located near the centromere regions, they may be difficult to detect by recombination data, because crossing over is often absent or severely reduced in these regions. If duplications can be obtained for these regions by choosing translocations with breakpoints near the centromeres, the duplicated genes may be revealed by their dosage effects.

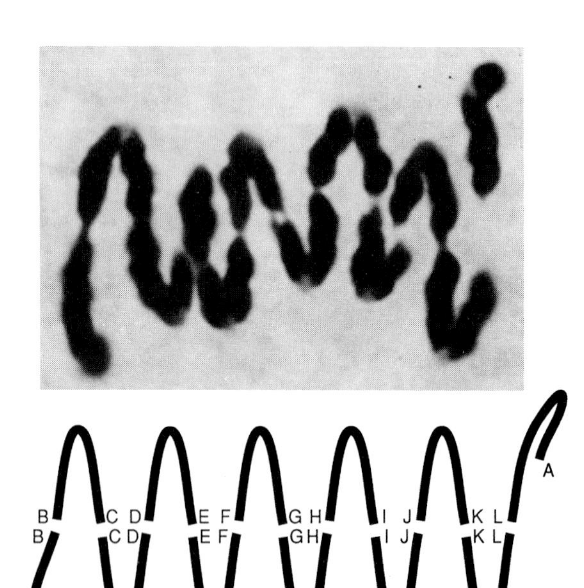

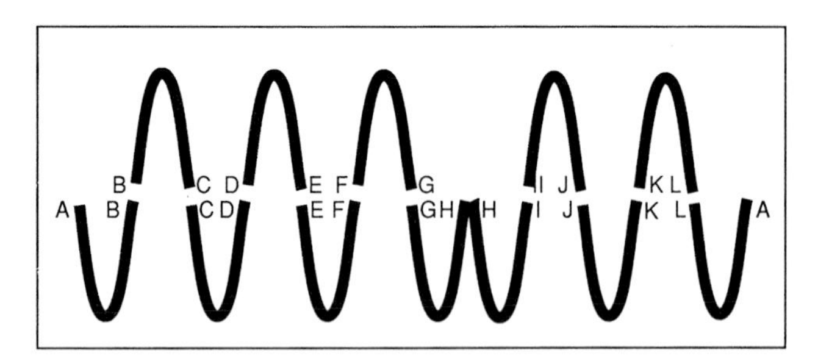

Fig. 9.33. A translocation ring-of-12 in *Rhoeo spathacea* (Sybenga, 1992). Photomicrograph of a metaphase I cell (top) showing alternate orientation of all the chromosomes in the chain. Some of the centromeres are visible as light areas. The two chromosomes on the right side are connected by a thin thread. Diagrammatic representations of (center) an alternate, balanced 6:6 segregation, and (bottom) adjacent segregation of the GH and HI chromosomes, to give a 7:5 segregation. The ends of each chromosome are designated by different letters.

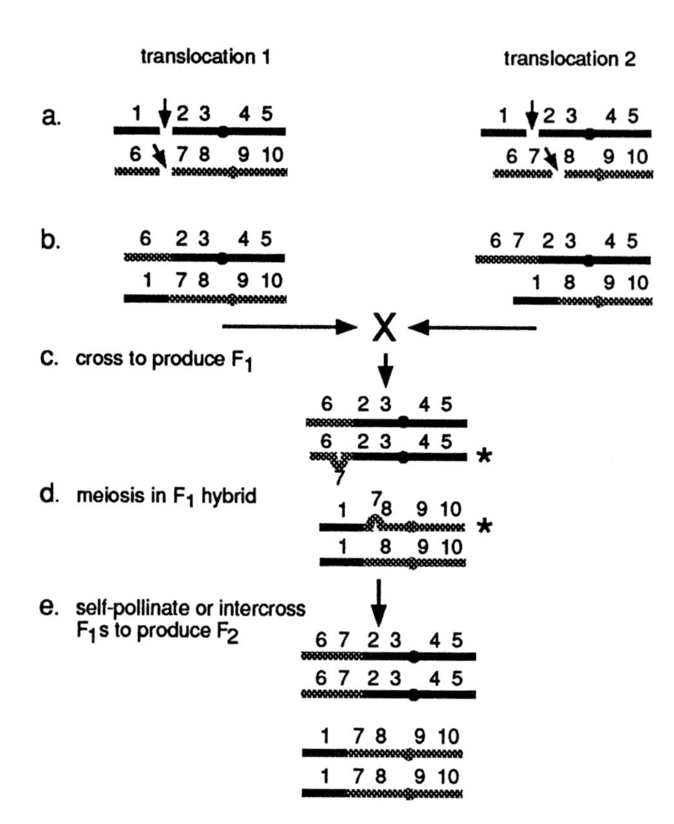

Fig. 9.34. The development of stable duplications of chromosome segments using reciprocal translocations (Hagberg and Hagberg, 1991). (a) Arrows mark the breakpoints between regions 1 and 2 of one chromosome in both translocations, and on opposite sides of region 7 in the other chromosome. (b) Gametes from the two translocations made homozygous after exchange of segments. (c) A cross between the two translocation lines. (d) Meiotic pairing in the F_1 hybrid. Region 7 is shown as a loop so that homologous pairing can occur. If the two chromosomes marked by asterisks enter the same two gametes, region 7 is effectively duplicated through its presence on two different chromosome pairs (e).

9.7.2 Use of Translocations to Control Insect Pests

Chemical insecticides have played a dominant role in controlling insect pests, particularly those that attack crop species, or those that carry organisms causing animal or human diseases. Concern for the possible harmful effects of some chemicals on the environment led to efforts to find safer methods of control. These alternatives, grouped under biological control, have included use of microorganisms or insects that attack and destroy specific pests, introduction of genes for insect resistance into crops, and use of induced gene mutations or chromosome aberrations to introduce sterility into the pest population.

As early as 1940, the suggestion was made that reciprocal translocations might be used as a method of controlling insect pests because, when heterozygous, they could introduce partial sterility into a pest population and thereby reduce its numbers. Because the frequency of translocations in natural populations is usually quite low, clastogenic agents are used to induce breaks and exchanges between chromosomes. Different types of translocations and procedures have been tried in efforts to improve the efficacy of this approach to pest control. Extensive studies on the Australian-sheep blowfly (*Lucilia cuprina*) have proven the theoretical feasibility of this approach, but have also demonstrated the difficulty in maintaining males carrying translocated chromosomes in the field.

The release of sterile males, produced by irradiation, has been successfully applied in the control of agronomically important pests such as the screw-worm. The principle applied in this case is the ability of the released males to compete with males in the natural population and thus reduce the overall production of progeny. This procedure requires the release of large numbers of sterile males and is economically feasible only for a major pest. Current research is examining the use of viruses that infect the particular host and express antibodies to proteins involved in sex-cell development.

BIBLIOGRAPHY

General

BURNHAM, C.R. 1962. Discussions in Cytogenetics. Burgess Publ. Co., Minneapolis, MN.

CARLSON, W.R. 1977. The cytogenetics of corn. In: Corn and Corn Improvement (Sprague, G.F., ed.). American Society Agronomy, Madison, WI. pp. 225–304.

GARBER, E.D. 1977. Cytogenetics. In: Cell Biology: A Comprehensive Treatise, Vol. 1, Genetic Mechanisms of Cells (Goldstein, L., Prescott, D. M., eds.). Academic Press, New York. pp. 235–278.

GUPTA, P.K., GUPTA S.N. 1991. Cytogenetics of chromosome interchanges in plants. In: Chromosome Engineering in Plants: Genetics, Breeding, Evolution Part A (Gupta, P.K., Tsuchiya, T., eds.). Elsevier, New York. pp. 87–112.

LEVITAN, M. 1988. Textbook of Human Genetics. Oxford University Press, New York.

SCHULZ-SCHAEFFER, J. 1980. Cytogenetics: Plants, Animals, Humans. Springer-Verlag, Berlin.

SYBENGA, J. 1992. Cytogenetics in Plant Breeding. Monographs on Theoretical and Applied Genetics 17. Springer-Verlag, Berlin.

THERMAN, E., SUSMAN, M. 1993. Human Chromosomes: Structure, Behavior and Effects, 3rd ed. Springer-Verlag, New York.

Section 9.1

WALPOLE, I.R., MULCAHY, M.T. 1991. Tyrosinase positive albinism with familial 46, XY, t(2;4)(q31.2; q31.22) balanced translocation. J. Med. Genet. 28: 482–484.

Section 9.2

BRANDHAM, P.E. 1969. Inversion heterozygosity and sub-chromatid exchange in *Agave stricta*. Chromosoma 26: 270–286.

CARSON, H.L. 1946. The selective elimination of inversion dicentric chromatids during meiosis in the eggs of *Sciara impatiens*. Genetics 31: 95–113.

CHANDLEY, A.C., MCBEATH, S., SPEED, R.M., YORSTON, L., HARGREAVE, T.B. 1987. Pericentric inversion in human chromosome 1 and the risk for male sterility. J. Med. Genet. 24: 325–334.

MCCLINTOCK, B. 1931. Cytological observations of deficiencies involving known genes, translocations and an inversion in *Zea mays*. Missouri Agr. Exp. Sta. Res. Bull. 163: 1–30.

MCCLINTOCK, B. 1938. The fusion of broken ends of sister half-chromatids following chromatid breakage at meiotic anaphases. Missouri Agr. Exp. Sta. Res. Bull. 290: 1–48.

MULLER, H.J. 1916. The mechanism of crossing over. Am. Nat. 50: 193–434, with skips.

STURTEVANT, A.H. 1921. A case of rearrangement of genes in *Drosophila*. Proc. Natl. Acad. Sci. Wash. 7: 181–183.

STURTEVANT, A.H., BEADLE, G.W. 1936. The relations of inversions in the X chromosome of *Drosophila melanogaster* to crossing over and disjunction. Genetics 21: 554–604.

SYBENGA, J. 1975. Meiotic configurations: a source of information for estimating genetic parameters. Monographs on Theoretical and Applied Genetics 1. Springer-Verlag, Berlin.

Section 9.3

BIRCHLER, J.A. 1991. Chromosome manipulations in maize. In: Chromosome Engineering in Plants: Genetics, Breeding, Evolution Part A (Gupta, P.K., Tsuchiya, T., eds.). Elsevier, Amsterdam. pp. 531–559.

BURNHAM, C.R., STOUT, J.T. 1983. Linkage and spore abortion in chromosomal interchanges in *Datura stramonium* L. Megaspore competition? Am. Nat. 121: 385–394.

CALDECOTT, R. S., SMITH, L. 1952. A study of X-ray induced chromosomal aberrations in barley. Cytologia 17: 224–242.

CHANDLEY, A.C., SPEED, R.M., MCBEATH, S., HARGREAVE, T.B. 1986. A human 9;20 reciprocal translocation associated with male infertility analyzed at prophase and metaphase I of meiosis. Cytogenet. Cell Genet. 41: 145–153.

ELDRIDGE, F.E. 1985. Cytogenetics of Livestock. AVI Publishing Co. Westport, CT.

ENDRIZZI, J.E. 1974. Alternate-1 and alternate-2 disjunctions in heterozygous reciprocal translocations. Genetics 77: 55–60.

HAN, J.-Y., CHOO, K.H., SHAFFER, L.G. 1994. Molecular cytogenetic characterization of 17 rob (13q 14q) Robertsonian translocations by FISH, narrowing the region containing the breakpoints. Am. J. Human Genet. 55: 960–967.

JACOBS, P.A., HASSOLD, T.J., HENRY, A. PETTAY, D., TAKAESU, N. 1987. Trisomy 13 ascertained in a survey of spontaneous abortions. J. Med. Genet. 24: 721–724.

JOHN, B., FREEMAN M. 1975. Causes and consequences of Robertsonian exchange. Chromosoma 52: 123–136.

LONGLEY, A.E. 1961. Breakage points for four corn translocation series and other corn chromosome aberrations. U.S. Dept. Agr. Res. Serv. ARS-34-16: 1–40.

RAKHA, F.A., ROBERTSON, D.S. 1970. A new technique for the production of A–B translocations and their use in genetic analysis. Genetics 65: 223–240.

ROBERTS, P.A. 1976. The genetics of chromosome aberration. In:

The Genetics and Biology of *Drosophila,* Vol. 1a (Ashburner, M., Novitski, E., eds.). Academic Press, New York. pp. 67–184.

Robertson, W.R.B. 1916. Chromosome studies. I. Taxonomic relationships shown in chromosomes of Tettigidae and other subfamilies of the Acrididae: V-shaped chromosomes and their significance in Acrididae, Locustidae, and Cryllidae: chromosomes and variation. J. Morph. 27: 179–331.

Roman, H. 1947. Mitotic nondisjunction in the case of interchanges involving the "B" type chromosome in maize. Genetics 32: 391–409.

Sears, E.R. 1969. Transfer of genes from wild relatives to wheat. In: Genetics Lectures, Vol. 1 (Bogart, R., ed.). Oregon University Press, Corvallis. pp. 107–120.

Simpson, P.R., Newman, M-A., Davies, D.R., Ellis, T.H.N., Matthews, P.M., Lee, D. 1990. Identification of translocations in pea by *in situ* hybridization with chromosome-specific DNA probes. Genome 33: 745–749.

Sybenga, J. 1984. The taxonomy of multivalent orientation: six modes of alternate or one? Can. J. Genet. Cytol. 26: 389–392.

Sybenga, J., Rickards, G.K. 1987. The orientation of multivalents at meiotic metaphase I: a workshop report. Genome 29: 612–620.

Section 9.4

Henikoff, S. 1990. Position–effect variegation after 60 years. Trends Genet. 6: 422–426.

Henikoff, S., Jackson, J.M., Talbert, P.B. 1995. Distance and pairing on the *brown*^{Dominant} heterochromatic element in *Drosophila*. Genetics 140: 1007–1017.

Section 9.5

Dobrovic, A., Peters, G.B., Ford, J.H. 1991. Molecular analysis of the Philadelphia chromosome. Chromosoma 100: 479–486.

Haluska, F.G., Tsujimoto, Y., Croce, C.M. 1987. Oncogene activation by chromosome translocation in human malignancy. Annu. Rev. Genet. 21: 321–345.

Rabbitts, T.H. 1985. The *c-myc* proto-oncogene: involvement in chromosomal abnormalities. Trends Genet. 1: 327–331.

Rabbitts, T.H., Boehm, T., and Mengle-Gaw, L. 1988. Chromosomal abnormalities in lymphoid tumours: mechanism and role in tumour pathogenesis. Trends Genet. 4: 300–304.

Rosson, D., Premkumar Reddy, E. 1988. Activation of the *abl* oncogene and its involvement in chromosomal translocations in human leukemia. Mut. Res. 195: 231–243.

Rowley, J.D. 1980. Chromosome abnormalities in human leukemia. Annu. Rev. Genet. 14: 17–39.

Stites, D.P., Terr, A.I. 1991. Basic and Clinical Immunology, 7th ed. Appleton and Lange, Norwalk, CT.

Section 9.6

Coyne, J. A., Aulard, S., Berry, A. 1991. Lack of underdominance in a naturally occurring pericentric inversiom in *Drosophila melanogaster* and its implications for chromosome evolution. Genetics 129: 791–802.

Harte, C. 1994. *Oenothera*: Contributions of a Plant to Biology. Monographs on Theoretical and Applied Genetics 20. Springer-Verlag, Berlin.

Hewitt, G.M. 1992. Population cytogenetics. Curr. Opinion Genet. Devel. 2: 844–849.

Wiens, D., Barlow, B.A. 1975. Permanent translocation heterozygosity and sex determination in East African mistletoe. Science 187: 1208–1209.

Section 9.7

Foster, G.G., Whitten, M.J. 1974. The development of genetic methods of controlling the Australian sheep blowfly, *Lucilia cuprina*. In: The Use of Genetics in Insect Control (Pal, R., Whitten, M.J., eds.). Elsevier/North Holland, Amsterdam. pp. 19–43.

Hagberg, A., Hagberg, P. 1991. Production and analysis of chromosome duplications in barley. In: Chromosome Engineering in Plants: Genetics, Breeding, Evolution Part A (Gupta, P.K., Tsuchiya, T., eds.). Elsevier, Amsterdam. pp. 401–410.

Robinson, A.S. 1976. Progress in the use of chromosomal translocations for the control of insect pests. Biol. Rev. 51: 1–24.

IV
Chromosome Polyploidy, Aneuploidy, and Haploidy

10

Multiples of Basic Chromosome Numbers—Polyploidy

- Polyploids have more than two basic chromosome sets and they occur mainly in plants.
- A major cause of spontaneous polyploidy in plants is unreduced gametes resulting from meiotic irregularities.
- Doubling chromosome numbers in somatic cells by means of colchicine is the universal method for inducing polyploidy in plants.
- The phenotypic consequences of polyploidy result from increased dosages of genes in each cell of all or some of the tissues of an organism, and they vary depending on whether the organism is autopolyploid or allopolyploid.
- The diploidlike pairing in many natural allopolyploids is due to the presence of genes that prevent pairing between partially homologous chromosomes.
- Gene segregations in polyploids depend on the relationships between the constituent genomes, and the meiotic behavior of the homologous or partially homologous chromosomes.
- Genome analysis is based on quantitative analyses of meiotic pairing, genetic compensation of substituted chromosomes, and homoeologous relationships between genomes.

Polyploidy is a potent force in the evolution of many plant species and in a few animal groups such as some species of fishes and amphibians. After ways of doubling the chromosome number were discovered, induced polyploids have had a significant role in examining the ancestries and interrelationships of species, and they have been tested in plant breeding programs. This chapter includes sections on the origin, reproductive behavior, genetics, and evolutionary aspects of polyploids.

10.1 DESIGNATING PLOIDY LEVELS AND SOMATIC VERSUS GAMETIC CHROMOSOME NUMBERS

The single-genome state, symbolized by x, with one dose of each chromosome, is the baseline for comparing different ploidy levels. It also is used for the somatic condition of haploids that are derived from diploids and are called monoploids or monohaploids (*see* Chapter 12). Ploidy levels above diploidy are polyploids and are designated according to the number of genomes in the somatic cells [i.e., triploidy ($3x$), tetraploidy ($4x$), pentaploidy ($5x$), hexaploidy ($6x$), etc.]. Haploids derived from polyploids are called polyhaploids. All of these ploidy levels contain the basic genome, in the case of monohaploids, or an exact multiple of it; therefore, they are characterized as euploids.

Different symbols are used to indicate the chromosome number in the somatic or body cells ($2n$) and the postmeiotic phase including the gametes (n). An unreduced gamete has the somatic chromosome number and is also symbolized by $2n$. Although some scientists have interchanged the designations of x and n, we use the terms n and $2n$ to define the euploid chromosome numbers in gametes and somatic cells, respectively, at any ploidy level.

10.1.1 Diploidy

Diploidy can be represented by $2n = 2x$, usually involving two doses of the same genome, in the somatic cells and $n = x$ in the gametes. In most organisms, it is the minimum ploidy level that provides normal functioning throughout the life of an individual. The presence of two homologous chromosomes of each type in the somatic cells during the active growth stages ensures cytological and genetical stabilities. Gene segregations are more straightforward in diploids than in polyploids that have most or all of their genes in doses above two. Diploidy is the normal ploidy level throughout much of the life cycle in humans, higher forms of animals, and some plants.

10.1.2 Polyploid Types in Relation to Genome Composition

Polyploids are classified both by the number of genomes and the genetic composition of the genomes (Fig. 10.1). If the same genome is multiplied three or more times, the condition is autopolyploidy (sometimes shortened to autoploidy). A $3x$ condition is an autotriploid, a $4x$ condition is an autotetraploid, and so on. If the basic genome of a species is designated by the letter A, an autotriploid is AAA and an autotetraploid is AAAA. Allopolyploidy (also called

a. Autopolyploidy

species 1 diploid (2x), AA —doubling→ autotetraploid (4x) AAAA x diploid (2x) AA
↓
autotriploid (3x) AAA

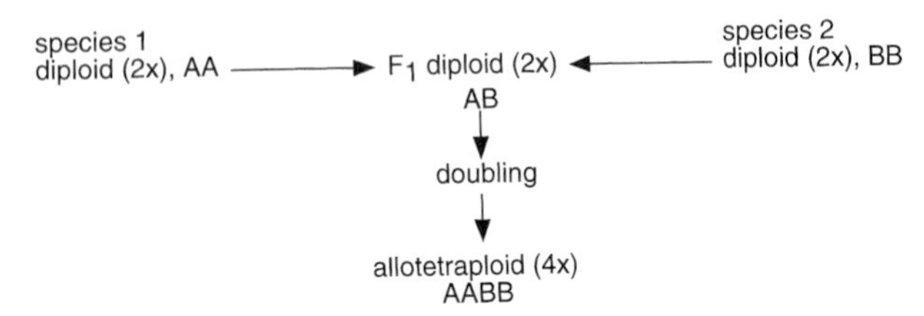

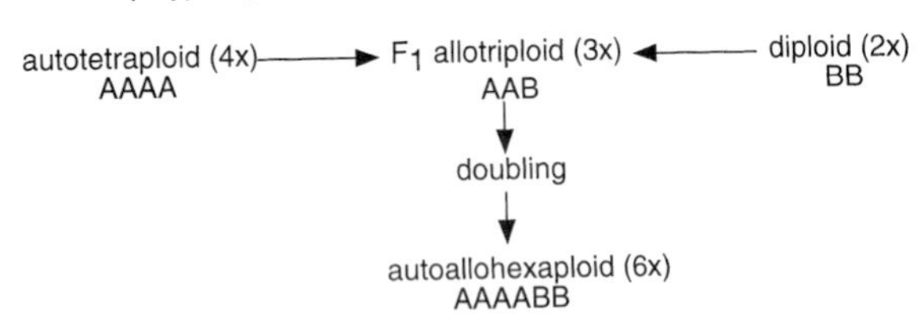

Fig. 10.1. The origins and genomic constitutions (capital letters) of different types of polyploids. In allopolyploids, the genomes from different parental species are often genetically related, but they are distinguished by different letters in this figure. The allotetraploid produced under (b) can also be called an amphidiploid.

alloploidy) arises from crosses between species or genera and includes at least two genomes from different sources. Allopolyploids from these sources have a broad range of variation in the amount of chromosome pairing and gene duplication between genomes from the parental sources. This variation is caused by several conditions, including the relationship between the parents and the evolutionary status of the allopolyploids. A former classification included true or genome allopolyploids and segmental allopolyploids based on the degree of relationship of the component genomes. Because this difference becomes indistinct at the level of DNA sequences, where all allopolyploids have some degree of homology, only the general term allopolyploidy is used in this book. Combinations of autopolyploidy and allopolyploidy can occur in plants above the tetraploid level either naturally or by experimental methods. The mixture is termed autoallopolyploidy (also autoalloploidy).

Another condition relating to allopolyploidy is amphiploidy (*amphi* is the Greek word for double), which is more often applied to induced than to spontaneous allopolyploidy. The usual experimental procedure is to cross the parental species or genera, then double the chromosome number in the highly sterile hybrids to produce the amphiploid and regain fertility. Amphiploidy applies to any level of allopolyploidy, but if the doubled hybrid is at the tetraploid level, the term amphidiploid (double diploid) is appropriate (*see* Fig. 10.1b).

10.2 POLYPLOIDY OCCURS IN WHOLE ORGANISMS OR PARTS OF ORGANISMS

Before considering the distribution of polyploidy among different types of organisms, we need to distinguish between whole-organism polyploidy and cell polyploidy, which involves certain tissues or organs. Whole-organism polyploidy commonly arises from sexual reproduction when unreduced gametes from one or both parents come together in the zygote. Cell polyploidy arises in the somatic cells at some stage in the life cycle of an individual and usually involves a small portion of the body, such as a specific organ or type of tissue. Cell polyploidy also occurs in cancer cells and in cells cultured *in vitro*.

Whenever a cell changes from the diploid to the polyploid state, adjustments must be made in cell metabolism, as well as in chromosome and gene interactions. Polyploidy is more likely to put a severe strain on cells that are dividing during the growth phase of an organism. Therefore, under natural conditions in a diploid organism, actively dividing cells are diploid. In the later stages of the life cycle, certain tissues or organs, in which cell division has ceased and differentiation is occurring, may change to a polyploid state. However, if the same organ or tissue is compared in different organisms, there is not always the same degree of ploidy. The predominant liver cells (hepatocytes) are mostly polyploid in adult mice and rats, but mostly diploid in humans and guinea pigs. When polyploidy does occur in a tissue or organ, it implies that the multiplication of chromosomes and their genes is needed for the differentiation or functioning of a specific part of the body. For example, in the Mediterranean flour moth, *Ephestia kühniella*, the sizes and locations of the wing scales relate to the ploidy level of the epidermal cells from which the scales project (Fig. 10.2).

Whole-organism polyploidy is more prevalent in the plant kingdom than in the animal kingdom, but there are too few cytological studies in some animal groups, such as amphibians and reptiles, for the true incidence to be known. Although polyploidy is the natural condition in from one-third to one-half of higher plant species for which chromosome numbers are available, it is largely confined to the angiosperms and is rare in the gymnosperms. Polyploidy is prevalent in ferns and mosses, but not in liverworts. It is found in most groups of algae and probably is more prevalent in fungi than the studies indicate because of the difficulty of getting accurate chromosome counts.

Although the normal condition in mammalian species, including humans, is diploidy, whole-organism polyploidy has the potential to occur. Triploidy and tetraploidy have been detected in studies of cattle and pig blastocysts, and of spontaneously aborted human embryos (Fig. 10.3). However, there are no reports of live-born offspring with overall polyploidy in any of these organisms, indicat-

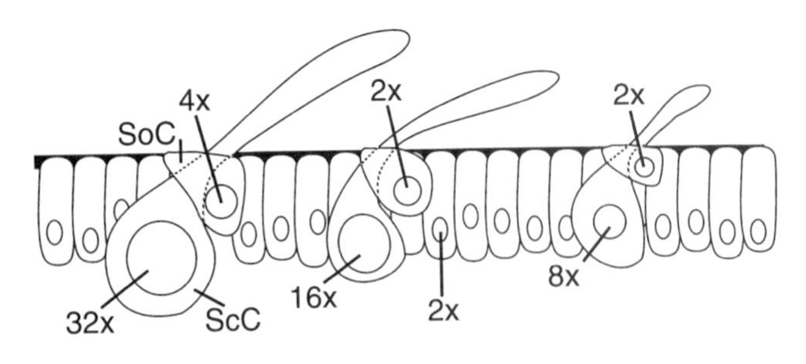

Fig. 10.2. A comparative diagram of the wing epidermis in the Mediterranean flour moth (*Ephestia kühniella*) to show the relationship between scale length and ploidy level of the scale-forming epidermal cells (ScC) and the socket cells (SoC). The uppermost and longest scale (left) is an outgrowth from a 32x cell, and the shortest scale (right) is from an 8x cell. The socket cells around the base of each scale are 4x with the longest scales and 2x with intermediate and short scales. The epidermal cells are 2x. (Redrawn from Brodsky and Uryvaeva, 1985.)

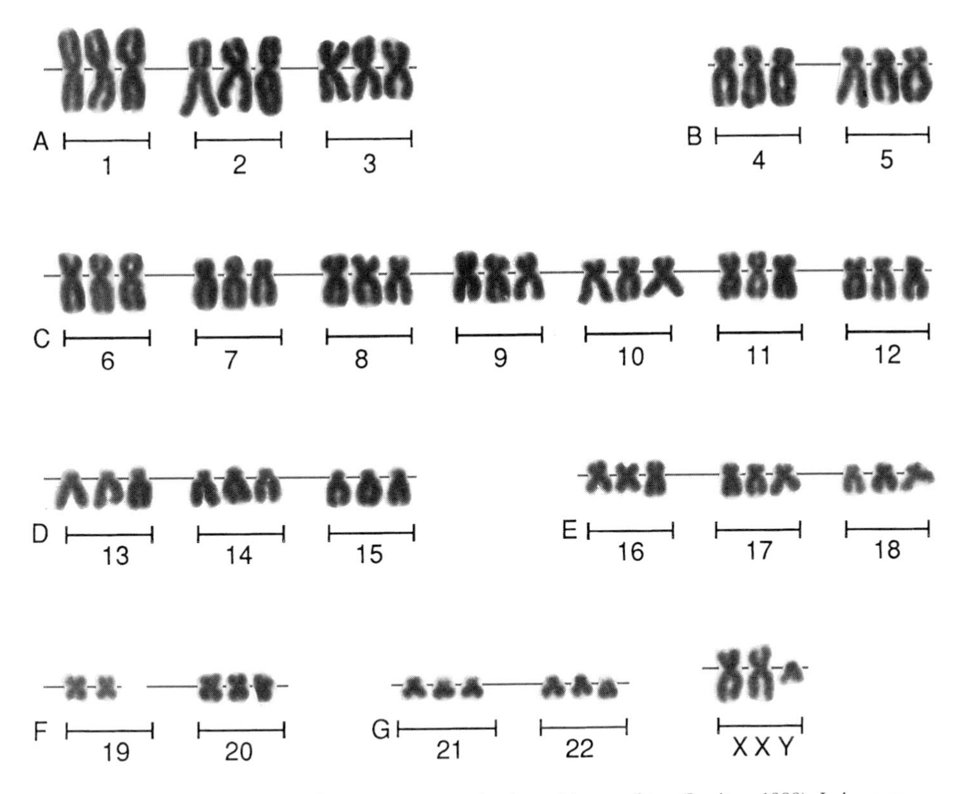

Fig. 10.3. A triploid karyotype from a spontaneously aborted human fetus (Levitan, 1988). In humans and other mammals, a mixture of polyploid and diploid cells is a normal condition in some tissues such as the liver, but it is not tolerated in the whole organism.

ing that the extra doses of chromosomes and genes cause lethality in the early embryonic stages. On the other hand, triploidy has been reported in adult chickens, and ploidy levels ranging from 3x to 8x occur in some fish species, and also in some amphibians and reptiles. Polyploidy is rare in insects and seems to be associated with a parthenogenetic mode of reproduction.

Explanations have been sought to explain why polyploidy is prevalent in plants but rare in animals. One suggestion is related to the fact that most animals are dioecious and have a Y chromosome with very little genetic activity, which is compensated by doubling of the transcription rate in the single X of males, or by inactivation of an X chromosome in females. Polyploidy disrupts the balance of X to autosomal gene products, normally maintained by dosage compensation, and is expected to result in lethality or sterility. On the other hand, most plants are not dioecious, and those that are, rarely have a degenerate sex chromosome.

10.3 WHOLE-ORGANISM POLYPLOIDY

Polyploid individuals arise spontaneously from diploids as a result of some cell-division error or the effect of some condition in the natural environment, and they are induced by applying a specific treatment or manipulation. The outcome is autopolyploidy or some type of allopolyploidy, depending on whether the diploid has one type of genome or two types as a result of hybridization. The timing of a polyploid-producing event in the life cycle determines whether the whole organism or only sectors become polyploid.

The spontaneous occurrence of whole-organism polyploidy due to chromosome doubling in somatic cells is thought to be a rare event. If the doubling takes place in the zygote, whole-organism polyploidy can result, but this causes death in the early embryonic stages of most higher animals, as indicated in the previous section. In a plant population, a newly polyploid individual has severe competition from the diploid majority and usually does not survive. Doubling in plant-meristematic cells results in sectors of polyploid tissue of varying sizes, depending on how early or late in the life cycle the doubling occurs. Unless the polyploid condition is carried into the reproductive organs, it will not be transmitted to the next generation.

The most likely natural cause of polyploidy in plants is the formation of unreduced ($2n$) gametes in a diploid individual. If a $2n$ gamete participates in fertilization by uniting with a $2n$ or n gamete, the zygote and all succeeding cells are polyploid. The advantage of $2n$ gametes over somatic doubling is that the doubling process occurs in mature plants after growth and differentiation have largely ceased. During seed development, there is time for the cells to adjust to the higher chromosome number, and the resulting offspring will be better able to compete with diploids, but certain gene combinations may give better adaptability in the polyploids. Some causes of unreduced gametes are as follows:

(a) Chromosome doubling in a premeiotic cell
(b) An incomplete type of first meiotic division known as first-division restitution
(c) An incomplete type of second meiotic division called second-division restitution
(d) Chromosome doubling after meiosis

(e) Development of a somatic cell in the ovule into a $2n$ gamete

Meiotic irregularities are the most frequent sources of unreduced gametes in plants. Normally, on the female side, a cell plate forms between the first and second meiotic divisions, and another plate at the end of meiosis, so that each of the four nuclear products is in a separate compartment. On the male side, the first cell division may or may not occur, depending on the species; if not, both cell divisions occur simultaneously at the end of meiosis. First-division restitution (FDR) means that the members of each homologous pair of chromosomes separate, but all the chromosomes stay within the membrane of one nucleus, called a restitution nucleus, at the end of the first meiotic division (Figs. 10.4 and 10.6). Second-division restitution (SDR) refers to an abnormal second meiotic division (Figs. 10.5 and 10.6). The amount of heterozygosity or homozygosity in the unreduced gametes depends on whether or not there is crossing over between genes and their centromeres, and whether there is first- or second-division restitution. With no crossing over between gene loci and their centromeres, all of the resulting gametes are heterozygous for these genes after FDR and all are homozygous after SDR. When crossing over occurs between a gene and its centromere, 50% of the gametes are heterozygous for that gene after FDR, but 100% are heterozygous after SDR (*see* the *B–b* gene pair in Fig. 10.6). Because heterozygosity is associated with heterosis, FDR is a desirable event in the absence of crossing over, but for genes farther from the centromeres, with crossover events more likely to occur, SDR events give maximum heterozygosity.

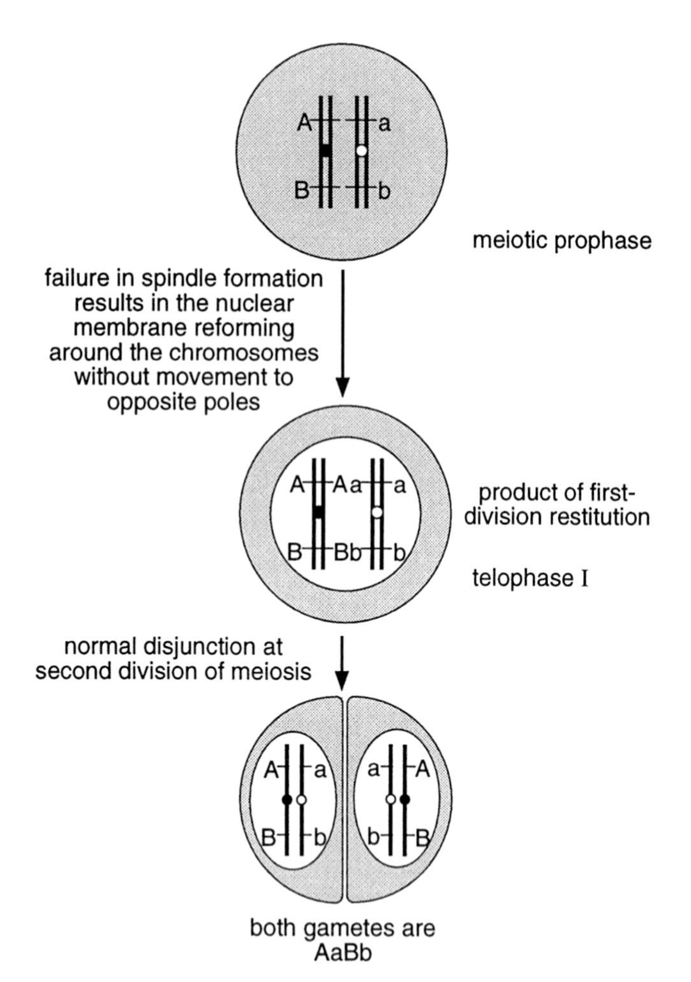

Fig. 10.4. A simplified diagram of meiosis to illustrate first-division restitution (FDR). Only one pair of chromosomes is shown with no crossing over between the centromere and the *A/a* or *B/b* loci. Although a cell plate does not develop between meiosis I and II, a plate does develop at the end of meiosis to give two diploid gametes (a dyad) in which nonsister centromeres are present and both loci are heterozygous. (Veilleux, 1985.)

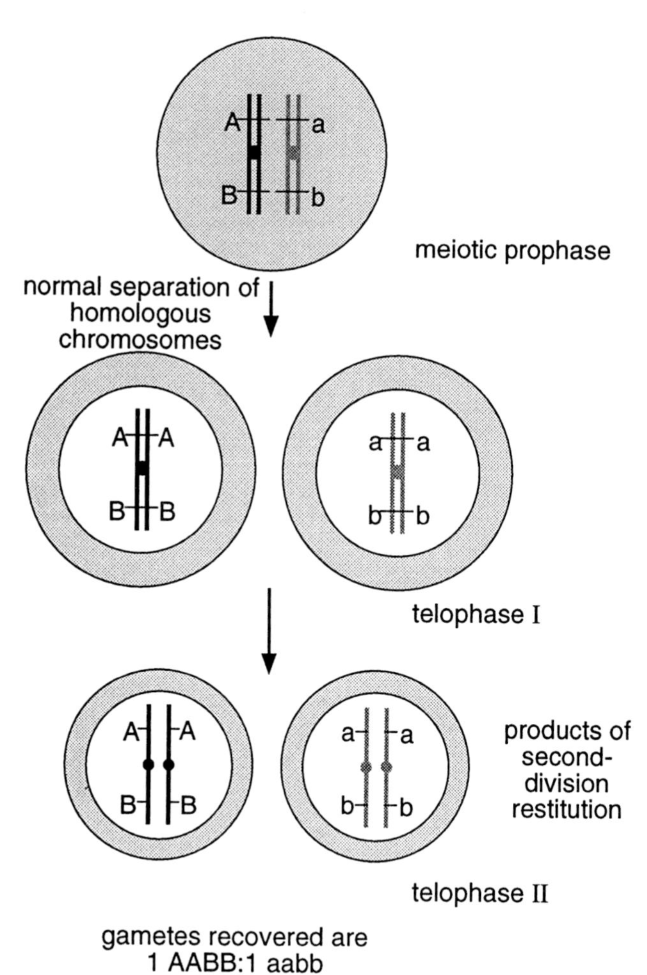

Fig. 10.5. A simplified diagram of meiosis illustrating second-division restitution (SDR). Again, only one pair of chromosomes is shown with no crossing over between the centromere and the *A/a* or *B/b* loci. In this case, meiosis I is normal and a cell plate forms between the two nuclei in the normal sequence. The separation of sister chromatids occurs during meiosis II, but no cell plate is formed at the end of meiosis, resulting in a dyad of 2n cells. In each cell, homologous chromosomes have sister centromeres although the chromatid arms distal to the *A/a* and *B/b* loci can be a mixture of crossover and noncrossover segments. The products of meiosis are 100% homozygous at both loci. (Veilleux, 1985).

The production of unreduced gametes is influenced by hybridization, the plant genotype, and environmental conditions. Hybrids from crosses between some species or genera have largely univalents at meiosis due to lack of intergenomic-chromosome pairing. FDR overcomes the problem of chromosome loss by keeping all the chromosomes in one nucleus at the end of meiosis I, and this results in dyads of functioning cells. Single-gene recessive mutations for lack of pairing (asynapsis) or premature separation of paired homologs (desynapsis, Fig. 10.7) occur in many plant species. They can cause irregular distributions of the chromosomes to the poles, and a high frequency of nonfunctioning male or female gametes. Therefore, the plants are prone to produce unreduced gametes by first-division segregation in order to overcome the high sterility.

In the microsporocytes of some plant species, a cell plate is not formed between the two meiotic divisions; instead, cell plates develop simultaneously in two directions at the end of meiosis II to produce a tetrad of reduced microspores. One condition that produces a dyad of unreduced microspores in these species is premature cytokinesis, the formation of a cell plate at telophase I or prophase II. Premature cytokinesis may be due to gene mutations or environmental conditions such as certain temperatures. Other genetic conditions with potentials for producing unreduced male gametes involve variations in spindle orientations during meiosis II. Normally, the two spindles are at right angles to each other or are positioned so that the four poles form a tetrahedron (Fig. 10.8a). Both fused (Fig. 10.8b) and parallel (Fig. 10.8c) spindles have been observed in potato (*Solanum*) species and hybrids, and alfalfa (*Medicago sativa*) haploids, which are at the diploid level. Cytokinesis occurs in only one plane at the end of meiosis, a dyad of unreduced microspores is produced, and the genetic effect is analogous to first-division segregation.

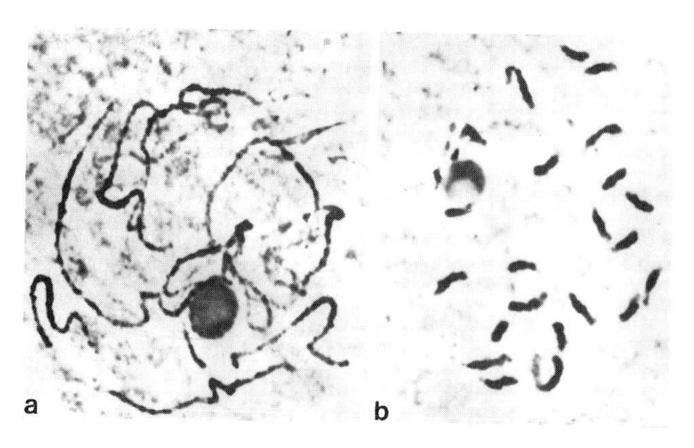

Fig. 10.7. Pachytene (a) and diakinesis (b) in a desynaptic mutant of diploid potato ($2x = 24$) (Ramanna, 1983). Although homologous chromosomes are paired at pachytene, most of the pairs separate prematurely and are univalents at diakinesis and later stages. Synaptic mutations, such as the one shown here, often stimulate first-division restitution, resulting in functioning $2n$ gametes rather than unbalanced gametes (Hermsen, 1984).

Parallel or fused spindles at meiosis II are not possible in megasporocytes because there is a linear arrangement of the cells after both meiotic divisions. However, asynaptic and desynaptic mutations, mentioned above, can produce unreduced female gametes by failure of cell division after the first meiotic division. Other mutations also can disturb the meiotic process such as the maize gene *elongate (el)*, which causes the chromosomes to uncoil (hence the name) during meiosis in the ovules. Only one of the two meiotic divisions takes place, resulting in a dyad of unreduced megaspores and, eventually, unreduced eggs. This gene has no effect on microsporogenesis.

Autotetraploids can arise from the spontaneous occurrence in diploids of unreduced gametes with two identical genomes. If union occurs between male and female unreduced gametes produced by the same plant, or by different plants with the same genome, autotetraploids can be produced directly. However, if an unreduced pollen grain originates from an irregular type of meiosis, it competes with the more numerous reduced pollen grains from normal meioses on the same plant. Therefore, a reduced sperm cell is more likely to fertilize an unreduced egg cell because of faster pollen-tube growth or a more favorable interaction with diploid-style tissue. The product of this mating is an autotriploid, which may occasionally produce an egg cell containing three identical genomes; this egg can unite with a reduced sperm cell, which contains the same genome from the triploid or a diploid, to give an autotetraploid.

Allotetraploids originate from crosses between diploid species with genomes from different sources, and the diploid hybrids are usually highly sterile due to lack of chromosome pairing. First-division restitution, followed by a normal second division, results in gametes containing one dose of each chromosome in the two genomes. Union of male and female gametes of this type restores a double dose of each chromosome and genome in an allotetraploid (*see* Fig. 10.1b).

Higher levels of polyploidy arise in nature by some of the same routes that produce tetraploids from diploids. For instance, natural crosses between an allotetraploid and a related diploid species with

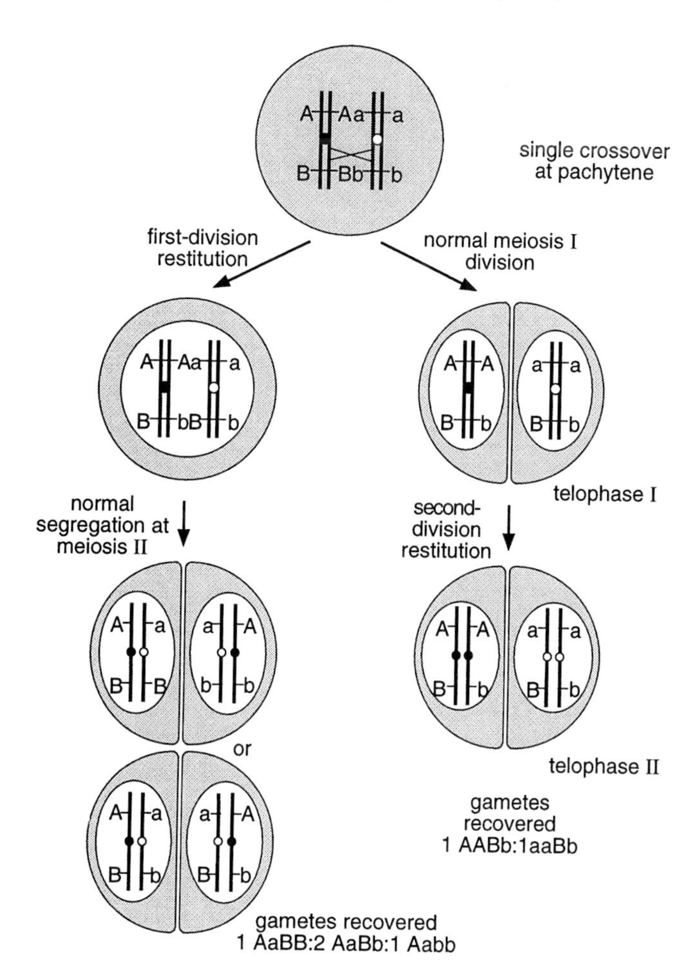

Fig. 10.6 Diagrams of meiotic stages, using one pair of chromosomes, to illustrate the results of FDR (left) and SDR (right) after a single crossover between the centromere and the *B/b* locus, but not between the centromere and the *A/a* locus. At the end of meiosis, the cells resulting from FDR are 100% *Aa*, 50% *Bb*, and 25% *BB* or *bb*. After SDR, they are 50% *AA* or *aa*, and 100% *Bb* (Veilleux, 1985). Both FDR and SDR are important causes of unreduced gametes in the potato (Peloquin et al., 1990).

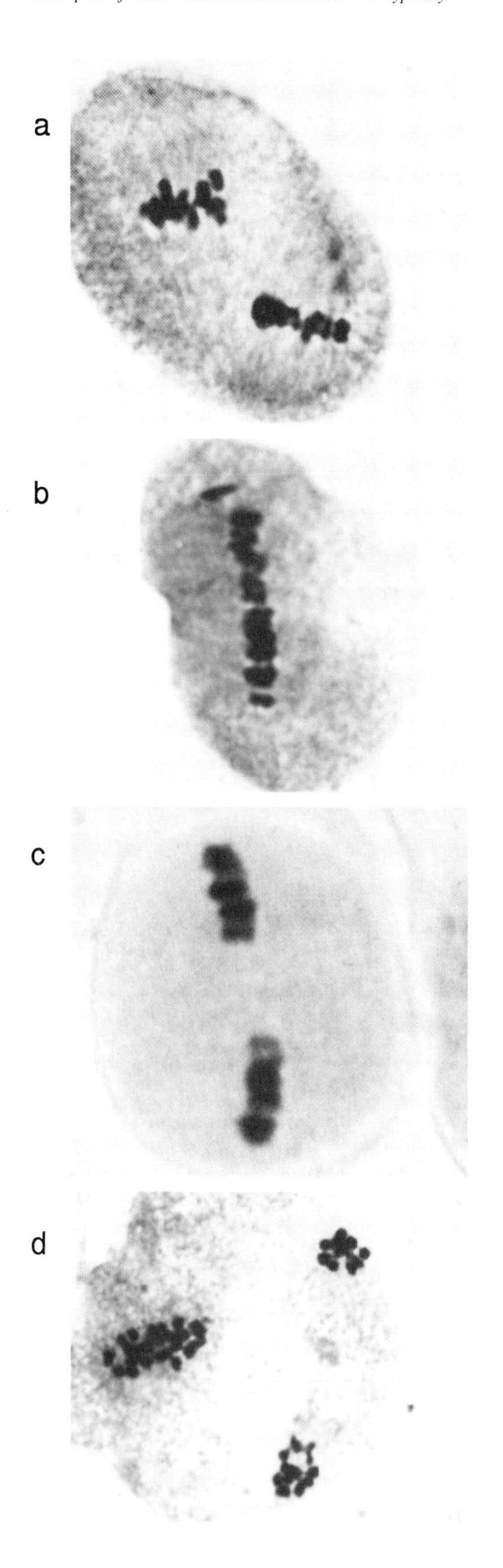

a third genome give allotriploid hybrids with a single dose of each genome and high sterility. Doubling of the triploid chromosome number somatically, or union of male and female gametes containing the three genomes, produces an allohexaploid with each genome in a double dose. Another possibility is doubling of the diploid chromosome number to give an autotetraploid, then introduction of a second genome from a diploid species, resulting in an autoallohexaploid (*see* Fig. 10.1c).

Polyploidy is associated with various forms of asexual reproduction in some genera or species of fishes, insects, reptiles, and plants. Polyploidy increases the dosage of gene loci, and thereby the chances for mutations that are favorable for the change from sexual to asexual reproduction. Hybridization increases the diversity of the gene pool by introducing heterozygosity. There is some evolutionary evidence in plant angiosperms that sexual polyploidy and hybridization precede apomixis (a general term used in plants to denote reproduction without fusion of gametes). However, in animals, hybridization and asexual reproduction likely would precede polyploidy. This difference between animals and flowering plants can be attributed to the more complex reproductive cycle in plants, so that more genetic changes must occur to bring about an asexual mode of reproduction. Both polyploidy and hybridization facilitate these changes, as pointed out above. In animals, the main type of asexual reproduction is parthenogenesis (i.e., the development of an unfertilized egg into an embryo and a subsequent individual). Only two genetic changes are needed in the oocyte to bring this about: elimination or circumvention of meiosis, and physiological or chemical changes in the oocyte to make it divide mitotically without stimulation from a sperm cell. A single-locus basis for parthenogenesis was reported in chickens, supporting this concept. If the asexual mode of reproduction can be established in diploid animals, it could be advantageous when polyploidy is introduced, by circumventing the sterility produced by meiotic irregularities in sexually reproducing polyploids.

Induced polyploidy results from various treatments or manipulations such as heat or cold shocks, which produce triploidy in fishes and amphibians. A temperature treatment applied to fish eggs in some cases suppresses the second meiotic division. As a result, the chromosomes that normally pass into the second polar body are retained in a restitution nucleus, and unreduced ova are produced. However, temperature treatments, particularly heat shocks, can lead to mortality and abnormalities. Heat shocks applied to plants around the time that the zygote is beginning to divide produce tetraploidy, apparently by preventing cell division. Temperature treatments in plants have been largely replaced by the alkaloid colchicine, as the most effective chromosome-doubling agent.

The discovery in the 1930s that colchicine (see Fig. 4.9, Chapter 4) could induce polyploidy led to a burst of experiments to study its effects in animals and plants. When colchicine is applied in a paste or solution to seeds or seedling meristems, it disorganizes

Fig. 10.8. Variations in spindle behavior during meiosis in potato (*Solanum* sp.). (a) Metaphase II showing two normal (nonparallel) spindles. (b) Fused spindles with the two chromosome groups together. (c) Parallel spindles at metaphase II. (d) An unreduced group of chromosomes at metaphase II and two reduced groups at anaphase II. (a, b, and d: Ramanna, 1974; c: Ramanna, 1979.)

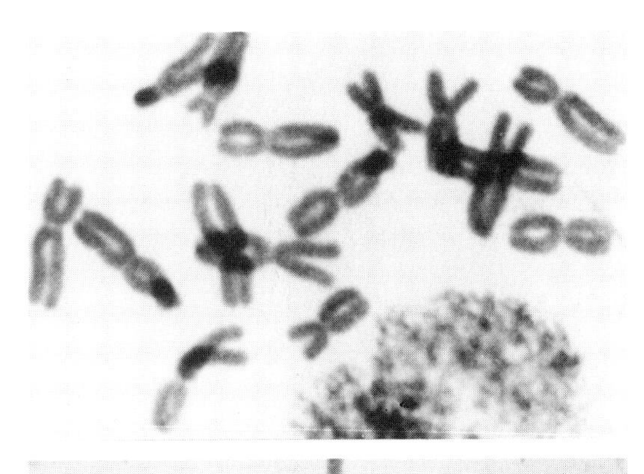

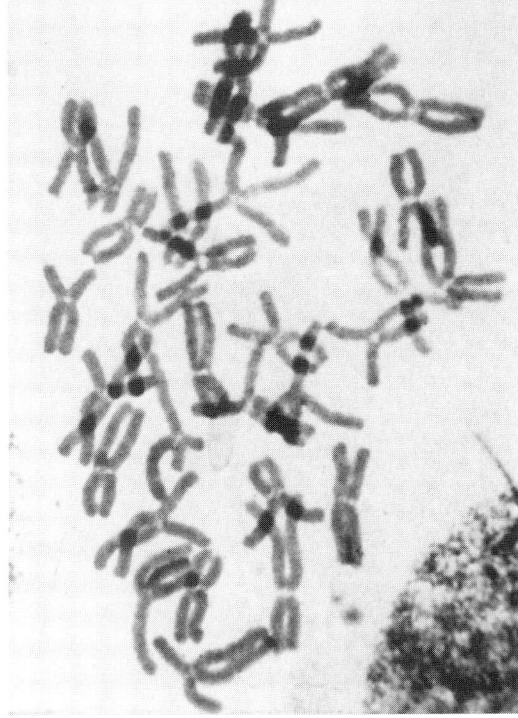

Fig. 10.9. The effects of colchicine on root-tip mitotic chromosomes of onion (*Allium cepa*, $2n = 2x = 16$) (O'Mara, 1939). **Top:** The 16 chromosomes showing lack of coiling and excessive contraction. They are not oriented in the cell because colchicine prevents spindle formation. **Bottom:** A near-tetraploid cell ($2n = 31$). This can result from the top cell if, after the chromosomes have separated, the sister chromatids remain in the same cell. In both cells, the sister chromatids are still attached at their centromeres. Some of the ends also show contact, but others are separated. These shapes are typical for colchicine-affected mitosis.

the spindles of cells that it contacts by binding to tubulin, the protein of the microtubules. Because colchicine prevents cell division but does not prevent chromosome or DNA replication, sister chromatids are retained in a restitution nucleus, which has a doubled chromosome number (Fig. 10.9). As long as colchicine remains in contact with the cell, the chromosomes continue to replicate without cell division, resulting in a higher level of polyploidy. Whenever colchicine is removed, the spindle function is restored in most plant species, and cell division proceeds normally. Colchicine has widespread effectiveness in higher plants and in some lower forms such as algae, but it is not generally effective in fungi. It has been applied to fertilized eggs or sperm of rabbits and pigs in attempts to induce polyploidy. Although there is evidence that chromosome doubling occurs, it is not possible to establish whole-organism polyploidy in these mammals.

Callus formation is another source of polyploidy in plants. In some species, particularly those in the family Solanaceae, a callus (outgrowth of cells) can form at the place where a stem is cut or injured. At least some of the cells in calluses have a doubled chromosome number, and a portion of the shoots from them are polyploid. Calluses also can form in tissue cultures of various plant parts if a callus-stimulating medium is used. If the plant parts involve somatic cells such as immature embryos, those callus cells that undergo chromosome doubling are polyploid.

Often, any one treatment does not act on its own, but interacts with another condition or treatment. The effectiveness of colchicine in doubling the chromosome number of grapes was enhanced by combining it with gibberellic acid. This combination was thought to increase the rate of cell division, making more cells available for colchicine action, and to cause cell elongation, making it easier for colchicine to enter the cells.

10.4 TISSUE OR ORGAN POLYPLOIDY

In diploid animals and plants, certain organs or tissues may become partially or entirely polyploid as a normal condition in the life cycle. This implies that there is some programmed irregularity in a mitotic cycle, resulting in polyploid cells. The most frequent type of irregularity is lack of cytokinesis (cell division) after a normal nuclear division. In a diploid organism, the result is a cell with tetraploidy distributed between two diploid nuclei (Fig. 10.10). Binucleate cells are a normal constituent of certain tissues in plants and animals; they are found in several parts of the mammalian body, including the liver (see also Fig. 4.16, Chapter 4), and particularly in some invertebrates. Another type of irregularity is failure of the later stages of mitosis due to lack of spindle formation or some other breakdown. Again, there is no cell division, and the cell has a single, tetraploid nucleus because the sister chromatids

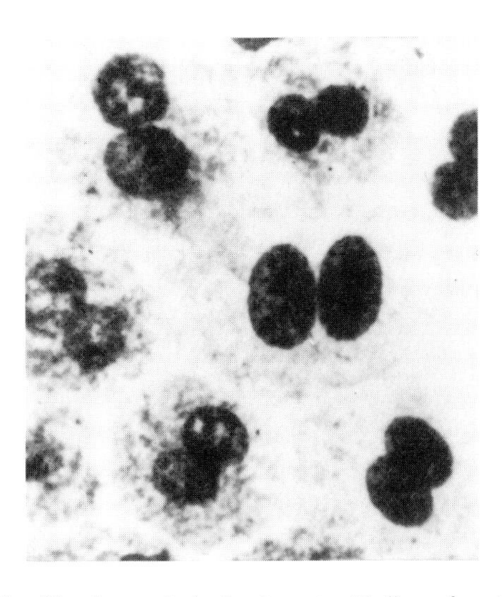

Fig. 10.10. Binucleate cells in the pigment epithelium of a rat retina stained with Feulgen (Brodsky and Uryvaeva, 1985.). The smaller paired nuclei are diploid, giving a tetraploid cell ($2x \times 2$), and the larger paired nuclei are tetraploid, giving octoploid cells ($4x \times 2$).

of each chromosome separate but cannot move to opposite poles. Failure of cytokinesis after the tetraploid nucleus divides results in octoploid cells with two tetraploid nuclei (*see* Fig. 10.10). Mononucleate, tetraploid cells can also be derived from binucleate, diploid cells by fusion of the two spindles and the two sets of chromosomes, with the formation of a single metaphase plate. Anaphase, telophase, and cell division proceed normally, resulting in two tetraploid cells with single nuclei. In certain plant and invertebrate cells, the two spindles of binucleate cells remain separate, but the two groups of daughter chromosomes at each pole are sufficiently close together during telophase that they are enclosed in one nuclear membrane, giving two mononucleate, tetraploid cells.

Endomitosis refers to situations where chromosomes replicate and sister chromatids separate without breakdown of the nuclear membrane. No nuclear or cell division occurs, giving a condition known as endopolyploidy. Repeated endomitoses in the same cell can give high ploidy levels, as has been found in differentiated tissues of certain insects, and in the tapetal layer of anthers in higher plants.

10.5 EFFECTS OF POLYPLOIDY ON PHENOTYPES AND INTERNAL TRAITS

The effects of polyploidy are most clearly detected in new, experimental polyploids, which can be compared with their diploid parents before they have had time to diverge because of mutations, or intergenomic recombination in the case of some allopolyploids. However, when diploid and neopolyploid (i.e., recent polyploid) races of the same plant species exist in natural populations, differences can be noted. Polyploidy of any type involves an increase in DNA amount per nucleus. When compared to the situation in diploids, gene dosage is increased at all loci in autopolyploids and at many loci in allopolyploids. The morphological, physiological, and chemical effects of polyploidy trace to these conditions within the cells.

The characteristics of polyploids have been determined mainly at the autotetraploid level. The initial effect of polyploidy is an increase in cell size to accommodate the greater chromosome number and DNA amount, but a decrease in the ratio of cell surface to cell volume. The increase in size of pollen and stomatal guard cells in plants is often used as an initial indicator of polyploidy (Fig. 10.11). The cell-division cycle is prolonged, resulting in a slower growth rate and later plant maturity (flowering and fruiting) than in diploids. In some plant species, the delayed development has been associated with decreases in growth-hormone levels in polyploid tissues and with lower rates of respiration. The final plant size may not be increased over that of the diploid because of fewer cell-division cycles, but there can be an increase in the size of flowers (*see* Fig. 10.11), fruits, and seeds, particularly in autotetraploids. Other typical traits of polyploids are less branching and changes in the shape and texture of leaves and flower parts. Polyploids have a higher water content, which may result from less transpiration, and a lower osmotic pressure except under low humidity. They are more sensitive to freezing temperatures because of their increased water content.

The activities of several enzymes involved in metabolic processes have been compared in tomato diploids and derived autotetraploids. For different enzymes, the activity increases, decreases, or remains unchanged in response to increased gene doses. Lack of consistency between gene dosage and enzyme activity could disturb metabolism or some other regulation mechanism and cause developmental abnormalities in the tetraploids.

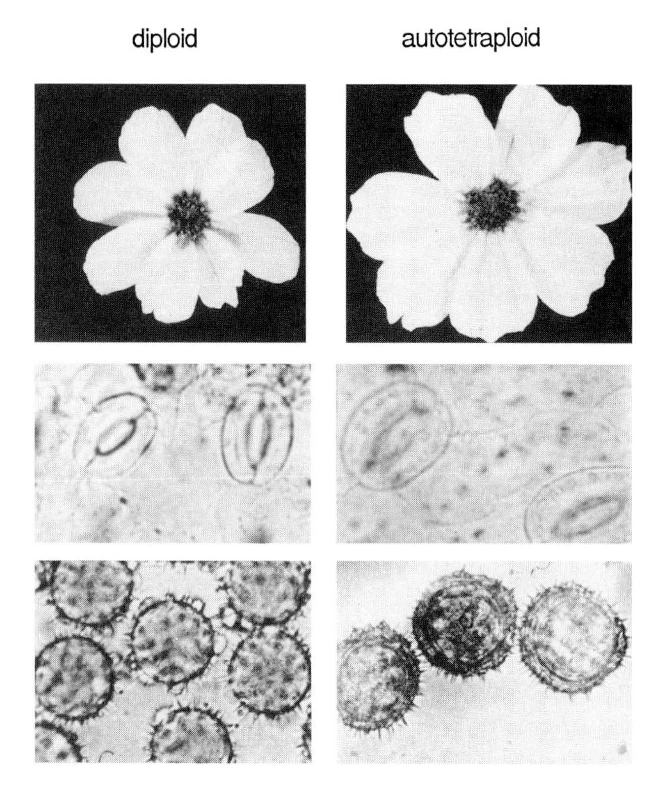

Fig. 10.11. Colchicine-induced autotetraploidy in *Cosmos bipinnatus* (Newcomer, 1941). From top to bottom: flower, stomata, and pollen grains of diploid (left) and autotetraploid (right). Induced autotetraploids usually have a mixture of favorable and unfavorable traits that must be weighed when considering their practical uses (Sybenga, 1992).

Polyploids contain higher concentrations of sugars, starch, and vitamin C than diploids, but lesser amounts of nitrogen and structural constituents such as cellulose. Studies on phenolic compounds and isozymes in polyploids have been useful in characterizing species or in helping to identify the parental species in natural polyploids, because different phenolic or isozyme patterns are often inherited as codominants. In such cases, patterns of two diploid parental species can appear in the F_1 hybrid or the derived allotetraploid. Sometimes, the parental patterns combine to produce a nonparental phenolic profile in the hybrid, making the parentage more difficult to detect. Some autopolyploids have nonparental phenolic or isozyme patterns when compared with the diploids from which they were derived. These differences between diploids and polyploids are explained by assuming that the pertinent genes are suppressed in the diploid parents, but chromosome doubling somehow restores their functions. The occurrence of nonparental patterns in autopolyploids indicates that this criterion must be combined with others when deciding whether a polyploid originated from interspecific hybridization.

10.6 MEIOTIC BEHAVIOR AND FERTILITY IN POLYPLOIDS

This section is based mainly on higher plants because of the availability of both natural and induced polyploids. In animals, viable polyploidy is an infrequent occurrence. By definition, autopolyploids should have identical multiples of each kind of chromosome, at least when they first arise. In both experimental and natural polyploids, structural changes over time can cause divergence among the homologous chromosomes of the multiplied genome, thereby affecting the kind of pairing during meiosis.

10.6.1 Meiotic Behavior of Autopolyploids

Autotetraploids have four doses of the same genome, which means that each chromosome in the genome is represented by four homologs. As synapsis is a specific two-by-two attraction between homologous chromosomes or segments, each group of four chromosomes cannot be paired as a unit throughout their lengths. Instead, there is pairing competition by segments within the group, and at least two chromosomes must change partners to maintain the two-by-two pairing relationship, and also the association of the four chromosomes as a quadrivalent (Fig. 10.12).

If chromosome arms pair at random, bivalents occur when there is pairing in both arms of two chromosomes instead of a switch in partners between one of these chromosomes and a third chromosome. Studies on induced autotetraploids in different plant species indicate that there are more bivalents and fewer quadrivalents at metaphase I than would be expected if partner exchange occurs without restrictions. There also is evidence that the proportion of bivalents to multivalents increases over a span of generations. The suggested explanations for the reduction in quadrivalents focus on the pairing mechanism during the zygotene and pachytene stages of meiosis and the locations of chiasmata.

In the silkworm, *Bombyx mori*, a study of synaptonemal complexes (SCs) in autotetraploid females, using the electron microscope, disclosed more quadrivalents at early synaptic stages than at later stages. This was explained by an adjustment in the SC so that it was extended between two homologous chromosomes and not between the other two homologs in the quadrivalent, which was resolved into two bivalents. However, if the pairing period was too short for the synaptic adjustment to occur, the frequency of quadrivalents remained high. Synaptic adjustment was observed in maize SCs but was not considered to be effective in reducing quadrivalent frequencies.

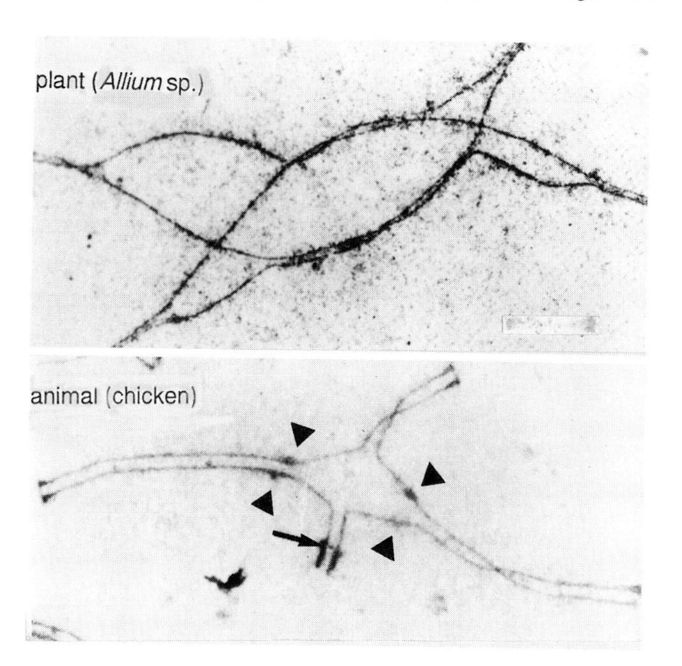

Fig. 10.12. Electron micrographs of quadrivalent formations in autotetraploid cells at pachytene. **Top:** A microsporocyte of tetraploid *Allium vineale*, with a quadrivalent of four chromosomes synapsed at the ends and in the middle (Loidl, 1986). There is lack of synapsis in two regions where the chromosomes change partners. **Bottom:** A tetraploid chicken oocyte showing a symmetrical, cross-shaped quadrivalent with non-pairing where the chromosomes change partners (Solari and Fechheimer, 1988.). The four centromeres are marked by arrowheads. The arrow marks telomeres that are folded back over the paired arms.

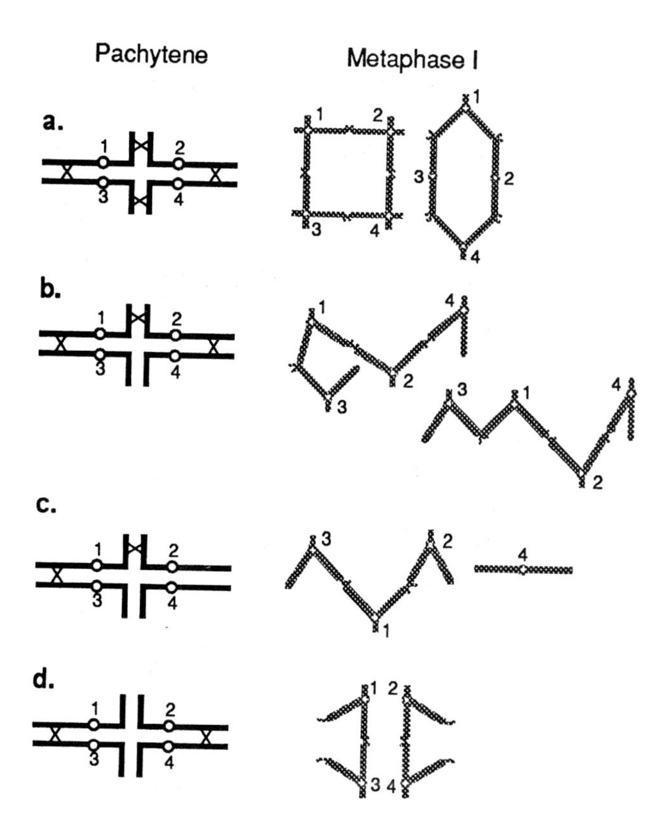

Fig. 10.13. Diagrams of quadrivalents (IVs) at pachytene in an autotetraploid (left) and some of the metaphase I configurations resulting from varying numbers of crossovers (right). Each chromosome is represented by one chromatid, and the centromeres are numbered; crossover sites are marked by X. (a) A crossover in each branch of a quadrivalent (left) gives closed IVs at metaphase I (right). The configuration on the left would be expected to give a 2:2 chromosome distribution because two chromosomes face each pole. In contrast, the configuration on the right will give a 1:1 distribution if the two nonoriented chromosomes lag behind, a 2:2 distribution if they go to opposite poles, a 1:2 distribution if one chromosome lags, and a 1:3 distribution if both go to the same pole. (b) A crossover in three of the arms will give an open quadrivalent. The left combination will give a 2:2 distribution; the right will give a 3:1 distribution as two adjacent chromosomes move to the same pole. (c) A crossover in two adjacent arms will give a trivalent (III) and a univalent. The distribution will be 2:1 if the univalent lags and is lost, 2:2 if it goes with chromosome 1, and 3:1 if it goes with 2 and 3. (d) A crossover in two nonadjacent branches will give two bivalents and a 2:2 segregation. (Sybenga, 1975.)

The number and locations of chiasmata in the pachytene configurations have a direct influence on whether quadrivalents or smaller configurations are seen at metaphase I (Fig. 10.13). Because the sites where the chromosomes change pairing partners are not fixed, as they are in reciprocal translocations, quadrivalents can vary in shape among cells depending on the locations of these sites. If partner exchange occurs near the ends of the arms, the

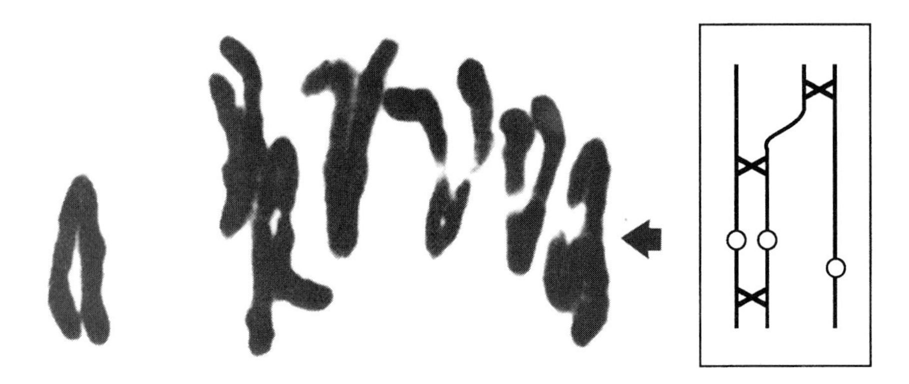

Fig. 10.14. **Left:** Photomicrograph of a metaphase I cell of an autotriploid *Triticum speltoides* ($3x = 21$). There are six V-shaped trivalents, and one "panhandle" trivalent (arrowhead) consisting of a closed bivalent and a third chromosome tied to one of the others after a switch of partners. (Photograph from Yen and Kimber, 1990). In the interpretive diagram for the trivalent (right), the two chromosomes on the left side form a closed bivalent structure after a crossover in each arm, then a partner switch in the upper arm and a second crossover ties the third chromosome like a panhandle to the other two. These complex trivalents are more likely to occur in autotriploids than in autotetraploids because after a crossover occurs in one arm between two chromosomes, the chances for that arm to pair and crossover with a third chromosome are greater when there is no fourth homolog competing for pairing, as is the case in tetraploids (Yen and Kimber, 1990). Although the photo shows only trivalents, bivalents and univalents also occur and must be included when quantifying pairing behavior (Jackson, 1991).

segments beyond an exchange site may be too short for crossing over and chiasmata to occur, and two rod-shaped bivalents result (*see* 10.13d). Another condition that can prevent chiasmata from forming is lack of pairing or nonhomologous pairing around the sites of partner exchanges. Chiasmata are sometimes localized in the centromere region such as in the natural autotetraploid *Allium porrum*, where partner switches must occur between chiasmata on opposite sides of the centromeres to retain quadrivalents. These events are rare, however, and, in their absence, quadrivalents are reduced to bivalents.

As the four chromosomes of each group in an autotetraploid are structurally identical, the gametes are balanced as long as there is a 2 : 2 distribution to the poles at anaphase I. Different metaphase I orientations that give balanced and unbalanced distributions are shown in Fig. 10.13. When these variations in behavior for one quadrivalent are extended to other quadrivalents that are present, the frequency of unbalanced gametes is increased. Because there are interactions among gene dosages on different chromosomes, changes in dosage caused by unequal chromosome distributions may produce nonfunctional gametes.

Autotriploids have three homologous chromosomes of each type, and their most common meiotic configurations are a trivalent (Fig. 10.14) or a bivalent plus univalent. The trivalents and univalents behave like those in autotetraploids during meiosis, but their higher frequencies in autotriploids increase the number of unbalanced gametes. The meiotic behavior of the trivalents that are associated with the trisomic states of individual chromosomes is given in Chapter 11 (Section 11.1).

With levels of induced autopolyploidy above tetraploidy, the configurations become more complex and cytological instability increases.

10.6.2 Fertility of Autopolyploids

When fertility is measured by seed set and/or normal pollen, it is usually lower in newly induced autotetraploids than in the parental diploids due to any one, or a combination, of cytological, genetical, and physiological conditions. Studies using different plant species have followed induced autotetraploids for a number of generations to detect a relationship between fertility and meiotic chromosome behavior. In maize, turnip (*Brassica campestris*), and grain amaranth (*Amaranthus* species), selection of plants with higher fertility over 10–19 generations was accompanied by a decrease in quadrivalents and an increase in bivalents. These results suggest that the decreased fertility in the newly induced autotetraploids was caused by the meiotic irregularities introduced by the multivalents and that the restoration of the diploid type of meiotic chromosome behavior, called diploidization, produced more functional gametes. There is support for this concept in some natural populations of plants and fishes, where inherent autopolyploids have bivalent pairing or are in the process of making this change. However, in some plant species, an increase in fertility is accompanied by no change, or an increase, in quadrivalent frequency, indicating that fertility is affected by genetical or physiological conditions.

Autotriploids typically have high sterility because of the preponderance of gametes with unbalanced chromosome numbers. Even if each group of three homologous chromosomes consistently forms a trivalent, there can be different orientations on the metaphase plate. For example, with each V-shaped trivalent (*see* Fig. 10.14), two chromosomes may be oriented toward either the upper pole or the lower pole, and the third chromosome to the opposite pole, and the gametes can have a range of chromosome numbers, with a few having one or two complete genomes. If bivalents plus univalents occur, the loss of univalents can contribute to this range. Gametes that deviate by more than one or two chromosomes from the reduced or unreduced euploid chromosome numbers are likely to be nonfunctional in plants because of the variation in dosage of different chromosomes.

Newly induced autopolyploids above tetraploids usually lack

the vigor and fertility to be maintained long enough to adapt to the higher level of ploidy. This is the situation with maize (*Zea mays*), which is normally diploid. On the other hand, plant species do exist in which high levels of autopolyploidy have evolved. In these cases, the plants are often perennials with some type of asexual propagation, so that seed propagation can be bypassed when high sterility is a problem. Ploidy levels up to 12 times the basic chromosome number occur in *Rubus* species (raspberries and blackberries) and in some groups of ferns (Pteridophytes).

10.6.3 Meiotic Behavior of Allopolyploids

As indicated in Section 10.1.2, all allopolyploids have at least two genomes from different sources in their background. The degree of genetic relationship between the genomes influences the kind of meiotic pairing. Allopolyploids that arise naturally or are successfully induced have some degree of relationship between genomes from different species, and they may have both bivalents and multivalents during meiosis, depending on the number and size of common chromosome segments. Those chromosomes having common segments across genomes, and the pairing of those segments, are called homoeologous. If intergenomic, homoeologous pairing and intragenomic, homologous pairing occur in the same nucleus, both multivalents and bivalents may be formed.

A number of naturally evolved allopolyploids have only bivalent pairing even though they have homoeologies between genomes. Genetic control of this pairing restriction was first discovered in common or bread wheat (*Triticum aestivum*), which is an allohexaploid with three related genomes. A major gene, symbolized *Ph1* for *pairing homoeologous*, restricts pairing to homologous chromosomes, with only bivalent formations in the hexaploid, and univalents in the haploids (Fig. 10.15a). The effect of this gene was detected by removing the chromosome or arm carrying the *Ph1* locus by cytogenetic manipulations, or by using X-rays or a chemical mutagen to delete or mutate the locus. In these situations, multivalents as well as bivalents were observed at meiosis in the hexaploids and haploids, with some univalents still occurring in the haploids (Fig. 10.15b). When a wild, diploid relative of wheat, *Triticum speltoides*, is crossed with common wheat, the action of *Ph1* is suppressed in the hybrid by the *speltoides* genome, and intergenomic pairing occurs.

Ph-like genes are thought to be widespread in plants and possibly in some amphibians, so there has been considerable interest in finding out how the *Ph1* gene functions in wheat. Characterization of the *Ph1* mutations presently available indicates that they are all deletions, and DNA sequences located in these deleted regions have been identified. Because point mutations for this important gene are not available, this limits detailed studies on its mechanism of action. Although a molecular explanation has not been established, two broad categories are possible.

1. The product of the *Ph1* gene may stimulate nuclear-attachment sites for the chromosome ends of homologous genomes so that pairing can occur. The homoeologous genomes would not be acted on in this way and would remain far enough apart so that pairing does not occur.
2. The product of the *Ph1* gene could modulate the fine tuning of the time in meiotic prophase when crossing over occurs. Early in this stage, homoeologous and nonhomologous, as well as homologous, associations of chromosomes can be readily observed, using three-dimensional reconstructions of sectioned nuclei analyzed by electron microscopy. As the cells progress to pachytene, all except the homologous associations disappear. If the *Ph1* gene product delays recombinational activity until well into pachytene, only homologous crossing over would occur and would result in only bivalents later in meiosis.

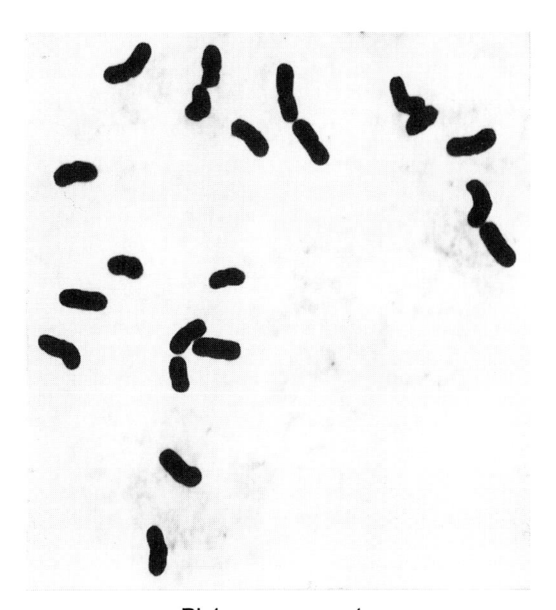

Ph1 gene present

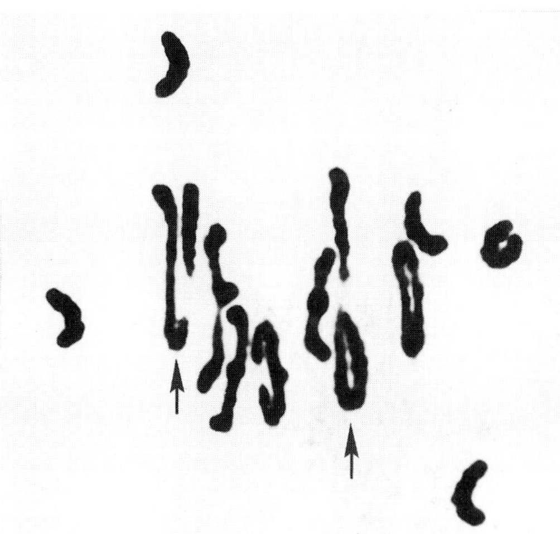

Ph1 gene absent

Fig. 10.15. Chromosome behavior in haploids of hexaploid wheat (*Triticum aestivum*), with and without the gene *Ph1*, located on chromosome 5B (Photographs from Jauhar et al., 1991). **Top:** Late metaphase I showing only univalents due to the presence of one dose of *Ph1* that prevents homoeologous pairing. **Bottom:** Metaphase I showing homoeologous pairing in the presence of the mutant allele *ph1*. There are two trivalents (marked by arrows), five bivalents, and five univalents. Meiotic pairing in wheat is controlled by genes on several chromosomes, with a balance between promoters and suppressors, but the *Ph* gene has a predominant effect (Kimber and Sears, 1987).

Variation in the dosage of *Ph1* may cause changes in the amount and kind of pairing. When six doses of *Ph1* are obtained by cytogenetic manipulations, even the close association between homologs is suppressed, and both homologs and homoeologs enter meiosis with a random distribution. It is evident that if evolutionary relationships between genomes are based on chromosome pairing observed in hybrids between two species under investigation, errors in interpretation can occur due to the unknown effects of a *Ph*-like gene.

10.6.4 Fertility of Allopolyploids

A major cause of reduced fertility in allopolyploids is homoeologous pairing during meiosis, because it produces multivalents, irregular chromosome distributions, and genetically unbalanced gametes. Allopolyploids without genetic restrictions on chromosome pairing have this problem. We might expect allopolyploids with a *Ph*-like suppression of homoeologous pairing to be as fertile as their diploid parents. However, most newly induced polyploids of this type have some reduction in fertility, indicating a lack of harmony when different genomes occur in the same cell, along with cytoplasm that came mainly from one species through the egg cell.

Another condition promoting sterility in allopolyploids is a high ploidy level. Several perennial grass genera ($x = 7$) in the tribe Triticeae have natural ploidy levels ranging from diploid to octoploid or higher, but two-thirds of the species are tetraploid, suggesting that this is an optimum ploidy level. Induced amphiploids at the octoploid level showed a continuous decrease in fertility over six generations. In such cases, the high chromosome number without pairing restrictions gives multivalents and unbalanced gametes. The established natural allopolyploids have survived because inherent mutations that restrict chromosome pairing to bivalents have become part of their genotype, or vegetative reproduction has eliminated the need to rely on seed production for perpetuation of the polyploid. With a combination of cytogenetic and molecular techniques, it may be possible to transfer *Ph1* or similar genes between species or genera, to lessen the problem of reduced fertility in induced amphiploids. The success of such transfers depends on how a gene of this type would function in foreign cells.

10.7 GENE SEGREGATIONS IN POLYPLOIDS

Diploids have disomic inheritance because they normally have two doses of each chromosome and each gene locus. The type of inheritance for polyploids may be disomic or polysomic (more than two doses of each chromosome and gene locus), depending on the type of polyploid and the genotype. The difference in genome composition between autopolyploids and allopolyploids is reflected in the way genes segregate in these two types of polyploids.

10.7.1 Allopolyploids

Established natural allopolyploids are thought to have been derived from hybridization between genetically related diploid species; therefore, the different genomes have many genes in common. In an allotetraploid, for example, each of the two parental genomes can have a gene contributing to the same function or phenotype. These genes show disomic segregation like that for duplicate genes arising from a duplication within a diploid genome. Any genes limited to one genome in an allopolyploid would also show disomic inheritance.

A good example of disomic inheritance in a natural allopolyploid is red versus white kernel color in wheat (Table 10.1). There exists one gene locus for kernel color in each of the three genomes

Table 10.1. Inheritance of Kernel Color in Allopolyploid Wheat with Triplicate Genes for Red Color, *R1* in Genome D, *R2* in Genome A, and *R3* in Genome B

	Crosses Involving Tetraploids (A and B Genomes)
Parents	*R2 R2 R3 R3* × *r2 r2 r3 r3* Dark Red White
F_1	*R2 r2 R3 r3* Medium red Selfed
F_2	9 *R2- R3-* 3 *R2- r3 r3* 3 *r2 r2 R3-* 1 *r2 r2 r3 r3* Different shades of red[a] to white

	Crosses Involving Hexaploids (A, B, and D Genomes)
Parents	*R1 R1 R2 R2 R3 R3* × *r1 r1 r2 r2 r3 r3* Dark red White
F_1	*R1 r1 R2 r2 R3 r3* Medium red Selfed
F_2	27 *R1- R2- R3-* 9 *R1- R2- r3r3* 9 *R1- r2r2 R3-* 9 *r1r1 R2- R3-* 3 *R1- r2r2 r3r3* 3 *r1r1 R2- r3r3* 3 *r1r1 r2r2 R3-* 1 *r1r1 r2r2 r3r3* Different shades of red[a] to white

[a] The intensity of red pigment depends on the number of dominant *R* alleles present.

(A, B, D) making up hexaploid wheat, and in each of the two genomes (A, B) making up tetraploid wheat. The lighter shades of red are difficult to distinguish from white by direct observation, but a chemical test detects the presence of any red pigment. Meiotic chromosome pairing is restricted to bivalents at both ploidy levels despite the presence of homoeologous chromosomes because of the pairing-restriction gene *Ph1*. The gene segregations are not, therefore, disturbed by irregular chromosome distributions. If allopolyploids have homoeologies between different genomes and no pairing-restriction genes, the proportion of multivalents and bivalents, and the behavior of the multivalents during meiosis influence gene segregations. Irregular segregations are also expected for odd-numbered ploidy levels such as allotriploids, with two doses of one genome and one dose of another, due to the unpredictable segregation behavior of the third genome.

10.7.2 Autopolyploids

Gene segregations in autopolyploids are more complex than in diploids or in allopolyploids with disomic inheritance, because of the replication of one genome. In autotetraploids, the following five genotypes are possible for alleles *R* and *r*, with the terms in parentheses referring to the dosage of the dominant allele: *RRRR* (quadruplex), *RRRr* (triplex), *RRrr* (duplex), *Rrrr* (simplex), or *rrrr* (nulliplex). The three heterozygous genotypes can be obtained by mutations or crosses between different genotypes. As the number of dominant alleles increases, the frequency of the *rrrr* genotype in an F_2 or backcross progeny decreases. The *RRRr* and *RRrr* genotypes are the most useful for distinguishing between diploid and polyploid segregations in F_2 or backcross progenies, because they give ratios of $R:r$ that deviate more markedly from diploid ratios than those from the *Rrrr* genotype (*see* Table 10.3).

The alleles at a single locus in an autotetraploid can be symbolized in a general way by *a, b, c,* and *d,* which can represent various combinations of dominant, recessive, or codominant alleles. Using these symbols, gametic formulas can be derived when there is no crossing over between the gene locus and the centromere in any of the four chromosomes due to close linkage (Fig. 10.16), or when the gene is far enough from the centromere for crossing over to occur in this region (Fig. 10.17). In the first situation, the alleles remain with their own centromeres through meiosis, and it is not possible for sister alleles to enter the same gamete; this is called chromosome segregation. In the second situation, sister alleles are separated into two chromosomes by crossing over, and if these chromosomes go to the same pole at anaphase I (e.g., *see* the 1 + 2, 3 + 4 distribution in Fig. 10.17), this is an equational distribution because both chromosomes have the same alleles. The random distribution of sister chromatids at anaphase II results in some gametes with sister alleles—a process called double reduction. This is also referred to as a type of chromatid segregation because the alleles segregate on a chromatid rather than a chromosome basis. The situation in Fig. 10.17, where there is a crossover between the gene and the centromere in each chromosome, is called maximum equational segregation, because it leads to the highest frequency of double reduction. Although maximum equational segregation presents the greatest contrast to chromosome segregation, it requires more specific conditions such as quadrivalent formation, and a crossover in each chromosome between the gene and the centromere. For these reasons, it is less likely to occur than a form of chromatid segregation requiring less stringent conditions.

The gametic formulas in Figs. 10.16 and 10.17 can be applied to the three heterozygous genotypes for the *R–r* locus by assigning the general symbol to each allele as follows.

R R R r	*R R r r*	*R r r r*
a b c d	*a b c d*	*a b c d*

For each of the three genotypes, the expected gametic frequencies for chromosome or maximum equational segregation can be derived by substituting *R* or *r* for *a, b, c,* or *d* in the formulas, and the results are given in Table 10.2.

When backcrosses of any of the three genotypes are made to the nulliplex genotype *rrrr*, the gametic frequencies in Table 10.2 become the genotypic frequencies of the backcross progenies when an *rr* gamete from the nulliplex is combined with each type of gamete. The F_2 segregations in Table 10.3 are obtained by self-pollinating the F_1 plants, which transmit the same kinds and frequencies of gametes through the male and female sides. The backcross and F_2 segregations for the triplex (*RRRr*) genotype distinguish tetrasomic inheritance from disomic inheritance (50% *r* for backcrosses, 25% *r* for F_2's) for both chromosome and maximum equational segregations, because of the absence or less frequent occurrence of the recessive phenotype. Both types of F_2 segregations and both types of backcross segregations of the duplex (*RRrr*) genotype are distinguishable from disomic segregations, so this genotype as well differentiates between a tetrasomic and a disomic segregation. Chromosome segregation for the simplex genotype

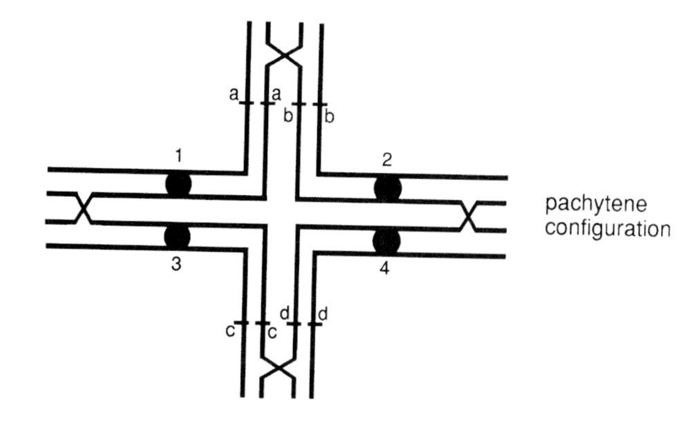

Anaphase I		Anaphase II		Gametes
1 + 2	aa + bb	a + b ⟷	a + b	ab + ab
↕	↕			
3 + 4	cc + dd	c + d ⟷	c + d	cd + cd
1 + 3	aa + cc	a + c ⟷	a + c	ac + ac
↕	↕			
2 + 4	bb + dd	b + d ⟷	b + d	bd + bd
1 + 4	aa + dd	a + d ⟷	a + d	ad + ad
↕	↕			
2 + 3	bb + cc	b + c ⟷	b + c	bc + bc

Gametic frequency: 2 ab + 2 cd + 2 ac + 2 bd + 2 ad + 2 bc

The other two pachytene pairing combinations are

1 (aa) – 3 (cc) and 1 (aa) – 4 (dd)
2 (bb) – 4 (dd) 2 (bb) – 3 (cc)

These combinations give the same kinds of gametes as the 1 – 2, 3 – 4 pairing shown above, and increase the frequency of each kind to 6, which can be reduced to 1, so the overall gametic formula for chromosome segregation is

1 ab + 1 cd + 1 ac + 1 bd + 1 ad + 1 bc

See text for use of this formula with any monogenic genotype.

Fig. 10.16. Diagram illustrating the derivation of gametic frequencies after chromosome segregation in an autotetraploid (after Burnham, 1962). The letters a, b, c, and d represent alleles of a single gene locus and there is no crossing over between the centromeres (numbered) and these loci. Randomness is assumed for the pachytene-pairing combinations, chromosome distributions at anaphase I, and chromatid distributions at anaphase II. Deviations from the gametic frequency shown are likely to occur if the four homologous chromosomes do not consistently form a quadrivalent.

Rrrr give the same frequency of recessives as a disomic segregation, and maximum equational segregation requires large progeny numbers to separate tetrasomic from disomic inheritance. The *RRRr* genotype gives some indication of the distance between the *R–r* locus and the centromere, because the recessive phenotype does not occur in backcross or F_2 progenies unless the gene is far enough from its centromere for crossing over to occur.

In natural populations, electrophoretic markers for various enzymes or DNA markers (*see* Chapter 21, Section 21.2) provide efficient ways of testing for tetrasomic versus disomic inheritance, because different alleles at a locus are codominant and therefore

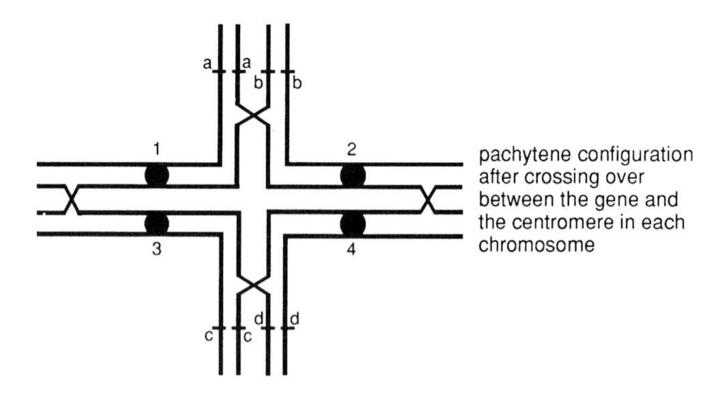

Anaphase I		Anaphase II			Gametes
1 + 2 ↕ 3 + 4	ab + ab ↕ cd + cd	a + a	←→ or	b + b	aa* + bb*
		a + b	←→	a + b	ab + ab
		c + c	←→ or	d + d	cc* + dd*
		c + d	←→	c + d	cd + cd
1 + 3 ↕ 2 + 4	ab + cd ↕ ab + cd	a + c	←→ or	b + d	ac + bd
		a + d	←→	b + c	ad + bc
		a + c	←→ or	b + d	ac + bd
		a + d	←→	b + c	ad + bc
1 + 4 ↕ 2 + 3	ab + cd ↕ ab + cd	This segregation has the same allelic combinations as the 1 + 3, 2 + 4 segregation, so each of its 8 gametes can be multiplied by 2.			

Gametic frequency: 1 aa* + 1 bb* + 1 cc* + 1 dd* + 2 ab + 2 cd + 4 ac + 4 bd + 4 ad + 4 bc

The other two pachytene pairing combinations after crossing over between each gene locus and its centromere are

1 (ac) – 3 (ac) and 1 (ad) – 4 (ad)
2 (bd) – 4 (bd) and 2 (bc) – 3 (bc)

Both of these combinations give the same types of gametes as the 1 – 2, 3 – 4 pairing but with different frequencies for some of them. The overall gametic formula for maximum chromosome segregation is

3 aa* + 3 bb* + 3 cc* + 3 dd* + 10 ab + 10 cd + 10 ac + 10 bd + 10 ad + 10 bc

See text for use of this formula with any monogenic genotype.

Fig. 10.17. Diagram illustrating the derivation of gametic frequencies after maximum equational segregation in an autotetraploid (redrawn from Burnham, 1962). The letters a, b, c, and d represent alleles of a single gene locus, and there is a crossover between these loci and the centromeres (numbered). Randomness is assumed for the pachytene-pairing combinations, chromosome distributions at anaphase I, and chromatid distributions at anaphase II. The gametes with an asterisk result from double reduction, which requires that after crossing over between the centromere and the gene marker in adjacent chromosomes, these chromosomes go to the same pole at anaphase I.

are expressed phenotypically. Some examples of tetrasomic segregation are given in Tables 10.4 and 10.5. The symbols *r*, *s*, *t*, and *u* for the different alleles at a locus were chosen to distinguish them from the general symbols *a*, *b*, *c*, and *d*, used in deriving the gametic formulas in Figs. 10.16 and 10.17. The three sets of data show good fits to the expected segregations for tetrasomic inheritance. Experiments such as these provide evidence for the occurrence of autotetraploidy in natural populations.

In autotriploids, with three doses of the same genome, a gene locus such as *R–r* can have any of the following four genotypes: *RRR* (triplex), *RRr* (duplex), *Rrr* (simplex), or *rrr* (nulliplex). The two heterozygous genotypes can be used to obtain gene segregations, and the theoretical expectations should be the same as for trisomics (*see* Chapter 13, Section 13.2). In reality, the trisomic segregations may be more reliable because autotriploids tend to have high sterility and a range of aneuploid progeny.

Discrepancies may occur between observed and expected segregations in autopolyploids because of variable conditions during meiosis such as the frequencies of multivalents, bivalents, and univalents, which can influence the proportion of balanced and unbalanced gametes, as well as fertility. Another uncertainty is that gene expressions in induced polyploids cannot always be predicted from their behavior in the diploid parents, because the increased gene dosage in autopolyploids sometimes changes the interactions between loci and also between alleles at a locus. A completely dominant allele in a diploid may become partially dominant or additive in a polyploid. These changes in gene expression complicate cytogenetic studies of characters in autopolyploids.

10.8 GENOME ANALYSES IN POLYPLOIDS

The chromosomal and genetical contents of genomes indicate the evolutionary pathways of extant species, and their taxonomic relationships. Knowledge of interspecific chromosome homoeologies facilitates the successful transfers of genes to crop species from their wild relatives by chromosome manipulations (*see* Chapter 15). The main focus of genome analysis has been on allopolyploids, which by definition, consist of more than one type of genome, but autopolyploids also can be investigated for the origin of their one type of genome, and possible divergence of the multiplied genome through mutation, chromosome rearrangement, or hybridization.

Genome analysis is best accomplished if each chromosome in the set can be identified. Telochromosomes can be distinguished from two-armed chromosomes by direct cytological observations, and their pairing relationships with other chromosomes can be determined. However, usually only one or two chromosomes of the set are introduced as telochromosomes in any one hybrid, so a series of crosses involving different telochromosomes must be made to test the complete set. Another direct way of assessing the pairing relationships of different genomes is by chromosome-banding pro-

Table 10.2. Expected Types and Frequencies of Gametes from Chromosome and Maximum Equational Segregations Involving a Single Heterozygous Locus with Two Alleles, *R* Dominant Over *r*, in an Autotetraploid

Genotype	Chromosome Segregation	% *rr*	Maximum Equational Segregation	% *rr*
RRRr	1*RR* + 1*Rr*	0.0	13*RR* + 10*Rr* + 1*rr*	4.2
RRrr	1*RR* + 4*Rr* + 1*rr*	16.7	2*RR* + 5*Rr* + 2*rr*	22.2
Rrrr	1*Rr* + 1*rr*	50.0	1*RR* + 10*Rr* + 13*rr*	54.2

Source: After Burnham (1962).

Table 10.3. Expected F_2 Segregations for a Single Locus with Two Alleles, *R* Dominant Over *r*, After Chromosome or Maximum Equational (Max. Equat.) Segregations in an Autotetraploid

		F_2 Genotypes						
Genotype	Type of Segregation	*RRRR*	*RRRr*	*RRrr*	*Rrrr*	*rrr*	Ratio (*R*:*r*)	% r
RRRr	Chromosome	1	2	1			all *R*	0.0
	Max. equat.	169	260	126	20	1	575:1	0.17
RRrr	Chromosome	1	8	18	8	1	35:1	2.8
	Max. equat.	4	20	33	20	4	77:4	5.19
Rrrr	Chromosome			1	2	1	3:1	25.0
	Max. equat.	1	20	126	260	169	407:169	29.03

Source: After Burnham (1962).

Table 10.4. Tetrasomic Inheritance, Based on Chromosome Segregation, for Two Alleles at a Locus for Phosphoglucomutase(*Pgm*), and for Three Alleles at a Locus for Phosphoglucoisomerase(*Pgi*), in a Natural Population of *Tolmiea menziesii*. The Symbols *r*, *s*, and *t* Represent Codominant Alleles

		Progeny					
Locus	Genotype of Parents	Genotypes	(Phenotypes)	Observed Segregation	Expected Ratio[a]	χ^2	*P*
Pgm	*rrss* × *rrss*	*rrrr*	(*r*)	3	1		
	(*abcd*) (*abcd*)	*rrrs*	(*rs*)	20	8		
		rrss	(*rs*)	38	18		
		rsss	(*rs*)	14	8		
		ssss	(*s*)	1	1		
						2.03	0.74
Pgi	*sttt* × *rttt*	*rstt*	(*rst*)	17	1		
	(*abcd*) (*abcd*)	*rttt*	(*rt*)	20	1		
		sttt	(*st*)	27	1		
		tttt	(*t*)	24	1		
0.46						2.64	0.46

[a] The expected ratios are derived by assigning the symbols *a*, *b*, *c*, and *d* to each parental genotype as shown, in order to substitute the alleles *r*, *s*, or *t* in the gametic formula for chromosome segregation in Fig. 10.16.

Source: Data from Soltis and Soltis (1988).

Table 10.5. Tetrasomic Inheritance, Based on Maximum Equational Segregation, for Four Alleles at the Locus (*Lap*) for Leucine-Aminopeptidase in a Cross Between Two Alfalfa Species, *Medicago sativa* and *M. falcata*

	F_1					
Parental Genotypes	Genotypes	(Phenotypes)	Observed Segregation	Expected Ratio[a]	χ^2	*P*
rrrr × *sstu*	*rrss*	(*rs*)	21	16		
(*abcd*) (*abcd*)	*rrst*	(*rst*)	37	20		
	rrsu	(*rsu*)	32	20		
	rrtu	(*rtu*)	16	10		
	rrtt	(*rtu*)	3	3		
	rruu	(*ru*)	2	3		
					3.9	0.60

Note: The symbols *r*, *s*, *t*, and *u* represent codominant alleles at the *Lap* locus.

[a] The expected ratios are derived by assigning the symbols *a*, *b*, *c*, and *d* to each parental genotype as shown, in order to substitute the alleles *r*, *s*, *t*, or *u* in the gametic formula for maximum equational segregation given in Fig. 10.17. The last two F_1 genotypes probably came from double reduction.

Source: Data from Quiros (1982).

cedures (*see* Chapter 6, Section 6.2), provided that most or all of the meiotic chromosomes have distinctive patterns. There is an increasing use of molecular techniques such as *in situ* hybridization (*see* Chapter 19) to detect repeated or single-copy DNA sequences in localized chromosome regions, and to relate them to intergenomic relationships. An example of this approach is given in the next section.

10.8.1 Identifying Diploid Ancestors of Allopolyploids

The first step is to locate diploid species that are likely progenitors of a naturally evolved allopolyploid species. This often involves expeditions to collect species from the area or center of origin for the taxonomic group to which the polyploid belongs. The potential ancestral species are selected for further investigations because they resemble the polyploid in one or more of the following ways: outward morphological traits, karyotype, and internal anatomical, biochemical, physiological, or molecular attributes.

One way of testing the source of different genomes in the allopolyploid is the genome-analyzer method. Hybrids are made between a selected diploid species and the allopolyploid, culturing the embryos on synthetic medium if needed to prevent abortion. Meiotic cells are observed in the hybrids to determine the extent of chromosome pairing. Each genome is present in a single dose, so pairing of the genome from the diploid species with one of the genomes from the polyploid indicates that the diploid contributed its genome to the allopolyploid. The same method can be used to trace the source of the second genome in an allotetraploid, or the second and third genomes in an allohexaploid. The genomes that are common to the polyploid and its diploid ancestor are given the same letter. Quantitative data on the types of chromosome pairing provide more details for the degree of relationship between the genomes.

A different approach is to make crosses between the selected diploid species, then double the chromosome number in the F_1 hybrids using colchicine. The naturally evolved allopolyploid can be compared with its re-created counterpart in a number of ways, and the degree of fertility in the hybrid between the synthetic and natural polyploids indicates the closeness of the relationship.

Both of these methods have been used in wheat and tobacco. The genome-analyzer method provided unequivocal evidence that the source of the A genome in tetraploid and hexaploid wheats was the diploid wheat *Triticum monococcum*, but there is a continuing search for the source of the B genome. The origin of the D genome in the hexaploid wheat *Triticum aestivum* was identified by crossing the tetraploid wheat *Triticum turgidum*, which has genomes A and B but lacks D, with the diploid species *Triticum tauschii*, which has the D genome. When the chromosome number in the hybrid was doubled to give a synthetic AABBDD hexaploid, it resembled the natural spelta group of hexaploid wheat and produced fertile hybrids when crossed both with this group and with bread wheat. There was less segregation in the progeny of the hybrids than would be expected if the cross had been between distantly related parents. With these clear results, *T. tauschii* was chosen as the contributor of the D genome to polyploid wheats.

When the naturally evolved allotetraploid tobacco, *Nicotiana tabacum*, with genomes S and T, was crossed with each of two diploid species, *N. sylvestris* (genome S) and *N. tomentosa* (genome T), each diploid genome paired with one genome in the tetraploid, indicating that they were the ancestors of the tetraploid. The synthetic allotetraploid, created by crossing the two diploid species and doubling the chromosome number of the hybrid, was like the natural allotetraploid *N. tabacum* in appearance and was male-fertile but female-sterile. When allotetraploids were derived using crosses between *N. sylvestris* and the diploid species *N. tomentosiformis* or *N. otophora*, they were completely fertile in both cases, indicating that either of the latter species could be involved in the ancestry of tobacco. Molecular cytogenetics has helped to solve this dilemma. When *N. tabacum* DNA was hybridized to biotinylated total genomic DNA from *N. tomentosiformis* and *N. otophora* using dot blot and *in situ* techniques (*see* Chapter 19), the results indicated that the T genome in *N. tabacum* may be the result of hybridization between the two diploid species. This idea was reinforced by molecular evidence for up to nine translocations between the S and T genomes in *N. tabacum*. A strong, uniform signal was given when biotinylated *N. sylvestris* DNA was hybridized to *N. tabacum* DNA using the same techniques, and thus confirmed the close relationship between the S genomes of *tabacum* and *sylvestris*.

10.8.2. Quantitative Tests of Genome Relationships in Polyploid Hybrids

Meiotic-chromosome pairing tests for DNA similarities throughout chromosome lengths, so considerable effort has been given to developing quantitative evaluations of observed sets of data. Mathematical models of chromosome pairing in triploid, tetraploid, or higher-level polyploid hybrids test the relative affinity (i.e., the degree of relationship) between different genomes. This approach is more objective than simply observing whether or not there is chromosome pairing between different genomes. In developing the models, certain assumptions are made such as the following: synapsis begins at the ends of the chromosome arms and not in other regions, therefore each arm of a chromosome acts independently in its synaptic behavior, and multivalents can occur; the tendency for synapsis and chiasma formation between homoeologous arms of two genomes is the same regardless of the arm and homoeologous group involved; and chiasma interference (*see* Chapter 21, Section 21.4.6 and Fig. 21.11) affects bivalents and multivalents equally. Although these assumptions are based on some factual evidence, they may not apply in all cases, but they are considered to be acceptable simplifications.

The mathematical models are based on observed frequencies of chromosome configurations at the diakinesis to metaphase I stages of meiosis. Using an allotriploid as an example, the configurations can include trivalents, ring or rod bivalents, and univalents. The frequencies of these configurations are used to calculate a mean arm-pairing frequency, c, which is the proportion of paired arms observed, compared to the maximum possible for a triploid. Rod bivalents have one paired arm, whereas ring bivalents and trivalents each have two paired arms. Therefore,

$$c = \frac{\text{rod bivalents} + 2(\text{ring bivalents} + \text{trivalents})}{2(\text{basic chromosome number})}.$$

Varying numbers of parameters, depending on the mathematical model, represent the relative affinities of different genomes. One model, developed for allotriploid hybrids, relates the meiotic behavior of chromosomes to two parameters, x and y, which are relative affinities between the more and less closely related genomes, respectively. Values of x range from 0.5 to 1.0, and $x + y = 1$, so there is one independent variable. If an allotriploid is formed from crosses between an autotetraploid with genomic composition AAAA and a diploid with genome B, the triploid will be AAB. Over time, mutations or chromosome rearrangements can cause the two A genomes to differ somewhat, in which case they can be designated A and A′, but they are still more closely related to each other than either is to genome B. Therefore, x is the relative affinity between A and A′, and y is the relative affinity between A and B or A′ and B, with a ratio of $x{:}y{:}y$. The calculated frequency of pairing between genomes A and A′ is $cx/(x + 2y)$ and between A and B or A′ and B, $cy/(x + 2y)$. Lack of pairing between arms of homoeologous groups is indicated by $1-c$. These formulas are used to obtain expected frequencies of different arm configurations, which are compared with an observed set of data. The best estimate of x (optimized x), which is facilitated using a computer program, is one that minimizes the differences between it and the calculated set of data, and this is attained by weighting the observed and calculated meiotic figures by the number of chromosomes in each type of configuration.

Other mathematical models use more than two parameters to describe meiotic behavior in triploid, tetraploid, or pentaploid hybrids. Some models are based on fewer assumptions and provide more variables than there are degrees of freedom in the set of data under study, resulting in ranges rather than single solutions for the parameters. Although this gives good agreement between observed and calculated frequencies of the various configurations for an optimized fit, it involves time-consuming computations. When the model described in the preceding paragraph was compared with two other models using the same sources of data for allotriploids in each case, essentially the same conclusions were reached concerning the relative affinities of different genomes.

10.8.3 Assigning Chromosomes to Genomes

In some cases, the chromosomes of one genome can be distinguished from those of a different genome because of a morphological feature that involves the whole genome. For example, in allotetraploid cotton, *Gossypium hirsutum*, the chromosomes of the A genome are about twice as large as those of the D genome. Another example occurs in hexaploid wheat, where the B genome as a group has more prominent C and N bands than either the A or D genomes (see Chapter 6, Fig. 6.8). Of course, additional information must be obtained before a group of chromosomes with distinctive features can be assigned to a specific genome, using a letter as well as a number for each chromosome (*see* next subsection).

When chromosome-group distinctions are absent, appropriate crosses are made to assign chromosomes to a genome on an individual basis, but there must be a way of distinguishing each chromosome from the others in the complement of allopolyploid species. If aneuploids such as monosomics or trisomics are available for each chromosome, they give decisive tests because the absence or addition of a chromosome is easily observed by cytology or, in some cases, by genetic effects (*see* Chapter 11, Section 11.5). The use of a set of monosomics to assign chromosomes to specific genomes is given in Chapter 14, Fig. 14.1. Telochromosomes are also good markers because they are distinctive at meiosis, either as univalents or when their one arm pairs with a homologous arm of a complete chromosome (see Chapter 11, Fig. 11.21).

10.8.4 Homoeologous Relationships Between Chromosomes of Different Genomes

Homoeologous chromosomes have some common segments, but they are not identical throughout their lengths, as are homologous chromosomes. Naturally occurring allopolyploids have intergenomic groups of homoeologous chromosomes, indicating that the different genomes are derived from a common ancestor. Cytological, cytogenetic, and biochemical methods have been used to detect these homoeologies.

Meiotic observations on haploids derived from allopolyploids provide information on homoeologous relationships because each chromosome lacks a homologous partner, so there is a greater chance for homoeologous pairing. If, however, a gene such as *Ph1* is present in one dose, it severely reduces homoeologous pairing, so haploids with a deletion or a mutant form of this locus must be used. A study on bread wheat haploids, with one dose of genomes A, B, and D, compared the amount of homoeologous pairing in the presence of one dose of the gene *Ph1* versus one dose of the mutant allele *ph1* (*see* haploids in Fig. 10.15). The percentage of the complement paired ranged from 6.45 to 14.05 in different haploids with *Ph1*. In *ph1* haploids, the pairing was much higher, ranging from 58.21 to 61.75 of the complement, and a haploid that was lacking chromosome 5B, which carries the *Ph1* locus, had comparable pairing (54.9%). Chiasma frequencies also were considerably higher when *Ph1* was not present. N-banding showed that 80% of the pairing was between A- and D-genome chromosomes, but there were some trivalents, which had a heavily banded B-genome chromosome attached at one end to more closely paired A and D chromosomes.

Individual chromosomes were not identified in the above study, but C- or N-banding can be used to distinguish whole chromosomes and, in some cases, each arm of a chromosome. Thus, when banding techniques are combined with a single dose of each genome and absence of genes that restrict homoeologous pairing, many intergenomic chromosome-arm relationships can be observed at meiosis. If homoeologous chromosomes pair closely enough in haploids for crossing over and chiasma formation to occur, the exchanges of segments amount to reciprocal translocations, which can be detected in the progenies of crosses between the haploids and the allopolyploids in which they occurred. By identifying the chromosomes in the translocations using chromosome banding or crosses with aneuploids, the homoeology of the chromosomes involved can be determined.

When an array of identified aneuploid types are available, homoeologous relationships can be tested by combining the nullisomic state for a chromosome in one genome with the tetrasomic state for a chromosome in a different genome. The nulli–tetra combinations are checked for the amount of phenotypic compensation that two extra doses of one chromosome provide for the missing chromosome. An intensive study was made in hexaploid wheat, in which many nulli–tetra combinations were developed in the cultivar Chinese Spring, using painstaking cytogenetic methods. Each nulli–tetra combination was compared with the normal condition—two doses of each chromosome—and with the nullisomic for the same chromosome that was missing in the nulli–tetra (Fig. 10.18). The criteria emphasized fertility, but also included plant and spike sizes, and other traits. The results of this study made it possible to classify wheat chromosomes into seven homoeologous groups, with three chromosomes in each group (one chromosome

Fig. 10.18. Spikes of Chinese Spring wheat showing nullisomic–tetrasomic compensation (top) and lack of compensation (bottom). N is a normal, disomic spike. The spikes labeled 3A, 3B, and 3D in the top row are nullisomics, and the two spikes to the right of each nullisomic are nullisomic for that chromosome and tetrasomic for one of the other two homoeologs. The spikes to the right of N in the bottom row are nullisomic–tetrasomic combinations—or nullisomic–trisomic for the rightmost spike—but involve nonhomoeologous chromosomes (Morris and Sears, 1967).

from each of the three genomes A, B, and D). There were differences among the groups in the degree of compensation, and in three groups, the compensation was poor when the nullisomic for one chromosome was combined with either of its homoeologs, (e.g., nulli-2B combined with tetra-2A or tetra-2D). With the accumulating evidence for the natural occurrence of reciprocal translocations in wheat, this type of chromosome aberration or small deletions can cause some differentiation among homoeologs, so that they lose some of their compensating effects.

The discovery that many genes in an allopolyploid are represented by a locus in each genome provides a genetic approach to detecting intergenomic homoeologies. The more of these homologous gene sets there are, the greater the extent of common areas among the chromosomes. The triplicate genes for kernel color in wheat are an example of a homologous set (*see* Table 10.1). Many homologous sets of structural gene loci for different enzymes span the wheat A, B, and D genomes, and they have also been located in homoeologous chromosome arms. Gluten, a nonenzymatic protein found in the endosperm of wheat kernels (see Fig. 21.7, Chapter 21) is essential for bread-making. Gliadin and glutenin, two of the important components of gluten, are each controlled by homologous sets of genes located in the homoeologous group 1 chromosomes and, in the case of gliadin, also in group 6. These genes are located on chromosomes by using a combination of identified monosomics and biochemical techniques, and they are localized to chromosome arms using telochromosomes (*see* Chapter 14, Sections 14.1 and 14.3). Strong evidence for homology of gene sets is provided if the genes are located at similar positions on the short or long arms of homoeologous chromosomes and if they are close to several loci that belong to known homologous sets.

Homoeologies between the genomes of natural allopolyploids and related species are important in tracing the ancestry of the polyploids or in transferring to them valuable genes from wild relatives. The homoeologies are detected by methods that are similar to those used within allopolyploids. For example, a hexaploid wheat variety (genomes A, B, and D) that was nullisomic for chromosome 5B and therefore lacked the pairing-restricting gene *Ph1*, was crossed with diploid rye, which has genome R. The hybrids were ABDR, lacking *Ph1*, and C-banding was used to identify chromosomes and their arms at meiosis. Whereas rye chromosomes could be distinguished from wheat chromosomes, only two of the seven rye chromosomes could be identified by their banding patterns. The pairing behavior of these two chromosomes with wheat chromosomes indicated that the long arm of 1R was homoeologous to the long arms of the wheat group 1 chromosomes, 1A, 1B, and 1D, and the short arm of 1R was probably homoeologous to the short arms of the same wheat group. The long arm of the other identifiable rye chromosome, 5R, was homoeologous to wheat 5AL, but partially homoeologous to the long arms of two chromosomes of wheat group 4. This and other lines of evidence suggested that during the evolution of the rye genome from the same ancestor as wheat, a translocation had occurred between 4RL and 5RL.

Tests for genetic compensation are effective in determining the amount of homoeology between chromosomes of an allopolyploid and a related species. Individual chromosomes of the related species, also called alien species, are substituted for individual chromosomes of the allopolyploid, and the amount of compensation is gauged by the vigor and fertility of the plants with the substitutions (*see* Chapter 15, Section 15.4.2). If telochromosomes are available for the alien chromosomes, the amount of compensation can be tested on an arm basis, substituting a pair of telochromosomes for one arm at a time.

In another genetic approach, an allopolyploid and its related species can be compared with respect to gene homologies and conservation of gene–chromosome arrangements (synteny) during the evolution of the polyploid and its relatives from a common ancestor. Structural chromosome changes, in particular reciprocal translocations, can disrupt the genetic relationships among chromosomes, but they can be detected by observations of meiotic pairing, by comparing related species for their gene arrangements, or by molecular techniques.

There is an extensive amount of information on the gene homologies of the grain crops barley, rye, and wheat, which belong to the tribe Triticeae within the family Poaceae. The hexaploid wheat cultivar Chinese Spring is used as a standard for cytogenetic comparisons with related species because its gene–synteny relationships are believed to be largely unchanged from those in the ancestral genome for the Triticeae. When the chromosomal locations of enzymes, proteins, and other traits are compared in wheat, barley, and rye, there is evidence of some homoeologies among chromosomes with the same numbers, particularly between barley and wheat. Although each rye chromosome has some genes that are homologous to those in the comparable chromosomes of barley and wheat, rye chromosomes have undergone more structural changes than barley or wheat chromosomes during their evolution.

This difference can be illustrated with chromosomes 4A, 4B, and 4D of wheat, 4H of barley, and 4R of rye. The short (S) arm of rye chromosome 4R carries genes for the enzymes phosphoglucomutase and alcohol dehydrogenase, which are homologous to genes on barley 4H and on homoeologous arms of the wheat group 4 chromosomes. However, some genes and molecular markers (cDNA clones) are located on 4R in rye but on one or more of the group 7 chromosomes in wheat. Conversely, the gene for acid phosphatase is located on 7RS in rye, but on chromosomes numbered 4 in wheat and barley. In chromosome-substitution experiments, 4RS gives good compensation for wheat group 4 chromosomes and 4RL (long arm) gives good compensation for the group 7 chromosomes of wheat. All these findings support the concept of a reciprocal translocation between 4RL and 7RS, so that some genes from 7RS are transferred to 4RL, and vice versa.

10.8.5 Species-Specific Translocations

During the evolution of different wheat groups, translocations that are specific for a group have occurred and have been detected by a combination of chromosome N-banding followed by *in situ* hybridization (*see* Chapter 19). The allotetraploid emmer group, *Triticum turgidum*, with genome formula AABB, has a translocation involving two chromosomes of the A genome (4A and 5A) and one chromosome of the B genome (7B). The interchanges between these chromosomes are thought to have occurred in a sequence, so the translocation is called cyclic. The tetraploid timopheevi group, *Triticum timopheevii*, with genome formula A^tA^tGG, has a different cyclic translocation involving chromosomes $6A^t$, 1G, and 4G. A hypothesis to explain why these translocations are species-specific is that when a new amphiploid occurs by combining and doubling two different genomes, it may be highly sterile because of an incompatibility between male nuclear genes and female nuclear or cytoplasmic genomes. This bottleneck is overcome by certain changes in the nuclear chromosomes such as cyclic translocations, which somehow restore fertility and harmony between nucleus and cytoplasm. The mechanism whereby this is achieved still has to be determined. The translocation difference between the emmer group and the timopheevi group indicates that two lines of descent were involved in the evolution of tetraploid wheats.

10.9 ADAPTABILITY OF NATURAL POLYPLOIDS

When polyploids are found in natural populations, questions arise about the attributes that enable them to compete with the diploid component and to adapt to specific environmental conditions. The potential is always present for polyploidy to originate from somatic doubling, or more likely from $2n$ gametes as a result of gene mutations affecting meiotic divisions. Some of the $2n$ gametes may have a higher number of adaptive gene combinations than others. Thus, some polyploids can survive and spread, while many are outcompeted by their diploid parents.

Conditions that favor polyploids in plant populations are a perennial rather than an annual habit of growth and an asexual mode of reproduction. With these traits, a polyploid has a longer time to become established while circumventing the sterility caused by irregular meioses. Apomixis can stabilize adaptive genotypes by transmitting the female genotype through unreduced eggs, thereby helping polyploids to colonize a suitable habitat. If the polyploids are limited to sexual reproduction, the presence of a *Ph*-like gene would help to normalize the polyploid meioses.

In both fish and insect natural populations, there is evidence that polyploidy has an adaptive role. In two fish families, Catostomidae (freshwater suckers) and Salmonidae (salmon, trout, etc.), all the members appear to be tetraploid based on cytological and genetical evidence. Members of the Catostomidae family and *Cyprinus carpio* (carp), a tetraploid in the Cyprinidae (minnow) family, seem to be very well adapted to different ecological areas because of their large size, fast growth rate, and long life. These traits are attributed to increased gene doses, genetic recombination, and especially heterozygosity. Polyploidy in insects is consistently associated with parthenogenesis, which in most cases involves suppression of meiosis. Although this condition prevents genetic recombination, both parthenogenesis and polyploidy cause a greater increase in heterozygosity than bisexual diploids because new mutations are retained. Genetic variability seems to make insects more efficient in occupying available niches; with parthenogenesis, one insect can establish a new population. Life spans of 2 or more years help polyploid insects to survive short-term environmental changes. Polyploidy increases genetic homeostasis (i.e., the stability of genetical and physiological states in variable environments) and has a buffering effect against deleterious mutations and chromosome aberrations. However, the accumulation of harmful mutations over a number of generations could cause loss of homeostasis and jeopardize the future of polyploid insect species.

10.10 PRODUCTION AND USES OF INDUCED POLYPLOIDS

Since the first use of colchicine in the 1930s to double chromosome numbers in plants, much has been learned about the optimum conditions when producing polyploids for breeding purposes. In addition to a perennial habit and asexual reproduction mentioned above for natural populations, other desirable traits of the diploids to be doubled are a low chromosome number, cross-pollination for the sexual mode of reproduction, and a short sexual-generation cycle. If the commercial product is a vegetative part of a plant such as tubers, leaves, or flowers, instead of seeds, the problem of a reduced seed set can be bypassed. At the same time, the larger size of some organs such as flowers, or the increased amount of certain chemical constituents such as sugar, give some commercial advantages to induced tetraploids.

A low chromosome number is considered to be one of the most important factors because most crop species have already reached their optimum ploidy level. In fact, some so-called diploid species may actually be polyploid at least in part of their genome and could have evolved from a true diploid ancestor. If diploids have a high chromosome number, the doubling process may add too much DNA to the cells, which would not function efficiently. Diploid species with a low chromosome number (e.g., maize, $n = 10$, or tomato, $n = 12$) can tolerate tetraploidy, but they lose vigor and fertility at the octoploid level.

Cross-pollination introduces new gene combinations and heterozygosity to autopolyploids, which have one genome multiplied. Successful induced autopolyploids such as ryegrass (*Lolium* species) and red clover (*Trifolium pratense*), as well as most of the natural autopolyploids, are cross-pollinating. This mode of pollination appears to be less essential in allopolyploids because they have built-in genetic diversity, with at least two genomes from different sources. Therefore, self-pollination in this group fixes desirable gene combinations.

Polyploidy breeding is a long-term program because the increase in chromosome number requires time for adaptation, both within the plants and in their interactions with the environment. A faster turnover of sexual generations accelerates testing of segregating genotypes for many traits, including adaptation. It also is important to start with a broad genetic base, using diploids with diverse genotypes to produce the polyploids. Genes in polyploids often react differently to environmental conditions than when they are in diploids, so a large number of polyploid genotypes should be available for tests.

BIBLIOGRAPHY

General

BRODSKY, V.YA., URYVAEVA, I.V. 1985. Genome Multiplication in Growth and Development. Biology of Polyploid and Polytene Cells. Cambridge University Press, Cambridge.

EIGSTI, O.J., DUSTIN, P., JR. 1955. Colchicine—in Agriculture, Medicine, Biology, and Chemistry. The Iowa State College Press, Ames.

ELDRIDGE, F.E. 1985. Cytogenetics of Livestock. AVI Publishing Co., Westport, CT.

GOTTSCHALK, W. 1978. Problems in polyploidy research. Nucleus 21: 99–112.

JACKSON, R.C. 1991. Cytogenetics of polyploids and their diploid progenitors. In: Chromosome Engineering in Plants: Genetics, Breeding, Evolution. Part A (Gupta, P.K., Tsuchiya, T., eds.). Elsevier, New York. pp. 159–180.

KIMBER, G. 1984. Evolutionary relationships and their influence on plant breeding. In: Gene Manipulation in Plant Improvement (Gustafson, P.J., ed.). 16th Stadler Genet. Symp. Series, Plenum Press, New York. pp. 281–293.

LEWIS, W.H. (ed.) 1980. Polyploidy: Biological Relevance. Basic Life Sciences Vol. 13. Int. Conf., Washington Univ., St. Louis, MO. Plenum Press, New York.

SYBENGA, J. 1992. Cytogenetics in Plant Breeding. Monographs on Theoretical and Applied Genetics 17. Springer-Verlag, Berlin.

Section 10.2

KÜHN, A. 1971. Lectures in Developmental Physiology. Springer-Verlag, Berlin.

LEVITAN, M. 1988. Textbook of Human Genetics. Oxford University Press, Oxford.

ORR, H.A. 1990. "Why polyploidy is rarer in animals than in plants" revisited. Am. Nat . 136: 759–770.

Section 10.3

HERMSEN, J.G.TH. 1984. Mechanisms and genetic implications of 2n-gamete formation. Iowa State J. Res. 58: 421–434.

IYER, C.P.A., RANDHAWA, G.S. 1965. Increasing colchicine effectiveness in woody plants with special reference to fruit crops. Euphytica 14: 293–295.

NEL, P.M. 1975. Crossing over and diploid egg formation in the elongate mutant of maize. Genetics 79: 435-450.

O'MARA, J.G. 1939. Observations on the immediate effects of colchicine. J. Hered. 30: 35–37.

PELOQUIN, S.J., YERK G.L., WERNER J. E. 1990. Ploidy manipulations in the potato. In: Chromosomes: Eukaryotic, Prokaryotic, and Viral. Vol. II (Adolph, K.W., ed.). CRC Press Inc., Boca Raton, FL. pp. 167–178.

RAMANNA, M.S. 1974. The origin of unreduced microspores due to aberrant cytokinesis in the meiocytes of potato and its genetic significance. Euphytica 23: 20–30.

RAMANNA, M.S. 1979. A re-examination of the mechanisms of 2n gamete formation in potato and its implications for breeding. Euphytica 28: 537–561.

RAMANNA, M.S. 1983. First division restitution gametes through fertile desynaptic mutants of potato. Euphytica 32: 337–350.

SHOFFNER, R.N. 1985. Bird cytogenetics. In: Cytogenetics of Livestock (Eldridge, F.E., ed.). AVI Publishing Co. Westport, CT. pp. 263–283.

SWARUP, H. 1959. Production of triploidy in *Gasterosteus aculeatus* (L.). J. Genet. 56: 129–142.

VEILLEUX, R. 1985. Diploid and polyploid gametes in crop plants: mechanisms of formation and utilization in plant breeding. In: Plant Breeding Reviews, Vol.3 (Janick, J., ed.). AVI Publishing Co., Westport, CT. pp. 253–288.

Section 10.4

MARSHAK,T.L., STROEVA, D.G. 1973. Cytophotometric investigation of the DNA content of the retinal pigment epithelium cells in postnatal development of rats: a cytophotometric study. Ontogenez 4: 472–475. (in Russian; English translation in Sov. J. Dev. Biol. 4: 472–475).

Section 10.5

NEWCOMER, E.H. 1941. A colchicine-induced tetraploid *Cosmos*. J. Hered. 32: 161–164.

TAL, M. 1980. Physiology of polyploids. In: Polyploidy: Biological Relevance. Basic Life Sciences Vol. 13 (Lewis, W.H. ed.). Plenum Press, New York. pp. 61–75.

Section 10.6

GILLIES, C.B. 1989. Chromosome pairing and fertility in polyploids. In: Fertility and Chromosome Pairing: Recent Studies in Plants and Animals (Gillies, C.B., ed.). CRC Press, Boca Raton, FL. pp. 137–176.

JAUHAR, P.P., RIERA-LIZARAZU, O., DEWEY, W.G., GILL, B.S., CRANE, C.F., BENNETT, J.H. 1991. Chromosome pairing relationships among the A, B, and D genomes of bread wheat. Theor. Appl. Genet. 82: 441–449.

KIMBER, G., SEARS E.R. 1987. Evolution in the genus *Triticum* and the origin of cultivated wheat. In: Wheat and Wheat Improvement, 2nd ed. (Heyne, E.G., ed.). American Society of Agronomy, Madison, WI. pp. 154–164.

LEIPOLDT, M., SCHMIDTKE, J. 1982. Gene expression in phylogenetically polyploid organisms. In: Genome Evolution (Dover, G.A., Flavell, R.B., eds.). Academic Press, New York. pp. 219–236.

LOIDL, J. 1986. Synaptonemal complex spreading in *Allium*. II. Tetraploid *A. vineale*. Can. J. Genet. Cytol. 28: 754–761.

PAL, M., PANDEY R.M. 1982. Decrease in quadrivalent frequency over a 10 year period in autotetraploids in two species of grain amaranths. Cytologia 47: 795–801.

RILEY, R. 1966. Genetics and the regulation of meiotic chromosome behaviour. Sci. Prog. Oxford 54: 193–207.

SEARS, E.R. 1977. Genetic control of chromosome pairing in wheat. Annu. Rev. Genet. 10: 31–51.

SOLARI, A.J., FECHHEIMER N.S. 1988. Quadrivalent formation in a tetraploid chicken oocyte. Genome 30: 900–902.

SYBENGA, J. 1975. Meiotic configurations. Springer-Verlag, Heidelberg.

YEN, Y., KIMBER, G. 1990. Production and meiotic analysis of autotriploid *Triticum speltoides* and *T. bicorne*. Theoret. Appl. Genet. 79: 525–528.

Section 10.7

BURNHAM, C.R. 1962. Discussions in Cytogenetics. Burgess Publishing Co., Minneapolis, MN.

KIMBER, G. 1971. The inheritance of red grain colour in wheat. Z. Pflanzen. 66:151–157.

QUIROS, C.F. 1982. Tetrasomic segregation for multiple alleles in alfalfa. Genetics 101: 117–127.

SOLTIS, D.E., SOLTIS, P.S. 1988. Electrophoretic evidence for tetrasomic segregation in *Tolmiea menziesii* (Saxifragaceae). Heredity 60: 375–382.

Section 10.8

BIETZ, J.A. 1987. Genetic and biochemical studies of nonenzymatic endosperm proteins. In: Wheat and Wheat Improvement, 2nd ed. (Heyne, E.G., ed.). American Society of Agronomy, Madison, WI. pp. 215–241.

CHAPMAN, C.G.D., KIMBER, G. 1992a. Developments in the meiotic analysis of hybrids. I. Review of theory and optimization in triploids. Heredity 68: 97–103.

CHAPMAN, C.G.D., KIMBER, G. 1992b. Developments in the meiotic analysis of hybrids. V. Second-order models for tetraploids and pentaploids. Heredity 68: 205–210.

CRANE, C.F., SLEPER. D.A. 1989. A model of chromosome association in triploids. Genome 32: 82–98.

ENDRIZZI, J.E. 1991. The origin of the allotetraploid species of *Gossypium*. In: Chromosome Engineering in Plants: Genetics, Breeding, Evolution, Part B (Tsuchiya, T., Gupta, P.K., eds.). Elsevier, Amsterdam. pp. 449–469.

HART, G.E. 1987. Genetic and biochemical studies of enzymes. In: Wheat and Wheat Improvement, 2nd ed. (Heyne, E.G., ed.). American Society of Agronomy, Madison, WI. pp. 199–214.

JIANG, J., GILL, B.S. 1994. Different species-specific chromosome translocations in *Triticum timopheevii* and *T. turgidum* support the diphyletic origin of polyploid wheats. Chromosome Res. 2: 59–64.

KENTON, A., PAROKONNY, A.S., GLEBA, Y.Y., BENNETT, M.D. 1993. Characterization of the *Nicotiana tabacum* L. genome by molecular cytogenetics. Mol. Gen. Genet. 240: 159–169.

MORRIS, R., SEARS, E.R. 1967. The cytogenetics of wheat and its relatives. In: Wheat and Wheat Improvement (Quisenberry, K.S., ed.). American Society of Agronomy, Madison, WI. pp. 19–87.

NARANJO, T., ROCA, A., GOICOECHEA, P.G., GIRALDEZ, R. 1987. Arm homoeology of wheat and rye chromosomes. Genome 29: 873–882.

SYBENGA, J. 1992. Cytogenetics in Plant Breeding. Monographs on Theoretical and Applied Genetics 17. Springer-Verlag, Berlin. pp. 271–300.

ZELLER, F.J., CERMEÑO, M-C. 1991. Chromosome manipulations in *Secale* (rye). In: Chromosome Engineering in Plants: Genetics, Breeding, Evolution. Part A. (Gupta, P.K., Tsuchiya, T., eds.). Elsevier, Amsterdam. pp. 313–333.

Section 10.9

DEWET, J.M.J. 1980. Origins of polyploids. In: Polyploidy: Biological Relevance. Basic Life Sciences Vol. 13 (Lewis, W.H., ed.). Plenum Press, New York. pp. 3–15.

LOKKI, J., SAUNA, A. 1980. Polyploidy in insect evolution. In: Polyploidy: Biological Relevance. Basic Life Sciences Vol. 13 (Lewis, W.H., ed.). Plenum Press, New York. pp. 277–312.

STEBBINS, G.L. 1980. Polyploidy in plants: unsolved problems and prospects. In: Polyploidy: Biological Relevance. Basic Life Sciences Vol. 13 (Lewis, W.H., ed.). Plenum Press, New York. pp. 495–520.

UYENO, T., SMITH, G. R. 1972. Tetraploid origin of the karyotype of catostomid fishes. Science 175: 644–646.

Section 10.10

DEWEY, D.R. 1980. Some applications and misapplications of induced polyploidy in plant breeding. In: Polyploidy: Biological Relevance. Basic Life Sciences Vol. 13 (Lewis, W.H., ed.). Plenum Press, New York. pp. 445–470.

11

Deviations from Basic Chromosome Numbers—Aneuploidy

- Aneuploidy can be caused by meiotic irregularities, chromosome aberrations, aging, or environmental stresses.
- The viability, fertility, and phenotypic effects of deleting or adding whole chromosomes varies among organisms.
- Aneuploidy for chromosome arms results from misdivision of univalent centromeres during meiosis.

In contrast to euploidy, aneuploidy refers to deviations of one or more chromosomes from the basic genome in monoploids, or from multiples of the basic genome in diploids and polyploids. The deviations can be additions or subtractions of individual chromosomes. Aneuploids are described symbolically by the somatic chromosome number because the gametic numbers vary. If $2n$ is used for the normal somatic chromosome number of a species, the addition of one chromosome to the somatic complement is designated by $2n + 1$, and the subtraction of a chromosome by $2n - 1$. These formulas apply to any ploidy level, but x can be used instead of n for a particular ploidy level. For example, in diploids, the formulas are $2x + 1$ and $2x - 1$.

Although the addition or subtraction of single chromosomes has severe effects in mammals and often causes lethality early in the life cycle, such changes are tolerated more readily in plants, particularly polyploids. However, if the doses of several chromosomes are changed, the imbalance in gene interactions cannot be tolerated even in the polyploid plant or its gametes. Aneuploidy is an important type of chromosome deviation because it is a major cause of spontaneous abortions in humans and other mammals. Certain aneuploids in plants are useful for gene-mapping and gene-dosage studies.

11.1 TYPES OF ANEUPLOIDS

The normal or disomic state of two doses per chromosome in somatic cells is the basis for comparing different types of aneuploids. The addition of a single chromosome to the $2n$ complement is a trisomic condition $2n + 1$. In a diploid species, this means that one chromosome (trisome) is represented three times in somatic cells; all the other chromosomes are in two doses. In plants, trisomics have been classified into different types, depending on the composition of the extra chromosome (Fig. 11.1). A tetrasomic ($2n + 2$) has two extra homologous chromosomes, so it has four doses of one chromosome compared to two doses of the other chromosomes.

A monosomic or $2n - 1$ condition is the loss of one chromosome from the $2n$ complement. In diploids or allopolyploids, the monosomic chromosome (monosome) and its genes are present once instead of twice in the somatic cells, but in autopolyploids, the genes are still present in two or more doses. The loss of both homologs of one type of chromosome constitutes a nullisomic condition. These and other types of subtraction aneuploids are illustrated in Fig. 11.2.

Some aneuploids involve dosage changes in more than one chromosome, such as double trisomics, which have an extra dose of two nonhomologous chromosomes, or double monosomics, which have one dose of two nonhomologous chromosomes.

11.2 OCCURRENCE AND VIABILITY OF ANEUPLOIDY

Aneuploidy can occur in any eukaryotic organism, but the viability of different aneuploid states depends on a number of factors, including the type of organism, the type of aneuploidy, the chromosome affected, and the background genotype and ploidy level.

In the plant kingdom, diploids usually tolerate the primary trisomic state, although the vigor can vary depending on which chromosome has an extra dose. Complete sets of primary trisomics exist in a number of diploid crop species, including barley, maize, and tomato. Tetrasomics can be obtained from primary trisomics if the male and female $n + 1$ gametes carry the same extra chromosome (Table 11.1), but they are rare in diploids because of the low transmission of $n + 1$ gametes through the male. The modified trisomics (secondary, tertiary, etc.) are viable in the relatively few plant species where efforts have been made to obtain them. Monosomics

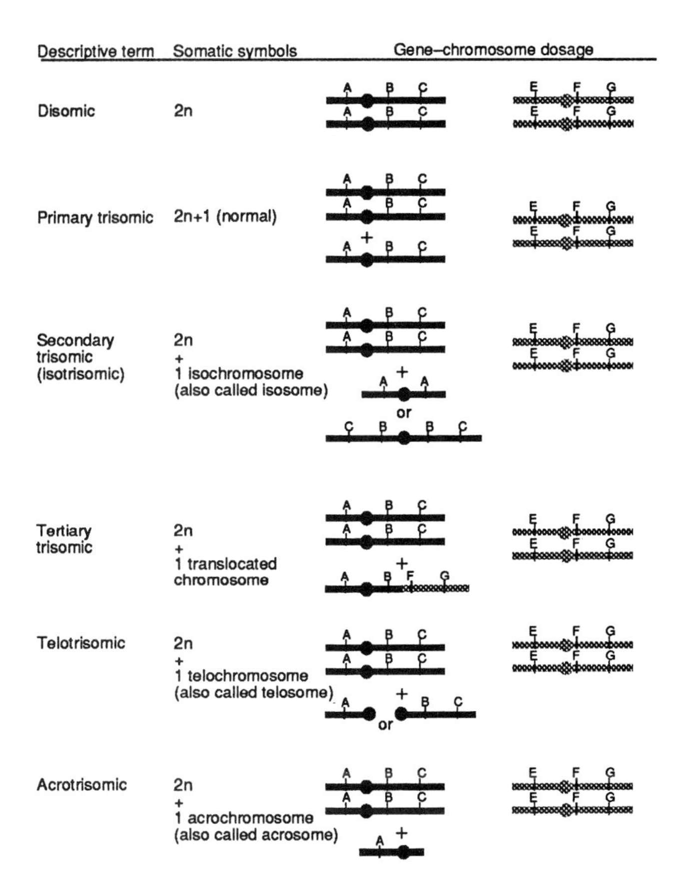

Fig. 11.1 Terms and diagrams illustrating the disomic and various addition-aneuploid states of one chromosome (e.g., A B C) in a diploid species. A second pair of chromosomes (E F G) represents the rest of the chromosome set. The symbol $2n$ is used here in a general sense for the somatic-chromosome number at any ploidy level, although x can be substituted for n when the ploidy level is known. Thus, in this figure, $2n = 2x$ for diploidy.

can be tolerated during the sporophytic stage of diploid plants, but the missing chromosome causes abortion of male and female gametes unless the loss occurs late in the development of pollen or eggs. Polyploids tolerate both trisomics and monosomics, but the latter are more useful because they have more distinctive phenotypic effects. Monosomic sets have been developed in several polyploid species starting with tetraploid tobacco (*Nicotiana tabacum*) and hexaploid wheat (*Triticum aestivum*).

Nullisomics come from the union of male and female $n - 1$ gametes with the same missing chromosome, usually by selfing a monosomic plant (Table 11.2). The frequencies of nullisomics for different chromosomes are usually low because of the competition between n and $n - 1$ pollen, or lethality in the early stages after fertilization. However, certain monosomics in hexaploid oats (*Avena sativa*), as well as certain monotelosomics and monoisosomics in hexaploid wheat, give higher frequencies of nullisomics. In these cases, there is less competition between $n - 1$ pollen and pollen with a complete chromosome set or with one or two doses of one arm. In tetraploid species, nullisomics cannot be recovered through sexual reproduction, but they were obtained indirectly in allotetraploid tobacco by culturing anthers containing $n - 1$ microspores. Haploid plants with a missing chromosome (nullihaploids) were obtained, and their chromosome number was doubled to get nullisomics.

The occurrence of aneuploidy in humans is widespread, but

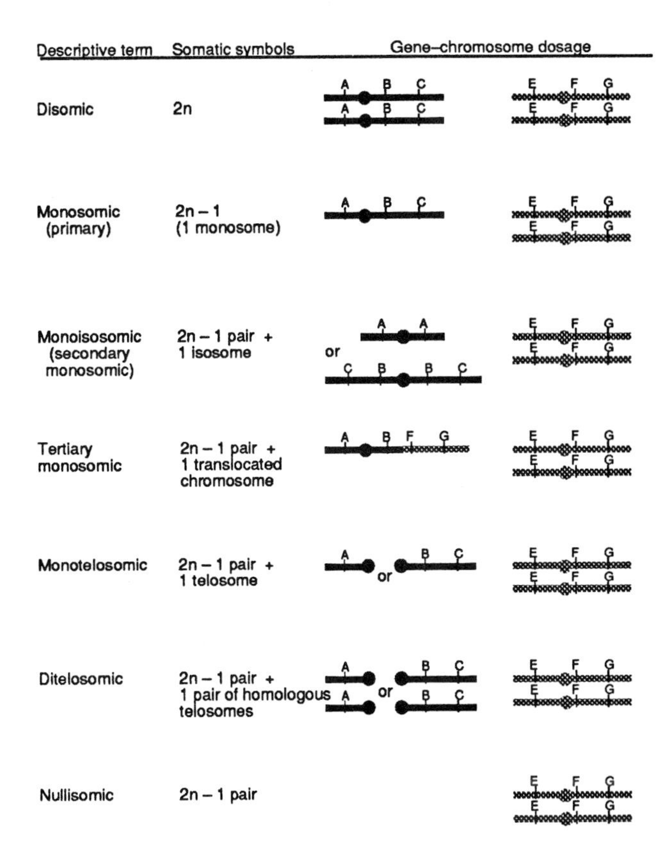

Fig. 11.2. Terms and diagrams illustrating the disomic and various subtraction–aneuploid states of one chromosome (e.g., A B C) in an allotetraploid species. A second pair of chromosomes (E F G) represents the rest of the chromosome set. See Fig. 11.1 for use of n and x. For the allotetraploids in this figure, $2n = 4x$.

most aneuploid states cause spontaneous abortions, often before the pregnancies are detected. For clinically recognized pregnancies, the frequencies of trisomy in extensive studies are 25% for spontaneous abortions, 4% for stillbirths, and less than 0.5% for live births. Trisomies for a few chromosomes are common, whereas those for others are rare. Chromosome 16 is involved in one-third of the trisomies causing spontaneous abortions. In live births, the sex chromosomes make up one-third of the trisomies, and chromosome 21 makes the greatest contribution to autosomal trisomy. The parental sources of monosomic conditions have been detected in spontaneous abortions by using restriction-fragment-length polymorphisms (RFLPs) (Fig. 11.3). Monosomy for any of the chromosomes except X does not occur in live births, presumably because of lethality very early in a pregnancy. However, partial monosomies

Table 11.1. Types of Progeny from the Self-Pollination of a Primary Trisomic ($2n + 1$); the Added Chromosome is Designated No. 5 as an Example

Female Gametes	Male Gametes	
	n	$n + 1$, No. 5
n	$2n$ Disomic for No. 5	$2n + 1$ Trisomic for No. 5
$n + 1$, No. 5	$2n + 1$ Trisomic for No. 5	$2n + 2$ Tetrasomic for No. 5

Table 11.2. Types of Progeny from the Self-Pollination of a Monosomic Plant ($2n - 1$); the Monosomic Chromosome is Designated No. 5 as an Example

Female Gametes	Male Gametes	
	n	$n - 1$, No. 5
n	$2n$ Disomic for No. 5	$2n - 1$ Monosomic for No. 5
$n - 1$, No. 5	$2n - 1$ Monosomic for No. 5	$2n - 2$ Nullisomic for No. 5

(deletions of chromosome segments), as well as partial trisomies (duplications of chromosome segments), for almost every human chromosome have been identified by banding techniques. The sizes of the excess or deficient segments usually do not exceed 5% of the total genome. There seems to be a concentration of partial aneuploidy in relatively few chromosome arms, and partial trisomies are more frequent than partial monosomies.

A mixture of aneuploid and diploid cells (mixoploidy) is another situation where monosomic cells could persist in a viable diploid body. This condition most likely results from nondisjunction in mitotic cells and seems to involve the sex chromosomes more often than the autosomes. From 10% to 20% of all sex-chromosome aneuploidies in humans are mixoploid conditions.

In other mammals, the aneuploid pattern in livestock is similar to that in humans, with sex-chromosome aneuploidy and mixoploidy predominating in animals surviving the gestation period. Studies on spontaneous aneuploid frequencies in mice have been confined mainly to the sex chromosomes using gene markers. Paternal X-chromosome losses are 5–10 times higher than maternal X-chromosome losses, both resulting in XO fertile females, and losses are also about 30 times as frequent as the expected complementary class, XXY (sterile males) with 1 maternal X and 1 paternal X. This indicates that most of the paternal X losses are not due to meiotic nondisjunction, otherwise the complementary classes should be equal in frequency.

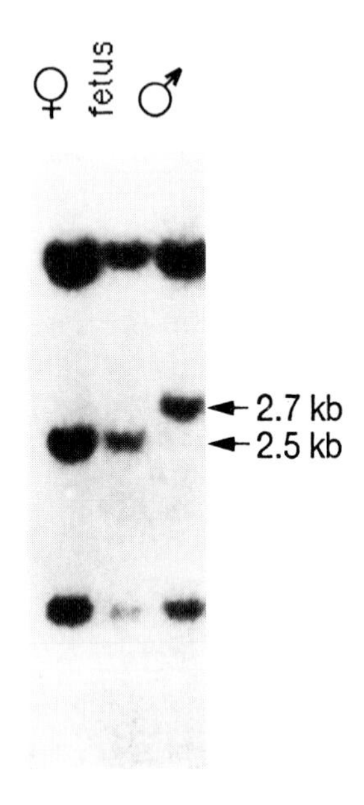

Fig. 11.3. X-chromosome RFLP DNA sequences in a spontaneously aborted human fetus monosomic for X (center), and its disomic parents. By comparing the bands located at 2.5 and 2.7 kb, it is evident that the single X chromosome of the fetus came from the mother. The monosomic condition was, therefore, caused by the loss of X or Y from the chromosomal complement of the father (Hassold, 1986).

In chickens (*Gallus domesticus*), trisomics for macrochromosomes (the large, identifiable chromosomes in the set; *see* Fig. 6.5, chapter 6 for avian karyotype) and mixoploids of euploid and aneuploid sectors have been observed in early-stage embryos, but most of these conditions cause embryonic death. Two exceptions include a viable and fertile chicken that was trisomic for a microchromosome, and a viable F_2 hybrid goose, which was shown to be trisomic for macrochromosome 1 (*see* Chapter 24, Section 24.14).

Among insects, *Drosophila melanogaster* has been more intensively investigated cytogenetically than any other species. The spontaneous aneuploid frequencies for each of the four chromosomes are low, whether originating in males or females. The frequencies of XO sterile males are consistently higher than those of XXY fertile females, similar to the results with mice given above, but the same sex-chromosome genotype has a different effect on sex expression and fertility in mice and *Drosophila*.

11.3 MODES OF ORIGIN OF ANEUPLOIDY

Most of the different types of aneuploids result from irregular chromosome distributions to the poles during cell divisions, so the conditions that produce these irregularities need to be considered. Gains and losses of chromosomes sometimes have a common origin.

11.3.1 Nondisjunction of the Centromeres

Occasional nondisjunction of the centromeres of sister chromatids during a mitotic division sends both chromatids to the same pole (Fig. 11.4a). One daughter cell will have a $2n + 1$ complement and the other a $2n - 1$ complement. If both types of cells divide, they can give rise to sectors of trisomic and monosomic tissues. Transmission of these aneuploid states depends on whether the sectors eventually involve reproductive structures. Nondisjunction can occur during either of the meiotic divisions. If a homologous pair of chromosomes fails to separate at anaphase I (Fig. 11.4b), both chromosomes go to the same pole, and there is no representative of that chromosome at the other pole. With the separation of sister chromatids during meiosis II, half of the cells are $n + 1$ and the other half are $n - 1$. Union of $n + 1$ and n gametes gives a trisomic, and union of $n - 1$ and n gametes gives a monosomic. Nondisjunction of sister chromatids of a chromosome during meiosis II also results in $n + 1$ and $n - 1$ gametes (Fig.11.4c).

11.3.2 Univalent State of Chromosomes

Conditions such as haploidy that cause univalents instead of bivalents to be present during the metaphase I and anaphase I stages of meiosis have the potential for producing aneuploids. If a univalent divides equationally during meiosis I, the sister chromatids go to opposite poles and are included in the nuclei. They cannot divide further during meiosis II, so they lag and are often excluded from the nuclei. A univalent may not divide during meiosis I and lags between the poles, but may be included in a nucleus because of its location. In this case, it divides equationally during meiosis II, and two of the four cells have one dose of this chromosome. A third possibility is that the univalent is not included in a nucleus at the end of meiosis I and remains in the cytoplasm. If this variability in behavior is extended to most or all of the chromosomes in a haploid complement, the resulting gametes have a range in chromosome number. A few gametes with one added or one missing chromosome can produce trisomics or monosomics, respectively, by uniting with a complete gamete. The gametes with one extra chromosome result from the inclusion of two unpaired homologs in the same nucleus at the end of meiosis I and equational separation of their sister chromatids during meiosis II, so that two of the four products have an extra dose of the chromosome. The loss of this chromosome from the other telophase I nucleus leads to deficient gametes.

Univalents are the main source of telochromosomes (telosomes) and isochromosomes (isosomes) because of occasional abnormal behavior (misdivision) of their centromeres during anaphase I or II (Fig. 11.5). Normally, a centromere divides lengthwise like the rest of the chromosome, and each chromatid has an equal amount of centromere material. A univalent centromere sometimes divides transversely, with each arm receiving varying amounts of centromere material depending on the dividing line. Because the centromere region consists of repetitive DNA sequences (*see* Chapter 20, Section 20.3), parts of it can be lost without interfering with its function, but if the transverse division is too unequal, only one arm has a functioning centromere. A linear centromere division and separation of the sister chromatids may precede misdivision, which can occur in one or both chromatids independently, and result in two or four telosomes. If misdivision occurs while the sister chromatids are still together, it produces two isosomes, each with two identical arms attached by part of the centromere. Once telosomes and isosomes are obtained, they can give rise to each other if they occur as univalents during meiosis. Misdivision of a telosome while its chromatids are together can give an isosome and a centromere fragment, which would probably be lost. Misdivision of an isosome after its chromatids have separated gives two telosomes for the same arm.

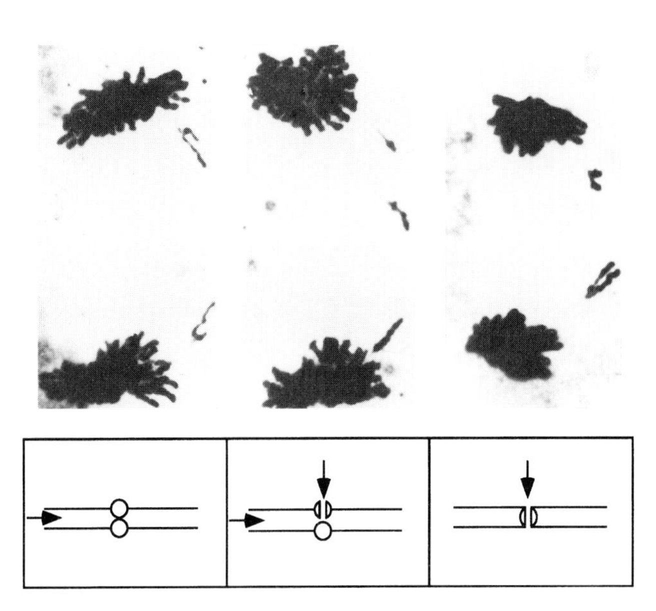

Fig. 11.5. Photographs and diagrams of normal division and misdivision of a wheat univalent at anaphase I. (Morris and Sears, 1967). **Left:** Normal centromere division, with sister chromatids moving to opposite poles. **Center:** Centromere misdivision in the upper chromatid after chromatid separation, resulting in two telochromosomes. There is no misdivision in the lower chromatid where the short arm overlaps the long arm. **Right:** Centromere misdivision before chromatid separation, resulting in two isochromosomes.

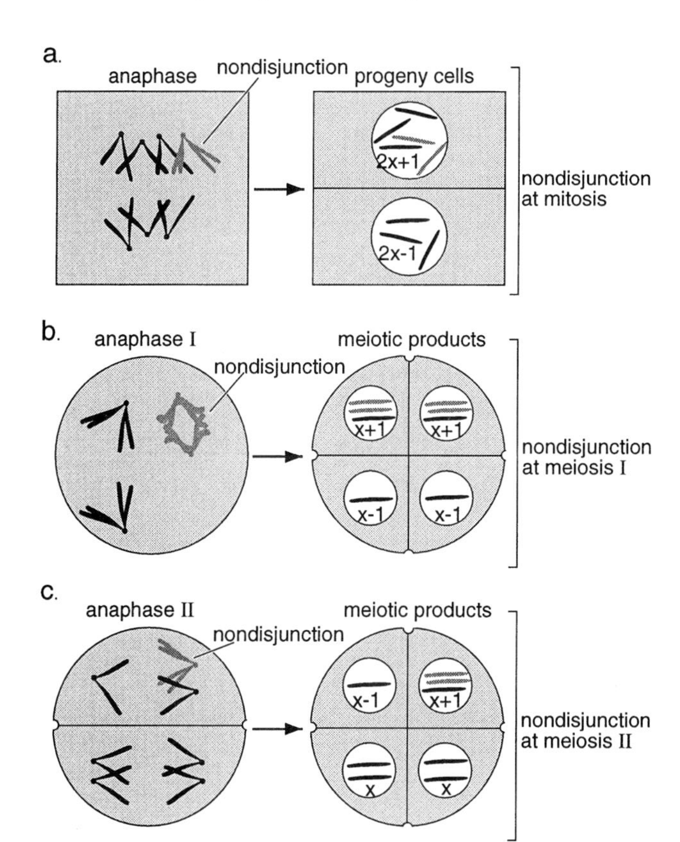

Fig. 11.4. Nondisjunction in diploid plant cells with 2x = 4. (a) mitotic anaphase—the sister chromatids of one of the chromosomes are at the upper pole, resulting in aneuploid states in the progeny cells. (b) Meiotic anaphase I—both members of one bivalent are at the upper pole (left), whereas the homologs of the other bivalent have separated. After sister-chromatid separation at anaphase II, the four products are aneuploid (right). (c) Meiotic anaphase II—the sister chromatids of one of the chromosomes have both moved to the pole on the right in the upper cell, resulting in aneuploid states for two of the four meiotic products.

11.4 CONDITIONS THAT CAUSE ANEUPLOIDY

Conditions that affect chromosome pairing, or that produce haploidy and triploidy, result in high sterility in the individuals where they occur, because many of the gametes are unbalanced due to added or missing chromosomes. The few gametes that are functional are a good source of aneuploids, and their sources are discussed in the following subsections.

11.4.1 Mutations in Genes Controlling Meiotic Pairing and Separation

Genes for asynapsis (no meiotic pairing) and desynapsis (premature separation of homologs) (*see* Fig. 10.7) result in univalents instead of bivalents at metaphase I. The irregular distribution of the univalents at anaphases I and II can result in a few viable gametes, with one or two chromosomes above or below the normal number for the species involved. When these gametes unite with gametes having the normal chromosome number, aneuploids are produced.

11.4.2 Occurrence of Haploid States

Monohaploids from diploids and polyhaploids from allopolyploids have high frequencies of univalents during meiosis because of the absence of homologous partners (*see* Chapter 12, Section 12.3). Both trisomics and monosomics are produced in plants by crossing haploids as females with their diploid or polyploid counterparts, because haploids are more likely to transmit aneuploid gametes through eggs than through pollen.

11.4.3 Crosses Involving Autotriploids

Autotriploids crossed as females with diploids are an efficient source of trisomics in plants, because many of the viable female gametes formed by the triploid have extra doses of one to three chromosomes as a result of trivalent formations and 2 : 1 separations during meiosis (*see* Chapter 10, Section 10.6). In the Solanaceae, the transmission of gametes with one to three extra chromosomes was 60% in the tomato (*Lycopersicon esculentum*) and close to 72% in *Datura stramonium*.

11.4.4 Crosses Between Related Species

Interspecific crosses between related species with different ploidy levels give some univalents at meiosis and are a potential source of monosomics (Fig. 11.6). The disadvantage of interspecific crosses is that they introduce heterozygosity, which makes it more difficult to distinguish the monosomic phenotypes. The solution is to transfer the monosomic states, through crosses, to a uniform genetic background such as a homozygous variety or line before comparing their effects.

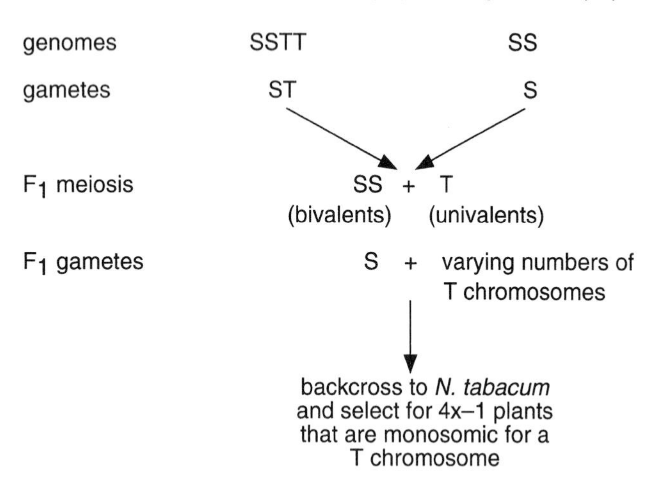

Fig. 11.6. An experimental procedure to obtain monosomics from interspecific crosses in *Nicotiana*. *N. tabacum* (tobacco) is an allotetraploid with an S genome from diploid *N. sylvestris*, and a T genome believed to come from a hybrid between two other diploid *Nicotiana* species. The cross shown is between the allotetraploid and the diploid source of the S genome, but a cross involving the T-genome diploid species could also be used. In order to compare the phenotypes of the different monosomics in a uniform genetic background, they were backcrossed several times to a particular line of *N. tabacum* (Clausen and Cameron, 1944).

11.4.5 Reciprocal Translocations with Irregular Chromosome Distributions During Meiosis

Heterozygous reciprocal translocations with a 3 : 1 distribution of chromosomes from a ring-of-four or chain-of-four during meiosis can produce $n + 1$ and $n - 1$ gametes. The type of trisomic or monosomic from these gametes depends on how the chromosomes in the translocation are aligned on the metaphase I plate (*see* Fig. 9.21c). If the centromeres of the two standard (nontranslocated) chromosomes are not oriented with respect to the poles and at anaphase I these chromosomes pass to one pole with one of the translocated chromosomes, the gametes resulting from this distribution have an extra, translocated chromosome. If such gametes unite with gametes having standard chromosomes, tertiary trisomics are produced (*see* Fig. 11.1). If the two nonoriented chromosomes are translocated and they pass to one pole with one of the standard chromosomes, the extra chromosome in the gametes is standard. The union of this type of gamete with a gamete having standard chromosomes gives an interchange trisomic (i.e., a trisomic state with a translocation background). In both situations, the nucleus at the other anaphase I pole receives only one chromosome from the ring or chain—a translocated chromosome in the first case and a standard chromosome in the second case. Such nuclei lack sizable amounts of chromatin, resulting in the abortion of gametes or zygotes in diploid organisms. However, tertiary monosomics with a single, translocated chromosome were obtained in tomato by irradiating mature pollen, which induced a break in the centromere regions of two nonhomologous chromosomes, followed by union of two of the four arms. The pollen functioned in fertilization because the male gametes contained the other two arms, which were lost during the zygotic cell division because of defective centromeres.

In compensating trisomics, two structurally modified chromosomes compensate for a missing standard chromosome. The compensations may consist of two translocated chromosomes from different sources, a translocated chromosome and a telosome, or other combinations that give the equivalent of the missing chromosome. Two types of compensating trisomics are shown in Fig. 11.7.

11.4.6 Robertsonian Translocations

Heterozygous Robertsonian translocations between two acrocentric chromosomes (*see* Chapter 9, Section 9.3) have the potential for producing trisomic states for most of one chromosome and monosomics for the other chromosome. These aneuploid conditions occur if the translocated chromosome goes with one of the normal chromosomes to the anaphase I pole (source of trisomics), whereas the other normal chromosome goes to the other pole (source of monosomics), and the resulting gametes unite with gametes having standard chromosomes (*see* Fig. 9.24).

11.4.7 Aging of Cells or Organisms

The frequencies of aneuploids in cell or organism populations may be increased by aging. In humans, a study of over 1300 sponta-

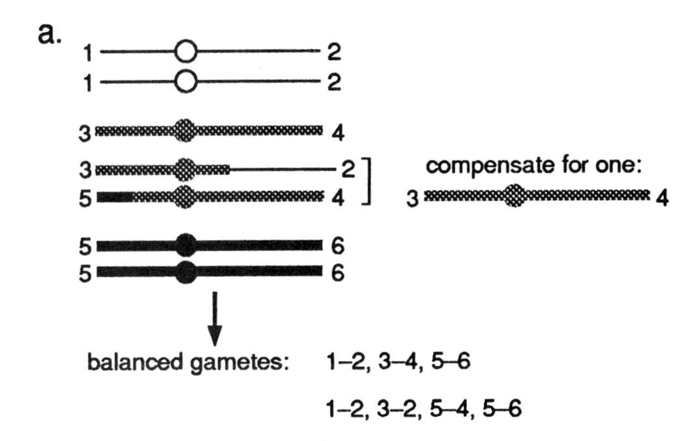

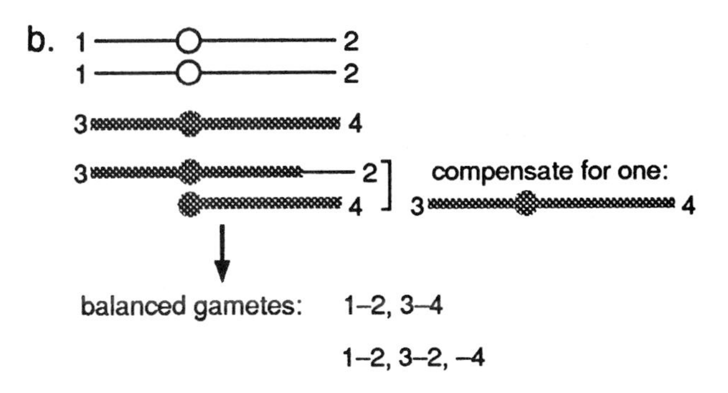

Fig. 11.7 Two types of compensating trisomics. The arms of each chromosome are marked with different numbers. (a) Two translocated chromosomes (3.2 and 5.4) from different translocations supply segments for a complete 3.4 chromosome. If the breakpoints in 3.4 are at different locations in the two translocated chromosomes as shown, the region between the breakpoints is trisomic, as are the distal parts of arms 2 and 5. (b) The translocated chromosome 3.2 and a telochromosome for arm 4 compensate for the missing chromosome 3.4. Parts of arms 2 and 4 are trisomic; the other parts of the three chromosomes are disomic. (Adapted from Sybenga, 1992.)

neous abortions showed that the mean maternal age was higher for trisomic conditions than for other types of chromosomal abnormalities. Among the trisomics, the pronounced effect of maternal age involved mainly acrocentric chromosomes (groups D and G). For trisomics surviving beyond birth, the frequencies for at least three chromosomes (numbers 13, 18, and 21) are influenced by the age of the mother. The most investigated condition is trisomy for chromosome 21, which causes a complex of abnormalities known as Down's syndrome after the person who first described the deviations. The mean age for mothers having children with Down's syndrome in 11 countries is 35.5 years compared to 28 years for mothers of children without the syndrome. The risk of a woman having a child with Down's syndrome therefore increases markedly in the later childbearing years. The main cause of trisomy 21 is thought to be nondisjunction of chromosome 21 during gametogenesis, and polymorphic DNA probes for this chromosome, as well as banding techniques, make it possible to trace the source of nondisjunction to meiosis I or II in the mother or father. Nondisjunction of chromosome 21 occurs more frequently in the mother, and one interpretation is that aging promotes some condition in the chromosomes or in the cell components that increases this type of chromosome behavior. Another suggestion is that older women do not have as many spontaneous abortions of trisomic 21 fetuses as younger women, therefore there are higher frequencies of live-born infants with Down's syndrome.

In *Drosophila melanogaster*, increases in frequencies of aneuploid offspring were obtained by a combination of maternal aging and treatments applied during the aging period such as a temperature of 10°C (control 25°C) or a minimal food medium. Oviposition is reduced or inhibited under these conditions, but it resumes after the treatments are ended. Separation of progenies into broods, based on the length of time after a treatment, made it possible to localize the origin of aneuploidy to meiotic or premeiotic cells. In both mice and Chinese hamsters, older females usually give higher frequencies of aneuploids in their offspring than younger females. The increase in mice can be attributed to nondisjunction of one or more chromosomes during meiosis I, based on observations of metaphase II in oocytes cultured *in vitro*.

11.4.8 Environmental Conditions

Two important environmental factors with the potential for inducing aneuploidy are radiations and chemicals. Experiments on yeast (*Saccharomyces cerevisiae*) show that ultraviolet (UV) and X radiations, when applied to cells that later undergo mitosis, increase the frequencies of aneuploids over those in untreated cells, and that UV is also effective when applied to premeiotic cells. In *D. melanogaster*, frequency of loss of an X chromosome depends on the stage of gametogenesis irradiated, as well as the X-ray dose, in both males and females. In studies of nondisjunction of the sex chromosomes, males were irradiated when most of the cells in the testes were primary spermatocytes. The small proportion of XXY female progeny was attributed to nondisjunction of X and Y in the damaged spermatocytes. In irradiated females, induced nondisjunction for chromosomes X and 4 depends on a specific stage of the oocytes when treated with X-rays. It should be pointed out that after irradiation, losses and gains of chromosomes may not trace to a common event because of other effects of the treatment. Whereas irradiation is able to induce aneuploidy in *Drosophila* and yeast, there is no decisive evidence for its effectiveness in Chinese hamsters after irradiation of females, or in mice after irradiation of young or old females, males, oocytes (*in vitro*), or spermatogonia. In humans, extensive chromosome studies have been made on groups at higher risk for irradiation effects because of their occupations, the regions where they lived, medical treatments, or catastrophic events such as exposure to radiations from atomic bombs in Japan during World War II. These studies indicate that at least for low dose levels and rates, there seems to be only a small risk of abnormalities from induced nondisjunction. However, irradiation can induce structural aberrations, and some of these can result in aneuploidy because of irregular distributions during meiosis.

Many chemicals with aneuploid-producing potentials are natural components of the environment, or are introduced by human activities. It has been challenging to develop assays that detect chemicals of this type, because some chemicals induce aneuploidy in some organisms but not in others. Another problem is that different chemicals can affect different targets or mechanisms in the cells, including the spindle, centrioles, centromeres, or chromosome activities during meiosis such as pairing and crossing over. Chemicals that interfere with spindle activities such as microtubule formation or spindle elongation can cause irregular chromosome distributions, resulting in aneuploidy. Diazepam induces aneuploidy in Chinese hamsters by inhibiting centriole separation. In maize, ethylene glycol causes equational separation of centromeres at anaphase I, and the sister chromatids are distributed independently during meiosis II, resulting in some aneuploids.

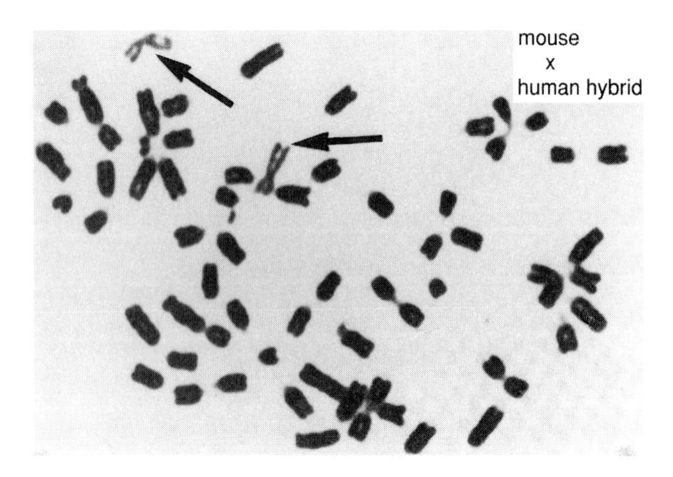

Fig. 11.8. Mitotic metaphase in a mouse–human hybrid cell after benomyl treatment. The human chromosome 2 was increased from one to two doses (arrows) by the treatment, and can be distinguished from the mouse chromosomes by lighter staining (Athwal and Sandhu, 1985).

Several species of fungi have been used to detect aneuploidy induced by chemicals because their chromosome organization and spindle formation are similar to those in higher eukaryotes, although some aspects of cell division are different. Other tests have included studies on the effects of chemicals on *in vitro* spindle–microtubule assembly, cultured cells from various organisms and tissues, *D. melanogaster*, and progenies of organisms exposed to chemicals.

In scoring cells for chromosome number, it is easier to detect chromosome gains than losses, because a chromosome loss attributed to a chemical may be due to the technique in squashing cells. This problem was solved by using a mouse–human hybrid cell culture containing one human chromosome 2 with a dominant gene for the enzyme xanthine–guanine phosphoribosyl transferase, transferred from *Escherichia coli*. Growth of the cells in a medium containing xanthine and mycophenolic acid retained chromosome 2, whereas growth in 6-thioguanine selected cells that had lost this chromosome. The human chromosome could be distinguished from mouse chromosomes by Giemsa-II or fluorescence banding. Three chemicals, benomyl, colcemid, and cyclophosphamide, caused both gains and losses of chromosome 2 when they were added to the medium in appropriate doses (Fig. 11.8). Another technique involving a human sex chromosome is the use of fluorescence after quinacrine staining to detect the number of Y chromosomes in spermatozoa from semen samples. This procedure showed that there were striking increases in cells with two Y chromosomes after the men were exposed to occupational or therapeutic chemicals.

11.4.9 Viral Infections

Viruses have been implicated in aneuploidy as well as other types of chromosomal aberrations. The barley-stripe-mosaic virus induces aneuploids in barley and wheat. One virus-induced condition is the persistence of nucleoli during mitosis, which makes it difficult for sister chromatids to separate, so that nondisjunction can occur and lead to aneuploidy. It is speculated that viruses may produce their effects by interfering with DNA or protein synthesis within cells, or that viral nucleic acid combines with cellular nucleic acid to disrupt cell division.

11.4.10 Physiological Stress

Most of the factors causing aneuploidy create stressful conditions for cells and disrupt the harmony of cellular processes. When mice were stressed by keeping them in small cages, there was an increase in aneuploid cells along with increases in hormone levels. Another stress-producing situation is cell culture *in vitro*, which can affect chromosomes as well as genes. The resulting variation has included monosomics and trisomics, which are thought to arise from mitotic errors.

11.5 EFFECTS OF ANEUPLOIDY ON PHENOTYPES AND INTERNAL TRAITS

In diploid organisms, both gains and losses of single chromosomes generally have pronounced effects on phenotypic traits due to associated disruptions in cellular processes. The severity of the effects varies with the organism and the individual chromosomes.

In humans, the phenotypic and other disorders associated with aneuploid conditions that survive beyond the gestation period have been described as syndromes, together with the names of the persons who first described the symptoms. The autosomal trisomies in this category include trisomy-13 or Patau's syndrome, trisomy-18 or Edwards' syndrome, and trisomy-21 or Down's syndrome (Fig. 11.9). Trisomy-13 and trisomy-18 have some common symptoms, including mental and developmental retardation, as well as some distinguishing features. The abnormalities in both cases are so severe that death usually occurs within a few months after birth. About one in six infants with trisomy-21 die during the first year, mainly due to congenital heart disease or respiratory infections, and the average life span is about 16 years. Males are sterile because of abnormal reproductive organs. At least some females are fertile and can transmit both n and $n + 1$ gametes to their offspring. From studies of persons with trisomy for part of chromosome 21, most of the symptoms associated with Down's syndrome appear when

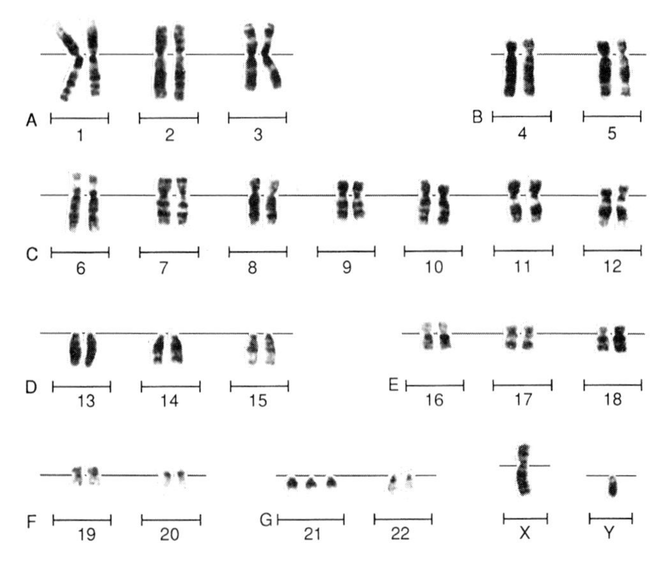

Fig. 11.9. Mitotic karyotype, stained with Giemsa, of a Down's syndrome human male with trisomy for chromosome 21 (Levitan, 1988).

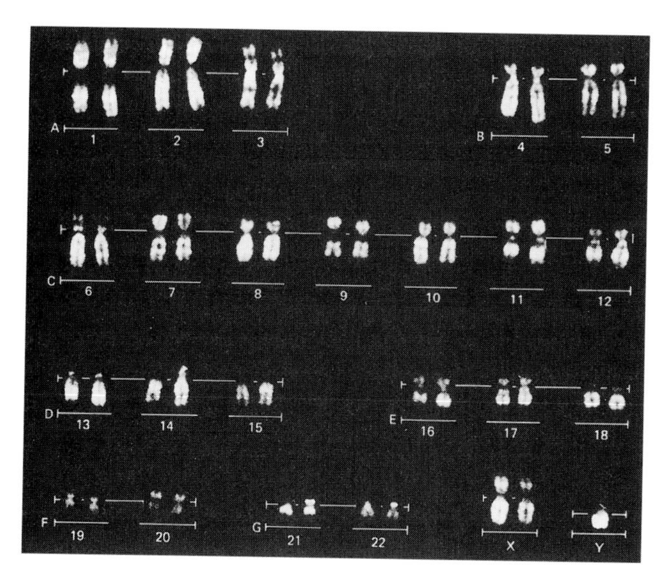

Fig. 11.10. Mitotic karyotype of a human male with Klinefelter's syndrome, showing trisomy for the sex chromosomes, XXY (Novitski, 1977).

the terminal part of q21 is trisomic. An increase in phosphofructokinase activity occurs in erythrocytes from persons with trisomy-21.

Trisomy for the sex chromosomes in humans consists of three genotypes, XXY, XYY, and XXX. Klinefelter's syndrome, which is based primarily on male reproductive abnormalities, is usually due to XXY trisomy (Fig. 11.10), but may occur with additional doses of X, as in tetrasomy (XXXY) or pentasomy (XXXXY). The phenotypic effects in different men can range from nearly normal to noticeable effects on body form, fertility, and such disorders as depressed thyroid function, chronic pulmonary problems, and diabetes mellitus. Mental retardation increases with the number of X chromosomes. Klinefelter's syndrome also results from mosaicism, most frequently the XY/XXY combination, which gives a lower frequency of men with reproductive and intelligence problems than overall XXY trisomy. XYY trisomic males have genital abnormalities and below-average intelligence. Some XYY individuals have normal psychological behavior; others may be aggressive against property. XXX trisomic females show variations ranging from a normal phenotype to abnormalities in sexual traits, reduction in intelligence, and behavior problems. Four or five doses of X intensify the symptoms.

In livestock, XXY and XXX trisomies in cattle and XXY trisomy in sheep have adverse effects on the reproductive organs. Trisomy for chromosome 17 in cattle is assayed by karyotype analysis of G-banded chromosomes and causes lethality due to a very short lower jaw and other defects. Animals with sectors of trisomic cells are sometimes viable and fertile.

Monosomy for the X chromosome (XO where O refers to a missing X or Y) has been found in humans (Turner's syndrome; Fig. 11.11) and in other mammals including horses (*Equus caballus*), swine (*Sus scrofa domestica*), and mice (*Mus musculus*). The XO phenotype in all these species is that of a sterile female, except in the mouse, where the female is fertile. The Turner's syndrome traits of XO human females have also been found in females with one X and an isochromosome for the long arm of X, thus giving a monosomic state for the short arm and a trisomic state for the long arm. Deletion of part of one X short arm gave

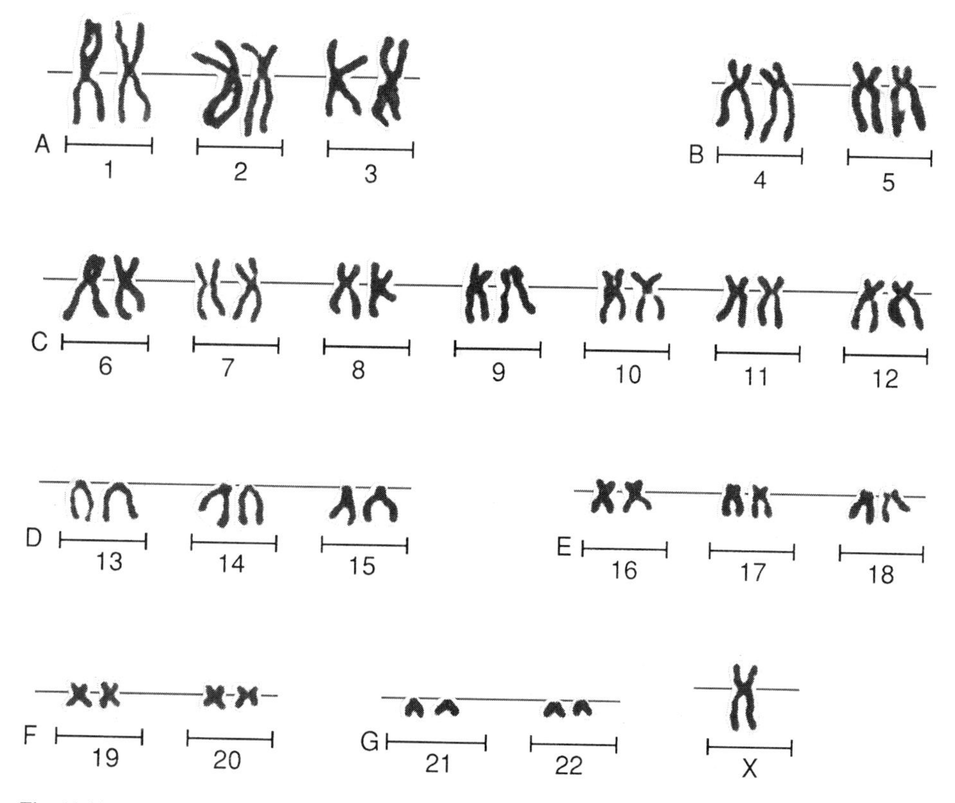

Fig. 11.11. Mitotic karyotype in a human with Turner's syndrome, showing monosomy for the sex chromosomes, XO (Novitski, 1977).

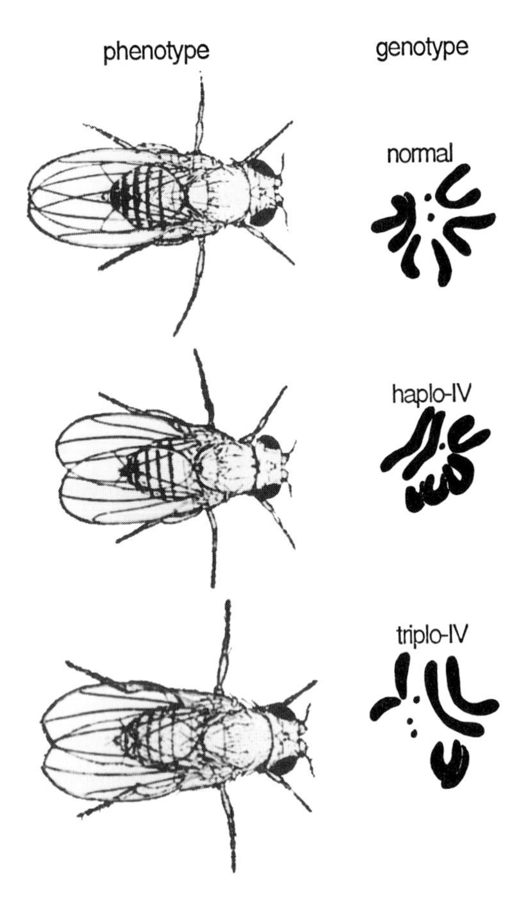

Fig. 11.12. Drawings of female flies of *Drosophila melanogaster* showing (top) a diploid ($2n = 2x = 8$), (center) a monosomic for chromosome 4 (haplo-IV), and (bottom) a trisomic for chromosome 4 (triplo-IV). Note the effect of aneuploidy on body size. In *Drosophila*, monosomy is tolerated only for 4, the smallest chromosome, and for the X chromosome (Morgan et al., 1925).

the complete Turner's syndrome expression, and deletion of part of one long arm gave a less severe expression of the syndrome. Thus, these partial monosomics have helped to localize the genes involved.

Monosomy for any of the autosomal chromosomes has not been reported in live births of mammals, so the loss of one of these chromosomes is presumably lethal early in the gestation period. *D. melanogaster* flies that are monosomic for the smallest chromosome, 4 (Fig. 11.12), develop slowly, may die, and are often sterile if they live, but the monosomic condition has been transmitted to progeny.

In plants, the phenotypic effects of trisomy are clearest in diploid species, because gene replications in polyploids often mask the dosage effects of an extra chromosome. The different trisomics of a set within a species are more easily differentiated from diploids and from each other when they are all in the same homozygous genetic background. Another requirement is favorable growing environments without competition from diploids, because trisomics generally develop more slowly.

Trisomic chromosomes affect many aspects of development and differentiation, both externally and internally, throughout the plant life cycle. In species with a range in chromosome length, longer chromosomes can have more pronounced trisomic effects than shorter chromosomes, including a greater reduction in fertility, as was found in rice (*Oryza sativa*). Observations of the primary trisomics in *Datura stramonium* and tomato (*Lycopersicon esculentum*) indicate that a specific trisomic condition modifies different organs in the same direction. For example, tomato trisomics 3 and 4 increase the length of leaves, stems, inflorescences, fruits, and seeds, whereas trisomics 7 and 10 decrease the lengths of most of these parts. Usually, certain key features are used to identify different trisomics. These include seedling traits in spinach (*Spinacia oleracea*) (Fig. 11.13), leaf traits in tomato, panicle types in sorghum (*Sorghum bicolor*), and seed-capsule sizes and shapes in *Datura* (Fig. 11.14). If the chromosomes involved in the trisomics cannot be identified by phenotypic effects or chromosome morphology, chromosome banding and *in situ* hybridization can be used to identify the trisomes and were applied to distinguish different primary trisomics in diploid wheat (*Triticum monococcum*).

When telotrisomics, isotrisomics, tertiary trisomics, or compensating trisomics are available, they can be compared phenotypically with the appropriate primary trisomics for the effects of extra doses of single arms or segments versus whole chromosomes. Isotrisomics, also called secondary trisomics, are tetrasomic for genes on the arm included in the isochromosome (*see* Fig. 11.01). In *Datura stramonium* and tomato, isotrisomics for opposite arms of a chromosome have divergent effects on several traits, whereas the phenotype of the primary trisomic for the same chromosome is intermediate between the two isotrisomics (Fig. 11.15). Tomato telotrisomics and isotrisomics for long arms resemble the pertinent primary trisomics, but those for the short arms are difficult to distinguish from disomics. If a tertiary trisomic has a translocated chromosome consisting of the short arm of one chromosome and the long arm of another, it resembles the primary trisomic for the chromosome contributing the long arm. These observations indicate that most of the genes producing trisomic effects are located on the long arms, at least in tomato. Compensating trisomics (*see* Section 11.4.5) have varying numbers of chromosome segments with three doses depending on their origin, so they resemble the related primary trisomic or one of the modified trisomics.

The effects of monosomy in plants vary according to the ploidy level of the species. In hexaploid species such as wheat (*Triticum aestivum*) and oats (*Avena sativa* and *A. byzantina*), if monosomics for each of the different chromosomes are grown in a favorable environment, most of them cannot be distinguished from each other or from disomics phenotypically. In these allopolyploids, genes on homoeologous chromosomes generally compensate for the reduced dosage of genes on the monosomes, and thereby produce a normal or nearly normal phenotype. A few genes lacking compensation have distinctive effects in reduced dosage and are useful for identifying the monosomics.

Monosomic features are clear enough at the tetraploid level to separate the monosomics from the disomics and from each other based on phenotype, but it is important to make comparisons in a uniform genetic background, as was mentioned for trisomics in diploids. Differences in various flower parts are useful in identifying tobacco (*Nicotiana tabacum*) monosomics, because flower traits are more stable in different environments than other characters. Monosomics of tetraploid cotton (*Gossypium hirsutum*) and tetraploid wheat (*Triticum turgidum*) are less vigorous and generally less fertile than tobacco monosomics. Cotton monosomics of the two component genomes, A and D, can be separated on the basis of boll traits.

Diploid plant species are most severely affected by monosomy. A few primary monosomics and a number of tertiary monosomics were obtained in tomato. All differed from disomics in traits affecting many parts of the plants, similar to those affected by trisomy, but with more extreme effects. The maize monosomics resulting from deletion-induced nondisjunction in the embryo sac (*see* Chapter 13, Section 13.4) were reduced in size compared to disomics, and each monosomic type could be distinguished by several quantitative traits.

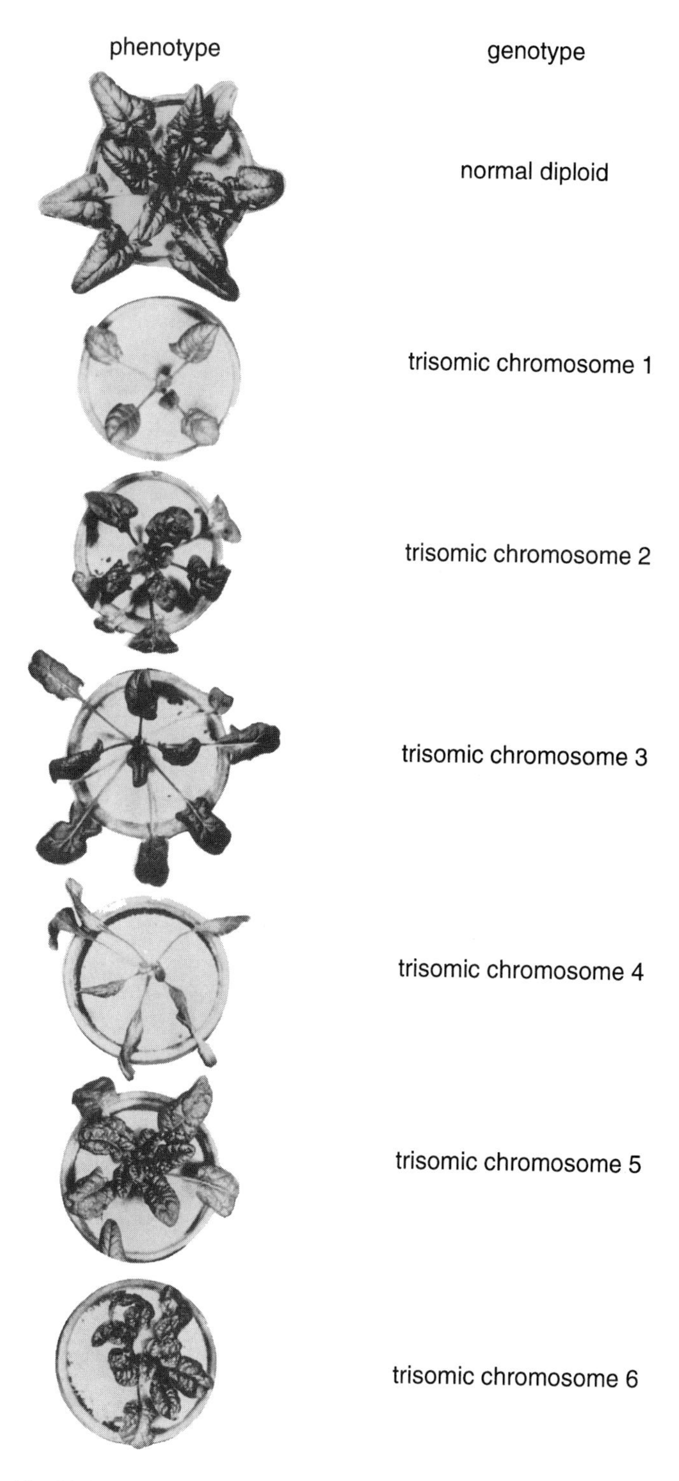

Fig. 11.13. Seedlings of spinach (*Spinacia oleracea*) grown in petri dishes to compare growth traits between the diploid (top) and the six primary trisomics (rearrangement of photo from Ellis and Janick, 1960).

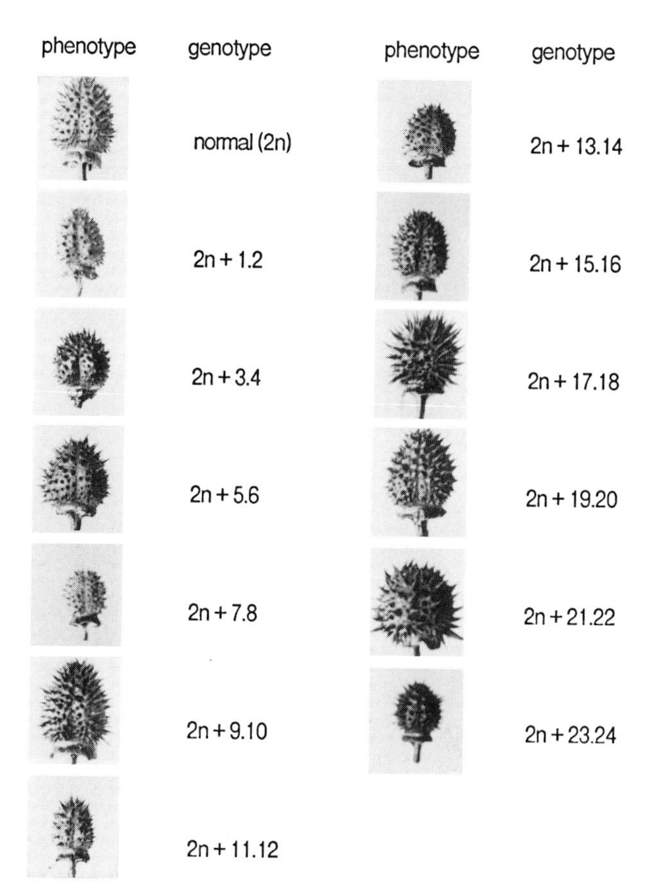

Fig. 11.14. Seed capsules of *Datura stramonium* ($2n = 2x = 24$) showing the normal diploid (top left) and the 12 primary trisomics. It is customary in this species to use numbers to designate the arms of each of the 12 chromosomes (rearrangement of photo from Avery et al., 1959).

11.6 MEIOTIC BEHAVIOR, FERTILITY, AND TRANSMISSION OF ANEUPLOIDS

The primary aneuploids, whether trisomics or monosomics, have an odd chromosome number, so we need to concentrate on the meiotic behavior of the extra chromosome in the case of trisomics, and on the single chromosome in the case of monosomics. As with polyploids, the studies on meiotic behavior in aneuploids have usually involved plant species because of the rarity of aneuploidy at the reproductive stage in animals.

11.6.1 Trisomics and Related Types

In the pachytene stage of meiosis, the trisome of a primary trisomic may form a trivalent with two-by-two pairing in different segments of the three chromosomes (Fig. 11.16), a bivalent and a univalent, or rarely three univalents. Even though there is consistent trivalent synapsis, there has to be a minimum of two chiasmata appropriately placed to maintain a trivalent until anaphase I. In species with a range in chromosome length such as maize and tomato, the longer chromosomes have higher frequencies of trivalents because they provide greater opportunity for chiasma formations. On the other hand, the chromosomes of barley and tobacco are relatively uniform in size, and the trisomics of a set do not differ significantly in trivalent frequencies.

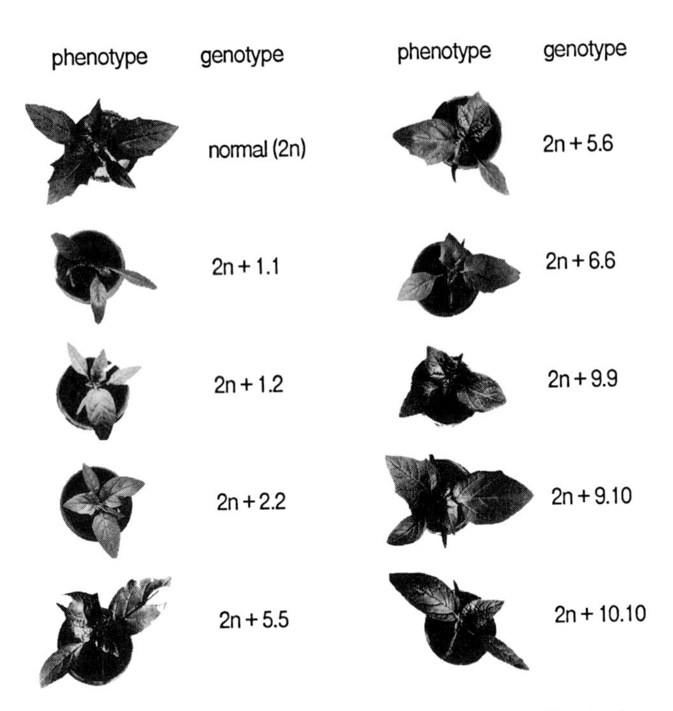

Fig. 11.15. Seedlings of *Datura stramonium* showing growth traits for the normal diploid (top left), three primary trisomics (e.g., $2n + 1.2$, 5.6, or 9.10) and several corresponding secondary trisomics or isotrisomics (e.g., $2n + 1.1$, etc.). Numbers designate the arms of each chromosome (rearrangement of photo from Avery et al., 1959).

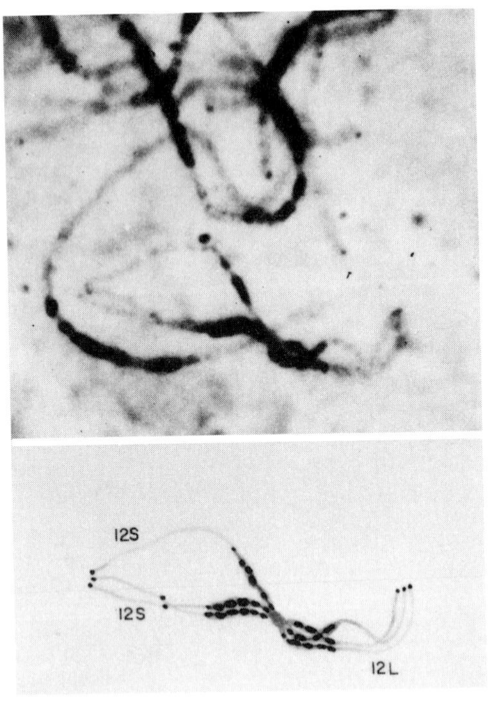

Fig. 11.16. Photograph (top) and diagram (bottom) of a trivalent configuration at pachytene in a primary trisomic for tomato chromosome 12 (S = short arm, L = long arm). Only two chromosomes are paired at any given point. The centromeres of the three chromosomes are grouped at the constriction between the two heterochromatic regions (Khush, 1973).

Fig. 11.17. Metaphase I in a barley ($2n = 2x = 14$) primary trisomic, showing a V-shaped trivalent and six bivalents (Tsuchiya, 1960).

The chainlike trivalent can be oriented in different ways on the metaphase I plate (Fig. 11.17). The V and Y shapes have each centromere oriented with respect to a pole, and there is a 2:1 distribution at anaphase I. Half of the meiotic products are $n + 1$ and the other half are n. Other configurations have one chromosome with a nonoriented centromere. This chromosome may pass with one of the other two to a pole, or it may lag like a univalent when it frequently remains in the cytoplasm. In the latter case, and also when the third chromosome is unpaired, the proportion of n gametes is increased. There may be a low frequency of gametes with a telosome or isosome instead of a complete extra chromosome because of univalent misdivisions during meiosis.

The fertility of primary trisomics varies depending on the genetic background in which they occur, and on the size and genetic composition of the trisome. The extra dose of genes on the trisome may interact unfavorably with the disomic doses of genes on the other chromosomes in some genotypes. Longer trisomes result in a higher proportion of $n + 1$ gametes with more genetic unbalance than shorter trisomes, causing some aborted pollen and ovules. Most primary trisomics have enough fertility to be maintained, but they segregate for trisomics and disomics.

The transmission rates of male and female $n + 1$ gametes can be tested by using primary trisomics as male and female parents in crosses with disomics. The male $n + 1$ gametes generally have zero to very low transmission rates, although a few trisomic types can overlap the rates through the female. Because a trisomic plant produces both n and $n + 1$ pollen, the viable aneuploid pollen often cannot compete with the euploid pollen because of slower pollen-tube growth or adverse physiological interactions with the disomic stigma or style. The female $n + 1$ gametes do not have competition because there is normally one egg cell per embryo sac, and although they usually have higher transmission rates than male $n + 1$ gametes, there are differences among trisomics as shown by the frequencies of trisomics in the progenies. In rice (*Oryza sativa*), the range is from about 15% to almost 44% for the 12 primary trisomics. These differences may be due to variations in loss of the extra chromosome during meiosis or in the amount of ovule or seed abortion.

The extra chromosome in acrotrisomics, telotrisomics, and isotrisomics forms a trivalent with its two homologs (Fig. 11.18) or remains unpaired as a univalent during meiosis. The frequency of trivalent formation, which depends on the extent of the pairing surface in these modified chromosomes, is influenced by the position of the breakpoint in acrotrisomics and the arm length in telotrisomics and isotrisomics. An isochromosome can synapse within itself because of its identical arms, forming an unusual ring univalent. Any of these trisomic types can yield n or $n + 1$ gametes. The extra chromosome in the aneuploid gametes can be the modified chromosome, or occasionally one of the two normal chromosomes if they go to the same pole at anaphase I. The fertility of these modified trisomics and the transmission of the extra chromosome depend on how much of the chromosome is present in three or four doses, and how much euchromatin and heterochromatin the extra segments contain.

Tertiary and interchange trisomics can have a variety of meiotic configurations, ranging from chains-of-five chromosomes (pentavalents) to univalents, depending on the lengths of translocated segments and the chiasma frequencies. In tomato, the most frequent configurations in 7 tertiary trisomics ($2n = 25$) were 11 bivalents + trivalent, 12 bivalents + univalent, and 10 bivalents + pentavalent (Fig. 11.19). These configurations result in a variety of gametic types, including n + a standard chromosome, which is a source of primary trisomics. Compensating trisomics also have various meiotic configurations, depending on whether they include translocated chromosomes, telochromosomes, or isochromosomes. In *Datura stramonium*, a diploid plant species, seven compensating trisomics, with compensation for a missing chromosome provided by two translocated chromosomes, were selfed or backcrossed as females to disomic plants. In each case, most of the progeny were disomics or the same type of compensating trisomic as the parent. These results indicated that most of the transmitted gametes from the compensating trisomics were n (all standard chromosomes) or $n + 1$, with the extra segments divided between two translocated chromosomes (*see* Fig. 11.7a).

Tetrasomics have four doses of one chromosome (*see* Table 11.1), and at meiosis, the four chromosomes can form a quadrivalent, a trivalent plus univalent, or two bivalents. The breeding behavior of a tetrasomic in *Datura stramonium* indicated that the transmitted gametes were mostly n or $n + 1$. A 3:1 distribution from the quadrivalent gives $n + 2$ (not functional) and n gametes, and a 2:2 distribution gives only $n + 1$ gametes. A 2:1 distribution from the trivalent gives $n + 1$ and n gametes provided the univalent is lost. The two bivalents give $n + 1$ gametes.

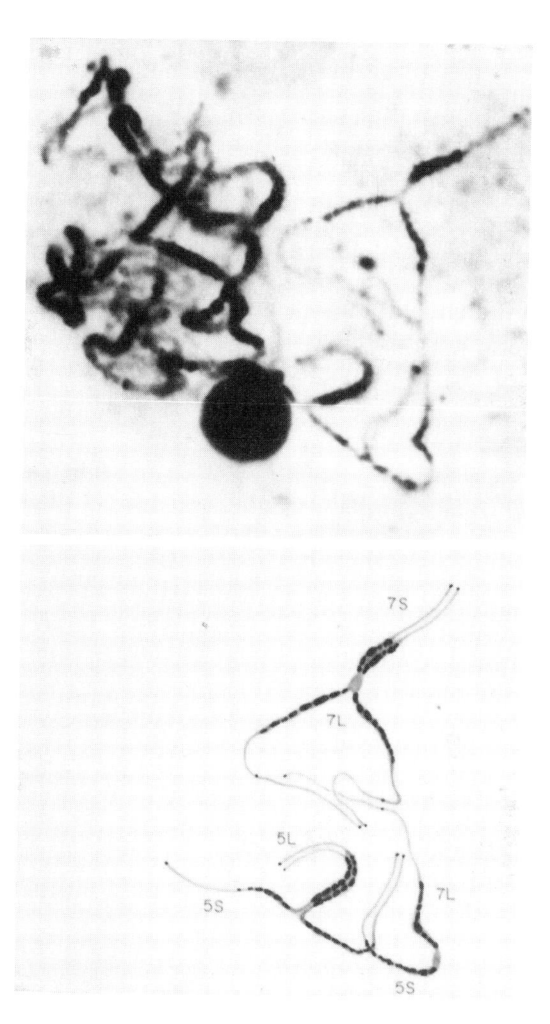

Fig. 11.19. Photograph (top) and diagram (bottom) of a pentavalent at pachytene in a tertiary trisomic tomato plant. The extra chromosome has parts of the short (S) arm of chromosome 5 and the long (L) arm of chromosome 7 and shows some pairing with homologous regions of those two chromosomes. This configuration will form a chain at metaphase I because of the one unpaired 5S arm (Khush, 1973).

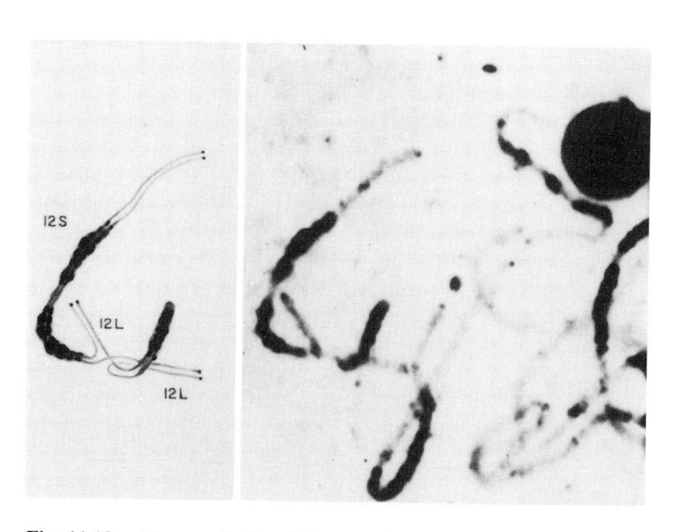

Fig. 11.18. Diagram (left) and photograph (right) of a trivalent at pachytene in a tomato secondary trisomic (isotrisomic), involving two doses of normal chromosome 12 and an isochromosome for the long (L) arm. The isochromosome is paired with the distal part of the long arm of both normal chromosomes and is paired on itself in the heterochromatic region. The centromeres are at the top of the heterochromatin in the isochromosome, and between the two heterochromatic regions in the normal chromosomes (Khush, 1973).

11.6.2 Monosomics and Related Types

The meiotic behavior of these deficient aneuploids has been observed in tetraploid and hexaploid plant species. In a monosomic individual, the monosome, which has no synaptic partner, usually occurs as a univalent during meiosis (Fig. 11.20). A much higher frequency of $n - 1$ gametes than of n gametes results from lagging of the monosome. In the allotetraploid *Nicotiana tabacum*, monosomics for two different chromosomes had the monosome included in a trivalent in about 25% of the cells. Presumably, the monosome was pairing with its homoeologous chromosome pair in the other genome in the absence of a gene-restricting pairing to homologs. The monosomic condition of certain chromosomes causes asynapsis or desynapsis in some of the other chromosomes and, therefore, a higher frequency of univalents than in disomics.

The meiotic behavior of monotelosomics and monoisosomics

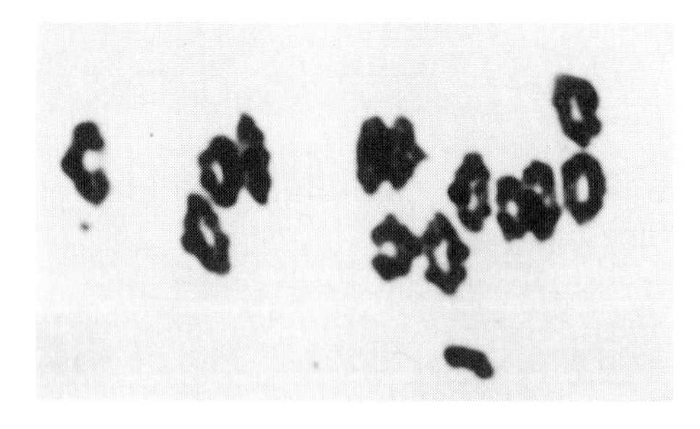

Fig. 11.20. Metaphase I in a monosomic of tetraploid wheat (*Triticum turgidum* var. *durum*) showing the unpaired univalent off the plate due to delayed movement (Mochizuki, 1968).

is similar to that of monosomics because the single telosome or isosome acts as a univalent and is often not included in the gametes due to lagging (Figs. 11.21a and 11.21c). In a ditelosomic, the homologous pair of telosomes forms a thin bivalent because of the missing arm (Fig. 11.21b). When a telosome is paired with a normal chromosome, a heteromorphic bivalent is formed (Fig. 11.21d). With both this condition and the ditelosomic, the modified bivalents move through meiosis at the same time as the normal bivalents, and the telosomes are included in the gametes. If the arms of the telosomes are too short to retain pairing by chiasma formations, these modified chromosomes remain as univalents and would be included in only a small portion of the gametes. Nullisomics are expected to have bivalent pairing, but some missing chromosome pairs cause partial or complete asynapsis. As a result, the gametes have varying numbers of chromosomes and often lack more than one chromosome.

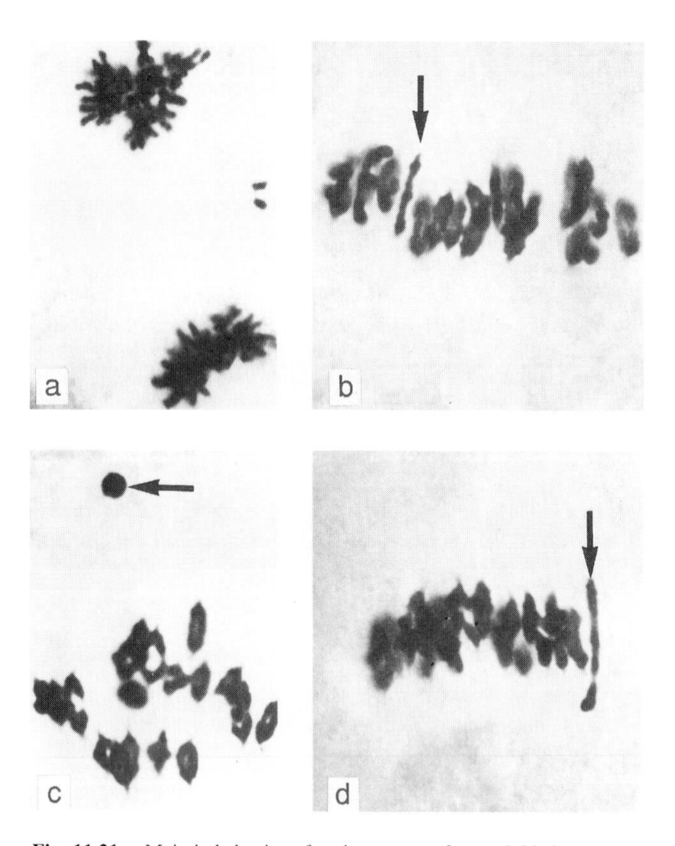

Fig. 11.21. Meiotic behavior of various types of aneuploids in hexaploid wheat (*Triticum aestivum*). (a) A dividing telochromosome at telophase I; (b) a pair of telochromosomes (arrow) at metaphase I; (c) a lagging isochromosome (arrow) at metaphase I; (d) a heteromorphic bivalent (arrow) consisting of a telochromosome (upper third) and a normal chromosome at a stage close to metaphase I. The centromeres are at the tip of the telochromosome and at the bend of the two-armed chromosome (Morris and Sears, 1967).

The fertility of monosomics depends on several factors, including ploidy level, the specific monosomic chromosome, genotype, and environmental conditions. Monosomics in tetraploid species have more sterility than those in hexaploid species when measured by pollen or ovule abortion, or seed fertility. There also is a difference in monosomic fertility among species at the same ploidy level. Monosomics of hexaploid wheat have normal fertility in a favorable environment, whereas those of hexaploid oats have reduced fertility and are more sensitive to environmental conditions. The differences among species are related to the amount of gene replication in the component genomes; species with more compensation for genes on the missing chromosome have better tolerance for environmental conditions.

The transmission rates of deficient $n - 1$ gametes of a monosomic are generally much higher through the female than through the male, as shown in reciprocal crosses between monosomics and disomics. In hexaploid species, the $n - 1$ eggs are viable and do not have competition from n eggs in the same embryo sac, so the frequency of monosomic progeny is similar to the frequency of $n - 1$ eggs. The $n - 1$ pollen grains are viable, but they have to compete with n pollen on the stigma and in the style. Even though there are fewer n pollen grains, their gametes usually predominate in fertilizing the egg cells. There is a wide range in the transmission rates of $n - 1$ gametes in different oat monosomics. For some monosomes, $n - 1$ gametes seem to participate in fertilization as often as n gametes, so that there is a high frequency of monosomics in the progenies of crosses between disomics and monosomics (used as males), or of nullisomics in progenies of selfed monosomics (*see* Table 11.2). Other oat monosomic chromosomes have intermediate to strong competition between the two kinds of pollen, resulting in lower or no transmission of $n - 1$ gametes. The variation in transmission rates reflects the effects of different missing chromosomes on pollen function.

In tetraploids, there is wide variation in the transmission frequency of different monosomic states through the female, with some missing monosomes causing high rates of $n - 1$ megaspore abortion. The 24 tobacco monosomics have a range in ovule abortion from about 16–87%, and a similar range in monosomic transmission through the female. However, for specific monosomics, the frequency of monosomic progeny is usually less than expected based on female fertility. The difference is due to the death of some monosomic progeny in the zygotic or early embryonic stage. Tetraploid species have low or no transmission of $n - 1$ gametes through the male due to a combination of pollen abortion, competition with n pollen, and death of monosomic zygotes or embryos. Nullisomics have not been obtained in progenies of selfed monosomics, and this is further evidence for the rare transmission of $n - 1$ gametes through the male.

The fertility and transmission behavior of monotelosomics, monoisosomics, and nullisomics have been studied most extensively in hexaploid wheat. The fertility of monotelosomics and monoisosomics varies depending on whether their arms carry genes affecting some aspect of fertility and whether there are genes for fertility in the missing arms. Some fertility genes have dosage effects, so that the monoisosomic is more fertile than the monotelosomic for the same chromosome arm. The transmission rates of female gametes with a telosome or isosome are similar to those with a monosome. Male gametes with a telosome or isosome have less advantage over $n - 1$ gametes than n gametes; as a result, monotelosomics and monoisosomics give a higher frequency of nullisomics in their progenies than do monosomics. Monotelosomics give some ditelosomics from transmission of the same telosome through both male and female gametes. Similarly, monoisosomics can produce diisosomics by transmission of the same isosome through both sides.

Only 2 of the 21 wheat nullisomics have fertility approaching that of disomics. Nine have male or female sterility and the rest have low to very low fertility. Those that do have enough fertility to produce seeds when selfed transmit both male and female $n - 1$ gametes to give mostly nullisomic progeny except for nullisomic 3B, which lacks a gene for synapsis and segregates for aneuploid conditions. Some nullisomics have a few offspring with a trisomic condition for a chromosome homoeologous to the missing chromosome, presumably as a result of a nondisjunction event during meiosis. These nullisomic–trisomic plants are more vigorous and fertile than their nullisomic sibs because of the compensating genes on the trisome.

BIBLIOGRAPHY

General

Hassold, T.J., Jacobs, P.A. 1984. Trisomy in man. Annu. Rev. Genet. 18: 69–97.

Khush, G.S. 1973. Cytogenetics of Aneuploids. Academic Press, New York.

Khush, G.S., Singh, R.J., Sur, S.C., Librojo, A.L. 1984. Primary trisomics of rice: origin, morphology, cytology, and use in linkage mapping. Genetics 107: 141–163.

Morris, R., Sears, E.R. 1967. The cytogenetics of wheat and its relatives. In: Wheat and Wheat Improvement (Quisenberry, K.S., ed.). American Society Agronomy, Madison, WI. pp. 19–87.

Sybenga, J. 1992. Cytogenetics in Plant Breeding. Monographs on Theoretical and Applied Genetics 17. Springer-Verlag, Berlin.

Therman, E., Susman, M. 1993. Human Chromosomes: Structure, Behavior, and Effects. 3rd ed. Springer-Verlag, New York.

Section 11.2

Hassold, T.J. 1986. Chromosome abnormalities in human reproductive wastage. Trends Genet. 2: 105–110.

Kurnit, D.M., Hoehn, H. 1979. Prenatal diagnosis of human genome variation. Annu. Rev. Genet. 13: 235–258.

Mattingly, C.F., Collins, G.B. 1974. The use of anther-derived haploids in *Nicotiana*. III. Isolation of nullisomics from monosomic lines. Chromosoma 46: 29–36.

Shoffner, R.N. 1985. Bird cytogenetics. In: Cytogenetics of Livestock (Eldridge, F.E. ed.). AVI Publishing Co., Westport, CT. pp. 263–283.

Section 11.3

Sankaranarayanan, K. 1979. The role of non-disjunction in aneuploidy in man: an overview. Mut. Res. 61: 1–28.

Section 11.4

Athwal, R.S., Sandhu, S.S. 1985. Use of a human × mouse hybrid cell line to detect aneuploidy induced by environmental chemicals. Mut. Res. 149: 73–81.

Bond, D.J. 1987. Mechanisms of aneuploid induction. Mut. Res. 181: 257–266.

Boué., A., Boué, J. 1974. Chromosome abnormalities and abortion. In: Physiology and Genetics of Reproduction (Coutinho, E.M., Fuchs, F., eds.). Basic Life Sciences Vol. 4, Part B. Plenum Press, New York. pp. 317–339.

Clausen, R.E., Cameron, D. R. 1944. Inheritance in *Nicotiana tabacum*. XVIII. Monosomic analysis. Genetics 29: 447–477.

Griffiths, A.J.F. 1982. Short-term tests for chemicals that promote aneuploidy. In: Chemical Mutagens, Principles and Methods for Their Detection, Vol. 7 (de Serres, F.J., Hollaender, A., eds.). Plenum Press, New York. pp. 189–210.

Khush, G.S., Rick, C.M. 1966. The origin, identification, and cytogenetic behavior of tomato monosomics. Chromosoma 18: 407–420.

Larkin, P.J. 1987. Somaclonal variation: history, method, and meaning. Iowa State J. Res. 61: 393–434.

Linde-Laursen, I.B., Siddiqui, K.A. 1973. Triploidy and aneuploidy in virus-infected wheat, *Triticum aestivum*. Hereditas 76: 152–154.

Nichols, W.W. 1970. Virus-induced chromosome abnormalities. Annu. Rev. Microbiol. 24: 479–500.

Oshimura, M., Barrett, J.C. 1986. Chemically induced aneuploidy in mammalian cells: mechanisms and biological significance in cancer. Environ. Mutagen. 8: 129–159.

Parry, J.M., Parry, E.M. 1987. Comparisons of tests for aneuploidy. Mut. Res. 181: 267–287.

Rick, C.M., Barton, D.W. 1954. Cytological and genetical identification of the primary trisomics of the tomato. Genetics 39: 640–666.

Sandfaer, J. 1973. Barley stripe mosaic virus and the frequency of triploids and aneuploids in barley. Genetics 73: 597–603.

Sankaranarayanan, K. 1982. Genetic Effects of Ionizing Radiation in Multicellular Eukaryotes and the Assessment of Genetic Radiation Hazards in Man. Elsevier Biomedical Press, Amsterdam.

Stewart, G.D., Hassold, T.J., Kurnit, D.M. 1988. Trisomy 21: molecular and cytogenetic studies of nondisjunction. Adv. Human Genet. 17: 99–140.

Section 11.5

Avery, A.G., Satina, S., Rietsema, J. 1959. Blakeslee: The Genus *Datura*. The Ronald Press Co., New York.

Brock, D.J.H., Mayo, O. 1972. The Biochemical Genetics of Man. Academic Press, New York.

Eldridge, F.E. 1985. Cytogenetics of Livestock. AVI Publishing Co., Westport, CT.

Ellis, J.R., Janick, J. 1960. The chromosomes of *Spinacea oleracea*. Am. J. Bot. 47: 210–214.

Epstein, C.J. 1988. Mechanisms of the effects of aneuploidy in mammals. Annu. Rev. Genet. 22: 51–75.

Friebe, B., Kim, N.-S., Kuspira, J., Gill, B.S. 1990. Genetic and cytogenetic analyses of the A genome of *Triticum monococcum*. VI. Production and identification of primary trisomics using the C-banding technique. Genome 33: 542–555.

Levitan, M. 1988. Textbook of Human Genetics, 3rd ed. Oxford University Press, New York.

Morgan, T.H., Bridges, C.B., Sturtevant, A.H. 1925. The genetics of *Drosophila*. Bibliog. Genet. 2: 1-262.

Novitski, E. 1977. Human Genetics. Macmillan Publishing Co., New York.

Weber, D.F. 1983. Monosomic analysis in diploid crop plants. In: Cytogenetics of Crop Plants (Swaminathan, M.S, Gupta, P.K., Sinha, U., eds.). Macmillan, New Delhi. pp. 351–378.

Section 11.6

Clausen, R.E., Cameron, D.R. 1944. Inheritance in *Nicotiana tabacum*. XVIII. Monosomic analysis. Genetics 29: 447–477.

Mochizuki, A. 1968. The monosomics of *durum* wheat. Proc. 3rd Int. Wheat Genet. Symp. Australian Academy of Science, Canberra. pp. 310–315.

Sears. E.R. 1954. The aneuploids of wheat. Missouri Agr. Exp. Sta. Res. Bull. 572: 1–58.

Sears, E R., Sears, L.M.S. 1979. The telocentric chromosomes of common wheat. Proc. 5th Int Wheat Genet. Symp. (Ramanujan, S., ed). IARI New Delhi. pp. 389–407.

Tsuchiya, K. 1960. Cytogenetic studies of trisomics in barley. Jpn. J. Bot. 17: 177–213.

12

The Cytogenetic Analysis of Haploids

- Some organisms can survive with half the usual somatic chromosome number.
- The smaller size of haploids is due to the reduction in the normal gene dosage of somatic cells.
- Haploids arise spontaneously from irregularities in sexual reproduction and are induced by various methods such as chromosome elimination in wide crosses or *in vitro* anther culture.
- Haploids are highly sterile because of the irregular meiotic behavior of unpaired chromosomes.
- Doubled-haploid plants are genetically homozygous and provide an important basis for producing new varieties with greater efficiency.

The term haploidy, in a whole-organism sense, refers to situations where the genome dosage in somatic cells is the same as in normal gametes of the organism. In most diploid higher organisms, the life cycle alternates between a prolonged $2n$ somatic phase of growth and differentiation, and a relatively short n phase of sexual reproduction, when the chromosome number is halved. For a specific genome in diploids, there is one dose in the gametes and two doses in the somatic cells, except for specialized tissues (*see* Chapter 10, Section 10.4). Haploidy in diploid organisms refers to an abnormal situation in which there is only one dose of a genome in the somatic cells. Vertebrate animals are not able to tolerate this condition beyond the early embryonic stage, so this chapter will be concerned almost entirely with haploids in plants.

The life cycle of higher plants is summarized in Fig. 12.1 to highlight the various sources of normal and abnormal haploid tissues or individuals. A haploid derived from a diploid is called a monoploid or a monohaploid, whereas a haploid derived from a polyploid is called a polyhaploid.

12.1 THE ORIGINS OF HAPLOIDS

12.1.1 Natural Occurrences of Haploids

Some organisms have adapted to a haploid state throughout much or all of their life spans. In the insect order Hymenoptera (ants, bees, and wasps), the males are haploid because they develop from unfertilized eggs. In many fungi, the n phase of the life cycle predominates, and the very short $2n$ phase occurs when two n nuclei with different mating-type alleles unite. The $2n$ fusion nucleus immediately undergoes meiosis to repeat the n phase. Fungi have evolved some compensating mechanisms for somatic haploidy such as coenocytic mycelia with numerous n nuclei, short generation times, and prolific spore production.

Spontaneous somatic-haploid states in higher plants are due to some type of apomixis (i.e., lack of fusion between egg and sperm) in contrast to amphimixis, where the egg and sperm unite (fertilization or syngamy). There are several types of apomixis depending on the behavior of the egg and sperm (*see* Fig. 12.1). The phenomenon of polyembryony, in which a seed has multiple embryos, is also a source of haploid plants.

Parthenogenesis refers to the development of the egg cell or some other cell in the embryo sac without fertilization. An unfertilized egg may divide in the absence of sperm (gynogenesis), or it may be stimulated to divide prematurely by the presence of sperm, but fertilization does not take place (pseudogamy). The deviation in timing of egg division may be due to failure of an inhibitor to prevent the egg from dividing before fertilization. With apogamy, a haploid embryo arises from some other cell in the embryo sac besides the egg, such as a synergid. These three variations in parthenogenesis result in maternal haploids. Androgenesis occurs when a sperm cell divides inside the embryo sac, possibly inside the egg cell, without syngamy because the egg nucleus has disintegrated due to gene mutation or environmental damage. Androgenesis gives rise to paternal haploids.

Semigamy also involves failure of syngamy although a sperm nucleus is inside the egg cell. Instead, the egg and sperm nuclei divide separately but usually synchronously. In the prairie lily, *Cooperia pedunculata*, the sperm nucleus divides in the basal part of the egg cell, and the descendant nuclei of the sperm nucleus are separated from the egg-nucleus descendants by cell walls. In cotton (*Gossypium barbadense*), semigamy is attributed to one dominant gene, and when marker genes are used for traits contributed by the egg or sperm, sectors involving maternal or paternal tissue appear on the plants. This provides further evidence that the descendant nuclei from sperm and egg segregate into different cells during at least some of the division cycles. Semigamy provides the opportu-

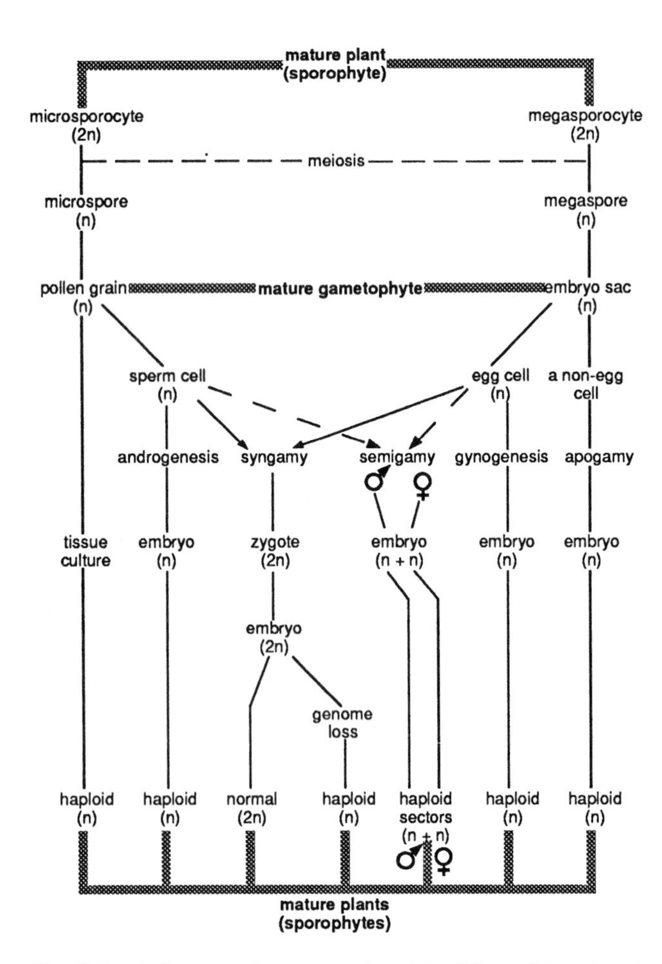

Fig. 12.1. A diagrammatic representation of the different life cycles of higher plants showing normal sexual reproduction and the possible deviations leading to haploidy. (Adapted from de Fossard, 1974.)

nity for obtaining maternal or paternal haploids if the sectors include reproductive tissue or if they can be propagated vegetatively.

The frequency of twin embryos, which are a potential source of haploids, seems to be under genetic control and is increased by wide crosses or by pollen irradiation. In a diploid species, twin embryos may be $2x/2x$, $2x/1x$, or $1x/1x$. The $2x$ embryos are normal diploids and the $1x$ embryos are monohaploids. In both diploid and polyploid species, most haploids seem to occur in one member of twins and are derived from an unfertilized embryo-sac cell other than the egg, whereas the $2n$ plant comes from a normal fertilization involving the egg cell. Relatively high frequencies of twins, with one haploid twin, have been reported in pepper (*Capsicum annuum*), rape (*Brassica napus*), flax (*Linum usitatissimum*), and in the wheat cultivar "Salmon."

12.1.2 The Induction of Haploids

The frequency of spontaneous haploids is usually too low to use them in genetic or breeding studies, so a number of different approaches have been used to increase their occurrence within plants. These have included temperature shocks applied just after pollination, irradiation of the embryo sac or pollen, delayed pollination, and application of certain chemicals to the stigmas after pollination. The purpose of each treatment is to inactivate or damage the male or female gamete so that fertilization cannot occur; then the unaffected gamete produces a haploid plant.

The frequency of apomictic haploids can be influenced by the genotype of one or both parents used in crosses. In maize, one genetic strain gives over 3% haploids with self-pollination, compared to the usual frequency of 0.1% maternal haploids and very rare paternal haploids. The increase in haploidy is transmitted with variable frequencies when different plants of the strain are used as male parents in crosses, but only maternal haploids are produced. In potato, selection of both female (seed) and male (pollinator) parents results in 30–100 haploids per 100 fruits. The technique of cutting off the upper portion of the seed parent and placing the cut stem in water in an air-conditioned greenhouse contributes to the high frequency of haploids.

Some single-gene mutations influence the frequency of haploids. The *indeterminate gametophyte (ig)* gene in maize causes abnormalities in the development of the embryo sac, resulting in several extra eggs and polar nuclei. When the female parent is homozygous for this gene, about 3% of the offspring are a mixture of maternal and paternal haploids (detected by using gene markers). The *hap* gene in barley produces 15–40% haploids when plants homozygous for the gene are self-pollinated or used as female parents in crosses, but no haploids occur when the plants are used as male parents. F_1 and F_2 plants heterozygous for *hap* also produce haploids, but in a lower frequency than homozygotes. It is thought that the *hap* gene may prevent fertilization of the egg nucleus or may stimulate it to divide prematurely. The maternal genotype seems to influence the formation and survival of haploid embryos.

Interspecific or intergeneric crosses have been used widely to promote haploidy because the stimulation of alien pollen often causes the egg cell to divide by pseudogamy. In potato, the most success has been obtained by pollinating the tetraploid species, *Solanum tuberosum*, with the diploid species, *Solanum phureja*. When the pollen tube reaches the embryo sac, the two sperm nuclei fuse and unite with the polar nuclei to produce the endosperm while the egg cell divides to produce a polyhaploid embryo.

Alien cytoplasm, introduced by wide crosses, also can induce haploidy. When hexaploid wheat or triticale (wheat–rye) genomes were combined with cytoplasms from several wild *Triticum* species, polyhaploids were obtained. The most effective wheat genotype was the strain "Salmon," which has already been mentioned in connection with twin embryos. This strain of wheat was derived from triticale, and it has a translocation between the wheat chromosome 1B and the homoeologous rye chromosome 1R (Fig. 12.2). The high frequency of haploids in "Salmon" is due to an interaction between the wild *Triticum* cytoplasm and a gene(s) on the 1B/1R chromosome. With this combination, it is thought that the egg cell often develops parthenogenetically to produce haploids while a synergid is fertilized by a male gamete, thus giving a high frequency of haploid–hexaploid twins. For each pair of twins, the haploid plant is larger than the hexaploid plant because an egg cell is larger than a synergid. The haploids are completely male sterile due to the wild *Triticum* cytoplasm, and possibly because of the loss of a fertility-restoring gene on the arm of 1B that is replaced by an arm from 1R.

Another way in which wide crosses can produce haploidy is by chromosome elimination of specific genomes. Crosses between barley (*Hordeum vulgare*) and a wild relative, *H. bulbosum*, followed by the rescue of the immature embryos on culture medium, result in haploid plants. The *bulbosum* chromosomes are eliminated during a number of cell-division cycles in the initial development

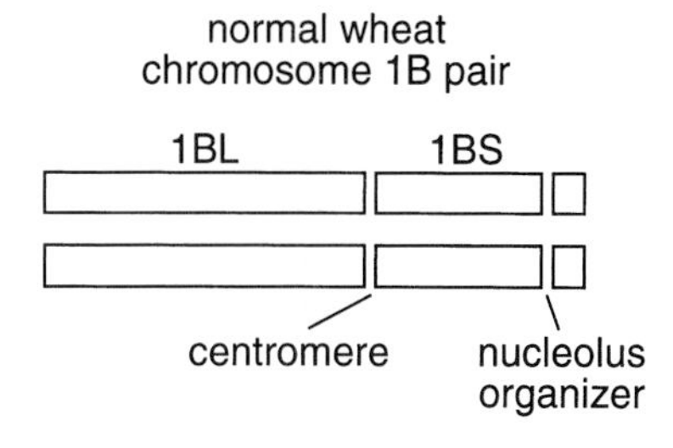

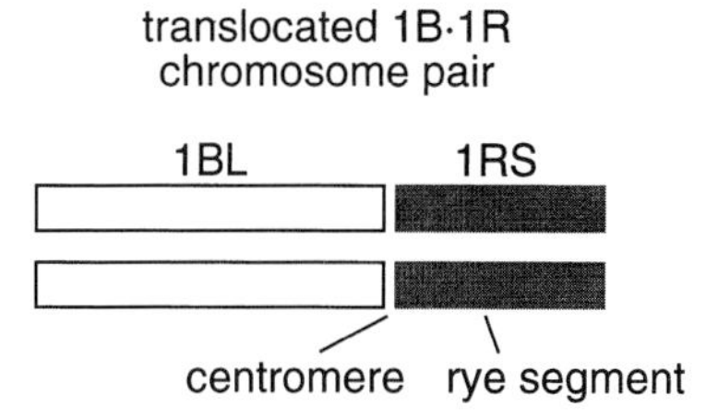

Fig. 12.2. Diagram showing the difference between a normal wheat chromosome 1B and the wheat-rye 1BL/1RS translocated chromosome found in the wheat strain "Salmon." The entire short arm and satellite of 1B has been replaced by a rye arm (adapted from Kobayashi and Tsunewaki, 1980). The 1RS arm includes the rye NOR, but the satellited constriction is suppressed when this chromosome is introduced into wheat.

of embryo and endosperm, and in less than 2 weeks after pollination, over 90% of dividing cells are haploid. Up to 60% of cultured embryos give mature plants and, of these, 90% are *vulgare* haploids and the rest are diploid hybrids. Both maternal and paternal *vulgare* haploids are obtainable from reciprocal crosses, because the *bulbosum* chromosomes are eliminated when this species is used as a male or female parent in crosses with *vulgare*. This suggests that elimination is not due to *vulgare* cytoplasm interacting with the *bulbosum* chromosomes, because it occurs when *bulbosum* is the female parent and introduces its own cytoplasm. Chromosome elimination is common in other interspecific *Hordeum* crosses, and in some of them, chromosomes are eliminated from specific tissues. For example, when *H. vulgare* is crossed with *H. marinum*, *vulgare* chromosomes are eliminated from the endosperm and *marinum* chromosomes are eliminated from the embryo. The elimination of different parental genomes from different tissues of the same seed has the potential for uncovering new aspects of the biology of chromosome behavior.

Suggestions for the selective mechanism underlying the elimination of the *bulbosum* genome have included suppression of nucleolar activity in this genome, or improper attachment of *bulbosum* chromosomes to the spindle during cell division, resulting in lagging chromosomes and other abnormalities. The *bulbosum* chromosomes in the *H. vulgare–H. bulbosum* hybrids are located in peripheral regions of mitotic cells, and they lag during cell divisions. Besides, their nucleolu-organizer and centromere-constriction regions are suppressed. These deviations from normal structure and behavior of the *bulbosum* chromosomes may be related to their interactions with the *vulgare* genome. Crosses between diploid or autotetraploid barley and autotetraploid *bulbosum* reveal that the number of *vulgare* genomes have to be equal to or greater than the number of *bulbosum* genomes for the consistent elimination of *bulbosum* chromosomes. Thus, the importance of genomic balance is evident when crosses include different ploidy levels. Chromosome-mapping studies using trisomic analysis indicate that genes influencing genomic balance are located on two *vulgare* chromosomes.

On the intergeneric level, haploid wheat plants are obtained by crossing hexaploid wheat with diploid and tetraploid *H. bulbosum*, with the tetraploid inducing a higher frequency of haploids. Crosses between wheat and maize (*Zea mays*), pearl millet (*Pennisetum americanum*), or *Tripsacum dactyloides* have also produced wheat haploids. The wheat × maize crosses have a major advantage over the wheat × *H. bulbosum* crosses, because maize pollen seems to be able to function in the presence of crossability genes (*Kr*) in wheat that greatly reduce the seed set when *H. bulbosum* pollen is used. The wheat × maize system also has an advantage over the anther-culture method because success in producing haploids from microspores is affected by the wheat genotypes used. In addition, cytological instability, such as aneuploidy, seems to be less frequent in haploids derived from wheat × maize crosses than in those derived from anther culture.

Anther culture on artificial nutrient media has been tested with many plant species as a haploid-producing method (Fig. 12.3). Anthers are $2n$ tissue, but the microspores within the anthers are n and the source of haploids. Isolated microspores, also called immature pollen, have produced haploids in some species, but they usually come from precultured anthers or have been cultured with aqueous extracts of cultured anthers. Substances in the anther wall seem to be needed for callus formation, and for the production of green rather than albino regenerated plants. The advantage of isolated-microspore culture over anther culture is that all the microspores are usually n, whereas anther-wall $2n$ cells can regenerate some $2n$ plants. The haploids resulting from anther or isolated-microspore cultures are called androgenic haploids because they consist of the male meiotic products, like those arising from androgenic apomixis.

A number of conditions affect the capability of microspores to produce haploids, whether by the anther or isolated-microspore

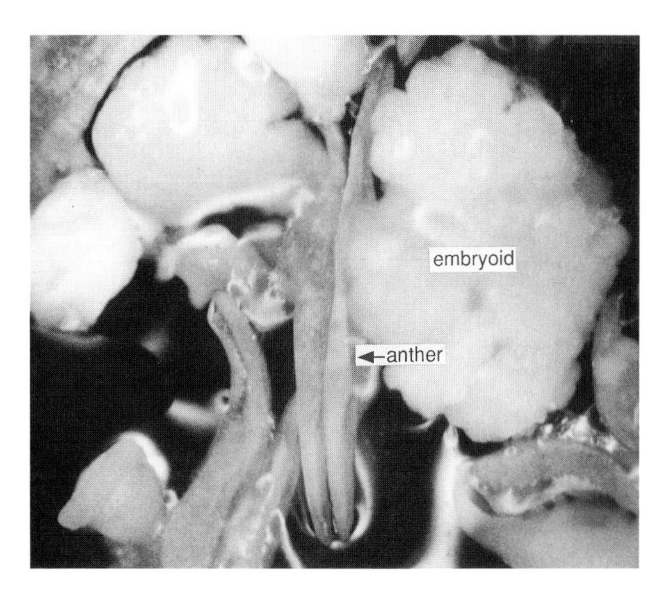

Fig. 12.3. The development of haploid plants from microspores via anther culture. Unpublished photograph kindly provided by Dr. N. Darvey, Sydney University, NSW, Australia.

method. The nuclear genotype of the plants from which the anthers or microspores are obtained has a strong influence on the induction of haploids, and, within a species, some cultivars are more responsive than others. The role of cytoplasmic genes is less certain, because there are differences between reciprocal crosses for microspore response in some experiments but not in others. Hybrids may produce more haploids than the parental cultivars, as in the case of *Brassica campestris*. The age of the donor plants and their growing conditions affect the embryogenic potential of anthers. Photoperiod, light intensity, temperature, and nutrition are thought to influence levels of hormones and inhibitory substances in the anther walls. Pretreatment of anthers or inflorescences with low temperatures or centrifugation before culturing them has enhanced the production of haploids in some species. The developmental stage of the microspores at the time of culturing is critical to success. The most favored stage in cereal crops is the uninucleate microspore (Fig. 12.4), but binucleate microspores, which have undergone the first mitotic division, or the earlier tetrad stage are suitable in other species. The culture requirements, including composition of different media, and the environmental conditions, particularly temperature, have to be determined for each species. Most of the species in the Gramineae produce haploid plants from microspores via the intermediate step of callus formation, which usually requires a different culture medium than that used for species predominantly in the Solanaceae, where haploid plants are produced directly from microspores. However, progress in culturing techniques makes it possible to obtain haploid plants by either pathway in wheat, a gramineous species.

There are several challenging situations during the process of regenerating plants from anther or microspore culture. As a rule, only a portion of the microspores respond even under favorable conditions. It has been tempting to associate the difference between embryogenic (producing embryos) and nonembryogenic microspores with two morphologically different kinds of pollen grains within an anther, and they have been described in several species.

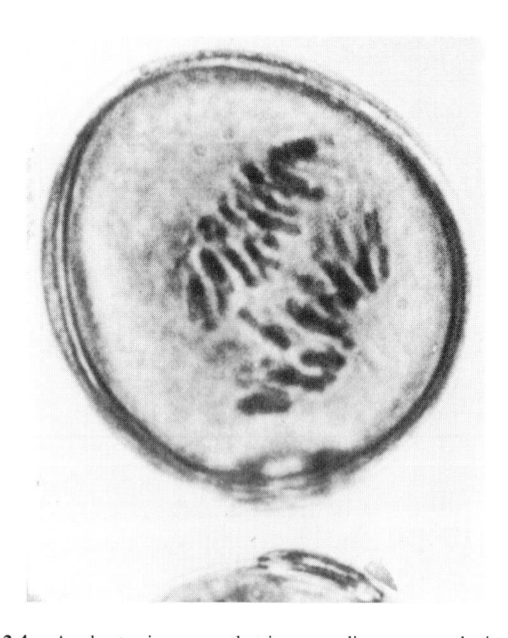

Fig. 12.4. A wheat microspore that is responding on a synthetic medium by undergoing the first mitotic division (anaphase stage) (Rybczynski et al., 1991). This microspore has the potential to differentiate into a haploid plant, but not all dividing microspores do so.

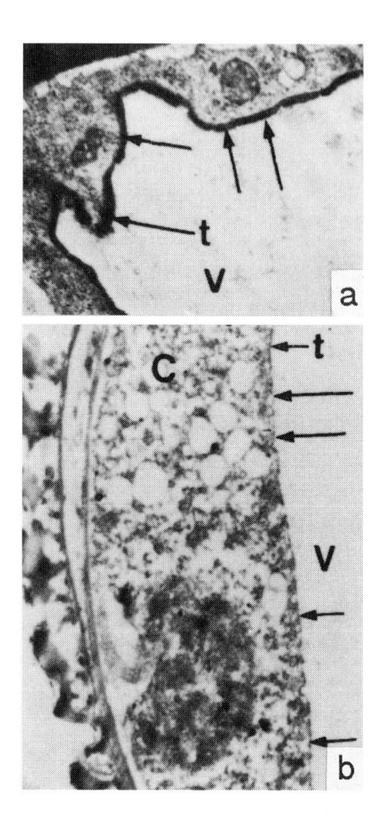

Fig. 12.5. Electron micrographs of portions of microspores in *Datura innoxia*. (a) Presence of tannin (arrows) on the tonoplast (t). (b) Absence of tannin from the tonoplast (t). V = vacuole; C = cytoplasm. (From Sangwan and Sangwan-Norreel, 1987.)

A critical test of this association was made in *Datura*, using a cytological marker that could differentiate two kinds of microspores as early as 12 h after putting them in culture. A tannin coating, which developed on the tonoplast of embryogenic microspores, was lacking on the tonoplast of nonembryogenic microspores (Fig. 12.5). The percentage of microspores with a coated tonoplast and the percentage that formed embryos were closely related.

Conditions arising in haploids derived from microspores are albino plants, chromosome numbers other than haploidy, changes in chromosome structure, and gene mutations. These changes are generally referred to as gametoclonal variation (*see* Chapter 7, Section 7.4.4), and it is important to check the regenerated plants cytologically as well as phenotypically. Albinism is detrimental to a haploid-inducing program, particularly in cereal crops, which can have high frequencies of albino plants. In wheat, albino plants have mininuclei in their cells, and deletions including up to 80% of the chloroplast genome. They also lack certain fractions of ribosomal RNA and protein. A physiological effect is implied by the fact that more albinos occur when cultures are exposed to higher temperatures during the stage when the microspores are converting from a gametophytic to a sporophytic pathway. In several cereal crops, the proportion of green plants is higher if microspores are excised before they have undergone division. The culturing conditions and the genotype also influence the frequency of green haploids. Nonhaploid chromosome numbers may be caused by irregular meioses that lead to aneuploidy and unreduced microspores, or by fusion of microspore nuclei. There also can be fusion between the vegetative and generative cells in the pollen grains before further division occurs. Some of the aneuploids and gene mutations resulting from haploid-inducing tissue culture have potential uses in gene-mapping studies (as discussed in Chapters 14 and 21).

The use of ovary, ovule, or pistil culture has been successful with a number of plant species, including those where anther culture is difficult. The female organ can be taken from male-sterile plants to produce haploids. Most gynogenetic haploids come from the egg cell, but there is evidence that synergid, and even antipodal, cells may produce haploids. The advantages of gynogenetic haploids over androgenetic haploids are that albino plants are rare and chromosome stability is high.

12.2 DETECTION AND PHENOTYPES OF HAPLOIDS

Gene markers for traits in various parts of the seeds or seedlings have been useful in detecting potential apomictic haploids in the early stages of the life cycle. The usual procedure is to cross a plant with one or more recessive gene markers as the female parent with a plant carrying the dominant alleles. Progeny showing the dominant phenotype are discarded because they should have both male and female genomes. Progeny with the recessive phenotype are potential haploids but need to be confirmed by root-tip chromosome counts. An example of a useful gene marker is the dominant gene *P* for the development of anthocyanin pigment in the hypocotyl of potato seedlings. When the tetraploid potato *Solanum tuberosum*, with four doses of the recessive allele *p*, is crossed as female with the diploid potato *S. phureja*, which has two doses of the dominant allele *P*, the progeny are screened as seedlings for pigmented hypocotyls, and nonpigmented seedlings are saved as potential maternal haploids.

Haploids in higher plants generally show a reduction in overall size, as well as in the size of cells and individual organs such as leaves and inflorescences (Fig. 12.6). Haploids from diploids show pronounced effects because they have only one dose of one genome. They cannot be maintained by sexual reproduction beyond their own generation because of gametic sterility, but there are a few examples where they have persisted for years by vegetative propagation. Haploids from allopolyploids also have one dose of each kind of genome, but usually there is considerable gene repetition across genomes, so the effects of reducing genome dosage are not as severe as in monohaploids. Haploids from autopolyploids have more than one dose of the one type of genome. For example, in autotetraploid potato, which has four similar genomes, the haploids have the same genome dosage as a diploid and are sometimes called dihaploids. Nevertheless, these haploids often have reduced vigor and fertility, which have been attributed to loss of interactions within and between loci that occur at the tetraploid level. Favorable interactions are retained in tetraploid potato by asexual propagation, but meiosis intervenes in the production of haploids, so crossing over and chromosome segregation can break up groups of favorable genes.

If monohaploids occur in cross-fertilizing species, which have many heterozygous loci, harmful recessive genes, which are normally masked by dominant alleles, have severe phenotypic effects. These effects occur less frequently in polyhaploids, which, in autopolyploids, can have a dominant allele to conceal a deleterious recessive allele at the same locus or, in allopolyploids, a dominant gene at a duplicate locus in one of the other genomes. It has been

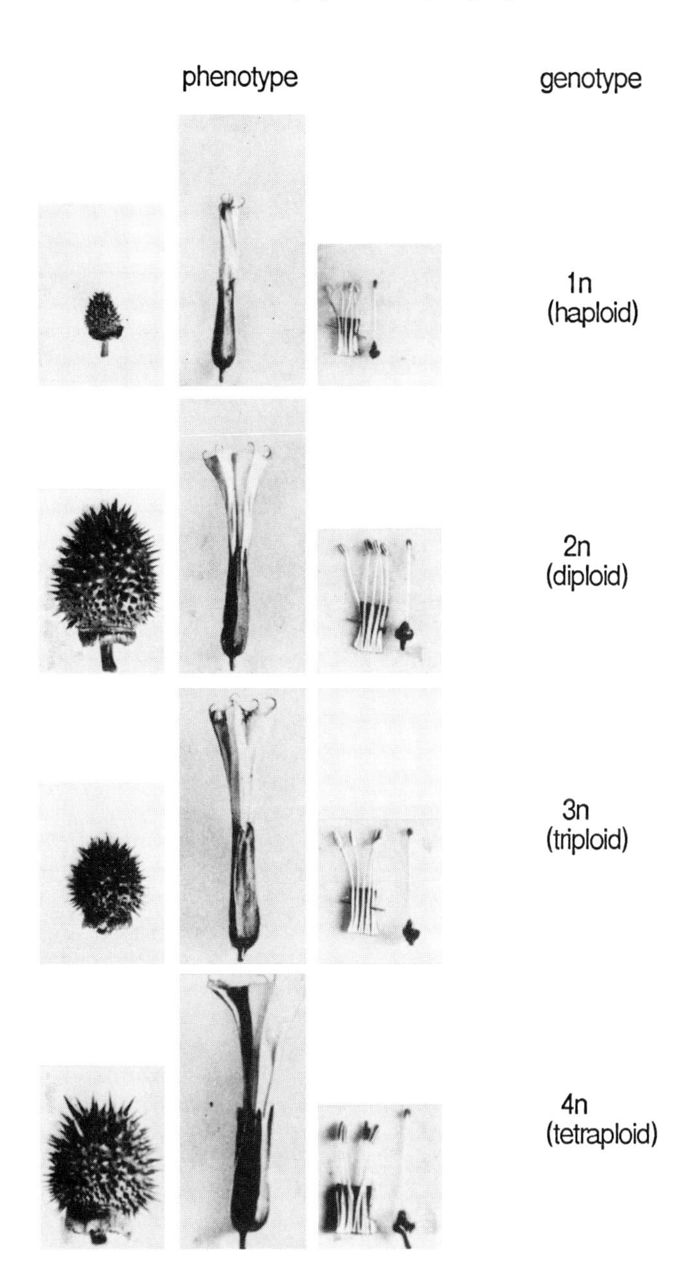

Fig. 12.6. The effect of ploidy level on the size of seed capsules (left), flowers (center), and stamens (right) in *Datura stramonium* (Jimson weed). From top to bottom: monohaploid, diploid, autotriploid, and autotetraploid. (Photographs from Avery, 1959.)

estimated that the process of obtaining a polyhaploid from an autotetraploid is equivalent to three generations of self-pollination in terms of increased homozygosity. The retention of harmful, recessive genes in self-fertilizing species is less likely, because the plants become homozygous for these genes and are eliminated from a population by competition from plants with the dominant alleles. Therefore, fewer unfavorable genes are uncovered with haploidy.

12.3 MEIOTIC BEHAVIOR AND FERTILITY IN HAPLOIDS

Variations in meiotic behavior relate to the type of haploid and can be affected by genes, chromosome structural changes such as duplications, and environmental factors such as temperature.

12.3.1 Monohaploids

A monohaploid has one dose of each type of chromosome; therefore, no chromosome pairing is expected except for a small amount involving duplicated regions, which can arise from chromosome rearrangements. Monohaploids of some plant species conform to these expectations, but those of other species have more extensive pairing. In several species, synaptonemal complexes are frequent in monohaploids, as shown for barley in Fig. 12.7. The ends of the synaptonemal complexes are aligned, as are the ends of paired chromosomes observed with the light microscope, indicating that the sites where pairing starts are not random. However, most of the associations are nonhomologous because they do not consistently involve the same chromosomes. Many associations break down into univalents by diakinesis and metaphase I, and the configurations that remain vary among cells. Univalents in the early stages of meiosis often show foldback pairing within themselves (Fig. 12.8a), and other configurations resemble bivalent or multivalent structures (Fig. 12.8b).

Metaphase I and anaphase I are hard to separate in microscopic observations on many monohaploids because of disorganized spindles and scattered univalents. In some species, the end of metaphase I is indicated by the separation of sister chromatids, except in the centromere regions. Unless a restitution nucleus containing all the chromosomes is formed, the univalents tend to lag between the poles at anaphase I, and varying numbers are included in the nuclei (Fig. 12.8c). Those not reaching the poles form micronuclei in the cytoplasm. Usually, a normal spindle forms during metaphase II, and those chromosomes that are in the nuclei are distributed irregularly to the poles. Restitution nuclei can occur during the second meiotic division (*see* Chapter 10, Fig. 10.5), but they usually lack one or more chromosomes unless all the chromosomes have undergone equational division during meiosis I, with sister chromatids going to opposite poles. This sequence can produce two n spores, as can a first-division restitution (*see* Chapter 10, Fig. 10.4), followed by separation of sister chromatids at anaphase II. Both first- and second-division restitutions are rare events, so most of the spores lack one or more chromosomes. Consequently, they vary in size and deviate in number from the normal tetrad.

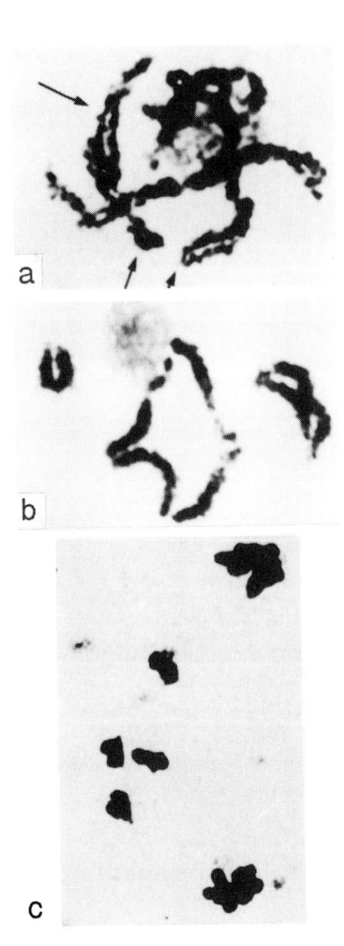

Fig. 12.8. Meiotic chromosome configurations in monohaploid barley. (a) Pairing within univalents (marked by arrows) at early diplotene. (b) Diakinesis structures resembling (from left to right) a univalent, a quadrivalent, and a bivalent. (c) Anaphase I with two lagging univalents after separation of sister chromatids. (Photomicrographs from Sadasivaiah, 1974.)

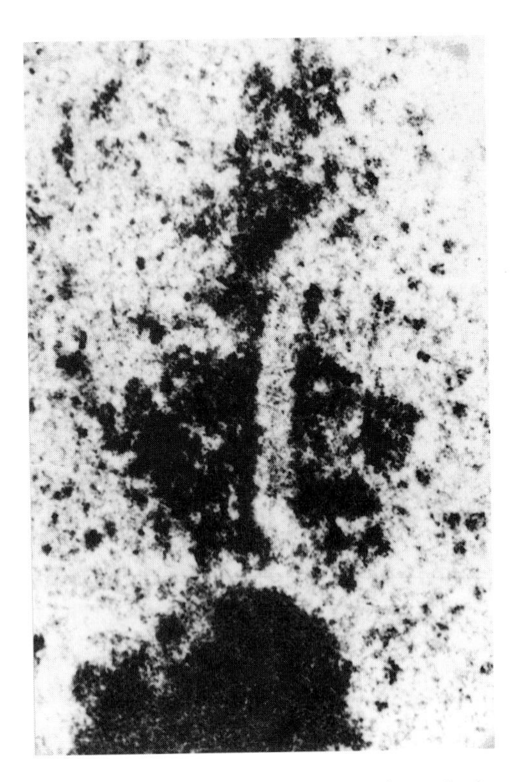

Fig. 12.7. Electron micrograph of a synaptonemal complex between nonhomologous chromosomes at the pachytene stage in a microsporocyte of monohaploid barley. (Photo from Sadasivaiah, 1974.)

Fertility is very low in monohaploids because most of the spores abort due to aneuploidy, but, occasionally, a sector of a plant shows a normal seed set as a result of chromosome doubling in a somatic cell (*see* Chapter 10, Section 10.3). On the other hand, an occasional normal seed is probably due to nuclear restitution during meiosis in a haploid plant, followed by a cross with its diploid counterpart. If the haploid is selfed, male and female gametes must come from separate meiotic restitution events if each is to contribute a complete genome. Seeds derived by these various means carry a doubled chromosome number in their embryos, and the resulting plants are known as doubled haploids.

12.3.2 Polyhaploids

Haploids derived from autotetraploids involve a reduction from the tetraploid to the diploid level. They have two doses of the same genome, so meiotic pairing is usually regular, with bivalent formation and normal distributions of the chromosomes during both meiotic divisions. The haploids of autotetraploid alfalfa and potato show normal meiotic behavior with a few exceptions, which may be due to the uncovering of recessive genes affecting some phase of meiosis. Such genes can contribute to the reduction in fertility, especially on the male side, where there may be complete sterility. The fertility can often be improved by crossing the haploids with diploid relatives either within the same species or in other species.

Most of the haploids from allopolyploids show some bivalent formations, indicating that there are some intergenomic genetic relationships and that the pairing is between homoeologous chromosomes (Fig. 12.9a). However, homoeologous pairing can be reduced by structural changes and also by genetic restrictions such as the *Ph* or similar genes, which limit pairing to homologous chromosomes. Therefore, caution is needed in interpreting genomic relationships from the amount of chromosome pairing. Some bivalents may be the result of stickiness or nonhomologous pairing rather than homoeologous attractions. None of the haploids from allopolyploids have enough meiotic pairing and regularity to have high fertility, but they may have enough female fertility to produce a few seeds when crossed with their polyploid sources. Polyhaploids, like monohaploids, can produce occasional n gametes as a result of the formation of restitution nuclei during meiosis.

12.3.3 Secondary Associations

In monohaploids and in polyhaploids from allopolyploids, loose attractions between univalents have been observed at metaphase I. These attractions may be end-to-end (e-e), side-by-side (s-s), or end-to-side (e-s) (*see* Figs. 12.9a and 12.9b). The chromosomes in these arrangements seem to be in contact without any evidence of chiasmata, and they do not have the shape of bivalents. Secondary associations may be attractions between heterochromatic regions with similar repeated-DNA sequences. Another suggestion is that side-by-side configurations are loose associations between homoeologous chromosomes. Chromosome-banding techniques can help in determining whether there is a consistency in the chromosomes that associate in this manner, thereby providing information on interchromosomal or intergenomic relationships.

12.4 USES OF HAPLOIDS IN PLANTS

12.4.1 Screening for Mutations Using Haploids

In diploids, and in both autopolyploids and allopolyploids, a newly arising, recessive mutation in a somatic cell usually occurs on only one of the homologous chromosomes. The mutation is concealed by a dominant allele on one or more of the other homologs, depending on the ploidy level and genome constitution. If the recessive mutation occurs in a gamete that participates in fertilization, the other gamete introduces a dominant allele. Therefore, in diploids and both types of polyploids, recessive mutations are not revealed until after one or more generations of inbreeding.

Monohaploids are the most suitable type of haploid to screen for recessive mutations because they have a single dose of one genome, so there are no covering dominant alleles unless there are intragenomic duplications. Polyhaploids are less suitable because of the replicated loci on homologous chromosomes in haploids

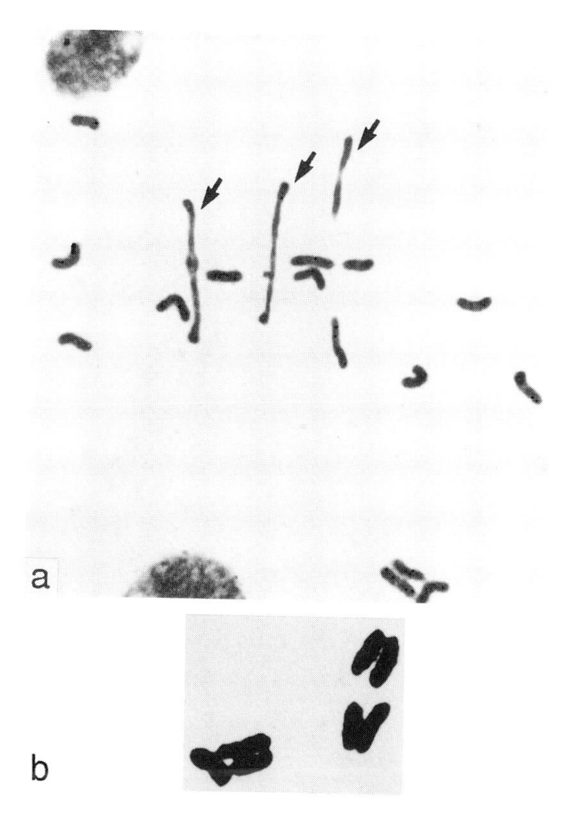

Fig. 12.9. Chromosome associations at metaphase I in haploids. (a) A polyhaploid in wheat showing 3 rod bivalents on the plate (the one on the right is broken into two pieces) and 15 scattered univalents. The three univalents in the lower right show end-to-end, side-by-side, and end-to-side associations. (Photomicrograph from Riley and Chapman, 1957.) (b) The seven univalents in monohaploid barley. The three univalents on the left show end-to-end, side-by-side, and end-to-side associations; the four univalents on the right show side-by-side associations. (Photomicrograph from Sadasivaiah, 1974.)

from autopolyploids, or on homoeologous chromosomes in those from allopolyploids. However, both monohaploids and polyhaploids can be used to detect dominant mutations, which are less frequent than recessive mutations.

The most efficient method for mutation screening is to introduce a treatment to haploid cell cultures such as microspores, callus tissue, or protoplasts. Treatments and their purposes include radiation and chemical mutagens to study mutation frequencies, weed-killing chemicals to isolate cells of crop plants that are resistant to the chemicals, and stress conditions such as a pathogen for a particular disease, to select disease-resistant cells, or an excess of salt, to select salt-tolerant cells. By exposing successive cycles of cells to increasing concentrations of a chemical such as salt, it may be possible to obtain cells with tolerance for levels of a chemical found in the growing environment of plants.

The next important step is to be able to regenerate plants from the selected cells and to double their chromosome number to get stable, homozygous genotypes, thereby perpetuating the mutations. The mutant plants can then be used in crosses with current crop varieties to transfer the mutant genes within or between species. If the genes have quantitative effects, the use of molecular markers, linked with the mutant genes, help in following the traits through the generations until a reconstituted variety is developed.

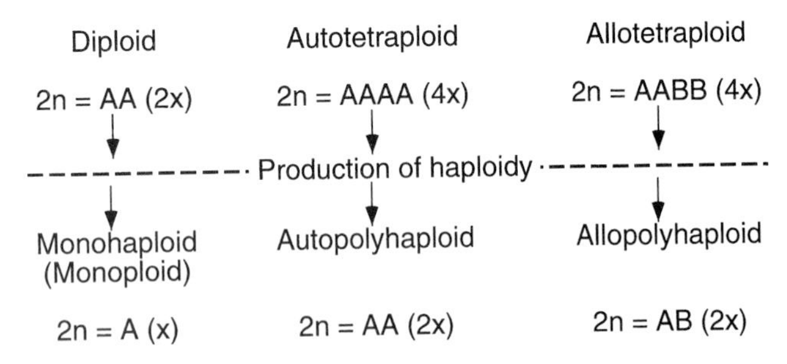

Fig. 12.10. The differences in origin and genome dosage (symbolized by A and B) of three kinds of haploids.

12.4.2 Uses of Haploids in Breeding and Genetic Research

Most haploids cannot be used directly in developing new crop varieties because of low vigor and high sterility due to the reduced genome number. An exception is the haploids from autotetraploids because they are at the diploid level with two doses of the same genome (Fig. 12.10). These haploids have distinct advantages over the tetraploids in gene-mapping studies and in transferring genes from wild diploid relatives, because they provide a simpler genetic situation. Two naturally evolved autotetraploid crops are the cultivated potato (*Solanum tuberosum*) and alfalfa or lucerne (*Medicago sativa*), both cross-fertilizing species, which depend on allelic variations at their gene loci to maintain vigor and fertility. The haploids have homologous pairs of chromosomes, which usually behave normally during meiosis. Vigor and fertility are often reduced because of fewer genetic combinations within or between loci, and the expression of deleterious, recessive genes when the haploids are maintained by self-fertilization. However, the severity of effects varies depending on the autotetraploid genotypes.

The most intensive use of haploids for genetic and breeding studies has been in potato, where they are derived from the tetraploid *S. tuberosum* in large numbers by various manipulations. Selection is made for the most vigorous and fertile plants, and improvement in these traits is sometimes obtained by intercrossing haploids from different tetraploid sources. The haploids are also crossed with wild, diploid relatives, which have the same or a similar type of genome, to introduce new and potentially valuable genes. The haploids do not have the potential of tetraploids for developing new varieties because of their more limited genetic variation, but they provide intermediate steps in transferring new genes to the tetraploid level. The transfers are facilitated by the formation of $2n$ gametes in the haploids as a result of first-division restitution (FDR) or second-division restitution (SDR) (*see* Chapter 10, Figs. 10.4–10.6). In potato, most of the $2n$ pollen results from FDR, whereas most of the $2n$ eggs result from SDR. The tetraploid level can be achieved by intercrosses at the diploid level between haploid lines, or between the haploids and diploid species, if some fertilizations unite $2n$ male and female gametes. Another approach is to cross a current tetraploid variety with haploids from different sources. When $2n$ gametes from the haploids meet $2n$ gametes formed by the tetraploids, the progeny are tetraploid with two genomes contributed by each parent. With selection of suitable parents, the derived tetraploid genotypes should be highly heterozygous and can be maintained by asexual propagation of tubers.

12.5 PRODUCTION AND USES OF DOUBLED HAPLOIDS IN PLANTS

Haploids from diploids have only one dose of one genome, and haploids from allopolyploids have one dose of each type of genome (*see* Fig. 12.10). These reduced doses cause disturbed meiotic divisions, resulting in sterility in most of the gametes, so the genomes must be doubled to restore vigor and fertility.

12.5.1 Methods for Doubling Haploids

Spontaneous genome doubling is desirable because it is less likely to produce changes in chromosomes and genes when compared with artificial treatments. Unfortunately, it happens too infrequently to obtain enough doubled haploids for research programs. Induction methods include the same types of treatments that are used to obtain polyploids from diploids (*see* Chapter 10, Section 10.3), except that some treatments can be applied to media used for culturing gametophytic cells. There are several requirements for a satisfactory induced-doubling method, including an efficient technique to handle large numbers of plants, seeds, or cultured cells over a wide range of species. The desired goal is homozygous, doubled-haploid plants with no genetic or physiological damage or changes. Overall, colchicine satisfies these needs more than other treatments, so it has been used most widely. It has been applied to seeds and seedlings, often producing $2n$ sectors, and has been added to anther-culture media. Some factors that may affect doubling in tissue culture are pollen stage when cultured, hormones, and low-temperature pretreatment. Because callus tissue, induced in appropriate culture media, tends to undergo chromosome doubling more than once, the regenerated plants with one doubling must be separated from those having a higher ploidy level. About 70% of the plants regenerated from anther culture in barley and rye are doubled haploids, and the rest are haploids or tetraploids.

12.5.2 Genetic Composition of Doubled Haploids

Doubled haploids from diploids or allopolyploids should be homozygous at all loci, because the haploids have one dose of each kind of chromosome before the doubling event. However, there can be genetic differences among doubled-haploid lines if the plants from which they were derived are heterozygous at a number of loci (Fig. 12.11). Another source of variation is the occurrence of

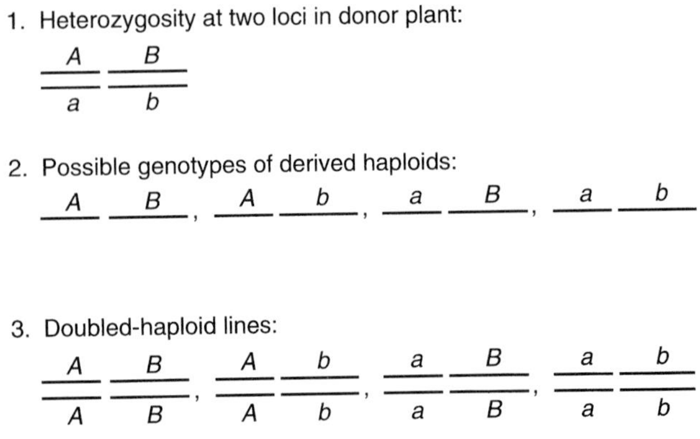

1. Heterozygosity at two loci in donor plant:

   A/a B/b

2. Possible genotypes of derived haploids:

   A B, A b, a B, a b

3. Doubled-haploid lines:

   A/A B/B, A/A b/b, a/a B/B, a/a b/b

Fig. 12.11. A diagrammatic representation of the effect of heterozygosity within a plant used for anther culture on genetic variability in the derived haploids and doubled-haploid lines. The two gene pairs used in this example are shown on separate chromosomes.

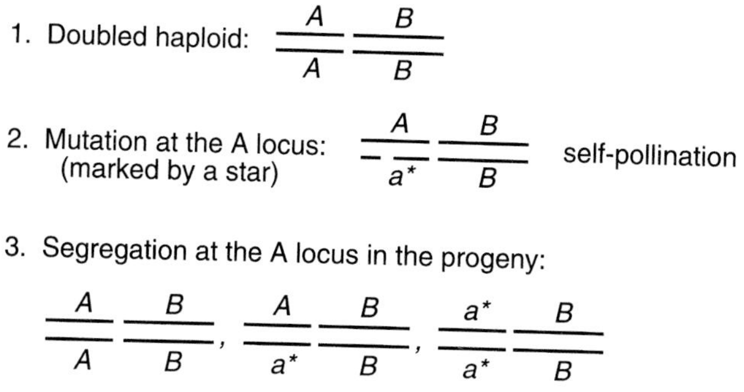

1. Doubled haploid: $\frac{A}{A}\ \frac{B}{B}$

2. Mutation at the A locus: (marked by a star) $\frac{A}{a^*}\ \frac{B}{B}$ self-pollination

3. Segregation at the A locus in the progeny:

$\frac{A}{A}\ \frac{B}{B}$, $\frac{A}{a^*}\ \frac{B}{B}$, $\frac{a^*}{a^*}\ \frac{B}{B}$

Fig. 12.12. The effect of a mutation in a somatic cell of a doubled haploid on segregation in its selfed progeny. If the mutation occurs during the early development of the doubled-haploid plant, it has a chance of being transmitted through the gametes.

mutations in the early stages of plant development after chromosome doubling (Fig. 12.12). Usually, only one of the alleles mutates at a particular locus on homologous chromosomes, so that the resulting plant is heterozygous at that locus, and segregation can occur in its progeny. Heterozygosity has been found to vary from 0% to 10% in doubled haploids of different species. Some segregating genotypes may have desirable characteristics to be investigated.

Genetic variations, as well as changes in chromosome structure and number, are frequent occurrences in doubled-haploid lines arising from *in vitro* culture. Some of the chromosome changes such as aneuploidy can be used in cytogenetic studies to relate genes to chromosomes. The genetic changes involve single-gene traits as well as polygenic traits such as height, maturity, and yield. The spectrum of variations produced by tissue culture is similar to the effects of applying mutagenic agents to plants. Although some of these changes may be useful for increasing variability when developing new crop varieties, they are a hindrance if they have deleterious effects. A complex of leaf abnormalities and decreased growth rates appeared in doubled haploids from cultured anthers of the diploid species *Nicotiana sylvestris*. Some doubled haploids transmitted these traits to all their selfed progenies, but others gave segregation in their progenies. There was increasing variation for these traits during five consecutive cycles of haploid induction, indicating that the new variation had arisen during anther culture. Molecular studies showed that the doubled haploids had an average of 10% more DNA than the original plant, with increases in the highly repeated sequences and in the number of inverted repeats. There were indications that regions rich in A–T and G–C were amplified. Wheat doubled haploids from anther culture differed from the parental cultivar in the organization of the nontranscribed spacer region of ribosomal DNA. The variation was inherited but was influenced by environmental and genetic factors.

There has been great interest in trying to define the causes of variation in doubled haploids from tissue culture, with the objective of being able to control the amount of stability or variability at will. Part of the variability is attributed to the age of the culture and the numerous mitotic cycles in culture, thus increasing the chances for mutations and chromosomal changes. The replication and segregation of chromosomes can be adversely affected, and this may contribute more to variability than the added number of cell divisions. Cultural conditions that may foster instability are the chemical composition of the medium and physical influences such as temperature.

It is important to be able to express rates of mutations or chromosomal changes on the basis of a cell-division cycle, because of variations in the number of cell divisions due to cultural or environmental conditions. This is often not possible because of the lack of data on the number of cell divisions. If most of the changes in genes and chromosomes occur during cell division, cultures that divide rapidly per time unit, or that are transferred often to fresh medium, are expected to generate more variability than cultures that divide, or are transferred, less frequently. Under cell-culture conditions, DNA replication and mitotic stages (except anaphase) take longer in unorganized cells, such as microspores, than in cells organized in meristems and can contribute to more instability. Mitotic abnormalities include spindle or nuclear fusions and endomitosis (*see* Chapter 10, Section 10.4). Heterochromatin may be more prone to breakage than euchromatin, based on evidence in oats and maize from culturing somatic cells. Genes may be suppressed by transposable elements, which may be activated by the stressful conditions of the culture regime (*see* Chapter 7, Section 7.5). The effect is a change in gene expression that resembles a mutation.

It is evident that when cells are removed from their natural environment within a plant body and cultured on an artificial medium, they are introduced to stressful conditions that can affect DNA, chromosome structure, and cell division. In spite of these adverse effects, it has been possible to obtain haploids and doubled haploids from cell cultures on a scale that is competitive with other methods.

12.5.3 Uses of Doubled Haploids

The prime advantage of using doubled haploids in developing new crop varieties is the achievement of homozygosity in one generation. The initial step in a breeding program involves crosses between varieties, lines, or species differing in important traits, in order to introduce variation through heterozygosity. The F_1 hybrids produce a range of gametes with different gene combinations, which reflect the genetic differences between the parents. A conventional plant-breeding program requires self-pollination of the F_1 hybrids and several succeeding generations before stable, relatively homozygous lines can be selected for increase and tests. If haploids are available from culturing anthers of F_1 plants, homozygous doubled haploids can be obtained in the equivalent of the F_2 generation. The only chance for genetic recombination is during meiosis in the F_1 hybrids. Therefore, enough doubled-haploid lines should be developed to include the range of variation represented by the F_1 gametes, so that lines with the most desirable traits can be selected. Some studies have shown a skewness toward one parent in the genotypes represented by the doubled haploids, indicating that these genotypes have a selective advantage in their passage through tissue culture.

Stable doubled haploids can be increased and tested several generations ahead of the conventionally developed lines (Fig. 12.13). This saving in time must be weighed against the expense

individuals with three genes homozygous:

comprise $(1/2)^3$ of population in doubled haploid derivatives
comprise $(1/4)^3$ of population in normal F2 generation

true-breeding lines recovered 1/8 in doubled haploid derivatives
true-breeding lines recovered 1/64 in normal F2 generation

Fig. 12.13. A statistical comparison of doubled-haploid and conventional-breeding methods as applied to spring wheat. (Adapted from Baenziger et al., 1984; Morrison and Evans, 1988.)

and expertise required to develop the doubled haploids. Their performance must be equal or superior to that of lines developed by conventional breeding methods to make this approach feasible. In tobacco and barley, some doubled-haploid lines were inferior to conventional lines in some decisive traits, whereas others equaled or surpassed the conventional lines. Chinese scientists, through concentrated efforts, have released new varieties of rice and wheat using the doubled-haploid pathway. A number of factors are involved in the successful application of this method, and one of the most important is the ability to regenerate a large number of haploid plants. They provide a wide base of genetic variation, upon which to draw, in developing and selecting the best doubled-haploid lines.

Doubled haploids have other potential uses besides the development of new varieties. They are a vehicle for incorporating in crop species homozygosity for valuable genes from related wild species. If the genomes of the crop and wild species have enough homology to pair during meiosis, the genes can be transferred to the crop chromosomes by crossing over in the F_1 hybrids, followed by extraction of doubled haploids with homozygous, recombined genes. Another mode of transfer takes advantage of the chromosomal instability in the haploid cells during tissue culture, with the possibility of breaks and exchanges between crop and alien chromosomes to produce translocations. Some exchanges may transfer segments of alien chromosomes, with needed genes, to the chromosomes of agricultural crops.

Doubled haploids are suitable for determining the mode of inheritance of complex quantitative traits, because their homozygous genotypes simplify the genetic segregations and statistical analyses from hybridizations. The studies can involve crosses between two doubled-haploid lines with different expressions of the traits under study, and collection of data from the F_2 or later generations. An alternative method involves tests of doubled-haploid lines developed from F_1 hybrids.

BIBLIOGRAPHY

General

CHU, C-C. 1982. Haploids in plant improvement. In: Plant Improvement and Somatic Cell Genetics (Vasil, I.K., Scowcroft, W.R., Frey, K.J., eds.). Academic Press, New York. pp. 129–158.

DE FOSSARD, R.A. 1974. Terminology in "haploid" work. In: Haploids in Higher Plants—Advances and Potential (Kasha, K.J., ed.). University of Guelph, Guelph, Canada. pp. 403–410.

HU, H., HUANG, B. 1987. Application of pollen-derived plants to crop improvement. Int. Rev. Cytol. 107: 293–313.

JENSEN, C.J. 1 974. Chromosome doubling techniques in haploids. In: Haploids in Higher Plants—Advances and Potential (Kasha, K.J., ed.). University of Guelph, Guelph, Canada. pp. 153–190.

KASHA, K.J. (ed.). 1974. Haploids in Higher Plants—Advances and Potential. University of Guelph, Guelph, Canada.

KASHA, K.J., SEGUIN-SWARTZ, G. 1983. Haploidy in crop improvement. In: Cytogenetics of Crop Plants (Swaminathan, M.S., Gupta, P.K., Sinha, U., eds.). Macmillan India Ltd., New Delhi. pp. 19–68.

MORRISON, R.A., EVANS, D. A. 1988. Haploid plants from tissue culture—new plant varieties in a shortened time frame. Bio/Technology 6: 684–690.

PELOQUIN, S,J., WERNER, J.E., YERK, G.L. 1991. The use of potato haploids in genetics and breeding. In: Chromosome Engineering in Plants: Genetics, Breeding, Evolution. Part B (Tsuchiya, T., Gupta, P.K., eds.). Elsevier, Amsterdam. pp. 79–92.

SHARP, W.R., REED, S.M., EVANS, D.A. 1984. Production and application of haploid plants. In: Crop Breeding: A Contemporary Basis (Vose, P.B., Blixt, S.G., eds.). Pergamon Press, New York. pp. 347–381.

SYBENGA, J. 1992. Cytogenetics in Plant Breeding. Monographs on Theoretical and Applied Genetics 17. Springer-Verlag, Berlin.

Section 12.1

CHASE, S.S. 1969. Monoploids and monoploid-derivatives of maize (*Zea mays* L.). Bot. Rev. 35: 117–167.

COE, E.H., JR. 1959. A line of maize with high haploid frequency. Am. Nat. 93: 381–382.

COE, G.E. 1953. Cytology of reproduction in *Cooperia pedunculata*. Am. J. Bot. 40: 335–343.

FINCH, R.A. 1983. Tissue-specific elimination of alternative whole parental genomes in one barley hybrid. Chromosoma 88: 386–393.

KASHA, K.J. 1974. Haploids from somatic cells. In: Haploids in Higher Plants—Advances and Potential (Kasha, K.J., ed.). University of Guelph, Guelph, Canada. pp. 67–87.

KASHA, K.J., ZIAUDDIN, A., CHO, U.-H. 1990. Haploids in cereal improvement: anther and microspore culture. In: Gene Manipulation in Plant Improvement II (Gustafson, J.P., ed.). 19th Stadler Genet. Symp., Plenum Press, New York. pp. 213–235.

KELLER, W.A., STRINGAM, G. R. 1978. Production and utilization of microspore-derived haploid plants. In: Frontiers of Plant Tissue Culture (Thorpe, T.A., ed.). University of Calgary Press, Calgary, Alberta, Canada. pp. 113–122.

KERR, W.E. 1969. Some aspects of the evolution of social bees (Apidae). Evol. Biol. 3: 119–175.

KISANA, N.S., NKONGOLO, K.K., QUICK, J.S., JOHNSON, D.L. 1993. Production of doubled haploids by anther culture and wheat × maize method in a wheat breeding programme. Plant Breed. 110: 96-102.

KOBAYASHI, M., TSUNEWAKI, K. 1980. Haploid induction and its genetic mechanism in alloplasmic common wheat. J. Hered. 71: 9–14.

LACADENA, J.-R. 1974. Spontaneous and induced parthenogenesis and androgenesis. In: Haploids in Higher Plants—Advances and Potential (Kasha, K.J., ed.). University of Guelph, Guelph, Canada. pp. 13–32.

LANGE, W. 1988. Cereal cytogenetics in retrospect. What came true of some cereal cytogeneticists' pipe dreams? Euphytica 39(Suppl.): 7–25.

LASHERMES, P., BECKERT, M. 1988. Genetic control of maternal haploidy in maize (*Zea mays* L.) and selection of haploid-inducing lines. Theoret. Appl. Genet. 76: 405–410.

LAURIE, D.A., REYMONDIE, S. 1991. High frequencies of fertilization and haploid seed production in crosses between commercial wheat varieties and maize. Plant Breed. 106: 182–189.

LORZ, H., BRETTSCHNEIDER, R., HARTKE, S., GILL, R., KRANZ, E., LANGRIDGE, P., STOLARZ, A., LAZZERI, P. 1990. In vitro manipulation of barley and other cereals. In: Gene Manipulation in

Plant Improvement II (Gustafson, J.P., ed.). 19th Stadler Genet. Symp. Plenum Press, New York. pp. 185–201.

Maniotis, J. 1980. Polyploidy in fungi. In: Polyploidy: Biological Relevance (Lewis, W.H., ed.). Plenum Press, New York. pp. 163–192.

Matzk, F., Mahn, A. 1994. Improved techniques for haploid production in wheat using chromosome elimination. Plant Breed. 113: 125–129.

Murray, B.E. 1985. Studies of haploid-diploid twins in flax (*Linum usitatissimum*) Can. J. Genet. Cytol 27: 371–379.

Peloquin, S.J., Yerk, G.L., Werner, J.E. 1990. Ploidy manipulations in the potato. In: Chromosomes: Eukaryotic, Prokaryotic, and Viral. Vol. II (Adolph, K.W., ed.). CRC Press, Boca Raton, FL. pp. 167–178.

Prakash, J., Giles, K.L. 1987. Induction and growth of androgenetic haploids. Int. Rev. Cytol. 107: 273–292.

Rybczynski, J.J., Simonson, R.L., Baenziger, P.S. 1991. Evidence for microspore embryogenesis in wheat anther culture. In Vitro Cell. Dev. Biol. 27P: 168–174.

San, L.H., Gelebart, P. 1986. Production of gynogenetic haploids. In: Cell Culture and Somatic Cell Genetics of Plants. Vol. 3. Plant Regeneration and Genetic Variability (Vasil, I.K., ed.). Academic Press, New York. pp. 305–322.

Sangwan, R.S., Sangwan-Norreel, B.S. 1987. Biochemical cytology of pollen embryogenesis. Int. Rev. Cytol. 107: 221–272.

Turcotte, E.L., Feaster, C.V. 1969. Semigametic production in Pima cotton. Crop Sci. 9: 653–655.

Vasil, I.K. (ed.) 1986. Cell Culture and Somatic Cell Genetics of Plants. Vol. 3. Plant Regeneration and Genetic Variability. Academic Press, Orlando, FL.

Section 12.2

Avery, A.G. 1959. Polyploidy. In: Blakeslee: The Genus *Datura* (Avery, A.G., Satina, S., Rietsema, J., eds.). The Ronald Press Co., New York. pp. 71–85.

Mendiburu, A.O., Peloquin, S.J., Mok, D.W.S. 1974. Potato breeding with haploids and 2*n* gametes. In: Haploids in Higher Plants—Advances and Potential (Kasha, K.J., ed.). University of Guelph, Guelph, Canada. pp. 249–258.

Peloquin, S.J., Hougas, R.W. 1960. Genetic variation among haploids of the common potato. Am. Potato J. 37: 289–297.

Rowe, P.R. 1974. Methods of producing haploids: parthenogenesis following interspecific hybridization. In: Haploids in Higher Plants—Advances and Potential (Kasha, K.J., ed.). University of Guelph, Guelph, Canada. pp. 43–52.

Sarkar, K.P. 1974. Genetic selection techniques for production of haploids in higher plants. In: Haploids in Higher Plants—Advances and Potential (Kasha, K.J., ed.). University of Guelph, Guelph, Canada. pp. 33–41.

Section 12.3

Hermsen, J.G.Th. 1984. Haploids as a tool in breeding polyploids. Iowa State J. Res. 58: 449–460.

Riley, R., Chapman, V. 1957. Haploids and polyhaploids in *Aegilops* and *Triticum*. Heredity 11: 195-207.

Sadasivaiah, R.S. 1974. Haploids in genetic and cytological research. In: Haploids in Higher Plants—Advances and Potential (Kasha, K.J., ed.). University of Guelph, Guelph, Canada. pp. 355–386.

Section 12.4

Sharp, W.R., Reed, S.M., Evans, D.A. 1984. Production and application of haploid plants. In: Crop Breeding: A Contemporary Basis (Vose, P.B., Blixt, S.G., eds.). Pergamon Press, New York. pp. 347–381.

Section 12.5

Baenziger, P.S., Kudirka, D.T., Schaeffer, G.W., Lazar, M.D. 1984. The significance of doubled haploid variation. In: Gene Manipulation in Plant Improvement (Gustafson, J.P., ed.). 16th Stadler Genet. Symp., Plenum Press, New York. pp. 385–414.

Choo, T.M., Reinbergs, E., Kasha, K.J. 1985. Use of haploids in breeding barley. Plant Breed. Rev. 3: 219–252.

De Maine, M.J., Jervis, L. 1989. The use of dihaploids in increasing the homozygosity of tetraploid potatoes. Euphytica 44: 37–42.

De Paepe, R., Prat, D., Huguet, T. 1983. Heritable nuclear DNA changes in doubled haploid plants obtained by pollen culture of *Nicotiana sylvestris*. Plant Sci. Lett. 28: 11–28.

Rode, A., Hartmann, C., Benslimane, A., Picard, E., Quetier, F. 1987. Gametoclonal variation detected in the nuclear ribosomal DNA from doubled haploid lines of a spring wheat (*Triticum aestivum* L., cv ''César''). Theoret. Appl. Genet. 74: 31–37.

V
Gene Locations, Genetic Manipulations, and Rearrangements

13

Locating Genes in Diploids Using Chromosome Aberrations

- Genes are assigned to specific chromosomes in diploids by using chromosome aberrations identified by their phenotypic effects, meiotic behavior, or banding techniques.
- Gene–chromosome associations are recognized by phenotypic segregations that deviate from normal when using aneuploids, or by linkages with breakpoints of structural chromosome aberrations.

One of the most important roles of cytogenetics is to discover the physical sites of genes within the karyotype of an organism. Information on genetic or linkage maps based on recombination data was first obtained while knowledge about the chromosomes was still in a primitive state. From the 1920s onward, starting with the classical organisms *Drosophila melanogaster* and maize (*Zea mays*), chromosomes of many organisms were identified by their physical characteristics (*see* Chapter 6). Another significant development was the stockpiling of structural and numerical aberrations from spontaneous or induced origins. Methods using these aberrations were developed to establish relationships between the linkage maps and the physical chromosomes, thereby producing cytogenetic maps. We shall discuss methods for gene mapping in diploids in this chapter, and methods for polyploids in Chapter 14. The fine-tuning of gene mapping by molecular techniques is discussed in Chapter 21.

Genes can be mapped in diploids using gains and, in some cases, losses of whole chromosomes or their segments, as well as rearrangements of segments. Aneuploids change gene dosage, and the association of a gene with an aneuploid condition is detected by a deviation from the normal phenotype or segregation pattern. Genes can be tested for linkage with the breakpoints of inversions and reciprocal translocations, and for association with deletions by pseudodominance, the appearance of recessive phenotypes in the F_1 from appropriate crosses. These different approaches will now be considered in more detail.

13.1 DIPLOID ANALYSIS USING TRISOMICS

The term disomic refers to two doses of each chromosome in a set such as in diploids, when compared to added or subtracted doses of one or more chromosomes. This chapter includes various types of trisomics, and these chromosome-dosage terms refer to individual plants as well as gene segregations.

Trisomy in its various forms is an efficient method for assigning genes to specific chromosomes in diploid plant species. The trisomic types for each chromosome usually have distinctive phenotypes involving such traits as height, leaf shape, inflorescence type, and maturity rate (*see* Chapter 11, Figs. 11.14 and 11.15). These are all quantitative characters, so direct observations of the effects of increasing the dosage of whole chromosomes, arms, or segments provide information on the inheritance of these traits. When these comparisons are made among trisomics of a set, they and the diploid controls should have the same genotypic background to avoid the possibility that some differences are due to background heterogeneity. Both trisomics and diploids should be grown in the same environment at the same time, because variations in such conditions as soil fertility or temperature may affect quantitative gene expression.

Trisomics also are very useful for locating qualitative genes that are introduced through crosses. Because of the distinctive trisomic phenotypes, large numbers of plants can be classified in segregating progenies without the need for cytology. The procedure is to make crosses between trisomics, identified for each chromosome in the set, and diploids that have contrasting alleles of the genes to be located. The F_1 trisomic plants are selfed to obtain F_2 segregations, or they are crossed with the diploid parent carrying the recessive alleles to get backcross-progeny segregations. The trisomic parents and trisomic F_1's are used as females in the crosses because the transmission rates of $n + 1$ gametes are usually much higher through the female than through the male. The trisomic ratios also depend on the number of genes controlling a trait and their modes of action or interaction, the genotypes of the parents, the behavior of the extra chromosome during meiosis, and the viability of trisomic progeny.

13.2 USE OF PRIMARY TRISOMICS

If a gene pair designated *A–a* is located on a trisomic chromosome, this is called the critical chromosome, and the cross involving this trisomic is the critical cross (Fig. 13.1). Crosses involving all the

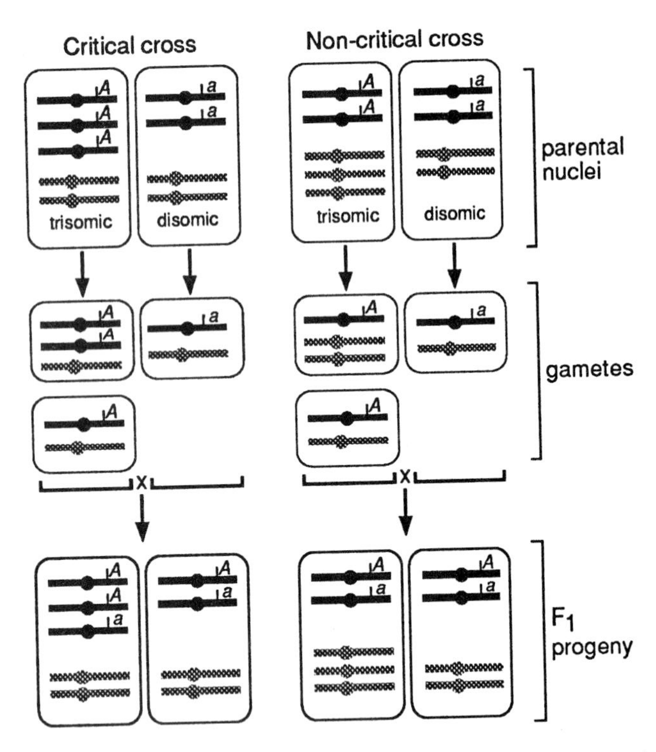

Fig. 13.1. Diagrams to show the difference between a critical cross and a noncritical cross, with respect to a gene locus *A–a*, when a trisomic is crossed with a disomic. In both cases, the parent used as female is the one on the left side because the transmission of the trisomic state is much higher through the female than through the male. Only two chromosomes of the genome are shown.

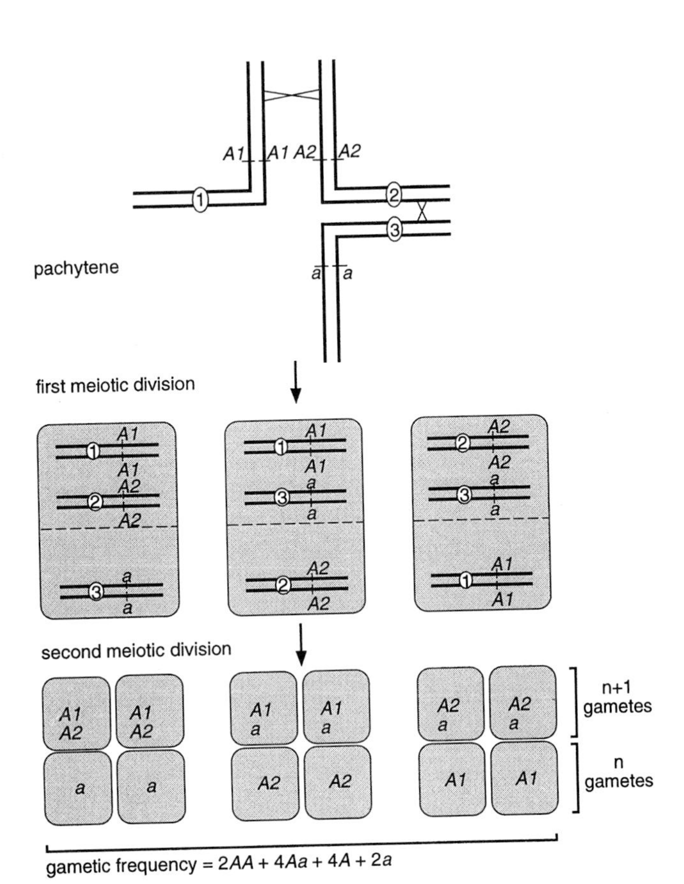

Fig. 13.2. Diagram of a trisomic pachytene configuration. The gene locus *A/a* is assumed to be completely linked with the centromere (numbered) in each chromosome. The dominant alleles *A1* and *A2* are distinguished by numbers to show how they segregate during meiosis. The anaphase I random segregations and the gametic array are shown. The other two random arrangements at pachytene, clockwise from the upper left, are *A1A1–aa–A2A2* and *A2A2–aa–A1A1*. Both orientations give the same kinds and frequencies of gametes as the one shown in this figure.

other trisomics in the set are noncritical crosses. The critical trisomic parent can have one of the following genotypes: *AAA*, *AAa*, *Aaa*, or *aaa*. The F_1 *AAa* genotype can be obtained by crossing trisomic *AAA* as female with disomic *aa*, so that an *AA* egg unites with an *a* sperm. The F_1 *Aaa* genotype can come from the reciprocal cross [i.e., the trisomic (*aaa*) supplies an *aa* egg and the disomic (*AA*) an *A* sperm]. If we assume that transmission of the $n + 1$ gamete is 25% through the female, which is within the range for plant trisomics (*see* Chapter 11, Section 11.6.1), the F_1's from these two types of crosses consist of 25% trisomics and 75% disomics. The trisomics are selected by phenotype or cytology and are self-pollinated or backcrossed to *aa*.

A pachytene trivalent configuration for an F_1 *AAa* genotype, with one of the three random-gene arrangements, is shown in Fig. 13.2. At least one chiasma must occur in one arm of each chromosome to retain the trivalent configuration until anaphase I. If the two dominant alleles are labeled *A1* and *A2*, the two chromosomes carrying them are shown paired in this region, and the a allele is unpaired. The amount of crossing over between the gene locus and the centromere in each of the three chromosomes influences the frequencies of n and $n + 1$ gametes. In Fig. 13.2, it is assumed that the *A–a* locus is close enough to the centromere that crossing over is rare, and no crossover products are recovered in a plant population of practical size. Because the two alleles in each $n + 1$ gamete are derived from different chromosomes, chromosome segregation is involved. Disregarding the numerical distinction between the *A* alleles, the overall gametic ratio is 2 *AA* + 4 *Aa* + 4 *A* + 2 *a*, which can be reduced to 1 *AA* + 2 *Aa* + 2 *A* + 1 *a*.

With female transmission rates of 25% $n + 1$ gametes and 75% n gametes, which is a 1 : 3 ratio, the transmission frequency can be obtained by multiplying the n gametes by 3.

The results of backcrossing and selfing the F_1 *AAa* trisomics with chromosome segregation of the gene are given in Table 13.1. In the backcross progeny, the genotypic frequency is the same

Table 13.1. Gametes and Progenies from F_1 Trisomic *AAa*, Backcrossed as Female to Disomic *aa* Male, or Selfed to Give an F_2 Progeny

F_1 Trisomic Eggs	Backcross to Disomic Sperm *a*	Self-Pollination Sperm 2 *A*	Self-Pollination Sperm 1 *a*
1 *AA*	1 *AAa*	2 *AAA*	1 *AAa*
2 *Aa*	2 *Aaa*	4 *AAa*	2 *Aaa*
6 *A* (2 × 3)	6 *Aa*	12 *AA*	6 *Aa*
3 *a* (1 × 3)	3 *aa*	6 *Aa*	3 *aa*

Note: It is assumed that there is no crossing over between the centromere and the gene and that the transmission of $n + 1$ gametes = 25% through the female and 0% through the male.

as the female gametic-transmission frequency. If trisomics can be distinguished from disomics by phenotypic traits, the trisomics have only the dominant phenotype, assuming that one dose of the A allele suffices. The disomic group segregates $2\ A:1\ a$ instead of $1\ A:1\ a$ in a noncritical backcross. If the trisomic phenotypes cannot be separated from the disomic phenotypes, the overall phenotypic segregation in the backcross progeny is $9\ A:3\ a$ or $3\ A:1\ a$, which is also different from a noncritical segregation. If we assume no transmission of $n + 1$ gametes through the male when the trisomic is selfed, the two kinds of n gametes compete equally and are transmitted in the same frequency as they are formed, which is $2\ A:1\ a$. Again, if trisomics are separable from disomics, all the trisomics have the dominant phenotype, and the disomics segregate $24\ A:3\ a$ or $8\ A:1\ a$. The overall segregation is $33\ A:3\ a$ or $11\ A:1\ a$. These segregations are compared with a $3\ A:1\ a$ noncritical segregation. If a low frequency of male $n + 1$ gametes is transmitted, they will not have much effect on the segregations, which can still be separated from noncritical segregations. A few tetrasomics may appear in the F_2 progenies from the union of female and male $n + 1$ gametes, but their phenotype is usually a more extreme form of the trisomic phenotype.

In all the noncritical crosses, the A–a gene pair segregates independently of the trisomic state. If enough progeny plants are grown, both the trisomic and disomic groups, as well as the combined groups, should have $1\ A:1\ a$ backcross segregations and $3\ A:1\ a$ F_2 segregations.

If crossing over occurs between the centromere and the gene locus in each chromosome, each allele must be considered on a chromatid basis. This situation is shown in Fig. 13.3, which has one of three possible gene arrangements. Three of the 12 $n + 1$ gametes (*A1A1*, *A2A2*, and *aa*) have sister alleles, a situation called double reduction, and the kind of chromatid segregation that produces this high frequency of gametes with sister alleles is described as maximum equational segregation (*see* Chapter 10, Section 10.7.2).

The expected segregations for the *AAa* genotype with maximum equational segregation are given in Table 13.2. The female n gametes have been multiplied by 3 (75% transmission) and, for the selfs, the male n gametes are reduced from $8\ A + 4\ a$ to $2\ A + 1\ a$. This table differs from Table 13.1 mainly in the occurrence of *aaa* in the trisomic group, although large progenies have to be grown to recover this phenotype, especially in the selfed progeny. The ratios from both backcrosses and selfs are distinguishable from

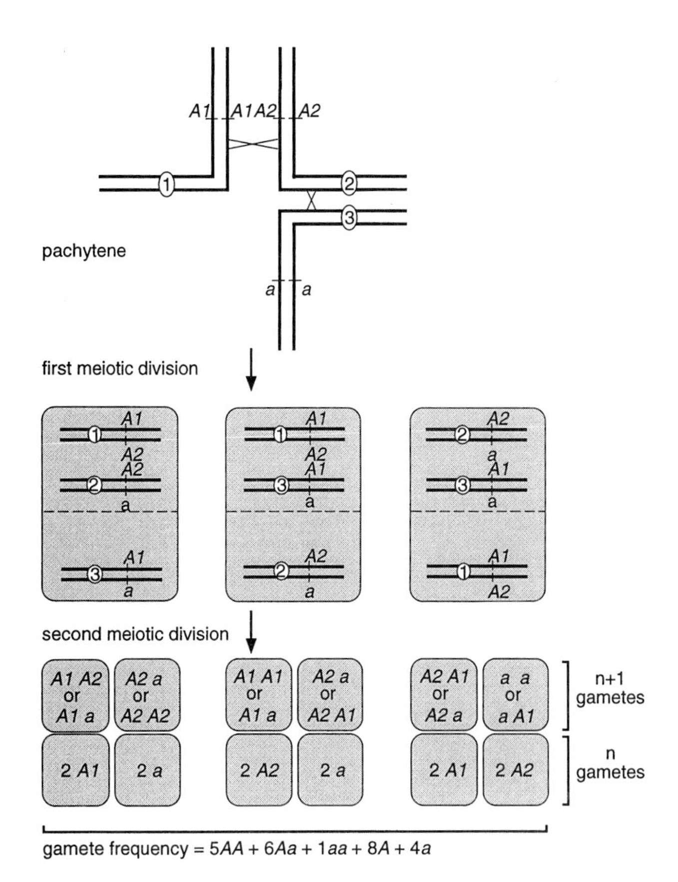

Fig. 13.3. Diagram of a trisomic pachytene configuration with crossing over between the centromere and the gene pair *A/a* in each chromosome. *See* Fig. 13.2 legend for the explanation of *A1* and *A2*. The random anaphase I segregations and the gametic array are shown. The frequencies of n gametes are multiplied by 2 to equal the frequencies of $n + 1$ gametes. The other two random arrangements at pachytene, clockwise from the upper left, are *A1A1–aa–A2A2* and *A2A2–aa–A1A1*. Both orientations give the same kinds and frequencies of gametes as the one shown in this figure.

noncritical ratios for the trisomic and disomic groups, as well as for the combined data.

The same procedures can be used for the other heterozygous trisomic genotype, *Aaa*. Comparisons of the segregations from the two genotypes (Table 13.3) indicate that *AAa* is the more efficient genotype for distinguishing between trisomic and disomic inheritance. Most of its segregations have greater deviations from disomic segregations than those from *Aaa*, so its progenies can be smaller in size. The *AAa* trisomic also gives some indication of the distance between a gene and its centromere, because there must be crossing over in this region to obtain any *aaa* genotypes. Trisomic progeny from *Aaa* include *aaa* both with and without crossing over between the gene and its centromere.

The segregations in Tables 13.1 through 13.3 are based on defined meiotic chromosome behavior in the F_1 trisomics such as consistent trivalent formations, crossing over in specific regions, and random chromosome distributions. There may be deviations from this pattern depending on which chromosome is trisomic. The transmission rates of $n + 1$ gametes vary considerably among different trisomics. Whereas the conditions used for these tables are useful in representing different aspects of theoretical expectations, some adjustments may be needed for comparisons with observed data. For instance, if the transmission frequencies of n and $n + 1$

Table 13.2. Gametes and Progenies from F_1 Trisomic *AAa*, Backcrossed as Female to Disomic *aa* Male, or Selfed to Give an F_2 Progeny

F_1 Trisomic Eggs	Backcross to Disomic Sperm *a*	Self-Pollination: Sperm 2 *A*	Self-Pollination: Sperm 1 *a*
5 *AA*	5 *AAa*	10 *AAA*	5 *AAa*
6 *Aa*	6 *Aaa*	12 *AAa*	6 *Aaa*
1 *aa*	1 *aaa*	2 *Aaa*	1 *aaa*
24 *A* (8 × 3)	24 *Aa*	48 *AA*	24 *Aa*
12 *a* (4 × 3)	12 *aa*	24 *Aa*	12 *aa*

Note: In this case, it is assumed that there is maximum crossing over between the centromere and the gene and that the transmission of $n + 1$ gametes = 25% through the female and 0% through the male.

Table 13.3. Trisomic Phenotypic Segregations for the *A/a* Locus Based on Female Transmission of $1(n + 1):3(n)$ Gametes Through the Female and No Male $n + 1$ Transmission

Type of Segregation and Genotype	F_1 Trisomic Backcrossed as Female to Disomic *aa*			F_1 Trisomic Selfed		
	$2n + 1$	$2n$	Total	$2n + 1$	$2n$	Total
Chromosome						
AAa	All *A*	2 *A*:1 *a*	3 *A*:1 *a*	All *A*	8 *A*:1 *a*	11 *A*:1 *a*
Aaa	2 *A*:1 *a*	1 *A*:2 *a*	5 *A*:7 *a*	7 *A*:2 *a*	5 *A*:4 *a*	11 *A*:7 *a*
Maximum equational						
AAa	11 *A*:1 *a*	2 *A*:1 *a*	35 *A*:13 *a*	35 *A*:1 *a*	8 *A*:1 *a*	131 *A*:13 *a*
Aaa	7 *A*:5 *a*	1 *A*:2 *a*	19 *A*:29 *a*	13 *A*:5 *a*	5 *A*:4 *a*	43 *A*:29 *a*

Note: Some of the ratios have been reduced as much as possible.

gametes through female and male are known for a particular trisome, those values should be used in calculating expected frequencies.

A useful observation in Table 13.3 is that the ratios of the disomic ($2n$) groups within a genotype are the same for chromosome and maximum equational segregations, but the frequencies of dominant and recessive phenotypes are reversed between the two genotypes. The disomic ratios also are the same with different transmission rates of the $n + 1$ gametes, but different from ratios resulting from crosses between diploids. Therefore, the disomics are an important component of a progeny in determining trisomic inheritance. If either of the two F_1 genotypes is used as male in backcrosses with *aa*, and no $n + 1$ gametes are transmitted, only disomics occur in the progenies, with a 2 *A*:1*a* ratio for *AAa* and a 1 *A*:2 *a* ratio for *Aaa*. These ratios are distinguishable from the 1 *A*:1 *a* diploid ratio if large enough progenies are tested.

We can see how trisomic analysis works in practice by using some data from a study in rice (Table 13.4). The gene symbols for *gold hull* are *gh1* and for *spotted leaf spl1* with the dominant alleles representing the normal phenotype in each case. The expected segregations are based on no transmission of $n + 1$ gametes through the male, and 33.3% through the female instead of 25% used in Tables 13.1 through 13.3. This change in frequency affects the segregation ratio for the total population (12.5:1 instead of 11:1 for chromosome segregation), but not the disomic portion (8:1). The actual transmission rates of the $n + 1$ gametes through the female in crosses of trisomic female by disomic male are 32.7% for trisomic 5 and 37.1% for trisomic 6, and through the male in the reciprocal cross, 1.6% for trisomic 5 and 5.6% for trisomic 6. Therefore, use of 33.3% transmission of $n + 1$ through the female and none through the male is close to the actual values. The chi-square tests give a good fit of the observed to the expected segregations. There are two recessive phenotypes in the F_2 trisomic group for trisomic 5, but none among a comparable number for trisomic 6. Assuming that the two recessive plants came from crossing over between the gene and the centromere, *gh1* is farther from its centromere on chromosome 5 than *spl1* is from its centromere on chromosome 6. All three sets of data for each trisomic are useful in locating the genes.

Trisomics and tetrasomics are useful for mapping restriction fragment-length polymorphisms (RFLPs) in diploid species when monosomics are not available. If a DNA fragment under study is on a trisomic or tetrasomic chromosome, the extra one or two doses of the fragment result in thicker bands than disomic doses when the fragments are hybridized with homologous probes. Once assigned to identified chromosomes, RFLPs can be used as genetic markers for exploring the genetic content of each chromosome. Their value is increased when their locations can be narrowed to chromosome arms or segments of arms. Telotrisomics or acrotrisomics show more intense bands if a DNA fragment is in an arm or a region that is trisomic. The band intensity is the same as for a disomic if the fragment is on the missing arm or missing segment of these modified trisomics because two doses of the normal chromosome are present.

13.3 USE OF TELOTRISOMICS, ACROTRISOMICS, AND TERTIARY TRISOMICS

These modified trisomics have three doses of genes in parts of the chromosome involved, and two doses in the remainder of the chromosome (*see* Chapter 11, Fig. 11.1). All are useful in localizing genes to arms or segments of arms.

13.3.1 Telotrisomics

These aneuploids, which have an extra dose of one chromosome arm, can be used for arm locations of genes. A gene gives a trisomic

Table 13.4. F_2 Segregations for Marker Genes in Rice from Self-Pollinating F_1 Trisomic Plants with Genotypes *Gh1 Gh1 gh1* or *Spl1 Spl1 spl1**

Trisomic Chromosome and Gene	$2n$ Progeny			$2n + 1$ Progeny		Total Progeny		
	Normal	Mutant	X^2 (8:1)	Normal	Mutant	Normal	Mutant	X^2 (12.5:1)
5, *gh1*	205	31	0.98	98	2	303	33	2.85
6, *spl1*	91	11	0.01	87	0	178	11	0.69

* *Note:* These genotypes are comparable to *AAa* in Figs. 13.2 and 13.3, and in Tables 13.1–13.3.

Source: Data from Khush et al. (1984).

segregation if on the same arm as the telochromosome (often abbreviated to telo), but a disomic segregation if it is on the opposite arm. If the telo and the two complete homologs form a modified trivalent during meiosis, the normal homologs usually go to opposite poles at anaphase I and the telo goes at random to either pole. This behavior results in equal frequencies of n and n + telo gametes. If the two complete chromosomes form a bivalent and the telo remains a univalent, most of the gametes are n because of frequent loss of the telo. The transmission rates of n and n + telo gametes through the female depend on the frequency of trivalent versus bivalent plus telo formations. Telochromosomes for longer arms pair with a homologous arm more frequently than those for shorter arms. The transmission of the n + telo through the male gametes is so low that it can be disregarded.

A pachytene trivalent involving a telochromosome is diagrammed in Fig. 13.4 for the F_1 genotype *AAa*, with no crossing over between the centromere and the gene locus in any of the three chromosomes. The gametic frequency can be reduced to 1 *AA* + 1 *Aa* + 1 *A* + 1 *a*. The segregations in Table 13.5 are based on consistent trivalent formations in the F_1 plants. If $2n$ + telo and $2n$ can be distinguished phenotypically, variations in transmission rate of the n + telo gametes do not affect backcross or F_2 segregations. These are 1 *A*:1 *a* for the disomics and all *A* for the telotrisomics in the backcross progenies, compared to 1 *A*:1 *a* from a diploid cross; and 3 *A*:1 *a* for the disomics and all *A* for the telotrisomics in the F_2 progenies, compared to 3 *A*:1 *a* from a diploid cross. If telotrisomics cannot be separated from disomics, the total ratio is influenced by the gametic transmission rates; with the assumptions given in Table 13.5, it is 5*A*:3 *a* in backcross progenies, and 13*A* : 3*a* in F_2 progenies. These ratios are close to those expected from a diploid cross, so the test is more efficient if telotrisomics and disomics can be separated, because smaller progeny numbers are needed.

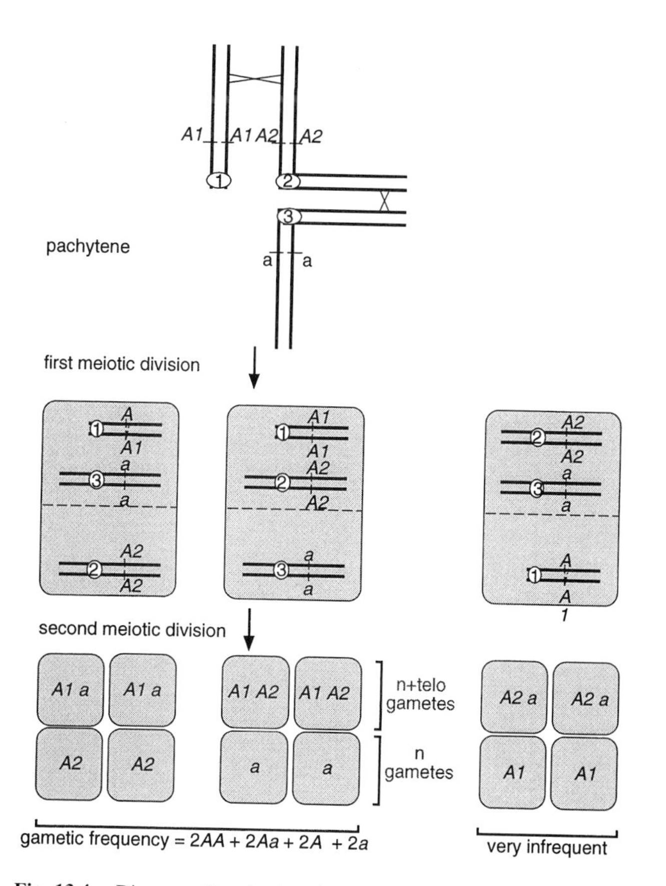

Fig. 13.4. Diagram of a telotrisomic pachytene configuration. It is assumed that the gene pair *A–a* is completely linked with the centromere (numbered) in each chromosome (*see* Fig. 13.2 legend for the explanation of *A1* and *A2*), and that the two complete chromosomes go to opposite poles in most anaphase I cells, with the telochromosome going to either pole. The gametic array is also shown. The other arrangement at pachytene, which gives the same gametic frequencies as the one shown in this figure, is *A1A1–aa–A2A2*.

Table 13.5. Gametes and Progenies from F_1 Telotrisomic *AAa* (see Fig. 13.4), Backcrossed as Female to Disomic *aa* Male, or Selfed to Give an F_2 Progeny

F_1 Telotrisomic Eggs	Backcross Disomic Sperm *a*	Self-Pollination Sperm	
		1 *A*	1 *a*
1 *AA*	1 *AAa*	1 *AAA*	1 *AAa*
1 *Aa*	1 *Aaa*	1 *AAa*	1 *Aaa*
3 *A* (3 × 1)	3 *Aa*	3 *AA*	3 *Aa*
3 *a* (3 × 1)	3 *aa*	3 *Aa*	3 *aa*

Note: It is assumed that there is no crossing over between the centromere and the gene, separation of the two normal chromosomes occurs anaphase I with the telochromosome going to either pole, and transmission of n + telo gametes = 25% through the female and 0% through the male. The extra chromosome in the trisomic progeny is a telochromosome.

Crossing over between the centromere and the gene in any or all of the three chromosomes gives different gametic frequencies and segregation ratios. One useful aspect of the effect of such crossovers will be given here. In Fig. 13.4, if chromosomes 1 and 3 are paired, and crossing over occurs between the telochromosome and chromosome 3 in the region between the centromeres and the *A–a* locus, the *a* allele can be transferred to the telochromosome. Then, if chromosomes 1 and 3 pass to the same anaphase I pole, an *aa* (double reduction) gamete can occur as a result of the anaphase II chromatid distribution. Then, segregation for the recessive phenotype can occur in the telotrisomic group in backcross or F_2 progenies. This indicates that the *A–a* locus is not completely linked with the centromere.

The use of telochromosomes makes it possible to localize the gene order within an arm. When two genes are being tested in the same study, and Gene 1 shows complete linkage with the centromere, whereas Gene 2 gives evidence of some crossing over in that region, the order of the genes can be determined, with Gene 1 being closer to the centromere than Gene 2.

13.3.2 Acrotrisomics and Tertiary Trisomics

The value of these aneuploids is that genes can be aligned with a physical marker in the extra chromosome; that is, the breakpoint that produced a deletion in the acrotrisomic and the breakpoint in the translocated chromosome in a tertiary trisomic.

Acrotrisomics are trisomic for all the genes in one arm and for those in the segment that is present in the opposite arm, but disomic for genes in the segment that is missing in the extra chromosome (*see* Chapter 11, Fig. 11.1). They are useful for gene locations in species such as barley, in which pachytene observations are very difficult because the chromosomes are not condensed enough. The meiotic behavior of an acrochromosome can resemble that of a telochromosome, but with a better chance for pairing with one of the normal homologs because part of the second arm is present. The calculations for the expected genetic segregations are similar to those for telotrisomics. Because genes in both the complete arm and the partial arm show trisomic segregations, they should be assigned to an arm before making the acrotrisomic analysis. Gene locations can be related to the physical chromosome by measuring the length of the shortened arm at a mitotic metaphase, then determining, by the type of segregation, whether the gene is somewhere in that segment or distal to it.

Some tertiary trisomics have extra doses of two complete arms from different chromosomes if the translocated chromosome originates from a break in the centromere region of two chromosomes, followed by fusion of two arms, one from each chromosome. Whereas this type of tertiary trisomic can be used to locate genes on either arm, additional tests are needed using telotrisomics to identify which of the two arms contains those genes. Other tertiary trisomics have a translocated chromosome with more than one arm of one chromosome and less than one arm of the other (*see* Chapter 11, Fig. 11.1). Genes from both chromosomes may give trisomic or disomic segregations depending on whether they are present or absent on the translocated chromosome. Their arm assignment should be determined by telosomic analysis before using tertiary trisomics.

13.4 DIPLOID ANALYSIS USING MONOSOMICS

The use of monosomics for mapping genes in diploids has been very limited because of $n - 1$ gametic abortion in plants, and $2n - 1$ zygotic or embryonic abortion in animals with the exception of monosomics for the tiny chromosome 4 in *Drosophila melanogaster*. Therefore, the effects of monosomics must be studied in the generation in which they occur.

In tomato, monosomics induced by mature-pollen irradiation were used to associate a linkage group with a specific chromosome. Plants carrying two recessive genes from the same linkage group were crossed with irradiated pollen, which carried the dominant alleles. Nine F_1 plants showing the recessive phenotype for both genes were examined cytologically and found to be monosomic for chromosome 11, one of the shortest chromosomes. The mature pollen was able to function after irradiation because there were no further nuclear divisions, but the treatment may have affected the function of the centromere region of chromosome 11, so that it lagged at the first mitotic division after fertilization and was excluded from a nucleus. Other plants were recessive for one of the two marker genes and had segmental deletions for chromosome 11.

A very effective way of producing monosomics for different chromosomes in maize involves the use of an irradiation-induced deletion (*r–X1*) which includes the *R* locus on chromosome 10. This deletion causes nondisjunction of sister chromatids, in any of the 10 chromosomes, at the second postmeiotic division of the female gametophyte, and about 18% of the female gametes lack a chromosome. Evidence for nondisjunction is the similar frequencies of monosomics and trisomics in the progenies of crosses using the deletion. Because the nondisjunction events occur late in the development of the female gametophyte, $n - 1$ eggs are functional. The *r–X1* deletion is not transmitted through the pollen, so it is maintained in the heterozygous state.

Maize plants homozygous for a recessive gene, for example, *aa*, which is to be assigned to a chromosome, are crossed as males on plants heterozygous for the deletion (*R/r–X1*) and homozygous *AA*. The F_1 plants showing the recessive phenotype (*a*) are checked cytologically and are nearly always monosomic. The single chromosome bearing *a* comes from the male parent, because this chromosome is missing in the egg cell from the female parent due to nondisjunction. Identification of the chromosome by pachytene observations or banding techniques aligns *A–a* with a specific chromosome. The test can be made more efficient by using male parents that have recessive marker genes for as many of the 10 chromosomes as possible, because the induction of nondisjunction in a particular chromosome by the *r–X1* deletion seems to be a random event. When these males with marker genes are crossed with females carrying the dominant alleles, the occurrence of the recessive phenotype for any of the marker genes usually indicates a monosomic state for the chromosome carrying the marker gene. If a recessive phenotype for a test gene accompanies the recessive phenotype for a marker gene, the test gene can be assigned to the marked chromosome. Occasionally, a segmental deletion including the test gene and the marker gene can occur, but it can be distinguished from a monosomic by the appearance of the plants (each monosomic type has a distinctive morphology) and by root-tip chromosome counts to determine if a whole chromosome is missing.

The *r–X1* deletion method can be used to locate mutant genes induced by a mutagenic agent. One approach is to pollinate *R/r–X1* plants with treated pollen carrying marker genes for different chromosomes. Specific monosomics are identified in the progeny by the marker-gene phenotypic effects. If new, recessive phenotypes appear in the monosomic plants, the mutant genes can be assigned to the pertinent chromosomes.

Monosomics induced by the *r–X1* deletion are valuable for locating protein genes that give products with different electrophoretic mobilities. The advantage of these genes is that different alleles are codominant; when two different alleles occur together, each one can be detected by the position and type of band (size and density) on the gel. One example is a locus with allelic variation for histone 1 subfractions (*H1a*). *R/r–X1* plants that had a fast-migrating H1 band were crossed as females with plants that had a slower-migrating band and a marker gene for chromosome 1, *leaf-brown midrib* (*bm2*). A monosomic plant in the progeny had brown midribs and only the slower-migrating band. Monosomics for eight other tested chromosomes had both types of bands, so it was clear that the *H1a* locus was on chromosome 1.

Restriction fragment-length polymorphisms (RFLPs) are valuable probes for mapping conventional genes because of their distribution throughout genomes, but they must be located within identified chromosomes before they can be used for this purpose. In maize, *r–X1*-induced monosomics were used to map a maize RFLP sequence. The F_1 monosomics came from a cross between a line designated MT, used as the male parent, and *R/r–X1*. Monosomics were obtained for all of the 10 maize chromosomes except Nos. 1 and 5, with the monosomic chromosomes coming from the male parent. A radioactive, cloned fragment of maize-genomic DNA was hybridized to membrane-bound DNA fragments from the two parents, the F_1 disomic plants, and each of the eight monosomic types. DNA fragments in the parents and their progeny with sequences homologous to the clone sequence were identified by autoradiography. The parents differed in the size of the sequence homologous to the clone, as shown by the different positions of the prominent band (Fig. 13.5). The F_1 disomic plants and seven of

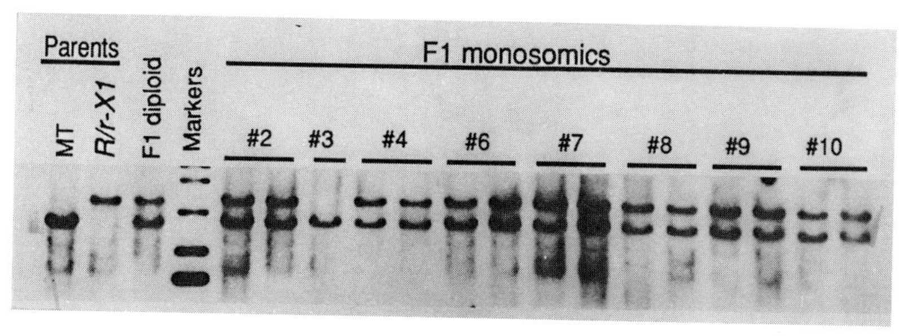

Fig. 13.5. Location of an RFLP site in maize using monosomics, generated by the *r–X1* deletion, and Southern blotting. Numbers 2 through 10 refer to chromosomes in the monosomic state. The monosomic for chromosome 1 is missing and is rarely produced when using the *r–X1* deletion. Monosomic 5 is also missing. The lane marked "markers" contains molecular-weight markers. The RFLP site is assigned to chromosome 3 because the band from the female parent, *R/r–X1*, is missing in monosomic 3, but is present in the diploid and the other seven monosomics. (From Weber, 1991).

the F_1 monosomics had the band from each parent because of co-dominance of the RFLP variants. The monosomic for chromosome 3 lacked this chromosome and its band from the female parent. Therefore, the cloned sequence can be assigned to chromosome 3 and can be used as a marker probe for that chromosome. By testing many cloned sequences from the maize genome by the monosomic method, about 400 DNA sequences were located on specific chromosomes, with some sequences on each chromosome but a greater number on the longer chromosomes.

13.5 DIPLOID ANALYSIS USING TRANSLOCATIONS AND INVERSIONS

Both reciprocal translocations and inversions are valuable cytogenetic tools for localizing genes on the physical chromosomes. They can be used to supplement information provided by aneuploid analyses or used on their own if aneuploids are not available. If a gene is first assigned to a chromosome, and perhaps to an arm, by the aneuploid method, but acrotrisomics or tertiary trisomics are not available, inversions or translocations with breakpoints in the identified chromosome or arm can be used to locate the gene within a physical segment. However, certain information about these structural aberrations must be available.

13.5.1 The Inverted or Translocated Chromosome Must be Identified

Early studies on chromosome aberrations used a genetic approach to identify aberrant chromosomes because the karyotypes were not well defined. A reversal in the normal order of marker genes in a linkage group indicated the presence of an inversion in that chromosome, and an exchange of genes between linkage groups suggested a reciprocal translocation. After the discovery of the salivary-gland chromosomes in *Drosophila melanogaster*, the natural banding pattern was used to discern rearrangements in this organism. The techniques developed to reveal banding patterns in human and plant chromosomes are helpful in detecting changes in linear patterns. In some cases, aberrant chromosomes are distinguishable at mitosis by changes in length and banding pattern (*see* Chapter 9, Fig. 9.1).

Several methods of identification using meiotic observations of aberrant chromosomes are available depending on the species. If the pachytene karyotype is identified, the chromosomes in the configuration formed by a heterozygous inversion or translocation can be identified by morphological characteristics (*see* Chapter 9, Fig. 9.4). If the species has only one pair of nucleolus-organizing chromosomes and one of them has an inverted or a translocated segment, the configuration is associated with the nucleolus quite consistently.

In plants, new, homozygous translocations can be crossed with a set of identified trisomics, used as females to increase the transmission of $n + 1$ gametes (*see* Chapter 11, Section 11.6), and diakinesis or metaphase I can be observed in the F_1's. If the trisome is one of the chromosomes in the translocation, the cells can have a chain of five, which includes the four chromosomes in the translocation and the extra dose of the trisomic chromosome. If the trisome is not one of those in the translocation, it may form a trivalent with its homologs while the translocation forms a ring- or chain-of-four.

If translocations with identified chromosomes are available, a tester set that checks for each chromosome in the complement can be used, and crosses are made between each tester and a new translocation. The meiotic configurations in the F_1 plants may be two separate rings-of-four if there are different chromosomes in a tester translocation and the new translocation, a ring-of-six if one chromosome is in common, and a ring-of-four or two bivalents if both chromosomes are in common.

13.5.2 The Breakpoints That Produce the Inversion or Translocation Must Be Determined Precisely in Order to Map the Genes Accurately

The sites can be narrowed to single bands in the salivary chromosomes of *Drosophila melanogaster* except when the breaks occur in heterochromatin around the centromeres, because these regions from different chromosomes fuse to form a chromocenter. The dye-induced banding in the mitotic chromosomes of humans, animals, and plants (*see* Chapter 6, Section 6.2) may be used if there are enough clear bands. The meiotic pachytene stage is very useful in plants if the chromosomes in the heterozygous inversion or translocation configuration can be followed from the centromeres to the ends of the arms (*see* Chapter 9, Figs. 9.4, 9.17, and 9.18). A configuration can be photographed if all parts are on one plane. Alternatively, the image of the configuration is projected by a camera lucida, a device with a prism placed over one eyepiece of the microscope, then traced on paper and interpreted. With either method, the location of a breakpoint is determined by measuring the distance from the centromere to the breakpoint (where the arms change partners), in relation to the length of the arm. For example, if a breakpoint is located at 0.25 on the short arm of a chromosome, it is one-quarter of the distance from the centromere to the end of the arm. It is painstaking work to locate the breakpoints for large numbers of translocations and inversions, but this has been done in maize, and many breakpoints distributed over the chromosome arms are available. In tomato, the breakpoints are more frequent in the centromere regions or in the heterochromatin around the centromeres than in the euchromatic regions.

13.6 GENERAL PROCEDURE USING INVERSIONS OR TRANSLOCATIONS TO LOCATE GENES

This general procedure applies to reciprocal translocations between A (essential) chromosomes of a genome, and a specific example is given in Section 13.6.1. Thr procedure is different for translocations between A and B (accessory) chromosomes, and it is discussed in Section 13.6.2.

The most efficient method is to cross plants that are homozygous for an inversion or translocation, and also for a gene under investigation, with plants that have the standard (normal) chromosomes and are homozygous for the contrasting allele. The gene to be located is designated the test gene to avoid confusion with a marker gene. If the chromosome and arm assignments of the test gene have already been determined using aneuploids, the inversion or translocation should have a breakpoint in the designated chromosome arm. If the chromosome assignment is unknown, the crosses should include an inversion or translocation for each chromosome in the set, with breakpoints covering both arms of each chromosome. The F_1 plants are backcrossed to the parent with standard chromosomes and homozygous for the recessive allele of the test-gene locus. In the backcross progenies, either a gene marker, closely linked with one of the inversion or translocation breakpoints, or pollen abortion is used to separate plants with the aberration from those without it, thus avoiding meiotic observations, which are more time-consuming. The segregations are analyzed to find out if the test gene shows linkage with pollen abortion or the gene marker, which in either case would indicate linkage with a breakpoint. The amount of aborted pollen should be high enough to identify plants that have a heterozygous aberration. Use of a gene marker requires less effort than pollen classification, and the most desirable gene markers control seed characters that can be classified before planting the seeds (*see* Section 13.6.1 for an example).

In self-pollinating plant species, where it is laborious to obtain large quantities of backcrossed seeds because of the need for emasculation (removal of anthers) before pollination, an alternative method is to grow F_2 progenies from selfed F_1's. This is more complex than backcrossing, because both male and female gametes consist of more than one kind, but it is possible to get information on test-gene locations using translocations or inversions. With quantitative traits, more decisive results are obtained by selfing F_2 plants that are homozygous for the normal or aberrant chromosomes and are selected either by a gene marker or on the basis of normal pollen. Further tests are needed to identify the F_3 lines that have either normal or aberrant chromosomes. A statistically significant difference between the two types of F_3 lines in the expression of the trait indicates linkage of one or more test genes with one or both of the translocation or inversion breakpoints. The differences are intensified in these F_3 lines because, in all the plants of each line, the regions near the breakpoints in both homologues come from one parent.

Linkage of a test gene with one translocation can involve the breakpoint in either of the two chromosomes. Unless the chromosome carrying the test gene has been identified by aneuploid analysis, the two chromosomes must be tested, one at a time, with different translocations. When a test gene is linked with the breakpoint in a certain chromosome, additional translocations involving the same chromosome, but with breakpoints marking different segments, must be used to determine on which side of the breakpoint the test gene is located. Another approach is to use inversion breakpoints as markers in the pertinent chromosome.

13.6.1 Use Of A–A Reciprocal Translocations

Most of the translocations used in cytogenetic studies on plants have involved exchanges between chromosomes of the normal complement, labeled A chromosomes to distinguish them from accessory or B chromosomes (*see* Chapter 6, Section 6.6). The illustration in Fig. 13.6 gives the procedure for locating a test gene,

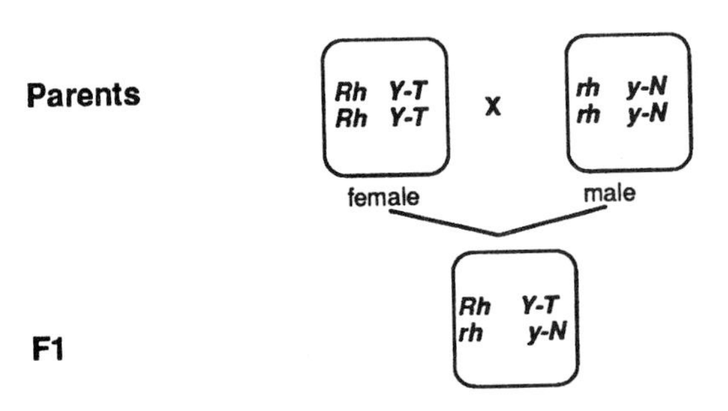

gametes from F1 — if Rh and Y-T are located extremely close to each other, only two types of gametes are produced:

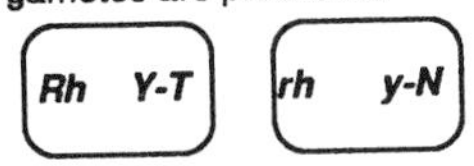

if Rh is located on a chromosome that is different from Y-T, or more than 50 map units away on the same chromosome, four types of gametes are produced in equal numbers:

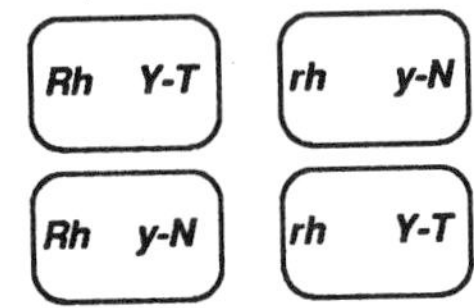

If *Rh* is located on the same chromosome as *Y-T*, four types of gametes are produced in proportions related to the distance between *Rh* and the breakpoint. The parental types (*Rh Y-T* and *rh y-N*) would be in excess over the recombinant types.

The different types of gametes that are produced can be assayed by backcrossing to the *rh rh* / *y-N y-N* parent.

Fig. 13.6. The use of a reciprocal translocation between chromosomes 6 and 7 in maize to test for the location of the gene *rh* (reduced height) versus *Rh* (normal height). T = the reciprocal translocation 6/7 chromosome, N = the normal 6 and 7 chromosomes. The locus for the alleles *Y* (yellow seeds) and *y* (white seeds) is so closely linked to the position of the breakpoint in chromosome 6 that yellow seeds (*Y*–T) serve as a marker for the translocation chromosomes and white seeds (*y*–N) for the normal chromosomes. The parents are homozygous for the genes and for the translocated or normal chromosomes.

Rh–rh in this example, using a translocation between chromosomes 6 and 7 in maize. It should be kept in mind that a complete study requires crosses involving a tester set of translocations that include each arm of every chromosome unless there is prior knowledge about the chromosome assignment of the test gene. If the *Y–y* gene pair on chromosome 6 of maize is used as a marker for each of the translocations, one of the translocated chromosomes must be 6. With no crossing over between *Y–y* and the breakpoint, the *Y* allele remains with translocated 6 and the *y* allele with normal 6 through the F_1 and backcross generations. The seeds from the backcross of the F_1 plants to the *rh rh y*–N *y*–N parent are separated into yellow and white groups, which are planted in separate rows. The plants from yellow seeds are heterozygous for the translocation (*Y*–T/*y*–N), and those from the white seeds have the standard chromosomes (*y*–N/*y*–N). The way in which the *Rh–rh* gene pair segregates in the paired rows indicates whether or not it is on one of the translocated chromosomes. If it is on 6, all the translocations will give similar results if the breakpoints in 6 are at about the same positions in the different translocations. If it is linked with the breakpoint of one of the other chromosomes, only the translocations involving that chromosome will show evidence of linkage.

If a gene to mark the translocation breakpoints is not available, the paired-row method cannot be used. Instead, a record has to be kept on individual plants in the backcross progenies with respect to the test-gene phenotype and the amount of pollen abortion. The data is grouped into those plants with around 50% pollen abortion, corresponding to the row from yellow seeds in the procedure given above, and those with normal pollen, corresponding to the row from white seeds. The linkage or independence of the test gene with respect to each translocation can be established.

The use of a gene marker such as *Y–y* to group the translocation-bearing plants and the normal-chromosome plants is an advantage in studies of quantitative characters, which are usually difficult to classify on a single-plant basis. With close linkage of the quantitative gene or genes with a breakpoint, the plants within a row from either *Y* or *y* seeds should have similar phenotypes except for a few possible recombinants. It is easier to detect subtle quantitative differences when groups of plants rather than individual plants are compared.

13.6.2 Use Of B–A Translocations

B–A translocations (*see* Chapter 9, Section 9.3) provide a very efficient method for mapping genes. The effectiveness of these translocations depends on the inherent, nondisjunctive property of B-sister chromatids during a specific stage in the life cycle, making it possible to locate genes in A-chromosome segments with confidence. The use of B–A translocations has been confined to maize, where nondisjunction of the B chromatids occurs at the division of the generative nucleus in the maturing pollen grain, resulting in one sperm cell with two B chromosomes and the other with none. When an A-chromosome segment bearing a dominant allele of a locus is attached to a B-centric segment, nondisjunction sends two doses of the A segment to one sperm cell and none to the other (Fig. 13.7). If the deficient (hypoploid) sperm unites with an egg bearing the recessive allele on a complete A chromosome, the F_1 plant shows the recessive phenotype because the A segment with the dominant allele is missing. Thus, the B–A translocations create a deletion method for locating genes. The clearest results are obtained if the nondisjunction frequency of the B centromeres is close to 100%, and this frequency is influenced by the distal part of the B chromosome, which is attached to the centric A segment and

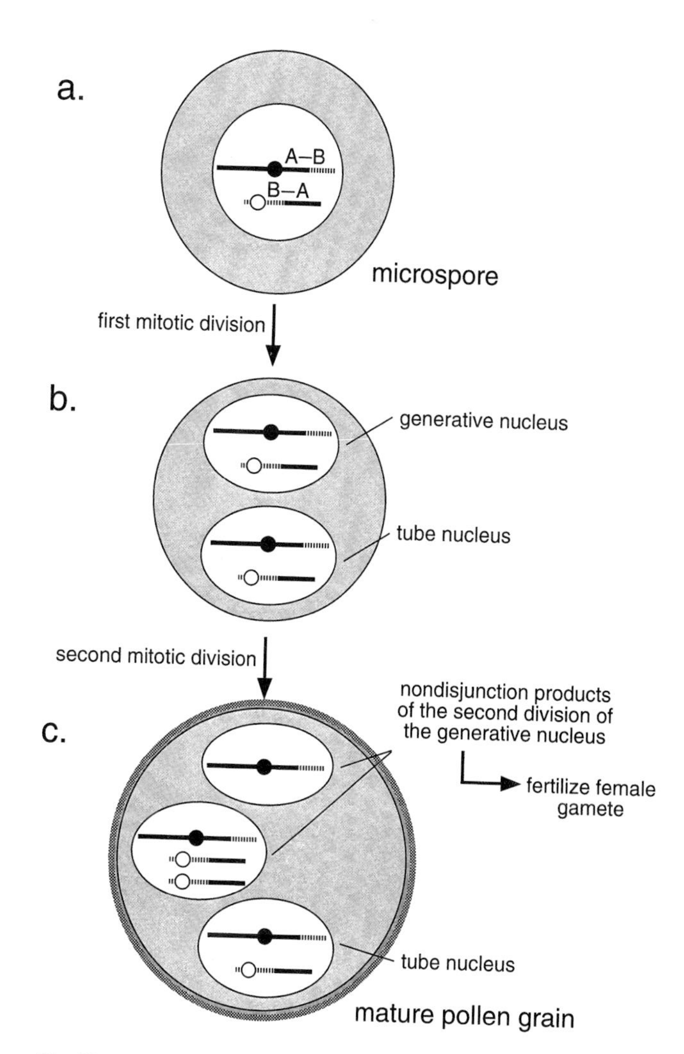

Fig. 13.7. The behavior of a B–A reciprocal translocation in a male gametophyte of maize. (a) The upper chromosome includes the centric segment (marked by a black circle) of an A chromosome and the acentric segment of a B chromosome (gray). The lower chromosome has an A acentric segment attached to a B centric segment. (b) Normal disjunction of both A and B centromeres during the first mitotic division following meiosis. (c) Nondisjunction of the B centromeres during mitosis in the generative nucleus results in a hypoploid sperm cell lacking the A segment that has been included in an extra dose in a hyperploid sperm cell. The tube nucleus does not divide further. (Adapted from Weber and Helentjaris, 1989.)

must be present in the generative nucleus. The postmeiotic behavior of the B chromosomes is normal on the female side in maize.

A set of B–A translocations involving almost all of the 20 chromosome arms has been developed in maize. Some of them were produced by irradiation-induced breaks and exchanges between A and B chromosomes. Others came from crosses between existing B–A and A–A translocations with a common A chromosome, which causes some pairing between the two types of translocations. Certain crossovers and segregations attach new segments of A chromosomes to the B-centric segments.

In order to locate a gene within a defined A segment, the accurate locations of the breakpoints in the A components of B–A translocations are obtained by pachytene measurements (*see* Section 13.5). If a gene has already been assigned to a specific A chromosome using trisomics, only B–A translocations involving segments of that A chromosome need to be used. If the A-chromosome assignment has not been made, a set of available B–A translocations should be used to test as many A segments as possible.

We can illustrate the use of B–A translocations in a study designed to localize a gene pair (*Hm*/*hm*) for resistance/susceptibility of maize to a race of the fungus *Helminthosporium carbonum*. The gene had already been assigned to chromosome 1, so two B–A translocations, with breakpoints in opposite arms of chromosome 1, were used to relate the *Hm*/*hm* locus to one of the breakpoints. Both translocation lines were resistant to infection by the fungus and were crossed as males with a susceptible source. The procedure for T B–1L, a translocation (T) with the breakpoint at about 0.12 in the long (L) arm of chromosome 1, is shown in Fig. 13.8. The expected results are diagrammed for the following two assumptions concerning the location of *Hm*: (1) between the centromere and the breakpoint, in which case *Hm* stays with its own centromere on the translocated 1^B chromosome, or (2) distal to the breakpoint, so that *Hm* goes with the B centromere on the B^1 translocated chromosome. If *Hm* stays with 1^B, each sperm cell has one dose of *Hm* from both normal disjunction and nondisjunction of the B

Testing for the location of *Hm* within chromosome 1

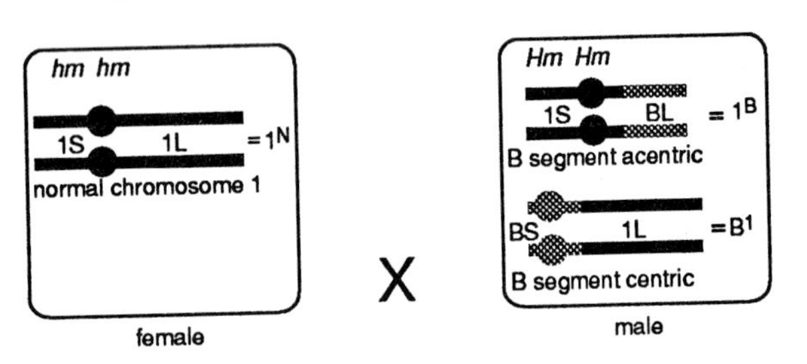

B^1 disjunction	F1 seeds: endosperm	F1 seeds: embryo	F1 seed size	F1 plant phenotype
normal disjunction	1^N 1^N 1^B B^1	1^N 1^B B^1	normal	normal
nondisjunction:				
1. B^1's to embryo	1^N 1^N 1^B	1^N 1^N 1^B B^1 B^1	small	normal
2. B^1's to endosperm	1^N 1^N 1^B B^1 B^1	1^N 1^B	large	short plants with narrow leaves due to missing chromosome 1 segment

If *Hm* is on 1^B, plants from embryos in both categories under nondisjunction are resistant. If *Hm* is on B^1, plants from embryos in category 1 are resistant whereas those from category 2 are susceptible. If B^1 chromatids disjoin normally in the generative nucleus, all the progeny plants are resistant.

Fig. 13.8. The use of a B–A reciprocal translocation to locate the gene pair *Hm/hm* in a segment of chromosome 1 in maize. The normal chromosome 1 is symbolized by 1^N and the translocated chromosomes by 1^B and B^1. The location of *Hm/hm* within chromosome 1 is not shown, as this would not be known at the beginning of the study, but alternative locations in 1^B or B^1 and expected results are given in the figure. The female parent contributes two doses of the 1^N chromosome to the $3n$ endosperm. Table 13.6 lists the results of an experimental study to determine this linkage.

Table 13.6. Segregations for Resistance and Susceptibility to *Helminthosporium carbonum* in Maize from a Cross of Susceptible Female × Resistant Male with a B–1L Translocation

	Normal Plant Phenotype		Abnormal Plant Phenotype	
Kernel Type	Resistant	Susceptible	Resistant	Susceptible
Large	160	0	0	344
Small	481	0	0	29

Source: Data from Roman and Ullstrup (1951).

centromeres, because the 1^B chromatids disjoin in the generative nucleus. If this type of sperm cell participates in fertilization, the resulting plants carry *Hm* and are resistant to the pathogen. If *Hm* is transferred to the B^1 chromosome, nondisjunction sends two doses of the gene to one sperm cell and none to the other within the same pollen grain. If the hypoploid sperm cells fertilize egg cells to produce embryos, the plants are susceptible. Regardless of the location of *Hm*, nondisjunction results in two types of seeds (small and large) depending on whether the endosperm came from the hypoploid sperm, which lacks the distal part of 1L, or the hyperploid sperm, which has an extra dose of the same segment.

The results of crosses involving T B–1L are given in Table 13.6. The data are separated by seed size and by plant appearance. The large seeds, giving plants with normal phenotype, came from normal B^1 disjunction, and the large seeds giving abnormal plants (shorter with narrower leaves) came from nondisjunction, with the hyperploid sperm contributing to the endosperm. The small seeds that gave rise to normal plants also came from nondisjunction, but the hypoploid sperm contributed to the endosperm. The unexpected group of 29 abnormal plants from small seeds illustrates that biological experiments sometimes have surprises. Possible explanations are that the small seed size for this group was not due to loss of B^1 from the endosperm, or that the sperm cells participating in double fertilization were both hypoploid and came from different pollen grains. The larger group of susceptible, abnormal plants from large seeds indicates that *Hm* is located distal to 0.12 in the long arm of chromosome 1. The location in the long arm was confirmed by the results from using the other translocation, with a break in the short arm of chromosome 1 close to the centromere. All the progeny from this cross were resistant, indicating that *Hm* was not on the short arm attached to the B centromere.

The small seeds, as well as the abnormal plants from large seeds, can be attributed to the single dose of the distal part of 1L and indicates that there are genes in this segment for endosperm development, plant height, and leaf width. Thus, a bonus of the B–A translocations is the possibility of locating genes in addition to those for which an experiment was undertaken.

B–A translocations are valuable tools for mapping isozyme and restriction fragment-length polymorphism (RFLP) loci. An example for an RFLP locus in maize is shown in Fig. 13.9. The same crossing procedure was used as shown in Fig. 13.8, and leaf samples were taken from the progeny for RFLP analysis. The results clearly indicate that the RFLP locus is located somewhere in the distal region beyond 0.70 in the long arm of chromosome 8 and was translocated to the centric part of the B chromosome. By analyzing large numbers of RFLP loci in this way, it is possible to identify the short and long arms of the RFLP map and to locate the centromeres on this map. The accurate placement of the B–A transloca-

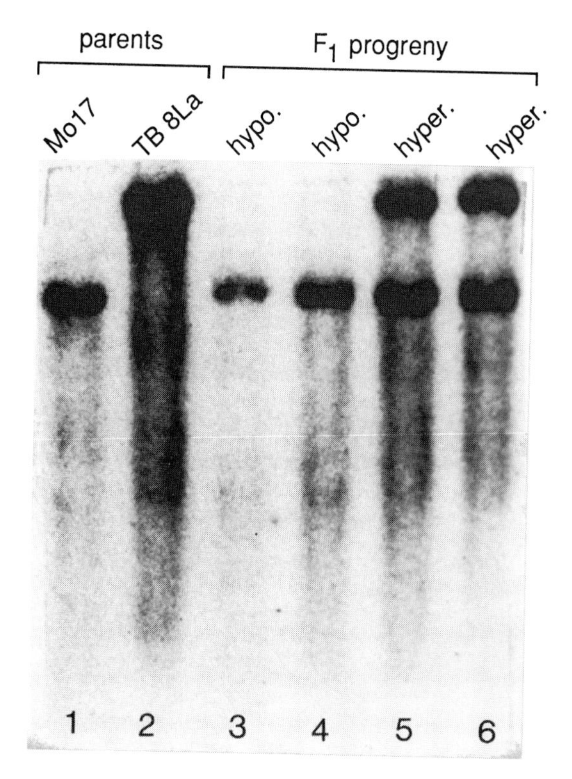

Fig. 13.9. The use of a B–A reciprocal translocation to locate a RFLP locus on maize chromosome 8. The parents Mo 17 (female) and TB–8La (male, breakpoint at 0.70 in the long arm of chromosome 8), have different RFLP alleles, as shown by the different band positions using a Southern blot. The hypoploid F_1 progeny, with one dose of the distal part of chromosome 8 from Mo 17, have only the Mo 17 band, whereas the hyperploid progeny, with three doses of the distal part of 8, have the bands from both parents. These results place the RFLP locus on the long arm of chromosome 8 distal to the breakpoint. (From Weber and Helentjaris, 1989.)

tion breakpoints makes it possible to correlate the RFLP map with both the physical map, based on cytology, and the genetic map, based on linkage studies.

13.6.3 Use of Inversions

It is more efficient to use inversions for gene locations after a chromosome assignment has been made using aneuploids or reciprocal translocations, then inversions for only one chromosome are needed. Inversion breakpoints can mark chromosome regions that are devoid of translocation breakpoints or gene markers. Very short inversions may not produce enough sterility to be detected, but a gene marker, closely linked with one of the breakpoints, can serve the same purpose as in the translocation method.

The procedure using inversions is the same as for translocations with or without a gene marker and is applicable to backcross or F_2 progenies. We can use the same genes as in Fig. 13.6, with the assumptions that the *Rh-rh* gene pair is on chromosome 6 of maize, and that the *Y–y* marker-gene pair is closely linked with an inversion breakpoint. The parental cross is *Rh Rh Y*(Inversion) *Y*(Inversion) (homozygous inversion in chromosome 6) X *rh rh y*–N *y*–N (standard 6), and the F_1 is backcrossed to *rh rh y*–N *y*–N. The yellow (*Y*) seeds produced on F_1 plants should be heterozygous for the inversion and the white (*y*) seeds, homozygous for the inversion. If *Rh–rh* is located more than 50 map units from both inversion breakpoints, there will be a 1 *Rh*:1 *rh* segregation in the rows from both yellow seeds and white seeds. If *Rh–rh* is close to one of the breakpoints or between the two breakpoints, it will show linkage with the inversion. Because most crossovers within or close to the breakpoints of a heterozygous inversion cause deletions as well as duplications of chromosome segments (*see* Chapter 9, Section 9.2.1), and result in gametic abortion, the progeny consists almost entirely of the parental type. Thus, plants in the row from *Y* seeds are normal height (*Rh*), and those in the row from *y* seeds are reduced in height (*rh*). It is not possible, in the test with one inversion, to decide which of the two breakpoints is closer to *Rh–rh*, therefore tests must be made using other inversions with one breakpoint in about the same location as one of the breakpoints in the first inversion, and the second breakpoint at a different location.

13.7 DIPLOID ANALYSIS USING DELETIONS

Deletions have been used in several diploid organisms to locate genes within the missing segments. The experimental procedure is to treat males in *Drosophila* or pollen in plants with a clastogenic agent to induce deletions in the sperm chromosomes. If the treatment is applied to essentially mature gametes, they can usually function in fertilization, and the deleted segment is lost in an early embryonic mitosis. The progeny may tolerate a heterozygous deletion in their body cells, although they often cannot transmit the deficient chromosome. Crosses are made between untreated females, homozygous for one or more recessive, test genes, and treated males or pollen carrying the dominant alleles. If one or more F_1 individuals show the recessive phenotype for one or more genes (pseudodominance), along with an induced deletion in the pertinent chromosome, the gene(s) is assumed to be located within the missing segment. Several overlapping deletions can be tested to narrow the site of the pseudodominant gene(s).

Deletion mapping has been especially efficient in *Drosophila melanogaster*, because a change in phenotype is paralleled by one or more missing bands in the salivary-gland chromosomes. Many of the dominant *Notch* mutations (mentioned in Chapter 20, Section 20.1.1 and Chapter 8, Section 8.6), are due to heterozygous deletions in the X chromosome that have a common missing band, 3C7. Therefore, the locus giving the normal phenotype is assumed to be in the vicinity of this band. Deletion mapping has been used in place of recombination mapping for the tiny fourth chromosome, which has an almost complete absence of crossing over in diploid females raised at standard temperatures. The use of overlapping deletions has made it possible to assign lethal genes to specific bands on the salivary chromosome 4. (*See* Chapter 20, Sections 20.1.5 and 20.2.1, for gene mapping in chromosome 2 heterochromatin using deletions.)

In plant species with clear pachytene karyotypes such as maize and tomato, the F_1 pseudodominant effect has been associated with deletions. Within the pertinent chromosome, the shortest deletion giving the pseudodominant effect is used for the approximate location of the gene. Some deletions may be too short to be detected by pachytene observations; in these cases, a change in F_1 phenotype can be interpreted as a gene mutation. A deletion rather than a gene mutation is indicated by reduced gametic transmission of the defect, especially through the male, and failure to make the defect homozygous. If there are genes very close to the one under investigation, a deletion may include them and cause some additional changes in phenotype. In maize, loss in anthocyanin activity at the *A* locus was accompanied by reductions in chlorophyll amount and somatic viability. These concurrent effects were attributed to deletions that eliminated several loci, including *A*. Some very short deletions require molecular procedures to be detected.

In humans, caution is needed in associating deletions with mutant phenotypes, which instead may be due to a transfer of the missing segment to another part of the genome. The disease chronic myelogenous leukemia (CML), which was first associated with an apparent deletion in the long arm of chromosome 22, was later attributed to a reciprocal translocation between 22 and 9 (*see* Chapter 9, Section 9.5). However, several abnormal syndromes have been associated with deletions on specific chromosomes by extensive studies on large numbers of individuals with the same or similar symptoms. An example is the cri-du-chat syndrome, which is caused by a deletion in the short arm of chromosome 5 (5p). The length of the deletion varies in different individuals, but most of the symptoms are due to a missing segment in region 1, band 5. The conclusion is that one or more genes in the missing segments control functions that are impaired or lacking in afflicted persons. Indirect evidence on human-gene locations also has been provided by deletion mapping. Genes that are heterozygous rather than hemizygous in the presence of a deletion are not located in the segment covered by the deletion.

BIBLIOGRAPHY

General

BURNHAM, C.R. 1966. Cytogenetics in plant improvement. In: Plant Breeding Symposia (Frey, K.J., ed.). Iowa State University, Ames. pp. 139–187.

GILL, B.S. 1983. Tomato cytogenetics—a search for new frontiers. In: Cytogenetics of Crop Plants (Swaminathan, M.S., Gupta, P.K., Sinha, U., eds). Macmillan India Inc., New Delhi. pp. 455–480.

ROBERTS, P.A. 1976. The genetics of chromosome aberration. In: The Genetics and Biology of *Drosophila*. Vol. 1a (Ashburner, M., Novitski, E., eds). Academic Press, New York. pp. 67–184.

SYBENGA, J. 1992. Cytogenetics in Plant Breeding. Monographs on Theoretical and Applied Genetics 17. Springer-Verlag, Berlin.

Section 13.1

HERMSEN, J.G. 1970. Basic information for the use of primary trisomics in genetics and breeding research. Euphytica 19: 125–140.

TSUCHIYA, T. 1991. Chromosome mapping by means of aneuploid analysis in barley. In: Chromosome Engineering in Plants: Genetics, Breeding, Evolution. Part A (Gupta, P.K., Tsuchiya, T., eds). Elsevier, Amsterdam. pp. 361–384.

Section 13.2

BURNHAM, C.R. 1962. Discussions in Cytogenetics. Burgess Publishing Co., Minneapolis, MN.

KHUSH, G.S. 1973. Cytogenetics of Aneuploids. Academic Press, New York.

KHUSH, G.S., SINGH, R.J., SUR, S.C., LIBROJO, A.L. 1984. Primary trisomics of rice: origin, morphology, cytology, and use in linkage mapping. Genetics 107: 141–163.

Section 13.3

SHAHLA, A., TSUCHIYA, T. 1987. Cytogenetic studies in barley chromosome 1 by means of telotrisomic, acrotrisomic and conventional analysis. Theoret. Appl. Genet. 75: 5–12.

Section 13.4

HOCHMAN, B. 1976. Fourth chromosome of *D. melanogaster*. In: The Genetics and Biology of *Drosophila*. Vol. 1b (Ashburner, M. Novitski, E., eds). Academic Press, New York. pp. 903–928.

WEBER, D.F. 1991. Monosomic analysis in maize and other diploid crop plants. In: Chromosome Engineering in Plants: Genetics, Breeding, Evolution. Part A (Gupta, P.K., Tsuchiya, T., eds). Elsevier, Amsterdam. pp. 181–209.

Section 13.6

RAKHA, F.A., ROBERTSON, D.S. 1970. A new technique for the production of A–B translocations and their use in genetic analysis. Genetics 65: 223–240.

ROMAN, H., ULLSTRUP, A.J. 1951. The use of A–B translocations to locate genes in maize. Agron. J. 43: 450–454.

WEBER, D., HELENTJARIS, T. 1989. Mapping RFLP loci in maize using B–A translocations. Genetics 121: 583–590.

Section 13.7

KHUSH, G.S., RICK, C.M. 1968. Cytogenetic analysis of the tomato genome by means of induced deficiencies. Chromosoma 23: 452–484.

LEVITAN, M. 1988. Textbook of Human Genetics. 3rd ed. Oxford University Press, New York.

STADLER, L.J., ROMAN, H. 1948. The effect of X-rays upon mutation of the gene *A* in maize. Genetics 33: 273–303.

THERMAN, E., SUSMAN, M. 1993. Human Chromosomes: Structure, Behavior, and Effects. 3rd ed. Springer-Verlag, New York.

14

Locating Genes in Polyploids Using Chromosome Aberrations

- Aneuploid chromosome stocks in polyploids provide a very efficient procedure for assigning genes and DNA sequences to specific chromosomes.
- Nullisomic lines can be scored at the DNA or phenotypic levels for the simple presence or absence of a DNA sequence or a gene.
- Monosomic lines usually need to be crossed and the segregation patterns analyzed before a gene can be assigned to a chromosome.
- Intervarietal-chromosome substitutions are valuable for assigning genes for quantitative traits to chromosomes.

Aneuploids, particularly monosomics and their derivatives, are the most useful type of chromosome aberration for gene-location studies in allopolyploid plant species, which have two doses of each kind of chromosome in two or more different genomes. Trisomics for most of the chromosomes in allopolyploids do not have the distinctive phenotypes that occur in diploids because of gene duplication between the different genomes. Their use in assigning genes to chromosomes must be based on backcross or F_2 segregations that deviate from disomic segregations, and larger progeny numbers are required when progenies cannot be separated phenotypically into trisomic and disomic plants. Primary trisomics are valuable in providing tetrasomics, which come from the union of two $n + 1$ gametes when the trisomics are selfed. Tetrasomics test the ability of two extra chromosomes to compensate for a missing (nullisomic) chromosome, and thereby indicate genetic relationships between chromosomes within or between genomes (*see* Chapter 10, Section 10.8.4; Fig. 10.18).

Monosomics provide an efficient method for gene–chromosome associations in allopolyploids. In allotetraploids, they usually have distinctive phenotypes, so that backcross or F_2 progenies can be grouped into monosomics and disomics, and gene segregations in each group can be checked. In some allotetraploid species such as durum wheat (*Triticum turgidum*), however, the monosomics are weak, with low fertility and low transmission of the $n - 1$ gametes, so other methods must be used. In allohexaploids, most monosomics have phenotypes that do not differ from disomic phenotypes because of more intergenomic gene duplication than in allotetraploids, but they do have attributes that give them advantages over the use of trisomics. At both the allotetraploid and allohexaploid levels, genes can be located in F_1 progenies when the dominant allele comes from the monosomic parent and the recessive allele from the disomic parent, and the recessive phenotype is expressed with one dose of the recessive allele (hemizygous-effective). If the crosses are made so that the F_1 monosomics have the dominant allele, reliable results are possible with small numbers of F_2 plants, because any recessive phenotypes are nullisomics and relatively weak, and any plants identified as disomics are homozygous for the dominant alleles.

Nullisomics have been obtained from monosomics, by the union of two $n - 1$ gametes, for all the chromosomes of hexaploid wheat (*Triticum aestivum*) and for some chromosomes of hexaploid oats (*Avena sativa* and *Avena byzantina*). Whereas the frequencies of nullisomics are usually low, most of them have distinctive phenotypes and can be used to locate genes by reduced dosage effects. Nullisomics are not available from monosomics in allotetraploids although tests show that $n - 1$ gametes may be transmitted through both egg and pollen. Lethality seems to occur in the zygotic or early embryonic stages.

Another important use of allopolyploid monosomics or their derivatives such as monotelosomics is in the development of chromosome substitution lines, whereby a chromosome pair from one source is substituted for a homologous or homoeologous pair from another source. Substitutions between varieties are called intervarietal or intercultivaral chromosome substitutions, and those between species or genera are called alien chromosome substitutions. Intervarietal substitution lines are particularly useful for studies of quantitative characters, because the data can be collected on groups of plants with the same genotype instead of on individual plants, as in F_2 or backcross monosomic studies. Alien substitutions are a means of introducing genes for valuable traits from related species or genera into crop plants (*see* Chapter 15).

Because autopolyploids have three or more doses of each chromosome, adding or subtracting single chromosomes gives aneuploid states that would probably not change the phenotype markedly and would give complex segregations. It is more expedient to derive diploids by haploidy (*see* Chapter 12) and to obtain trisomics at this ploidy level for cytogenetic analyses. In alfalfa or lucerne (*Medicago sativa*), for example, natural autotetraploids were crossed with derived diploid lines to get autotriploids, which were then backcrossed with the derived diploids. Five out of eight possible trisomics were obtained at the diploid level by the union of $n + 1$ gametes from the triploids with n gametes from the diploids. Several genes were then located using crosses between trisomics and diploids.

14.1 POLYPLOID ANALYSIS USING MONOSOMICS

The first requirement for monosomic studies is to have available a set of identified monosomics for each chromosome in a species. When monosomics are first discovered, the identities of the monosomic chromosomes (monosomes) are generally unknown. The procedure for their identification can be illustrated using the allotetraploid *Nicotiana tabacum*, which has genomes from two diploid species with 12 chromosomes in each genome (Fig. 14.1). A different approach is possible in allotetraploid cotton (*Gossypium hirsutum*), where monosomics can usually be assigned to one of the two component genomes, A and D, by the size of the monosome, because the A genome has larger chromosomes than the D genome. These assignments are confirmed by crossing the monosomics with reciprocal translocations that have identified chromosomes, then checking whether the translocation configurations in the monosomic progeny lack one chromosome. By combining the results of crosses with different translocations, the missing chromosome can be identified. Where DNA and/or protein markers are available for the species they provide an efficient means of chromosome identification.

<u>Materials for test</u>:

Nicotiana sylvestris, S genome, 2n = 2x = 24
Nicotiana tomentosa, T genome, 2n = 2x = 24
Nicotiana tabacum, S and T genomes, 2n = 4x = 48
Nicotiana tabacum monosomics, 2n–1 = 47

<u>Cross 1</u>:
N. tabacum monosomic X *N. sylvestris*

Gametes: n = 24 (12S + 12T) n = 12 (S)
n–1 = 23 (12S + 11T)

Meiosis in F_1 monosomic plants:
35 chromosomes : 12 S bivalents + 11 T univalents

<u>Cross 2</u>:
N. tabacum monosomic X *N. tomentosa*

Gametes: n = 24 (12S + 12T) n = 12 (T)
n–1 = 23 (12S + 11T)

Meiosis in F_1 monosomic plants:
35 chromosomes : 12 S univalents + 11 T bivalents + 1 T univalent

Fig. 14.1. A procedure to assign monosomic chromosomes to one of the two genomes in *Nicotiana tabacum* ($2n = 4x = 48$). In the crosses shown, the monosomic is for a T-genome chromosome. If an S-genome chromosome is monosomic, the F_1 meiotic configurations are reversed; that is, 11 bivalents + 13 univalents from the cross with *N. sylvestris*, and 12 bivalents + 11 univalents from the cross with *N. tomentosa*. The latter species was used to assign monosomics to the T genome, but molecular–cytogenetic studies indicate that the T genome of *N. tabacum* was derived from hybridization between two other diploid species related to *N. tomentosa* (*see* Chapter 10, Section 10.8.1).

Table 14.1. Types and Frequencies of Progeny from Self-Pollinating a Monosomic Plant, Assuming that $n - 1$ Gametes have 75% Transmission Through the Female and 5% Transmission Through the Male

	Sperm	
Eggs	95% n	5% $n - 1$
25% n	24% $2n$ Disomics	1% $(2n - 1)$ Monosomics
75% $n - 1$	71% $(2n - 1)$ Monosomics	4% $(2n - 2)$ Nullisomics

The next phase is to relate the monosomics to individual chromosomes. The first sets to be developed, in cotton, tobacco, and wheat, were assigned arbitrary numbers or letters in the order in which they were obtained. In wheat, after intergenomic, homoeologous relationships had been resolved using nullisomic–tetrasomic combinations, the chromosomes were renumbered so that homoeologous chromosomes were given the same number followed by the letter of the genome to which they belonged; for example, 1A, 1B, and 1D. With this knowledge, the development of chromosome-banding techniques could be used routinely to identify monosomic chromosomes with distinctive banding patterns.

The success of monosomic analysis depends on the transmission rates of $n - 1$ gametes, which are much higher for eggs than for sperm (*see* Chapter 11, Section 11.6.2). Rates of transmission can vary for both male and female gametes depending on the species and the specific monosome. For illustration, we shall use representative transmission rates of 25% n:75% $n - 1$ through the female, and 95% n:5% $n - 1$ through the male. Based on these transmission rates, the progeny from a selfed monosomic consists of 24% disomics, 72% monosomics, and 4% nullisomics (Table 14.1). If the monosomic is crossed as female with a disomic, the progeny segregates 25% disomics and 75% monosomics, a reflection of the n and $n - 1$ egg frequencies, because the sperm are all n. In the reciprocal cross, the progeny are largely disomic because of the low transmission rate of $n - 1$ gametes through the male. Therefore, monosomics should be used as the female parent in crosses with disomics in order to recover a high frequency of monosomics.

We shall outline how monosomics can be used to locate a gene pair, *G–g*, using the transmission rates shown in Table 14.1. Other assumptions are that *G* in one dose is hemizygous-effective because it is completely dominant over *g*, which is an inactive recessive allele and produces the same phenotype whether it is homozygous (*gg*), hemizygous (*g* –), or missing altogether (– –). The term "critical cross" refers to a cross where the pertinent gene is on the monosomic chromosome, whereas crosses involving all the other monosomics in the set are noncritical. The results vary depending on whether the disomic parent carries the recessive or dominant alleles.

A set of identified monosomics is crossed with a disomic that differs from the monosomic lines at the *G–g* locus, and three types of crosses are shown in Fig. 14.2. On the left side, under critical crosses, the monosomic parent has the dominant allele *G* on the monosome, and the disomic parent has the recessive allele *g*. In the F_1, monosomics have the recessive phenotype, disomics have the dominant phenotype, and the gene can be located on the monosomic chromosome in this generation. If F_2 progenies from F_1 monosomics and disomics are grown to confirm the finding, progenies from the monosomics have only the recessive phenotype while progenies from the disomics segregate 3 G:1 g. The F_2 progenies from the noncritical crosses also segregate 3 G:1 g because the gene is independent of the monosomic condition. On the right side under critical crosses, the allelic arrangement is reversed; G is with the disomic parent and g is with the monosomic parent. In the F_1, both monosomics and disomics have the dominant phenotype, and the gene cannot be located in this generation. In the F_2 from F_1 monosomics, nullisomics are the only plants expected to show the recessive phenotype with the assumption that the missing locus gives the same phenotype as gg. If the nullisomic frequency is low, as we have postulated, there is no problem in distinguishing between critical (largely G) and noncritical (3 G:1 g) ratios. The frequency of nullisomics from different monosomes is known to vary and, in a few cases, may be high enough to obscure the difference between critical and noncritical segregations. The critical segregations may be affected by other chromosome abnormalities. For example, misdivision events in the F_1 monosomes may produce F_2 plants with a telochromosome or isochromosome for the arm not carrying the G–g locus. These plants add to the recessive group because the arm carrying the locus is missing. Therefore, at least some of the plants with the recessive phenotype in the potential critical progeny should be checked cytologically to confirm the nullisomic state. The use of F_2 monosomic analysis to locate the gene *Pm3*, which gives resistance to powdery mildew, caused by *Erisyphe graminis* in hexaploid wheat is shown in Table 14.2.

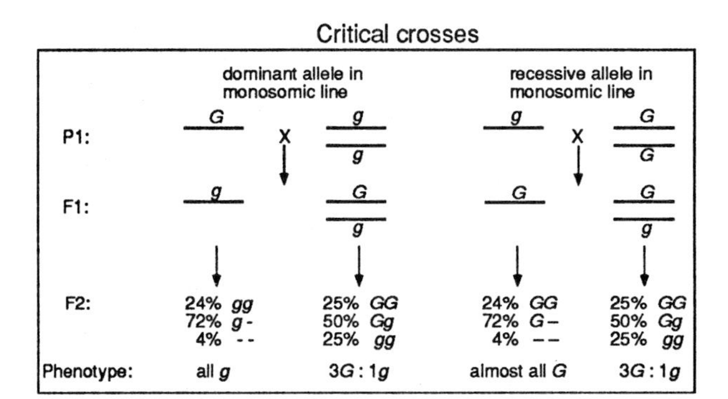

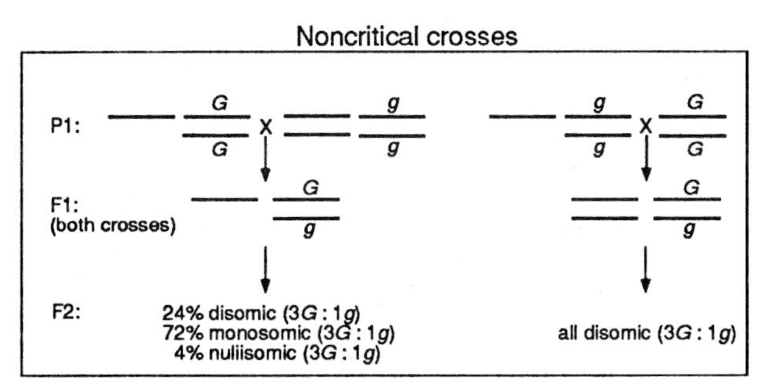

Fig. 14.2. Procedure for assigning a hemizygous-effective gene to a specific chromosome using monosomics as female parents in the initial crosses. The symbols G–g are allelic states of a gene locus. In critical crosses, the monosomic chromosome carries the gene; in noncritical crosses, the gene is not on the monosomic chromosome. The F_1 monosomics and disomics are selfed to get the F_2. The hyphens in the F_2 frequencies indicate missing chromosomes. The gametic frequencies from Table 14.1 were used in this figure.

Table 14.2. Segregations for Seedling Reaction to Powdery Mildew in the F_2 Progenies of F_1 Monosomic Plants from Crosses Between the 21 Monosomics in the Susceptible Variety "Chinese Spring" (*pm3/pm3*), and Disomics of a Resistant Source (*Pm3/Pm3*)

Chromosome Tested	Resistant	Susceptible	Total	χ^2 (3:1)	*P*-Value
1A	224	15	239	44.69	<0.001
1B	119	40	159	0.002	0.99–0.95
1D	51	15	66	0.18	0.95–0.50
2A	43	15	58	0.02	0.95–0.50
2B	60	16	76	0.63	0.50–0.20
2D	62	18	80	0.27	0.95–0.50
3A	52	19	71	0.12	0.95–0.50
3B	48	16	64	0.00	1.00
3D	44	19	63	0.89	0.50–0.20
4A	78	27	105	0.03	0.95–0.50
4B	43	20	63	1.53	0.50–0.20
4D	54	20	74	0.16	0.95–0.50
5A	51	21	72	0.67	0.50–0.20
5B	56	16	72	0.30	0.95–0.50
5D	17	8	25	0.65	0.50–0.20
6A	69	16	85	1.73	0.20–0.10
6B	33	7	40	1.20	0.50–0.20
6D	65	14	79	2.23	0.20–0.10
7A	49	12	61	0.01	0.95–0.50
7B	38	18	56	1.52	0.50–0.20
7D	44	18	62	0.54	0.50–0.20
All but 1A	1076	355	1431	0.03	0.95–0.50

Note: The significant deviation from a 3:1 segregation for chromosome 1A indicates that the gene pair *Pm3/pm3* is located on that chromosome.

Source: Data from McIntosh and Baker (1968).

In self-fertilizing species such as wheat, F_2 progenies are used more than backcrosses because of the labor involved in emasculating anthers. If, however, the F_1 G– monosomics from the critical cross on the right in Fig. 14.2 are backcrossed as females to gg disomics, the progenies segregate 75% g–:25% Gg, a ratio of 3 recessive:1 dominant instead of 1 Gg:1 gg in a noncritical cross. If a few plants pen progeny from critical backcrosses are checked cytologically, the recessive plants are monosomic and the dominant plants are disomic. If a reciprocal backcross is made using the F_1 monosomics as males, 95% of the backcross progeny show the dominant phenotype and are disomics, and 5% are monosomics with the recessive phenotype.

Other recessive genes in hexaploid wheat require two doses to express the recessive phenotype. These genes are called hemizygous-ineffective because one dose is not enough for expression. The dosage effect is due to the influence of modifying genes from the other wheat genomes, so it is likely that hemizygous-ineffective genes occur in allopolyploids of other species. The monosomic and nullisomic states of such genes give phenotypes closer to those expressed by the dominant alleles. A monosomic analysis of this type of gene is shown in Fig. 14.3. The recessive allele *virescent* (v) (turning green) in two doses causes chlorophyll deficiency in the early stages of plant growth, but in one and zero doses, it has no or very little effect on chlorophyll production. The disomic parent in the critical cross should have the recessive alleles because one dose of the v allele is not detectable in the monosomic parent. Both F_1 monosomics and disomics show the dominant phenotype, and in the F_2, the disomics are the only group showing the recessive phenotype. However, because this F_2 ratio is very close to 3 dominant:1 recessive, it cannot

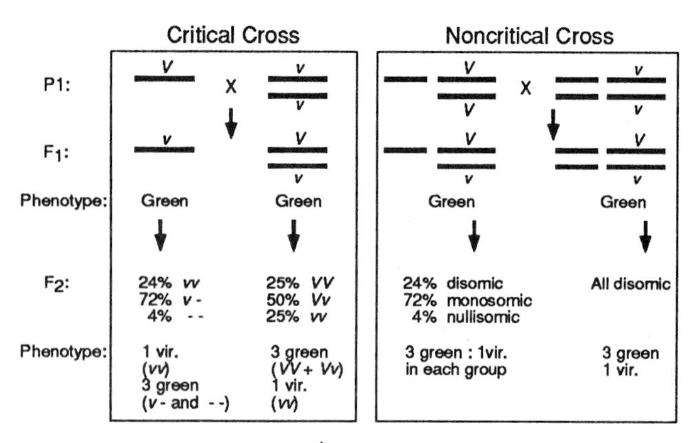

Although the segregation is 3 green:1 virescent in both the critical and noncritical progenies, cytological checking shows that all the virescent plants are disomic in the critical progeny from F_1 monosomics, but only about 25% are disomic in noncritical progenies from F_1 monosomics.

Fig. 14.3. Procedure for assigning a hemizygous-ineffective gene to a specific chromosome using monosomics. *VV* or *Vv* = green plant, *vv* = *virescent*, *v*– or –– = green. The F_1 monosomics and disomics are selfed to get the F_2. The hyphens in the F_2 frequencies indicate missing chromosomes. The gametic frequencies from Table 14.1 were used in this figure.

be distinguished from the noncritical 3 : 1 ratio. Cytological studies of plants with the recessive phenotype are required to show that all of those in the critical progeny are disomics, whereas only about 25% of those in the noncritical progenies are disomics.

With some characters such as a disease reaction, it is not always possible to determine the critical chromosome(s) from backcross or F_2 segregations, because the expression of the character may not be consistently clear when individual plants are classified. In such cases, F_3 progenies usually overcome this problem because, in a critical cross, the F_2 disonics from F_1 monosomics are homozygous for the gene(s) controlling the trait, and all the plants within a F_3 progeny should have a uniform phenotype. Chromosome counts and probability estimates are needed to confirm that a uniform F_3 progeny came from a F_2 disomic derived from a critical cross rather than a noncritical cross. After the F_3 critical progenies have been identified by uniformity of phenotype, a few reserve seeds from each pertinent F_2 plant are germinated and root-tip chromosome counts are made. If three or four consecutive F_3 seedlings from the same F_2 plant are all disomic, that F_2 plant is determined to be disomic because of the low probability of obtaining this frequency from a F_2 monosomic. Furthermore, if three or four F_3 uniform progenies from the same cross come from different F_2 disomic plants, there is confidence that a critical chromosome is involved, because in a noncritical cross, 75% of an F_2 progeny segregate in F_3 or are homozygous for the contrasting phenotype. The finding that at least two F_3 seedlings from an F_2 plant are monosomic indicates that it must be monosomic. The most likely explanation for a F_2 monosomic plant giving a uniform F_3 progeny is that a noncritical cross is involved, and the F_2 plant is homozygous for the gene of interest such as *GG*, which is on a chromosome other than the monosome.

14.2 POLYPLOID ANALYSIS USING NULLISOMICS

Because nullisomics lack both homologs of a chromosome pair, their phenotypic effects are the result of the absence of genes, and contrasting alleles do not have to be present as in monosomic analyses. Most nullisomics have distinctive phenotypes in comparison with disomics, and both qualitative and quantitative genes with distinct presence–absence effects can be located on specific chromosomes. The use of nullisomics is largely limited to allohexaploids because they do not occur in the progenies of monosomics from allotetraploids and have to be obtained by special manipulations (*see* Chapter 11, Section 11.2). Although nullisomics can be produced for some or all chromosomes depending on the species, they usually cannot be maintained as stable lines because of male or female sterility. Nevertheless, many genes have been located in the hexaploid wheat variety "Chinese Spring" by comparisons between disomics and nullisomics in the progenies of monosomics. For instance, nullisomics for chromosome 2A (nulli-2A) are about half the normal height and have fewer tillers, shorter and wider leaves, and larger culms. They are also delayed in maturity and female-sterile. These phenotypic deviations from disomics indicate that there are genes for height, tillering, leaf and culm size, maturity, and fertility somewhere on chromosome 2A. Other chromosomes that can be investigated through their absence have given comparable results. From such studies, it is evident that nullisomic analysis is a very effective way of assigning genes to chromosomes and should be used whenever possible.

14.3 POLYPLOID ANALYSIS USING TELOCHROMOSOMES AND ISOCHROMOSOMES

The origin of telochromosomes (telos) and isochromosomes (isos) from centromere misdivisions of monosomes has been described in Chapter 11, Section 11.3.2. One arm of a chromosome pair is present in one dose in monotelosomics (monotelos), and in two doses in ditelosomics (ditelos) and monoisosomics (monoisos). These aneuploids can be used for direct phenotypic observations, in comparisons with nullisomics and disomics, to determine the arm locations of genes. They have the advantage over nullisomics in that the presence of one arm gives some of them enough vigor and fertility to be maintained. Monotelosomics can also be used in crosses to estimate the physical distance between a gene and its centromere.

14.3.1 Chromosome-Arm Locations of Genes Based on Dosage Effects

If genes have been assigned to an identified chromosome by observing the nullisomic effects, monotelos and monoisos of the particular chromosome can then be used to localize the genes to either arm. This method is effectively one-arm nullisomy, so allelic variation is not needed. The effect of nulli-2A on female fertility was mentioned in the previous section. Both a monotelosomic and a monoisosomic are available for the short (S) arm, and their presence shows that a gene or genes for fertility are located on 2AS. There is a dosage effect, with the monoiso having more fertility than the monotelo. Conversely, the monoiso for 2AL is female-sterile like the nullisomic, indicating that there are no fertility genes on this arm. If aneuploids are available for only one arm of a chromosome, a gene may be located on the opposite arm by inference. In the case of female fertility, if the aneuploids for 2AS are not available, the result with monoiso-2AL indicates that there must be a gene or genes for fertility on 2AS. Some traits may be controlled by

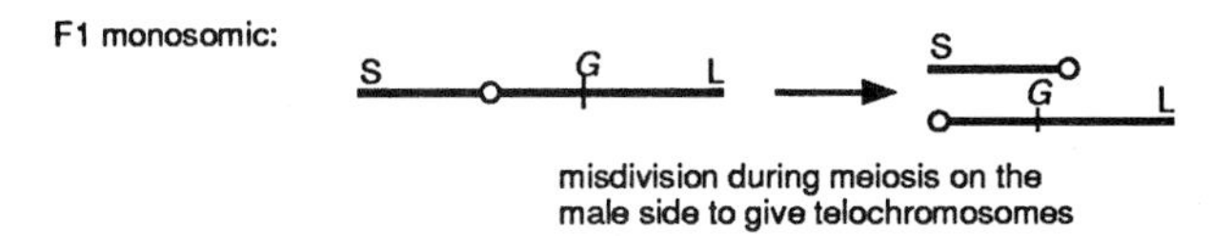

female gametes	male gametes with a telochromosome: S	male gametes with a telochromosome: G L
S G L	*S / S G L phenotype: G	* G L / S G L phenotype: G
chromosome missing	S phenotype: g (due to missing *G*)	G L phenotype: G

* at meiosis I, this chromosome pair is observed as a heteromorphic bivalent consisting of the telochromosome and the complete chromosome (see section 11.6.2,Chapter 11)

Fig. 14.4. Misdivision of a monosome during microsporogenesis in a F_1 monosomic plant giving a telochromosome for each arm, and the transmission of each telochromosome to the F_2 progeny by selfing the F_1 monosomic. Male gametes with a complete chromosome or a missing chromosome can also occur (not shown) to give disomics, monosomics, or nullisomics when combined with the pertinent female gametes. The gene pair *G–g* is assumed to be hemizygous-effective and on the long arm. S = short arm, L = long arm.

genes on both arms, in which cases, one or two doses of one arm, but absence of the opposite arm, are indistinguishable from the nullisomic state.

Occasionally, monotelosomic or monoisosomic plants occur in F_2 progenies as a result of misdivision of the monosome during meiosis in the F_1 monosomics. In the example shown in Fig. 14.4, the monosome has the dominant allele *G* on the long arm. As a result of misdivision, some F_2 plants are expected to have a normal homolog along with the telo, forming a heteromorphic bivalent during meiosis; these plants are called monotelodisomics. If the telo is for the long arm, the monotelosomics have the dominant phenotype *G* like the monotelodisomics, but if the telo is for the short arm, it lacks *G*, and the recessive phenotype of the monotelosomics contrasts with the dominant phenotype of the monotelodisomics, which have one or two *G* [one or two Gs.]. The monotelosomics should be weaker plants than the monotelodisomics, which have a normal dose of one arm and one dose of the second arm, and cytological observations of somatic or meiotic dividing cells can confirm the presence of a telochromosome. It should be tested to make sure that it is derived from the monosome, because pairing disturbances during meiosis can cause univalent states of other chromosomes and the possibility of centromere misdivision. For the test, the monotelosomic plants lacking *G* in Fig. 14.4 are crossed with the ditelosomics, if available, for each arm of the chromosome that was monosomic in the F_1. If observations of meiosis in the progenies show pairing between the new telo and one of the identified telos, *G* is assigned to the long arm of the monosome used in this study. Lack of pairing indicates that the new telo was derived from some chromosome other than the monosome.

14.3.2 Mapping Gene–Centromere Distances Using Telochromosomes

A gene should be assigned to a chromosome, and preferably to a chromosome arm, before attempting to determine the distance

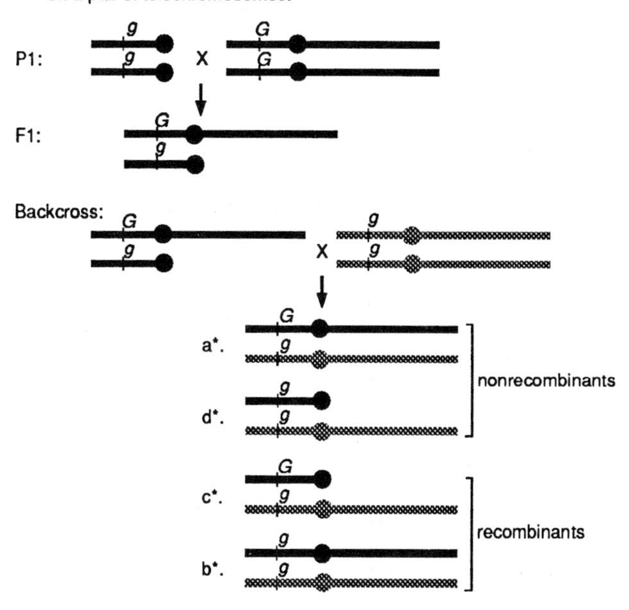

a*. G g
d*. g g
nonrecombinants
c*. G g
b*. g g
recombinants

If *G–g* is not on the telochromosome, it is not possible to score recombination between the gene and the centromere

Fig. 14.5. The use of a ditelosomic to map the gene pair *G–g* to one arm of a chromosome, and to determine the distance between the centromere and the gene locus. The nonrecombination genotypes are *Gg* disomic (a) and *gg* monotelodisomic (d). The recombination genotypes, which result from a crossover between the centromeres and the *G–g* alleles in the F_1 monotelodisomic, are *Gg* monotelodisomic (c) and *gg* disomic (b). * The letters a, b, c, d, correspond to the observed frequencies in Table 14.3.

between the gene and the centromere. Allelic variations in backcross or F_2 progenies are needed to detect recombination in this region and to determine how far the gene is from the centromere. Therefore, the gene to be tested must be on the arm represented by the telochromosome if crossovers are to occur between the centromere and the gene locus. The expected results in the backcross progeny from the F_1 *Gg* monotelodisomic are shown in Fig. 14.5 and are listed in Table 14.3. It is important to check all the backcross progeny cytologically to have low standard errors for the recombination values.

Table 14.3. Expected and Observed Frequencies of Genotypes from Backcrossing the F_1 *Gg* Monotelodisomic Shown in Fig. 14.5 to a *gg* Disomic

	Disomic		Monotelodisomic	
Genotype	Expected Frequency[a]	Observed Frequency[a]	Expected Frequency[a]	Observed Frequency[a]
Gg	$0.5(1-p)$	a	$0.5p$	c
gg	$0.5p$	b	$0.5(1-p)$	d

[a] p = the recombination fraction and $(1-p)$ = the nonrecombination fraction, each distributed between two genotypes. The letters a, b, c, and d are the observed frequencies of particular gene combinations. The respective gene combinations are illustrated in Fig. 14.5 using the a, b, c, d designation.

When data are obtained for the four observed classes, the recombination frequency can be calculated from the formula

$$p = \frac{b + c}{N}$$

where N is the total progeny number. The standard error of the recombination value is obtained from the formula

$$s_p = \left(\frac{p(1 - p)}{N}\right)^{1/2}.$$

Finally, the transmission rate of the telochromosome can be obtained from

$$\frac{c + d}{N}.$$

If no recombinant classes are obtained, the gene may be closely linked with the centromere. Another cause, particularly with telos of short arms, is a general failure of pairing between a telo and the corresponding arm of the complete chromosome. Cytological observations to determine the frequency of asynapsis are needed to indicate whether this is a problem.

An alternative procedure is to use F_2 progenies from selfed F_1 monotelodisomics. Estimates of male and female transmission rates of the telochromosome are needed to obtain expected frequencies of the chromosome types and phenotypes in the F_2 progenies. These frequencies and the observed data are used with statistical methods to derive a recombination value.

Recombination values from backcross and F_2 progenies involving telochromosomes result from crossing over between a gene and its centromere, which can be regarded as a physical chromosome marker. Thus, a recombination value translates into a physical distance between the gene and the centromere. The accuracy of this distance depends on the consistency of pairing between the telo and the complete chromosome. A considerable degree of asynapsis reduces the amount of recombination and makes the distance seem shorter than it actually is. Another factor is how accurately the chromosome types and phenotypes are classified, especially if a gene is close to its centromere and recombinant classes are rare. There is evidence in both cotton and wheat that even if there is pairing between the telo and the complete chromosome, crossing over is reduced near the centromeres of heteromorphic bivalents. In spite of these problems, gene mapping using telochromosomes is possible in polyploids, where the use of translocations or inversions is not generally feasible because of genetic duplication and triplication.

14.4 POLYPLOID ANALYSIS USING INTERVARIETAL CHROMOSOME SUBSTITUTIONS

14.4.1 Development of Intervarietal Chromosome Substitution Lines

Although intervarietal chromosome substitutions are especially suitable for locating genes governing quantitative traits, they can also be used for analyses of qualitative inheritance. The development of substitution lines involves the replacement of a pair of chromosomes in one variety, the recipient, by the homologous pair from another variety, the donor (Fig. 14.6). There must be clearly detectable allelic differences between the recipient and donor varieties for the characters under study, so that individual donor chromosomes have the potential to produce measurable changes in the recipient variety. For a complete genetic analysis, substitution lines need to be developed for each chromosome in the complement.

A set of monosomics and/or their derivatives such as nullisomics and monotelosomics, must be available in the recipient variety to supply $n - 1$ gametes for the crosses and backcrosses. This

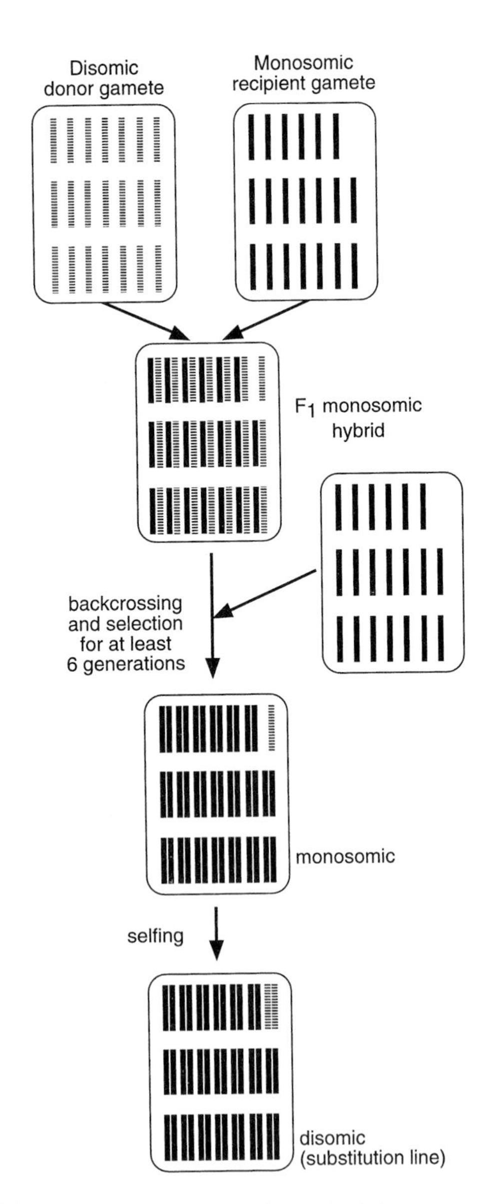

Fig. 14.6. Diagram to show how an intervarietal chromosome substitution line is developed. Using hexaploid wheat as an example, the donor gamete comes from a disomic variety and has a complete set of 21 chromosomes; the recipient gamete lacks one chromosome and comes from a monosomic or a monosomic derivative such as a monotelosomic. The disomic substitution line at the end of the backcrosses has a pair of chromosomes from the donor variety substituted for the homologous pair in the recipient variety. (Adapted from Morris and Sears, 1967.) *See* Fig. 14.7 for further details.

ensures that a donor chromosome is unpaired at meiosis and, therefore, genetically unchanged, barring mutation, in the F_1 and throughout the backcrosses. Nullisomics are an efficient type of aneuploid because they produce only $n - 1$ gametes if their meiotic behavior is regular, but, even in hexaploid wheat, only a few nullisomic types are fertile enough to use. If fertile, they sometimes have a trisomic or tetrasomic condition for a homoeologous chromosome to compensate for the missing chromosome pair, so they must be checked cytologically.

Monotelosomics and monoisosomics prevent the selection of monosomics with the recipient monosome, which is a possibility if monosomics of the recipient variety are used in the backcrosses. This is illustrated in Table 14.1, where selfing a monosomic gives the same distribution of progeny types as backcrossing a monosomic as male to the recipient monosomic. The monosomic progeny have two sources for their n gametes, the major source being the sperm and the minor source, the egg. Unless these two gametic sources can be distinguished by phenotype or chromosome morphology, there is a low probability that a monosomic plant chosen for a backcross received its monosome from the recipient parent through the egg cell (*see* the 1% group in Table 14.1). From this point onward, there would not be a substitution. If a recipient monotelosomic or monoisosomic instead of a monosomic is used as the female parent for the backcrosses (*see* Fig. 14.7), the group of $2n - 1$ progeny with a very low frequency is monotelosomic or monoisosomic, whereas the predominant $2n - 1$ group is monosomic for a donor monosome. These two groups can be distinguished by decreased vigor of the infrequent group and by the difference in the morphology of the unpaired chromosome at meiosis. The monosomics can then be selected for further backcrosses.

If monosomics are the only aneuploids available in the recipient variety, they should be selfed between each backcross (*see* Fig. 14.7). The disomics in the selfed progenies are homozygous for the donor chromosome and, when backcrossed to the recipient monosomic, produce only the correct monosomics. Whereas this adds an extra generation to each backcross cycle and prolongs the development of the substitution lines, it has the benefit of preserving the integrity of the substitutions. After six or more backcrosses to the recipient parent and provided that chromosome pairing and crossing over have been regular, the genes on the chromosomes other than the monosome should come almost entirely from the recipient. The final step is to self the monosomics from the final backcross and in their progenies select the disomics, which have a pair of donor chromosomes with their genes in a homozygous state. All the other chromosomes should have mostly homozygous gene pairs from the recipient, and they constitute the genetic background. When the disomics are selfed to increase seed supplies, the resulting substitution lines can be used for quantitative genetic studies and are known as near-isogenic lines (NILs).

The monosomic state for any chromosome can be transferred, for example, from variety A to variety B by using the monosomics as female parents in crosses with disomics of variety B, followed by at least six backcrosses of the monosomic progeny as females to variety B to recover its genetic background. In the F_1 and each backcross generation, the monosomes with their genes come from variety B, but the cytoplasm is derived mainly from variety A. The crosses are made in this direction because the transmission of $n - 1$ gametes through the male is usually so low that a large progeny is required to recover one monosomic. If the transferred monosomics are then used as females in developing the intervarietal substitution lines, these lines also have cytoplasm derived mainly from

REC = recipient variety, DON = donor variety

P_1: Monosomic set * in REC X DON disomic
2n–1 2n

Desired Gametes: n–1 n

↓

F_1: Select monosomics with DON monosome for backcrosses.

Source of n–1 gametes:

(1)

BC_1: REC Nullisomics or / REC Monotelosomics or / REC Monoisosomics] X F_1 monosomics

↓

BC_2: Select monosomic progeny and backcross to REC (n–1) gametes.

$BC_3 - BC_6$: Repeat the procedure for BC_2
(or more)

(2)

BC_1: REC Monosomics X Disomics from F_1 monosomics selfed

↓

Select monosomic progeny and self to get disomics.

BC_2: REC Monosomics X Disomics

↓

Select monosomic progeny and self to get disomics.

$BC_3 - BC_6$: Repeat the procedure for BC_2.
(or more)

For both (1) and (2), select monosomics after the last backcross and self to get disomics. Seed increases of these disomics will constitute the substitution lines; each line in the set has a different DON chromosome substituted for a homologous REC chromosome in REC genetic background (see Section 14.4.1).

* Nullisomics, monotelosomics, or monoisosomics may be used if available.

Fig. 14.7. A procedure to develop intervarietal chromosome substitution lines using monosomics or monosomic derivatives in the backcrosses. The plants used as females in the crosses and backcrosses are shown on the left side. For each substitution line developed, the aneuploids must be for the same chromosome in each generation.

variety A. Substitutions of cytoplasm between varieties is less likely to have phenotypic effects than substitutions between species or genera (*see* Chapter 15), but there is a possibility of a cytoplasmic–nuclear interaction for some genes when significant results are interpreted. This situation can be avoided at either the beginning or end of transferring the monosomic states. For instance, at the beginning of the transfer, variety A monosomics can be crossed as males with variety B. If there are no phenotypic differences between monosomics and disomics, a large seedling population must be checked to recover the low frequencies of monosomics from $n - 1$ male transmission. These monosomics, which have the cytoplasm and monosome from variety B, are then backcrossed as females to variety B to get a higher frequency of monosomics while retaining B cytoplasm.

Monotelosomics and monoisosomics can be derived from the monosomics established in variety B by centromere misdivisions of the monosomes, or they can be transferred from a variety in which they are established. The transfer of monosomics or their derivative aneuploids from one variety to another makes it possible to extend the genetic backgrounds in which donor chromosomes are tested. It is also possible to make reciprocal substitutions between two varieties with monosomic sets, whereby homologous chromosomes are exchanged.

With both one-way substitutions and reciprocal substitutions, each substitution line should be developed in duplicate to check the uniformity of its genetic background. If the duplicates are grown and tested in the same environment, there should be no significant differences between them for the characters measured. If there are, it is likely that there is genetic heterogeneity in the nonsubstituted chromosomes. The substituted chromosomes in the duplicate lines should be identical except for rare, spontaneous mutations or heterozygosity in the donor variety.

Although many substitution lines have been developed in hexaploid wheat (*Triticum aestivum*), a different approach has been needed in tetraploid wheat (*Triticum turgidum*), because the vigor, fertility, and $n - 1$ gametic transmission rates of tetraploid monosomics are low. Tetraploid wheat has two genomes, A and B, and each chromosome in both genomes is homoeologous with one chromosome in the D genome of hexaploid wheat. A complete set of D-genome disomic substitution lines has been developed in a tetraploid variety using cytogenetic manipulations. Each of the substitution lines has an A or B chromosome pair replaced by the homoeologous D-genome pair. Most of these D(A) and D(B) substitutions provide enough genetic compensation that they can be maintained. These lines provide the means to replace single chromosomes of the recipient variety with homologous chromosomes from other tetraploid wheat varieties (*see* Section 14.4.3).

14.4.2 Genetic Parameters for Chromosome Substitution Lines and Their Hybrids

The behavior of substitution lines, or of hybrids from crosses between lines, may be described by three types of genetic parameters. In explaining these types, let us label one chromosome C in variety I, and its homolog c in variety II, with the letters also referring to allelic differences between the genes on this pair of chromosomes. The state of the genes on all the other chromosomes of the substitution lines can be represented by G in variety I and g in variety II. With two doses of each chromosome in an allopolyploid, the parental varieties are $CCGG$ and $ccgg$, and the reciprocal substitution lines are $CCgg$ and $ccGG$. Variety I is the donor of chromosome C to the recipient variety II in the first substitution line, and the situation is reversed in the second line.

Additive effects refer to the genetic differences between a donor chromosome and the homologous recipient chromosome such as the difference between CC and cc. Between-chromosome interactions are interactions between a donor chromosome and one or more nonhomologous recipient chromosomes in the substitution line, or between two different donor chromosomes in hybrids from crosses between two substitution lines. Examples of the first type are the interactions between CC and gg in $CCgg$, or between cc and GG in $ccGG$; an example of the second type is the interaction between C or c and D or d, another substituted chromosome, in a hybrid from a cross between the two substitution lines. Within-chromosome interactions refer to interactions within a locus such as dominant versus recessive alleles, and interactions between loci on the same chromosome such as recombination due to crossing over.

These three parameters are not always independent of each other. For instance, a donor chromosome may have an additive effect in a substitution line, but it may also interact with the recipient-background chromosomes. Although these two parameters can be separated in tests of one-way substitutions, the results are sometimes misleading. They are most precisely separated in a reciprocal substitution series, where two-way tests can be conducted on homologous, substituted chromosomes for additive effects and for interactions with the recipient-background chromosomes. The additive effects of the two genetic backgrounds can also be determined.

14.4.3 Cytogenetic Studies Using Intervarietal Chromosome Substitution Lines

The advantage of substitution lines for quantitative-inheritance studies is that, provided enough backcrosses have been made, all the plants within a line should have the same homozygous genotype. Therefore, the large amounts of seed that can be obtained for each substitution line are an advantage when investigating quantitative traits.

A considerable amount of genetic information can be obtained from the substitution lines without further crosses. One-way or reciprocal substitutions, with duplicate lines, and their two parental varieties, can be grown in statistically designed arrangements, with several replications and preferably several environments over locations and years. The data collected are analyzed statistically, usually for a number of quantitative traits, to detect whether there are significant differences between any of the substitution lines and the recipient variety and whether the differences are in the direction of the donor variety. If duplicate lines of a substitution differ significantly in the same direction from the recipient variety, the interpretation is that there are one or more genes on the substituted chromosome with enough effect to be distinguished from their alleles on the recipient chromosome. A lack of significant differences between the substitution line and the recipient parent may be due to one of the following situations:

1. The genes are in the same allelic state on the donor chromosome as they are on the recipient homologous chromosome.
2. There is an absence of genes on the donor chromosome for the traits involved.
3. The genes on the donor chromosome must interact with genes on other donor chromosomes that are not in the substitution line.
4. Interactions between genes on the donor chromosome and genes in the recipient background prevent the donor genes from being expressed.

If there are significant differences between duplicates for any substitution, heterogeneity in the genetic background is implicated, and it is usually difficult to decide which duplicate is giving the true effect.

The value of reciprocal sets of chromosome substitution lines in locating genes for quantitative traits is evident in Table 14.4. Seventeen of the 21 wheat chromosomes have significant effects on 1 or more traits in 1 or both of the reciprocal sets, and there are inverse effects for 7 chromosomes when reciprocal comparisons are made. For example, when ''Wichita'' chromo-

Table 14.4. Significant Effects ($P < 0.05$) of Reciprocal Chromosome Substitutions Between the Wheat Varieties "Cheyenne" and "Wichita"

Trait	Negative Effect on Trait	Positive Effect on Trait
"Wichita" Chromosomes in "Cheyenne" Background		
Grain yield	3B	3A, 6A
Seeds per culm	None detected	3B
Seed weight	3B	3A, 5A, 6A, 2D, 3D, 4D
Culms per sq. meter	3B	None detected
Grain test weight	6B, 7B	3A, 3B
Plant height	3A, 6A, 3D	None detected
Anthesis date	3A, 6A, 3D	None detected
"Cheyenne" Chromosomes in "Wichita" Background		
Grain yield	3A, 6A	None detected
Seeds per culm	1A, 5A, 7A, 7D	2B, 3B, 1D
Seed weight	2B, 3D, 4D	2A, 6A
Culms per sq. meter	3A, 6A, 3B	None detected
Grain test weight	3A, 2B, 3D	7A, 6D
Plant height	4A, 3B	2B, 3D
Anthesis date	None detected	3D

Note: The letters A, B, and D refer to the three genomes of hexaploid wheat. The study included reciprocal sets of substitution lines representing each of the 21 wheat chromosomes, with duplicate lines for each chromosome substitution.

Source: Data from Berke et al. (1992).

some 3A is placed in a "Cheyenne" background, it increases grain yield over that of the recipient variety "Cheyenne," whereas "Cheyenne" chromosome 3A in "Wichita" background decreases yield in comparison with the yield of the recipient variety "Wichita."

In tetraploid wheat, the D-genome disomic substitution lines described in Section 14.4.1 can be used without further crossing to determine the genetic effects of missing A- or B-genome chromosomes (nullisomic effects) of the variety in which the lines were developed and of the added D-genome chromosomes. The substitution lines can also be crossed with a different tetraploid variety, which contributes to the F_1 plants a chromosome from the A or B genome that is homologous with the missing chromosome in the substitution line (Fig. 14.8). If there are recessive, hemizygous-effective genes on this chromosome, which is in the monosomic state in the F_1, they should be detected in the phenotype if the homoeologous D-genome chromosome does not have an overriding effect.

Intervarietal chromosome substitutions in tetraploid wheat are used to locate genes for valuable traits such as gluten proteins, which need to be strong to produce good quality pasta products. In the example shown in Fig. 14.8, the 1D(1B) substitution line is in the genetic background of the variety "Langdon," which has weak gluten. The substitution of chromosome 1B from "Edmore," a strong-gluten variety, indicates that this chromosome contributes in a major way to the gluten properties of "Edmore." This finding was made by comparing the different gliadin proteins, which are components of gluten and can be detected with polyacrylamide gel electrophoresis (*see also* Chapter 21, Section 21.2.5).

14.4.4 Crosses Involving Substitution Lines

When the substitution lines are tested for gene–chromosome associations, the chromosome is the unit and the genes on the substituted chromosomes should all be homozygous. After an association has been established between a particular chromosome and a quantitative character, the substitution line involving that chromosome can be used in crosses to detect recombination and to estimate the number of genes controlling the character (Fig. 14.9). The substitution line, the recipient variety in which the chromosome is substituted, and the F_1 hybrid are each crossed as males to the recipient variety monosomic in order to obtain nonrecombinant (RR and DD) or recombinant (RD and DR) chromosomes in monosomic progeny. As shown in Fig. 14.9, preliminary tests can be made on the monosomics in the progenies of these crosses. The amount of variation that can be attributed to nonrecombinant chromosomes alone is estimated by the parental products ($P_1 + P_2$), and the amount of variation due to both nonrecombinant and recombinant chromosomes can be estimated from the F_1 products. Significant differences between these two products can be attributed to recombination. If a larger population of plants is needed for testing, the monosomics from the three types of crosses are selfed, and then disomics are selected in the progenies and developed into lines by seed increase. The disomics derived from the recipient variety (P_1) and the substitution line (P_2) should be homozygous for the respective nonrecombinant chromosome, whereas those derived from the F_1 hybrid are homozygous for either a nonrecombinant or recombinant chromosome. Significant differences between the disomic lines derived from the two parents and those derived from the F_1 hybrid indicate that recombination occurred in the F_1 hybrids and that at least two linked genes control the character. On the other hand, lack of any differences indicate that there was no

Plants used in example:

tetraploid wheat, Langdon, 2n = 4x = 28 [A and B genomes, 7" in each genome]

tetraploid wheat, Edmore, 2n = 4x = 28 [A and B genomes, 7" in each genome]

1D (1B) Langdon, a tetraploid wheat in which a pair of 1D chromosomes from hexaploid wheat substitute for the Langdon 1B chromosome pair
2n = 4x = 28, [7 A" + 6 B" + 1D"]

P1: 1D (1B) Langdon X Edmore

Gametes: 7 [A] + 6 [B] + 1D 7 [A] + 6 [B] + 1B*

F1: 7 [A]" + 6 [B]" + 1B*' + 1D'

BC1 – BC6 (or more):

Cross F1 and also the backcross progeny plants that have 13" + 2', to 1D (1B) Langdon until the genetic background of Langdon is considered to be recovered.

After the last backcross, self plants with 13" + 2', which should have the same chromosome constitution as the F1. In their progeny, select plants with 14", which can be 7 [A]" + 6 [B]" + 1D", or 7 [A]" + 6 [B]" + 1B*". The latter type is a substitution line with Edmore 1B* replacing Langdon 1B. It can be distinguished from the 1D substitution by crossing with Langdon or Edmore and checking meiosis in the progeny. The 1D substitution gives 13" + 2', whereas the 1B* substitution gives 14".

Increase seed supplies of the 1B* substitution line by selfing.

*** The asterisk marks the Edmore 1B chromosome, which is to be substituted in Langdon background.**

Fig. 14.8. The use of a 1D(1B) disomic substitution line in the tetraploid wheat cultivar "Langdon," to substitute chromosome 1B from a second cultivar, "Edmore." Genome designations are inside brackets. Double primes denote a bivalent; a single prime denotes a univalent.

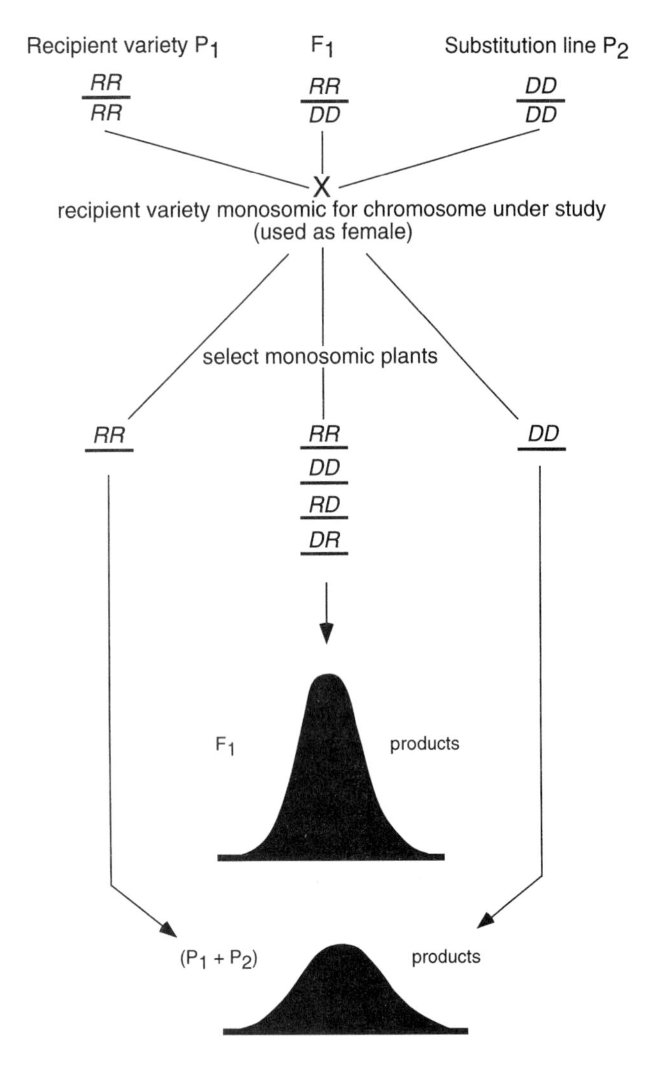

Fig. 14.9. The development of chromosome recombinant lines from crosses between a recipient variety (P_1) and a chromosome substitution line (P_2) to test for the number of genes controlling a trait and their linkage relationships. Only the substituted chromosome and its homolog are shown. The nonrecombinant chromosome from the recipient variety is designated R, the nonrecombinant chromosome from the donor variety, D, and recombinant chromosomes, RD and DR. (Adapted from Law, 1966b.)

recombination in the F_1 and that the character is controlled by one gene or closely linked genes.

Gene locations can be narrowed to chromosome arms using the same procedure as in Fig. 14.9 except that the initial crosses are made between the substitution line and recipient ditelosomics for each arm of the pertinent chromosome. In the F_1 hybrids from crosses involving either ditelo, the complete chromosome pairs with a telo and crossing over is restricted to one arm. If the genes for the character under investigation are on the arm represented by the telo, there can be a difference between the F_1 products and the parental products because of crossing over within the arm. If there is no difference between the two groups, the genes are either on the opposite arm or are so close to the centromere in the existing arm that no crossing over occurs between the centromere and the genes. Results of crosses with the ditelo for the opposite arm help in resolving these alternative possibilities.

14.5 EARLY DETECTION OF GENETIC DIFFERENCES BETWEEN DONOR AND RECIPIENT HOMOLOGOUS CHROMOSOMES

Although the intervarietal chromosome substitution lines described in the previous section have great value, they take a long time and much labor to develop, and they test only for genes in the two varieties involved in their development. Several methods have evolved to overcome these drawbacks, and one of them, the backcross reciprocal–monosomic method, is outlined in Fig. 14.10. It requires a monosomic set in only one variety to be used in crosses with any number of other varieties to determine the gene contents of their chromosomes. Because hemizygous or homozygous chromosomes are tested against segregating backgrounds in the early generations after the initial crosses, these chromosomes need to have noticeable genetic effects to be detected.

The reciprocal sets of progenies can be compared in Generation 3 (G3), where they consist of monosomics and disomics in about a 3:1 ratio. The disomics, with the test chromosome from both parents, can decrease the differences between the reciprocal sets, but there still can be detectable differences due to the tester monosome versus the varietal monosome. For some traits, the reduced-dosage effect due to the monosomic state of the test or varietal chromosome may obscure allelic differences between reciprocal sets. In such cases, disomics from selfing monosomics in G3 can be tested in G4. They are homozygous for the chromosome that came through the monosomic pathway from either the monosomic parent or from variety A (*see* Fig. 14.10).

Significant differences between the reciprocal sets in G3 or later generations can be attributed to allelic differences for genes on the varietal chromosome being tested, because the genetic background, while not homogeneous, should have similar amounts of heterozygosity and homozygosity. When chromosomes with major effects are identified, the performance of the relevant substitution lines can be predicted. This information expedites the transfer of these

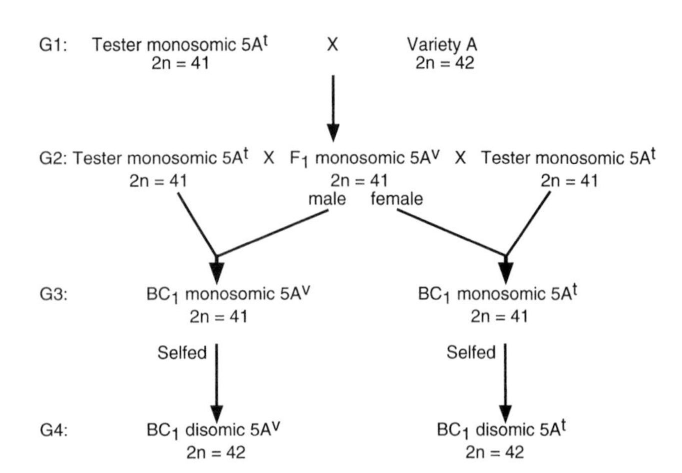

Fig. 14.10. Procedure for the backcross reciprocal–monosomic method to test substituted chromosomes for their effects in early generations. G = generation; BC = backcross; $5A^t$ = a chromosome from the tester monosomic; $5A^v$ = a chromosome from variety A. The tester monosomic can be crossed with as many varieties as are manageable. (Adapted from Snape and Law, 1980.)

chromosomes to other varieties and the incorporation of desirable genes in a plant-breeding program.

14.6 CHROMOSOME IRREGULARITIES ASSOCIATED WITH MONOSOMICS AND SUBSTITUTION LINES

14.6.1 Monosomic Shift

This condition results from nonpairing or premature separation of two homologous chromosomes during meiosis in a monosomic plant. If, for example, these two chromosomes are 1A and the monosome is 1B, at metaphase I the 1A chromosomes are univalents along with monosome 1B. By chance, 1B may be included in a telophase I nucleus, whereas the two 1A chromosomes are excluded. The $n - 1$ gametes derived from this nucleus are deficient for 1A instead of 1B, and progeny plants from a union between $n - 1$ and n gametes are monosomic for 1A instead of 1B.

There are several ways of checking the identity of monosomics. If there are distinctive phenotypic effects, which occur more frequently in tetraploids than in hexaploids, monosomic shift can be detected immediately. In the absence of phenotypic differences, the monosomes can be checked cytologically for morphology or banding pattern (*see* Chapter 6, Section 6.2). If they still cannot be identified, the monosomic plants are crossed with double ditelosomics, which have a pair of telochromosomes for each arm of the designated monosome and are more vigorous than ditelosomics (Fig. 14.11).

Monosomic is for 1B, no monosomic shift:

	Monosomic 1B	X	Double ditelosomic for 1B
Cross:	Monosomic 1B, $2n = 41$	X	Double ditelosomic for 1B, $2n = 44$
Meiosis:	$20'' + 1B'$		$20'' + 1B^{S''} + 1B^{L''}$
Gametes:	$n = 20 + 1B$		$n = 20 + 1B^S + 1B^L$
	$n - 1 = 20$		

Progeny: The $2n - 1$ plants at meiosis have $20''$ + two unpaired telochromosomes.

Monosomic is for 1A, monosomic shift has occurred, but the test is for 1B:

	Monosomic 1A	X	Double ditelosomic for 1B
Cross:	Monosomic 1A, $2n = 41$	X	Double ditelosomic for 1B, $2n = 44$
Meiosis:	$19'' + 1B'' + 1A'$		$19'' + 1A'' + 1B^{S''} + 1B^{L''}$
Gametes:	$n = 19 + 1B + 1A$		$n = 19 + 1A + 1B^S + 1B^L$
	$n - 1 = 19 + 1B$		

Progeny: The $2n - 1$ plants at meiosis have $19'' + 1A'$ + a trivalent made up of $1B + 1B^S + 1B^L$. Instead of a trivalent, there may be a heteromorphic bivalent, most likely consisting of $1B + 1B^L$, and an unpaired $1B^S$, or 1B and the two telochromosomes may all be unpaired.

All these configurations are distinguishable from the one shown above for the correct monosomic.

The 2n plants give the same configurations in both situations, namely, $20''$ + a trivalent ($1B + 1B^S + 1B^L$ or variations as given above), so they do not detect monosomic shift.

Fig. 14.11. The use of a double ditelosomic to check the identity of a monosomic plant using hexaploid wheat ($2n = 6x = 42$) and chromosomes 1A and 1B as examples. $1B^S$ = a telochromosome for the short arm of 1B; $1B^L$ = a telochromosome for the long arm of 1B; double primes denote a bivalent; a single prime denotes a univalent. The two situations are based on the assumption that the monosomic should be chromosome 1B.

Monosomic shifts are more likely to occur in the early generations of monosomic or substitution-line development, before cells have become adapted to the introduction of different genomes and possibly different cytoplasms. Although it is essential to test the identity of monosomics in these early stages, they should be checked periodically in later backcross generations, and also during their maintenance. In the development of substitution lines, monosomics from the last backcross must definitely be checked before disomics are selected from their selfed progenies.

14.6.2 Misdivision of Monosomes

The telochromosomes or isochromosomes produced by centromere misdivisions in lagging monosomes can affect monosomic analyses. For instance, failure to associate a gene with a chromosome may be due to the absence of the arm carrying the gene, and misdivision in the monosomics during the backcross program can interfere with the substitutions of whole chromosomes between varieties. Usually, monotelosomics and monoisosomics are less vigorous and less fertile than monosomics, so that careful observations and cytological checking should avoid this problem. At least in wheat, some chromosomes are more prone to misdivision than others, so particular attention should be given to these chromosomes.

14.6.3 Reciprocal Translocations

Varieties of a self-fertilizing species may differ by a reciprocal translocation, which is homozygous in the variety in which it occurs, but heterozygous in hybrids from crosses with varieties lacking it. The effect of such a translocation on monosomic transfers or the development of substitution lines using monosomics depends on whether one of the chromosomes involved in the translocation is monosomic. If it is not, the monosome and the translocation are transmitted independently. On the other hand, if a monosomic chromosome is one of those involved in the translocation, this relationship can be detected as a chain-of-three instead of a ring- or chain-of-four at meiosis in the F_1 hybrids. In the example shown in Fig. 14.12, variety B is disomic and homozygous for a translocation and variety A has normal chromosomes but is monosomic for one of the chromosomes in the translocation of variety B. The desirable situation in the F_1 hybrids and the backcrosses to either parent is to have consistent chain-of-three formation, with the gametes receiving either the two translocated chromosomes (n gametes) or the normal chromosome ($n - 1$ gametes) (*see* column 2 in Fig. 14.12). The n gametes are needed for backcrosses to variety A, the monosomic parent, when substituting a chromosome from variety B for its homolog in variety A, and the $n - 1$ gametes are needed for backcrosses to variety B to transfer the monosomic state from variety A. At the end of the backcross program, the monosomic chromosome in variety B and the substituted chromosome in variety A are divided between two centromeres by the translocation, the same as in the F_1 diagram in Fig. 14.12. In either case, this chromosome carries the genes of variety B because it had no pairing partner in the F_1 and backcross generations, and the monosomic or substitution line involving this chromosome can be used in cytogenetic studies.

There can be problems with the overlapping of a translocation and a monosomic state. The chain-of-three may not form consistently, depending on the position of the breakpoints, so that a biva-

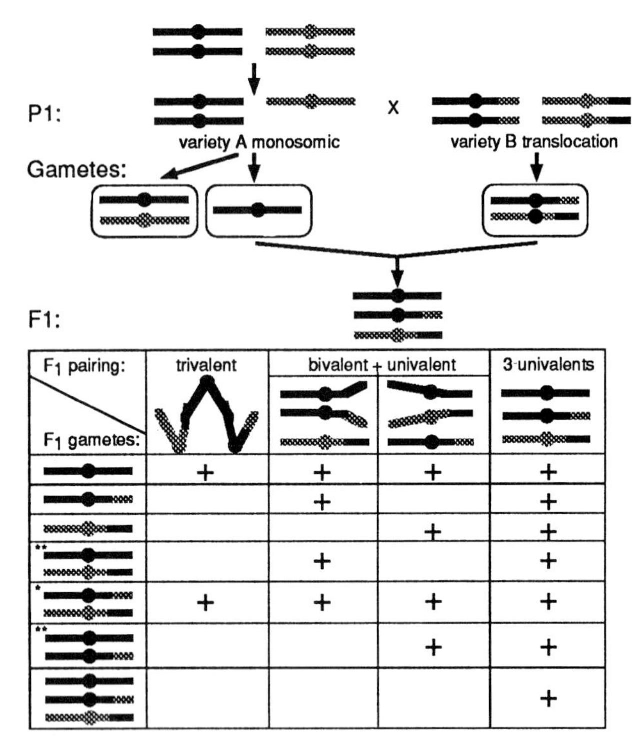

* When backcrossed to variety A monosomic, results in the same constitution as the F1. Chain of three regularly formed if translocated segments are long.

** Backcrosses to variety A (n - 1 gametes) may also have chain of three but less regularly.

Fig. 14.12. Diagram showing a cross between a monosomic of variety A and a disomic in variety B, which has a reciprocal translocation involving the chromosome that is monosomic in variety A. The circles represent centromeres. The " + " symbol denotes F_1 gametes that can result from the different F_1 pairing configurations. The gray chromosome is monosomic in variety A and is involved in a reciprocal translocation with a nonhomologous chromosome (black) in variety B. (Adapted from Morris and Sears, 1967).

lent and a univalent, or even three univalents, occur in some meiotic cells (*see* Fig. 14.12). Because these variations can give gametes with undesirable as well as desirable chromosome constitutions, careful screening is needed in each generation. Another problem relates to the amount of pairing and subsequent crossing over between the nontranslocated chromosome from variety A and its homologous segments in the translocated chromosomes from variety B. The amount of pairing is influenced by the locations of the breakpoints, and, in at least some meiotic cells, there is nonpairing around the breakpoints, where the exchange of segments occurs (*see* Chapter 9, Section 9.3.1). Nonpairing of segments or whole chromosomes prevents crossing over, which is required to recover the genotype of variety B for the segments of the nontranslocated chromosomes while simultaneously incorporating the monosomic condition. Crossing over also is needed to recover the genotype of variety A while substituting the chromosome segments, separated by the translocation, from B into A.

14.7 GENETIC HETEROGENEITY

It is important to have uniform genetic backgrounds in monosomic lines or their derivatives and in intervarietal chromosome substitution lines when using them for cytogenetic studies. Some genetic variation exists among plants within all varieties produced from intercrossing different parental lines and can also be present in cytogenetic lines. A way of reducing heterogeneity within a variety is to select single plants to maintain the line or use it in a cross. However, when narrowing the genetic scope in this way, the selected plants should have the genetic characteristics considered typical for the variety.

The number of backcrosses needed to recover essentially all of the genetic background of the recipient variety can be calculated on a theoretical basis but may not be the same in practice because of chromosome irregularities, which can restrict crossing over between chromosomes constituting the genetic background. It has been estimated that a chromosome substitution line derived from six backcrosses has about a 1% probability of carrying a donor gene on other than the substituted chromosome. When this probability is extended to all the substitution lines in a set, the chances of any one of these lines carrying such a gene increases substantially.

As was mentioned in describing the development of substitution lines, duplicate lines for each substituted chromosome are helpful in disclosing heterogeneity. The use of the same substitution series for studies of many traits may reveal heterogeneity for some traits but not for others. When genetic control of a trait has been associated with a substituted chromosome, other approaches such as recombinant lines or DNA sequencing can be used to substantiate this finding.

BIBLIOGRAPHY

General

KHUSH, G.S. 1973. Cytogenetics of Aneuploids. Academic Press, New York.

LAW, C.N., SNAPE, J.W., WORLAND, A.J. 1981. Intraspecific chromosome manipulation. Phil. Trans. Roy. Soc. London Ser. B 292: 509–518.

MCCOY, T.J., BINGHAM, E.T. 1988. Cytology and cytogenetics of alfalfa. In: Alfalfa and Alfalfa Improvement (Hanson, A.A., ed.). American Society of Agronomy, Madison, WI. pp. 737–776.

MCINTOSH, R.A. 1987. Gene location and mapping in hexaploid wheat. In: Wheat and Wheat Improvement, 2nd ed. (Heyne, E.G., ed.). American Society of Agronomy, Madison, WI. pp. 269–287.

MORRIS, R., SEARS, E.R. 1967. The cytogenetics of wheat and its relatives. In: Wheat and Wheat Improvement, 2nd ed. (Heyne, E.G., ed.). American Society of Agronomy, Madison, WI. pp. 19–87.

SEARS, E.R. 1954. The aneuploids of wheat. Missouri Agr. Exp. Sta. Res. Bull. 572, Columbia, MO.

SEARS, E.R. 1975. The wheats and their relatives. In: Handbook of Genetics, Vol. 2 (King, R.C., ed.). Plenum Press, New York. pp. 59–91.

SYBENGA, J. 1992. Cytogenetics in Plant Breeding. Monographs on Theoretical and Applied Genetics 17. Springer-Verlag, Berlin.

Section 14.1

MCINTOSH, R.A., BAKER, E.P. 1968. Chromosome location and linkage studies involving the *Pm3* locus for powdery mildew resistance in wheat. Proc. Linn. Soc. NSW 93: 232–238.

Section 14.3

Sears, E.R., Sears, L.M.S. 1979. The telocentric chromosomes of common wheat. In: Proc. Fifth Int. Wheat Genet. Symp. (Ramanujam, S., ed.). Indian Soc. Genet. Plant Breed., IARI, New Delhi. pp. 387–407.

Section 14.4

Berke, T.G., Baenziger, P.S., Morris, R. 1992. Chromosomal location of wheat quantitative trait loci affecting agronomic performance of seven traits, using reciprocal chromosome substitutions. Crop Sci. 32: 621–627.

Joppa, L.R. 1987. Aneuploid analysis in tetraploid wheat. In: Wheat and Wheat Improvement, 2nd ed. (Heyne, E.G., ed.). American Society of Agronomy, Madison, WI. pp. 255–267.

Joppa, L.R., Khan, K., Williams, N.D. 1983. Chromosomal location of genes for gliadin polypeptides in durum wheat *Triticum turgidum* L. Theoret. Appl. Genet. 64: 289–293.

Law, C.N. 1966a. Biometrical analysis using chromosome substitutions within a species. In: Chromosome Manipulations and Plant Genetics (Riley, R, Lewis, K.R., eds.). Plenum Press, New York. pp. 59–85.

Law, C.N. 1966b. The location of genetic factors affecting a quantitative character in wheat. Genetics 53: 487–498.

Section 14.5

Snape, J.W., Law, C.N. 1980. The detection of homologous chromosome variation in wheat using backcross reciprocal monosomic lines. Heredity 45: 187–200.

Sections 14.6 and 14.7

Person, C. 1956. Some aspects of monosomic wheat breeding. Can. J. Bot. 34: 60–70.

15

Gene Transfers by Chromosome Manipulations

- Genetic variation is essential for the development of new plant varieties and animal breeds, and this can often be achieved by introducing chromosomes from relatively distant species by hybridization.
- In plants, new species have been generated by the formation of amphiploids.
- Additions or substitutions of alien chromosomes are important steps in gene transfers to a host genome.
- Reciprocal translocations, or mutations that allow homoeologous pairing in polyploids, provide the basis for transferring foreign-chromosome segments to the host chromosomes.
- Reliable means for identifying particular chromosome segments are essential for carrying out chromosome manipulations.

Chromosome and gene transfers occur naturally or by manipulation between varieties or subspecies within a species, between species, or between genera. The closer the relationship between individuals used in crosses, the greater the likelihood of obtaining fertile hybrids, because the chromosomes from the two parents are homologous and pair at meiosis. If two crop varieties within a species are crossed, standard breeding procedures are used to transfer genes from one variety to the other. The F_1 hybrids are backcrossed for several generations to the variety receiving the genes, or they are advanced to the F_2 and later generations by selfing or intercrossing plants. Selections for the desired gene effects are made in each generation by phenotype or by biochemical or molecular tests, depending on the traits. Intervarietal chromosome substitutions or reciprocal monosomic crosses (*see* Chapter 14) can be used for transfers of specific chromosomes with genes that would improve other varieties.

The movements of whole chromosomes, or segments, with pertinent genes into crop species from other species or genera, whether wild or domesticated, are called alien transfers. The genome relationships between a crop species and a donor species may be close enough that fertile hybrids are produced. In these cases, the breeding procedures are the same as those outlined for intervarietal crosses, with the alien-gene transfers depending on normal crossing-over events. The cultivated tomato, *Lycopersicon esculentum*, is crossable with eight wild species in the same genus, although for some of the crosses the hybrid embryos have to be excised and grown on synthetic medium. The hybrid plants have enough vigor and fertility to be used in obtaining backcross or F_2 progenies. When the relationship between a wild species or genus and a crop is more distant, these wide crosses require complex manipulations, including embryo-rescue techniques, induced polyploidy, and the use of aneuploids, to achieve the goal of gene transfers. For this reason, most of the procedures included in this chapter are limited to plants. Methods that bypass the sexual route and that can be applied to a wide range of organisms and ploidy levels are discussed in Chapter 23.

15.1 THE REASON FOR GENE TRANSFERS

The need for gene transfers depends on the extant genetic variability in a crop, as well as the availability of wild species or genera. Many valuable genes residing within a crop species are easier to manipulate than alien genes, especially when multilocus, quantitative traits are involved. Furthermore, alien genes for desirable traits are often closely linked with genes having undesirable effects and if they cannot be separated by recombination, other methods such as translocated segments must be used.

On the other hand, the amount of genetic diversity in domesticated species of plants is generally more limited than in their wild relatives, which have had a much longer time to evolve and adapt to the natural environment. The surviving genotypes among wild species often include genes for resistance to harsh conditions such as diseases and insects, heat and cold, drought, and other stresses. Cultivated crops have been developed by human selection of a relatively narrow segment of the available gene pool, so they may not have the genetic buffers to withstand unexpected adversities. Popular varieties are often planted over wide areas because of desirable agronomic traits such as high yield or superior quality of a product. Their relatively limited genetic diversity makes them vulnerable to new races of pathogens and insects, which are continually evolving in response to environmental conditions. For this reason, the most prevalent interspecific or intergeneric transfers have involved resistance to specific diseases or insects, which often are controlled by one to a few major genes.

The efficiency of transfers from wild to cultivated sources can be increased by thorough genetic evaluations of the wild species in a process called prebreeding. After collections of these species have been made in their native habitats, the potentially valuable characters are investigated for their mode of inheritance. Attempts are then made to combine the genes for these characters in breeding lines, because they often are distributed among accessions collected at different times by different scientists. These prebreeding materials are evaluated for their adaptation to the environmental conditions in which the pertinent crops are grown. After these steps have been taken, the improved genotypes of the wild species are ready to be crossed with the crop genotypes.

15.2 INTERSPECIFIC AND INTERGENERIC CROSSES

When sexual reproduction is the channel for transmission of alien genes, the first step is a cross between two individuals. Interspecific and intergeneric crosses have been made in plants and animals, including birds and insects, with varying degrees of success. Male sterility is more prevalent than female sterility among hybrid offspring in both plants and animals. Reciprocal crosses are made when possible, because the chances for viable and fertile progeny may be greater in one direction than the other; in plants, phenotypic differences between reciprocal progenies are quite common.

The results of wide crosses in animals can be illustrated with two mammalian genera. Nonviable or sterile hybrids are common when interspecific crosses are made within the genus *Equus* (horses, asses, and zebras), and within the genus *Bos* (cattle). Karyotypic differences between species and physiological disturbances are thought to cause these offspring failures. The only successful *Equus* cross is between the domestic horse (*Equus caballus*, $2n = 64$) and the now rare Przewalski horse from Mongolia (*Equus przewalski*, $2n = 66$). Both male and female hybrids are fertile, with $2n = 65$ (Fig. 15.1). Depending on which species is ancestral to the other, the difference in chromosome number in the two species may be due to either centric fusion of two acrocentric chromosomes (Robertsonian translocation) to give the domestic horse karyotype, or centric fission of metacentric chromosomes to give two acrocentric chromosomes (four in the somatic cells), as in the Przewalski horse karyotype.

In domestic cattle, both male and female hybrids from crosses between *Bos taurus* and *Bos indicus* are fertile and have been used to derive some important cattle breeds. The intergeneric cross between *Bos taurus* and *Bison bison* (American buffalo) produces sterile males, but the females are fertile and have been used in backcrosses to both parents. The only significant difference between the karyotype of *Bos taurus* and those of the other two species seems to be the centromere position in the Y chromosome.

In plants, scientists are encouraged by the knowledge that natural allopolyploids evolved from spontaneous crosses between species or genera. However, in trying to make experimental crosses, they have often encountered barriers before or after fertilization that keep species intact. In sexual hybridizations, the advantage of plants over animals is that various chromosome manipulations and

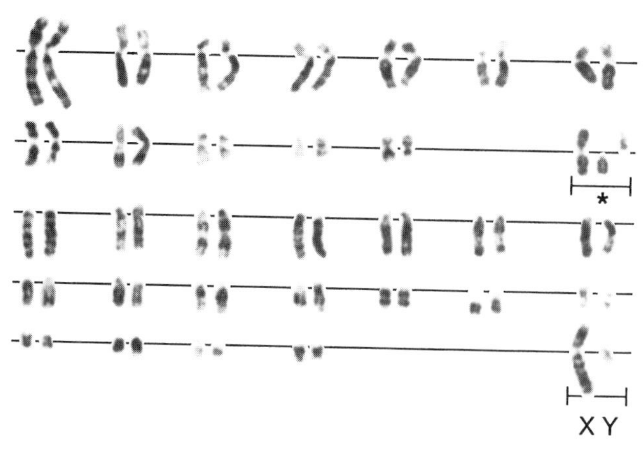

Fig. 15.1. A G-banded karyotype of a male hybrid ($2n = 65$) between the Przewalski horse ($2n = 66$) and the domestic horse ($2n = 64$). (From Short et al., 1974.) The asterisk marks a metacentric chromosome and two small, acrocentric chromosomes. These three chromosomes form a trivalent during meiosis, indicating homologies among them, while the other chromosomes pair as bivalents. There is a 2:1 separation of the chromosomes in the trivalent, giving $n + 1$ and n gametes, and both types produce normal-appearing offspring in crosses with the domestic horse. The solid line marks centromeres.

in vitro methods can be applied to overcome some of the prefertilization and postfertilization barriers.

The ease or difficulty in achieving fertilization with wide crosses in plants seems to be associated with physiological factors such as pollen–pistil incompatibilities, or mechanical factors such as time of flower abscission or size and shape of reproductive cells, rather than the degree of relationship between the parents. Lack of fertilization may be due to failure of pollen to germinate on a foreign stigma, or to restricted or blocked pollen-tube growth in the style, so that the tubes do not reach the embryo sac. Hormones can be used to prevent premature flower abscission, and boric acid or other substances can be applied to stigmas to increase pollen germination. Pollen-tube growth in the style may be stimulated by temperature changes or growth regulators. An incompatible reaction between pollen tubes and styles may be bypassed by excising the incompatible stylar region, or by injecting a suspension of pollen grains into the ovary.

Another successful approach for many incompatible interspecific crosses is known as the mentor effect. Pollen from one species may have an incompatible reaction in another species either on the surface of the stigma or after its tubes have penetrated the style. This incompatible pollen may be activated by mixing it with compatible pollen that has been damaged by some treatment such as ionizing radiation, a cycle of freezing and thawing, or methanol, so that it cannot participate in fertilization. The treated pollen is called mentor pollen because it supplies substances, most likely proteins, that recognize the stigmatic cells, so that the incompatible pollen can be stimulated to germinate, and its sperm cells can participate in fertilization. An interspecific cross was made in the tree genus *Populus* by obtaining mentor pollen from *P. deltoides* (black cottonwood) using one of the treatments mentioned above, then mixing it with the incompatible pollen from *P. alba* (white poplar). Proteins extracted from the walls of *P. deltoides* pollen grains had a mentor effect on *P. alba* pollen.

In vitro pollination and fertilization are useful when flowers drop from a plant prematurely or when the pollen tubes cannot reach the ovary. Intact pistils, ovules with some of the placental tissue attached, and isolated ovules have all been cultured on artificial media, and dusted with pollen. The success of in vitro methods in producing viable embryos and plants depends on a number of factors, including the genotypes of the parents, the culture medium, the female tissue used (with isolated ovules giving the most difficulty), and methods of collecting and applying pollen. Viable embryos, which in some cases developed into plants, have been obtained in several interspecific crosses involving the genera *Gossypium*, *Melandrium*, *Nicotiana*, *Petunia*, and *Zea*.

A postfertilization complication in wide crosses is failure of the embryo, or more commonly the endosperm, to develop, but the frequencies vary depending on the type of cross. When hexaploid wheat is crossed with maize or sorghum, most of the fertilized ovules have only an embryo, whereas both embryo and endosperm develop after crosses between hexaploid wheat and pearl millet, or between rye and maize. In all of these crosses, elimination of the maize, sorghum, or pearl-millet chromosomes occurs either during the zygotic division, or during early divisions of the embryo, leaving a haploid complement of wheat or rye chromosomes. In order to retain a hybrid state, a way has to be found to prevent genome elimination.

The breakdown of the embryo and/or endosperm is due to various factors depending on the degree of relationship between the parental species or genera. Incompatibility between genomes from different sources or differences in ploidy levels can disturb harmonious genetical and physiological interactions involving embryo, endosperm, and maternal tissues. Even if an embryo has the potential to develop, it is deprived of nutrients and will eventually abort if the endosperm degenerates. In such situations, in vitro culture plays an important role in rescuing hybrid embryos. The developmental stage of the embryo when transferred to a synthetic medium depends on when the endosperm deteriorates. From fertilization up to what is called the heart stage (based on its shape), the embryo depends on the endosperm for nutrition. If it has to be rescued during this critical period, the chances for success are increased if it is placed on the surface of normal endosperm tissue, which is cultured in vitro. Another method is to culture ovules, ovaries, or even intact flowers or spikelets (in grasses), provided that these tissues do not inhibit embryo development. This method avoids the danger of damaging the embryo by dissection, especially if it is very small when it is transferred. If transfer can be delayed to a more mature stage, the embryo is easier to culture on its own.

One solution for prefertilization and postfertilization barriers is to use additional species that are more compatible with the donor of desirable genes than the recipient species and that serve as genetic bridges in the gene-transfer process. In *Solanum*, resistance to late blight (*Phytophthora infestans*) cannot be transferred directly from the diploid species *S. bulbocastanum* to tetraploid cultivated potato *S. tuberosum*, because of the lack of offspring. The tetraploid species *S. acaule* is used as a bridging species because it can be crossed as female with the resistant diploid species. The sterile, triploid hybrids can be converted by colchicine treatment to hexaploids, which are difficult to cross with *S. solanum*, so they are crossed with a second bridging species, diploid *S. phureja*. The hexaploids contribute $3x$ gametes and the diploids $1x$ gametes to produce tetraploid, triple-species hybrids. Although many pollinations need to be made between these triple hybrids and *S. solanum*, some tetraploid progeny can be obtained. The effort is worthwhile because some of these tetraploids have strong resistance to late blight, as well as cytological stability and female fertility.

Another way of overcoming the failure of initial crosses is to double the chromosome number of one or both parental species, depending on their ploidy level. If both are diploids, raising one or both to a tetraploid state often facilitates interspecific crosses. If one species is diploid and the other tetraploid or hexaploid, inducing tetraploidy in the diploid often makes the crosses with the natural polyploids possible. This type of genetic bridge has been successful in a wide range of crop species.

15.3 TRANSFERS AT THE GENOME LEVEL (AMPHIPLOIDS)

Plant hybrids from interspecific or intergeneric crosses often are viable but highly sterile due to various types of incompatibilities. The sterility may be due to genetic incompatibility between parental genomes, sometimes with indirect physiological effects. If the hybrids have at least some female fertility, several generations of backcrosses can be made to the parental species that is to receive the desirable gene or genes, with selection for higher fertility in the progenies. Cytological incompatibility in the hybrids is expressed as little or no chromosome pairing during meiosis due to lack of homology between the parental genomes, or gene-controlled prevention of homoeologous pairing if there are some segments in common. A solution to this dilemma is chromosome doubling to produce two doses of each genome, which should restore cytological stability unless there is an inherent, unfavorable genetic or cytoplasmic effect.

Many naturally evolved allopolyploid plant species are basically amphiploids, which arose through spontaneous chromosome doubling in interspecific or intergeneric hybrids. When the ancestors of some natural amphiploids are traced, it is found that the original crosses were between diploid species. In some species such as cotton, tobacco, and durum wheat, the crops remained at the tetraploid level in subsequent evolution. In others such as hexaploid oats and wheat, crosses between the tetraploids and a third diploid species, followed by chromosome doubling, produced amphiploidy at the hexaploid level. An important factor in maintaining genome integrity is the absence of pairing between genomes from different sources. This situation occurs if there is no homology or homoeology (partial homology) between genomes, or if there are gene mutations that allow only homologous pairing. Homoeologous pairing can cause sterility because of multivalent formations, leading to gametes with varying chromosome numbers. Homoeologies between genomes in the successful natural amphiploids have been accompanied by pairing-restriction genes such as the *Ph* gene in wheat (*see* Chapter 10, Section 10.6.3), so that only homologous pairing occurs.

The conditions that produce stable, fertile natural amphiploids are equally important for induced amphiploids. The concept that the initial crosses should involve species or genera with low chromosome numbers has been demonstrated in perennial grasses. When the amphiploids are above the optimum ploidy level for the species or genera involved, vigor and fertility decline in further generations of maintenance. Some synthetic amphiploids with intergenomic homoeologies are restricted to homologous pairing by gene control and are fertile, whereas others lacking the genetic control have problems with unbalanced gametes and sterility. In the latter cases, attempts can be made to introduce a pairing-restriction gene into a parental species or genus by mutation or transfer from another source, and to test its expression in the amphiploid. The sterility problem can also be solved if the amphiploid has some type of vegetative propagation.

The most intensively investigated amphiploids have been the triticales (*X Triticosecale*), which combine hexaploid or tetraploid wheat with diploid rye. These wheat–rye crosses do not meet two of the desired conditions mentioned above, because one of the parents is polyploid, and both parents as well as the amphiploids reproduce by seeds only. A major problem with the amphiploids is a high frequency of shriveled seeds. Nevertheless, the keen interest in creating a new crop with the desirable traits of wheat (high yield along with protein and baking quality) and rye (drought tolerance, disease resistance, adaptation to poor soils, and amount of lysine) has stimulated a tremendous research effort in triticale breeding and cytogenetics.

The triticale amphiploids are octoploids or hexaploids, depending on the ploidy level of the wheat parent. Both types of amphiploids have cytological irregularities, including univalents at meiosis and chromatin bridges in dividing endosperm nuclei (Fig. 15.2). Studies have been made not only on the amphiploids but also on lines derived from crosses between amphiploids at the same or different ploidy levels, or from crosses between triticale and wheat. The main cause of the cytological abnormalities seems to be the difference in rate of nuclear development in rye and wheat. The durations of meiosis and endosperm-nuclear divisions are longer in rye than in wheat, and this difference in division rates is thought to cause cytological instability. This is probably one of several features in the genetic background and environment that contribute to the abnormalities seen in triticale endosperm development; nevertheless, triticale varieties have been produced after extensive crossing and selection.

Another amphiploid that is being used successfully in agriculture was derived from a cross between two diploid species, *Lolium perenne* (perennial ryegrass) and *Lolium multiflorum* (Italian ryegrass). The rapid early growth and high nutritive quality of Italian ryegrass were combined with the persistency of perennial ryegrass. The amphidiploid has some quadrivalent configurations at meiosis, and the diploid hybrid from which it came has consistent bivalent pairing, indicating that the genomes of the two parental species are very similar. The amphidiploid can be propagated vegetatively, so any reduction in fertility due to multivalent formations is not a problem.

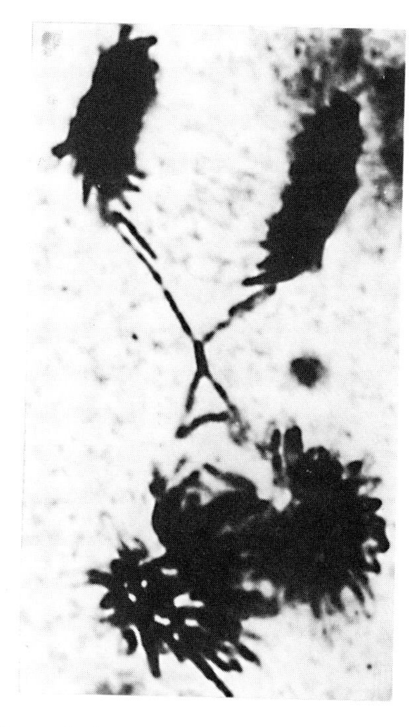

Fig. 15.2. Chromatin bridges between different nuclear chromosome groups at anaphase in the coenocytic endosperm of a triticale line. (From Gustafson and Bennett, 1982.) Such abnormalities are believed to contribute to defects in the endosperm.

Most induced amphiploids are, like triticale, not necessarily the end of the gene-transfer process. When amphiploids come from crosses between cultivated varieties and their wild relatives, the latter usually have a number of undesirable traits along with the desirable ones. Therefore, the amount of alien genetic material must be reduced, with selection for the desirable effects.

15.4 TRANSFERS OF INDIVIDUAL CHROMOSOMES (ADDITION AND SUBSTITUTION LINES)

Alien-addition lines have the full chromosome complement of the host species or genus with the addition of a chromosome from another species or genus. The added chromosome can be present in one dose (monosomic addition) or two doses (disomic addition). Alien-substitution lines have a chromosome pair in the host species or genus replaced by a chromosome from another species or genus in one dose (monosomic substitution) or two doses (disomic substitution). The monosomic state of the added or substituted chromosome is not as stable as the disomic state because of the irregular behavior of the alien univalent during meiosis. The usual sequence is to develop addition lines first, then use them to derive substitution lines. If an interspecific or intergeneric F_1 hybrid has some fertility, the addition and substitution lines can be derived by backcrossing the hybrid to the host parent. If the hybrid is sterile, the amphiploid approach must be used.

15.4.1 Alien-Addition Lines

An amphiploid has the full complement of chromosomes from two parental species or genera, so the best way to reduce the complement of the one parent is to backcross the amphiploid to the other parent. The procedure is illustrated in Fig. 15.3 by the development of rye-chromosome additions in wheat, using the triticale amphiploids discussed in the last section. The occurrence of any of the rye chromosomes in the monosomic-addition plants is a random event and depends on the meiotic behavior of the rye univalents. However, complete sets of rye monosomic-addition lines have been obtained when a number of backcross-progeny plants are checked. It may be difficult to obtain disomic-addition plants if the alien chromosome reduces the ability of pollen bearing it to compete with pollen carrying only host chromosomes. The transmission rates of different alien chromosomes vary from very low to preferentially high, depending on their genetic composition in relation to the host chromosomes.

When transmission rates of alien chromosomes are low, alternative methods may increase the chances of recovering disomic-addition lines. One solution is to obtain haploids with the added alien chromosome by culturing anthers from monosomic addition plants,

(1) 6x wheat (female) x 2x rye (male)
(2n = 42, AABBDD) (2n = 14, RR)

(2) F_1 sterile hybrid
2n = 21 wheat (ABD) + 7 rye

(3) Double chromosome number with colchicine.

(4) Amphiploid obtained is:
2n = 42 wheat (AABBDD) + 14 rye (RR)

(5) Backcross to wheat.

(6) Backcross progeny at meiosis I:
21 wheat bivalents (AABBDD) + 7 rye univalents (R)

Make further backcrosses to wheat until progeny plants with 21 wheat bivalents and only a single rye univalent are obtained. These are monosomic additions.

(7) Self the monosomic addition plants and select progeny plants with 21 wheat bivalents and 1 rye bivalent (disomic additions).

Fig. 15.3. The experimental development of alien-chromosome addition lines, using diploid rye as the alien species and hexaploid wheat as the host. The rye genome is represented by R and the three wheat genomes by A, B, and D.

or by crossing these plants with *Hordeum bulbosum* if feasible (*see* Chapter 12, Section 12.1.2). The chromosome number of the haploids is then doubled with colchicine to get a disomic addition. Another method is to self plants with two or three alien chromosomes added to the host complement. These can occur after the second backcross to wheat using the plants in Fig. 15.3(6). In practice, a higher frequency of disomic additions for single rye chromosomes is obtained from plants with two or three rye chromosomes than from those with only one rye chromosome. A third possibility is to cross the plants shown in Fig. 15.3(6) as males (assuming they are male-fertile) with the amphiploid indicated in Fig. 15.3(4). Each rye univalent has a much lower chance of being transmitted through the male gametes than through the female gametes, so the likelihood of transmitting single rather than multiple rye chromosomes is increased. In one study, over one-third of the transmitted male gametes had only one rye chromosome, which was combined with its homolog from the amphiploid to give progeny with a disomic addition. The other six rye chromosomes from the amphiploid are in single doses in these progeny plants, and they are lost as univalents in a generation or two of selfing.

Disomic additions are more stable than monosomic additions because the alien chromosome is present in two doses, but they also have a tendency for instability, with some nonpairing of the alien homologs in a foreign nuclear environment, and loss of the alien chromosome from some gametes. Male gametes with only host chromosomes may have a competitive edge over those with the alien chromosome; hence, some of the progeny from selfing are not disomic additions.

The phenotypic effects of an added chromosome depend on the genes it carries and how they function in the host-genetic background. When each of the 12 chromosomes of *Oryza officinalis* are added in single doses to the complement of *O. sativa* (Asian cultivated rice), the monosomic-addition plants have distinctive morphological traits and resemble the primary trisomics of *O. sativa* (Fig. 15.4). Therefore, each *O. officinalis* chromosome has at least some genes in common with genes on a chromosome of *O. sativa*, indicating a close genetic relationship, although pairing homology is very low. Disomic additions of four *Elymus trachycaulus* chromosomes were obtained in hexaploid wheat (Fig. 15.5), and genes for vigor and fertility were located on an arm of one of the added chromosomes by comparing the effects of monotelosomic additions for the two arms. The telochromosomes resulted from misdivision of the *Elymus* univalent at meiosis in the monosomic-addition line.

Addition lines are useful in dissecting the genotype of an alien species or genus by isolating each chromosome from all the others in the background of a cultivated relative. Genes for valuable traits such as disease resistance can be assigned to specific alien chromosomes by testing which addition line(s) in a set conveys resistance. If the genes have dosage effects, there are differences between monosomic and disomic additions. The genetical and cytological effects of the added alien chromosomes help to determine the degree of relationship between the parental species or genera that were used in developing the addition lines.

Disomic-addition lines cannot be used as new varieties because of their tendency for instability and some undesirable genetic effects. However, they can be used to develop alien-chromosome substitutions or as a step in the transfer of chromosome segments.

15.4.2 Alien-Substitution Lines

When an alien-chromosome pair replaces a host-chromosome pair, the chromosome number of the host species is not changed, but the complement lacks the genes on the absent host-chromosome

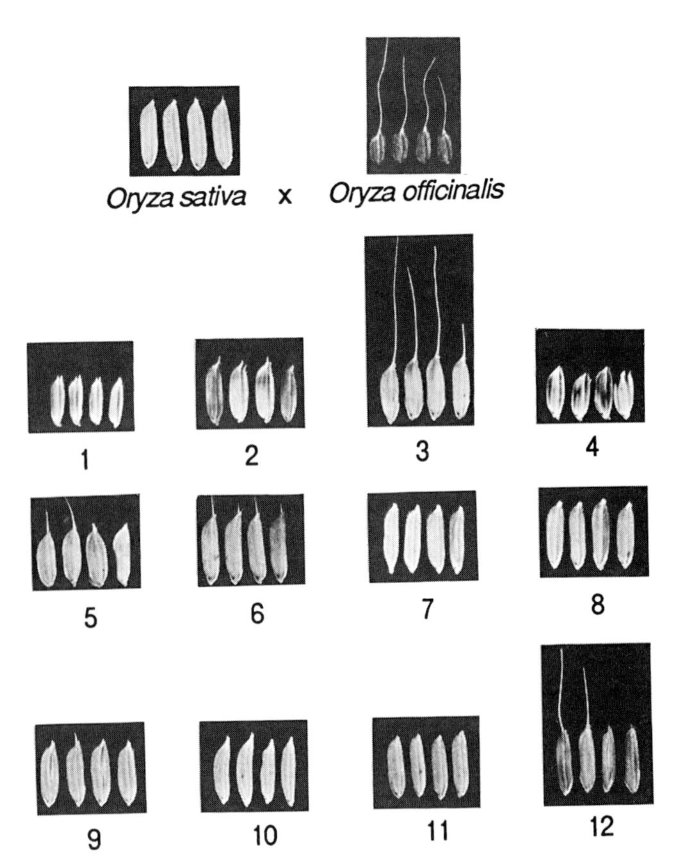

Fig. 15.4. Samples of the grain of *Oryza sativa*, *O. officinalis*, and the 12 monosomic-addition lines consisting of individual *O. officinalis* chromosomes added to the *O. sativa* complement. (From Jena and Khush, 1989.) Compare the monosomic-addition lines with each other and with the parents for variation in grain size and awn length.

Fig. 15.5. Spike morphology of (a) diploid *Elymus trachycaulus*, (b) the hexaploid wheat cultivar "Chinese Spring," (c) their F_1 hybrid, and (d)–(j) various addition lines, consisting of individual *E. trachycaulus* chromosomes added to the wheat complement. (d), (g), (h), and (i) are disomic additions involving four different *E. trachycaulus* chromosomes; (f) is a monotelosomic addition; and (e) and (j) are ditelosomic additions. (From Morris et al., 1990.)

pair. Therefore, the successful substitutions from the standpoint of plant vigor and fertility are those in which the substituted alien chromosome is homoeologous with a host chromosome, so that it provides similar genetic effects.

Some alien substitutions occur spontaneously after wide crosses and may not be detected until some generations later, such as in the development of some European wheat varieties. In selecting for resistance to several diseases, the wheat breeders unconsciously retained a homoeologous rye–wheat substitution (the rye chromosome 1R pair replacing the wheat chromosome 1B pair), which probably occurred after triticale lines were crossed with wheat in the 1930s. The substitution must have resulted from the presence of two wheat 1B univalents (instead of a bivalent) and a rye 1R univalent during meiosis I. This situation made it possible for a gamete to include rye 1R while lacking wheat 1B. The union of two such gametes gives plants with the normal chromosome number for hexaploid wheat, but with a disomic substitution and disease resistance from rye.

Another example of a spontaneous substitution occurred during a program to transfer a dominant gene for tobacco-mosaic-virus resistance to tetraploid tobacco (*Nicotiana tabacum*) from a diploid species (*N. glutinosa*). The hexaploid amphiploid from the interspecific cross was backcrossed to *N. tabacum*. After a second backcross followed by selfing, a line was obtained with the same chromosome number as *N. tabacum* and with homozygosity for resistance. The substitution of a *N. glutinosa* chromosome pair for a *N. tabacum* chromosome pair was detected by crossing the line with *N. tabacum* and, in the progeny, noting lack of pairing between the *N. glutinosa* monosome and the *N. tabacum* monosome that it replaced. There must have been some common genes between these two chromosomes for the line to be maintained.

When the chromosome homoeologies between a host species and an alien species are known, specific alien chromosomes can be substituted for the appropriate host chromosomes, provided identified addition lines and a set of monosomics are available in the host species. The procedure is shown in Fig. 15.6, using as an example the substitution of rye chromosome 1R for wheat chromosome 1B. If there is enough genetic overlap between an alien-chromosome pair and the replaced host-chromosome pair, the substitution line should have normal vigor. These homoeologous substitution lines are more stable than addition lines in chromosome behavior, and therefore are fertile more consistently. In the case of hexaploid wheat, which has three chromosomes from different genomes in each of its seven homoeologous groups, an alien chromosome of a related species usually can substitute for any of the three homoeologous wheat chromosomes.

Alien-substitution lines can have one or more distinctive traits determined by genes carried on the alien chromosome. The phenotypic effects of these genes can be compared in an addition line, which has a complete complement of host chromosomes, and a substitution line, which has the same alien chromosome and lacks

(1) Wheat monosomic 1B (female) X rye 1R disomic addition in wheat (male)

At meiosis:
The wheat monosomic 1B line forms 20" + 1B'
The rye 1R disomic addition in wheat forms 21" +1R"

(2) The pertinent gametes are:

n–1 (lacking 1B) from wheat mono 1B and
n+1 (21 wheat + 1R rye) from the addition line.

(3) Select F_1 plants with 20" + 2'. They should have

20 wheat bivalents + 1B wheat univalent + 1R rye univalent.

Self, or backcross (as male) to wheat monosomic 1B.

(4) Select progeny with 20 wheat bivalents + 1 rye univalent (1R) at meiosis monosomic substitution line) and self pollinate. The univalent can be identified as 1R by its C-banding pattern, or by its genetic effects.

(5) Select progeny with 20 wheat bivalents + 1R rye bivalent to form the disomic substitution line. Wheat 1B is missing and has been replaced by 1R.

Fig. 15.6. The experimental development of an alien-chromosome substitution line, using rye 1R as the alien chromosome and wheat 1B as the replaced host chromosome. Single prime denotes a univalent; double primes denote a bivalent.

one pair of host chromosomes. This comparison tests the effects of the presence or absence of genes on the host chromosome that is replaced by the alien chromosome. Another important point is how the alien genes interact with the nuclear and cytoplasmic genes of the host species. If the female and male parents in the initial crosses are as shown in Figs. 15.3 and 15.6, the derived addition and substitution lines have wheat (the host species) cytoplasm.

From a practical standpoint, alien-substitution lines are seldom suitable to be developed as commercial varieties because, like addition lines, the alien-chromosome pair introduces a mixture of favorable and unfavorable genes, while not replacing some of the essential genes on the missing host chromosome. The commercial use of the spontaneous substitution lines in wheat and tobacco mentioned above are exceptions rather than common occurrences. However, the likelihood of designing a substitution line with potential use can be increased if the alien chromosome replaces a homoeologous host chromosome that bears the least important genes for plant functions. Alternatively, it is possible to choose a genetic background that compensates for, or masks, the undesirable effects of genes on the alien chromosome.

The presence of undesirable genes on substituted alien chromosomes often requires the incorporation of alien segments rather than whole chromosomes into the host genotype. Alien additions and substitutions are intermediate steps in achieving this objective.

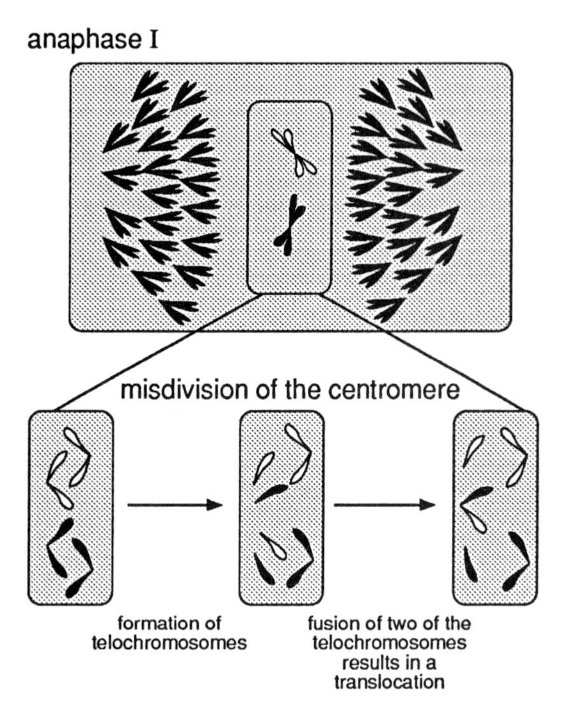

Fig. 15.7. The transfer by translocation of an alien-chromosome arm to a host chromosome by centromere fusion after misdivision in one chromatid of both the host (black) and alien (gray) monosomes during anaphase I of meiosis. The constrictions in the middle of the two-armed chromosomes and the pointed ends of the telochromosomes mark the centromere regions.

15.5 TRANSFERS OF ALIEN CHROMOSOME ARMS

There are several ways of reducing the amount of alien chromatin from a whole chromosome to an arm or a segment of an arm. These reductions are sometimes spontaneous, but their frequency can be increased by various methods. Alien-chromosome arms need to be homoeologous with the host arms they replace, otherwise too many host genes are missing in derived individuals. One way of initiating arm transfers is by spontaneous arm fusions after centromere misdivisions during meiosis, when both an alien chromosome and a host chromosome are in a monosomic state. This situation results from crossing an alien-substitution line with the host species. In the presence of a pairing-restriction gene, the two monosomes do not pair and usually lag as univalents between the anaphase I poles (Fig. 15.7). Occasionally, within the same cell, both monosomes misdivide after chromatid separation, resulting in four separated arms (telochromosomes). If an arm of the host chromosome lies near an arm of the alien chromosome at the time of misdivision, they can fuse in the centromere region. When these double-monosomic plants are crossed with the host species, the newly synthesized chromosome may be transmitted by chance inclusion in a gamete, which unites with a gamete from the host species. During meiosis in the progeny from such a union, one arm of the partially alien chromosome can pair with the homologous arm of the host chromosome. The alien arm remains intact because of nonpairing with the host arm. By selfing the plants involved, the alien arm can be made homozygous to give a disomic arm substitution. A number of alien-arm substitutions in triticale × wheat progenies are thought to have arisen from the misdivision process.

15.6 TRANSFERS OF ALIEN-ARM SEGMENTS

Arm substitutions may still introduce undesirable alien genes, so the alternative is to find ways of transferring smaller alien segments, which still include the gene or genes of interest. These transfers can be achieved by homoeologous pairing and crossing over between an alien chromosome and a host chromosome, or by translocations involving the alien chromosome.

15.6.1 Homoeologous Pairing

Several conditions are necessary to induce the type of homoeologous exchange that transfers alien segments to host chromosomes. In the plants developed for this purpose, the alien chromosome and a homoeologous host chromosome should be in the monosomic state, in order to foster an attraction between them in the absence of homologs. Any genes that inhibit homoeologous pairing should be removed, mutated to an inactive form, or suppressed. The *Ph* pairing control in hexaploid wheat has been manipulated in inventive ways to promote homoeologous pairing, so we shall use it to illustrate the methods. Lack of homoeologous pairing is due mainly to the strong gene *Ph1* on the long arm of chromosome 5B, but there are several other genes with weaker suppressive effects, including a locus designated *Ph2* on the short arm of 3D. In this discussion, keep in mind that monosomic states in hexaploid wheat have no effect on vigor and fertility in favorable growing conditions, but nullisomic states for most chromosomes have severe effects.

The *Ph1* effect on homoeologous pairing is suppressed when wheat is crossed with certain related diploid species such as *Triticum speltoides*. This method was used to transfer a dominant gene, *Yr8*, for resistance to stripe (yellow) rust, caused by *Puccinia strii-*

formis, from chromosome 2M of *Triticum comosum* to the homoeologous wheat chromosome 2D. A monosomic-addition line, with chromosome 2M added to the wheat complement, was crossed with *T. speltoides*. The selected hybrids had chromosome 2M, as well as the complete wheat and *T. speltoides* complements, in a single dose, so conditions were favorable for homoeologous pairing to take place. After several backcrosses to wheat, with selection of resistant plants in each generation, one resistant plant was obtained with the same chromosome number as wheat ($2n = 42$) and with bivalent pairing at meiosis. Further tests showed that this plant and its progeny had a chromosome consisting of the short arm of 2M, the proximal part of its long arm including *Yr8*, and part of one arm of the homoeologous wheat chromosome 2D. Although conditions were suitable for the origin of this transfer by homoeologous crossing over, a spontaneous translocation is also a possibility, because another study showed that chromosome 2M was prone to break in the long arm. A line homozygous for this chromosome was called "Compair" and was stable enough to be very useful as a source of stripe-rust resistance in wheat-breeding programs.

Another approach to inducing homoeologous pairing is to obtain a nullisomic state for chromosome 5B, which removes the *Ph1* gene. This can be done by crossing monosomic 5B with an alien-substitution line, and selfing the monosomic 5B progeny, which are also monosomic for the alien chromosome and the homoeologous host chromosome. A search is made by chromosome analyses in the F_2 for the few 5B nullisomics that have both the alien chromosome and the host homoeolog as monosomes. Although these nullisomic plants are weak and male sterile due to the absence of 5B genes, the female gametes can be rescued by crossing with normal pollen. One or two added doses of the homoeologous chromosome 5D can compensate to a large extent for the missing 5B genes, but this entails a cross with a nullisomic 5B–tetrasomic 5D line. As this procedure requires several generations of crosses and selfings with extensive cytological checking, it has not been widely applied.

A third way of promoting homoeologous pairing is to induce mutations to inactive alleles at the *Ph* loci. Both X-rays and a mutagenic chemical, ethyl methanesulfonate (EMS), have been used to produce inactive mutants, designated *ph*. At least some of the mutants involve deletions of varying lengths, but all include a *Ph* locus. Suppose that instead of using *Triticum speltoides* to transfer the gene *Yr8* for stripe-rust resistance (in the example discussed above), we use the *ph1* mutation on chromosome 5B. We have an alien-substitution line available in which chromosome 2M from *Triticum comosum* is substituted for 2D of wheat. In order to replace *Ph1* 5B with *ph1* 5B in the substitution line, the substitution line is first made monosomic for its 5B chromosome by crossing with a monosomic 5B line as female (Fig. 15.8). The desired F_1 plants are monosomic for *Ph1* 5B and 2M from the substitution line, and for 2D from the monosomic 5B line. When these triply monosomic plants are crossed as females with a homozygous *ph1* line, most of the progeny are monosomic for *ph1* 5B and 2D (male transmission), and some of these plants are also monosomic for 2M (female transmission). The presence of 2M can be detected by testing the plants for stripe-rust resistance, the presence of 2D and 5B by chromosome counts, and the presence of *ph1* rather than *Ph1* by observing meiotic multivalent formations, which indicate homoeologous pairing. Thus, conditions are favorable for pairing and exchange between 2M and 2D, both of which are in single dose. These plants and successive progenies are crossed with wheat, selecting in each generation for stripe-rust resistance, until the normal chromosome number for wheat is restored. Then, if only bivalent pairing occurs, it indicates that the pertinent 2M segment has been transferred to a wheat chromosome, most probably 2D.

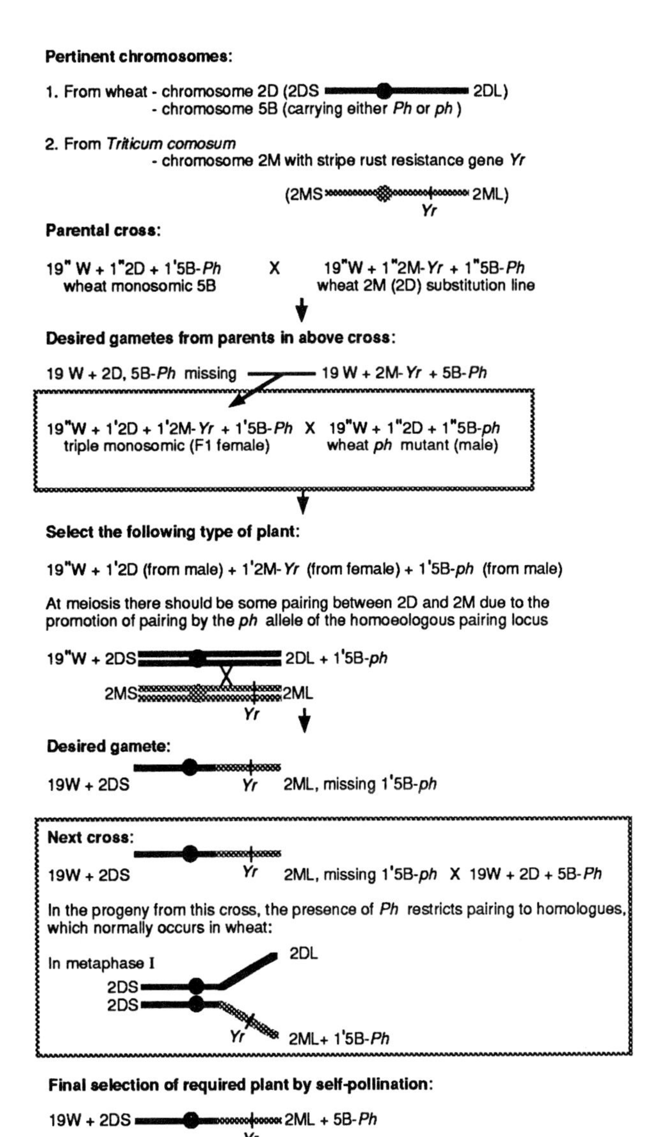

Fig. 15.8. An experimental application of the induced mutation *ph1* (shown as *ph* in the figure) to allow pairing and crossing over between chromosome 2M from *Triticum comosum* and the homoeologous wheat chromosome 2D. The objective is to transfer to 2D the 2M segment carrying the gene *Yr8* (shown as *Yr*) conferring resistance to stripe rust to wheat. Single primes = univalent, double primes = bivalents.

The choice of *ph* mutants on 5B or 3D depends on how much homoeologous pairing is desired, because the mutants on chromosome 5B have a more widespread effect than the 3D mutants, which restrict pairing to closely related homoeologs. A combination of *ph1* and *ph2* mutants may give a higher level of pairing than the *ph1* mutant by itself. Lower levels of homoeologous pairing may be obtained by introducing extra doses of genes that promote pairing.

15.6.2 Translocations

Some alien chromosomes are unable to pair with their wheat homoeologs even if the *Ph* restriction is removed or inactivated. If they do pair, synapsis may be restricted to the distal regions of the chromosomes, so that valuable alien genes located in the proximal regions are not transferred to the wheat chromosomes. In such cases, induced translocations provide the means for making the transfers. Alien-addition or alien-substitution lines ensure that any alien transfers involve specific chromosomes with desirable genes.

The methods that produce chromosome breaks with the potential for translocation exchanges were described in Chapter 7, Section 7.4. When irradiation is used, the chromosome breaks often occur at random, and the reciprocal exchanges usually involve terminal segments. An alien segment that is transferred to the homoeologous arm of a host chromosome is usually able to compensate for the missing host segment. If it is transferred to a nonhomoeologous host arm and the translocated chromosome is made homozygous, the loss of host genes may affect plant development adversely. At the same time, the alien segment provides an extra dose of the genes that are already present on the homoeologous host chromosomes. Depending on the genes involved, the extra doses have beneficial, neutral, or deleterious effects.

The ideal type of transfer involves the insertion of an alien segment containing the desirable genes into a host chromosome without loss of host genes. This requires one break in the host chromosome and two breaks in the alien chromosome, with the desirable genes between the breaks (Fig. 15.9c). The excised alien segment has unstable ends, which can unite with the unstable ends of the host chromosome at the point of breakage. If the inserted alien segment is quite short, it should not interfere with the pairing of the homologous host chromosomes, particularly if the chromosome with the insertion is made homozygous by selfing. The probability of getting this type of transfer is very low because of the sequence of events that has to occur (*see* Figs. 15.9a and 15.9b for other transfer types). It might be worthwhile to search for insertions in plants regenerated from tissue culture, which has produced many translocations. The tissue-culture route would be even more productive if it promoted transfers between homoeologous chromosomes, but more research is needed to resolve this point. Even if the breaks and exchanges are random, the high frequency of breaks increases the chances of getting an insertion type of transfer.

The first alien transfer to a crop species via an irradiation-induced translocation conveyed a dominant gene (*Lr9*) for leaf-rust resistance from the wild species *Triticum umbellulatum* to wheat. This study is a classic for logic and careful planning. A monosomic-addition line was used in which the alien chromosome was present as an isochromosome, with two doses of the arm carrying *Lr9* (Fig. 15.10). By using an isochromosome instead of a normal chromosome, which had *Lr9* on only one arm, the chances for irradiation-induced transfers of the gene were increased. In an attempt to increase the proportion of gametes with favorable translocations, X-rays were applied to the plants before meiosis. An intact isochromosome, and probably the centric (nontranslocated) part of the alien chromosome with a wheat segment attached, behave as univalents and are not included in most of the nuclei by the end of meiosis.

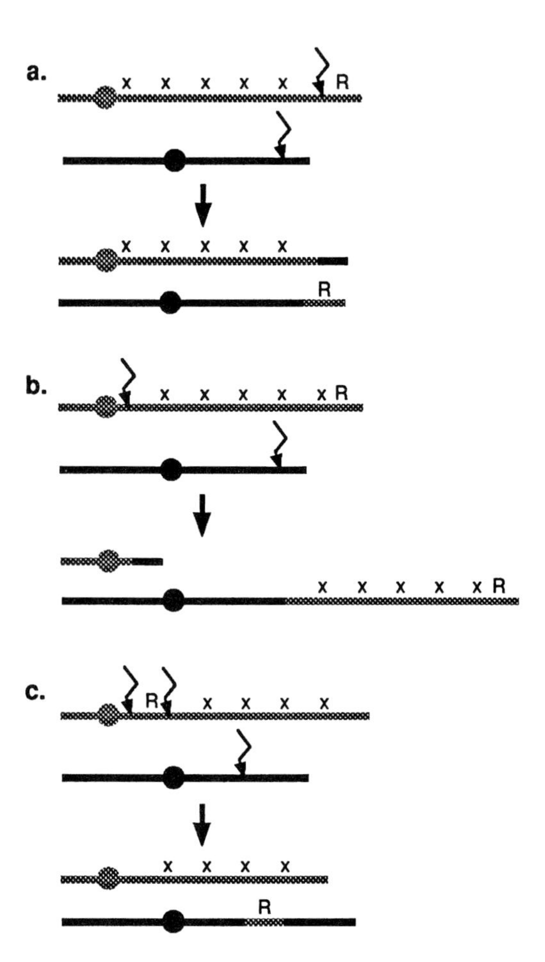

Fig. 15.9. The use of induced translocations to transfer a gene for disease resistance (*R*) from an alien chromosome (gray) to a host chromosome (black). Because the location of *R* and the exchanges shown in (a) are in the distal regions, *R* is transferred without the deleterious genes, marked x. In (b) the deleterious genes are transferred with *R* because both *R* and the breakpoint are near the centromere in the alien chromosome. In both (a) and (b), some host genes are lost through transfer to the alien chromosome. The transfer shown in (c) requires three breaks to insert the alien segment with *R* in the host chromosome without loss of host genes. In this diagram, the broken ends of the alien chromosome have rejoined. (Redrawn from Morris and Sears, 1967.)

The centric part of the wheat chromosome, with an alien segment attached, should pair with the normal wheat homolog throughout much of its length, so it is not lost. Provided that essential wheat genes are not lost with the other translocated chromosome, the retained, translocated chromosome can be transmitted through the pollen.

Crosses were made by applying pollen from the irradiated plants to untreated wheat plants without the alien chromosome. The *Lr9* gene was traced through succeeding generations by screening progenies for resistance to the pathogen *Puccinia recondita*. Resistant plants were checked cytologically to select those that had an alien segment attached to a wheat chromosome while discarding those with an intact isochromosome. Out of 17 different translocations with an alien segment carrying *Lr9*, 1

An alien chromosome addition to wheat:

21" wheat chromosomes + U —Lr—Lr— U isochromosome (U) from *Triticum umbellulatum*

These plants were irradiated before meiosis to induce translocations between the isochromosome and wheat chromosomes. One translocation that was recovered involved wheat chromosome 6B:

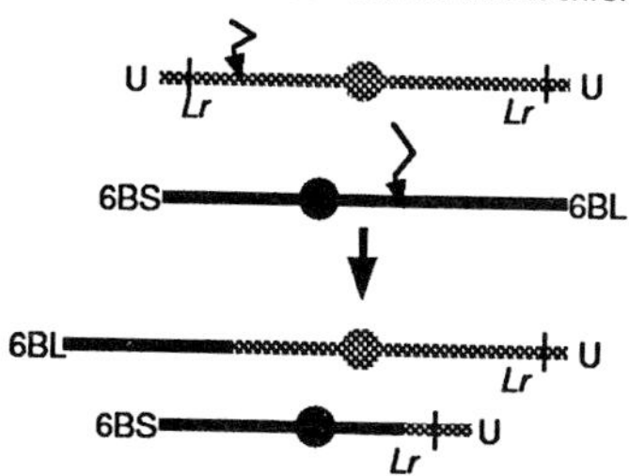

The two translocation chromosomes can form a "ring-of-three" by pairing with each other and one of the other untranslocated chromosomes such as the normal 6B:

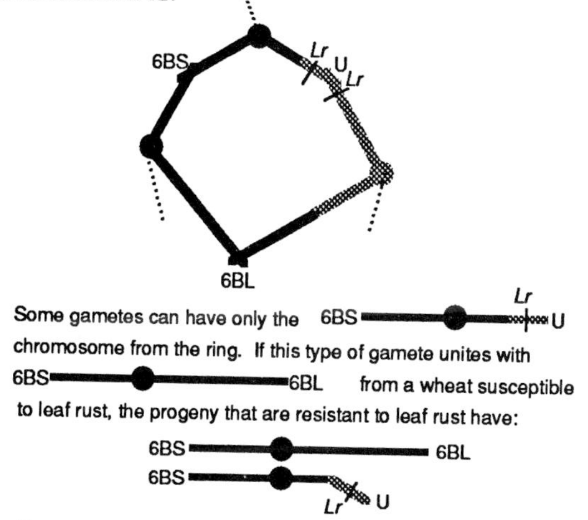

Some gametes can have only the 6BS —Lr— U chromosome from the ring. If this type of gamete unites with 6BS——6BL from a wheat susceptible to leaf rust, the progeny that are resistant to leaf rust have:

6BS ——— 6BL
6BS ——Lr— U

These two chromosomes have a good chance of pairing in the homologous regions. Self-pollinating these plants and selecting for true-breeding leaf rust resistance will result in the following homozygous wheat line:

Fig. 15.10. The use of an induced translocation to transfer the gene *Lr9* from the wild species *Triticum umbellulatum* (U genome) to chromosome 6B of hexaploid wheat. (Based on research by Sears, 1956.)

was at first thought to be an insertion as described above. However, further study showed that a reciprocal exchange had occurred between a fairly large segment of the alien chromosome and a terminal segment of a homoeologous wheat chromosome. This line was named "Transfer;" it was useful as a source of leaf-rust resistance until new races of the pathogen overcame the effectiveness of *Lr9*.

Since this pioneering study, labor-saving methods have been devised that reduce the need for extensive cytological observations while increasing the chances of obtaining alien-gene transfers. These include irradiation of disomic addition or substitution lines, which have two doses of an alien chromosome and therefore a better chance of exchanges with a host chromosome than when a single alien chromosome is present, and irradiation of seeds, which is more convenient than treatment of pollen or plants.

15.7 METHODS FOR DETECTING ALIEN CHROMATIN

The detection of alien chromatin in a host background includes genetical, cytological, and molecular approaches. The method used depends on the availability of effective alien-gene markers, the features of alien chromosomes that set them apart from host chromosomes, and the adaptability of molecular techniques to the chromosomes under investigation.

Desirable alien genes such as *Lr9* in the preceding section can be used to follow an alien segment through the generations after an interspecific or intergeneric cross. Morphological markers are most useful if they have been located on specific chromosomes and have a distinct phenotypic effect when combined with the host genotype. A good example of this type of marker is a gene on rye chromosome 5R for pubescent peduncle (hairy neck), which shows clear expression in a wheat background.

Certain biochemical markers such as isozymes, which are different forms of a single enzyme, are particularly useful for detecting alien chromatin. All organisms need many enzymes for a variety of functions, so each chromosome is likely to have one or more enzyme loci. When a genotype is heterozygous for one of these loci, the alleles controlling different isozymes are usually codominant, so that both an alien allele and a host allele are expressed. When crosses are made between distantly related species or genera, the presence in the progeny of isozymes from both parents confirms hybridization. Isozymes can serve as markers for alien chromosomes in addition or substitution lines and for some segmental transfers produced by crossing over or translocations.

Cytological methods for detecting alien chromatin are most efficient if they can be applied to mitotic rather than meiotic chromosomes. Occasionally, alien chromosomes are distinguishable from host chromosomes in a mitotic–metaphase spread because of size, arm ratio, or nucleolus-organizer constrictions. If the alien chromosome is introduced as a telosome into a two-armed host complement, it can easily be identified, but the arm would have to carry the desired genes. Cytological checks have been greatly facilitated by the development of chromosome-banding procedures (*see* Chapter 6, Section 6.2), which are used to distinguish genomes from different sources, or individual alien chromosomes in addition or substitution lines. Segmental transfers can be detected by mitotic banding if the segments have a different banding pattern from that of the host chromosomes. If the alien-segment banding is not distinctive, meiotic observations on the pairing configurations can be combined with the use of gene markers.

The molecular method of *in situ* hybridization (described in Chapter 19) makes it possible to identify transfers of alien chromosomes or segments to host species more efficiently and, in the case of segmental transfers, with greater precision than by other methods. The use of biotin rather than isotopes to label the DNA or RNA probes greatly reduces the time needed for the test and localizes the hybridization sites more accurately. An example of the use of biotin-labeled probes is to identify alien segments transferred to host chromosomes by translocations or crossing over. Wheat–rye lines with radiation-induced translocations and resistance to Hessian fly, *Mayetiola destructor*, were analyzed by in situ hybridization, using total-genomic rye DNA and two highly repetitive rye DNA sequences as probes. This molecular procedure made it possible to locate the translocation breakpoints precisely and to determine the exact sizes of the transferred rye segments (Fig. 15.11). The insertion of a small rye segment containing the gene for Hessian fly resistance into a wheat chromosome was also detected and is the most desirable type of translocation because none of the host chromatin is lost (*see* Fig. 15.11, lower panel).

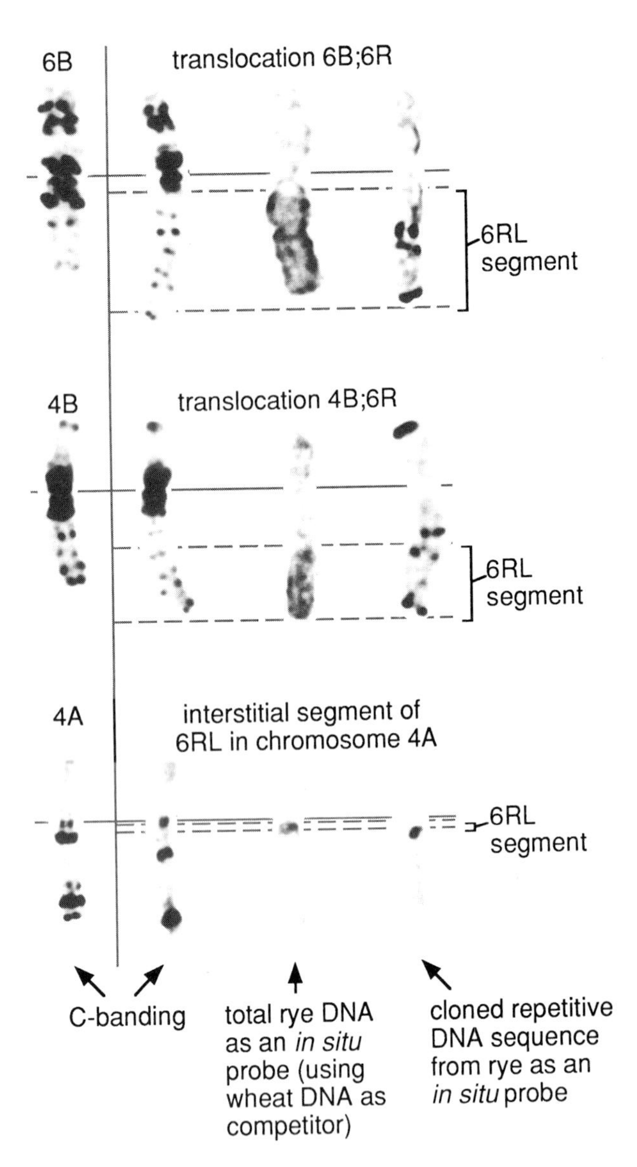

Fig. 15.11. The application of C-banding and *in situ* hybridization to detect transfers of rye (R) chromosome segments to wheat chromosomes by translocation, as well as wheat–wheat transfers. *Left*: C-banded wheat chromosomes. *Right*: translocated wheat/rye chromosomes detected using, respectively, C-banding, rye total-genomic DNA, and two highly repetitive rye DNA sequences (pSc119 and pSc74), singly or combined, as probes. The solid horizontal lines mark the centromeres, and the dashed lines mark the breakpoints. The upper two rows involve a terminal transfer of most of 6RL to wheat 6BL or 4BL; the last row shows the insertion (TC) of a small segment of 6RL into wheat 4AL. (From Mukai et al., 1993.)

BIBLIOGRAPHY

General

Dewey, D.R. 1980. Some applications and misapplications of induced polyploidy to plant breeding. In: Polyploidy: Biological Relevance (Lewis, W.H., ed.). Plenum Press, New York. pp. 445–470.

Feldman, M. 1988. Cytogenetic and molecular approaches to alien gene transfer in wheat. In: Proc. 7th Int. Wheat Genet. Symp. (Miller, T.E., Koebner, R.M.D., eds.). Institute of Plant Science Research, Cambridge. pp. 23–32.

Hart, G.E., Tuleen, N.A. 1983. Introduction and characterization of alien genetic material. In: Isozymes in Plant Genetics and Breeding, Part A (Tanksley, S.D., Orton, T.J., eds.). Elsevier Science Publishers B.V., Amsterdam. pp. 339–362.

Hermsen, J.G.Th. 1984. Some fundamental considerations on interspecific hybridization. Iowa State J. Res. 58: 461–474.

Morris, R., Sears, E.R. 1967. The cytogenetics of wheat and its relatives. In: Wheat and Wheat Improvement (Quisenberry, K.S., ed.). American Society of Agronomy, Madison, WI. pp. 19–87.

Rick, C.M. 1982. The potential of exotic germplasm for tomato improvement. In: Plant Improvement and Somatic Cell Genetics (Vasil, I.K., Scowcroft, W.R., Frey, K.J., eds.). Academic Press, New York. pp. 1–28.

Riley, R., Law, C.N. 1984. Chromosome manipulation in plant breeding: progress and prospects. In: Gene Manipulation in Plant Improvement (Gustafson, J.P., ed.). Plenum Press, New York. pp. 301–322.

Section 15.1

Duvick, D.N. 1990. The romance of plant breeding and other myths. In: Gene Manipulation in Plant Improvement II (Gustafson, J.P., ed.). Plenum Press, New York. pp. 39–54.

Goodman, R.M., Hauptli, H., Crossway, A., Knauf, V.C. 1987. Gene transfer in crop improvement. Science 236: 48–54.

Simmonds, N.W. 1984. Gene manipulation and plant breeding. In: Gene Manipulation in Plant Improvement (Gustafson, J.P., ed.). Plenum Press, New York. pp. 637–654.

Snape, J.W., Law, C.N., Worland, A.J., Parker, B.B. 1990. Targeting genes for genetic manipulation in plant species. In: Gene Manipulation in Plant Improvement II (Gustafson, J.P., ed.). Plenum Press, New York. pp. 55–76.

Section 15.2

Collins, G.B., Taylor, N.L., DeVerna, J.W. 1984. In vitro approaches to interspecific hybridization and chromosome manipulation in crop plants. In: Gene Manipulation in Plant Improvement II (Gustafson, J.P., ed.). Plenum Press, New York. pp. 323–383.

Eldridge, F.E. 1985. Cytogenetics of Livestock. AVI Publishing Co., Westport, CT.

Hermsen, J.G.Th., Ramanna, M.S. 1973. Double-bridge hybrids of *Solanum bulbocastanum* and cultivars of *Solanum tuberosum*. Euphytica 22: 557–566.

Knox, R.B., Gaget, M., Dumas, C. 1987. Mentor pollen techniques. Int. Rev. Cytol. 107: 315–332.

Laurie, D.A., Donoughue, L.S.O., Bennett, M.D. 1990. Wheat × maize and other wide sexual hybrids: their potential for genetic manipulation and crop improvement. . In: Gene Manipulation in Plant Improvement II (Gustafson, J.P., ed.). Plenum Press, New York. pp. 95–126.

Raghavan, V. 1986. Variability through wide crosses and embryo rescue. In: Cell Culture and Somatic Cell Genetics of Plants, Vol. 3. Plant Regeneration and Genetic Variability (Vasil, I.K., ed.). Academic Press, Orlando, FL. pp. 613–633.

Short, R.V., Chandley, A.C., Jones, R.C., Allen, W.R. 1974. Meiosis in interspecific equine hybrids. II. The Przewalski horse/domestic horse hybrid. Cytogenet. Cell Genet. 13: 465–478.

Sprague, J.J. 1982. Combining genomes by conventional means. In: Plant Improvement and Somatic Cell Genetics (Vasil, I.K., Scowcroft, W.R., Frey, K.J., eds.). Academic Press, New York. pp. 99–118.

Section 15.3

Gustafson, J.P. 1983. Cytogenetics of triticale. In: Cytogenetics of Crop Plants (Swaminathan, M.S., Gupta, P.K., Sinha, U., eds.). Macmillan India Ltd., New Delhi. pp. 225–250.

Gustafson, J.P., Bennett, M.D. 1982. The effect of telomeric heterochromatin from *Secale cereale* L. on triticale (× *Triticosecale*). I. The influence of the loss of several blocks of the telomeric heterochromatin on early endosperm development and kernel characteristics at maturity. Can. J. Genet. Cytol. 24: 83–92.

Hadley, H.H., Openshaw, S.J. 1980. Interspecific and intergeneric hybridization. In: Hybridization of Crop Plants (Fehr, W.R., Hadley, H.H., eds.). American Society of Agronomy, Madison, WI. pp. 133–159.

Jauhar, P.P. 1983. Some aspects of cytogenetics of the *Festuca–Lolium* complex. In: Cytogenetics of Crop Plants (Swaminathan, M.S., Gupta, P.K., Sinha, U., eds.). Macmillan India Ltd., New Delhi. pp. 309–350.

May, C.E. 1990. Triticale × wheat hybrids. In: Wheat. Biotechnology in Agriculture and Forestry 13 (Bajaj, Y.P.S., ed.). Springer-Verlag, Berlin. pp. 229–249.

Section 15.4

Feldman, M., Sears, E.R. 1981. The wild gene resources of wheat. Sci. Am. 244: 102–112.

Gerstel. D.U. 1943. Inheritance in *Nicotiana tabacum*. XVII. Cytogenetical analysis of glutinosa-type resistance to mosaic disease. Genetics 28: 533–536.

Jena, K.K., Khush, G.S. 1989. Monosomic alien addition lines of rice: production, morphology, cytology, and breeding behavior. Genome 32: 449–455.

Knott, D. R. 1988. Transferring alien genes to wheat. In: Wheat and Wheat Improvement, 2nd ed. (Heyne, E.G., ed.). American Society of Agronomy, Madison, WI. pp. 462–471.

Law, C.N. 1981. Chromosomes and plant breeding. Chromosomes Today 7: 194–205.

Lukaszewski, A.J. 1988. A comparison of several approaches in the development of disomic alien addition lines of wheat. In: Proc. 7th Int. Wheat Genet. Symp. (Miller, T.E., Koebner, R.M.D., eds.). Institute of Plant Science Research, Cambridge. pp. 363–367.

Morris, K.D.L., Raupp, W.J., Gill, B.S. 1990. Isolation of H^t genome chromosome additions from polyploid *Elymus trachycaulus* (S^t S^t H^t H^t) into common wheat (*Triticum aestivum*). Genome 33: 16–22.

Swaminathan, M.S., Gupta, P.K. 1983. Improvement of crop plants—emerging possibilities. In: Cytogenetics of Crop Plants (Swaminathan, M.S., Gupta, P.K., Sinha, U., eds.). Macmillan India Ltd., New Delhi. pp. 1–18.

Zeller, F.J. 1973. 1B/1R wheat-rye chromosome substitutions and translocations. In: Proc. 4th Int. Wheat Genet. Symp. (Sears, E.R., Sears, L.M.S., eds.). Columbia, MO. pp. 209–221.

Section 15.5

Lukaszewski, A.J., Gustafson, J.P. 1983. Translocations and modifications of chromosomes in triticale × wheat hybrids. Theoret. Appl. Genet. 64: 239–248.

Sears, E.R. 1972. Chromosome engineering in wheat. Stadler Symp. Series 4: 23–38.

Sears, E.R. 1984. Mutations in wheat that raise the level of meiotic chromosome pairing. In: Gene Manipulation in Plant Improvement (Gustafson, J.P., ed.). Plenum Press, New York. pp. 295–300.

Section 15.6

Driscoll, C.J., Jensen, N.F. 1963. A genetic method for detecting intergeneric translocations. Genetics 48: 459–468.

Miller, T.E., Reader, S.M., Singh, D. 1988. Spontaneous non-Robertsonian translocations between wheat chromosomes and an alien chromosome. In: Proc. 7th Int. Wheat Genet. Symp. (Miller, T.E., Koebner, R.M.D., eds.). Institute of Plant Science Research, Cambridge. pp. 387–390.

Riley, R., Chapman, V., Johnson, R. 1968. The incorporation of alien disease resistance in wheat by genetic interference with the regulation of meiotic chromosome synapsis. Genet. Res. Camb. 12: 199–219.

Sears. E.R. 1956. The transfer of leaf rust resistance from *Aegilops umbellulata* to wheat. Brookhaven Symp. in Biol. No. 9. Genetics in Plant Breeding. pp. 1–22.

Section 15.7

Appels, R., Moran, L.B. 1984. Molecular analysis of alien chromatin introduced into wheat. In: Gene Manipulation in Plant Improvement (Gustafson, J.P., ed.). Plenum Press, New York. pp. 529–557.

Flavell, R.B. 1982. Recognition and modification of crop plant genotypes using techniques of molecular biology. In: Plant Improvement and Somatic Cell Genetics (Vasil, I.K., Scowcroft,

W.R., Frey, K.J., eds.). Academic Press, New York. pp. 277–291.

Hutchinson, J., Flavell, R.B., Jones, J. 1981. Physical mapping of plant chromosomes by *in situ* hybridization. In: Genetic Engineering: Principles and Methods, Vol. 3 (Setlow, J.K., Hollaender, A., eds.). Plenum Press, New York. pp. 207–222.

Mukai, Y., Friebe, B., Hatchett, J.H., Yamamoto, M., Gill, B.S. 1993. Molecular cytogenetic analysis of radiation-induced wheat–rye terminal and intercalary chromosomal translocations and the detection of rye chromatin specifying resistance to Hessian fly. Chromosoma 102: 88–95.

Rayburn, A.L., Gill, B.S. 1985. Use of biotin-labeled probes to map specific DNA sequences on wheat chromosomes. J. Hered. 76: 78–81.

VI
DNA Organization, Structure, Replication

16

Organization of DNA Sequences in Protosomes and Chromosomes

- The amounts of nuclear DNA vary greatly between organisms even though (80–100) × 10^6 base pairs of DNA are theoretically sufficient to define a basic set of genes necessary for eukaryotic life-forms.
- Under suitable experimental conditions, denatured DNA spontaneously reforms a double helix, based on A–T and G–C base pairing.
- Repetitive DNA sequences are a major source of variation in DNA amount and contribute to modulating gene activity.
- Many repetitive sequences are dispersed throughout the genome and were transposable or retrotransposable elements at some stage in their evolutionary history.
- Some regions of DNA that affect the expression of a gene can be thousands of base pairs away from the respective gene.

Most prokaryotic and eukaryotic life-forms have their genetic information encoded in DNA. In prokaryotes, the bulk of the genetic information is usually present in a very large, circular protosome, with lesser amounts in small, circular molecules, plasmids, which often encode genes for resistance to chemicals. In eukaryotes, genetic information is located in the nucleus, mitochondria, and chloroplasts. The nuclear genome is organized in linear molecules, which are present in chromosomes, whereas the mitochondria and chloroplast genomes are small circles of DNA in cytoplasmic organelles. Small, plasmidlike molecules have also been found in some eukaryotes.

16.1 THE VARIABLE AMOUNT OF DNA AND THE ARRANGEMENT OF SEQUENCES IN ORGANISMS

In eukaryotic life-forms, the size and number of chromosomes, thus their nuclear DNA content, vary enormously (Fig. 16.1). For example, the nucleus of a single diploid cell of a wheat plant contains 11.2 meters (m) of DNA, corresponding to a mass of approximately 36.2 picograms (pg), or 32 billion (32 × 10^9) nucleotide base pairs (bp), and arranged in 21 pairs of chromosomes. In the same terms, a human haploid gametic cell contains 2.1 m of DNA, equal to approximately 3.1 pg or 3 billion base pairs, arranged in a haploid complement of 22 chromosomes plus the X or Y sex chromosome.

The amount of coding DNA required for eukaryotic life-forms is most likely in the range (80–100) × 10^6 bp, or (80–100) × 10^3 kilobase pairs (kb), as judged by "model" organisms such as the plant *Arabidopsis thaliana*, as well as large-scale sequencing projects on the genes that are expressed in humans. These amounts often represent less than 1% of the total amount of DNA in an organism, and it is evident that the DNA in the chromosomes is not simply a "set of genes on a string." Research has shown that in addition to coding DNA, one strand of which is committed to coding for RNA and protein products for the cell, there exist large tracts of noncoding DNA, which do not code for any specific product. The tracts of noncoding DNA provide the environment that regulates the expression and transmission of genes, and variation in the amount of this class of DNA is mostly responsible for the large differences in DNA content (C-value) among eukaryotes. The high content of noncoding DNA in many organisms makes it important to understand the organization of DNA sequences in the genome.

16.2 CODING AND NONCODING DNA SEQUENCES

The DNA molecule is composed of two strands, each of which contains an array of bases held in place by sugar–phosphate backbones (Fig. 16.2). In the simplest instance, a gene (one of many thousand in an organism) may consist of an array of approximately 1600 bases, which include, on the average, 400 of each of 4 bases, Adenine, Cytosine, Guanine, and Thymine. In more general terms, a gene can be defined as a combination of DNA segments that together comprise a unit, which has an effect on the appearance, or phenotype, of an organism by coding

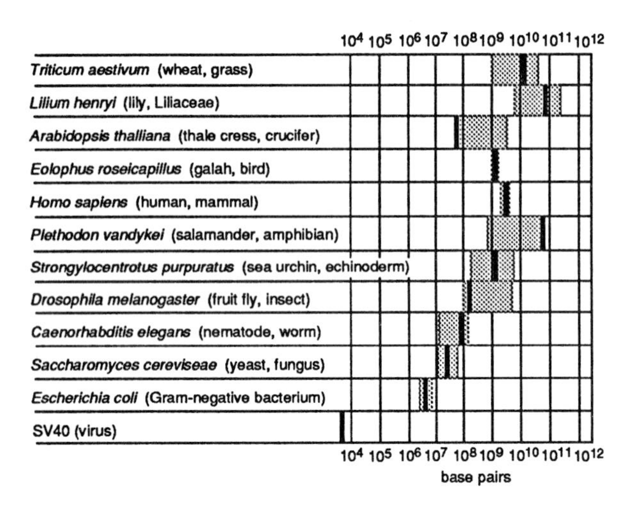

Fig. 16.1. Variation in DNA content of different life-forms. The approximate DNA contents of the species listed are indicated by black bars (Sparrow and Nauman, 1976; Bennett and Smith, 1976; Lewin, 1990). The range of DNA contents in organisms belonging to the same broad classification as the individual species is indicated by a box. For example, *Triticum aestivum* (wheat) has a DNA content of 16×10^9 base pairs (bp), whereas the broad category of organisms to which wheat belongs—the Gramineae or grasses—have a range of DNA contents from 10^9 to 6×10^{10} bp. In converting cytologically determined DNA contents (usually in pg) to numbers of base pairs, 9.13×10^8 bp are assumed to be present in 1 pg of DNA.

The nucleotide

base
sugar
phosphate

The bases

Adenine (A)
Cytosine (C)
Guanine (G)
Thymine (T)

The DNA sequence

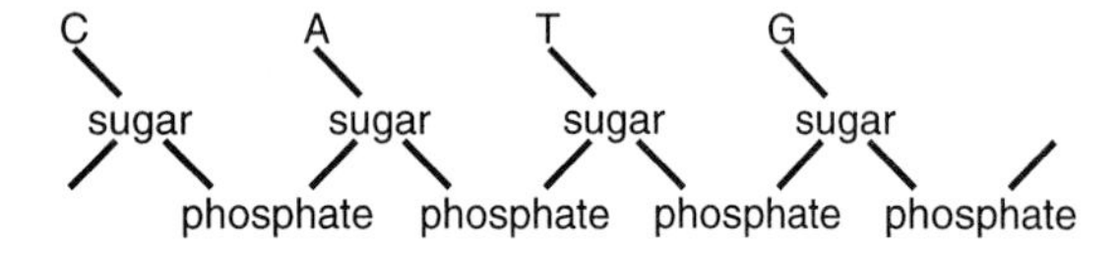

The double-stranded DNA molecule

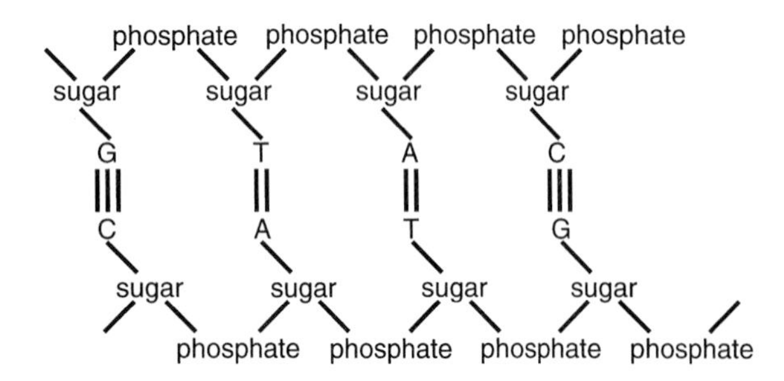

Fig. 16.2. A basic representation of the DNA molecule, emphasizing the presence of two sugar–phosphate strands held together by hydrogen-bonding between complementary base pairs.

for either a RNA or protein product. The genetic information in an organism is defined by the unique order of A, C, G, and T bases in the linear arrays that are present in the DNA molecules. The order of base pairs, or base sequence, in the DNA molecule is the critical aspect of genetic information. Considering the fact that organisms can have millions of base pairs, very large numbers of different sequences, 1000–2000 bp long, are possible.

The standard flow in the decoding of information present in double-stranded DNA includes the transcription of single-stranded messenger RNA, mRNA, from one of the DNA strands, followed by translation of a protein from the mRNA template. The triplet code defines the three consecutive bases in RNA that are required to specify a single amino acid. It can be seen from Fig 16.3 that any segment of DNA can create two different RNA molecules, one from each strand of the duplex. Each of the RNA molecules has the potential for creating three different proteins depending on the reading frame of translation, which determines the set of triplets to be decoded.

The way in which prokaryotes arrange coding and noncoding DNA sequences provides a framework for understanding the organization of eukaryotic DNA. In the following sections, the DNA sequences of a virus, SV40, and a bacterium, *Escherichia coli*, are considered, followed by a consideration of eukaryotic DNA organization.

16.3 DNA OF THE VIRUS SV40 ENCODES FOR OVERLAPPING GENETIC INFORMATION AND CONTAINS REPETITIVE SEQUENCES

The simian virus SV40 was one of the first life-forms to have its DNA completely sequenced. The DNA is in the form of a circle, 5243 bp long, and analysis of the coding DNA indicates a highly efficient use of the genetic information. The processes that decode the information in the SV40 genome exploit the overlapping nature of the genetic information (Fig. 16.4); this feature is not normally found in eukaryotes. Further modification of the decoding of the genetic information into protein is achieved by removing parts of the transcribed RNA molecule by splicing, prior to translation into protein. The large T-antigen protein is formed in this manner.

Approximately 15% of the SV40 DNA is noncoding, and a large proportion of this is located near the origin of replication. Initiation of DNA synthesis requires complex structural changes and, for SV40, the critical step is the binding of the large T-antigen protein to the origin of replication (*see* Fig. 16.4). The region of the SV40 genome containing the origin of replication is not transcribed and features a 27-bp sequence, with the bases arranged in a special form of symmetry called an inverted repeat (Fig. 16.5), which is the core of the T-antigen binding site. In addition, several repetitive sequences from 9 to 56 bp in length can be identified, and these are nonessential for viability of the virus. The SV40 virus, in which regions of repetitive DNA were deleted, survived in laboratory-mutation experiments. Repetitive and inverted-repeat sequences are common in the noncoding regions of eukaryotic DNA.

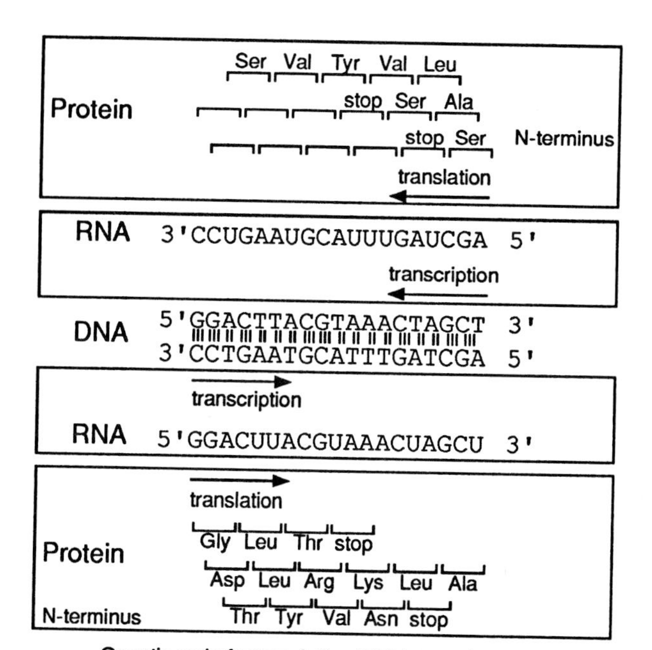

Genetic code for translating RNA into amino acids

first base (5')	second base U	C	A	G	third base (3')
U	Phe	Ser	Tyr	Cys	U
	Phe	Ser	Tyr	Cys	C
	Leu	Ser	Stop	Stop	A
	Leu	Ser	Stop	Trp	G
C	Leu	Pro	His	Arg	U
	Leu	Pro	His	Arg	C
	Leu	Pro	Gln	Arg	A
	Leu	Pro	Gln	Arg	G
A	Ileu	Thr	Asn	Ser	U
	Ileu	Thr	Asn	Ser	C
	Ileu	Thr	Lys	Arg	A
	Met[1]	Thr	Lys	Arg	G
G	Val	Ala	Asp	Gly	U
	Val	Ala	Asp	Gly	C
	Val	Ala	Glu	Gly	A
	Val[2]	Ala	Glu	Gly	G

[1] AUG is the most common initiator codon
[2] GUG can sometimes code for Methionine and initiate a protein

Fig. 16.3. The genetic code. Decoding of the information in DNA can occur from either strand and, for each strand, in three different phases (top half). Triplets of bases in the nucleic-acid sequence (codons) code for a single amino acid, whereas different codons can code for the same amino acid. This is referred to as degeneracy of the genetic code, which results from the fact that there are only 20 or so essential amino acids (bottom half) while there are 64 possible codon triplets. Organisms can differ in their preference for certain triplets coding for a particular amino acid, and some triplets can have different coding capacities depending on their position in the sequence. For example, the AUG codon can denote the start of translation and also code for methionine; the UUG and GUG codons, which normally code for leucine and valine, respectively, can also initiate translation by coding for the incorporation of *N*-formyl-methionine; and the UGA codon, which normally codes for termination of translation, also codes for tryptophan in mitochondria and mycoplasms, as well as for the incorporation of selenocysteine. (From Matthews and van Holde, 1990.) The single-letter code for amino acids is Ala (A), Arg (R), Asp (D), Asn (N), Cys (C), Glu (E), Gln (Q), Gly (G), His (H), Ile (I), Leu (L), Lys (K), Met (M), Phe (F), Pro (P), Ser (S), Thr (T), Trp (W), Tyr (Y), and Val (V).

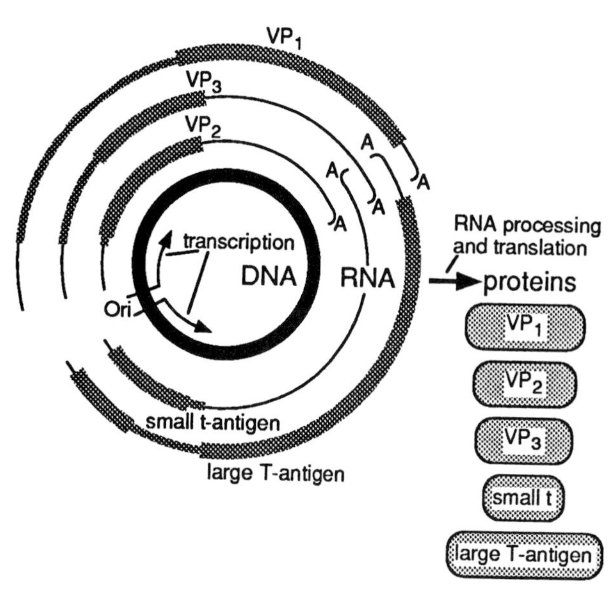

Fig. 16.4. Simian virus 40 (SV40) DNA. The central circle represents the DNA molecule. Initiation of transcription is in the region where the origin of replication, *Ori*, is located, and the directions of transcription into RNA are indicated by the arrows. The messenger RNAs that are produced are shown as lines of three differing widths. The thinnest line indicates RNA that is transcribed and terminated with a run of A residues (denoted as A), analogous to normal eukaryotic messenger RNA. The medium-thick line indicates the portion of the RNA molecule that is translated to give polypeptide molecules, which, in turn, are posttranslationally modified to give the final proteins. The regions of RNA sequences that correspond to the final proteins (VP_1, etc.) are indicated by the thickest lines. (From Fiers et al., 1978.)

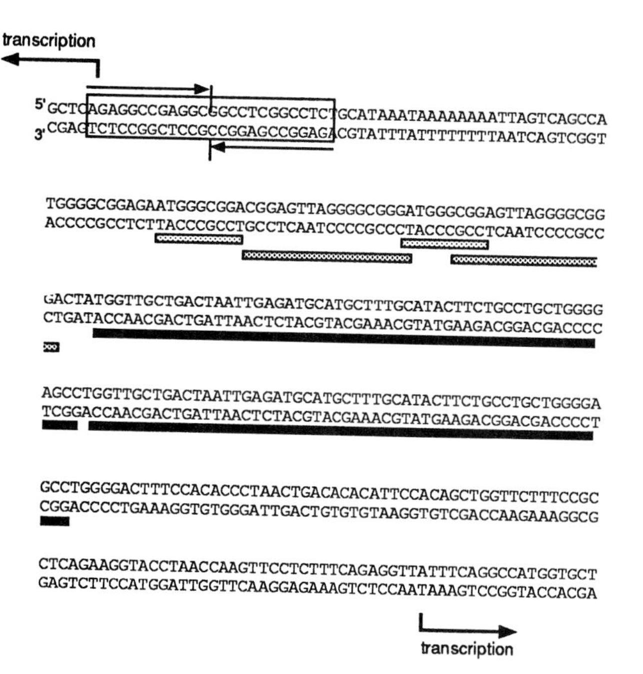

Fig. 16.5. The nucleotide sequence that defines the replicative origin of DNA in the noncoding DNA of SV40 (Fiers et al., 1978; Reddy et al., 1978). The boxes underneath the sequence indicate repetitive sequences. The box enclosing some of the sequence indicates an inverted repeat or dyad-symmetry structure, a feature of binding sites for highly specific DNA-binding proteins. The sequence is defined on both sides by regions of the genome that are transcribed.

16.4 BACTERIAL DNA IS CHARACTERIZED BY THE OCCURRENCE OF GROUPS OF GENES UNDER A COMMON CONTROL

The bacterium *Escherichia coli* has an extensive history of genetic-linkage and molecular/biochemical studies. The amount of DNA present in the central nucleoid region of an *E. coli* cell has been determined both chemically and by direct observation of the circular protosome. Careful breakage of bacterial cells releases intact protosomes, which can then be prepared for examination by the electron microscope. If the DNA is made radioactive by feeding the bacterium radioactive precursors, light microscopy can be used to examine the image of the spread, radioactive DNA exposed by means of autoradiography (Fig. 16.6). Different types of measurements indicate that the standard *E. coli* genome contains approximately 4.2×10^6 bp (See also Fig. 24.2, Chapter 24), with some variation among different strains (*see* Fig. 16.1).

16.4.1 The Operon Organization of Genes

The entire protasome of *E. coli* DNA, has been sequenced. In this bacterium, groups of genes are usually under a common control and are transcribed as a single RNA molecule. Such operons have formed a major focus for studying the control of gene expression; those that have been well studied include the lactose, tryptophan, histidine, L-arabinose, maltose, galactose, proline-utilization, histidine-utilization, and D-serine deaminase operons. Analysis of the noncoding DNA sequences, preceding the coding DNA of the operons, has defined the nucleotide sequences that are essential for regulating the expression of the associated group of genes. Studies of these regions provided the foundations for similar studies in eukaryotic DNA.

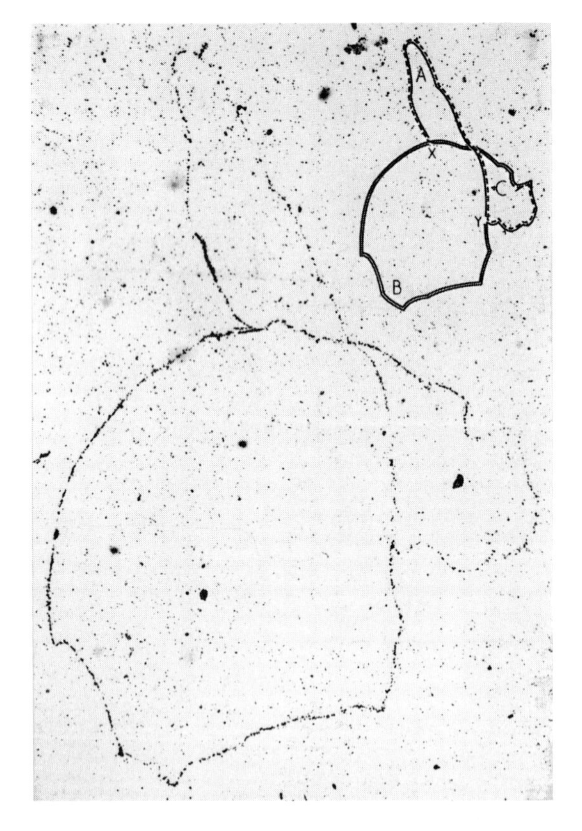

Fig. 16.6. An autoradiograph of an *E. coli* protosome undergoing replication. (From Cairns, 1963.) The DNA was extracted from bacterial cells that were initially labeled with ^{3}H-thymidine, and then gently broken open and allowed to spread on a glass slide. The distribution of radioactivity was observed by covering the glass slide with a photographic emulsion. DNA replication is occurring at X and Y to give two new protosomes (A and C), each of which contains one strand of the original protosomal DNA (B).

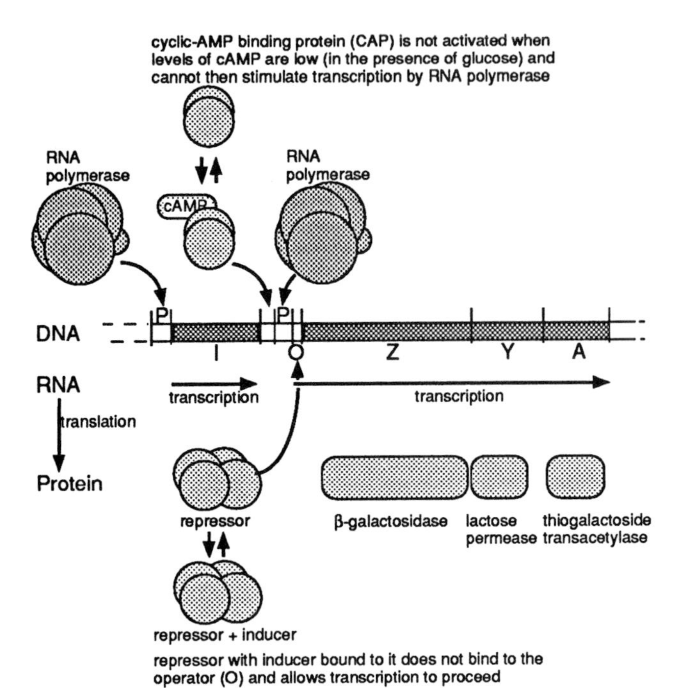

Fig. 16.7. The *lac* operon of *E. coli* (Beckwith, 1987) and its expression. Positive control is based on the stimulation of transcription by the CAP protein, and negative control is based on repressing or preventing transcription. The *lacI* gene codes for a monomer that forms a tetrameric protein, which binds to the operator region, O, 5′-upstream from the Z gene in the *lac* operon, thereby repressing transcription. X-ray crystallography has shown that the tetrameric repressor consists of two dyad-symmetric dimers with the DNA-binding domain forming a deep V-shaped cleft in between (Friedman et al., 1995). The presence of an inducer such as isopropylthiogalactoside (IPTG) or β-galactosyl glycerol (Egel, 1988) prevents the binding of the repressor protein to the operator by binding to the protein and causing a conformation change, which blocks the repressor–operator recognition site.

The lactose operon or *lac* operon, for example, is composed of three nonoverlapping genes, *Z, Y,* and *A*, preceded by a short, noncoding region of DNA that acts as a receptor site for a repressor protein (Fig. 16.7). The idea of a single repressor controlling a group of three closely linked genes was developed from studies on different classes of mutations that affected the induction of expression of the *lac*-operon genes. The repressor protein binds to a site on the DNA, known as the operator, which is located between the promoter, the site where the RNA polymerase binds to start transcription, and the start of the gene to be transcribed. The binding of the repressor protein physically stops the progress of RNA polymerase toward the gene.

The positive control that determines *lac*-operon activity was revealed by genetically analyzing suppression of *lac*-operon in-

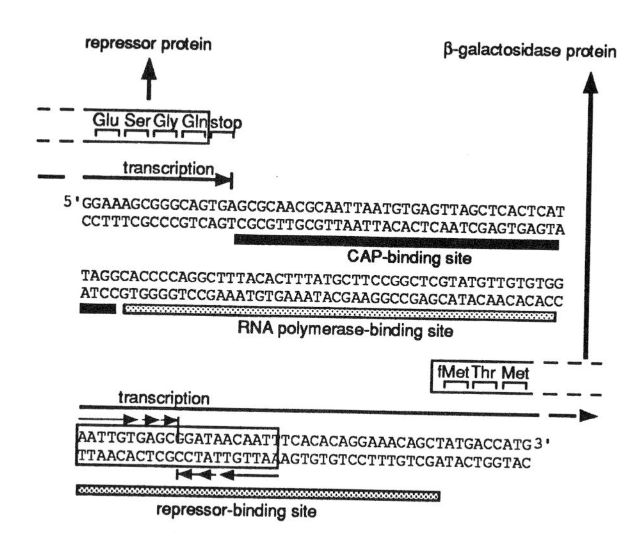

Fig. 16.8. The noncoding DNA region of the *lac* operon. The location of the binding sites for the repressor, CAP, and RNA polymerase proteins are indicated. As in Fig. 16.5, the box encloses the inverted repeat or dyad-symmetry region, which is characteristic of DNA-binding sites for proteins with a high degree of specificity for binding to DNA. This inverted symmetry is evident only in the repressor-binding site. Transcription, when it occurs, is indicated by the arrow. The translation-stop codon UGA in the repressor messenger RNA and the translation-start codon AUG for the β-galactosidase protein of the *lac* operon are shown.

ducibility by glucose, an effect known as catabolite repression. In this case, a site near the promoter was shown to bind a catabolite-activated protein, CAP, which greatly facilitates the binding of RNA polymerase to its promoter, in preparation for transcription. The sequences of noncoding DNA preceding the *lac* operon that are important in the positive and negative controls of transcription are shown in Fig. 16.8. The sequence that binds the repressor protein demonstrates the same kind of special symmetry that was noted in the DNA sequence at the SV40 origin of replication, namely an inverted-repeat structure, or dyad symmetry (*see* box in Fig. 16.8).

It is evident that clear principles underlying the positive and negative control of gene expression have been established from the study of bacterial operons. Many of these principles apply to the control of eukaryotic genes, as discussed later.

16.4.2 Transcription Attenuation

When RNA polymerase binds to a promoter, a number of conformational changes occur in the complex, as assayed by its contact with the DNA template. Before RNA polymerase moves into a productive transcription mode, it stalls at the promoter and produces short RNA molecules no longer than nine bases long. When it goes into a productive transcription mode, its progress along the DNA molecules is believed to be uneven and, in some cases, can be related to the translation of the messenger RNA that is being synthesized. The interaction between transcription and translation has been named "transcription attenuation." Attenuation is the premature termination of transcription speculated to be due to the presence of a loop structure in newly synthesized messenger RNA (Fig. 16.9). This loop is

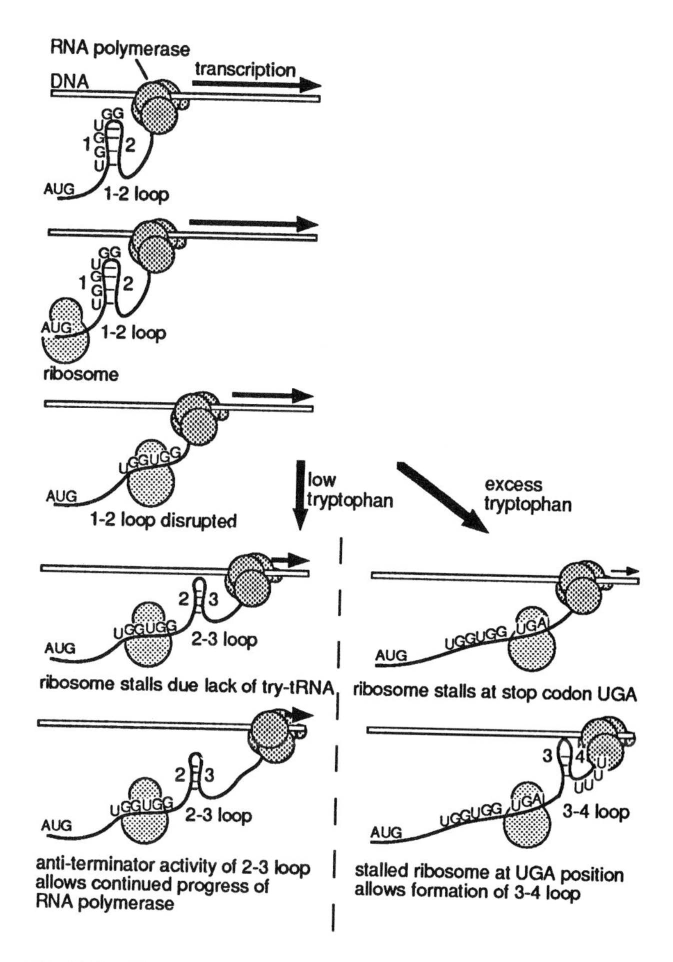

Fig. 16.9. The structure and mechanism of action of the noncoding/coding DNA junction region of the *tryp* operon of *E. coli* and its effect on transcription attenuation (Yanofsky and Crawford, 1987; Landrick and Yanofsky, 1987). Prior to transcribing the tryptophan operon to give the mRNA that manufactures the enzymes required for tryptophan synthesis, RNA polymerase tends to pause during transcription of the leader region that immediately precedes the first gene of the *tryp* operon. This pause is postulated to lead to the formation of a 1–2 loop (top), which includes two repeated tryptophan-tRNA codons (UGG). At the same time, a ribosome attaches to the AUG initiator codon. If tryptophan is in limited supply (left), the ribosome begins to move along the messenger RNA, disrupting the 1–2 loop but stalling at the UGG codons due to a limiting supply of tryptophan-tRNA. Continuing transcription by the RNA polymerase then synthesizes another region of messenger RNA that forms a 2–3 loop, which acts as an antiterminator and allows further transcription by the RNA polymerase. Under these conditions, messenger RNA continues to be formed, from which the enzymes for tryptophan biosynthesis are synthesized and tryptophan is produced. Conversely, if tryptophan is in excess with a subsequent overabundance of tryptophan-tRNA (right), the ribosome does not stall at the UGG codons, halting instead at the UGA stop codon further along the leader region. Under these conditions, a 3–4 loop forms in close juxtaposition to the UUUU tract, and together they form a signal for transcription termination.

similar to the normal loop structure that precedes a set of four or more U bases and is recognized by a protein factor named *rho* to terminate transcription. The attenuation mode of control arose from a study of the effects of amino-acid analogs on the synthesis of the respective amino-acid biosynthetic enzymes. The control is thought to be a dynamic process involving both transcription and translation of a region between the promoter site for RNA polymerase and the first gene of the operon, and the relative timing of these processes is critical (*see* Fig. 16.9).

16.5 REPETITIVE DNA SEQUENCES COMPRISE UP TO 2% OF THE *E. COLI* PROTOSOME

In experiments to determine the renaturation kinetics of denatured DNA, the DNA from *E. coli* has long provided a standard for characterizing DNA from other organisms. The two strands of a DNA molecule can be separated, or denatured, and specifically reassociated, or renatured, to reconstitute the original base-pairing. The renaturation of the single strands of *E. coli* DNA into a double-stranded DNA molecule follows normal physical laws, in a renaturation curve largely typical of a bimolecular reaction (Fig. 16.10). This observation is expected from the presence of single copies of particular base sequences approximately 1000–2000 bp in length. The renaturation curve depends on the temperature and ionic conditions of the solution in which the reaction is occurring, as well as the size of the DNA fragments participating in the reaction. The renaturation reaction can be described by

$$\frac{C}{C_0} = \frac{1}{1 + kC_0t}$$

where the concentration of double-stranded DNA (C), measured in moles/liter, present in the reaction after t seconds is dependent on the initial concentration of DNA (C_0) and the number of different

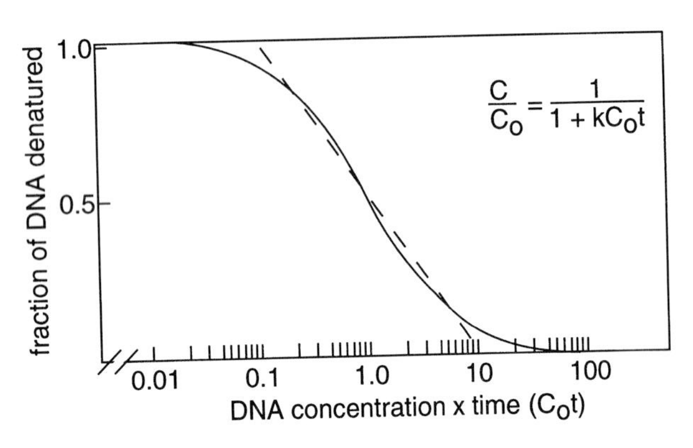

Fig. 16.10. A renaturation curve of denatured DNA from *E. coli*. The proportion of DNA that renatures over time can be determined by measuring (a) the absorbance of the DNA in solution, (b) the differential binding of double-stranded DNA to hydroxyapatite, or (c) its differential sensitivity to enzymes that degrade single-stranded DNA. The dashed line indicates the symmetry of the renaturation curve and is used to define the halfway point for the renaturation process as calculated from the formula, where C_0 is the starting concentration of DNA, C is the concentration of DNA denatured at time t where t is the renaturation time (s), and k = a reaction constant. An approximate, but convenient, measure of the C_0t value of a particular renaturing reaction is C_0t = ($OD_{260}/2$) time (in h) at the optimum temperature of $T_m - 25°C$ in 0.12 × SSC. In this calculation, OD_{260} is the optical density of the DNA solution determined at a wavelength of 260 nm, T_m is the melting point of the double-stranded DNA (i.e., the temperature at which half the DNA strands are dissociated), and SSC is the standard saline citrate. (From Britten et al., 1974.)

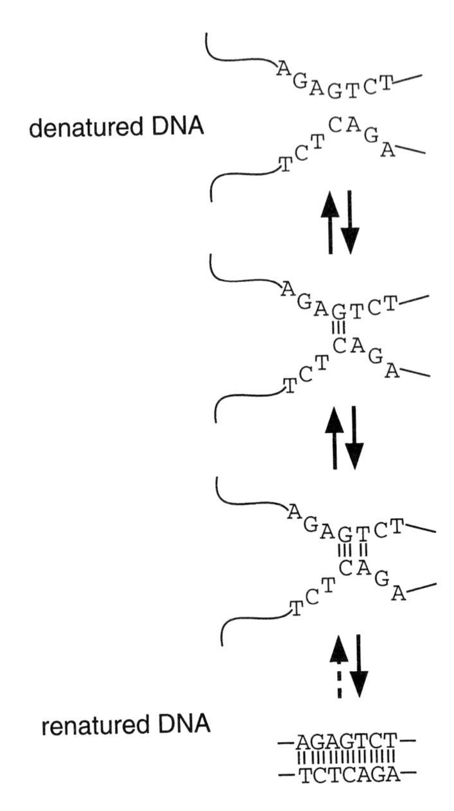

Fig. 16.11. The renaturation of DNA strands. The diagram shows strands of DNA in the various stages of renaturation and emphasizes the reversibility of the denaturation/renaturation reaction. (From Wetmur and Davidson, 1968.) This model is particularly important in understanding the behavior of short primers in the polymerase chain reaction (PCR; *see* Chapter 19), because the DNA polymerases used in PCR operate at temperatures where equilibrium differences between partially formed hybrid partner strands are likely to be significant.

sequences present in the DNA (complexity). The value of the rate constant k is proportional to the complexity, or number of base pairs, of the DNA sample. The relative values of k for different bacterial DNA samples can be readily determined by measuring the renaturation kinetics. The size of the *E. coli* genome, determined by independent means, can then be used to convert the k values into genome sizes. The measurement of initial rates of renaturation is important in using k values to determine genome size, because the rate-limiting step in renaturation is the initial contact between the two renaturing strands (Fig. 16.11).

Curves of the type shown in Fig. 16.10 can also be obtained if single- and double-stranded DNA fractions are separated at various times in the renaturation reaction, using columns of crystalline calcium phosphate, hydroxyapatite, which adsorbs DNA, and solutions of sodium phosphate can be used to elute the DNA. A higher concentration of sodium phosphate is required to remove double-stranded DNA from hydroxyapatite than if the DNA is single stranded, and this is particularly useful in preparing large amounts of DNA renatured to a given $C_0\,t$ value. Experiments of this type have shown that 0.5–2% of *E. coli* DNA consists of sequences that do not follow the renaturation kinetics shown in Fig. 16.10, and these sequences are repetitive in nature.

Renaturation and molecular–genetic experiments have demonstrated that the *E. coli* genome contains both coding and noncoding repetitive sequences. Coding sequences that contribute to the pres-

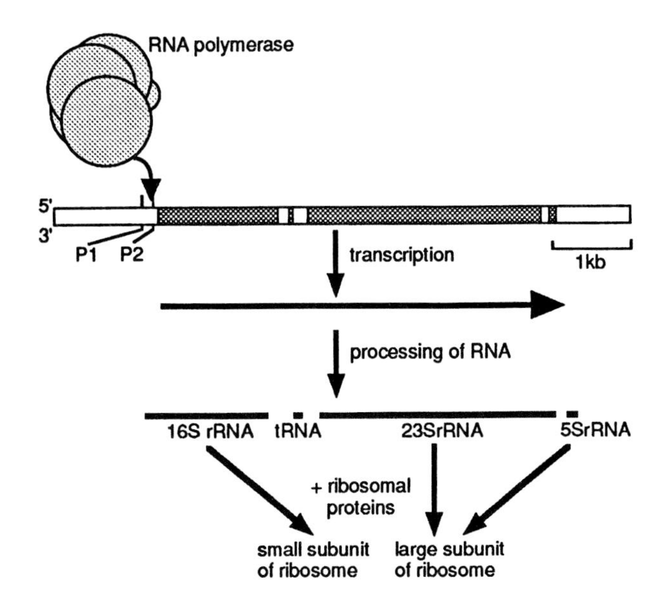

Fig. 16.12. The ribosomal RNA (*rrn*) operon of *E. coli*. Processing of the polycistronic rRNA by a specific RNA-degrading enzyme that is capable of recognizing specific secondary structures yields the 16S (small), 23S (large), and 5S RNA molecules, which then combine to form ribosomes. (From Reiter et al., 1990.) Some tRNA molecules are also released. The operon is preceded by a TATA sequence in the noncoding region, 30 bp before the start of transcription, and is essential for efficient transcription; this TATA "box" or its equivalent is a highly conserved feature of transcription-initiation complexes and is also required by eukaryotic genes for efficient transcription [*see* histone genes (Fig. 16.23), and globin genes (Fig. 16.25).

ence of repeated sequences include ribosomal RNA gene sequences (Fig. 16.12), coding for the RNA components required for the assembly of ribosomes, and genes coding for proteins involved in biosynthetic activity, or for responses to environmental changes. The ribosomal RNA gene operon is repeated seven times and is dispersed throughout the *E. coli* protosome. Environmental selective pressure can lead to mutations in the DNA resulting from amplification of a small section of the protosome carrying the gene of interest. Resistance to the antibiotic ampicillin, for example, is directly proportional to the number of copies of the gene specifying β-lactamase, a secreted enzyme that degrades ampicillin. Mutants of *E. coli* with up to 40 copies of the gene as integral parts of the protosome have been isolated.

Repetitive–extragenic–palindromic (REP) sequences, 36 bp long (Fig. 16.13), constitute another family of repetitive sequences that is widely distributed in the protosome. Although the REP sequences are found in most of the bacterial operons examined, their function has not been established. They occur as 1–4 copies as either direct or inverted repeats.

Some classes of short repetitive sequences function as target sites for recombination. Short, inverted-repeat sequences flanking a gene, for example, are required for site-specific recombination, which is associated with the inversion of the gene and subsequent change in its expression. For example, a site-specific inversion or phase variation, with 26-bp target sequences, is responsible for the expression of alternative flagellin or pili proteins in *Salmonella aureus* and *E. coli*. This type of rearrangement is one example of genetic reassortment that occurs in higher organisms either by site-specific recombination, unequal crossing over, or specialized enzymatic reactions.

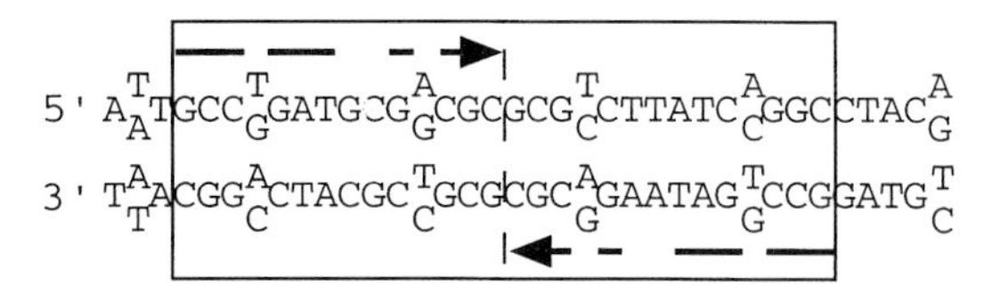

Fig. 16.13. The consensus–nucleotide sequence of a family of bacterial repetitive–extragenic–palindromic (REP) units. REP sequences display an inverted-repeat structure that allows a stem loop to form in a single strand of, for example, an RNA molecule transcribed from such a sequence. The sequence shown is a consensus sequence for the family of REP units that occur in bacteria. Where bases occur in approximately equal frequency, the two alternatives are shown (Gilson et al., 1984; Stern et al., 1984).

Transposable elements are also associated with repetitive sequences, but these particular families of DNA sequences are discussed in detail in Chapter 7.

16.6 SMALL, CIRCULAR DNA MOLECULES (PLASMIDS) IN E. COLI CODE FOR 1–3 GENES

In natural populations of bacteria, plasmids are small, extraprotosomal circles of DNA that contain genes coding for enzymes or proteins, and conferring resistance to antibiotic or toxic metal ions. Some of the plasmid molecules of *E. coli* have been completely sequenced and manipulated, by molecular biologists during the course of constructing vectors suitable for DNA cloning. Because plasmids form the basis for a range of techniques used in molecular cytogenetics, their properties are discussed here from the standpoint of their use as cloning vectors. The commonly used plasmid pUC19 is 2686 bp in length (Fig. 16.14).

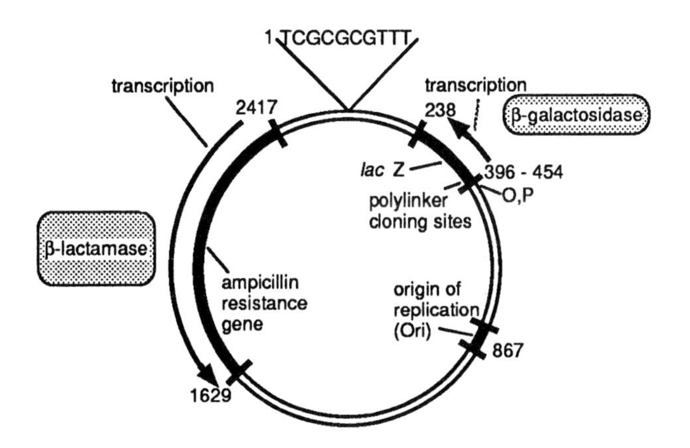

Fig. 16.14. An overview of the plasmid pUC19. The diagram identifies the two genes (β-lactamase and β-galactosidase) encoded by the plasmid, and the position of the polylinker region that was introduced to clone foreign DNA sequences. The numbers indicate nucleotide positions within the molecule, and O and P signify the operator and promoter regions, respectively, of the *lac* operon (as described in Fig. 16.7). The directions of transcription are indicated by arrows, and the respective translated protein products are shown in boxes. (From Yanisch-Perron et al., 1985.)

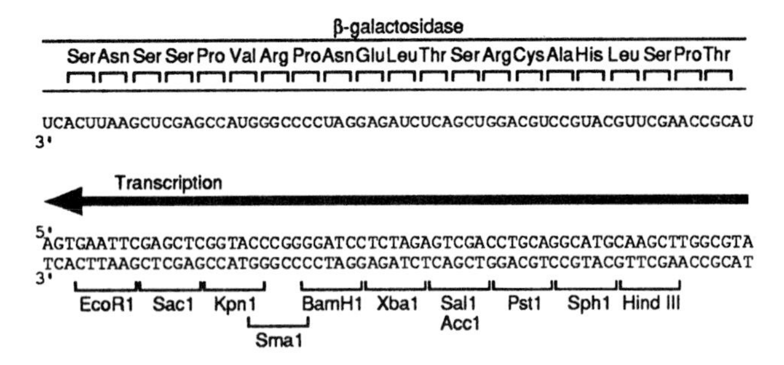

Fig. 16.15. The nucleotide sequence of the polylinker inserted into pUC19. The positions of different restriction-enzyme cleavage sites are indicated (bottom), as well as the transcription product from this region (top). The successful insertion of foreign DNA into the polylinker can be distinguished by chemical color reactions (*see* Fig. 19.18), which verify that the production of the enzyme β-galactosidase has been disrupted owing to the insertion of foreign DNA into the gene sequence. (*See* Fig. 16.16 legend for derivations of the names of some restriction enzymes.)

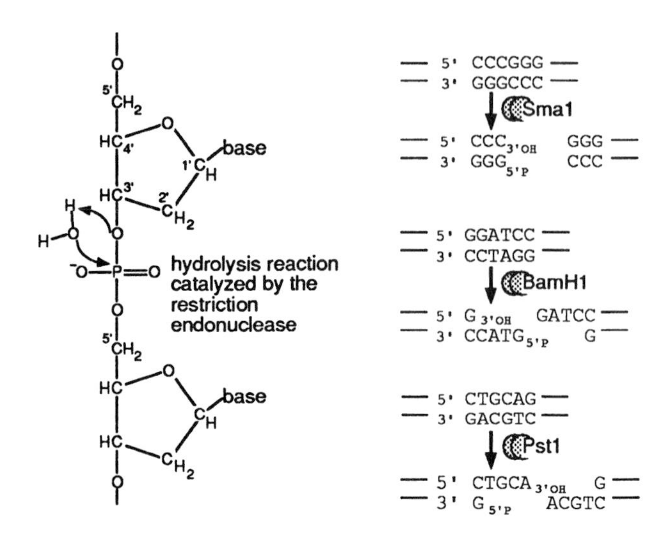

Fig. 16.16. The three different ways in which common restriction endonucleases cleave DNA. For simplicity, the products of the reactions are shown with the liberated 3′-OH and 5′-phosphate groups on only the left end of the divided molecule; where one of the DNA strands protrudes with a 5′-P, it is called a "5′-overhang," and in the case of a protruding 3′-OH DNA strand, it is called a "3′-overhang." The hydrolysis reaction leading to the cleavage of the sugar–phosphate backbone of a DNA strand is shown on the left-hand side. The names of the enzymes (type II in this figure) derive from the organism from which they were isolated: Sma1 from the bacterium *Serratia marcescens;* BamH1 from *Bacillus amyloliquefaciens;* and Pst1 from the bacterium *Providencia stuarti.*

An important region of the plasmids inserted by molecular biologists is the cloning or polylinker region (Fig. 16.15). The DNA sequences in the polylinker occur only once in the entire plasmid and are the recognition sites for the enzymes, restriction endonucleases, that cleave the DNA at specific recognition sites. Thousands of bacteria and Archaea, as well as some viruses that infect fungi, have been examined for enzymes that specifically cleave DNA and a wide range of DNA sequences, 4–8 bp in length, can be specifically cleaved by particular enzymes. These enzymes are generally classified as restriction endonucleases and can be divided into three types: I, II, and III. The types I and III restriction endonucleases have both endonuclease and methylase activities. Although they bind to specific nucleotide sequences, they cleave DNA at nonspecific sites a certain number of nucleotide base pairs distant from the recognition sequence. Type II restriction endonucleases are composed of dimeric proteins containing identical subunits. These usually bind to DNA sequences with characteristic inverted-repeat or dyad symmetries (right-hand side), although some such as MboII, recognize sequences with no dyad symmetry, 5′-GAAGA. Other type II restriction endonucleases also cleave DNA at specific sites away from the recognition sequence. Restriction enzymes cleave the sugar–phosphate backbone of both strands of the DNA helix.

As shown in Fig. 16.16, type II restriction endonucleases hydrolyze the sugar–phosphate backbone of their respective recognition sites in three different ways with respect to the relative orientations of the released 3′-hydroxyl and 5′-phosphates of each strand of the DNA helix. The ends of the DNA molecules can be "blunt-ended," or can have 2–4 bases as a single-stranded section of DNA carrying either a terminal 5′-phosphate (5′-overhang), or a terminal 3′-hydroxyl (3′-overhang). For example, a plasmid cleaved with the restriction endonuclease BamH1, produces the 5′-overhang sequence 5′-GATC at one end of the molecule and CTAG-5′ at the other end of the molecule. These ends are referred to as "sticky ends" in that they are complementary to each other and will reform a double-stranded molecule, following the same principles as described earlier for the renaturation of denatured DNA. The sticky ends produced by cleavage of any prokaryotic or eukaryotic DNA molecule with a specific restriction endonuclease are identical, and thus form the basis for joining different molecules together. When the sticky ends are held together by complementary base-pairing, the enzyme DNA ligase reforms the respective sugar–phosphate backbones of the strands of the DNA helix. The entire plasmid molecule, with a segment of foreign DNA now inserted into it, is then reintroduced into *E. coli* by transformation, and the bacterial cells carrying the plasmid are clonally propagated. Selection for transformed cells carrying a plasmid is conveniently carried out by growing them in a medium containing the antibiotic ampicillin. The *β*-lactamase gene on the plasmid (*see* Fig. 16.14) confers resistance to ampicillin by allowing the *E. coli* cells with the plasmid to produce the enzyme that degrades the antibiotic. Because a single bacterium generally receives only one plasmid molecule in the initial transformation, any one colony of *E. coli* carries only one segment of foreign, cloned DNA maintained in the respective plasmid.

16.7 REPETITIVE SEQUENCES ARE A MAJOR FEATURE IN EUKARYOTIC NUCLEAR DNA

Repetition of DNA sequences in the eukaryotic genome is the rule rather than the exception. Sequences potentially coding for protein products, but not necessarily active in this respect, are often repeated. In other cases, well-understood genes such as those coding for ribosomal RNA, histone, globin, seed-storage protein, and the small subunit of ribulose bisphosphate decarboxylase are repeated many times. In addition, regions within genes that actively code for protein products can also display repetition of sequence motifs 15–30 bp long. In proteins such as elastin and collagen in animals, seed-storage proteins in plants, and an avirulence protein in bacteria, there exist repetitive sections in the final polypeptide chain. The function of many other repetitive sequences is not clear, but as molecular studies on the genomes of a wide range of organisms are carried out, the database of repetitive-DNA-sequence families that have known functions continues to grow.

In early experiments monitoring the renaturation of denatured DNA, it was predicted that denatured calf-thymus DNA would renature at a slower rate than that of *E. coli* DNA (*see* Fig. 16.10) because of the increased DNA content of the eukaryotic nucleus. The greater content of DNA was expected to have a dilution effect on any single sequence in the genome because of the presence of an increased variety of sequences, and this would reduce the chance of correct partner strands finding each other. Instead, the renaturation turned out to be much more complex (Fig. 16.17). A significant proportion of the DNA renatured very rapidly and was deduced to be composed of repetitive DNA sequences that were present up to thousands of times within the genome. The presence of repetitive DNA raised the concentration of a particular subset of sequences, so that the overall renaturation kinetics failed to follow the pattern for a simple bimolecular reaction. The complex curve of eukaryotic DNA renaturation can be described by a series of simple curves superimposed upon each other, as shown in Fig. 16.17. Each of the individual curves represents a particular family of DNA sequences, present at their own particular concentration of C_0 and renaturing at a rate defined by C_0t.

Renaturation curves of this type are affected by the length of the DNA molecules used in the experiments. This effect is particularly evident when a solid matrix such as hydroxyapatite is used to separate single-stranded DNA from double-stranded DNA. The DNA in a chromosome is a single, large molecule (Fig. 16.18), and it is clear that if molecules of this length could be analyzed in a renaturation-rate experiment, any one molecule would acquire a double-stranded region at a rate determined by its most highly repeated sequence family. Such molecules could be isolated on hydroxyapatite columns, and the sequences neighboring the repetitive-sequence family and present as single-stranded "tails" could be made available for study. In practical terms, it is extremely difficult to handle chromosome-size DNA molecules, and the usual isolation procedures yield double-stranded DNA approximately 50 kb long. Due to occasional breaks in either DNA strand, this corresponds to a single-strand length of 5–10 kb when the DNA is denatured. However, even with these shorter DNA molecules, there exists a length dependency for the isolation of certain sequences as single-stranded tails linked to repetitive-DNA-sequence families.

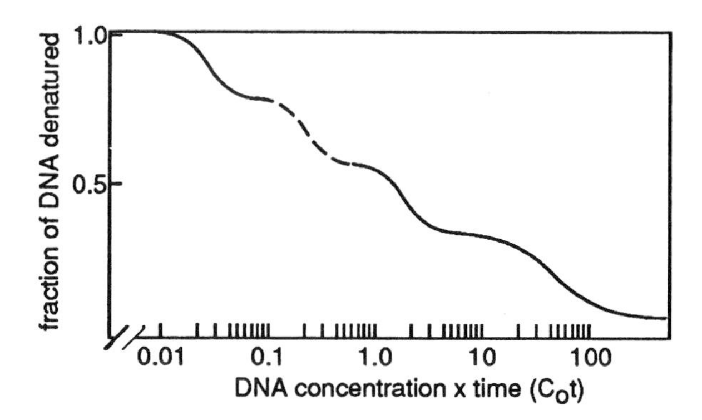

Fig. 16.17. A simplified diagrammatic version of a complex renaturation curve obtained when denatured eukaryotic DNA is renatured. Each individual down-slope one of which is highlighted as a dotted line is the result of a second-order reaction occurring for a given family of sequences within the complex renaturation reaction of a mixture of different sequence families. (From Britten and Kohne, 1968.)

When DNA in the double-stranded, 50-kb class is studied in denaturation/renaturation experiments, many repetitive-DNA-sequence families are found to behave independently of each other. The implication of these results is that some families of repetitive DNA exist in long, tandem arrays. Several of these sequence categories correspond to classical cytogenetic landmarks and are described in Chapter 20. Some repetitive-sequence classes, however, can only be obtained in a purified form when the DNA is broken down to smaller lengths of approximately 500 bp, because in longer DNA segments, they remain linked to a wide range of other sequences. Repetitive sequences in this category are called dispersed repetitive sequences.

16.7.1 Repetitive-Sequence Families Are Characterized by Episodes of Amplification and Deletion During Evolution

During the course of evolution of modern-day species, deletion–amplification events within tandem arrays of repetitive sequences can lead to a small subgroup of sequence variants dominating the array in a given species. The actual mechanisms underlying this process include unequal crossing over between tandem arrays, and extra-chromosomal replication (*see* Chapter 20), as well as slippage during replication (*see* Chapter 18). The net result of these processes is that tandem arrays of sequences can show "species-specificity" because certain, relatively rare sequence variants within an array in one species can become major components of similar arrays in a closely related species. The family of DNA sequences that constitute the various tandem arrays are usually ancient in evolutionary terms, although the modulation in numbers of particular sequence variants can occur in a relatively short period of time.

Slippage during replication can also create short, tandem arrays of DNA sequences that are 2–5 base pairs in length and dispersed throughout the genome. These tandem arrays are usually referred to as microsatellites (repetitive unit 2–5 bp in length) and can vary in length from 10 units upward; minisatellites (repetitive unit 30–110 bp in length) often contain only 15–20 repetitive units. These tandem arrays of sequences provide excellent genetic markers for distinguishing individuals within a species, as discussed in Chapter 20.

16.7.2 Dispersed Repetitive Sequences Fall into Two Broad Classes

Repeated DNA sequences that are dispersed throughout the genome are generally considered to have been transposable elements at some time in their evolutionary history and can represent major portions of chromosomal DNA. Long, interspersed elements occur with either long, terminal, repetitive-DNA sequences (LTRs, Fig. 16.19) or without LTRs. Short, interspersed elements (SINEs) often

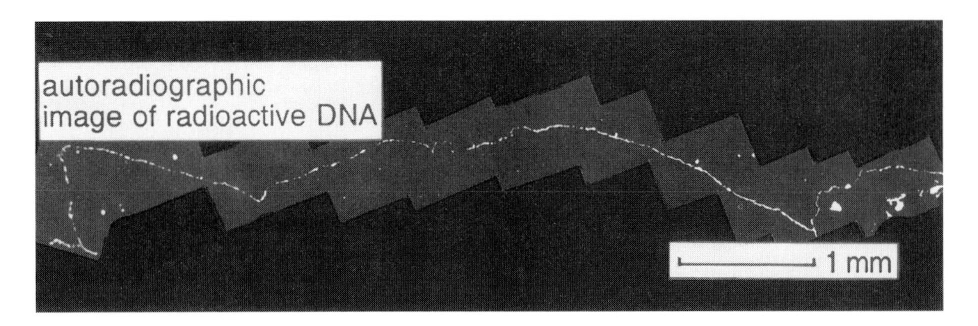

Fig. 16.18. An autoradiographic image of a chromosomal DNA molecule isolated from tissue-culture cells of *Drosophila melanogaster* that were labeled with ^{3}H-thymidine and lysed gently to release the DNA onto a glass slide. The autoradiographic image was then photographed using dark-field light microscopy. (From Kavenoff and Zimm, 1973.)

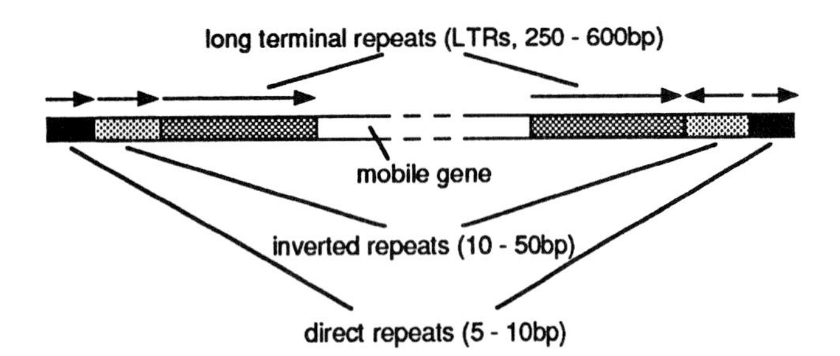

Fig. 16.19. A diagrammatic representation of the structure of a long, interspersed, repeat-DNA sequence. The long terminal repeats or LTRs are of the direct-repeat type.

have characteristic poly-A/poly-T tracts as part of their structure (Fig. 16.20).

Particularly well-characterized examples of these interspersed repetitive elements are the family of *Ty* elements of yeast and the *Alu* sequences in humans. The structures of both families of sequences indicate that they are capable of movement to other parts of the genome via an RNA intermediate, so they act as retrotransposons. There are usually 30–40 copies of the *Ty* element in the yeast genome and 300,000 or more copies of the *Alu* sequence in the human genome. These figures are representative of the copy number of long, interspersed repetitive elements and SINEs, respectively, that are present in most eukaryotic organisms.

The retrotransposon nature of the *Ty* elements in yeast has been demonstrated using a plasmid carrying an inserted *Ty* element in combination with a galactose-enhanced transcriptional activator. In the presence of galactose, transcription of the *Ty* element increases and, concomitantly, the frequency of movement of the *Ty* element also increases (Fig. 16.21). When the DNA structure of the initial *Ty* element was altered by including an intron, a special section of DNA that is transcribed as part of a gene and later removed during subsequent processing of the RNA, a very striking result was obtained. The induction of *Ty* element transcription by galactose caused movement of the *Ty* element to other regions of the genome, but the newly integrated elements *lacked* the intron (*see* Fig. 16.21). This is exactly the result expected if retrotransposition of the *Ty* element occurred via a RNA intermediate, as this step provides the opportunity for the intron to be eliminated from

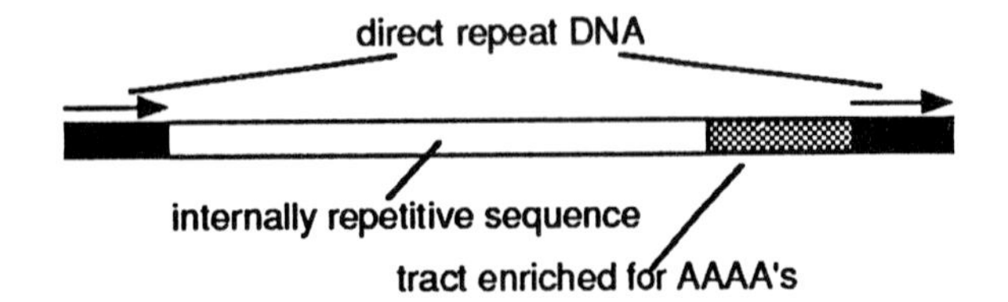

Fig. 16.20. A summary of the structure of a short, interspersed, repeat-DNA sequence. The example shown is based on the human *Alu* sequence.

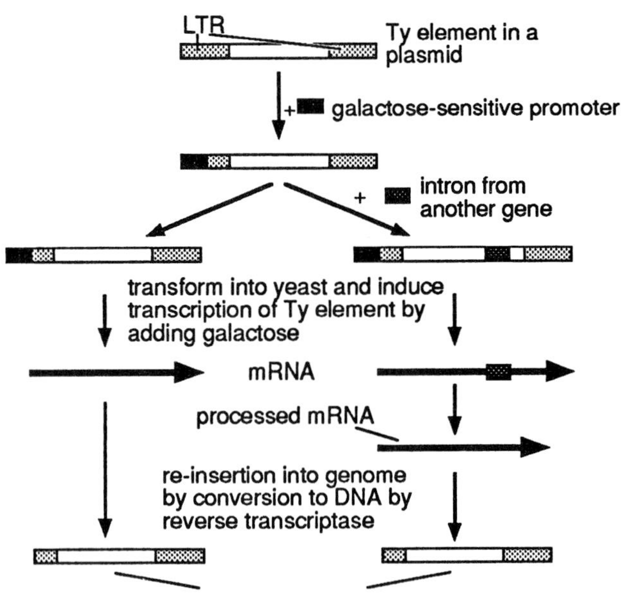

Fig. 16.21. The mechanism of movement of a *Ty* element via a RNA intermediate. Proof of this came from an experiment using an element that had been previously modified by the insertion of an intron. Introns are usually removed, or spliced, from mRNA by standard enzymatic processes. The observation that *Ty* elements that had moved to different parts of the genome had lost their introns was consistent with the scheme shown for *Ty* element movement (Boeke et al., 1985). The process of reverse transcription, first found to occur in the RNA retroviruses associated with certain cancers, makes a DNA copy from an RNA template.

Table 16.1. Fully Sequenced Retrotransposable Elements in Plant Genomes

Element	Size (kb)	No. of Copies	Host Species
Ta1	5.2	1–3	*Arabidopsis thaliana*
Tnt1	5.3	>100	*Nicotiana tabacum*
Tst1	5.1	1	*Solanum tuberosum*
del1	9.3	>13,000	*Lilium henryi*
IFG	5.9	~10,000	*Pinus lambertiana*
Cin1	0.69	1000	*Zea mays*
Bs1	3.2	1–5	*Zea mays*
Cin4	>6.8	25–50	*Zea mays*
del2	4.5	240,000	*Lilium speciosum*

Source: Data from Smyth (1991).

the element by standard excision processes in the cell. Interestingly, the movement of the *Ty* element causes the accumulation of mutations within the element, suggesting that reverse transcriptase, the enzyme responsible for synthesizing DNA from an RNA template (*see* Fig 16.21), is prone to introducing errors.

The main structural evidence suggesting that the human SINEs called *Alu* sequences are retroposonlike elements is the presence of a region of A residues at the 3′-end of the sequence (*see* Fig. 16.20). Such regions are characteristic of many eukaryotic messenger RNAs and are added after transcription. Their presence in *Alu* sequences is consistent with the involvement of an RNA intermediate in processes leading to the dispersion of these sequences throughout the genome.

Several dispersed sequences in plants have been described (Table 16.1), and their sequence structure fits into the general pattern of sequence organization described for the yeast and human elements. Not all movement of genetic material in eukaryotes occurs via RNA. When the movement of DNA sequences occurs, the mobile DNA elements are called transposable elements, which were first discovered in association with chromosome breakage events in maize, as discussed in more detail in Chapter 7.

16.8 REPETITIVE SEQUENCES IN AND AROUND EUKARYOTIC GENES

Although eukaryotic gene expression does not usually follow the operon model established for prokaryotes, a striking exception is the multigene RNA produced as the first step in the formation of ribosomal RNA. Because the region containing the ribosomal RNA genes is particularly significant in cytogenetic studies, it is treated separately in Chapter 20.

Many of the principles underlying the control of prokaryotic gene expression do apply to eukaryotes. The noncoding DNA near the eukaryotic gene contains numerous regions involved in positive and negative controls of gene expression. The genes coding for histone and hemoglobin proteins are discussed here as examples.

16.8.1 The Histone Genes

The histone proteins are crucial for the assembly of chromosomes and the packaging of DNA into more compact forms. The genes coding for histones are usually arranged in multigene families, with any single repetitive unit of the family often containing the genes for all five histones. Although the detailed arrangement of histone–gene families varies enormously between different organisms, the synthesis of histones is generally coordinated with the synthesis of DNA. The structure of a DNA replication-dependent histone–gene locus from yeast is shown in Fig. 16.22. The locus contains two genes, H2A and H2B, which are transcribed from opposite strands of the DNA double helix and therefore in opposite directions. The noncoding regions before the start of each gene, in the region between the two genes, control the increased transcription of the genes in the S phase of the cell cycle. The enzyme that transcribes the genes is RNA polymerase II, and TATA is a key nucleotide sequence in the promoter region that is recognized by this enzyme. The TATA sequence is part of the promoter region usually located approximately 30 bp before the 5'pr start of the gene; it is referred to as being 5′-upstream from the gene. Approximately 400 bp 5′-upstream from each of the histone genes in this locus is a 16-bp sequence, which is repeated three times (*see* Fig. 16.22). These sequences are involved in coordinating the transcription of the genes, starting late in G1 of the cell cycle and reaching a peak early in the S phase. Deletion of the sequences results in the loss of the cell-cycle-activated transcription, thus implying that the repetitive sequences are part of a positive control mechanism. In this same region, there also exists a 19-bp sequence with an inverted-repeat, or dyad-symmetry, structure. When this sequence is deleted, continuous transcription occurs, independent of the cell-cycle stage, and thus implies the involvement of a negative regulation of histone transcription. Not all organisms have the 16- and 19-bp controlling regions near the histone genes, but they do have other repetitive sequences whose function remains to be determined.

Studies on the control of transcription of the sea-urchin H4 histone has demonstrated that in addition to the core-promoter sequence TATA, several sequence elements in the region 60–140 bp, 5′-upstream from the start of the gene, exert a positive control function on transcription (Fig. 16.23). This region also includes a sequence with a negative control function. It is thus clear that the final transcription of histone genes by RNA polymerase II is subject to controls that are analogous to those described in bacteria. An additional level of control operating on the histone genes and other eukaryotic genes is the control of messenger RNA (mRNA) stability. In the S phase of the cell cycle, the half-life of histone mRNA

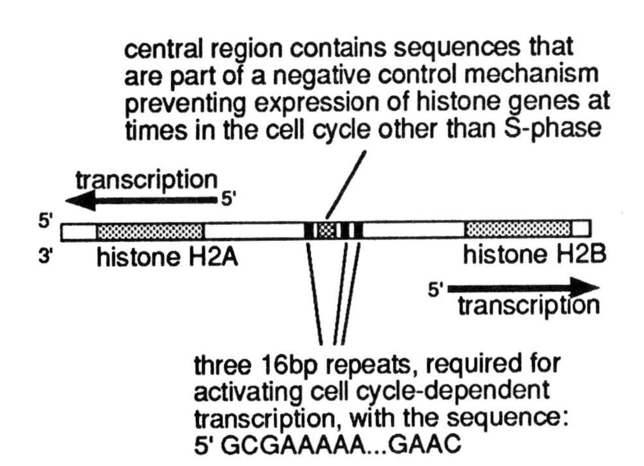

Fig. 16.22. The structure of the yeast-histone locus carrying the gene sequences for H2A and H2B proteins. The products of transcription are shown as arrows indicating the direction of synthesis. Initiation of transcription occurs 50–100 bp before the start of the gene and termination occurs approximately 200 bp after the end of the gene in an A-rich region of DNA. [The diagram is based on Osley et al. (1986) and Briggs et al. (1989).]

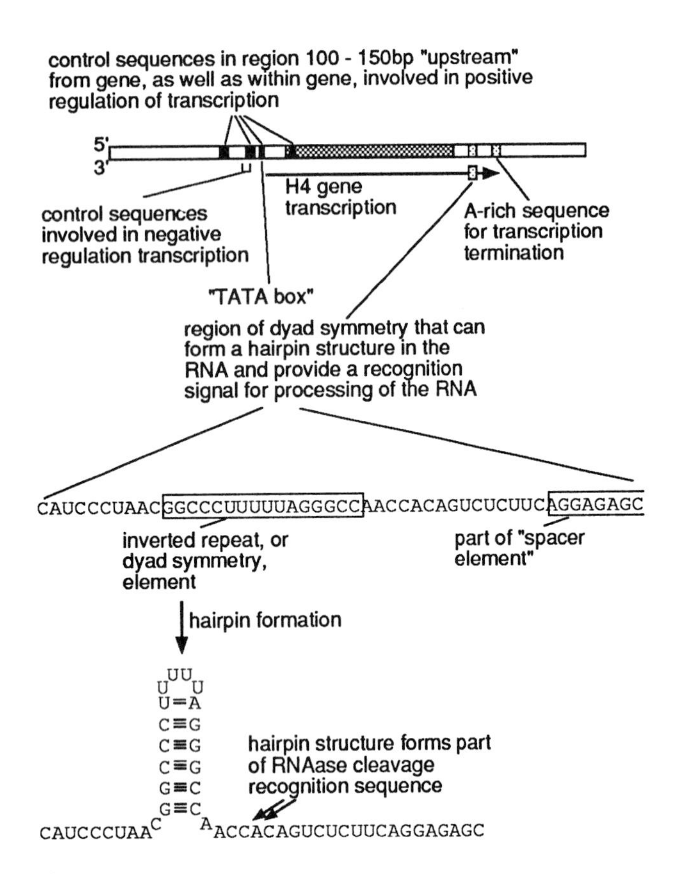

Fig. 16.23. Structure of an idealized histone H4 gene. The figure summarizes information about the control of H4 expression in sea urchin and mouse. The "spacer element" in the 3′ region of the mRNA binds a small nuclear RNA (snRNA) called U7, which is important in the RNA processing steps. (Seiler-Tuyns and Paterson., 1987; Schrumperli, 1988; Tung et al., 1990; Hoffmann and Birnstiel, 1990.)

is 30–60 min, whereas in the G2 phase, it drops to 10–15 min. This cell-cycle-regulated destruction of histone mRNA helps to coordinate histone and DNA syntheses. A DNA sequence that is central in this control occurs in the region 3′-downstream from the end of the gene (*see* Fig. 16.23), although the exact mechanism that utilizes this sequence in the control of mRNA stability is still under investigation.

A DNA region near the histone genes is involved both in the control of gene expression and in the attachment of DNA to a nuclear scaffold. A number of independent lines of evidence indicate the existence of a protein-based scaffold within the nucleus that contributes to maintaining a degree of order in the folding of chromatin. The earliest evidence for the existence of such regions came from the study of the histone genes in *Drosophila* (Fig. 16.24). The scaffold-attachment regions are approximately 200 bp long and contain repetitive elements that have DNA sequences related to the binding site for the DNA-binding protein topoisomerase II. This protein is a prominent component of the nuclear scaffold, as discussed further in Chapter 17.

16.8.2 The Hemoglobin Genes

Another family of genes that has been significant in developing our understanding of the function of noncoding sequences near genes, code for the globin proteins that make up the various hemoglobins present in the red-blood cells in humans (*see also* Chapter

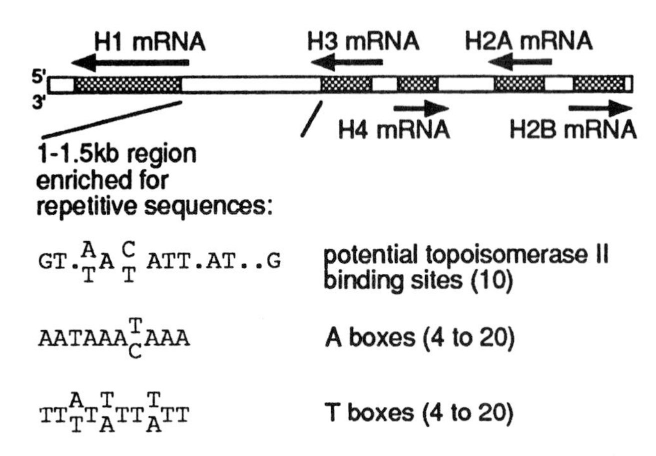

Fig. 16.24. Nuclear-scaffold attachment sites in the *Drosophila* histone–gene family. The 1.0–1.5-kb region between the H1 and H3 histone genes is considered to be the point of attachment of the histone-family DNA to the nuclear scaffold. This region is particularly enriched for the three repetitive sequences indicated. (Gasser and Laemmli., 1987; Kremer and Hennig, 1990.)

21, Section 21.2.5). Defects in the expression of one or more of these genes are associated with blood-cell disorders, and their study has a long history of research at the genetic, biochemical, and molecular levels. An overview of the β-gene cluster is shown in Fig. 16.25. The β-globin gene has been particularly well studied and the ATA sequence, 30 bp before the start of transcription, is typical of the core promoter required for RNA polymerase II transcription of eukaryotic genes. The CCAAT sequence, located a further 40 bp, 5′-upstream from the ATA sequence, also defines a positive control region regulating gene expression. The region extending 60 bp, 5′-upstream from the start of transcription, contains sequences that define the induction of β-globin mRNA synthesis in murine erythroleukemia cells stimulated to undergo erythropoiesis (i.e., cell differentiation into red blood cells).

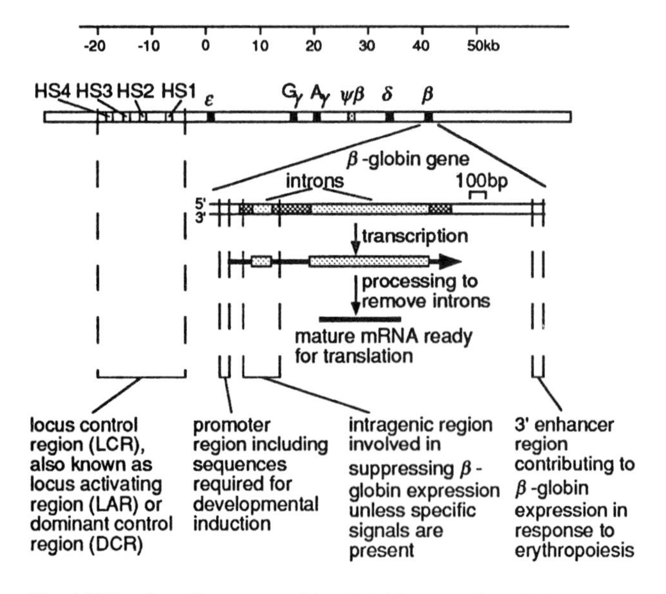

Fig. 16.25. Overall structure of the β-globin-gene cluster in humans. [Based on Van der Ploeg et al. (1980); Wright et al. (1984); Tuan et al. (1985); Trudel and Constantini (1987); Townes and Behringer (1990).] (*See also* Chapter 17, Section 17.5.3).

Interestingly, in the absence of a signal that commits a given cell to erythropoiesis, a region within the gene itself inhibits the transcription of β-globin. In response to developmental signals leading to erythropoiesis, other sequences controlling transcription are also found 3′-downstream from the gene.

From the point of view of the overall arrangement of DNA sequences within chromosomes, it is significant that the expression of the entire cluster of globin genes also relies on sequences located 50–65 kb, 5′-upstream from the β-globin gene. Initial indications for such a region came from the analyses of patients suffering from a particular form of red blood cell disorder, $\delta\beta^0$-thalassemia. Biochemical analyses indicated that β-globin was not produced, although DNA studies indicated that the entire gene plus 5′- and 3′-controlling sequences were present. Subsequent research revealed that a region of repetitive DNA sequences, normally present approximately 6 kb, 5′-upstream from the ϵ-globin gene (50–65 kb, 5′-upstream from the β-globin gene, *see* Fig 16.25), were missing in the $\delta\beta^0$-thalassemia disorder. These repetitive sequences are located in chromatin that changes its structure markedly in response to the process of erythropoiesis (*see also* Chapter 17, Fig. 17.30). The structural changes are measured biochemically using the DNA-degrading enzyme DNAase I, as discussed in Chapter 17, Section 17.5.3, and are called hypersensitive sites, or HS. Although the regions upstream from the ϵ-globin gene do not appear to be structurally analogous to the nuclear-scaffold attachment sites discussed above, they may be functionally analogous in defining the chromosomal domain of the hemoglobin family of genes. The region has been called various names including dominant-control region (DCR), locus-activating region (LAR), and locus-control region (LCR).

Table 16.2. Processes That Modify RNA

Process	Significance
Processing of precursor RNA molecules into smaller RNA molecules by specific RNase enzymes	A single RNA molecule provides several functional products.
Splicing of RNA in special ribonuclear protein structures	Specific, small nuclear RNAs can bring together different RNA sequences, which are not necessarily part of a single transcript, to create a new RNA molecule for translation into protein.
Alternative splicing of RNA	Different proteins can be made from the same initial RNA molecule.
Self-splicing of RNA	Removal of intron sequences without the requirement of proteins.
Capping of 5′-end of RNA and A-tailing of 3′-end of RNA	Posttranscriptional modifications determining the stability of messenger RNA.
RNA editing	Alters mainly C's to U's and can change the amino-acid sequence of the protein that is translated from the modified RNA; discovered in plant mitochondria, where it may be involved in male sterility.

Note: Differential stability of RNA is an important determinant in gene expression but is not summarized in the table. Observations on transgenic plants suggest that excess RNA production from an introduced gene may stimulate loss of all RNA produced from both the introduced gene and the normal gene resident in the host, leading to increased turnover of the RNA and gene silencing (de Carvalho et al., 1992; Matzke and Matzke, 1995; Stam et al., 1997). Although the phenomenon is not clearly understood, it demonstrates that interactions involving RNA are crucial in determining the final level of gene expression in an organism.

16.9 RNA PROCESSING IS AN INTEGRAL ASPECT OF UNDERSTANDING THE WAY DNA SEQUENCES ARE ORGANIZED

It is evident throughout this chapter that an understanding of the way DNA sequences are organized is difficult to separate from the processes that modify the RNA transcript derived from that DNA. Table 16.2 summarizes the various RNA modification processes that have been discovered to date. The discovery of RNA editing in mitochondria and chloroplasts is relatively recent and is discussed in Chapter 22 in relation to nuclear–cytoplasmic interactions that have important consequences for the phenotype of an organism.

BIBLIOGRAPHY

General

LEWIN, B. 1990. Genes IV. Oxford University Press, New York.

MATTHEWS, C.K., VAN HOLDE, K.E. 1990. Biochemistry. Benjamin/Cummings, Redwood City, CA.

NEIDHARDT, F.C. (ed.-in-chief). 1987. *Escherichia coli* and *Salmonella typhimurium:* Cellular and Molecular Biology. 2 Vols. American Society of Microbiology, Washington, DC.

Section 16.1

BENNETT, M.D., SMITH, J.B. 1976. Nuclear DNA amounts in angiosperms. Proc. Roy. Soc. London Ser. B 274: 227–274.

SPARROW, A.H., NAUMAN, A.F. 1976. Evolution of genome size by DNA doublings. Science 192: 524–529.

Section 16.2

CATTANEO, R. 1989. How "hidden" reading frames are expressed. Trends Biol. Sci. 14: 165–167.

EICK, D., WEDEL, A., HEUMANN, H. 1994. From initiation to elongation: comparison of transcription by prokaryotic and eukaryotic RNA polymerases. Trends Genet. 10: 292–296.

Section 16.3

FIERS, W., CONTRERAS, R., HAEGEMAN, G., ROGIERS, R., VAN DE VOORDE, A., VAN HEUVERSWYN, H., VAN HERREWGHE, J., VOLCKAERT, G., YSEBAERT, M. 1978. Complete nucleotide sequence of SV40 DNA. Nature 273: 113–120.

REDDY, V.B., THIMMAPPAYA, B., DHAR, R., SUBRAMANIAN, K.N., ZAIN, B.S., PAN, J., GHOSH, P.K., CELMA, M.L., WEISSMAN, S.M. 1978. The genome of simian virus 40. Science 200: 494–502.

Section 16.4

Beckwith, J. 1987. The lactose operon. In: *Escherichia coli* and *Salmonella typhimurium*: Cellular and Molecular Biology Vol. 2 (Neidhardt, F.C., ed.-in-chief). American Society of Microbiology, Washington, DC. pp. 1444–1452.

Cairns, J. 1963. The chromosome of *Escherichia coli*. Cold Spring Harb. Symp. Quant. Biol. 28: 43–46.

Charlier, D., Roovers, M., Van Vliet, F., Boyen, A., Cunin, R., Nakamura, Y., Glansdorff, N., Pierard, A. 1992. Arginine regulon of *Escherichia coli* K-12: A study of repressor–operator interactions and of in vitro binding affinities versus in vivo repression. J. Mol. Biol. 226: 367–386.

Egel, R. 1988. The "lac" operon: an irrelevant paradox? Trends Genetics 4: 31.

Friedman, A.M., Fischmann, T.O., Steitz, T.A. 1995. Crystal structure of *lac* repressor core tetramer and its implications for DNA looping. Science 268: 1721–1726.

Landrick, R., Stewart, J., Lee, D.N. 1990. Amino acid changes in conserved regions of the β-subunit of *Escherichia coli* RNA polymerase alter transcription pausing and termination. Genes Dev. 4: 1623–1636.

Landrick, R., Yanofsky, C. 1987. Transcription attentuation. In: *Escherichia coli* and *Salmonella typhimurium:* Cellular and Molecular Biology Vol. 2 (Neidhardt, F.C., ed.-in-chief). American Society of Microbiology, Washington, DC. pp. 1276–1301.

Reiter, W-D, Hudepohl, U., Zillig, W. 1990. Mutational analysis of an archaebacterial promoter: Essential role of a TATA box for transcriptional efficiency and start-site selection in vitro. Proc. Natl. Acad. Sci. USA 87: 9509–9513.

Rznikoff, W.S., Siegele, D.A., Cowing, D.W., Gross, C.A. 1985. The regulation of transcription initiation in bacteria. Annu. Rev. Genet. 19: 355–387.

Wetmur, J.G., Davidson, N. 1968. Kinetics of renaturation of DNA. J. Molec. Biol. 31: 349–370.

Yanofsky, C., Crawford, I.P. 1987. The tryptophan operon. In: *Escherichia coli* and *Salmonella typhimurium*: Cellular and Molecular Biology Vol. 2 (Neidhardt, F.C., ed.-in-chief). American Society of Microbiology, Washington, DC. pp. 1453–1472.

Section 16.5

Britten, R.J., Graham, D.E., Neufeld., B.R. 1974. Analysis of repeating DNA sequences by reassociation. Methods Enzymol. 29: 363–419.

Gilson, E., Clement, J-M., Brutlag, D., Hofnung, M. 1984. A family of dispersed repetitive extragenic palindromic DNA sequences in *E. coli*. EMBO J. 3: 1417–1421.

Herbers, K., Conrads-Strauch, J., Bonas, U. 1992. Race-specificity of plant resistance to bacterial spot disease determined by repetitive motifs in a bacterial avirulence protein. Nature 356: 172–174.

Johnson, R.C., Simon, M.I. 1985. Hin-mediated site-specific recombination requires two 26 bp recombination sites and a 60 bp recombinational enhancer. Cell 41: 781–791.

Kingsman, A.J., Chater, K.F., Kingsman, S.M. (eds.) 1988. Transposition. Society for General Microbiology Symposium 43. Cambridge University Press, Cambridge.

Stern, M.J., Ames, G. F-L., Smith, N.H., Robinson, E.C., Higgins, C.F. 1984. Repetitive extragenic palindromic sequences: a major component of the bacterial genome. Cell 37: 1015–1026.

Wetmur, J.G., Davidson, N. 1968. Kinetics of renaturation of DNA. J. Molec. Biol. 31: 349–370.

Section 16.6

Yanisch-Perron, C., Viera, J., Messing, J. 1985. Improved M13 phage cloning vectors and host strains: nucleotide sequences of the M13mp18 and pUC19 vectors. Gene 33: 103–119.

Section 16.7

Boeke, J.D., Garfinkel, D.J., Styles, C.A., Fink, G.R. 1985. *Ty* elements transpose through an RNA intermediate. Cell 40: 491–500.

Britten, R.J., Kohne, D.E. 1968. Repeated sequences in DNA. Science 161: 529–540.

Heslop-Harrison, J.S. 1991. The molecular cytogenetics of plants. J. Cell Sci. 100: 15–21.

Kavenoff, R., Zimm, B.H. 1973. Chromosome-sized DNA molecules from *Drosophila*. Chromosoma 41: 1–27.

Klar, A.J.S. 1990. Regulation of fission yeast mating-type interconversion by chromosome imprinting. Development supplement. (Monk, M., Susani, A., eds. The Company of Biologists Ltd, Cambridge England, pp. 3–8.

Smyth, D.R. 1991. Dispersed repeats in plant genomes. Chromosoma 100: 355–359.

Section 16.8

Behringer, R.R., Hammer, R.E., Brinster, R.L., Palmiter, R.D., Townes, T.M. 1987. Two 3′ sequences direct adult erythroid-specific expression of human β-globin genes in transgenic mice. Proc. Natl. Acad. Sci. USA 84: 7056–7060.

Briggs, D., Jackson, D., Whitelaw, E., Proudfoot, N.J. 1989. Direct demonstration of termination signals for RNA polymerase II from the sea urchin H2A histone gene. Nucleic Acids Res. 20: 8061–8071.

Charnay, P., Treisman, R., Mellon, P., Chao, M., Axel, R., Maniatis, T. 1984. Differences in human α- and β-globin gene expression in mouse erythroleukemia cells: the role of intragenic sequences. Cell 38: 251–263.

de Carvalho, F., Gheysen, G., Kushnir, S., Van Montagu, M., Inze, D., Castresana, C. 1992. Suppression of β-1,3-glucanase transgene expression in homozygous plants. EMBO J. 11: 2595–2602.

Forrester, W.C., Thompson, C., Elder, J.T., Groudine, M. 1986. A developmentally stable chromatin structure in the human β-globin gene cluster. Proc. Natl. Acad. Sci. USA. 83: 1359–1363.

Gasser, S.M., Laemmli, U.K. 1987. A glimpse at chromosomal order. Trends Genet. 3: 16–22.

Hoffmann, I., Birnstiel, M.L. 1990. Cell cycle-dependent regulation of histone precursor mRNA processing by modulation of U7 snRNA accessibility. Nature 346: 665–668.

Kollias, G., Hurst, J., deBoer, E., Grosveld, F. 1987. The human β-globin gene contains a downstream developmental specific enhancer. Nucleic Acids Res. 15: 5739–5747.

Kremer, H., Hennig, W. 1990. Isolation and characterization of a *Drosphila hydei* histone DNA repeat unit. Nucleic Acids Res. 18: 1573–1580.

MAXSON, R., COHN, R., KEDES, L. 1983. Expression and organization of histone genes. Annu. Rev. Genet. 17: 239–277.

OSLEY , M.A., GOULD, J., KIM, S., KANE, M., HEREFORD, L. 1986. Identification of sequences in a yeast histone promoter involved in periodic transcription. Cell 45: 537–544.

SCHRUMPERLI, D. 1988. Multilevel regulation of replication-dependent histone genes. Trend Genet. 4: 187–191.

SEILER-TUYNS, A., PATERSON, B.M. 1987. Cell cycle regulation of a mouse histone H4 gene requires the H4 promoter. Molec. Cell. Biol. 7: 1048–1054.

SPIKER, S. 1988. Histone variants and high mobility group non-histone chromosomal proteins of higher plants: their potential for forming a chromatin structure that is either poised for transcription or transcriptionally inert. Physiol. Plant. 74: 200–213.

TOWNES, T.M., BEHRINGER, R.R. 1990. Human globin locus activating region (LAR) role in temporal control. Trends Genet. 6: 219–223.

TRAINOR, C.D., STAMLER, S.J., ENGEL, J.D. 1987. Erythroid-specific transcription of the chicken histone H5 gene is directed by a 3′ enhancer. Nature 328: 827–830.

TRUDEL, M., CONSTANTINI, F. 1987. A 3′ enhancer contributes to the stage-specific expression of the human β-globin gene. Genes Dev. 1: 954–961.

TUAN, D., SOLOMON, W., LI, Q., LONDON, I.M. 1985. The "β-like-globin" gene domain in human erythroid cells. Proc. Natl. Acad. Sci. USA 82: 6384–6388.

TUNG, L., LEE, I.J., RICE, H.L., WEINBERG, E.S. 1990. Positive and negative transcriptional regulatory elements in the early H4 histone gene of the sea urchin, *Strongylocentrotus purpuratus.* Nucleic Acids Res. 24: 7339–7348.

VAN DER PLOEG, L.H.T., KONINGS, A., OORT, M., ROOS, D., BERNINI, L., FLAVELL, R.A. 1980. γ-β Thalassaemia studies showing that deletion of the γ- and δ-genes influences β-globin gene expression. Nature 283: 637–642.

VOGT, P. 1992. Code domains in tandem repetitive DNA sequence structures. Chromosoma 101: 585–589.

WRIGHT, S., ROSENTHAL, A., FLAVELL, R., GROSVELD, F. 1984. DNA sequences required for regulated expression of β-globin genes in murine erythroleukemia cells. Cell 38: 265–273.

Section 16.9

COVELLO, P.S., GRAY, M.W. 1993. On the evolution of RNA editing. Trends Genet. 9: 265–268.

DE CARVALHO, F., GHEYSEN, G., KUSHNIR, S., VAN MONTAGU, M., INZE, D., CASTRESANA, C. 1992. Suppression of β-1,3-glucanase transgene expression in homozygous plants. EMBO J. 11: 2595–2602.

GALL, J.G. 1991. Spliceosomes and snurposomes. Science 252: 1499–1500.

MATZKE, M.A., MATZKE, A.J.M. 1995. Homology-dependent gene silencing in transgenic plants: what does it tell us? Trends Genet. 11:1–3.

PRING, D., BRENNICKE, A., SCHUSTER, W. 1993. RNA editing gives a new meaning to the genetic information in mitochondria and chloroplasts. Plant Molec. Biol. 21: 1163–1170.

SARAO, R., GUPTA, S.K., AULD, V.J., DUNN, R.J. 1991. Developmentally regulated alternative RNA splicing of rat brain sodium channel mRNAs. Nucleic Acids Res. 19: 5673–5679.

STAM, M., MOL, J.N.M., KOOSTER, J.M. 1997. The silence of genes in transgenic plants. Ann. of Bot. (Lond) 79: 3–12.

17

Variable Structure and Folding of DNA

- The three-dimensional, double-helix structure of DNA has numerous variations depending on the sequence of internal base pairs and the presence of interacting external molecules.
- Methylation of DNA is a chemical modification of the cytosine and adenosine residues that results in changes in the control of gene expression and DNA imprinting.
- Proteins interacting with DNA to control gene expression have characteristic structures and distort the double helix upon binding.
- The folding of DNA into chromosomes involves the formation of nucleosomes, which consist of a core of four different histones with DNA wrapped around the outside.
- Further folding involves the ordering of nucleosomes into fibers and chromatin domains by the formation of loops, which are attached at their bases to a nuclear scaffold.
- Single chromosomes tend to form domains within the nucleus and have nuclear-membrane attachment sites.

The DNA inside a eukaryotic nucleus is highly compacted and adopts numerous variations in its three-dimensional, double-helical structure as a result of the different molecules that interact with it. The three-dimensional structures of many different forms of DNA have been solved by X-ray crystallography, and they provide a sound foundation for studying the variations in structure that can occur in the DNA double helix.

17.1 THE STRUCTURE OF DNA

The DNA molecule is composed of two strands, each consisting of a deoxyribose sugar–phosphate backbone with purine and pyrimidine bases protruding from them. The helical structure of DNA provides the means for the negatively charged sugar–phosphate backbone to be located externally, in contact with water molecules, and the relatively hydrophobic bases to be stacked internally. The forces holding the two strands together derive from the hydrogen (H) bonds between the bases. The usual configuration of H bonds is that seen in the so-called Watson–Crick base pairing (Fig. 17.1). The coiling of the two sugar–phosphate backbone strands of the DNA molecule into a helical structure means that they cannot be separated unless the coil is unwound. In the classical B form of DNA, there are 10 base pairs per turn of the helix, and in this form, as well as numerous related forms, the double helix is in the direction of a right-handed twist (Fig. 17.2).

17.1.1 Polarity of the Sugar–Phosphate Backbone Is a Key Feature of DNA

Chapter 16 described the decoding of information from the DNA molecule (*see* Fig. 16.2) and indicated that each strand had either a 5′–3′ or 3′–5′ direction (polarity). The structural detail of this polarity is shown in more detail in Fig. 17.3. The polarity of the sugar–phosphate backbone is most easily considered as a consequence of the way it is formed. A nucleoside-triphosphate molecule, with a phosphate group attached to the 5′ carbon atom of the deoxyribose sugar, is joined to the hydroxyl group on the 3′ carbon atom of the deoxyribose moiety of an existing oligonucleotide. Both DNA and RNA are synthesized in this way, and the direction of synthesis is said to be in the 5′ to 3′ direction because the oligonucleotide, to which incoming nucleoside triphosphates are added, has a free 5′-phosphate or a special structure called a cap. In the process of the transcription of DNA into RNA, the DNA is "read" in the 3′ to 5′ direction. The main difference between sugar–phosphate backbones of DNA and RNA is that RNA features ribose sugars (OH groups at both the 2′ and 3′ positions of the sugar residue—*see* Fig. 17.3), and this means that the molecule does not form double-stranded structures, as does DNA. Although RNA is single stranded, it does fold to form secondary structures that are stabilized by H-bonding.

17.1.2 Variation in the Structure of DNA

The view of the DNA molecule that is usually portrayed is the B form (*see* Fig. 17.2). This form was deduced from X-ray diffraction studies on fibers of highly purified DNA and from biochemical studies on the composition of DNA from different organisms. The "standard" helix displays a number of characteristics such as the even stacking of the base pairs, as they contribute to formation of the helix, and both a major and a minor groove (Fig. 17.4). How-

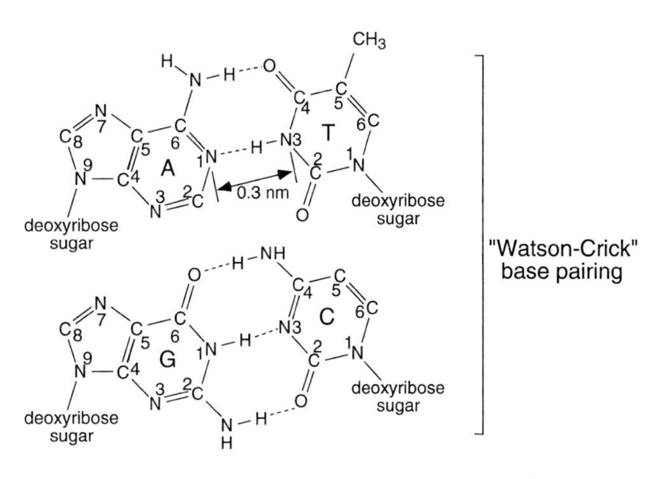

Fig. 17.1. Chemical structure of the base pairs that occur in DNA. A = adenine, T = thymine, G = guanine, C = cytosine. In the standard Watson–Crick base-pair configuration, there are two H bonds between the A and T residues and three between the G and C residues. The configuration is named after Watson and Crick (1953).

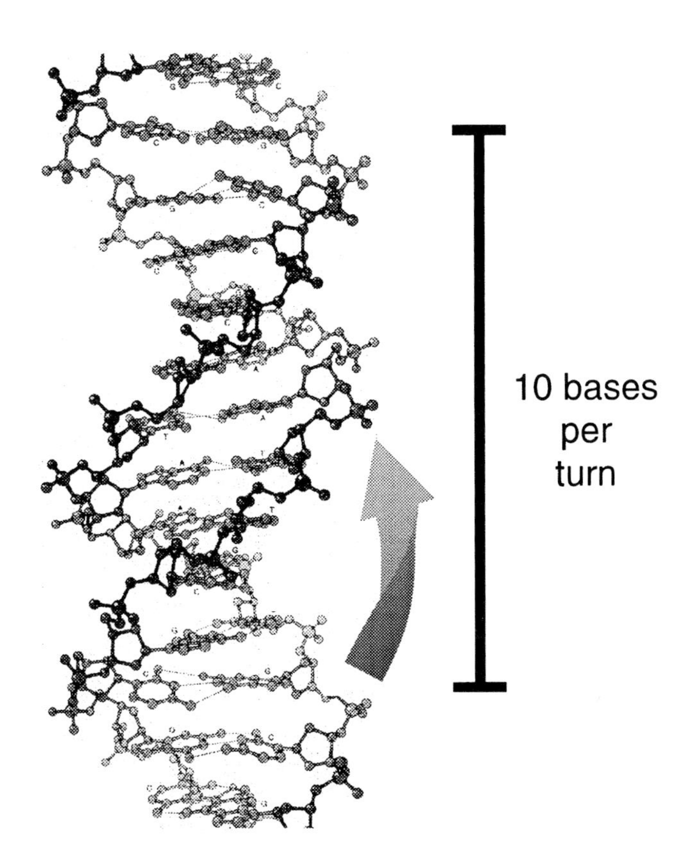

Fig. 17.2. Model of the classical B form of DNA. The arrow indicates a right-handed twist. (Reproduced with permission from I. Geis: this model originally appeared in Dickerson, 1983).

Fig. 17.3. The 5′–3′ direction of the sugar–phosphate backbone is a crucial feature of the biological interactions that involve DNA. The different directions of the backbones are particularly evident when the orientations of the deoxyribose sugar and O atom are compared in the two DNA strands.

ever, the DNA in living organisms does not always adopt the ''standard'' B helix structure, and in order to understand the basis for the numerous structural variations that have been found, the following parameters of the DNA molecule need to be defined:

Helical twist (T): The orientation of one nucleotide base pair relative to its neighbor with respect to the helical twist of the DNA molecule. The value varies from 36° in the classical B form (Fig. 17.5) to 32.7° in the A form (Fig. 17.6). DNAs isolated from the genomes of most organisms have, on average, a helical twist of $T = 34°$. The value of T determines the helical rotation of the DNA molecule, and in the classical B form, the $T = 36°$ value generates 10 base pairs per helical turn of the molecule. The more common value of $T = 34°$ gives the DNA molecule a helical turn every 10.6 base pairs.

Roll (R): Describes the deviation of neighboring base pairs from a perfectly parallel arrangement (Fig. 17.7). In the classical B form of DNA, $R = 0°$. If the roll angle is 12°, as in the A form of DNA, the DNA molecule is still straight at each step in the

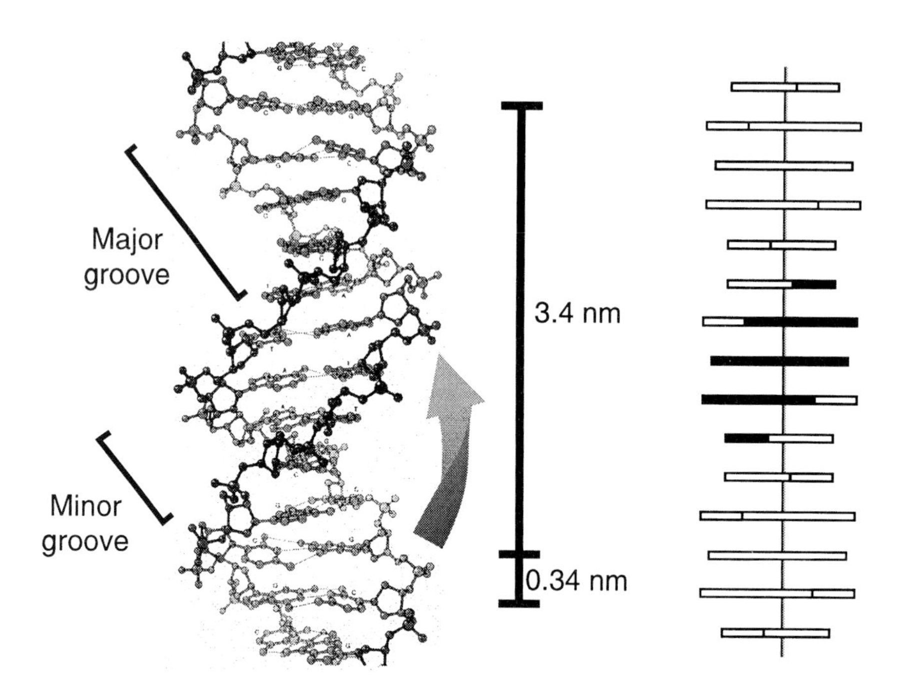

Fig. 17.4. The minor and major grooves of B-form DNA. The left part of the diagram is based on the drawing by I. Geis (Fig. 17.2), and the arrow indicates a right-handed twist. The right part of the diagram is a simplified view of the DNA molecule by representing a base pair as a rectangular block, with the edge facing the minor groove shown in black. (Calladine and Drew, 1992; Dickerson, 1983.)

helix, but the base pairs exhibit a 20° incline with respect to the helical axis.

Slide (S): The distance by which base pairs are pulled away from the central axis of the helix (Fig. 17.8). In the B form of DNA, $S = 0$, whereas in the A form of DNA, $S = 0.15$ nm (1.5 Å). DNA molecules with a low roll often have a low slide, whereas those with a high roll often have a high slide.

''Propeller twist'' within a base pair: This is a deviation in the relative orientation of the two members of a base pair from a coplanar arrangement (Fig. 17.9). This modified conformation is thought to allow improved stacking of bases in certain situations. ''Propeller twist'' tends to be more pronounced in A.T pairs than in G.C pairs.

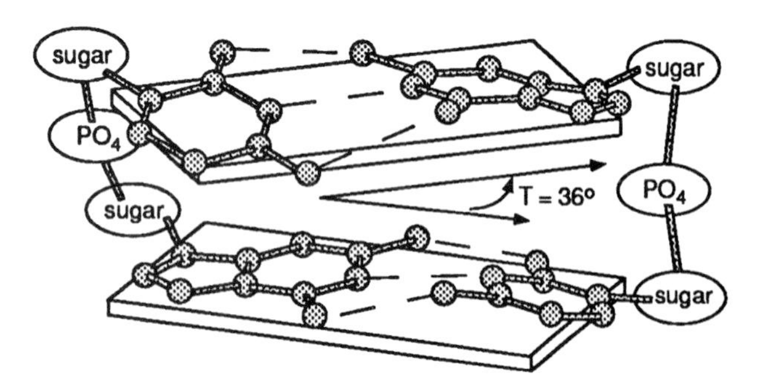

Fig. 17.5. The helical twist T of the DNA molecule is the degree of turn in the DNA helix between each base pair. The rectangular block superimposed underneath each base pair assists in following the orientation of the base pair as a unit. (From Calladine and Drew, 1992).

The parameters defined above serve as a basis for considering the mobility in the conformation of the DNA, within the constraints imposed by the sugar–phosphate backbone, when it interacts with other molecules. The extremes among the range of conformations of the DNA molecule are displayed by the A and B forms discussed above, as well as the Z form (Fig. 17.10); the key features of these conformations are summarized in Table 17.1. The particular conformation favored by a DNA molecule depends on the environment and the range of molecules interacting with it. The final structure adopted by the DNA helix is also dependent on the sequence of base pairs within the molecule; therefore, DNA should not be envisaged as a stiff and rigid structure.

17.1.3 Curvature of DNA Can Result from a Particular Sequence of Base Pairs

A DNA sequence containing a run of adenines in one strand and thymidines in the complementary strand, dA.dT, has $T = 36°$

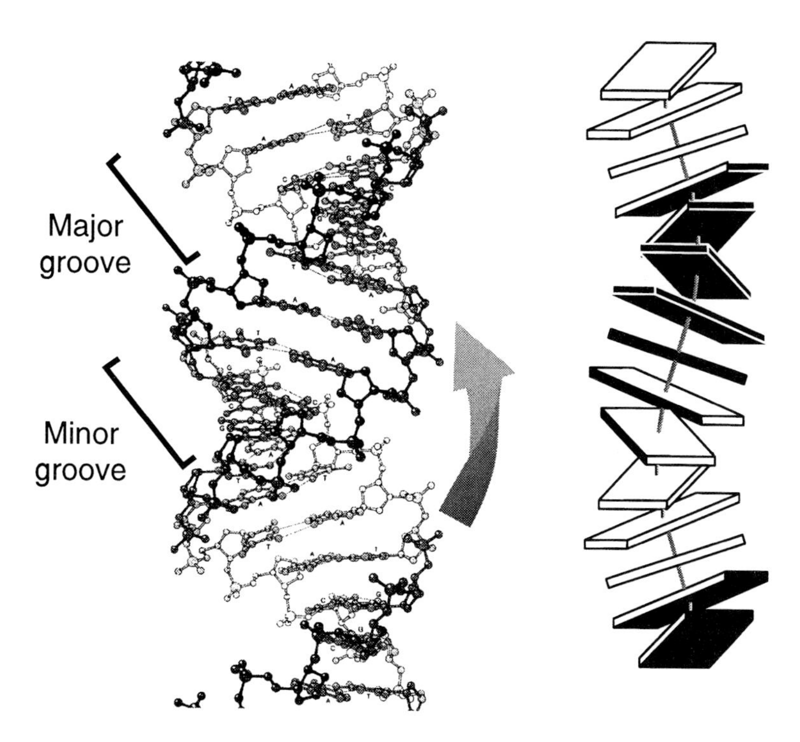

Fig. 17.6. The A form of DNA. This structure is characteristic of the shape of DNA molecules at lower humidities of approximately 75%. The left part of the diagram is based on a drawing by I. Geis (used with permission), and the right part of the diagram illustrates a simplified view of the DNA molecule by representing a base pair as a rectangular block, with the edge facing the minor groove shown in black. (From Calladine and Drew, 1992.)

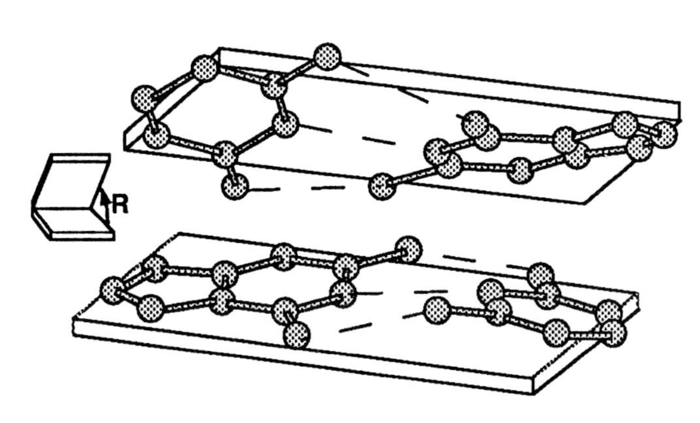

Fig. 17.7. The roll function *R* between consecutive base pairs. The rectangular block superimposed on each base pair assists in following the orientation of the base pair as a unit. (From Calladine and Drew, 1992.)

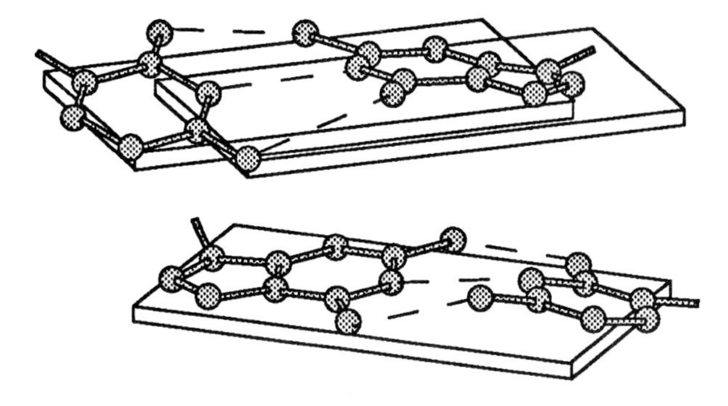

Fig. 17.8. The slide function *S* indicating the degree of slippage between consecutive base pairs. The rectangular block superimposed beneath each base pair assists in following the orientation of the base pair as a unit. (From Calladine and Drew, 1992.) The upper base pair has moved to the left relative to the lower base pair, and its original position is indicated by the rightmost rectangular block.

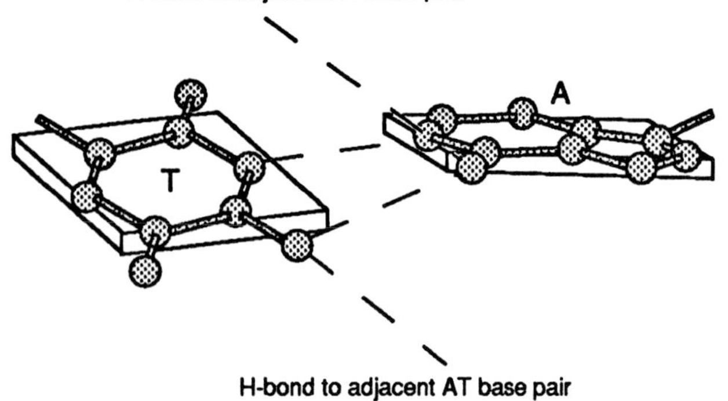

Fig. 17.9. The "propeller twist" within an A.T nucleotide pair as induced by an adjacent A.T base pair. Blocks are placed underneath the individual base pairs to illustrate the twist of one base pair relative to the other (Dickerson, 1983; Aymami et al., 1989; Drew et al., 1990).

Table 17.1. Characteristics of Some DNA Conformations

Variable	B Form	A Form	Z Form
Helical twist (*T*)	36°	32.7°	45°/15°
Base pairs/helical turn	10	11	12
Distance between bp	0.34 nm	0.26 nm	0.57 nm
Helix direction	Right hand	Right hand	Left hand

and thus a helical repeat of 10.0 base pairs. This is significantly less than the 10.6 base pairs for mixed-sequence DNA. When the A.T pairs take on a "propeller twist," H bonds in the major groove connect the nitrogen atom, attached to carbon #6 of each adenine, to the oxygen atom, attached to carbon #4 of the partner thymine, as well as the thymine in the base pair below (Fig. 17.11). The between-base pair H-bonding of dA.dT means that the occurrence of three consecutive A or T residues in a strand of DNA, once every helical repeat, causes the DNA to adopt a curved rather than a straight helix (Fig. 17.12). As the packaging of DNA into chromosomes and the processes of transcription inevitably involve some sort of folding, the natural tendency of certain DNA sequences to curve is a significant consideration. The structural modifications in DNA that result from the presence of three consecutive A or T residues, once every helical repeat in DNA, are generally considered to have significant effects on the activity of regulatory sequences in the noncoding DNA near a gene.

17.2 STRUCTURAL MODIFICATION OF DNA BY METHYLATION

DNA methylation occurs in prokaryotic and eukaryotic organisms. The methylation of both adenine and cytosine residues (Fig. 17.13) has been observed and is an important variable in gene expression

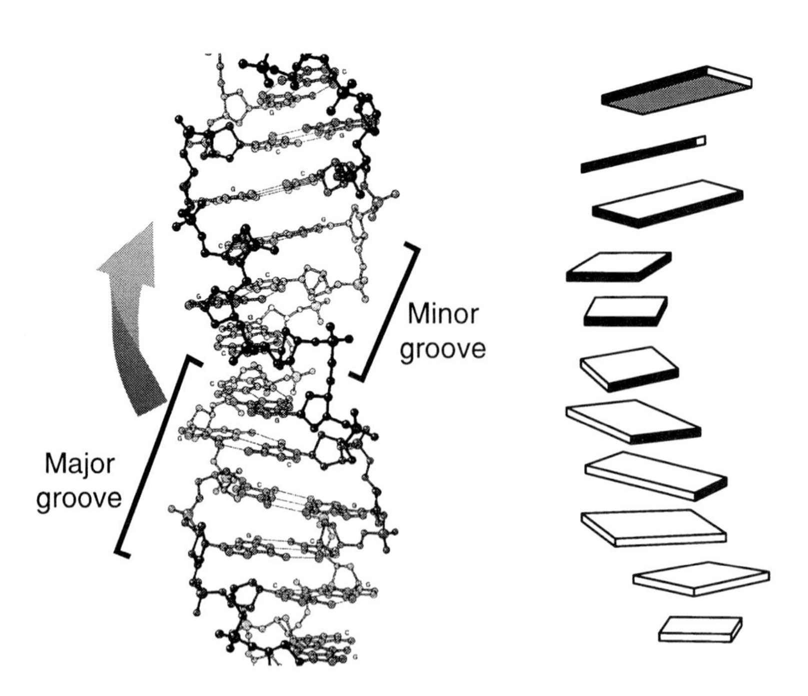

Fig. 17.10. The Z form of DNA. The left part of the diagram is based on a drawing by I. Geis (used with permission); the arrow indicates a left-handed twist. The right part of the diagram illustrates a simplified view of the DNA molecule by representing a base pair as a rectangular block with the edge facing the minor groove shown in black. (From Calladine and Drew, 1992.)

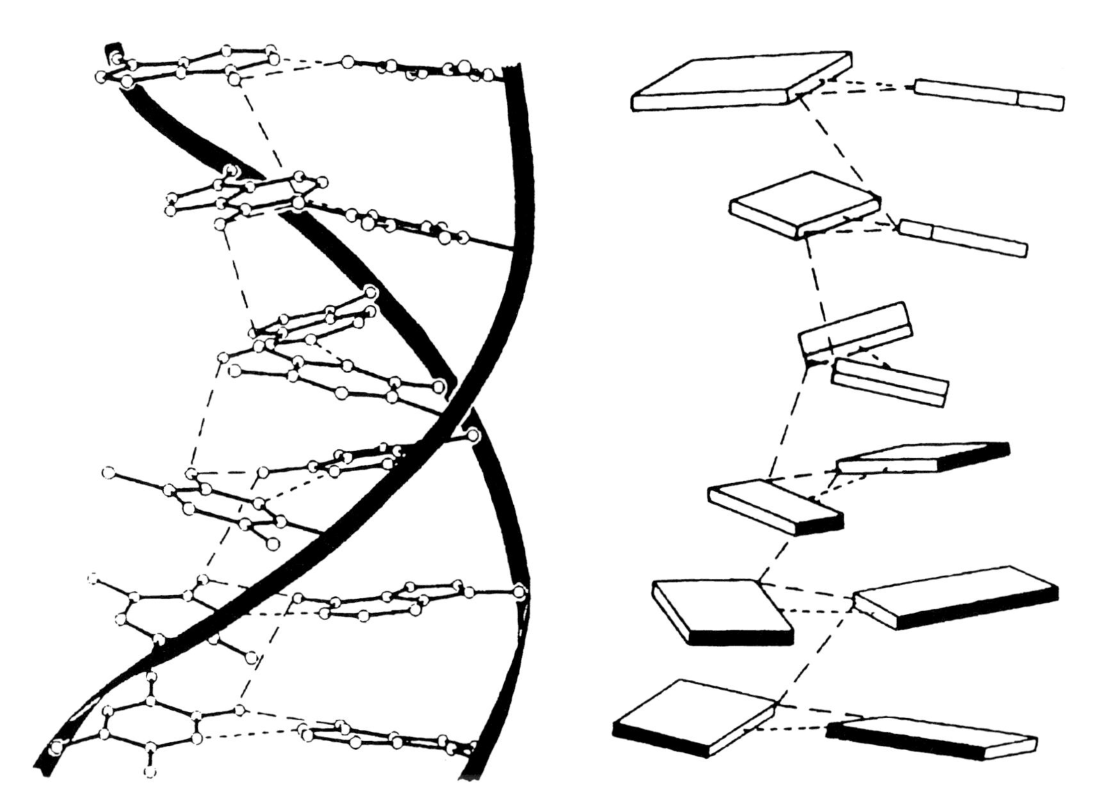

Fig. 17.11. Hydrogen-bonding between consecutive base pairs in special sequences of DNA such as d(AAAAAA).d(TTTTTT), where only A.T base pairs are present with A's in one strand and T's in the other. The propeller-twist conformation of the successive A.T base pair (*see* Fig. 17.9) makes it feasible that interbase pair H bonds can form (indicated by long dashed lines), resulting in a more rigid structure (Dickerson, 1983; Aymami et al., 1989: Drew et al., 1990). The simplified view of the structure on the right uses blocks to represent individual bases.

from a particular DNA region. Methyl transferases methylate DNA using S-adenosylmethionine as a source of the methyl group. In *E. coli*, GATC sequences are the sites of action of an adenine–methylating enzyme, Dam methylase (*Dam*). The importance of this type of methylation has been demonstrated in studies of the movement of the Tn group of transposable elements, which are DNA sequences capable of movement from one region of the genome to another by virtue of the enzymatic action of transposase (*see* Chapter 7, Section 7.5.1), as well as other host functions involved in DNA synthesis and repair. The Tn elements encode a transposase gene, which needs to be transcribed and translated so that movement of a Tn element can be catalyzed by transposase. Using structural studies of DNA to monitor its methylation status, and mutations that remove the *Dam* methylase activity from the host, it has been shown that methylation of GATC sequences in the promoter region of the transposase gene significantly reduces its transcription. The resulting reduced production of transposase-protein molecules means that movement of the transposable element is inhibited. Similar observations apply to the transposable elements characterized in maize (*Ac*, *Spm*, and *Mu*).

Methylation of DNA at cytosine residues is carried out in the sequences CG and CNG (where N can be either C, G, A or T) and is particularly prominent in plants, although it does occur in other organisms. This modification of DNA can be reduced by the incorporation of 5-azacytosine during replicative cycles of the cell, because the 5-azacytosine moieties incorporated into DNA cannot be methylated by the normal methyltransferase enzymes. The presence of 5-azacytosine during developmental stages of an organism can also directly inhibit the methylation process. Demethylation of genomic DNA, induced by 5-azacytosine, causes the transcription of a number of different genes; a well-characterized example is the analysis of the GUS gene (*see* Chapter 19, Fig. 19.19) expression in tobacco tissue-culture cells. Methylation was correlated with inactivity, and demethylation, induced by 5-azacytosine, was sufficient to activate expression.

In the replication of DNA and subsequent mitotic divisions, the pattern of methylation is usually transmitted with high fidelity to daughter cells and is associated with the phenomenon of genomic imprinting, which controls the particular pattern of gene expression depending on whether it is derived from the maternal or paternal parent. Abnormal methylation patterns in certain regions of the genome can arise as a result of mutation, and they have been associated with specific diseases in humans; for example, increased methylation at the fragile X site is associated with mental retardation (*see* Chapter 7, Section 7.4.5). In fungi, the induction of methylation of repeated gene regions (methylation induced premeiotically or MIP), leading to reversible gene silencing, has been well studied. The phenomenon was first demonstrated by the transformation of *Ascobolus* with a 5.7-kb duplication of a DNA segment carrying a $met2^+$ gene (introduced from a plasmid with the $met2^+$ wild-type allele). The $met2^+$ gene was recovered by homologous recombination with the equivalent region of the recipient strain, which had a mutant $met2^-$ allele (required methionine for growth). The duplication was present in the haploid nuclei of the vegetative growth stage, and expression of the $met2^+$ allele provided the basis for selection of the transformants. After sexual reproduction, the *met2*$^{+}$ allele was not expressed in 90% of the progeny even though DNA analyses demonstrated that it was still present, and its lack of expression was correlated with increased levels of cytosine methylation. Subsequent selection for revertants to the *met2*$^{+}$ phenotype was correlated with reduced levels of cytosine methylation. It has been speculated that the MIP phenomenon is closely related to *r*epeat-*i*nduced-*p*oint-mutation (RIP), which results in relatively high levels of mutations in repetitive DNA sequences. The increased levels of mutation are thought to be due to deamination of 5 methyl cytosine (*see* Chapter 22, Fig. 22.11). Both phenomena are speculated to involve the transient association of identical or near-identical sequences to form structures that are substrates for enzymes causing the respective DNA changes.

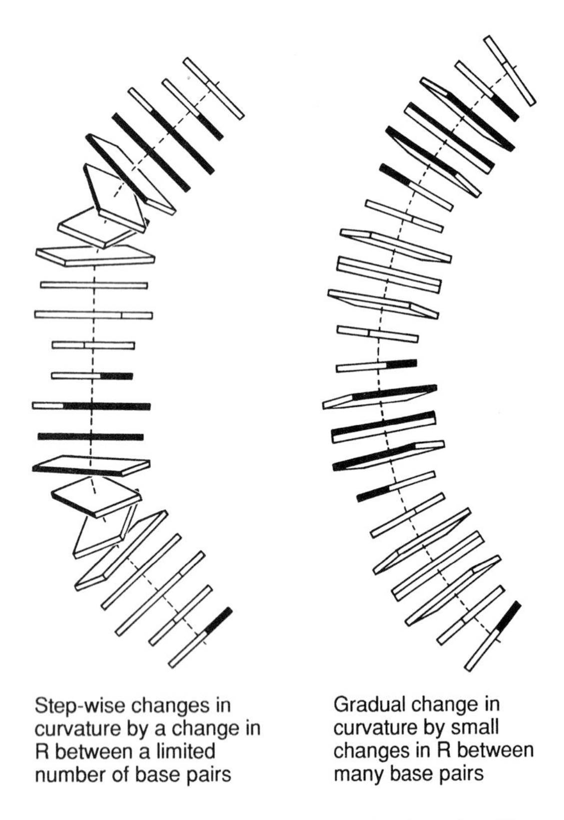

Fig. 17.12. A diagram of a curved DNA molecule produced by representing base pairs as rectangular blocks with the edges facing the minor groove shown in black. (From Calladine and Drew, 1992; drawing kindly provided by Dr. C.R. Calladine.) The left panel indicates curvature resulting from stepwise changes in roll, R, affecting only a limited number of base pairs; the right panel shows a similar curvature achieved by small changes in *R* between all base pairs. The degree to which a particular segment of DNA is inherently curved depends on its sequence: for example, d(GGCCNAAAAN)$_n$, where N is any nucleotide and the subscript *n* indicates the number of repeats, will undergo curvature when several repeats occur in tandem. Because the bending of DNA occurs when proteins bind to DNA, the inherent curvature of a DNA sequence can have significant consequences for the control of gene expression (Collis et al., 1989). The curvature may decrease the affinity of DNA-binding proteins if the attachment of these proteins causes a distortion or change in DNA conformation that opposes the inherent curvature of DNA. For example, the TATA-binding protein causes a distortion of the DNA molecule in the position at which it binds to the molecule (*see* Section 17.4.4); similarly, the site-specific recombinase *- resolvase causes a sharp bend of 60° in the DNA toward the major groove upon binding (Yang and Steitz, 1995).

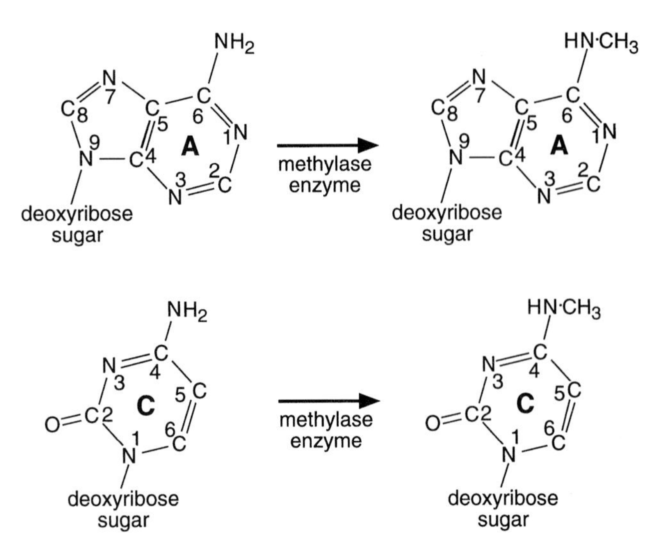

Fig. 17.13. The mode of action of methylation of bases in DNA. Methylation is a structural modification of DNA that occurs after replication and has important implications in the control of gene expression (Shen and Maniatis, 1980; Matzke and Matzke, 1991; Pfeifer et al., 1990; Takahashi et al., 1994). The enzymatic methylation of cytosine (C) or adenine (A) involves deformation of the local helix structure, and the rate at which this occurs is suggested to be affected by the G + C environment (Bestor et al., 1992). Plants have particularly high levels of methylation, with as much as 82 % of the C residues in CG sequences within DNA being affected (Wagner and Capesius, 1981; Gruenbaum et al., 1981). Mutants that reduce methylation display a number of developmental defects (Kakutani et al., 1995).

17.3 INTERACTION OF SMALL MOLECULES WITH DNA

17.3.1 Water

The importance of determining the nature of the interaction between DNA and water stems from the fact that water comprises 70–90% of most forms of life and is the continuous phase common to all living organisms. The DNA molecule has many sites at which water molecules can interact through H-bonding. As a result, the wide range of molecules that interact with DNA in living systems also need to interact with the ordered networks of water molecules that are present in the major and minor grooves of the DNA helix.

The sites on DNA that interact with water include the oxygen atoms of the sugar–phosphate backbone as well as functional groups, on the bases, where oxygen and nitrogen atoms protrude into the major and minor grooves. X-ray diffraction techniques have been used to study crystals of the A and B forms of DNA, from 4 to 12 bp in length. These studies have shown that DNA helices are coated with a layer of strongly bound water one molecule deep.

In one B form of DNA that has been analyzed, a chain of water molecules zigzags along the minor groove to form a "spine of hydration." In the major groove, most functional groups are hydrated, and the water molecules form discontinuous clusters with interstrand and intrastrand links. The DNA-associated water structures, however, vary according to the conformation and sequence of the DNA segment because other B-form oligonucleotides, as well as A forms of DNA, do not show the "spine of hydration" in the minor groove.

17.3.2 Spermine, Ions, and Atoms

Relatively little information is available on the positions occupied by a range of other small molecules such as spermine or the ions Ag^+, Mg^{2+}, and Hg^{2+}, within the DNA helix. This lack of information results from variation in the positions of these moieties in different molecules of crystallized DNA. The positions of platinum atoms (used as anticancer agents) in the major groove have been determined because their positions in the DNA helix are stable.

One example of where it was possible to position a spermine molecule within the DNA helix is shown in Fig. 17.14. Spermine is important in stabilizing the chromatin structure in nuclei, and it is seen to span the major groove, with ionic links to the oxygen atoms of the phosphate groups in the sugar–phosphate backbone. Hydrogen-bonding between spermine and the base pairs in the DNA helix, either directly or via water molecules, also occurs.

17.3.3 Dye and Antibiotic Molecules

When dyes and antibiotics bind to the double-stranded DNA molecule, their properties are usually changed with respect to their interaction with light. This has been exploited in cytological work where dyes and antibiotics, in this context called fluorochromes,

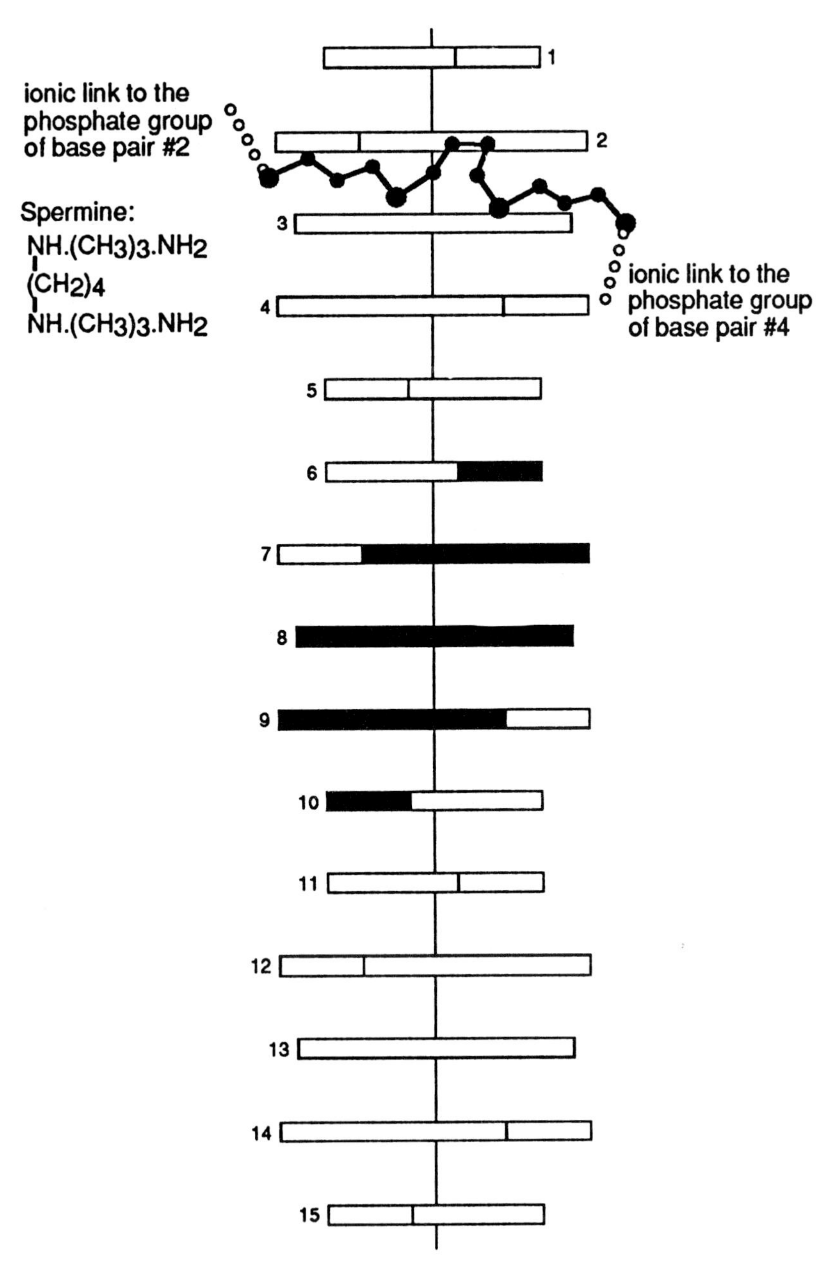

Fig. 17.14. The position of a spermine molecule in DNA (Drew and Dickerson, 1981). Spermine is a biologically important cation, which can neutralize the highly negatively charged sugar–phosphate backbone of each strand of the DNA helix, and induce conformational changes in the DNA (Feuerstein et al., 1989).

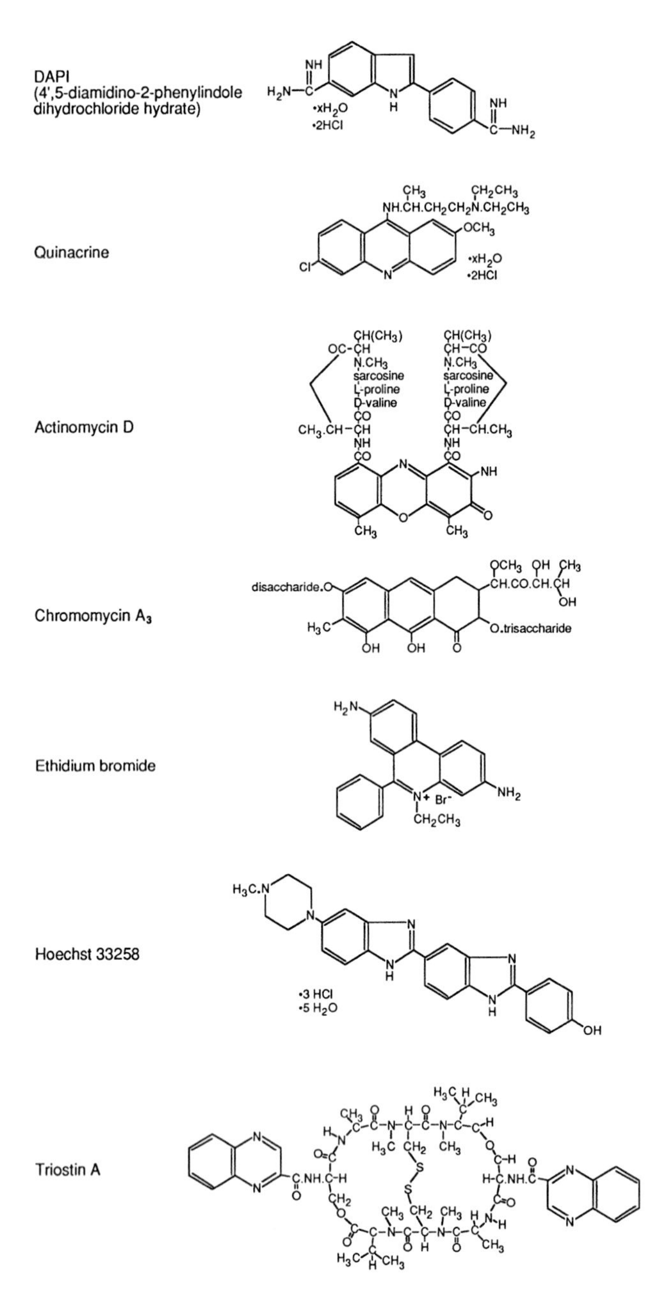

Fig. 17.15. A selection of fluorochromes that interact with DNA. The central three planar rings present in many of the molecules usually are intercalated between consecutive base pairs and distort the helix as they bind to the DNA molecule.

are commonly used to detect DNA. Fluorochromes show a characteristic fluorescence after they are bound to DNA and irradiated with light of particular wavelengths. The structures of a selection of fluorochromes are shown in Fig. 17.15. It is evident that they often share the feature of a planar set of rings, which can fit into the hydrophobic environment between the base pairs of the double helix. Not all fluorochromes are intercalated between the base pairs; some such as Hoechst dye #33258 simply bind intimately into the minor groove. The traditional stain used to detect DNA, namely Feulgen or Schiff's reagent, normally reacts chemically with acid-hydrolyzed DNA to form a purple compound.

Because many RNA molecules have secondary structures, most likely consisting of A-form-related helices, the fluorochromes used to detect DNA can also detect RNA, although the wavelengths of the incident light used to illuminate the target, as well as the emitted light, differ from those used to detect DNA.

A number of complexes between small DNA molecules and anticancer drugs or antibiotics have been crystallized to determine their molecular structure. The work provides an insight into the variety of structures adopted by DNA in complexing with larger molecules and has indicated how peptide moieties interact with the DNA helix. Extreme distortion of the DNA helix is seen in the crystal structure of the anticancer drug triostin-A, bound to a small DNA molecule with the sequence GCGTACGC (Fig 17.16). In this figure, two triostin-A molecules are bound to the DNA and, as a result, the two chains are only loosely intertwined rather than forming a normal helical structure. The peptide portion of the triostin-A molecule interacts intimately with the bases of DNA while the two planar rings are intercalated between the base pairs. The

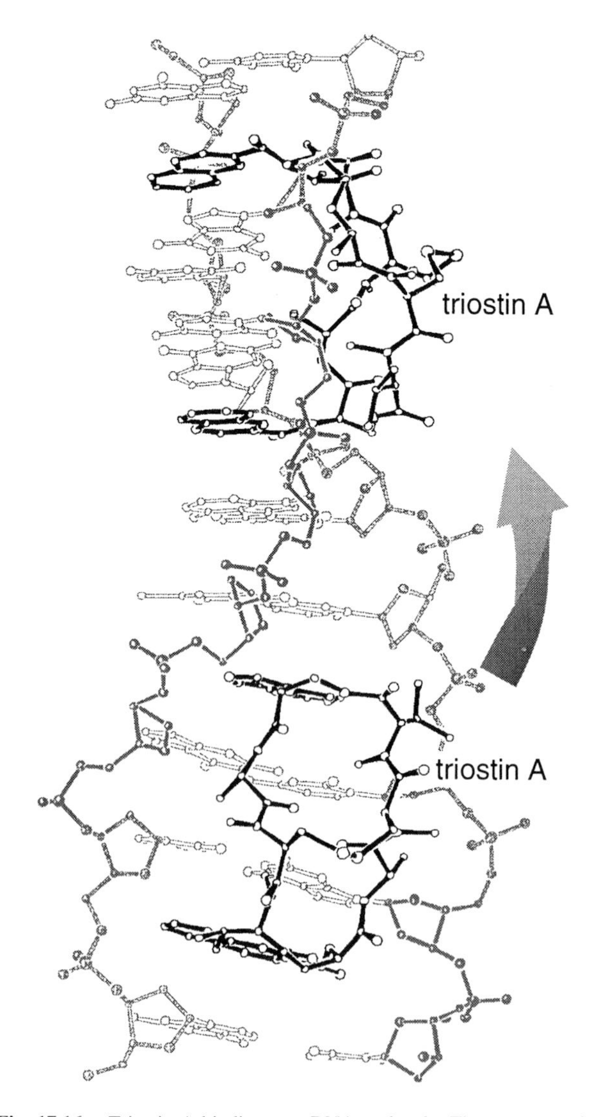

Fig. 17.16. Triostin-A binding to a DNA molecule. The structure of the antibiotic is shown in Fig. 17.15. The intercalation of the two sets of planar rings/molecule severely distorts the DNA helix (Quigley et al., 1986), resulting in some unusual base pairing, called Hoogsteen pairing (Fig. 17.17).

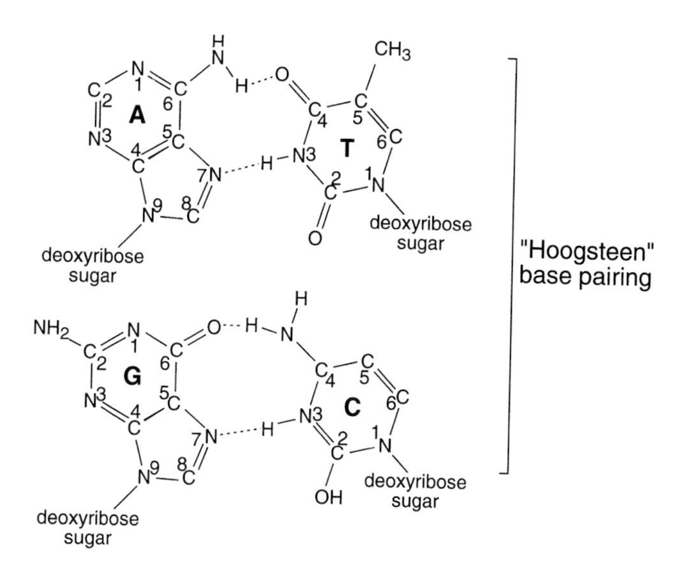

Fig. 17.17. Hoogsteen H-bonding between base pairs. The Hoogsteen base-pair configuration (Hoogsteen, 1959), where two H bonds occur between both the A and T, and the G and C residues (Portugal 1989), causes severe deformation of the DNA helix.

configuration of the base-pairing is such that Hoogsteen base-pairing (Fig. 17.17) is observed rather than the usual Watson–Crick base-pairing.

Molecules and ions that form complexes with DNA change the buoyant density of DNA in cesium salts. Because antibiotics such as actinomycin-D and netropsin, as well as the fluorochromes listed in Fig. 17.15, do not bind randomly to DNA, the extent of change in buoyant density is dependent on the sequence characteristics of the DNA. This sequence-dependent change was widely used to purify tandem arrays of sequences, as well as histone and ribosomal RNA genes, before DNA cloning technology was developed.

17.4 PROTEIN MOLECULES INTERACT WITH DNA TO DETERMINE ITS FOLDING AND TRANSCRIPTION PROPERTIES

Using X-ray diffraction, the classical examples of specific DNA–protein complexes that have been studied involve bacterial proteins such as the restriction endonuclease EcoR1, binding to a small DNA molecule with the base sequence GAATTC (Fig. 17.18). A wide range of proteins that interact with DNA have been studied at the biochemical level to determine the amino acids that interact with DNA. These studies, along with X-ray diffraction analyses, have defined domains or modules of protein that are important in forming a DNA–protein complex. Although the amino acids in DNA-binding domains may vary between different proteins, the three-dimensional geometry remains constant and forms the basis for defining modules of structure that are used to characterize proteins binding to DNA.

17.4.1 Helix-Turn-Helix Modules

The prokaryotic proteins involved in highly specific interactions with DNA often show a characteristic geometry in their structure called helix-turn-helix. The domain of the helix-turn-helix protein that interacts with the DNA contains two α-helices, separated by a β-turn, with one of the α-helices being responsible for the highly specific interaction. The α-helix that interacts specifically with the DNA does so in the major groove. The DNA sequence to which the protein binds can sometimes have an inverted-repeat structure, in which case the respective protein usually binds as a dimer.

17.4.2 Zn Finger Modules

One of the more extensively studied eukaryotic protein–DNA complexes is the binding of transcription factor IIIA (TFIIIA) to the DNA coding for 5S ribosomal RNA. The 5SrRNA is stored in cytoplasm in two forms of ribonucleo/protein particles (RNPs); one is a 7S RNP with the RNA complexed with TFIIIA and the other is a 42S RNP with the RNA complexed with transfer RNA (tRNA) and two nonribosomal proteins. The entire sequence of the 5SrRNA gene plus adjoining, noncoding DNA (5SDNA) has been determined for many plants and animals, and the 5SrRNA genes were among the first to be physically localized to specific sites in the chromosomal complement of the frog *Xenopus laevis*. The entire sequence of the gene coding for *X. laevis* TFIIIA has also been determined.

Detailed in vitro studies of the transcription of the 120-bp 5SrRNA molecule from the 5SDNA template revealed that at least three proteins, TFIIIA, TFIIIB, and TFIIIC, are necessary to form the transcription-initiation complex, which includes the enzyme RNA polymerase, in preparation for efficient transcription. RNA polymerase III is responsible for transcribing 5SDNA.

The cloning of a complementary DNA or cDNA copy of the messenger RNA for *X. laevis* TFIIIA provided the starting material for the production of modified TFIIIA proteins. Deletions in the TFIIIA cDNA were created biochemically, and the modified cDNA

Fig. 17.18. The X-ray crystallographic structure (low resolution) of a restriction endonuclease (EcoR1) bound to DNA containing the site GAATTC that is usually cut by the enzyme; the grey DNA molecule is enclosed by the endonuclease. The structures of endonucleases bound to their DNA-binding sites have been determined for Eco R1 (McClarin et al., 1986; Kim et al., 1990), EcoRV (Winkler et al., 1990), BamH1 (Newman et al., 1995), and Pvu II (Cheng et al., 1994). Other DNA/protein complexes studied at this level of resolution include λ 434 repressor binding to its respective operator, CAP protein binding to promoter DNA sequences, DNA gyrase binding to DNA, and *cer* recombinase binding to its recognition sequence.

was then transcribed and translated in cell-free extracts to provide new TFIIIA proteins. By studying the properties of these modified proteins in transcription assays, a large proportion of the protein, including the region near the N-terminus, was found to comprise the DNA-binding domain. This domain consists of 9 segments, with each segment including approximately 30 amino acids and a zinc ion, so that each segment is referred to as a Zn finger (Fig. 17.19). The Zn-finger structure is a common feature to several eukaryotic DNA-binding proteins so far studied, in contrast to the helix-turn-helix structure, which is more common among the prokaryotic DNA-binding proteins.

The positioning of cysteine and histidine amino-acid pairs to form a site of interaction with the Zn ion is the dominating feature of Zn fingers. Substitution of histidine by asparagine leads to structural disruption of the Zn finger and allows the function of individual Zn fingers in TFIIIA to be examined. From mutational analyses

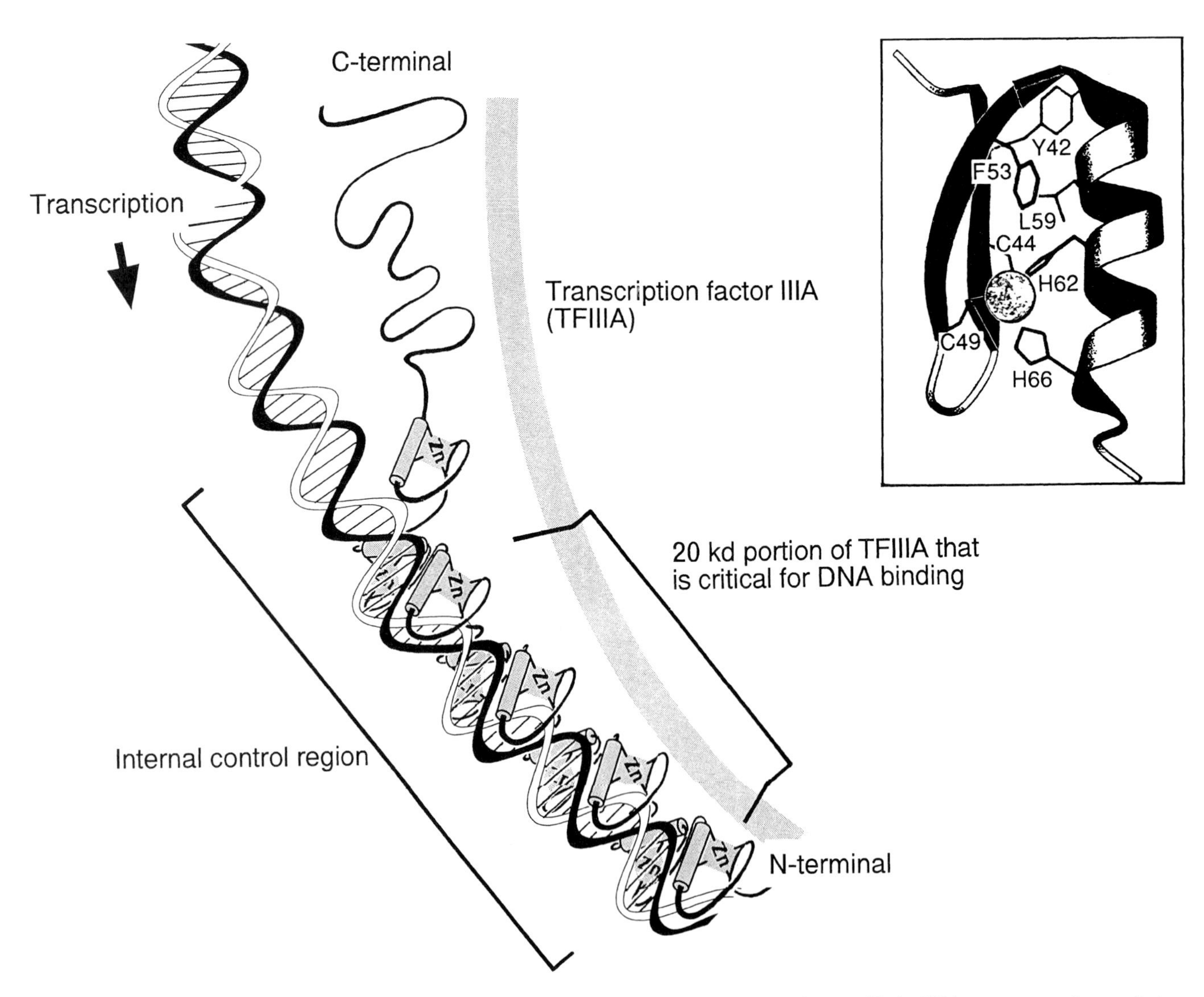

Fig. 17.19. The transcription factor IIIA (TFIIIA)–5SDNA complex, showing the interaction of Zn fingers with the DNA sequences at the so-called internal-control region. [Based on Smith et al. (1984); Brown et al. (1985); Miller et al. (1985); Tyler (1987); Sharp and Garcia (1988); Churchill et al. (1990); Keller et al. (1990); Kim et al. (1990); Pavletich and Pabo (1991); Sands and Bogenhagen (1991).] The Zn fingers are numbered from 1 to 9 starting from the N-terminal. Mutational studies in which the Zn-finger domains of the C-terminal region are disrupted reduce the activity of TFIIIA in stimulating transcription, indicating that Zn fingers are probably required for higher-order interactions in establishing the transcription complex. Disruption of any one of the six Zn fingers in the N-terminal region (marked in the diagram as the portion critical for DNA binding) reduces DNA binding but does not alter its ability to stimulate transcription (Vrana et al., 1988; Del Rio and Setzer, 1993; Hansen et al., 1993). Two important functions of TFIIIA, together with another factor TFIIIC, are the positioning of the TFIIIB protein, which acts as a transcription-initiation factor (Kassavetis et al., 1990; Paule, 1991), and the prevention of nonspecific repression by the packaging of DNA into nucleosomes (Stunkel et al., 1995). The inset shows details of a Zn finger. The protein structure is depicted using a standard ribbon type of presentation, with the Zn atom shown as a gray sphere. The amino acids considered particularly significant for this structure include phenylalanine (F53), tyrosine (Y42), leucine (L59), two cysteines (C49, C44), and two histidines (H62, H66) (Luisi, 1995).

of this type, it is clear that each of the nine Zn fingers is important in the efficient DNA binding/stimulation of transcription of TFIIIA, and that the binding is cooperative. For example, the binding of finger 4 is a prerequisite for fingers 5–9 to bind. Only Zn fingers 7–9, however, are critical for the assembly of the other components into the final transcriptional complex. The TFIIIA protein is a good model for a feature of transcription factors that appears to be widespread, namely that the DNA-binding domain is separate from the domain involved in actually stimulating transcription. In the case of TFIIIA, structural disruption of Zn fingers 1–6 affects only the ability of the protein to bind to the internal-control sequence, whereas structural disruption of Zn fingers 7–9 results in the loss of transcription stimulatory activity. The latter region is therefore involved in higher-order interactions and also binds to the internal-control DNA sequence.

17.4.3 Leucine-Zipper Modules

Some eukaryotic DNA-binding proteins contain a leucine-zipper structure (Fig. 17.20), considered to be a major point of interaction between these proteins, which are often comprised of two subunits. The dimeric form of the proteins is reflected in the inverted-repeat structure of the DNA sequences with which they interact. An important consequence of the dimeric nature of these proteins is that a wide range of heterodimeric molecules can be formed from a limited pool of different protein subunits, interacting via a common leucine-zipper module in their structures. In addition, identical protein subunits can combine to form homodimers. As the various protein subunits can have different DNA-sequence-binding specificities, dimeric DNA-binding proteins can clearly provide for a great deal of variability in the type of DNA sequence that is bound. The DNA-binding sites for these proteins are involved in the formation of transcription-initiation complexes of genes, so variations in the relative concentration of the different interacting protein subunits would be expected to vary the control over the expression of certain genes. The mechanism by which this differential control is thought to operate is that the range of DNA-binding proteins compete for involvement, as transcription cofactors, in the formation of specific transcription-initiation complexes. In this way, the efficiency of transcription by a RNA polymerase of a particular gene can be modulated as part of the cascade of differential gene activity leading to cell differentiation. The blockage of the binding of specific transcriptional activators is also an important determinant in defining the final suite of genes expressed in a particular cell type; this blockage may be due to nucleosome formation (discussed later in this chapter) or the presence of specific repressor-type molecules.

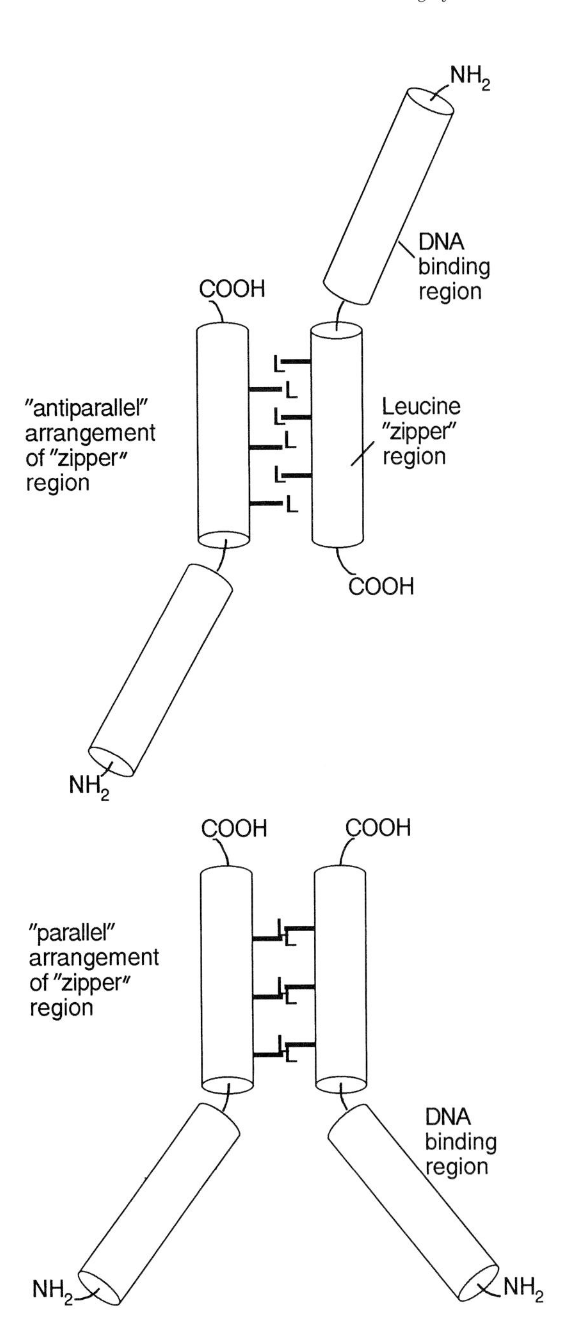

Fig. 17.20. The leucine-zipper module that characterizes eukaryotic dimeric DNA-binding proteins and that is the site of interaction between the dimeric protein subunits. The leucine-zipper module, the helix-turn-helix (*see* Section 17.4.1), and Zn-finger motifs are evolutionally conserved features of DNA-binding proteins that are becoming widely recognized (Klug and Rhodes, 1987; Lee et al., 1989; Struhl, 1989; Jones, 1990; Harrison and Aggarwal, 1990; McKnight, 1991).

17.4.4 Distortion of the DNA Helix by TATA-Box Binding Protein

Most genes that code for proteins in eukaryotes are transcribed by the enzyme RNA polymerase II. The way in which this enzyme forms a transcription-initiation complex with other protein factors is analogous to what was described for the RNA polymerase III-initiation complex formed to synthesize 5S RNA (*see* Section 17.4.2). One of the first steps in the formation of the RNA polymerase II transcription-initiation complex is the binding of a transcription factor IID, TF IID, to a recognition sequence approximately 30 bp, 5′-upstream from the start of a gene. The recognition sequence is TATA, which is extremely widespread in eukaryotes as a signal to form RNA polymerase II transcription-initiation complexes. The subunit of TFIID that is responsible for binding to the TATA box is the evolutionally highly conserved TATA-box binding protein or TBP. The X-ray diffraction analysis of this protein demonstrates a severe distortion of the DNA helix, which results from the protein interacting in the minor groove and forcing the helix into an 80° bend, or kink (*see* Fig. 17.12). The kink is produced by large roll angles (R) induced between base pairs of the DNA helix.

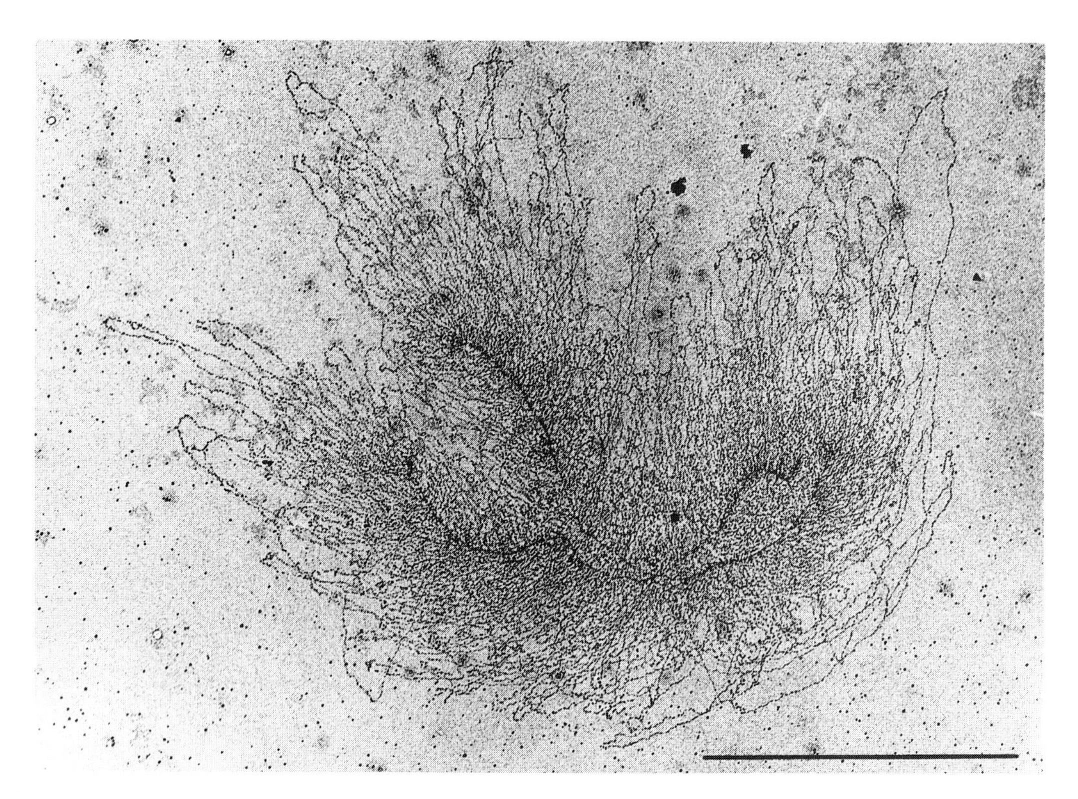

Fig. 17.21. Electron-microscopic evidence for the nucleosome structure of chromatin in a chromosome. Analysis of figures such as this indicate that chromatin is composed of chromosomal DNA complexed with protein to form arrays of beadlike structures, arranged in large loops attached to a matrix, as is clearly evident in the photograph (Weith and Traut, 1980; photograph kindly supplied by Dr. W. Traut). The bar represents 1 μm.

17.5 FOLDING OF DNA INTO CHROMATIN: THE NUCLEOSOME

In the early 1970s, biochemical studies and electron microscopy indicated that DNA in the chromatin of eukaryotic nuclei was folded into regular units. Biochemical evidence for these units came from investigating a problem of DNA degradation during the incubation of isolated nuclei to assay DNA polymerase activity, an enzyme required for the replication and repair of DNA. When DNA was isolated from nuclei after incubation in the assay and analyzed using electrophoretic procedures, it was evident that degradation of the DNA was occurring in discrete steps of approximately 200 base pairs (*see* Chapter 2, Fig. 2.10). The investigators suggested that this reflected the presence of a regular folding of DNA, moderated by proteins, and resulting in a high relative sensitivity of the DNA helix to DNA-degrading enzymes, DNases. Electron-microscopic studies also provided evidence for a primary folding of DNA into beadlike structures which are visible after gentle disruption of chromosomes (Fig. 17.21), and which are units of approximately 200 base pairs around a core of histone proteins. The combination of a histone core and associated DNA comprises the nucleosome.

17.5.1 Histone Proteins

The intimacy of histone–DNA interactions is reflected by the fact that histone synthesis is closely coordinated with the replication of DNA in the S phase of the cell cycle (*see* Chapter 18). There are five main types of histones, of which H2A, H2B, H3, and H4 form the core of the nucleosome, and H1 binds to the nucleosome-linker regions. Minor structural variants of H1 and the other histones can be distinguished. The importance of histones in the basic arrangement of chromosomes is reflected by their evolutionally highly conserved amino-acid sequences. The primary amino-acid sequences, taken together with physical studies, indicate that the H2A, H2B, H3, and H4 histones have two domains. The N-terminal portion of the histone polypeptide, near the terminal NH_2 group, is composed of an excess of positively charged amino acids, particularly lysine and arginine, forming a distinctive hydrophilic ''tail,'' which reacts readily with water. The C-terminal portion of the polypeptide, near the terminal COOH group, is composed largely of uncharged amino acids, which react poorly with water and are hydrophobic. These hydrophobic domains form a variety of secondary structures. Although the amino-acid sequences of both domains are highly conserved in evolution, only the C-terminal domain is critical for the formation of the nucleosome, as discussed below. Within the nucleosome, the four histone molecules form two pairs, H3–H4 and H2A–H2B, which, in turn, make up a complex, comprised of $(H3–H4)_2$ plus $(H2A–H2B)_2$, forming the protein core of the nucleosome (Fig. 17.22). The $(H3–H4)_2$ tetramer is more stable than the $(H2A–H2B)_2$ tetramer. The histones interact with the minor groove in DNA and with each other via their hydrophobic C-terminal portions. The hydrophilic tails are free to interact with the negatively charged portions of other proteins, as well as the sugar–phosphate backbones of nearby DNA molecules.

The dispensable nature of the N-terminal region of the histones for nucleosome formation has been inferred from genetic studies in yeast. In these experiments, the two normal chromosomal copies of the histone gene under study (H4) were inactivated, and the cells were rescued by introducing the respective histone gene on a plasmid. When this plasmid–histone gene had parts of the N-terminal domain missing or modified, the yeast cells were still viable. Interestingly, deletions in the N-terminal domain of H4 can also activate the silent mating loci of yeast, thus emphasizing that the biology of chromosome structure and its alteration is still an area of intense research. The silent mating loci contain sequences that can interact, by a process called conversion, with genes encoding mating and sporulation functions to facilitate a change in the mating type of a yeast cell. These silent mating loci, and genes placed near them, are usually completely repressed with respect to transcriptional activity. Similar deletions in the other three histones do not have these effects.

DNA folds around the histone core to form a left-handed superhelix of almost two turns (*see* Fig. 17.22), and this folding involves a major distortion of the DNA helix. The curvature of the

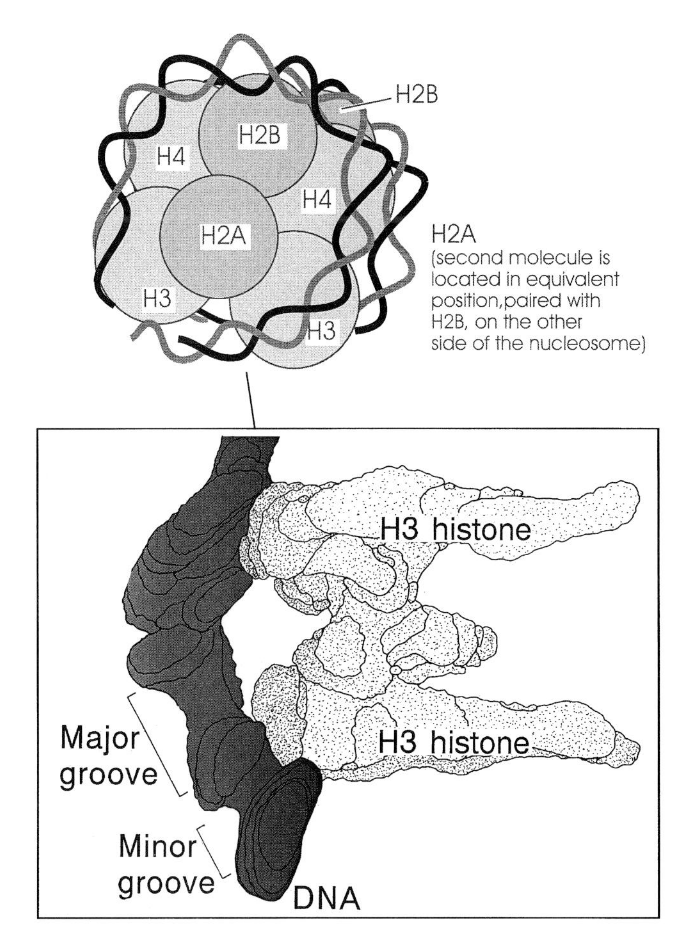

Fig. 17.22. The structure of a nucleosome as determined by X-ray crystallography. The upper panel is a diagrammatic interpretation of nucleosome structure. The lower panel shows the electron-density map (0.7-nm resolution) of the H3 histone dimers and the respective DNA segment, to indicate in more detail the protein–DNA contacts. (From Richmond et al., 1984). In assembling a nucleosome, an $(H3–H4)_2$ tetramer is the first to bind to the DNA, followed by two H2A–H2B dimers, and, finally, by H1 proteins (not shown) (Wolffe, 1994). The incorporation of DNA sequences into nucleosomes acts as a general negative control on gene expression (Han and Grunstein, 1988; Hirschhorn et al., 1992; Felsenfeld, 1992; Roth et al., 1992).

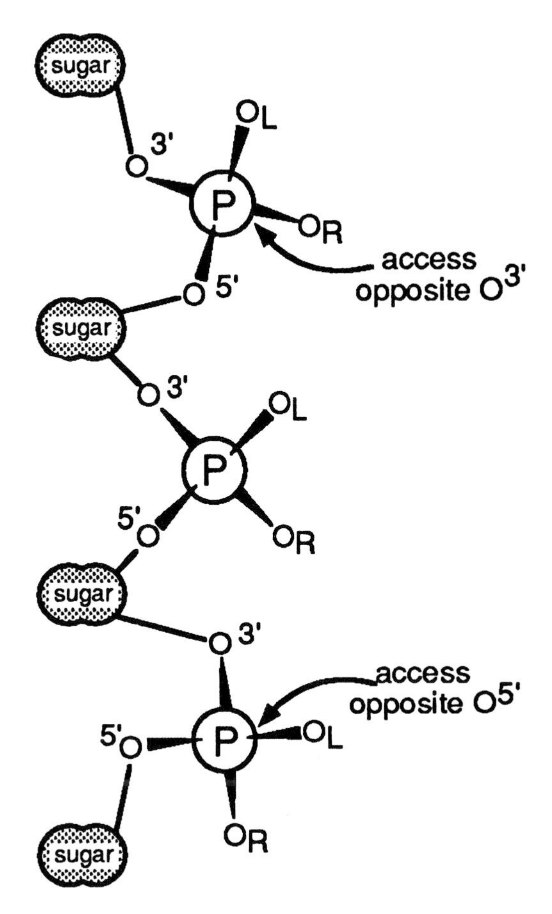

Fig. 17.23. The DNase sensitivity of DNA as a function of its folding around the nucleosome and exposure of active sites for nuclease access. The diagram indicates that the conformation of the DNA determines the relative sensitivity of sites in the sugar–phosphate backbone to nucleases such as micrococcal nuclease, DNase I and II. Analysis of the DNA products resulting from limited action by these enzymes indicates the parts of the DNA double helix that face outward, away from the parts that interact with the nucleosomal protein (Drew and Travers, 1985). The protection of DNA from digestion by nucleases is a key technique in determining the positions where proteins bind to the DNA.

DNA helix means that the sugar–phosphate backbone on the outside of the curve, at particular points, is relatively more exposed and therefore more susceptible to cleavage by an enzyme such as DNase I (Fig. 17.23). Experiments with DNase I have accurately described the winding of the DNA helix and, in combination with X-ray studies, have provided data to elucidate the structural details of nucleosomes. Regions of chromatin that are particularly sensitive to DNase I are referred to as hypersensitive sites (HS).

A fifth histone, H1, is also involved in the nucleosome complex, but not as part of the core particle. The H1 class of histone is generally considered to bind to the DNA between nucleosomes (Fig. 17.24), and it is possible that the DNA molecule may continue to coil around this protein. The H1 histone and a specialized H5 form specific to avian erythrocytes are speculated to be involved in the arrangement of nucleosomes into larger complexes (Fig. 17.25), as part of the process for compacting the eukaryotic DNA into chromosomes and repressing transcription. Higher forms of coiling are implied to occur in order to pack the long strands of DNA into nuclei. Cytological evidence for coiling in mitotic chro-

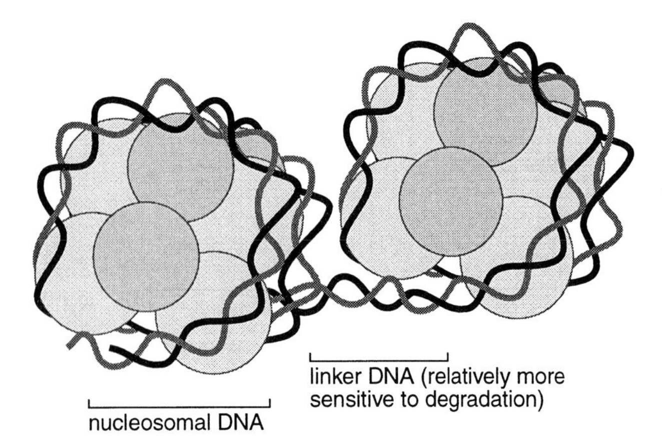

Fig. 17.24. A diagrammatic representation of the linker DNA between nucleosomes. The positioning of nucleosomes leaves some DNA that is not directly bound to histones, called linker DNA. This unbound DNA is more sensitive to cleavage by enzymes such as DNase I and it is also a region of the genome that is more accessible to specific binding by proteins that control levels of DNA methylation, the formation of transcription complexes, chromatin conformation, and initiation of DNA replication. Although the positioning of nucleosomes on the DNA is generally considered to be random, the presence of DNA-binding proteins attached to specific sites on the DNA can result in an array of nucleosomes that are arranged at specific locations near this binding site (Kornberg and Stryer, 1988; Thoma and Zatchej, 1988; Hayes et al., 1990).

mosomes was obtained in the late 1960s by preparing chromosomes in relatively low-salt solutions (Fig. 17.26) and is consistent with a model of several levels of coiling, or spiraling, to pack DNA strands.

17.5.2 Nonhistone Proteins Are Also Important in the Organization of DNA into Chromosomes

Proteins that are intimately associated with chromatin, and thus not readily lost when chromatin is isolated for biochemical studies, are extremely heterogeneous in nature. The nuclear proteins not defined as histones are called nonhistone proteins, and those that are intimately associated with the chromatin complex, at some stage in the cell cycle, include the proteins involved in the formation of selected transcription-initiation and transcription-termination complexes and DNA replication. The phosphorylation of nuclear proteins, including histones, by protein kinases is critical in inducing conformational changes in the chromatin. The modification of nonhistones and histones by the addition of acetyl, methyl, and ADP-ribosyl groups (Fig. 17.27) also requires the presence of specific enzymes. One class of nonhistone proteins that are reproducible components of the nuclear chromatin is the so-called high-mobility group, HMG. Two proteins in particular, HMG 14 and HMG 17, have been highly purified from calf thymus and, in biochemical experiments, have a high binding affinity for nucleosomes. HMG proteins act to disrupt the compaction of DNA induced by the histones and may therefore be involved in the derepression of chromatin for transcription in vivo; the HMG 1(Y) protein has been implicated in the assembly of transcription-initiation complexes. The abundance of the HMG class of protein has also led to the suggestion that they may function to facilitate the formation of nucleosomes by bending DNA in preparation for folding into the compact nucleosome structure.

Another group of nonhistone proteins that were initially characterized by their enzymatic activity but were later shown to be significant components of chromatin at certain stages in the cell cycle are the DNA topoisomerases. Their enzymatic activities are characterized by their ability to alter the overall shape, or topology, of DNA. Topoisomerase II is particularly important in untangling newly replicated DNA molecules, and temperature-sensitive mutations in yeast have demonstrated that topoisomerase II is vital for life. It also is a major component of the protein ''scaffold,'' or nuclear matrix, to which defined sections of the DNA molecule attach in vivo. Immunostaining with antibodies to topoisomerase II has demonstrated that this protein is a major component of the axial core of mitotic chromosomes. The location of topoisomerase II is not, however, limited to the central chromosome axis, and at

Fig. 17.25. The solenoid arrangement of nucleosomes. The solenoid has a left-hand helical turn and the vertical distance between nucleosomes is 110 Å (Widom and Klug, 1985). The H1 histone that binds to the linker DNA is one of the main proteins involved in forming this higher-order folding of chromatin (Zlatanova and van Holde, 1992).

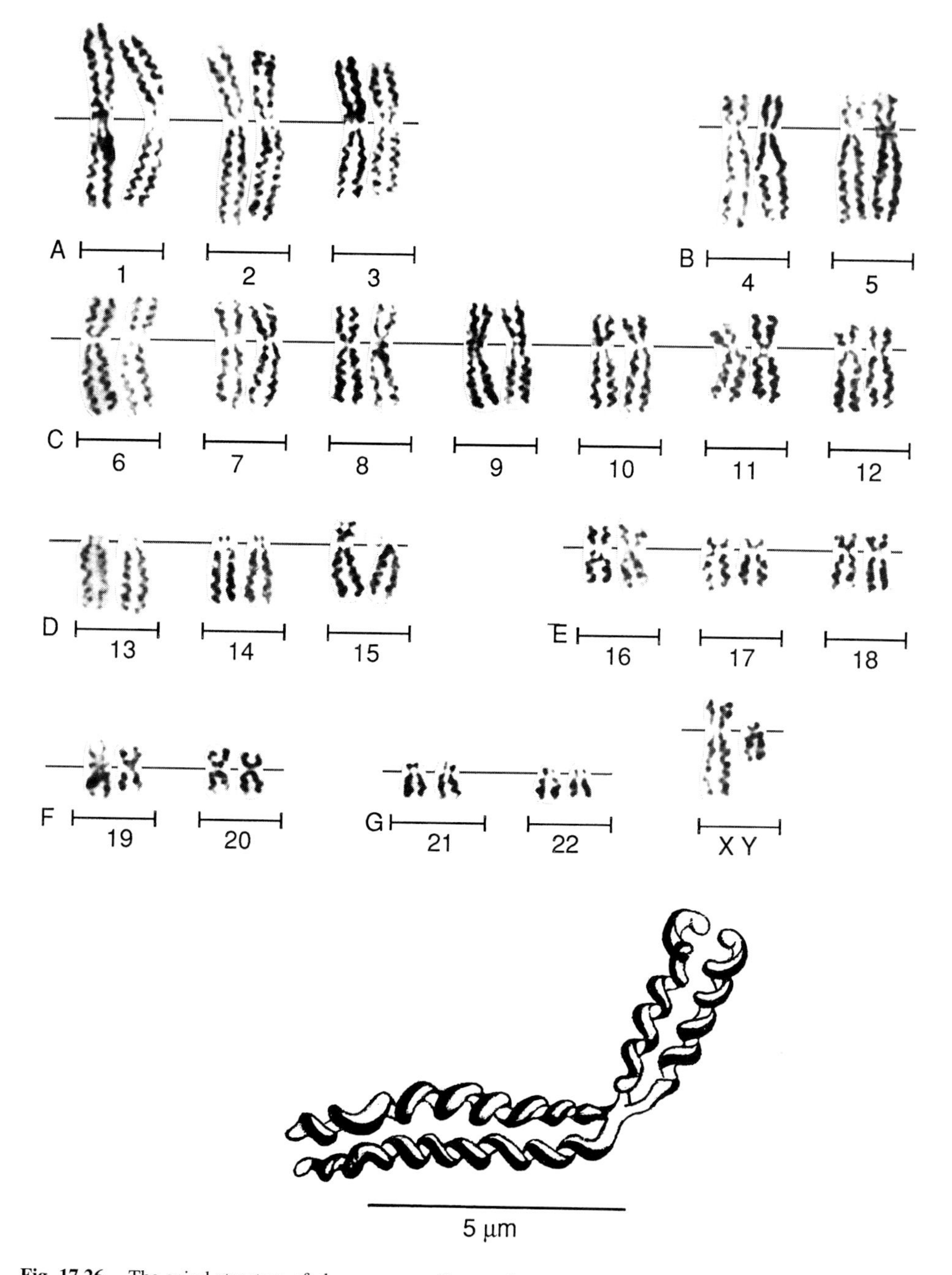

Fig. 17.26. The spiral structure of chromosomes. Human chromosomes arranged in a standard-karyotype format (Ohnuki, 1968) after preparation for analysis under relatively low-salt conditions. White blood cells stimulated to divide (*see* Chapter 6) were resuspended in a 4:1:2:2 mixture of KCl (0.055 *M*), KSCN (0.062 *M*), $NaNO_3$ (0.055 *M*), and CH_3COONa (0.055 *M*), at room temperature for 90–120 min, before preparing slide preparations for microscopic examination. This particular mixture of salts uncoiled the chromosomes sufficiently to display the spiral structure. The drawing reproduces the interpretation of the cytological observations in Ohnuki (1968).

interphase, this molecule has been shown to occupy well-defined regions of the nucleus.

The eukaryotic topoisomerase I cleaves one of the DNA strands and binds to the 3′-phosphate via a tyrosine residue. The complex then rotates to release torsional stress (*see* Fig. 17.31) and reforms the sugar–phosphate backbone upon removal of the topoisomerase I protein. Topoisomerase II cleaves both strands in a staggered manner (Fig. 17.28), with both 5′-ends becoming covalently bound to the protein. Rotation of the complex, or untangling of double-stranded DNA molecules, is followed by the reassembly of the sugar–phosphate backbone and removal of the protein. Repetitive DNA-sequence motifs in the noncoding regions near genes have been shown to have sequence similarities to the natural binding site of topoisomerase II. These regions are considered to be central in the overall folding of DNA in the nucleus and, as discussed in Chapter 16, in modulating the activity of groups of genes.

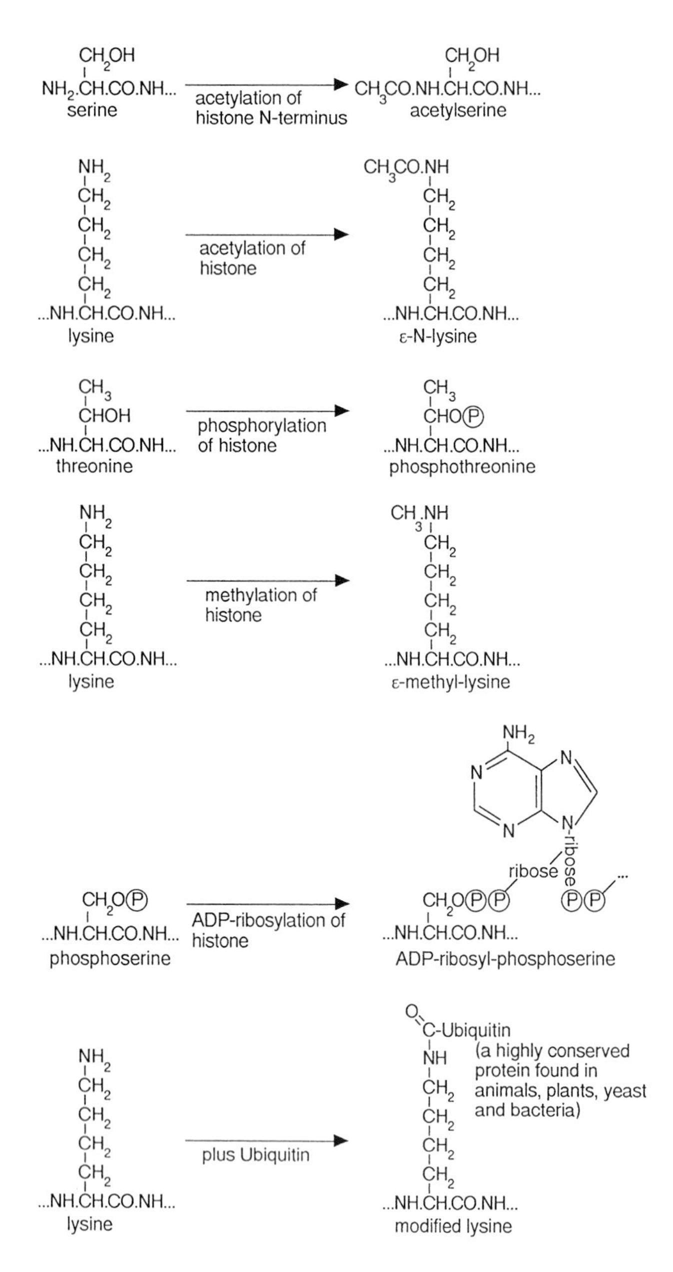

Fig. 17.27. The chemical modifications that can occur to histones. These modifications are not unique to histones and are widely utilized in other cellular processes to modify the properties and activity of proteins (Hayaishi and Ueda, 1977; Finley et al., 1989; Hill et al., 1991; Loidl, 1994).

17.5.3 Chromatin Domains

The chromatin domain consists of loops of the 30-nm fiber attached to a nuclear scaffold (Fig. 17.29). The classical studies on lampbrush chromosomes (*see* Chapter 2, Fig. 2.9) demonstrated loops varying in size from 3 to 300 kb of DNA and had a major influence on the concept that chromatin domains are important in the control of gene activity. Detailed biochemical studies on the hemoglobin locus have resulted in attractive models that argue for the importance of changing chromatin domains in the developmental timing of gene expression. One model, illustrated in Fig. 17.30, argues that when critical-control regions, HS1, HS2, HS3, or HS4, are physically close to the promoter of a particular hemoglobin gene, the respective gene is activated for transcription. The HS regions have the biochemical characteristic of hypersensitivity to DNase digestion as a result of a structural modification to the chromatin that leads to a reduced level of protection of the respective DNA against DNase; normally, this protection is provided

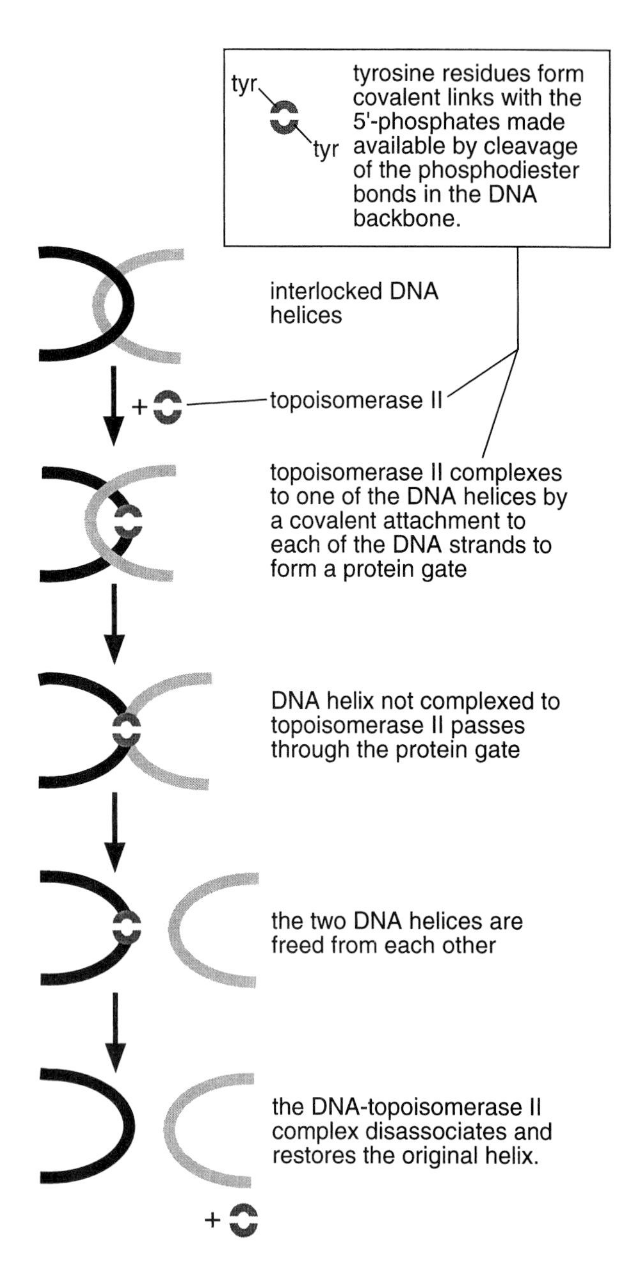

Fig. 17.28. A molecular pathway showing the mode of action of cleavage of DNA by topoisomerase II and the resolution of interlocked double helices. The diagram is based on the work of Wang (1985) and Huang et al. (1992). This activity is considered a crucial feature of topoisomerase II as an integral chromosome-scaffold protein, because chromosome pairing and separation, as well as protein binding to DNA and transcription, are expected to continually introduce strains that need to be relieved within the chromatin complex. DNA topoisomerase II undergoes cell-cycle-dependent alterations in amount and stability, with much of the protein being degraded upon cessation of mitosis. These changes are consistent with this protein being a major structural component of condensed chromosomes (Boy de la Tour and Laemmli, 1988; Heck et al., 1988; Käs and Laemmli, 1992).

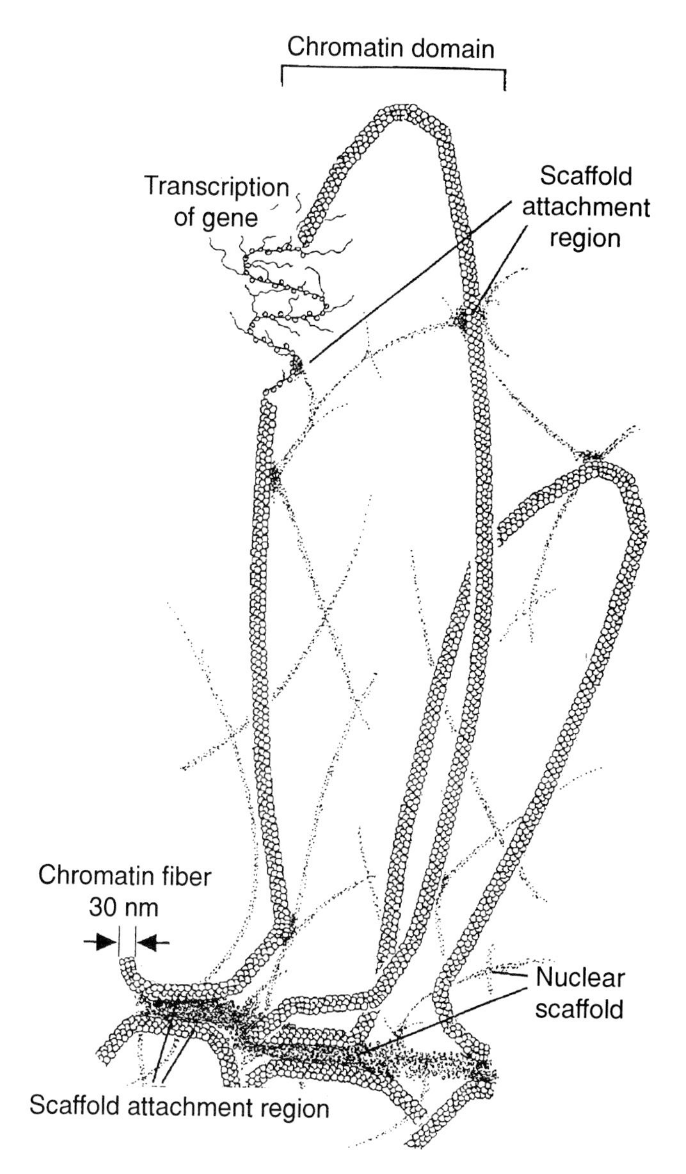

Fig. 17.29. A diagrammatic representation of the chromatin domain. Although the nuclear scaffold is shown as a network of fibers, the detailed structure is unclear because many cytological techniques use acid fixatives that can cause major changes to biological structures (Hill et al., 1991). The diagram is based on cytological observations of chromosome structure in spermatozoa (Ward et al., 1989; Zalensky et al., 1995), lampbrush chromosomes (Gall and Callan, 1962), electron-microscopic observations on spread chromosomes (*see* Fig. 17.21), and biochemical studies on the existence of scaffold-attachment sites (Gasser and Laemmli, 1986; Slatter et al., 1991; van Driel et al., 1991). A functional definition of a chromatin domain is that it corresponds to a region that can be moved to any place in the genome without showing any effect on gene expression (Kellum and Schedl, 1991). The expression of genes introduced into a genome by transformation techniques (*see* Chapter 23) is sensitive to where they "land" within the genome, and the definition implies that there exists a clear limit to the DNA sequences that can affect gene expression. Furthermore, the definition implies that the sequences affecting gene expression lie within the loops illustrated and that specialized structures may exist at the base of the loops to insulste the genes within the loop from the influence of sequences elsewhere in the genome. The expression of some genes such as the rRNA gene unit (*see* Chapter 20, Section 20.2) is not affected by their position within the genome (Karpen et al., 1989).

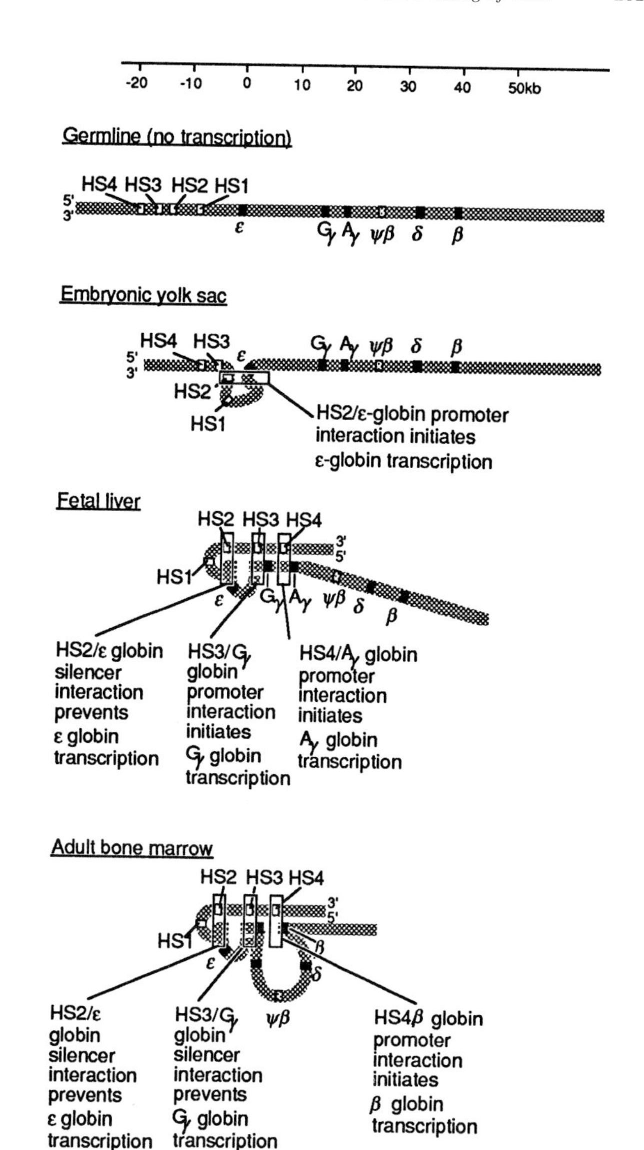

Fig. 17.30. The hypothetical long-range interactions in the chromatin complex that modify specific globin genes (ϵ, Gγ, Aγ, $\psi\beta$, δ, and β) at particular stages of development. The principle is that by suitable folding of the chromatin domain, specific locus-control regions (HS1–HS4) are spatially relocated to positions immediately adjacent to the promoters of certain globin genes, resulting in transcription of the respective genes (Townes and Behringer, 1990). The three-dimensional folding of the chromatin is argued to be an important variable in the control of gene expression (Forrester et al., 1987; Shih et al., 1990; Watt et al., 1990; Talbot and Grosveld, 1991; Vyas et al., 1992; Reitman et al., 1993; Wijgerde et al., 1995). The production of transgenic mice carrying the entire 70-kb region of the globin complex confirms that most of the control sequences are resident in this DNA segment, although additional regulatory sequences may exist outside this region (Gourdon et al., 1994). The type of long-range interactions suggested for the activation of specific globin genes is also considered important for the control of bacteriophage λ transcription (Hochschild and Ptashne, 1988).

by intimate association between DNA and nucleosomes. The HS regions are defined as "open" chromatin configurations and are associated with gene activity. At different stages in development, the model in Fig. 17.30 postulates that different "active" chromatin domains are formed and, as a result, different hemoglobin genes are transcribed. The HS regions are part of the locus-control region (LCR; *see* Chapter 16, Fig. 16.25) and although details of the model will evolve as more data accumulates, it provides a focus for the concept that dynamic, long-range interactions occur in chromatin to modify gene expression.

17.6 STRUCTURAL CHANGES IN CHROMATIN DURING GENE EXPRESSION

The intimate DNA–protein complexes formed in order to initiate efficient transcription can conflict with, or be enhanced by, the folding of the DNA molecule into nucleosomes and chromatin fibers. The formation of a transcription-initiation complex, and subsequent transcription, involve changes in conformation of the DNA molecule and present special challenges at a conceptual level.

The following terms are used to describe the topology of a DNA molecule:

Superhelix: Derived from a straight double helix by curvature of the helix and a change in the T (helical twist) value between successive base pairs. The wrapping of DNA about a histone core to form a nucleosome is an example of a superhelix, and its curvature can be calculated from the average R value between base pairs. A DNA superhelix, comprising 145 bp around a histone core, contains 1.8 left-handed turns and results in an average deflection of the double-helical axis by 47° for each turn.

Supercoil: The change in shape of a helical DNA when its ends are fixed and a part of the helix undergoes local unwinding or increased winding (i.e., a local change in the T value). The local unwinding may result from transcription of the DNA (Fig. 17.31), and occurs because torsional stresses in the DNA molecule need to be dispersed.

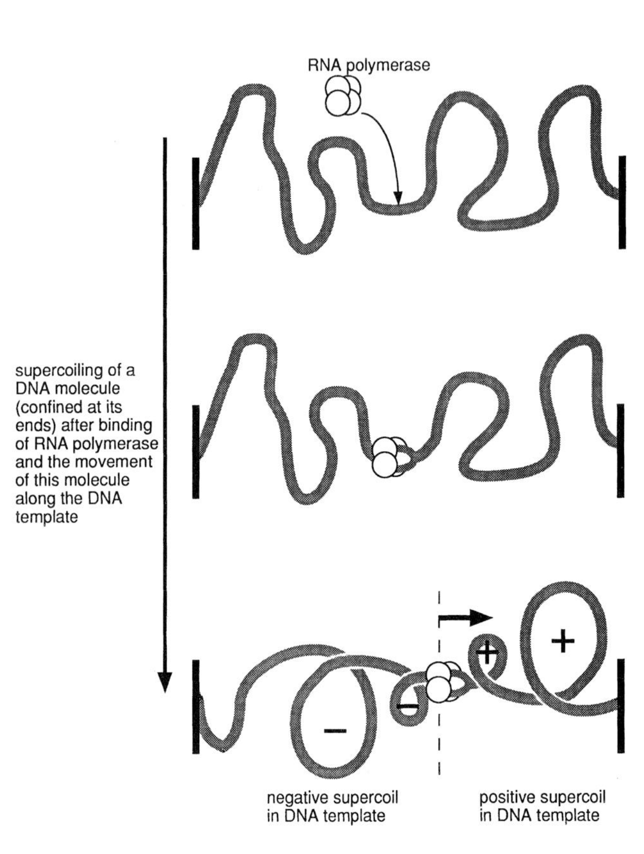

Fig. 17.31. Changes in DNA topology after distortion in one part of a constrained DNA helix. The example shows the binding and progressive movement of RNA polymerase along a DNA double helix that is fixed at both ends. This DNA transcription involves a local unwinding of the DNA helix as the polymerase progresses from left to right, resulting in an induced torsion that can be relieved only by the formation of a positive supercoil in front and a negative supercoil behind the point of attachment (Liu and Wang, 1987; Drlica, 1990). Similar problems arise when any protein binds to a DNA helix that is constrained by attachment to a larger structure such as a nuclear scaffold. The potential of a particular DNA sequence to take up the Z-DNA conformation (with a left-handed instead of a right-handed helical turn; *see* Fig. 17.10) could also relieve some of the stress that leads to negative supercoiling (Hill, 1990).

The right-hand helix of the DNA molecule is considered to have positive winding, and the supercoil that forms to compensate for the unwinding of the helix is said to be negative. Conversely, an increase in winding of the helix induces a positive supercoil. The induction of supercoiling is a reflection of torsional stress, produced by a process such as transcription (*see* Fig. 17.31) or replication, and proteins with topoisomerase activity release this stress by transiently cleaving the DNA molecule and then rejoining it. In eukaryotes, the cleavage of DNA by topoisomerase II has been argued to occur in nucleosomal linker regions and unfolded chromatin, close to the AT-rich regions that characterize topoisomerase II-binding sites.

The folding of DNA, particularly the noncoding DNA preceding a gene, into a nucleosome structure can be considered a general negative control of the initiation of transcription. In addition to repression by nucleosome formation, however, examples of specific repressions analogous to those found in operon-transcription control in prokaryotes can also be found in eukaryotes (for example: *see* control of histone synthesis in Chapter 16, Section 16.8).

Specific derepression, or positive control, of the DNA region involved in forming the transcription-initiation complex is required for transcription of a gene to proceed. There are at least two ways (not mutually exclusive) in which this positive control can occur. It is possible that one of several proteins, essential for the transcription-initiation complex, becomes stably bound to its particular DNA-recognition site and excludes histone binding. The protein would have to remain in position even during DNA replication. Initiation of transcription can then take place when the other required proteins become available. A variation on this all-or-none competition between histones and transcription-initiation complex factors is the possibility that they can both occupy the same DNA segment. Some transcription-initiation complex factors have been shown to bind nucleosomes and it is thus possible that in some instances the histones are an intimate part of the transcription-initiation complex.

Another way of resolving the competition between histone binding and specific activator-protein or repressor-protein binding to

DNA is for part of the transcription-initiation complex to form after DNA replication, before the nucleosome structure has time to establish itself. This dynamic competition between nucleosome formation and specific DNA-binding proteins may be particularly important early in the process of establishing differential gene expression in development. Although a relatively small number of transcription factors establish the complex patterns of gene expression in eukaryotes, the required variation is likely provided by the combination of these transcription factors acting as specific DNA-binding proteins and also inducing structural changes in the DNA, bringing together other proteins bound to particular sites.

Once transcription starts, the nucleosomes do not inhibit RNA polymerase movement, and major structural changes can occur in the chromatin, as highlighted by the formation of puffs in polytene chromosomes (Fig. 17.32). Polytene puffs are classical cytogenetic landmarks (*see* also Chapter 2, Fig. 2.7).

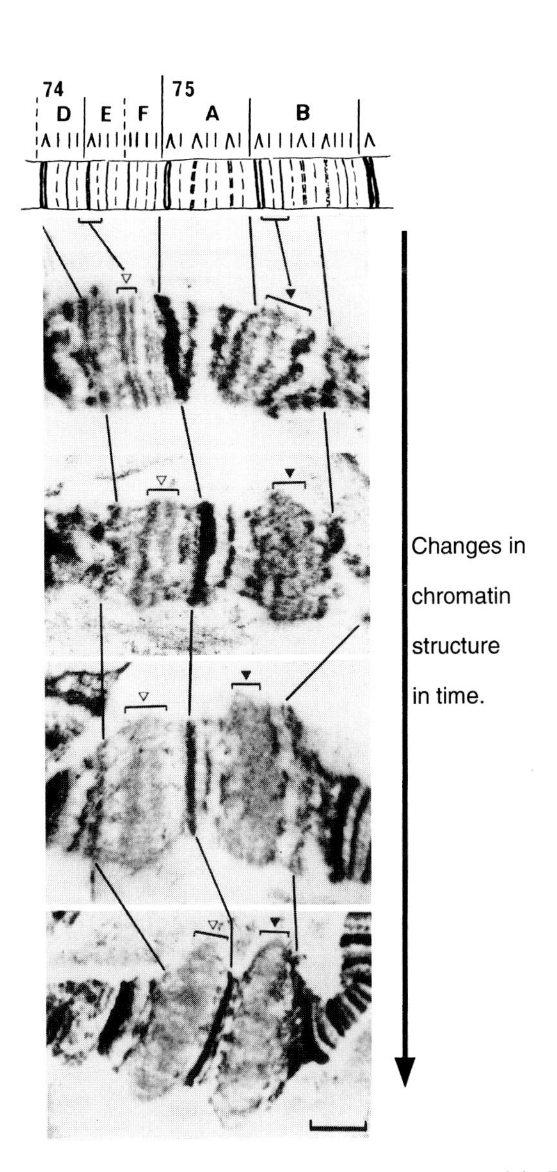

Fig. 17.32. The development of chromosome puffs in *Drosophila*. The puffs that occur in polytene chromosomes are not only cytogenetic landmarks but also reflect the major changes in chromatin structure that occur when a gene is transcribed (Semeshin et al., 1985, Ashburner, 1990).

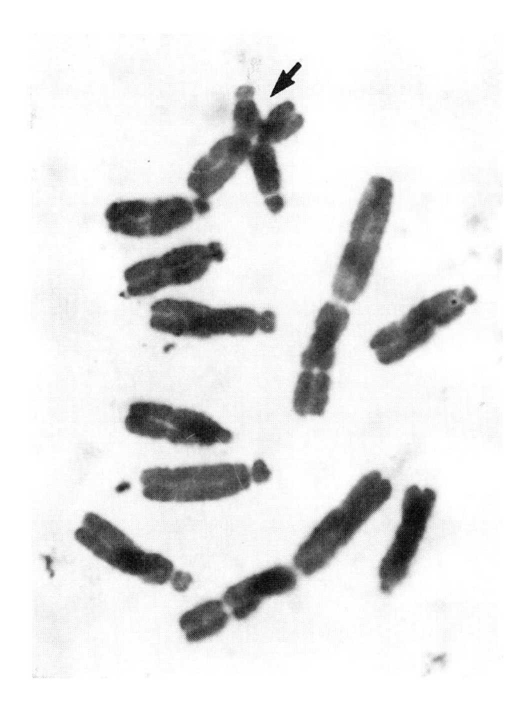

Fig. 17.33. Mitotic chromosomes from a root-tip cell nucleus of a bean plant showing a pair of homologous chromosomes apparently undergoing mitotic crossing over (arrow). (Photograph kindly supplied by Dr. B. Friebe.) The close association of homologous chromosomes in metaphase preparations from some organisms reflects a degree of order regarding the distribution of chromosomes within a nucleus. The biological consequences of disruption of the relationships between chromosomal DNA sequences in the nucleus are significant in phenomena such as position-effect variegation and homology-dependent gene silencing (*see* Chapter 9, Section 9.4.1 and Table 16.2, Chapter 16 respectively).

17.7 FOLDING OF DNA-PROTEIN COMPLEXES ON A BROADER SCALE WITHIN THE INTERPHASE NUCLEUS

In addition to the chromatin domains previously discussed, it is clear that chromosome domains also exist within the eukaryotic nucleus. Some of the earliest studies in cytology suggested that chromosomes occupy distinct domains within the nucleus. One manifestation of this is that homologous chromosome pairs can be intimately associated in mitotic nuclei (*see* also Chapter 4, Fig. 4.6) and may even undergo what is referred to as somatic crossing over (Fig. 17.33). Recent studies, using the technique of in situ hybridization, have demonstrated that when chromosomes decondense after mitosis, they do not intermingle but, instead, retain clear divisions between individual chromosomes (Fig. 17.34) and between groups of chromosomes. In human sperm nuclei, the centromeres form a cluster (or chromocenter), and the ends of any one chromosome are located near each other at the nuclear membrane. Attachment of specific chromosome regions to sites embedded in the nuclear lamina may contribute toward the maintenance of these domains. In vitro studies with cell-free extracts from *Xenopus* eggs have shown that the pool of soluble proteins present in these extracts can fold introduced DNA into chromatin, after which four different types of nuclear-lamina proteins, in the size range 60–70 kd, are assembled onto the chromatin. Finally, nuclear-membrane proteins are added in a process that does not require ATP to provide

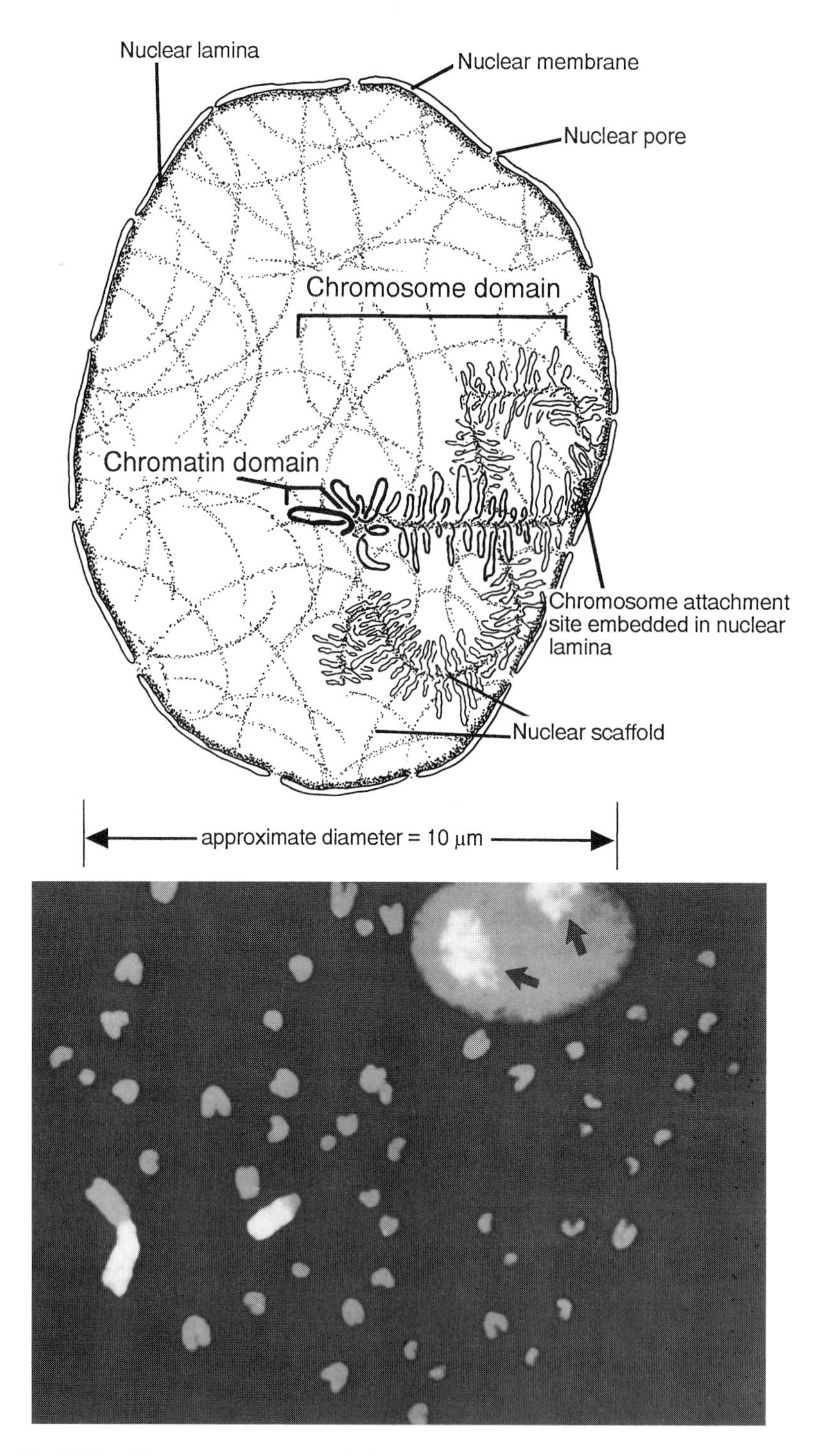

Fig. 17.34. Chromosome and chromatin domains within a nucleus. The diagram (upper) indicates the attachment of a chromosome to the nuclear lamina within a specific chromosomal domain, and the attachment of chromatin loops to the nuclear scaffold in chromatin domains. The diagram is based on Appels (1989), Hilliker and Appels (1989), Leitch et al. (1990), van Driel et al. (1991), Broccoli and Cooke (1994), and Zalensky et al. (1995). Evidence for such arrays has been obtained using *in situ* hybridization technology (*see* Chapter 19) in which chromosome-specific DNA probes are hybridized to nuclear preparations to determine the location of the respective chromosome among all of the others present in the nucleus. The fluorescent probe can then be detected using confocal microscopy (*see* Chapter 3, Section 3.1.5) to determine the position of the labeled chromosome in the nucleus. As an example, the photograph (lower) shows the distribution of DNA sequences that are specific to the X and Y chromosomes (shown in the lower left of the photograph) of a rodent (*Microtus*). It is evident that in the nucleus (at the top of the photograph), these chromosomes (marked by arrows) remain in distinct regions (Modi, 1993). In the case of crosses between different species, it is also common for the chromosomes derived from each species to remain clustered together within the hybrid nucleus.

the energy for the assembly, but may, instead, derive energy from the dephosphorylation of proteins involved in the process. The strict order of events leading to the formation of nuclei in the cell-free extracts suggests that chromosome-attachment sites may be of critical importance in the re-formation of nuclei after mitosis.

In special cases such as in the somatic cells of *Drosophila*, homologous chromosomes can associate and lead to two chromosomes forming a common domain. The heterochromatic regions of nonhomologous chromosomes can form aggregates or chromocenters, which also alter the nature of chromosome domains. Nucleoli, the sites of ribosomal RNA synthesis in eukaryotic nuclei, are major cytological features of interphase nuclei, and nucleoli from different chromosomes can fuse and thus also modify particular chromosome domains. The molecular description of specific cytological features of chromosomes such as euchromatic bands in polytene chromosomes, heterochromatin, and nucleoli are treated in detail in Chapter 20. They are mentioned here in the context of the broad organization and folding of DNA within a nucleus, because evidence is accumulating that in certain situations, these aspects may be important in gene expression. Chromosome domains as such do not appear to be of major importance to viability and fertility, as judged from experiments carried out in *Drosophila*, but more subtle changes in gene expression do occur as a result of changes in these domains, as is discussed further in Chapter 20.

BIBLIOGRAPHY

General

CALLADINE, C.R., DREW, H.R. 1992. Understanding DNA: the Molecule and How It Works. Academic Press, London.

Section 17.1

AYMAMI, J., COLL, M., FREDERICK, C.A., WANG, A.H.-J., RICH, A. 1989. The propeller DNA conformation of polydA.polydT. Nucleic Acids Res. 17: 3229–3245.

COLLIS, C.M., MOLLOY, P.L., BOTH, G.W., DREW, H.R. 1989. Influence of the sequence-dependent flexure of DNA on transcription in *E. coli*. Nucleic Acids Res. 17: 9447–9469.

DICKERSON, R.E. 1983. The DNA helix and how it is read. Sci. Am. 249: 94–111.

DREW, H.R., MCCALL, M.J., CALLADINE, C.R. 1990. New approaches to DNA in the crystal and in solution. In: DNA Topology and Its Biological Implications (Cozzarelli, N.R., Wang, J.C., eds.). Cold Spring Harbor Laboratory Press, Cold Spring Harbor, NY. pp 1–56.

MANZINI, G., XODO, L.E., GASPAROTTO, D., QUADRIFOGLIO, F., VAN DER MAREL, G., VAN BOOM, J. 1990. Triple helix formation by oligopurine–oligopyrimidine DNA fragments: electrophoretic and thermodynamic behavior. J. Molec. Biol. 213: 833–843.

MCCLELLAN, J., A., LILLEY, D. 1991. Structural alteration in alternating adenine-thymine sequences in positively supercoiled DNA. J. Molec. Biol. 219: 145–149.

WANG, A. H-J., QUIGLEY, G.J., KOLPAK, F.J., CRAWFORD, J.L., VAN BOOM, J.H., VAN DER MAREL, G., RICH, A. 1979. Molecular structure of a left-handed double helical DNA fragment at atomic resolution. Nature 282: 680–686.

WATSON, J.D., CRICK, F.H.C. 1953. A structure for deoxyribose nucleic acid. Nature 171: 737–738.

Section 17.2

BERG, D. 1989. Transposon Tn5. In: Mobile DNA (Berg, D.E., Howe, M.M., eds.). American Society for Microbiology, Washington DC. pp. 185–210.

BESTOR, T.H., GUNDERSEN, G., KOLSTØ, A-B., PYRDZ, H. 1992. CpG islands in mammalian gene promoters are inherently resistant to *de novo* methylation. Genet. Anal. Tech. Anal. 9: 48–53.

FINNEGAN, E.J., BRETTELL, R.I.S., DENNIS, E.S. 1993. The role of DNA methylation in the regulation of plant gene expression. In: DNA Methylation: Molecular Biology and Biological Significance (Jost, J.P., Saluz, H.P., eds.). Birkhäuser Verlag, Basel. pp. 218–261.

FLAVELL, R.B., O'DELL, M. 1990. Variation and inheritance of cytosine methylation patterns in wheat at the high molecular weight glutenin and ribosomal RNA gene loci. Development Supplement. (Monk, M., Surani, A., eds.) The Company of Biologists Ltd., Cambridge, England. pp. 15–20.

GRUENBAUM, Y., NAVEH-MANY, T., CEDAR, H., RAZIN, A. 1981. Sequence specificity of methylation in higher plants. Nature 292: 860–862.

KAKUTANI, T., JEDDELOH, J.A., RICHARDS, E.J. 1995. Characterization of an *Arabidopsis thaliana* DNA hypomethylation mutant. Nucleic Acids Res. 23: 130–137.

MATZKE, M.A., MATZKE, A.J.M. 1991. Differential inactivation and methylation of transgene in plants by two suppressor loci containing homologous sequences. Plant Molec. Biol. 16: 821–830.

OHLSSON, R., BARLOW, D., SURANI, A. 1994. Impressions of imprints. Trends Genet. 10: 415–417.

OLINS, A.L., OLINS, D.E. 1974. Spheroid chromatin units (ν bodies). Science 183: 330–332.

PFEIFER, G.P., TANGUAY, R.L., STEIGERWALD, S.D., RIGGS, A.D. 1990. In vivo footprint and methylation analysis by PCR-aided genomic sequencing: comparison of active and inactive X chromosomal DNA at the CpG island and promoter of human PGK-1. Genes Dev. 4: 1277–1287.

RICHARDS, E.J. 1997. DNA methylation and plant development. Trends in Genetics 13: 319–323.

ROSSIGNOL, J.-L., FAUGERON, G. 1994. Gene inactivation triggered by recognition between DNA repeats. Experientia 50: 307–317.

Selker, E. U. 1997. Epigenetic phenomena in filamentous fungi. Trends in Genetics 13: 296–301.

SHEN, C-K. J., MANIATIS, T. 1980. Tissue-specific DNA methylation in a cluster of rabbit β-like globin genes. Proc. Natl. Acad. Sci. USA 77: 6634–6638.

SURANI, M.A. 1994. Genomic imprinting: control of gene expression by epigenetic inheritance. Cell Biol. 6: 390–395.

TAKAHASHI, Y., MITANI, K., KUWABARA, K., HAYASHI, T., NIWA, M., MIYASHITA, N., MORIWAKI, K., KOMINAMI, R. 1994. Methylation imprinting was observed of mouse mo-2 macrosatellite on the pseudoautosomal region but not on chromosome 9. Chromosoma 103: 450–458.

TRASLER, J.M., ALCIVAR, A.A., HAKE, L., BESTOR, T., HECHT, N. 1992. DNA methytransferase is developmentally expressed in replicating and non-replicating male germ cells. Nucleic Acids Res. 20: 2541.

WAGNER, I., CAPESIUS, I. 1981. Determination of 5-methyl cytosine from plant DNA by high performance liquid chromatography. Biochim. Biophys. Acta 654: 52–56.

Section 17.3

DAWSON, R.M., ELLIOTT, D.C., ELLIOTT, W.H., JONES, K.M. 1986. Data for biochemical research. Oxford Science Publications/ Clarendon Press, Oxford.

DREW, H.R., DICKERSON, R.E. 1981. Structure of a β-DNA dodecamer: Geometry of hydration. J. Molec. Biol. 151: 535–556.

EISENSTEIN, M., FROLOW, F., SHAKKED, Z., RABINOVICH, D. 1990. The structure and hydration of the A-DNA fragment d(GGGTACCC) at room temperature and low temperature. Nucleic Acids Res. 18: 3185–3194.

FEUERSTEIN, B.G., PATTABIRAMAN, N., MARTON, L. 1989. Molecular dynamics of spermine–DNA interactions: sequence specificity and DNA bending for a simple ligand. Nucleic Acids Res. 17: 6883–6891.

HOOGSTEEN, K. 1959. Structure of a crystal containing a hydrogen-bonded complex of 1-methylthymine and 9-methyladenine. Acta Crystallogr. 12: 822–823.

KENNARD, O., CRUSE, W.B.T., NACHMAN, J., PRANGE, T., SHAKKED, Z., RABINOVICH, D. 1986. Ordered water structure in an A-DNA octamer at 1.7 Å resolution. J. Biomolec. Struct. Dynam. 3: 623–647.

PORTUGAL, J. 1989. Do Hoogsteen base pairs occur in DNA? Trends Biochem. Sci. 14: 127–130.

QUIGLEY, G.J., UGHETTO, G., VAN DER MARCEL, G.A., VAN BOOM, J.H., WANG, A.H-J., RICH, A. 1986. Non-Watson–Crick G.C and A.T base pairs in a DNA–antibiotic complex. Science 232: 1255–1258.

TAKAHARA, P.M., ROSENZWEIG, A.C., FREDERICK, C.A., LIPPARD, S.J. 1995. Crystal structure of double-stranded DNA containing the major adduct of the anticancer drug cisplatin. Nature 377: 649–651.

Section 17.4

BERG, J.M. 1990. Zinc fingers and other metal-binding domains. J. Biol. Chem. 265: 6513–6516.

BERG, J.M. 1992. Sp1 and the subfamily of zinc finger proteins with guanine-rich binding sites. Proc. Natl. Acad. Sci. USA 89: 11109–11110.

BROWN, R.S., SANDER, C., ARGOS, P. 1985. The primary structure of transcription factor TFIIIA has 12 consecutive repeats. FEBS Lett. 186: 271–274.

BURATOWSKI, S. 1994. The basics of basal transcription by RNA polymerase II. Cell 77: 1–3.

CHENG, X., BALENDIRAN, K., SCHILDKRAUT, I., ANDERSON, J.E. 1994. Structure of PvuII endonuclease with cognate DNA. EMBO J. 13: 3927–3935.

CHURCHILL, M.E.A., TULLIUS, T.D., KLUG, A. 1990. Mode of interaction of the zinc finger protein TFIIIA with a 5S RNA gene of *Xenopus*. Proc. Natl. Acad. Sci. USA 87: 5528–5532.

CORMACK, B.P., STRUHL, K. 1992. The TATA-binding protein is required for transcription by all three nuclear RNA polymerases in yeast cells. Cell 69: 685–696.

DEL RIO, S., SETZER, D.R. 1993. The role of zinc fingers in transcriptional activation by transcription factor IIIA. Proc. Natl. Acad. Sci. USA 90: 168–172.

GREENBLATT, J. 1992. Riding high on the TATA box. Nature 360: 16–17.

HANSEN, P.K., CHRISTENSEN, J.H., NYBORG, J., LILLELUND, O., THOGERSEN, H.C. 1993. Dissection of the DNA-binding domain of *Xenopus laevis* TFIIIA. J. Molec. Biol. 233: 191–202.

HARRISON, S.C., AGGARWAL, A.K. 1990. DNA recognition by proteins with the helix-turn-helix motif. Ann. Rev. Biochem. 59: 933–969.

JONES, N. 1990. Transcriptional regulation by dimerization: two sides to an incestuous relationship. Cell 61: 9–11.

KADONAGA, J.T., COUREY, A.J., LADIKA, J., TJIAN, R. 1988. Distinct regions of Sp1 modulate DNA binding and transcriptional activation. Science 242: 1566–1570.

KASSAVETIS, G.A., BRAUN, B.R., NGUYEN, L.H., GEIDUSCHEK, E.P. 1990. *S. cerevisiae* TFIIIB is the transcription factor proper of RNA polymerase III, while TFIIIA and TFIIIC are assembly factors. Cell 60: 235–245.

KEEGAN, L., GILL, G., PTASHNE, M. 1986. Separation of DNA binding from the transcription-activating function of a eukaryotic regulatory protein. Science 231: 699–704.

KELLER, H.J., YOU, Q., ROMANIUK, P.J., GOTTESFELD, J.M. 1990. Additional intragenic promoter elements of the *Xenopus* 5SRNA genes upstream from the TFIIIA-binding site. Molec. Cell. Biol. 10: 5166–5176.

KIM, J.L., NIKOLOV, D.B., BURLEY, S.K. 1993. Co-crystal structure of TBP recognizing the minor groove of a TATA element. Nature 365: 520–527.

KIM, S.H., DARBY, M.K., JOHO, K.E., BROWN, D.D. 1990. The characterization of the TFIIIA synthesized in somatic cells of *Xenopus laevis*. Genes Dev. 4: 1602–1610.

KIM, Y., GRABLE, J.C., GREENE, P.J., ROSENBERG, J.M. 1990. Refinement of endonuclease crystal structure: a revised protein chain tracing. Science 249: 1307–1309.

KIM, Y., GEIGER, J.H., HAHN, S., SIGLER, P.B. 1993. Crystal structure of a yeast TBP/TATA-box complex. Nature 365: 512–520.

KLUG, A. 1993. Opening the gateway. Nature 365: 486–487.

KLUG, A., RHODES, D. 1987. "Zinc fingers:" a novel protein motif for nucleic acid recognition. Trends Biol. Sci. 12: 464–469.

LEE, M.S., GIPPERT, G.P., SOMAN, K.V., CASE, D.A., WRIGHT, P.E. 1989. Three-dimensional solution structure of a single zinc finger DNA-binding domain. Science 245: 635–637.

LUISI, B. 1995. Zinc standard for economy. Nature 356: 379–380.

MCCLARIN, J.A., FREDERICK, C.A., WANG, B-C., GREENE, P., BOYER, H,W., GRABLE, J., ROSENBERG, J.M. 1986. Structure of the DNA–EcoR1 endonuclease recognition complex at 3 Å resolution. Science 234: 1526–1541.

MCKNIGHT, S.L. 1991. Molecular zippers in gene regulation. Sci. Am. 264: 32–39.

MILLER, J., MCLACHLAN, A.D., KLUG, A. 1985. Repetitive zinc-binding domains in the protein transcription factor IIIA from *Xenopus* oocytes. EMBO J. 4: 1609–1614.

MUKAI, Y., ENDO, T.R., GILL, B.S. 1990. Physical mapping of the 5S rRNA multigene family in common wheat. J. Hered. 81: 290–295.

MURDOCH, K., ALLISON, L.A. 1996. A role for ribosomal protein L5 in the nuclear import of 5S rRNA in *Xenopus* oocytes. Exp. Cell Res. 227: 332–343

NEWMAN, M., STRZELECKA, T., DORNER, L.F., SCHILDKRAUT, I., AGGARWAL, A.K. 1995. Structure of Bam H1 endonuclease bound to DNA: partial folding and unfolding on DNA binding. Science 269: 656–669.

NIKOLOV, D.B., HU, S-H., LIN, J., GASCH, A., HOFFMANN, A., HORIKOSHI, M., CHUA, N-C., ROEDER, R.G., BURLEY, S.K. 1992. Crystal structure of TFIID TATA-box binding protein. Nature 360: 40–45.

PARDUE, M.L., BROWN, D.D., BIRNSTIEL, M.L. 1973. Location of the genes for 5S ribosomal RNA in *Xenopus laevis*. Chromosoma 42: 191–203.

PAULE, M.R. 1991. In search of the single factor. Nature 344: 819–820.

PAVLETICH, N.P., PABO, C.O. 1991. Zinc finger-DNA recognition: crystal structure of a Zif268–DNA complex at 2.1 Å. Science 252: 809–817.

SANDS, M.S., BOGENHAGEN, D.F. 1991. The carboxyterminal zinc fingers of TFIIIA interact with the tip of helix V of 5S RNA in the 7S ribonucleoprotein particle. Nucleic Acids Res. 19: 1791–1801.

SHARP, S.J., GARCIA, A.D. 1988. Transcription of the *Drosophila melanogaster* 5SRNA gene requires an upstream promoter and four intragenic sequence elements. Molec. Cell Biol. 8: 1266–1274.

SMITH, D.R., JACKSON, I.J., BROWN, D.D. 1984. Domains of the positive transcription factor specific for the *Xenopus* 5S RNA gene. Cell 37: 645–652.

SORENSEN, P.D., FREDERIKSEN, S. 1991. Characterization of human 5S rRNA genes. Nucleic Acids Res. 19: 4147-4151.

STRUHL, K. 1989. Helix-turn-helix, zinc-finger, and leucine-zipper motifs for eukaryotic transcriptional proteins. Trends Biochem. Sci. 14: 137–140.

STUNKEL, W., KOBER, I., KAUER, M., TAIMOR, G., SEIFART, K.H. 1995. Human TFIIIA alone is sufficient to prevent nucleosomal repression of a homologous 5S gene. Nucleic Acids Res. 23: 109–116.

TJIAN, R., MANIATIS, T. 1994. Transcription activation: a complex puzzle with few easy pieces. Cell 77: 5–8.

TRAVERS, A.A., NER, S.S., CHURCHILL, M.E.A. 1994. DNA chaperones: a solution to a persistence problem. Cell 77: 167–169.

TYLER, B. 1987. Transcription of *Neurospora crassa* 5S rRNA genes requires a TATA box and three internal elements. J. Molec. Biol. 196: 801–811.

VRANA, K.E., CHURCHILL, M. E. A., TULLIUS, T.D., BROWN, D.D. 1988. Mapping functional regions of transcription factor TFIIIA. Molec. Cell. Biol. 8: 1684–1696.

WINKLER, F.K., BANNER, D.W., OEFNER, C., TSERNOGLOU, D., BROWN, R.S., HEATHMAN, S.P., BRYAN, R.K., MARTIN, P.D., PETRATOS, K., WILSON, K.S. 1993. The crystal structure of EcoRV endonuclease and of its complexes with cognate and non-cognate DNA fragments. EMBO J. 12: 1781–1795.

YANG, W., STEITZ, T.A. 1995. Crystal structure of the site-specific recombinase $\gamma\delta$ resolvase complexed with a 34 bp cleavage site. Cell 82: 193–207.

Section 17.5

BAVYKIN, S.G., USACHENKO, S.I., ZALENSKY, A.O., MIRZABEKOV, A.D. 1990. Structure of nucleosomes and organization of internucleosomal DNA in chromatin. J. Molec. Biol. 212: 495–511.

BOY DE LA TOUR, LAEMMLI, U.K. 1988. The metaphase scaffold is helically folded: sister chromatids have predominantly opposite helical handedness. Cell 55: 937–944.

DREW, H.R., TRAVERS, A.A. 1985. DNA bending and its relation to nucleosome positioning. J. Molec. Biol. 186: 773–790.

DURRIN, L.K., MANN, R.K., KAYNE, P.S., GRUNSTEIN, M. 1991. Yeast histone H4 N-terminal sequence is required for promoter activation in vivo. Cell 65: 1023–1031.

FELSENFELD, G. 1992. Chromatin as an essential part of the transcriptional mechanism. Nature 355: 219–224.

FINLEY, D., BARTEL, B., VARSHAVSKY, A. 1989. The tails of ubiquitin precursors are ribosomal proteins whose fusion to ubiquitin facilitates ribosome biogenesis. Nature 338: 394–401.

FORRESTER, W.C., TAKEGAWA, S., PAPAYANNOPOULOU, T., STAMATOYANNOPOULOS, G., GROUDINE, M. 1987. Evidence for a locus activation region: the formation of developmentally stable hypersensitive sites in globin-expressing hybrids. Nucleic Acids Res. 15: 10159–10177.

GALL, J.G., CALLAN, H.G. 1962. H^3 uridine incorporation in lampbrush chromosomes. Proc Natl Acad Sci USA 48: 562–570.

GASSER, S.M., LAEMMLI, U.K. 1986. Cohabitation of scaffold binding regions with upstream/enhancer elements of three developmentally regulated genes of *D. melanogaster*. Cell 48: 521–530.

GOURDON, G., SHARPE, J.A., WELLS, D., WOOD, W.G., HIGGS, D.R. 1994. Analysis of a 70 kb segment of DNA containing the human ϵ and α-globin genes linked to their regulatory element (HS-40) in transgenic mice. Nucleic Acids Res. 22: 4139–4147.

HAN, M., GRUNSTEIN, M. 1988. Nucleosome loss activates yeast downstream promoters in vivo. Cell 55: 1137-1145.

HAYAISHI, O., UEDA, K. 1977. Poly(ADP-ribose) and ADP-ribosylation of proteins. Annu. Rev. Biochem. 46: 95–116.

HAYES, J.J., TULLIUS, T.D., WOLFFE, A.P. 1990. The structure of DNA in a nucleosome. Proc. Natl. Acad. Sci. USA 87: 7405–7409.

HECK, M.M.S., HITTELMAN, W.N., EARNSHAW, W.C. 1988. Differential expression of DNA topoisomerases I and II during the eukaryotic cell cycle. Proc. Natl. Acad. Sci. USA 85: 1086–1090.

HEWISH, D.R., BURGOYNE, L. 1973. Chromatin sub-structure. The digestion of chromatin DNA at regularly spaced sites by a nuclear deoxyribonuclease. Biochem. Biophys. Res. Commun. 52: 504–510.

HILL, C.S., RIMMER, J.M., GREEN, B.N., FINCH, J.T., THOMAS, J.O. 1991. Histone-DNA interactions and their modulation by phosphorylation of –Ser–Pro–X–Lys/Arg– motifs. EMBO J. 10: 1939–1948.

HIRSCHHORN, J.N., BROWN, S.A., CLARK, C.D., WINSTON, F. 1992. Evidence that SNF2/SW12 and SNF5 activate transcription in yeast by altering chromatin structure. Genes Dev. 6: 2288–2298.

HOCHSCHILD, A., PTASHNE, M. 1988. Interaction at a distance between λ repressors disrupts gene activation. Nature 336: 353–357.

HUANG, H-W., JUANG, J-K., LIU, H-J. 1992. The recognition of DNA cleavage sites by porcine spleen topoisomerase II. Nucleic Acids Res. 20: 467–473.

JACOBSEN, K., LAURSEN, N.B., JENSEN, E.O., MARCKER, A., POULSEN, C., MARCKER, K.A. 1990. HMG1-like proteins from leaf and nodule nuclei interact with different AT motifs in soybean nodulin promoters. Plant Cell 2: 85–94.

KARPEN, G.H., SCHAEFER, J.E., LAIRD, C.D. 1989. A *Drosophila* rRNA gene located in euchromatin is active in transcription and nucleolus formation. Genes Dev. 3: 1745–1763.

KÄS, E., LAEMMLI, U.K. 1992. In vivo topoisomerase II cleavage of the *Drosophila* histone and satellite III repeats: DNA sequences and structural characteristics. EMBO J. 11: 705.

KELLUM, R., SCHEDL, P. 1991. A position-effect assay for boundaries of higher order chromosomal domains. Cell 64: 941–950.

KLEINSCHIMDT, J.A., SEITER, A., ZENTGRAF, H. 1990. Nucleosome assembly in vitro: separate histone transfer and synergistic interaction of native histone complexes purified from nuclei of *Xenopus laevis* oocytes. EMBO J. 9: 1309–1318.

KORNBERG, R., STRYER, L. 1988. Statistical distributions of nucleosomes: nonrandom locations by a stochastic mechanism. Nucleic Acids Res. 16: 6677–6690.

LOIDL, P. 1994. Histone acetylation: facts and questions. Chromosoma 103: 441–449.

OHNUKI, Y. 1968. Structure of chromosomes. I. Morphological studies of the spiral structure of human somatic chromosomes. Chromosoma 25: 402–428.

REITMAN, M., LEE, E., WESTPHAL, H., FELSENFELD, G. 1993. An enhancer/locus control region is not sufficient to open chromatin. Molec. Cell. Biol. 13: 3990.

RICHMOND, T.J., FINCH, J.T., RUSHTON, B., RHODES, D., KLUG, A. 1984. Structure of the nucleosome core particle at 7D resolution. Nature 311: 532–537.

ROTH, S.Y., SHIMIZU, M., JOHNSON, L., GRUNSTEIN, M., SIMPSON, R.T. 1992. Stable nucleosome positioning and complete repression by the yeast α2 repressor are disrupted by amino-terminal mutations in histone H4. Genes Dev. 6: 411–425.

SAPP, M., WORCEL, A. 1990. Purification and mechanism of action of a nucleosomal assembly factor from *Xenopus* oocytes. J. Biol. Chem. 265: 9357–9365.

SESSA, G., RUBERTI, I. 1990. Assembly of correctly spaced chromatin in a nuclear extract from *Xenopus laevis* oocytes. Nucleic Acids Res. 18: 5449–5455.

SHIH, D.M., WALL, R.J., SHAPIRO, S.G. 1990. Developmentally regulated and erythroid-specific expression of the human embryonic β-globin gene in transgenic mice. Nucleic Acids Res. 18: 5465–5472.

SLATTER, R., DUPREE, P., GRAY, J.C. 1991. A scaffold-associated DNA region is located downstream of the pea plastocyanin gene. Plant Cell 3: 1239–1250.

SVAREN, J., CHALKLEY, R. 1990. The structure and assembly of active chromatin. Trends Genet. 6: 52–56.

TALBOT, D., GROSVELD, F. 1991. The 5′HS2 of the globin locus control region enhances transcription through the interaction of a multimeric complex binding at two functionally distinct NF-E2 binding sites. EMBO J. 10: 1391–1398.

THOMA, F., ZATCHEJ, M. 1988. Chromatin folding modulates nucleosome positioning in yeast minichromosomes. Cell 55: 945–953.

TOWNES, T.M., BEHRINGER, R.R. 1990. Human globin locus activating region (LAR): role in temporal control. Trends Genet. 6: 219–223.

VYAS, P., VICKERS, M.A., SIMMONS, D.L., AYYUB, H., CRADDOCK, C.F., HIGGS, D.R. 1992. Cis-acting sequences regulating expression of the human α-globin cluster lie within constitutively open chromatin. Cell 69: 781–793.

WANG, J.C. 1985. DNA topoisomerases. Annu. Rev. Biochem. 54: 665–697.

WARD, S.W., PARTIN, A.W., COFFEY, D.S. 1989. DNA loop domains in mammalian spermatozoa. Chromosoma 98: 155–159.

WATT, P., LAMB, P., SQUIRE, L., PROUDFOOT, N. 1990. A factor binding GATAAG confers tissue specificity on the promoter of the human globin gene. Nucleic Acids Res. 18: 1339–1350.

WEITH, A., TRAUT, W. 1980. Synaptonemal complexes with associated chromatin in a moth, *Ephestia kuehniella* Z. Chromosoma 78: 275–291.

WIDOM, J., KLUG, A. 1985. Structure of the 300 A chromatin filament: X-ray diffraction from oriented samples. Cell 43: 207–213.

WIJERDE, M., GROSVELD, F., FRASER, P. 1995. Transcription complex stability and chromatin dynamics in vivo. Nature 377: 209–213.

WOLFFE, A. P. 1994. Transcription: in tune with the histones. Cell 77: 13–16.

ZALENSKY, A.O., ALLEN, M.J., KOBAYASHI, A., ZALENSKAYA, I.A., BALHORN, R., BRADBURY, E.M. 1995. Well-defined genome architecture in the human sperm nucleus. Chromosoma 103: 577–590.

ZLATANOVA, J., VAN HOLDE, K. 1992. Histone 1 and transcription: still an enigma. J. Cell Sci. 103: 889–895.

Section 17.6

ASHBURNER, M. 1990. Puffs, genes, and hormones revisited. Cell 61: 1–3.

BROWN, D.D. 1984. The role of stable complexes that repress and activate eukaryotic genes. Cell 37: 359–365.

DRLICA, K. 1990. Bacterial topoisomerases and the control of DNA supercoiling. Trends Genet. 6: 433–437.

ENGEL, J.D. 1993. Developmental regulation of human β-globin gene transcription: a switch of loyalties. Trends Genet. 9: 304–309.

FELSENFELD, G. 1996. Chromatin unfolds. Cell 86: 13–19.

HILL, R. 1990. Z-DNA: a prodome for the 1990s. J. Cell Sci. 99: 675–680.

LIU, L.F., WANG, J.C. 1987. Supercoiling of the DNA template during transcription. Proc. Natl. Acad. Sci. USA 84: 7024–7027.

Prioleau, M-N., Huet, J., Sentenac, A., Mechali, M. 1994. Competition between chromatin and transcription complex assembly regulates gene expression during early development. Cell 77: 439–449.

Semeshin, V.F., Baricheva, E.M., Belyaeva, E.S., Zhimulev, I.F. 1985. Electron microscopical analysis of *Drosophila* polytene chromosomes II. Development of complex puffs. Chromosoma 91: 210–233.

Swedlow, J.R., Sedat, J.W., Agard, D.A. 1993. Multiple chromosomal populations of topoisomerase II detected by time-lapse, three-dimensional wide-field microscopy. Cell 73: 97–108.

Section 17.7

Appels, R. 1989. Three dimensional arrangements of chromatin and chromosomes: old concepts and new techniques. J. Cell Sci. 92: 325–328.

Bode, J., Kohwi, Y., Dickinson, L., Joh, T., Klehr, D., Mielke, C., Kohwi-Shigematsu, T. 1992. Biological significance of unwinding capability of nuclear matrix-associating DNAs. Science 255: 195–197.

Broccoli, D., Cooke, H.J. 1994. Effect of telomeres on the interphase location of adjacent regions of the human X chromosome. Exp. Cell Res. 212: 308–313.

Collins, C.M., Molloy, P.L., Both, C.W., Drew, H.R. 1989. Influence of the sequence-dependent flexure of DNA on transcription in *E. coli*. Nucleic Acids Res. 17: 9447–9468.

Hilliker, A.J., Appels, R. 1989. The arrangement of interphase chromosomes: structural and functional aspects. Exp. Cell Res. 185: 297–318.

Kim, Y., Grable, J.C., Love, R., Greene, P.J., Rosenberg, J.M. 1994. Refinement of EcoRI endonuclease crystal structure: A revised protein chain tracing. Science 249: 1307–1309.

Leitch, A.R., Mosgoller, W., Schwarzacher, T., Bennett, M.D., Heslop-Harrison, J.S. 1990. Genomic *in situ* hybridization to sectioned nuclei shows chromosome domains in grass hybrids. J. Cell Sci. 95: 335–341.

Modi, W.S. 1993. Rapid, localized amplification of a unique satellite DNA family in the rodent *Microtus chrotorrhinus*. Chromosoma 102: 484–490.

Newman, M., Strelecka, T., Dorner, L.F., Schildkraut, I., Aggarwal, A.K. 1994. Structure of restriction endonuclease BamHI and its relationship to EcoRI. Nature 368: 660–664.

Newport, J.W., Forbes, D.J. 1987. The nucleus: structure, function and dynamics. Annu. Rev. Biochem. 56: 535-565.

Philpott, A., Leno, G.H. 1992. Nucleoplasmin remodels sperm chromatin in *Xenopus* egg extracts. Cell 69: 759-767.

van Driel, R., Humbel, B., de Jong, L. 1991. The nucleus: a black box being opened. J. Cell Biochem. 47: 311–316.

18

Replication of Protosomes and Chromosomes

- The replication of DNA is semiconservative, involving a single origin of replication in prokaryotes and multiple origins of replication in eukaryotes.
- Synthesis of new DNA occurs by copying an existing template through the addition of 5′-nucleoside monophosphate units to the 3′OH moiety of a primer sequence.
- Many proteins are involved in the replication process, and the enzymatic activities are the same in prokaryotes and eukaryotes.
- The high fidelity of DNA replication is the result of several different postreplicative editing activities that remove errors.
- Any single DNA segment is usually replicated only once per cell cycle, and the assembly of chromatin after DNA replication in eukaryotes is a critical period for competition between transcription-control factors and histones.
- Developmentally regulated deviations from the "once only per cell cycle" rule for DNA replication can lead to amplification of regions of the genome.

Chromosomes usually replicate in the S phase of the eukaryotic cell cycle, when DNA and histone syntheses occur to provide the components required for the assembly of new chromosomes. Although many features of chromosome replication are still not understood, much has been learned about the complex processes that underlie it by a combination of molecular and cytogenetic experiments. The DNA component of chromosomes is replicated by using a preexisting parental DNA strand as a template for the synthesis of an identical but complementary DNA strand. Each parental strand becomes incorporated into a new, and separate, DNA double helix in a process referred to as semiconservative replication. The basic protein components of chromosomes, histones, are assembled onto newly formed DNA from a pool that includes both newly synthesized histones and histones displaced from DNA during the course of replication. The replication of the protein component of chromosomes is therefore not semiconservative.

18.1 SEMICONSERVATIVE DNA REPLICATION

Shortly after the publication of the structure of the DNA molecule, the semiconservative nature of DNA replication in both prokaryotes and eukaryotes was proven experimentally. Semiconservative replication of DNA has important genetical consequences, because it is a simple mechanism for ensuring that accurate copies of DNA are passed from parent to progeny. At the biochemical level, the process clarified experimental predictions that were important to confirm before the concept was generally accepted. For example, semiconservative replication predicts that a radioactive, or otherwise physically marked, precursor for DNA synthesis is incorporated into the newly synthesized DNA and that after one round of replication, both DNA molecules will be labeled. After a second round of replication in the absence of labeled substrate, two of the four DNA molecules will be labeled, and two will be unlabeled. Alternatively, a second round of replication in the presence of labeled substrate results in two DNA molecules with both strands labeled and two with only one strand labeled.

18.1.1 Protosomes

The early experiments on DNA replication in protosomes used ^{15}N-NH_4Cl to produce DNA, in living bacteria, that was distinguishable from normal DNA in equilibrium cesium-chloride gradients (Fig. 18.1). In this way, inhibitors, radioactivity, or other agents that may have induced artifacts were avoided. The data indicated that hybrid $^{15}N/^{14}N$-labeled DNA molecules were produced in *Escherichia coli* grown on medium containing ^{15}N-NH_4Cl and then transferred to medium containing ^{14}N-NH_4Cl. The production of hybrid molecules was expected from semiconservative, but not from conservative, replication of the DNA molecule, as illustrated in Fig. 18.1.

18.1.2 Chromosomes

The classical experiments to monitor DNA replication in eukaryotes used ^3H-thymidine as a precursor for DNA by adding it to a medium in which *Vicia faba* root-tip cells were actively dividing. Technically, the experiments relied on the following knowledge:

1. ^3H-thymidine is a specific probe for DNA synthesis but not for RNA or protein synthesis,

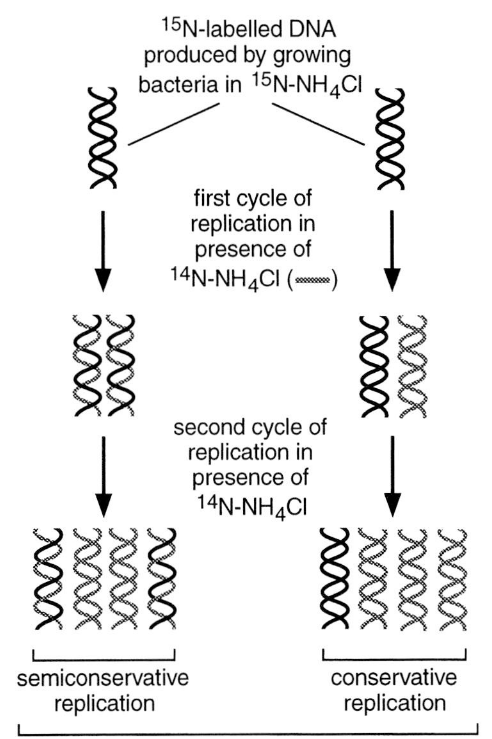

Fig. 18.1. Proof of the semiconservative nature of DNA replication in prokaryotes. In these experiments, designed to investigate the nature of DNA replication in bacteria (Meselson and Stahl, 1958), $^{15}N-NH_4Cl$ was used to produce radioactively labeled DNA. This DNA could be distinguished from normal DNA because of the fact that the incorporation of ^{15}N-labeled bases increased its buoyant density in equilibrium cesium-chloride gradients. Semiconservative replication predicted that after one cycle of replication, one strand of every DNA helix would be newly synthesized, resulting in double-stranded DNA molecules, all of which had hybrid densities, as was detected. In contrast, conservative replication, where both strands of a double helix would be newly synthesized, would have yielded one double-stranded DNA molecule of high density (^{15}N) and one of low density (^{14}N). A second cycle of replication in the presence of ^{14}N then produces DNA products that easily distinguish between semiconservative and conservative replications.

2. The low energy of the β particles emitted from decaying ^3H atoms results in a short path length, allowing the point of detection of the radioactivity to be related to the actual position of the source of the radioactivity.
3. The drug colchicine inhibits cell division but not the process of chromosome duplication.
4. Covering chromosome preparations with a thin layer of photographic emulsion allows the radioactivity to be detected and relates the positions of silver grains in the emulsion to the physical shape of the chromosomes.

The results of these experiments are illustrated in Fig. 18.2. The distribution of radioactivity in the first and second mitotic divisions, following the addition of ^3H-thymidine, can be satisfactorily explained by semiconservative replication of DNA and the existence of a single DNA molecule extending the full length of the chromatid. In the second mitotic division after the administration of ^3H-thymidine, most chromosomes showed only a single chromatid labeled with radioactivity.

A complication in these experiments was that occasionally blocks of radioactivity appeared in otherwise unlabeled chromatids. This phenomenon, which is due to sister-chromatid exchange, was studied in greater detail in later experiments, labeling with 5-bromo-2′-deoxyuridine (BrdU), instead of ^3H-thymidine because

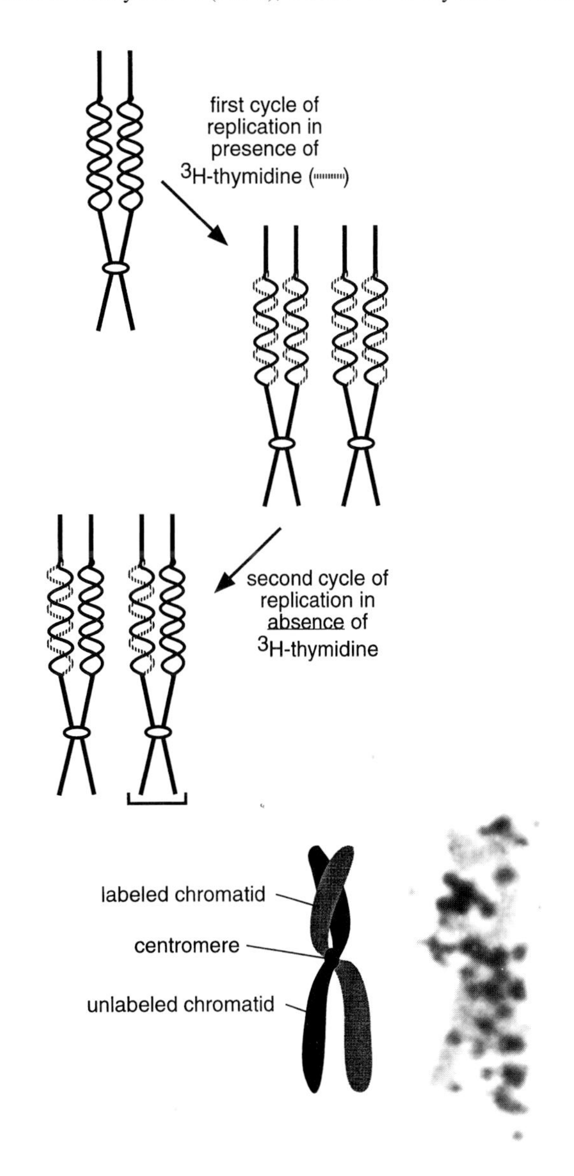

Fig. 18.2. Semiconservative DNA replication in eukaryotes. (From Taylor et al., 1957.) Following the addition of ^3H-thymidine to plant root tips in which the cells were actively dividing, chromosomes that were undergoing first and second mitotic divisions became radioactively labeled. The root-tip cells that were of most interest were those that were in G1 at the time of incubation with the ^3H-thymidine. These cells produced chromosomes with both chromatids labeled (top half). After an ensuing second cycle of replication in the absence of ^3H-thymidine, these cells produced chromosomes with one chromatid labeled and one not (bottom half and also shown in photograph). The distribution of radioactivity in the chromosomes provided the proof for semiconservative replication in eukaryotes. Colchicine (*see* Chapter 4, Fig. 4.9) was used to block cell division (but not DNA synthesis), so that cells in the second cycle of replication were clearly identifiable because they had double the chromosome number.

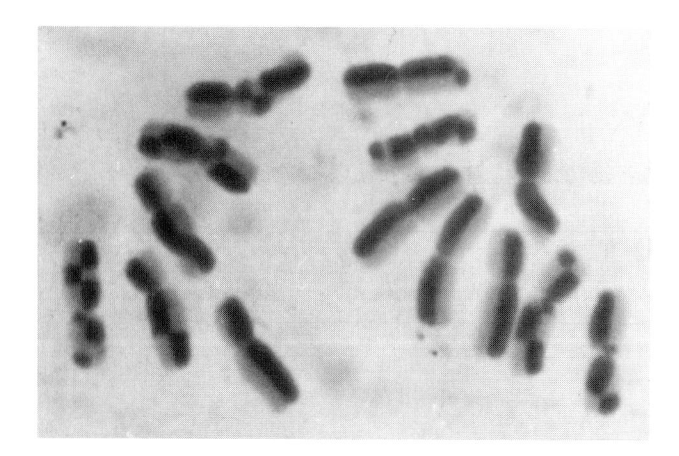

Fig. 18.3. Sister-chromatid exchanges in *Secale africanum* (Friebe 1980) in at least 8 out of 14 mitotic chromosomes after labeling and incubation with 5-bromo-2′-deoxyuridine (BrdU). In such experiments, it was found that if the chromosomes that had incorporated BrdU were subsequently stained with a fluorescent dye (Hoechst 33258), exposed to light, and then stained with Giemsa, they showed a staining intensity that depended on the level of incorporation of BrdU (Wolff and Perry, 1974). The darkly staining chromatids contain DNA with only one of the two chains of the helix containing BrdU, whereas the lightly staining chromatids have DNA with both strands containing BrdU.

of the higher resolution afforded by the BrdU technique (Fig. 18.3). In these experiments, the BrdU is present in two mitotic divisions and, as a result, creates DNA molecules that have one or both strands of the DNA helix labeled with BrdU. The two types of DNA molecules are readily distinguished by staining with a fluorescent dye because of the different quenching levels of the emitted light. This difference can be further accentuated by Giemsa staining. The data obtained in these experiments demonstrated that reciprocal exchanges of sections of chromatids can occur. Although sister-chromatid exchanges (SCEs) are, in part, induced by the incorporation of unusual bases into DNA during an experiment, the phenomenon is not entirely an artifact induced by the repair responses of the cell. This can be proven by cytogenetic experiments in the absence of any labeling procedures. Ring chromosomes occur naturally in both plant and animal cells, and when sister-chromatid exchange occurs in these chromosomes, a dicentric chromosome is formed (Fig. 18.4). The finding of such dicentric chromosomes in dividing cells that carried ring chromosomes is evidence that sister-chromatid exchange can occur in the absence of any chemical stress.

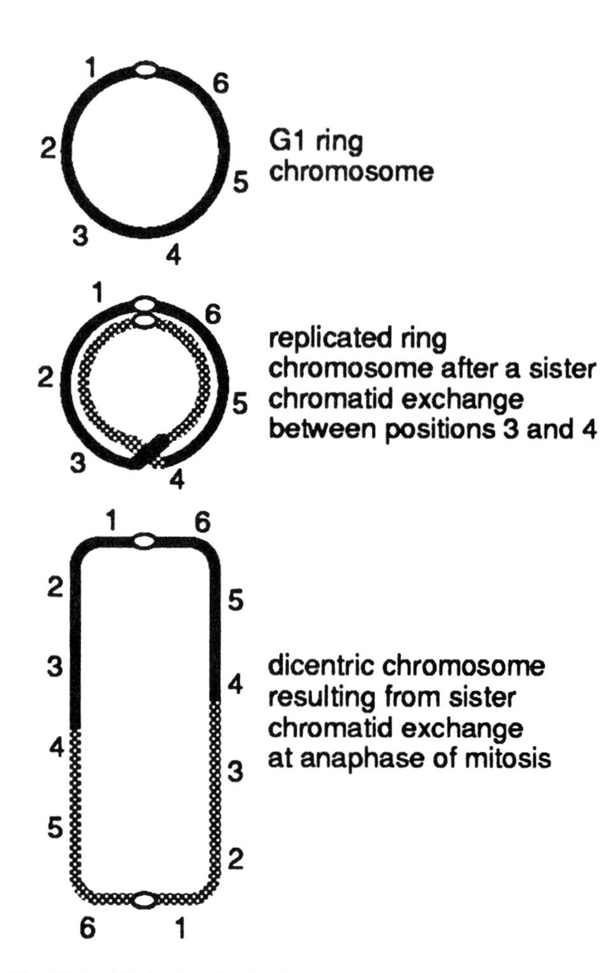

Fig. 18.4. Mechanism for the formation of dicentric chromosomes from ring chromosomes. When sister-chromatid exchange occurs between replicating ring chromosomes, a double-sized dicentric chromosome is formed (McClintock, 1940). Labeling experiments (*see* Fig. 18.3) have confirmed McClintock's analysis. More recent experiments have also demonstrated that the experimental labeling of chromosomes with BrdU and ^{3}H-thymidine actually increases the frequency of sister-chromatid exchange (Peacock, 1979).

18.2 DNA SYNTHESIS IN PROKARYOTES

The DNA sequences where DNA replication is initiated are origins of replication (*ori*) and have been identified in some organisms. The origins of replication are characterized by the presence of repetitive sequences (Fig. 18.5; *see also* Chapter 20 Fig. 20.5).

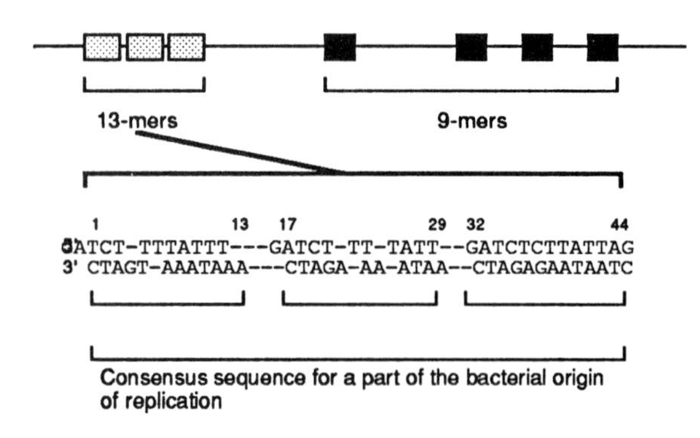

Fig. 18.5. The structure of a bacterial origin of replication. The illustrated consensus sequence is highly conserved among bacteria including the marine bacterium *Vibrio harveyi* (Zyskind et al., 1983; McMacken et al., 1987; Diffley and Stillman, 1990). Bacterial origins of replication are characterized by the presence of simple repetitive sequences, and the protein DnaA (*see* Table 18.1) binds to DNA with these sequences. The timing of initiation is thought to be dependent on the concentration of a form of the DnaA protein to which ATP has been bound (Donachie, 1993).

18.2.1 The Proteins Involved

There is usually only a single origin of replication in a protosome, and DNA replication proceeds bidirectionally from this point (Fig. 18.6). Specific proteins recognize origins of replication and,

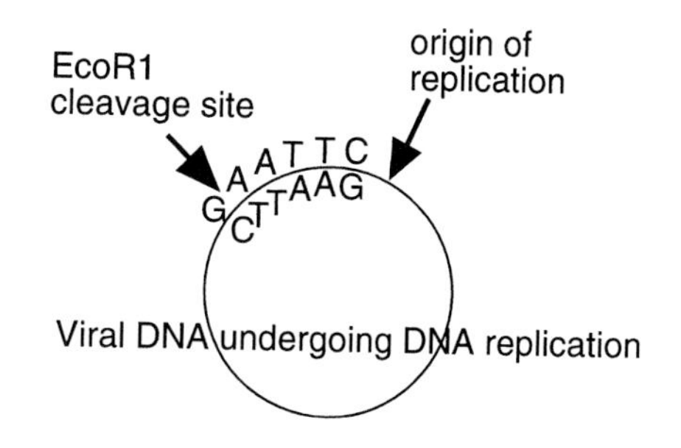

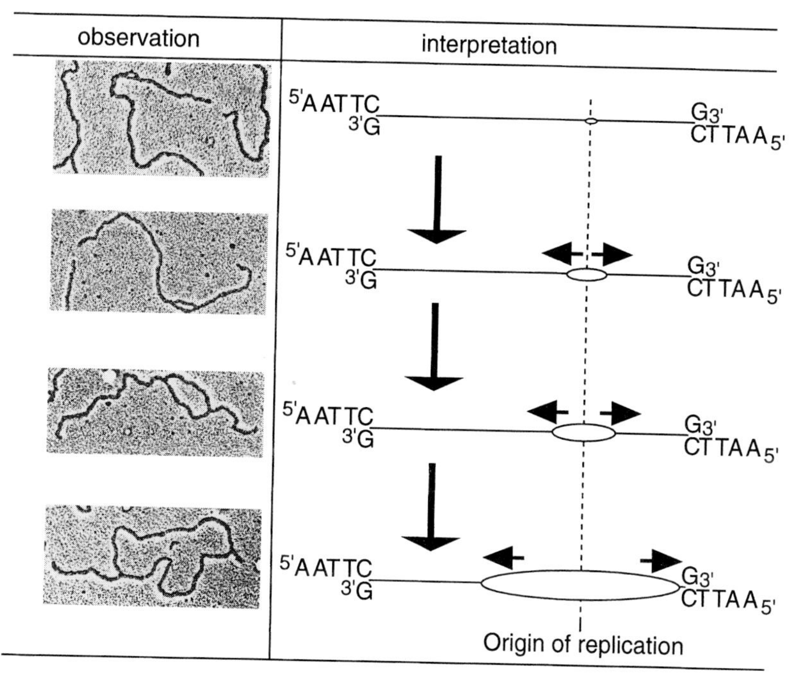

Fig. 18.6. The bidirectional replication of protosomal DNA from a single origin of replication. This was clearly demonstrated by cleaving the molecules at a single restriction-endonuclease site (top part of figure), and then examining replicating molecules by electron microscopy (left-hand side, lower part of figure; Fareed et al., 1972). In the lower right part of the figure, it can be seen how the asymmetric position of the origin of replication (vertical dotted line) relative to the restriction site allows the unambiguous orientation of the DNA molecules and thus demonstrates that the progression of the replicating DNA is toward both ends of the molecule.

in conjunction with other proteins, cause the changes in helical structure that allow the replication machinery access to the base pairs. The replication of the *Escherichia coli* protosome requires the unfolding of DNA, unwinding of the DNA helix, protection of the single strands of DNA, fidelity of DNA synthesis, a progressive and high rate of movement of the replication fork, and coordination of the synthesis from both DNA strands. Under ideal conditions of growth, the entire protosome in *E. coli* is replicated in 20 min; therefore, it can be calculated that DNA synthesis must proceed at the rate of approximately 200,000 bp/min. The various proteins involved in these steps are summarized in Table 18.1.

It has been noted earlier (Chapter 17, Section 17.1.1) that the sugar–phosphate backbone of the DNA helix has a clear polarity. The synthesis of the sugar–phosphate backbone can occur only in one direction, namely the 5′-phosphate of the incoming nucleoside triphosphate is added to the 3′-OH moiety of the oligonucleotide that is in the process of being extended. This oligonucleotide is therefore said to grow in the 5′ to 3′ direction, as it is the 3′-end that is constantly being increased in length. The template strand with the sugar–phosphate backbone in the 3′–5′ polarity is called the leading strand, because it provides the basis for a relatively uncomplicated synthesis of the new DNA strand in the complementary 5′ to 3′ direction.

The parental DNA strand with a 5′–3′ polarity is called the lagging strand and cannot be simply copied in the same manner as the leading strand. This would require the 3′-OH of the incoming nucleoside triphosphate to be added to the 5′-phosphate of the growing oligonucleotide; the enzymes available for this mode of DNA replication do not exist. The result is that the lagging strand is copied in a discontinuous manner, with the actual synthesis occurring in the normal 5′–3′ manner, but in a direction that is opposite to the overall direction of movement of the replication fork (Fig. 18.7). The enzyme machinery copying the lagging strand periodically translocates forward, in the direction of the replication fork, to begin copying the segment of parental DNA exposed by the moving replication fork. This process of replicating the lagging

Table 18.1. Proteins in DNA Synthesis

Protein	Primary Function
	Initiation of Replication
DnaA	Recognizes and binds to *ori* (origins of replication).
DnaB, DnaC	Form an aggregate with the DnaA–*ori* complex.
Helicase	Responsible for unwinding the DNA helix near the *ori*–protein complex.
HU	Histone-like protein binds to DNA–protein complex at *ori*. This protein is active as a heterodimer and binds to the minor groove to induce bending of DNA.
Single-strand-binding protein (SSB)	Binds to the single-stranded DNA resulting from helicase activity in order to block reassociation of strands into a duplex.
Primase	Synthesizes RNA primers at *ori* in preparation for DNA replication.
	DNA Chain Elongation
DNA polymerase III (holoenzyme)	An aggregate of at least seven proteins that binds to the initiation complex to replicate the leading strand continuously and the lagging strand discontinuously. The copying of the lagging strand is in the form of a series of short DNA sequences, or Okazaki fragments, which are subsequently joined together by DNA ligase. The role of the enzyme complex is to catalyze the formation of phosphodiester bonds between 3′-OH termini and the 5′-P of incoming nucleoside triphosphates that are complementary to the template base. One of the protein subunits of DNA polymerase III (subunit ϵ) is responsible for the 3′–5′ proofreading exonuclease activity. The helicase and primase activites in the initiation complex, on the lagging strand, are believed to migrate ahead of the DNA polymerase III complex on the leading strand (see Fig. 18.7).
DNA gyrase (topoisomerase II type activity)	Acts as a swivel during DNA replication to separate the highly intertwined parental strands of DNA. The α-subunit carries out the nicking–closing activity to facilitate untanglement of strands, whereas the β-subunit provides the ATPase activity. DNA gyrase can also induce supercoiling of DNA, which is considered to be of importance in the formation of the initiation complex.
DNA polymerase I	Removes RNA primers (5′–3′ exonuclease activity) and fills in gaps. Also has a proofreading function by degrading and resynthesizing newly formed DNA.
DNA ligase	Seals the breaks in the sugar–phosphate backbone and joins Okazaki fragments.
	Termination of DNA Replication
dnaT	Terminates DNA replication. DNA gyrase activity is required for the final resolution of the DNA strands.

DNA strand requires repeated initiation of DNA synthesis, hence the continued requirement for primase, to produce RNA primer in the replicating-fork complex. The initial series of small DNA fragments that are synthesized from the lagging DNA strand are called Okazaki fragments. Coordination of leading- and lagging-strand syntheses is considered to depend on interactions between the helicase and primase proteins, and the DNA polymerase III subunits involved in leading-strand and lagging-strand syntheses.

18.2.2 Fidelity of Replication

The DNA polymerase III–protein complex carrying out the replication of DNA has a certain probability of making errors, because the selection of bases complementary to the template strand, during the process of DNA synthesis, is a complicated procedure. The physical differences between the various possible pairs of bases are extremely small, and the overall accuracy of DNA replication most likely depends on the following three steps:

1. G.C and A.T base pairs are selected on the basis of small differences in geometry and thermal stability that differentiate them from the other possible combinations of base pairs. In the presence of the correct G.C and A.T base pairs, the DNA polymerase is thought to undergo a conformational change that facilitates the formation of a new phosphodiester bond.
2. Editing, or proofreading, of the newly synthesized DNA by enzymes that can degrade DNA in the 3′–5′ direction (3′–5′ exonuclease activity), followed by resynthesis of the DNA strand, can remove errors that occurred during the initial replication (*see* Fig. 18.7).
3. Post-replicative editing, or DNA repair, can remove base mismatches formed during replication, as well as bases damaged by the action of ultraviolet (UV) light or chemicals. This activity is part of a complex series of activities, called the SOS response, that come into play when a damaged base causes a block in DNA replication. These repair activities are discussed in detail in Chapter 7, Section 7.2.3.

The large differences in spontaneous mutation rates between different organisms most likely reflect differences in the efficiency of protein machinery involved in DNA proofreading and post-replicative editing (Table 18.2).

18.3 DNA REPLICATION IN CHROMOSOMES

The overall process of DNA replication in eukaryotes is considered to be similar to that of the well-characterized *E.coli* DNA replication described above. Although distinct differences such as the occurrence of multiple origins of replication exist, the enzymology of DNA replication in eukaryotes includes the same activities that have been studied in *E. coli*.

18.3.1 Origins of Replication

Eukaryotes are characterized by multiple origins of replication in the chromosomes. Although the initiation of DNA replication is not understood in as much detail as in prokaryotes, it is closely

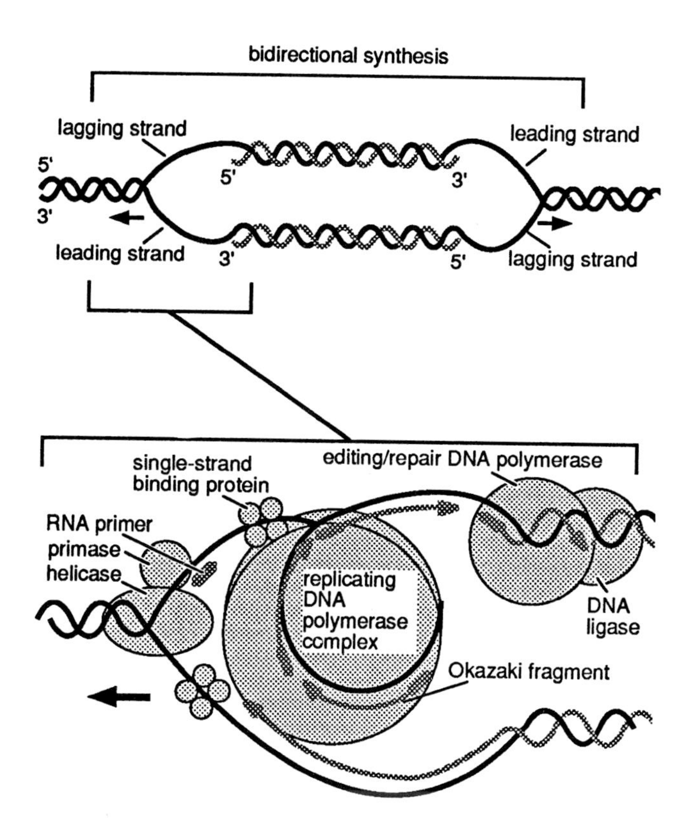

Fig. 18.7. The bidirectional replication of DNA at a replication fork. The diagram (based on McMacken et al., 1987; Brambill and Kornberg, 1988) summarizes the directions of chain growth (top) and the enzymes involved in DNA synthesis (bottom). Numerous challenges are overcome during the replication of DNA, including the release of torsion in the helix (*see* Fig. 17.31), the coupling of replication to other metabolic processes in the cell (Schmidt and Migeon, 1990; Rivier and Rine, 1992), and the attainment of an extremely low error rate (Echols and Goodman, 1991) while still maintaining a high level of efficiency (Debyser et al., 1994).

tied to the cell cycle. The cyclin/cdc complexes, formed during G1 and S phases (*see* Chapter 4, Section 4.2.1, and Fig. 4.18) to determine the phosphorylation/dephosphorylation events controlling the progression of cells through the cell cycle, also have a major control over DNA replication. The protein Cdc6, for example, has been implicated in yeast as a key factor, interacting with the origin of replication complex, to determine the frequency of initiation of DNA replication. The human DNA α-polymerase is also phosphorylated in a cell-cycle-dependent manner, with hyperphosphorylation occurring in mitosis. The replication of virus SV40 DNA in mammalian cells is particularly well characterized and utilizes the host-replication machinery. An essential protein required for initiation of SV40 DNA replication is T-antigen, encoded by SV40, and

Table 18.2. Mutation Frequency per Base Pair per Replication

Qβ RNA Primed RNA Synthesis	DNA Replication in T4 and λ Bacteriophages	Reversion Frequency in *E. coli* Tryptophan Gene	Mutation Frequency in *Drosophila*
10^{-3}–10^{-4}	10^{-8}	5×10^{-10}	5×10^{-11}

Source: Data taken from J. Langridge (1991).

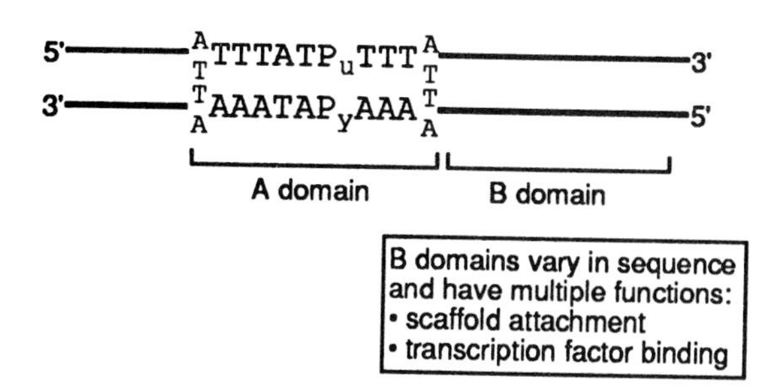

Fig. 18.8. Autonomously replicating sequences (ARS elements) from yeast. These DNA sequences provide sites for the initiation of DNA synthesis and contain a common A.T-rich, 12-bp DNA sequence in the A domain (Diffley and Stillman, 1990, Heintz et al., 1992). The B domain is not highly conserved in sequence but is involved in a range of other activities such as the interaction between the initiation of DNA replication and transcription (Rivier and Rine, 1992; Liu and Alberts, 1995).

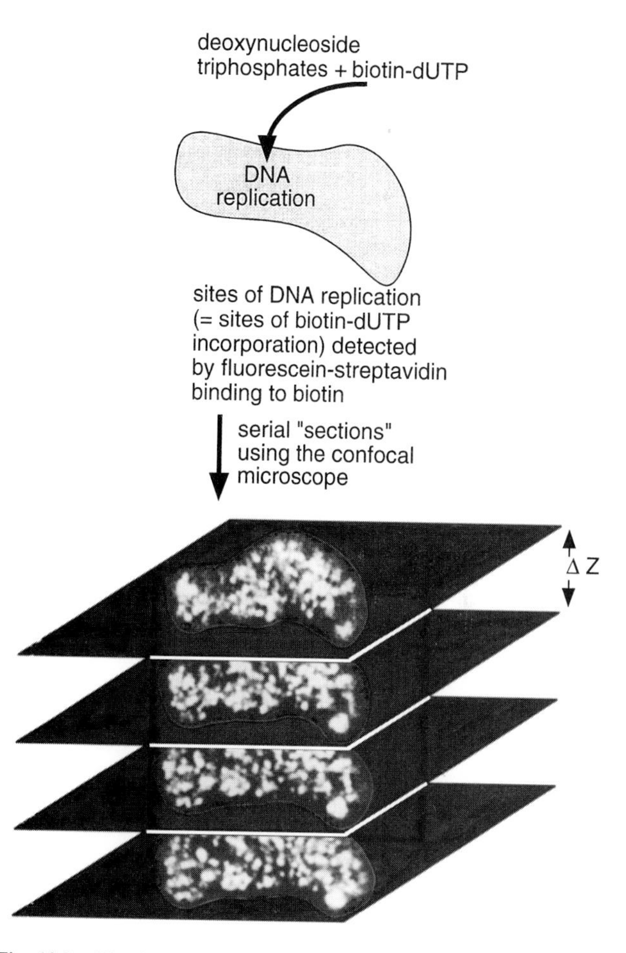

Fig. 18.9. The detection of multiple DNA-initiation sites in eukaryotic cells. The initiation sites can be demonstrated cytologically by providing cells with biotin-dUTP (for structure, *see* Fig. 19.4, Chapter 19) for a brief time (top), and then assaying the sites of incorporation with fluorescein–streptavidin, which binds very tightly to the biotin, giving a complex that can be observed with fluorescence microscopy. Optical sectioning using the confocal microscope (bottom) allows individual sites to be mapped in three dimensions relative to other landmarks in the nucleus. (Modified from Mills et al., 1989.) The different overall rates of DNA replication in different tissues and in different parts of the genome are generally due to variation in the frequency of DNA replication initiation rather than variation in the inherent rate of DNA replication.

this protein must be activated by phosphorylation at its threonine 124 by the host's $p34^{cdc2}$. Phosphorylation of the serine 120 and 123 residues in the T-antigen can, however, inhibit its activity.

DNA sequences that provide sites for the initiation of DNA synthesis have been isolated from yeast and named autonomously replicating sequences or ARS elements. These ARS elements contain a common A.T-rich, 12-bp DNA sequence (Fig. 18.8), and they provide an avenue for unraveling the details of DNA replication in eukaryotes by isolating proteins that bind to these sequences. The A.T-rich sequence, or A domain, is flanked by a region, the B domain, that varies between different origins of replication even though it is required for the initiation of replication. The B domain is thought to have multiple functions, including binding to nuclear-skeleton proteins (*see* Fig. 18.11) and transcription-activating factors. The latter function relates the involvement of transcription-activating factors to the temporal coordination of both replication and transcription during the cell cycle; in an extreme case, regions of chromosomes that are poorly transcribed (called heterochromatin, *see* Chapter 20, Section 20.1) are replicated late in the S phase. In eukaryotes, DNA synthesis is bidirectional, and the multiple points of initiation can be demonstrated cytologically by providing cells with BrdU for brief periods, then assaying the sites of incorporation with fluorescently labeled antibodies to BrdU (Fig. 18.9). The DNA replication machinery accepts the BrdU 5′-triphosphate that is formed in the cell, as an alternative to thymidine 5′-triphosphate. Multiple sites can be seen and their pattern of distribution depends on the particular stage of the S phase that is labeled. Independent experiments, which have measured the rate of DNA synthesis in eukaryotic nuclei and the time taken to replicate the entire genome, predict that many more replication forks than the ones visualized cytologically must exist. This discrepancy suggests that cytological observations are not differentiating clustered regions of DNA undergoing replication. Experiments in which replicating DNA has been observed after the incorporation of a pulse of ^{3}H-thymidine, followed by autoradiography of the isolated DNA, also indicate that origins of replication are close together (Fig. 18.10).

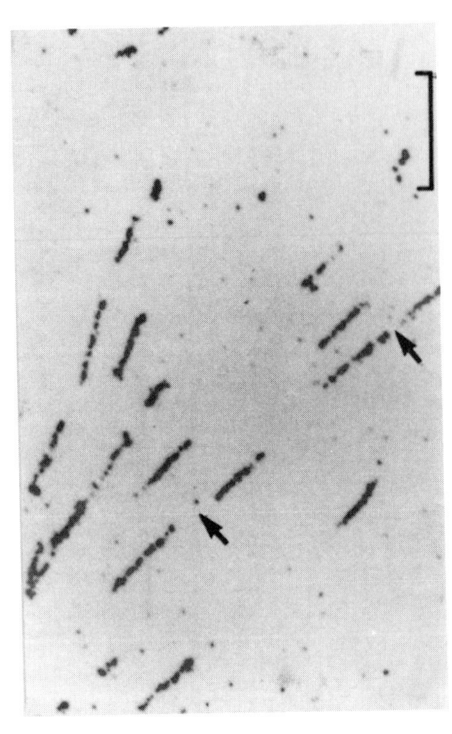

Fig. 18.10. The bidirectional nature of DNA replication in plant chromosomes. Root-tip cells of cereal rye were given a pulse of ^{3}H-thymidine, followed by a short period of growth in unlabeled thymidine. The nuclei were then gently lysed on to glass slides and autoradiographed (Huberman and Riggs, 1968). In the photograph (from Francis and Bennett, 1982), the labeled, dark strands are regions of DNA that incorporated ^{3}H-thymidine. These are commonly separated by shorter, unlabeled regions of DNA that subsequently replicated in the presence of unlabeled thymidine. The origins of replication are interpreted to be in the positions indicated by the arrows. The bar indicates 25 μm.

Table 18.3. Proteins in Eukaryotic DNA Replication

Protein Location	Function	Chromosome
α-DNA polymerase 180 kDa protein	DNA polymerase activity plus 3′–5′ exonuclease proofreading	Short arm of human X (Xp21.3–Xp22.1)
70 kDa protein		
49 kDa protein	RNA primase	On mouse #9
58 kDa protein	RNA primase	On mouse #1
β-DNA polymerase 40 kDa protein (335 amino acids)	DNA repair	On human #8
γ-DNA polymerase 144 kDa protein	Replication of mitochondrial DNA	
δ-DNA polymerase 125 kDa protein	Leading-strand DNA synthesis	
ϵ-DNA polymerase group of proteins 132–200 kDa	Not yet defined	
Proliferating cell nuclear antigen (PCNA or cyclin) 32 kDa protein	Binds to replication complex to coordinate DNA replication to cell cycle	
Replication factor A (RF-A)	Single-strand-binding (SSB) protein required for the initiation of DNA replication	
Replication factor C (RF-C)	Required to facilitate lagging-strand synthesis	
Helicase Topoisomerase II	Unwinding of DNA helix and disentanglement of DNA strands	

18.3.2 Proteins Involved in Chromosomal DNA Replication

The autonomously replicating sequences, ARSs, consist of A and B elements (*see* Fig. 18.8) and it is these DNA sequence elements that are recognized, and bound, by the origin-recognition complex (ORC). The ORC complex is comprised of at least six protein subunits, some of which include the proteins of α DNA polymerase and the single-strand-binding (SSB) protein. Following the formation of the ORC, DNA replication proceeds using the proteins listed in Table 18.3. Studies to date have shown that eukaryotic DNA synthesis requires an SSB protein, similar to the prokaryotic DNA replication requirement for SSB (*see* Table 18.1); in eukaryotes, this protein is subject to control by phosphorylation. The eukaryotic SSB is also referred to as replication factor-A (RF-A; Table 18.3). The completion of eukaryotic DNA replication is closely tied to the synthesis of histones, which may be analogous to the involvement of the HU protein in prokaryotic DNA synthesis.

The replication of certain DNA sequences such as long arrays of tandemly repetitive sequences late in the S phase after other sequences have been replicated, may result from a low frequency of origin-of-replication sites in these regions of the genome. In addition, because a nucleoskeleton is involved in the replication process, with its own properties becoming modified as the cell progresses through the S phase, it may influence the timing of replication of certain classes of DNA sequences.

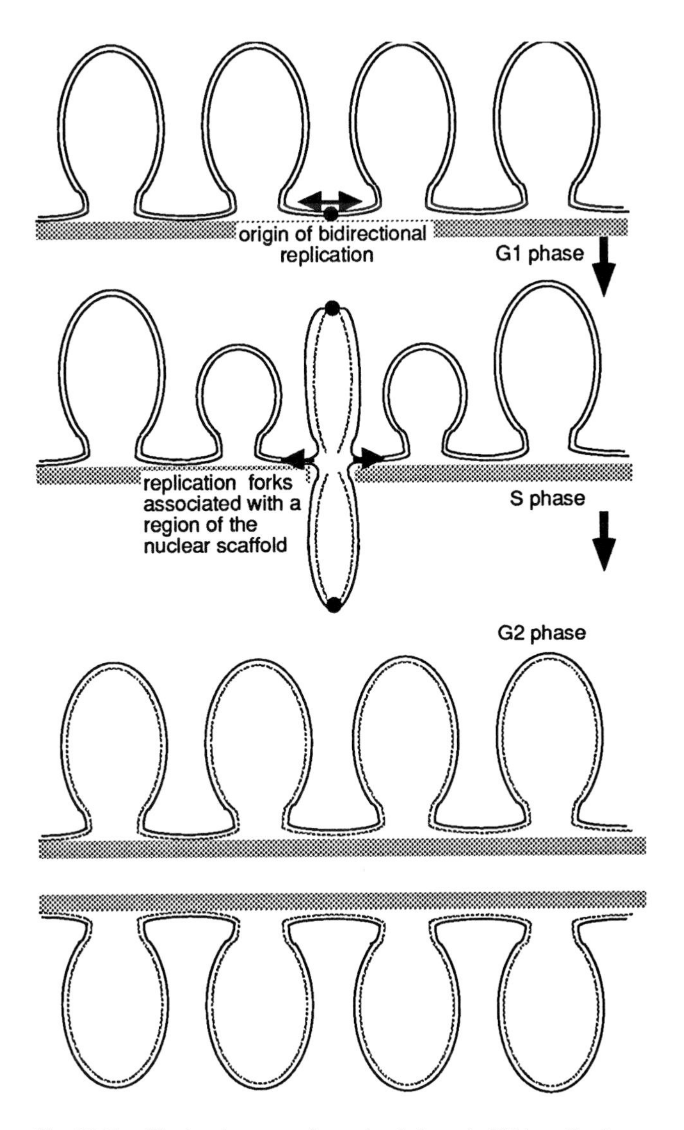

Fig. 18.11. The involvement of a nucleoskeleton in DNA replication. The existence of some type of DNA-protein-nucleoskeleton attachment to help stabilize the replicating chromosomal DNA is generally accepted, even though it may only be transient. (Based on Cook, 1991.)

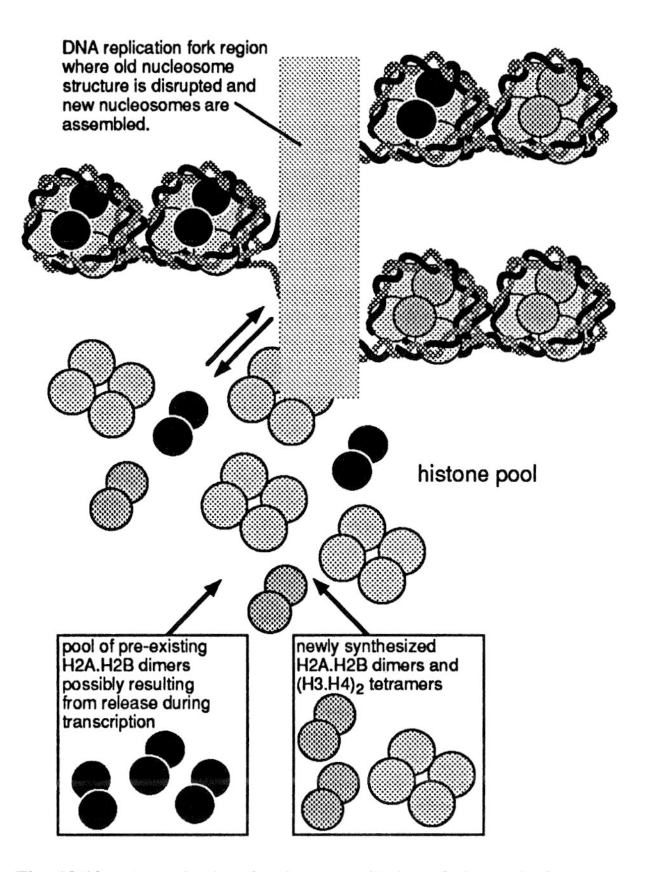

Fig. 18.12. A mechanism for the reconstitution of chromatin from freshly replicated DNA and histones. The histone addition does not distinguish between "old" histone molecules, released prior to DNA replication, and newly synthesized histone molecules. The H3/H4 histone pairs form a pool of tetramers, $(H3/H4)_2$ (center) and the H2A and H2B histone proteins pair to form a pool of dimers (lower), both of which combine with the newly synthesized DNA molecules (upper). (Based on Dilworth and Dingwall, 1988.)

It is significant that any single DNA segment is replicated only once per cell cycle and, although the mechanism for differentiating newly synthesized DNA from unreplicated DNA is not understood, the involvement of other proteins such as those of the nucleoskeleton may be important. The existence of some type of DNA–protein–nucleoskeleton attachment is generally accepted, even though it may only be a transient one, and its involvement in DNA replication is illustrated in Fig. 18.11. In special, developmentally regulated situations, the "once only per cell cycle" rule for replication of genomic DNA is broken and can lead to amplification of defined regions of the genome. Examples of DNA amplification are discussed in Section 18.4.

18.3.3 The Assembly of Newly Formed Chromosomes

As newly synthesized DNA is formed, it is complexed with either newly synthesized or preexisting histone molecules. The pool of preexisting histones is augmented by both the replicative and transcriptional processes, as both can cause some displacement of histones from the DNA. In dividing cells, the pool of free histones is extremely small because of the demand to form nucleosomes with newly synthesized DNA. The association of histones with newly made DNA molecules is illustrated in Fig. 18.12. It seems likely that other protein factors are involved in the assembly of the final nucleosomal complex, but their exact role remains to be determined. One such protein, nucleoplasmin, has been characterized biochemically and is required for in vitro experiments, in which the DNA–histone complex is reconstituted from its individual components. The assembly of chromatin from newly synthesized DNA is most likely a critical period for establishing new DNA–protein interactions that are important for determining gene expression during development of an organism. As discussed in Chapter 17, Section 17.5.1, nucleosome assembly generally represses gene activity. The postreplicative period of chromatin assembly provides an opportunity for DNA-binding proteins involved in controlling gene expression, to compete with nucleosome assembly for binding to critical regions of DNA.

18.3.4 The Synthesis of Histones in the S Phase of the Cell Cycle

The bulk of the histone H3, H4, H2A, and H2B molecules are synthesized during the S phase but are not the result of any single controlling step. The various means by which the increased synthesis is attained are summarized in Table 18.4. The replication-dependent histones are formed in the cytoplasm, using mRNA that does not have poly-A at its 3′-end (except for fungi and ciliates), and are rapidly transported to the nucleus. The genes coding for these histones are characterized by the absence of introns and the presence of a sequence in the 3′-downstream region that can adopt a highly conserved hairpin structure when transcribed into RNA (*see* Chapter 16, Fig. 16.23). This stem–loop structure is a critical signal in determining the stability of the messenger RNA in relation to the cell cycle.

Table 18.4. Control of Histone Synthesis in Relation to the Cell Cycle

Controls	Action
Transcriptional	The complex array of genes interacts with control regions 5′-upstream to create both positive and negative controls on transcription.
Post-transcriptional	
Processing of histone mRNA	A precursor RNA is processed by enzymes that use a small nuclear RNA molecule as a cofactor for specific cleavage.
Stability of histone mRNA	A hairpin in the 3′-downstream region of the mRNA binds a protein factor that controls stability of the mRNA.
Translation of histone mRNA	The process of translation initiates degradation of the mRNA through removal of the hairpin.
Effect of excess histones	Via a feedback mechanism—the presence of excess histone preferentially stimulates histone mRNA degradation.

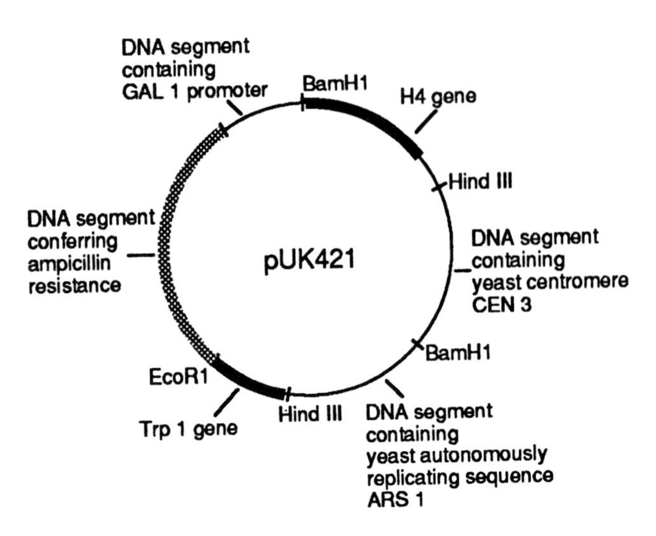

Fig. 18.13. The plasmid pUK421 containing a cloned yeast H4 gene sequence. The significance of the presence of H4 in chromatin has been investigated by deleting the yeast genes that code for H4 protein and then reintroducing the protein in a plasmid carrying an H4 gene. The expression of the plasmid-encoded H4 gene is under the control of a GAL1 promoter, so that the H4 gene is activated in galactose-containing medium and repressed in glucose-containing medium. This makes it possible to switch H4 synthesis on or off as required (Kim et al., 1988).

The genetic manipulation of histone synthesis in yeast cells has provided a basis for relating histone synthesis to the physiology of the entire cell. Although yeast normally contains two unlinked copies of the H4 histone gene, strains have been developed that replace these genes with the marker genes HIS3 (histidine-requiring) and LEU2 (leucine-requiring). The H4 protein is then supplied by a plasmid carrying an H4 gene (Fig. 18.13). As it is possible to synchronize yeast cells at the G1 phase of the mitotic cycle, glucose repression can be used to inhibit H4 synthesis and demonstrate, using flow cytophotometry, that H4 depletion leads to an accumulation of cells in the G2 phase (Fig. 18.14). The association of H4 depletion with lethality was determined by testing the ability of cells to recover from H4 depletion when plated on galactose-containing medium. By relating the recovery ability of the population of cells to the time spent in the H4-depleted state, the period leading to lethality is shown to occur when nucleosomes are assembled on newly replicated DNA in the S phase. Direct assays for nucleosome structure in chromatin, under conditions of no H4-histone synthesis, indicate that a general loss of nucleosomes occurs and can lead to unusually high levels of expression of certain loci that are normally repressed.

18.3.5 Synthesis of Histones Can Occur Outside the Limits of the S Phase

The range of histone molecules that are synthesized at a time when no DNA replication is occurring include minor sequence variants of H1, H2A, and H2B. A tissue-specific histone found in avian erythrocytes, H5, is in the H1 class of histone and is also made at low levels outside the S phase of the cell cycle. These histones replace chromosomally bound histones, which are degraded during chromatin structural changes that accompany transcriptional processes.

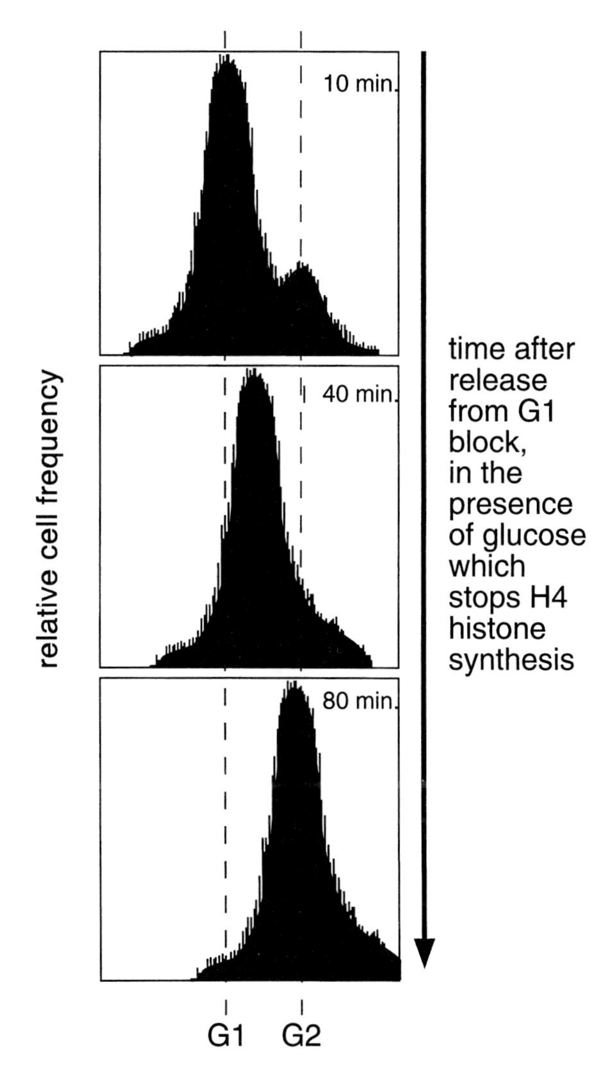

Fig. 18.14. The effects of glucose repression on yeast containing the pUK421 plasmid. It is possible to synchronize yeast cells at the G1 phase of the mitotic cycle, and glucose repression can then be used to inhibit H4 synthesis by the plasmid (*see* Fig. 18.13). Using flow cytometry, Kim et al. (1988) demonstrated that H4 depletion leads to an accumulation of cells in the G2 phase.

18.4 DNA SYNTHESIS OUTSIDE THE NORMAL CELL CYCLE IS LIMITED TO SPECIFIC REGIONS OF THE GENOME

Developmentally regulated deviations from the ''once only per cell cycle'' rule for the replication of genomic DNA lead to amplification of regions containing certain genes or, on a larger scale, to the overreplication of large sections of the genome, giving rise to polytene chromosomes.

The amplification in *Drosophila* embryos of the genes coding for chorion proteins on chromosome 3 (at position 66D) and the X chromosome (at position 7F) has been well studied. The tandemly arranged genes encode eggshell structural proteins and are amplified by frequent initiation of DNA replication and relatively slow fork movement in the chromosome region carrying the gene array, plus at least 50 kb of flanking DNA (Fig. 18.15). Sequences located upstream from the chorion gene cluster, cis-acting elements [named amplification-control elements (ACEs)] and amplification enhancers (AERs), control the level of amplification. The origins of replication responsible for the amplification of the chorion-gene domain on chromosome 3, containing four tandemly arranged chorion genes, are localized to a small region within the domain and resemble the yeast autonomously replicating sequences (ARSs) in being A.T-rich.

Amplification of specific regions of the genome has also been demonstrated by selection pressure for a gene conferring drug resistance to mammalian cells in tissue culture. The region amplified is usually larger than the average replicon in mammalian cells (100 kb) and can be several thousand kilobases in length. The most thoroughly studied example of this phenomenon is the amplification of the dihydrofolate reductase (DHFR) gene in response to selection pressure by growing cells in the presence of methotrexate. The replication origin involved has amplification-promoting activity. The amplification results in the formation of homogeneously staining regions (HSRs), and further biochemical studies are required to elaborate the factors that define the amplified region of the genome.

The formation of episomes, 200–600 kb, and of very small chromosomes, called double minutes (DMs), have also been correlated with DNA amplification in tissue-culture cells and tumors

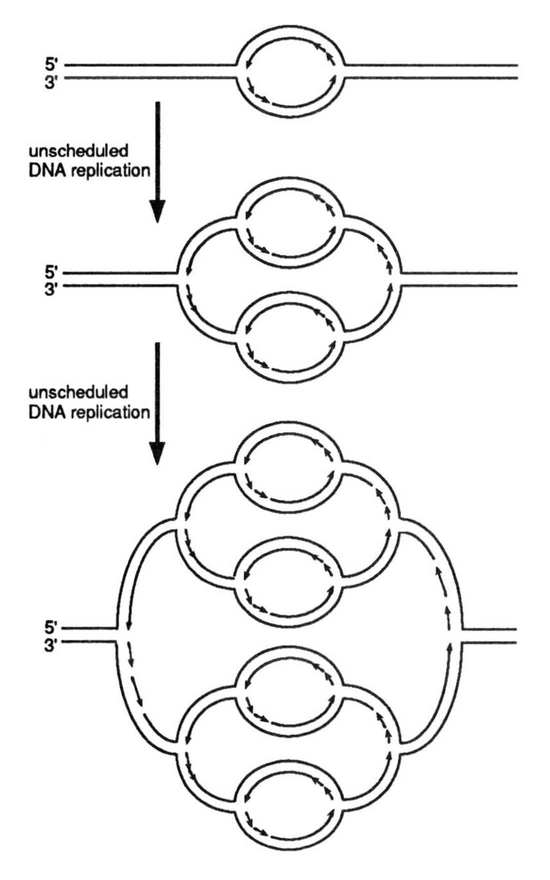

Fig. 18.15. The ''onion-skin'' model for the localized amplification of DNA. The model postulates a large increase in the amplification of DNA replication within a region of newly replicated DNA and creates a structure that resembles the layers in an onion, hence the name of the model Several rounds of extra initiation events lead to the local amplification of specific DNA sequences. This mechanism is believed to account for the amplification of chorion genes in *Drosophila* (Delidakis and Kafatos, 1989; Heck and Spradling, 1990; Wintersberger, 1994).

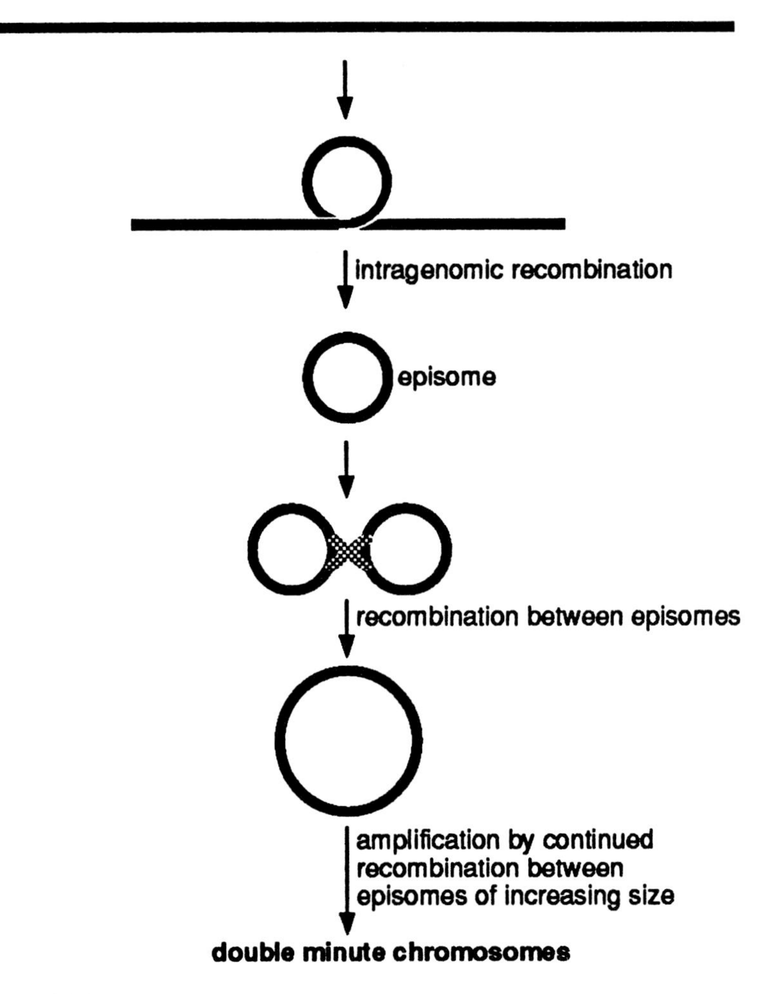

Fig. 18.16. The extrachromosomal amplification of specific DNA sequences. This mechanism has been postulated to account for the large increases in number of ribosomal RNA genes that have been demonstrated in the oocytes of many organisms (Kubrakiewicz and Billinsky, 1995). Not only does this mechanism account for recombination between episomic circles, it is also postulated to account for the occurrence of double-minute chromosomes that arise in response to exposure to drugs (reviewed in Wintersberger, 1994).

(Fig. 18.16). The actual mechanism of DNA amplification in these situations is different from the processes characterized for the chorion-gene cluster in that numerous cell cycles are required before the amplification is evident. It is possible that bridge–breakage–fusion cycles, initiated by sister-chromatid exchange (SCE) (*see* Chapter 7, Section 7.3), are the primary causes for DNA amplification and associated formations of episomes and double-minute chromosomes. The SCEs may be induced by drugs that are used in selection experiments.

The formation of the macronucleus in *Tetrahymena* spp. and other ciliated protozoa also involves a developmentally controlled amplification of DNA, without accompanying nuclear or cell division. The development of the macronucleus, which is characteristic of somatic tissues, involves the excision of DNA segments, the addition of telomere sequences, joining of two segments to form a palindrome, and replicative amplification. The DNA segments that undergo replicative amplification have a single origin of replication and appear to function as minichromosomes.

BIBLIOGRAPHY

Section 18.1

Friebe, B. 1980. Comparison of sister chromatid exchange and chiasma formation in the genus *Secale*. Microscopica Acta 83: 103–111.

McClintock, B. 1940. The association of mutants with homozygous deficiencies in *Zea mays*. Genetics 25: 542-571.

Meselson, M., Stahl, F.W. 1958. The replication of DNA in *Escherichia coli*. Proc. Natl. Acad. Sci. USA 44: 671–682.

Peacock, W.J. 1979. Strandedness of chromosomes and segregation of replication products. Cell Biol. 2: 363-387.

Taylor, J.H., Woods, P., Hughes, W.L. 1957. The organization and duplication of chromosomes as revealed by autoradiographic studies using tritium-labeled thymidine. Proc. Natl. Acad. Sci. USA 43: 122–128.

Wolff, S., Perry, P. 1974. Differential Giemsa staining of sister chromatids and the study of sister chromatid exchanges without autoradiography. Chromosoma 48: 341–353.

Section 18.2

Borowiec, J.A., Dean, F.B., Bullock, P.A., Hurwitz, J. 1990. Binding and unwinding—how T antigen engages the SV40 origin of DNA replication. Cell 60: 181–184.

Brambill, D., Kornberg, A. 1988. A model for the initiation at origins of DNA replication. Cell 54: 915–918.

Crooke, E., Thresher, R., Hwang, D.S., Griffith, J., Kornberg, A. 1993. Replicatively active complexes of DnaA protein and the *Escherichia coli* chromosomal origin observed in the electron microscope. J. Molec. Biol. 233: 16–24.

Debyser, Z., Tabor, S., Richardson, C.C. 1994. Coordination of leading and lagging strand DNA synthesis at the replication fork of bacteriophage T7. Cell 77: 157–166.

Diffley, J.F.X., Stillman, B. 1990. The initiation of chromosomal DNA replication in eukaryotes. Trends Genet. 6: 427–432.

Donachie, W.D. 1993. The cell cycle of *Escherichia coli*. Annu. Rev. Microbiol. 47: 199–230.

Echols, H., Goodman, M.F. 1991. Fidelity mechanisms in DNA replication. Annu. Rev. Biochem. 60: 477–512.

Erickson, H.P. 1995. FtsZ, a prokaryotic homolog of tubulin? Cell 80: 367–370.

Fareed, G.C., Garon, C.F., Salzman, N.P. 1972. Origin and direction of replication of simian virus 40 deoxyribonucleic acid replication. J. Virol. 10: 484–491.

Kornberg, A. 1988. DNA replication. J. Biol. Chem. 263: 1–4.

Langridge, J. 1991. Molecular Genetics and Comparative Evolution. Research Studies Press Ltd., Taunton, Somerset, UK.

Liu, B., Alberts, B.M. 1995. Head-on collision between a DNA replicating apparatus and RNA polymerase transcription complex. Science 267: 1131–1136.

Matic, I., Rayssiguier, C., Radman, M. 1995. Interspecies gene exchange in bacteria: the role of SOS and mismatch repair systems in evolution of species. Cell 80: 507–515.

McMacken, R., Silver, L., Georgopoulos, C. 1987. DNA replication. In: *Escherichia coli* and *Salmonella typhimurium:* Cellular and Molecular Biology (Neidhardt, F.C. ed.-in-chief). American Society for Microbiology, Washington, DC. pp. 564–612.

Messer, W., Weigel, C. 1997. DNA-A initiator-also a transcription factor. Mol. Microbiol. 24: 1–6.

Rivier, D.H., Rine, J. 1992. An origin of DNA replication and a transcriptional silencer require a common element. Science 256: 659–663.

Schmidt, M., Migeon, B.R. 1990. Asynchronous replication of homologous loci on human active and inactive X chromosomes. Proc Natl. Acad. Sci. USA 87: 3685–3689.

Zahn, K., Blattner, F.R. 1987. Direct evidence for DNA bending at the lambda replication origin. Science 236: 416–422.

Zyskind, J.W., Cleary, J.M., Brusilow, W.S.A., Harding, N.E., Smith, D.W. 1983. Chromosomal replication origin from the marine bacterium *Vibrio harveyi* functions in *Escherichia coli ori C* consensus sequence. Proc. Natl. Acad. Sci. USA 80: 1164–1168.

Section 18.3

Cook, P. 1991. The nucleoskeleton and the topology of replication. Cell 66: 627–635.

Dilworth, S.M., Dingwall, C. 1988. Chromatin assembly in vitro and in vivo. Bioessays 9: 44–49.

Durrin, L.K., Mann, R.K., Kayne, P.S., Grunstein, M. 1991. Yeast histone N-terminal sequence is required for promoter activation in vivo. Cell 65: 1023–1031.

Fairman, M.P. 1990. DNA polymerase δ/PCNA: actions and interactions. J. Cell Sci. 95: 1–4.

Francis, D., Bennett, M.D. 1982. Replicon size and mean rate of DNA synthesis in rye (*Secale cereale* L.) cv Petkus Spring. Chromosoma 86: 115–122.

Heck, M.M.S., Spradling, A.C. 1990. Multiple replication origins are used during *Drosophila* chorion gene amplification. J. Cell Biol. 110: 903–914.

Heintz, N.H., Dailey, L., Held, P., Heintz, N. 1992. Eukaryotic replication origins as promoters of bidirectional DNA synthesis. Trends Genet. 8: 376–381.

Huberman, J.A., Riggs, A.D. 1968. On the mechanism of DNA replication in mammalian chromosomes. J. Molec. Biol. 32: 327–341.

Kim, U-J., Han, M., Kayne, P., Grunstein, M. 1988. Effects of histone H4 depletion on the cell cycle and transcription of *Saccharomyces cerevisiae*. EMBO J. 7: 2211–2219.

Mills, A.D., Blow, J.J., White, J.G. Amos,W.B., Wilcock, D., Laskey, R.A. 1989. Replication occurs at discrete foci spaced throughout nuclei replicating in vitro. J. Cell Sci. 94: 471–477.

Osley, M.A. 1991. The regulation of histone synthesis. Annu. Rev. Biochem. 60: 827–861.

Pont, G., Degroote, F., Picard, G. 1988. Illegitimate recombination in the histone multigene family generates circular DNAs in *Drosophila* embryos. Nucleic Acids Res. 16: 8817–8833.

Svaren, J. Chalkley, R. 1990. The structure and assembly of active chromatin. Trends Genet. 6: 52–56.

Travers, A.A., Ner, S.S., Churchill, M.E.A. 1994. DNA chaperones: a solution to a persistence problem. Cell 77: 167–169.

Wang, T S.-F. 1991. Eukaryotic DNA polymerases. Annu. Rev. Biochem. 60: 512–552.

Section 18.4

Delidakis, C., Kafatos, F.C. 1989. Amplification enhancers and replication origins in the autosomal chorion gene cluster of *Drosophila.* EMBO J. 8: 891–901.

Kapler, G.M., Orias, E., Blackburn, E.H. 1994. *Tetrahymena thermophila* mutants in the developmentally programmed maturation and maintenance of the rDNA minichromosome. Genetics 137: 455–466.

KUBRAKIEWICZ, J., BILLINSKY, S. 1995. Extrachromosomal amplification of rDNA in oocytes of *Hemerobius* spp. (Insecta, Neuroptera). Chromosoma 103: 606–612.

STOLZENBURG, F., GERWIG, R., DINKL, E., GRUMMT, F. 1994. Structural homologies and functional similarities between mammalian origins of replication and amplification promoting sequences. Chromosoma 103: 209–214.

WINTERSBERGER, E. 1994. DNA amplification: new insights into its mechanism. Chromosoma 103: 73–81.

VII
In Situ Biochemical Reactions, Chromosome Landmarks, Genetic Mapping

19

Exploring Chromosomes by *In Situ* Biochemical Reactions

- Reactions carried out on chromosomes in situ involve a primary reaction that defines the specificity of the reaction, and a secondary reaction that provides the means for detecting the product of the primary reaction.
- The chemicals used in the preparation of chromosomes can determine the specificity of the primary reaction.
- Antibodies that detect specific features of chromosomes often provide the basis for a primary reaction.
- The hybridization of a nucleic-acid probe to its complementary sequence in chromosomal DNA is widely used in primary reactions to determine the sequence organization in chromosomal DNA.
- The polymerase chain reaction (PCR) provides a highly sensitive primary reaction for characterizing chromosome structure.
- Secondary reactions utilize enzymes that cause the precipitation of a dye at the site of the primary reaction, or bind a fluorescent-labeled antibody to a component of the primary reaction.

Procedures for staining biological materials provide the means for enhancing contrast and facilitating microscopic observations. Specificity in the staining reaction has the advantage of allowing contrast to be enhanced only in those structures that are of interest to the observer. Molecular biology has greatly expanded the number of procedures available to observe specific DNA sequences and proteins associated with chromatin. The analysis of chromosome structure by in situ reactions generally involves at least two steps, and in this chapter the principles underlying these reactions are discussed.

19.1 PRINCIPLES UNDERLYING IN SITU REACTIONS

Most in situ reactions have a primary and a secondary step (Fig. 19.1), and the interpretation of data from such reactions requires the limitations of each step to be taken into consideration. A key variable in the development of *in situ* reactions to assay the components of chromosomes is the preparation of cytological materials for study. The chemical fixation of tissues can have drastic effects on the interpretation of the data obtained.

19.1.1 Preparation of Tissues for Study

The common procedure for preparing tissues for chromosomal observations is to fix cells in an ethanol/acetic acid mix (ratio 3 : 1), followed by 45% acetic acid. Although this procedure is still commonly used, a classical example of the problems associated with the interpretation of cytological observations on material prepared in this way is the demonstration of Z-DNA in salivary-gland chromosomes in the early 1980s. The Z-DNA conformation is quite different from the right-handed B-DNA type (*see* Chapter 17, Fig. 17.10), so in order to determine whether this type of conformation exists in vivo, antibodies that specifically assay Z-DNA and distinguish it from the usual right-handed B-DNA, were allowed to bind to polytene chromosomes from *Drosophila melanogaster* salivary glands. The experiments demonstrated unambiguous reaction of the antibody to polytene chromosomes and suggested that Z-DNA is widespread within the insect genome. However, subsequent studies using chromosomes isolated under physiological conditions, without acidic fixatives, proved that only after treatment with 45% acetic acid was the Z-DNA antibody binding observed (Fig. 19.2). Furthermore, the binding was evident first in the interband region and then, after longer exposure to 45% acetic acid, more generally distributed. The treatment of the acetic-acid-modified chromosomal DNA with topoisomerase I removed Z-DNA antibody binding and suggested that the Z-DNA conformation was the result of torsional stress induced in the DNA molecule after removal of the histones by the acid.

Although the results described above indicate that DNA is a dynamic molecule and is capable of switching to unusual conformations, the major impact of the studies has been to emphasize the caution that needs to be exercised in interpreting cytological information.

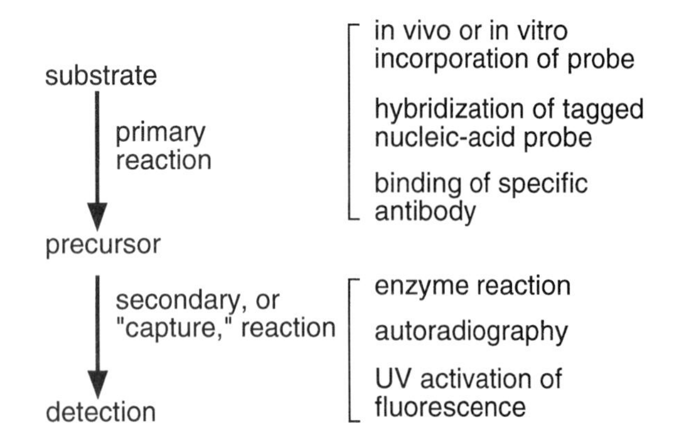

Fig. 19.1. A summary diagram to explain the two basic steps in the *in situ* detection of specific chromosomal DNA sequences or protein components. (Based on Holt and O'Sullivan, 1958.)

19.1.2. Specificity in the Primary Reaction

Antibody probes interact specifically with either the nucleic-acid or the protein components of chromosomes and provide a common source of molecules for the primary in situ reaction. The nature of antibody interactions with chromosomal antigens is complex and involves H-bonding, ionic attractions, and hydrophobic

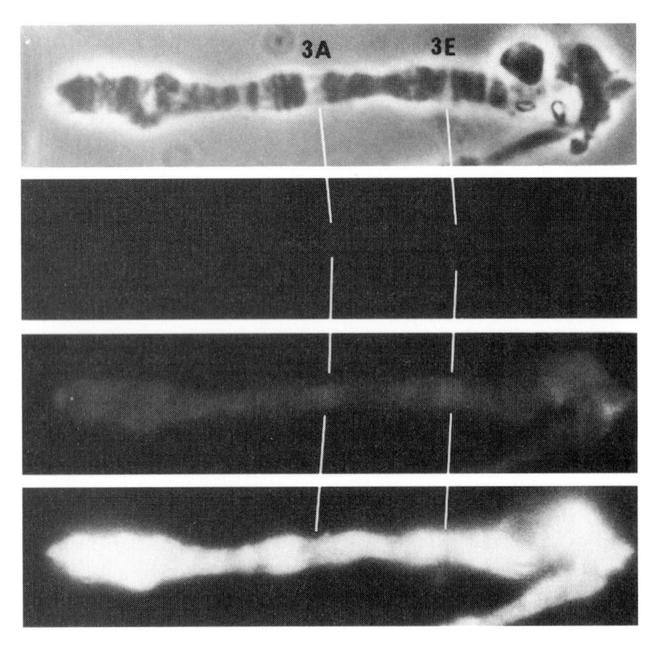

Fig. 19.2. Polytene chromosomes of *Drosophila melanogaster* isolated by microdissection and fixed in formaldehyde rather than traditional acetic/ethanol fixatives. (From Hill and Stollar, 1983.) The top panel shows the phase-contrast image (3A and 3E refer to salivary-gland chromosome bands); the panel below it shows the fluorescent image after staining with an antibody specific for Z-DNA. The absence of a staining reaction contrasts with the increased staining observed after treatment of the chromosome preparation with 45% acetic acid (two lower panels). Clearly, the formation of Z-DNA and its appearance using a Z-DNA antibody-binding reaction results from the way in which the chromosomes were prepared. Even so, it is believed that this reflects the potential of certain sequences to adopt the Z-DNA configuration, in order to release torsional stress created in the molecule during the course of processes such as transcription (Hill, 1991).

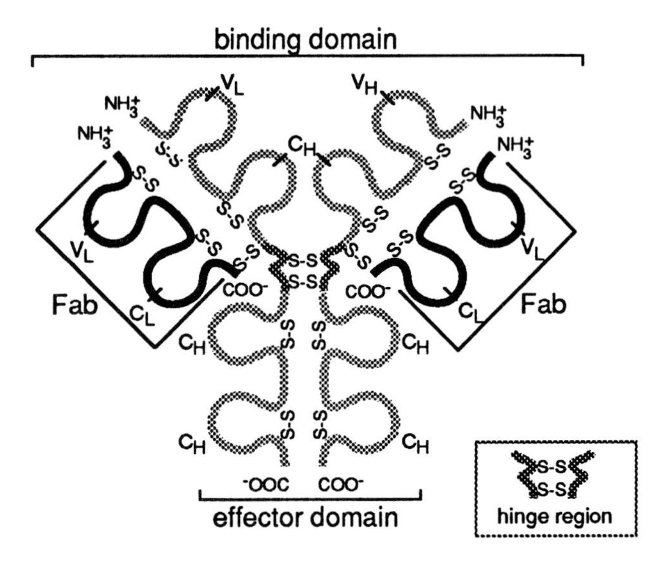

Fig. 19.3. Diagrammatic representation of an immunoglobulin molecule, which comprises heavy (H) and light (L) chains, held together by sulfur bonds, with both chains containing constant (C_H and C_L) and variable (V_H and V_L) regions. The H chain distinguishes the five main classes of immunoglobulin, namely IgM, IgD, IgG, IgE, IgA. The L chains occur in two types called k and l. The hinge region in the H chain defines the two basic domains, referred to as the binding and effector domains.

interactions. Many cytological-detection procedures are based on the unique attributes of immunoglobulin molecules that result from their normal biological functions in animals, where the binding domains are responsible for the specific interaction with an antigen (Fig. 19.3). The effector domains provide the point of interaction with the cells responsible for destroying the antigen and also provide the signals for distributing the antibodies to specific parts of the body. When the immunoglobulin molecules are used in cytogenetics, the binding domain is directed against an antigen such as digoxigenin, which can be incorporated into DNA via a nucleoside triphosphate, or against a specific protein that has been purified from cells. Naturally occurring antibodies against nuclear components in the serum of patients suffering from diseases such as rheumatoid arthritis and lupus erythematosus also provide valuable probes for chromosomal and nuclear structures.

The direct binding of DNA and RNA probes to complementary DNA sequences in chromosome preparations, referred to as in situ hybridization, has been widely used. The specificity of binding derives directly from the base-pairing that occurs when complementary strands interact via H-bonding (*see* Chapter 16, Fig. 16.11).

19.1.3. The Secondary or "Capture" Reaction to Detect the Results of the Primary Reaction

The purpose of the secondary reaction is to enhance the contrast, for microscopical observation, of the molecules used in the primary in situ reaction. Generally, this involves one or more of the following:

1. Detection of the radioactivity, or fluorescence, that may have been introduced as an integral part of the molecule in the primary step.

2. An enzymatic reaction, which targets the molecule involved in the primary step and results in the deposition of a dye.
3. An antibody-binding reaction, which detects some feature of the molecule used in the primary reaction.

The radioactive labeling of the molecule used in a primary in situ reaction usually utilizes ^3H. The secondary reaction used to detect this radioactivity is the deposition of silver grains in a photographic emulsion overlaying the chromosome preparation. The short path length of a weak source of radioactivity such as ^3H, means that the silver grains are deposited close to the origin of the emission, and this allows the location of the primary molecule to be determined relative to microscopical features of chromosomes. Other isotopes such as ^{125}I and ^{35}S, have been used in in situ reactions, but they suffer from the fact that the β emissions are more energetic than those from ^3H, so their point of origin is more difficult to ascertain.

The labeling of the molecule used in the primary reaction with an entity that is capable of fluorescing is more versatile than radioactive labeling, because a range of colors can be distinguished in a single preparation. The secondary reaction in detecting fluorescently labeled probes involves absorption of relatively short-wavelength light by the probe and reemission of this energy as light of a characteristic wavelength in the visible range. Absorption of light energy by the fluorescent probe can lead to its breakdown, referred to as photobleaching, and reduces the efficiency with which the probe can be observed. In addition, molecules located adjacent to the probe may absorb the energy reemitted by it, effectively quenching the fluorescence and preventing the observation of the probe location in the chromosome or nucleus preparation.

A typical set of secondary reactions, nonradioactive techniques, is illustrated for the detection of nucleic acid or antibody probes labeled with biotin (Fig. 19.4). The following series of reactions result in the deposition of dye at the location of the original probe:

Binding of streptavidin. Streptavidin is a glycoprotein composed of four subunits, each with a binding site for biotin, and forms a complex with biotin that is essentially irreversible.

Binding of biotinylated alkaline phosphatase complex. The spare binding sites on the streptavidin protein are utilized.

Detection of alkaline phosphatase. The cytochemical detection of enzyme activity has been particularly well studied. Ideally, the substrate for the enzyme should be converted to a ''stain precursor'' form in a single step, then converted rapidly to an insoluble form by a ''capture'' reaction to minimize diffusion from the original site of enzyme activity. For example, alkaline phosphatase will hydrolyze a colorless indoxyl-based substrate to liberate indoxyl molecules, which can subsequently be oxidized, in a capture reaction, to an insoluble indigoid dye (Fig. 19.5). The colored indigoid dye is then observed microscopically.

The binding of labeled antibodies to specific sites on the molecule used in the primary in situ reaction is also a common basis for secondary reactions, leading to its detection in the microscopic preparation. An example using the digoxigenin molecule is illustrated in Fig. 19.6. The antibodies used in these procedures often have the effector domain removed to produce immunoglobulins called Fab fragments (*see* Fig. 19.3). The immunoglobulins used in the analysis of chromosomes have enzymes such as alkaline

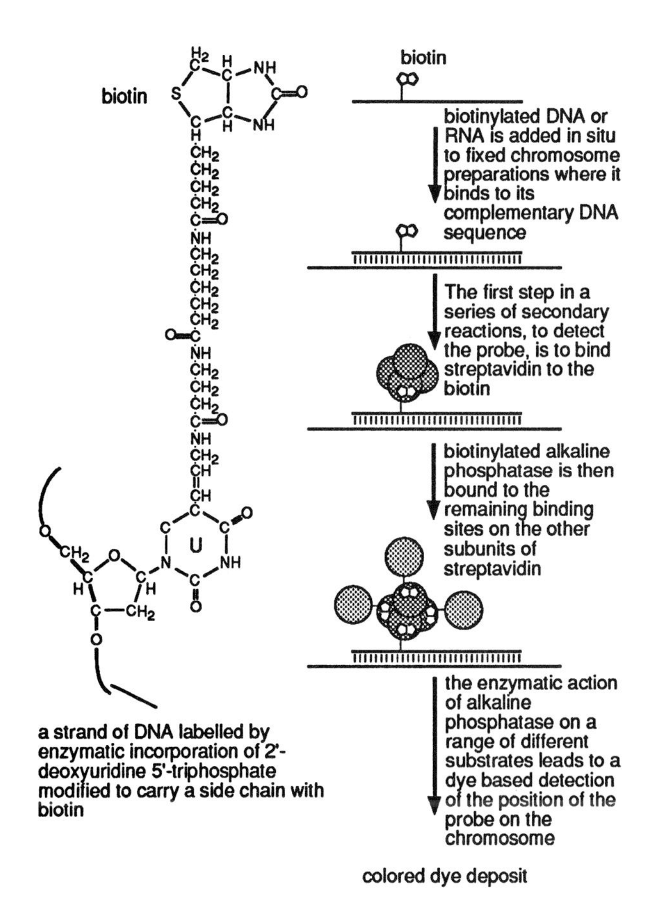

Fig. 19.4. Steps in the detection of a biotin-labeled probe localized *in situ* to chromosome preparations. *Left*: the chemical structure of the probe. *Right*: the secondary reactions to detect the biotin in the probe make use of the very specific interaction between biotin and streptavidin, as well as the ability of alkaline phosphatase to initiate the formation of a colored precipitate (*see* Fig. 19.5). The use of nonradioactively labeled probes provided a major step in high-resolution, in situ-hybridization reactions to locate DNA sequences in specific regions of chromosomes (Manuelidis et al., 1982). The temperature at which the hybridization is carried out is usually 15–25°C below the T_m or melting point of the DNA duplex. A guide to the T_m of a DNA duplex can be obtained from the following formula $T_m = 81.5 + 16.6(\log S) + 0.41(GC) - 0.72(F) - 1(M) - 500/L$, where S is the molar-salt concentration, GC is the % GC base content of the sequence, F is the % formamide, M is the % mismatch, and L is the length of the probe in nucleotides. The use of formamide in hybridizations lowers the temperature at which reactions need to be carried out.

phosphatase, or moieties such as colloidal gold, or a fluorescent dye, bound to them in order to detect their position in the cell by in situ reactions. Fluorescent molecules attached to immunoglobulins for the purpose of detection include rhodamine isothiocyanate, Texas red, fluorescein isothiocyanate (FITC), and phycoerythrin. The binding of these molecules to immunoglobulins can affect their binding to the antigen. Chemical reactions that bind proteins to immunoglobulins use either gluteraldehyde or a more complex bifunctional reagent such as 3-maleimidobensoyl-*N*-hydroxysuccinimide ester (Fig. 19.7). Where loss of antigen binding occurs, alternate empirical procedures have to be devised for labeling the immunoglobulin.

The binding of colloidal gold to immunoglobulins allows them to be observed in the electron microscope. The technique has pro-

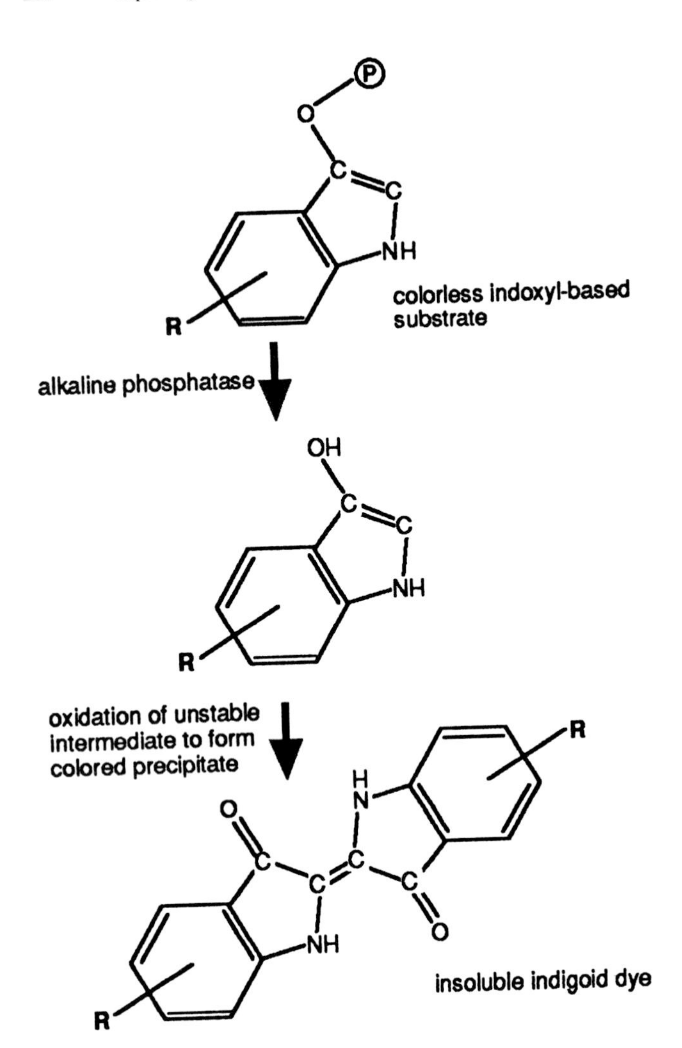

Fig. 19.5. The chemistry of formation of an insoluble dye complex. (From Holt and Sadler, 1958.) In this instance, the indoxyl-based substrate is bound to a phosphate moiety, and alkaline phosphatase is the enzyme acting on the substrate, resulting in the formation of an insoluble dye. The original substrate can be attached to a variety of chemical entities and thus can be used to assay a range of enzyme activities (*see* Fig. 19.18).

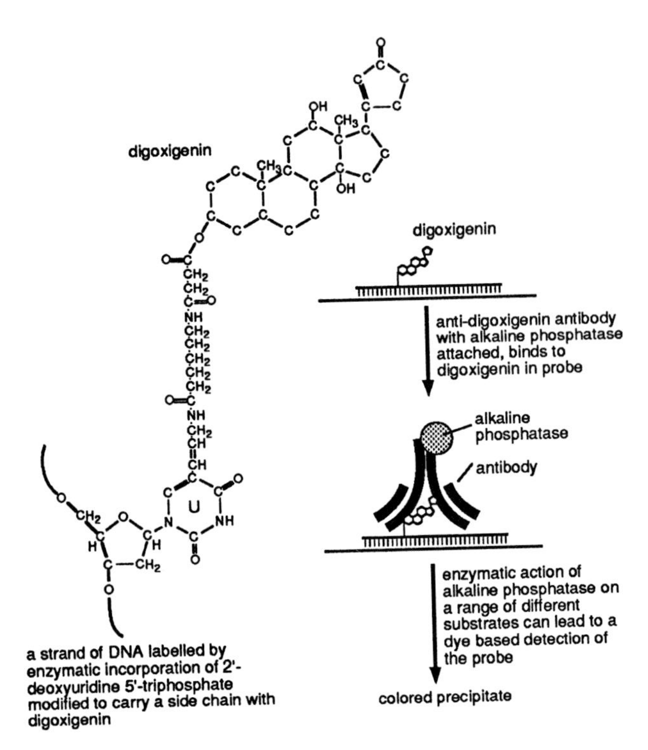

Fig. 19.6. Steps in the detection of a digoxigenin-labeled probe localized *in situ* to chromosome preparations. The chemical structure of the probe is shown on the left. The secondary reaction (right) uses an antibody labeled with either alkaline phosphatase, as shown in the diagram, or a fluorescent dye to locate the position of the digoxigenin-labeled probe (Boeringher Mannheim 1994 Biochemicals Catalog).

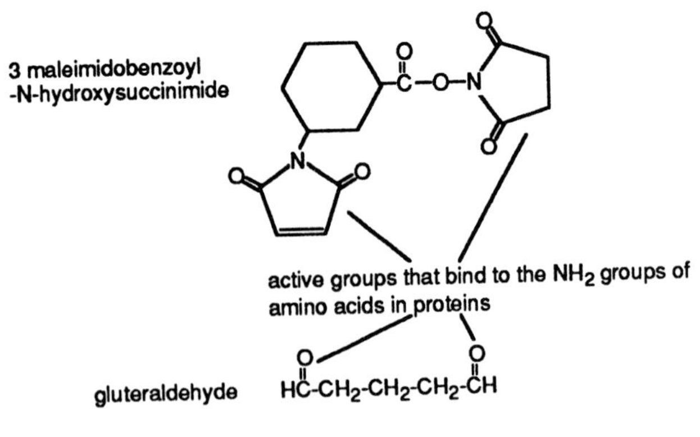

Fig. 19.7. A chemical agent (3-maleimidobenzoyl-*N*-hydroxysuccinimide) used to cross-link proteins and its method of interaction with gluteraldehyde. It has been found that the cross-linked positions within proteins are unable to be controlled, so that active sites may be damaged by the chemical reactions that link, for example, alkaline phosphatase to an antibody. In such cases, alternative cross-linking agents and reaction conditions are needed.

vided high-resolution mapping of the positions of, for example, nuclear proteins such as RNA polymerase, within the nucleus, as well as the path of DNA in the synaptonemal complex found in the pachytene stage of meiosis (Fig. 19.8). Colloidal-gold particles can be fractionated into uniform-size classes, which are distinguishable under the electron microscope and thus allow double-labeling experiments.

19.2 SOURCES OF RADIOACTIVE AND NONRADIOACTIVE DNA AND RNA PROBES

The ability to hybridize well-defined, labeled DNA and RNA molecules to chromosome preparations, following the same principles of DNA renaturation described in Chapter 16, Sections 16.5 and

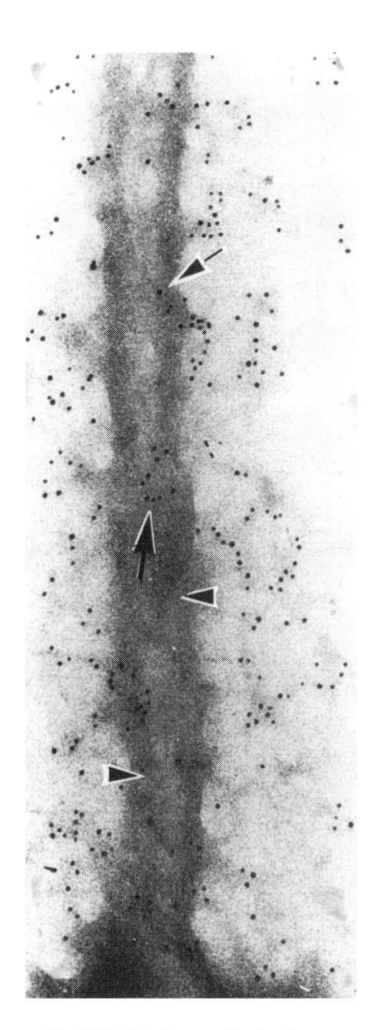

Fig. 19.8. A specific DNA-binding antibody (from a mouse) bound to a preparation of a synaptonemal complex from pachytene cells. The position of the mouse anti-DNA antibody was located using a secondary reaction with a goat antimouse antibody, labeled with colloidal gold to facilitate observation under the electron microscope (Vázquez Nin et al., 1993). The arrows with tails indicate colloidal gold particles; arrows without tails indicate unlabeled regions of the synaptonemal complex.

16.7, has enabled specific nucleic-acid sequences to be localized within the nucleus and cytoplasm. Radioactive labeling using ^{3}H-2′ deoxythymidine 5′-triphosphate (for DNA) or ^{3}H-uridine 5′-triphosphate (for RNA), with the *E.coli* enzymes DNA and RNA polymerase, respectively, were used in early in situ hybridization experiments to localize repetitive DNA sequences within chromosomes (Fig. 19.9). Direct radiolabeling of nucleic acid with ^{125}I was also important in localizing DNA sequences complementary to RNA molecules such as 5S, 5.8S,18S, and 26S RNA, the structural components of ribosomes.

The different classes of DNA probes for in situ hybridization are based on the range of sequence classes found within the genomes of eukaryotes (*see* Chapter 16). In addition to specific gene sequences, the following range of repetitive sequences have been utilized as probes to characterize chromosomes:

1. Tandemly repeated sequences that hybridize to chromosomes and give a "banding" pattern to distinguish different chromosomes
2. Telomeric repeats hybridizing to the ends of chromosomes
3. Chromosome-specific, tandemly repeated, sequences that hybridize at or near the centromere
4. Tandemly repeated genes such as ribosomal and 5S DNA sequences, to provide for the identification of classical landmarks such as the nucleolus-organizer region (*see* Chapter 20)
5. Dispersed repetitive sequences that are limited to a single chromosome, or subset of chromosomes, and allow the "painting" of whole chromosomes.

The combination of probes that have different fluorescent colors (discussed below) allows a high-resolution analysis of whole chromosomes for diagnostic or mapping purposes and has been termed "whole-chromosome hybridization."

19.2.1. In Situ Hybridization with Fluorescently Labeled DNA (FISH)

Techniques for the nonradioactive labeling of nucleic acids (*see* Figs. 19.4 and 19.6) have expanded in situ hybridization to allow the use of the confocal microscope and the associated computer-image-enhancement technology (*see* Chapter 3). The advantage of

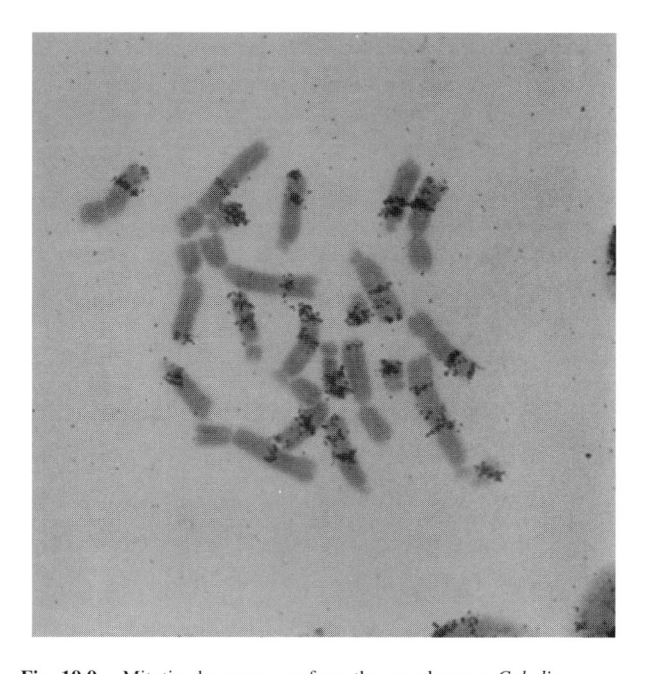

Fig. 19.9. Mitotic chromosomes from the grasshopper, *Caledia captiva,* subjected to *in situ* hybridization with a ^{3}H-labeled, repetitive-sequence probe. The chromosomal locations of the repeated DNA sequences were determined by covering the preparation with a film of photographic emulsion and, after a few days in the dark, photographically developing the slide to detect the deposition of silver grains resulting from the exposure to radioactivity. Consequently, in this detection system, the primary reaction is the annealing of the radioactive RNA probe, enzymatically synthesized from a cloned DNA sequence, to denatured, single-stranded chromosomal DNA. The secondary reaction is the photographic-detection system. Denaturation of the chromosomal DNA in situ can be carried out by heating (the usual technique), acid denaturation, or enzymatically (van Dekken et al., 1988). (Photograph kindly provided by D. Shaw and N. Contreras).

the confocal microscope is that it can be used to optically "section" nuclear preparations to locate complementary DNA sequences in a three-dimensional configuration (Fig. 19.10).

A particularly valuable aspect of alternate methods of labeling nucleic acids is the possibility of double- (or multi-) labeling experiments, where two or more different sequences can be observed simultaneously in the same nucleus or chromosome preparation. The highly sensitive nature of fluorescent-detection techniques, as well as the possibility of amplifying the signal (Fig. 19.11), makes the detection of single-gene sequences possible. To detect a specific gene, the DNA probe often comprises a DNA segment 15–20 kb long, combined with a large excess of genomic DNA to block any hybridization due to repetitive-sequence families that may neighbor the gene of interest.

19.2.2. In Situ Hybridization to DNA Fibers (FIBER-FISH)

An extension of the FISH technique to DNA fibers, rather than mitotic or meiotic chromosomes, has provided an increase in resolution for mapping DNA sequences relative to each other at the 1–500-kb level: to map two sequences relative to each other in mitotic chromosomes, they need to be separated by at least 1 Mb (1000 kb).

Fibers of DNA or chromatin can be prepared by lysis of nuclei on a glass slide and then allowing the lysate to spread over the surface of the slide. The standard FISH technique, using two or more DNA probes labeled with different fluorochromes, can then

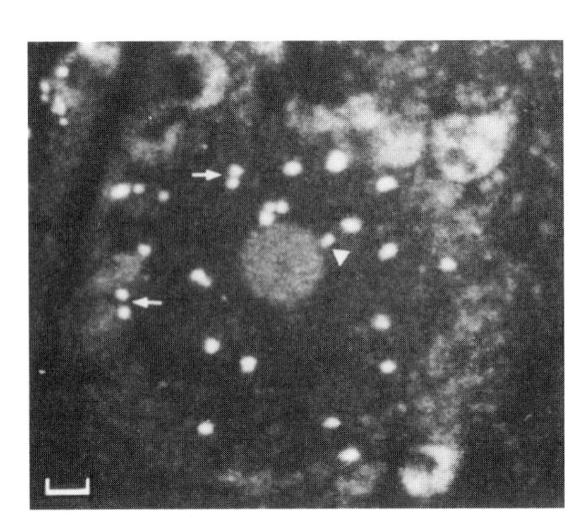

Fig. 19.10. The detection of chromosomal telomeres using in situ hybridization. In this nucleus of a tomato cell, 28 telomeres are labeled and all are arranged around the nuclear periphery except for one adjacent to the nucleolus (arrowhead). Some are arranged in pairs (arrows). Bar = 2 μm. The probes used to detect telomeric sequences comprise either sequences isolated from genomic DNA or synthetic DNA segments manufactured to match consensus sequences (Rawlins et al., 1991; Schwarzacher and Heslop-Harrison, 1992; Mukai et al., 1993). The significant improvement in the use of fluorescently labeled probes is that several probes labeled with different colors can be used in a single reaction (Leitch et al., 1991; Mukai et al., 1993). Commonly used colors include fluorescein (yellow–green), hydroxycoumarin (blue), and resorufin (red). Mixtures of labeled precursors can also be incorporated into a single probe to produce a new color (Smith et al., 1992). The fluorescent molecule is generally incorporated into the probe by its attachment to dUTP (*see* Fig. 19.4) or digoxigenin (*see* Fig. 19.6).

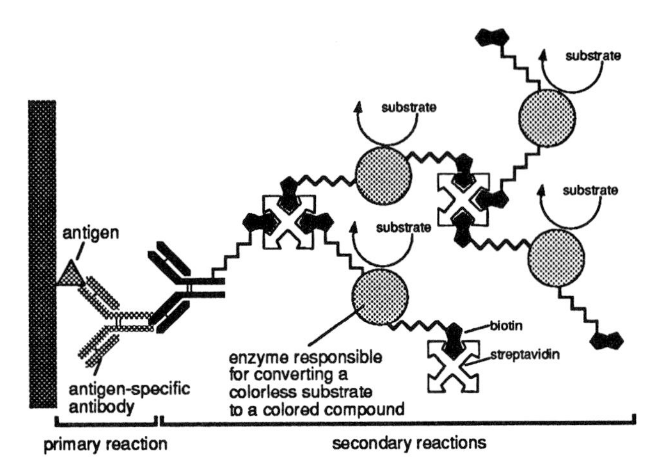

Fig. 19.11. A diagrammatic representation of a primary reaction and its amplification by a secondary reaction, designed to enhance observation of the primary site of attachment. In this case, the primary reaction is the binding of an antigen-specific antibody from, for example, a rabbit, to the site of interest that contains the original antigen. The secondary reaction then involves the binding of biotinylated antirabbit antibody from, for example, a mouse, to the product of the primary reaction. The subsequent binding of streptavidin to biotin then provides the basis for the further binding of a biotinylated enzyme such as alkaline phosphatase, to carry out the color reaction for visual detection of the complex. (Modified from Skerritt and Appels, 1995.) There are numerous variations on the scheme shown and, although it is possible to use streptavidin in further rounds of amplification using an intermediate molecule with two biotin groups (Lichter et al., 1990; Dominguez-Steglich et al., 1992), this is not usually successful because nonspecific adsorption effects come into play, and the background fluorescence becomes too great. If DIG-labeled probes are used (*see* Fig. 19.6), a fluorescein-labeled sheep anti-DIG antibody is used to carry out the first phase in the secondary-detection step. Amplification of the signal is achieved by the addition of fluorescein-labeled rabbit antibody directed against the heavy chain of sheep antibodies (Kallioniemi et al., 1992). An amplification procedure based on peroxidase-mediated deposition of biotin- or fluorochrome-tyramides has also been described (Raap et al., 1995). In this case, the first layer of the secondary reaction is an immunoglobulin or streptavidin molecule conjugated to peroxidase enzyme, which subsequently reacts with biotin- or fluorochrome-tyramides to deposit biotin or fluorochrome molecules near the site of attachment of the enzyme. In the case of biotin, the deposited molecules are then detected by incubation with fluorochrome-labeled streptavidin; in the case of fluorochrome-tyramides, by immediate-fluorescence microscopy.

be applied to such a preparation to provide information on the relative positioning of the DNA sequences assayed by the respective probes. In Fig. 9.12, fibers from *Arabidopsis* nuclei are illustrated after probing with cosmid clones derived from chromosome 4 to reveal the chromatin strands that contain the sequences in these clones; the cosmid is a cloning vector-based bacteriophage lambda and capable of accepting DNA segments approximately 50 kb long. When multiple colors are displayed, the fibers provide the basis for the unambiguous ordering of different sequences along their length.

19.3 IN SITU REACTIONS TO MODIFY ACCESSIBLE REGIONS OF DNA

The enzymes that modify DNA such as DNA methylases and DNA terminal transferase can be used in in situ reactions to add distinctive groups to regions of DNA that are exposed. An example is

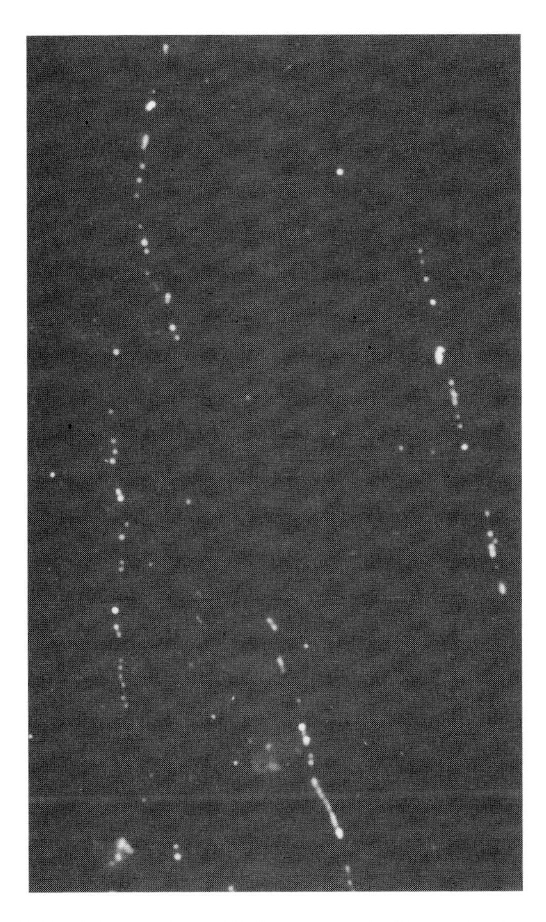

Fig. 19.12. The detection of DNA or chromatin fibers using the fiber-FISH technique. The nuclei from *Arabidopsis* tissue were lysed, using a buffer containing EDTA and SDS, and spread on a glass slide, then hybridized with DNA probes to assay sequences that derive from chromosome 4 (Franz et al., 1996). The technique is used to position DNA sequences relative to each other by hybridizing two or more probes labeled with different fluorochromes, to the same preparation; the source of the DNA to which the probes are hybridized can be either lysed nuclei or DNA preparations from bacterial/yeast cloning vectors carrying large segments of eukaryotic genomic DNA (Heiskanen et al., 1996; Franz et al., 1996). (Photo courtesy of J.H. de Jong.)

shown in Fig. 19.13 where bromo-deoxyuridine, BrdU, is added to the exposed 3′-OH ends of DNA molecules in a tissue section prepared for electron microscopy. The BrdU is then detected using an anti-BrdU antibody labeled with colloidal gold.

19.4 *IN SITU* BIOCHEMICAL REACTIONS FOR ASSAYING SPECIFIC ENZYME ACTIVITIES

Specific enzymes such as isozymes and RNA polymerase can be assayed in cells after they have been fixed for microscopic examination. Numerous enzyme activities can be assayed using the same principles described for alkaline phosphatase. Changes in the structure of chromatin often provide the basis for renewed gene expression, and the ability to assay RNA polymerase directly on cells that have been fixed can provide a useful localization of the potential for transcriptional activity. Early studies used ^{3}H-labeled nucleoside triphosphates to detect RNA polymerase activity (Fig. 19.14). The use of digoxigenin or biotin-labeled nucleoside triphosphates allow fluorescent antibodies to detect sites of RNA polymerase activity and may provide greater levels of sensitivity.

One of the most exciting applications of the knowledge about DNA replication is the development of the polymerase chain reaction (PCR), which can amplify regions of the genome without prior cloning. The PCR reaction, first described in 1985, has revolutionized the way in which molecular biology is carried out. The procedure enables small amounts of specific DNA, which may be mixed with large amounts of contaminating DNA, to be amplified approximately a million-fold.

Polymerase chain reaction is an in vitro procedure for the enzymatic synthesis of DNA, using two oligonucleotide primers that hybridize to opposite strands of the parental DNA template and flank the region of interest (Fig. 19.15). A repetitive series of cycles involving template denaturation, primer annealing, and the extension of the annealed primers by DNA polymerase results in the exponential accumulation of a specific fragment whose termini are defined by the 5′-ends of the primers. The amplification is dramatic

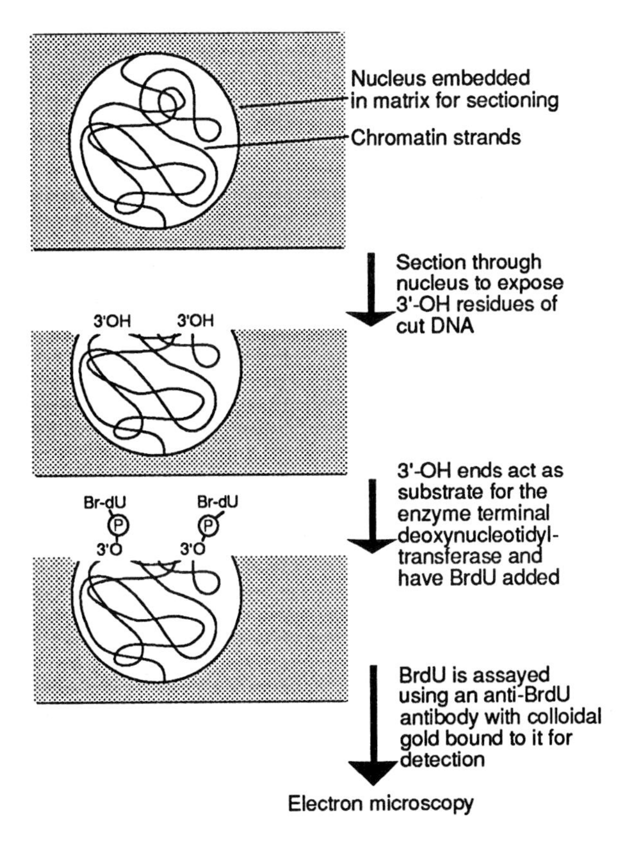

Fig. 19.13. The detection of DNA strands in the nucleus using thin sections and electron microscopy. Although the primary reaction (i.e., the labeling of the 3′-hydroxy ends of the cut DNA with BrdU-labeled terminal deoxynucleotidyl transferase) is highly specialized (Thiry, 1992), it is used to illustrate the wide variety of chemical reactions that can be used to target specific entities in cytological preparations.

because the extension products of one cycle serve as templates for the following reactions, and thus the number of target copies, for the reaction, double at every cycle. After 20 cycles of PCR, a million-fold (2^{20}) amplification of a particular DNA sequence is achieved. The key factor in the widespread use of PCR was the introduction, in 1986, of the thermostable DNA polymerase (Taq polymerase) from *Thermus aquaticus*. The commercial supply of the enzyme meant that the reaction components (template, primers, Taq polymerase, nucleoside triphosphates, and buffer) could all be mixed and subjected to temperature cycling. The kinetics of this reaction can be monitored directly if small amounts of the DNA-binding dye, ethidium bromide, are included in the reaction mix (Fig. 19.16). The reaction is not exponential over the entire range of cycles, as expected from the above theoretical calculation, presumably because some Taq polymerase activity is lost in each heating cycle of the amplification reaction. However, the amplification reaction is still occurring after 50 cycles and, as expected, is dependent on the number of starting-template DNA molecules.

The direct impact of PCR on cytogenetic studies of chromosomes is that the reaction can be carried out in situ on metaphase chromosomes immobilized for microscopic observation. This allows an extremely wide range of sequences to be physically mapped on chromosomes. In this procedure, short oligonucleotides are hybridized to their complementary sites in chromosomes and nuclei. This method provides suitable substrates for the enzyme Taq polymerase to synthesize a complementary sequence in situ, which can be labeled and detected as discussed above (Fig. 19.17). The advantage of this technique is that a long segment (1000 nucleotides or longer) of labeled DNA is synthesized at the place where the oligonucleotides are localized; therefore, the procedure has the potential for being more sensitive than the use of externally added probes. The single-strand length of externally added probes is usually not greater than 200 nucleotides, and this limits the amount of label that can be deposited at a complementary site in the chromatin.

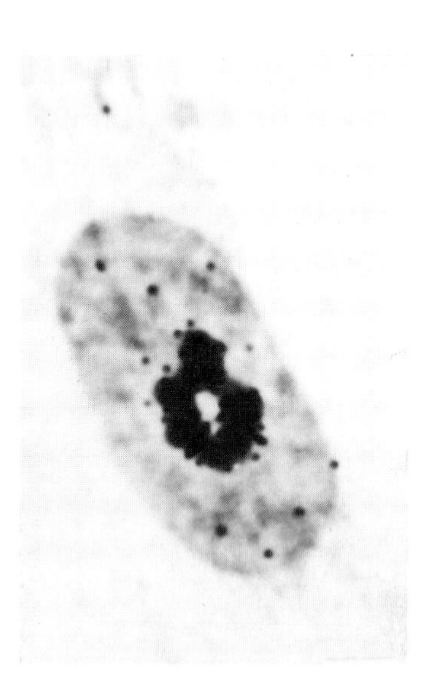

Fig. 19.14. The detection of endogenous RNA polymerase activity in a cultured mammalian (HeLa) cell, which was fixed for microscopic observation before incubation with ^{3}H-nucleoside triphosphates. RNA polymerase enzymes then incorporated the labeled triphosphates into acid-insoluble molecules (i.e., RNA). The incorporated radioactivity was detected by covering the preparation with a thin film of emulsion and detecting the deposition of silver grains resulting from the exposure to radioactivity. (From Moore and Ringertz, 1973.)

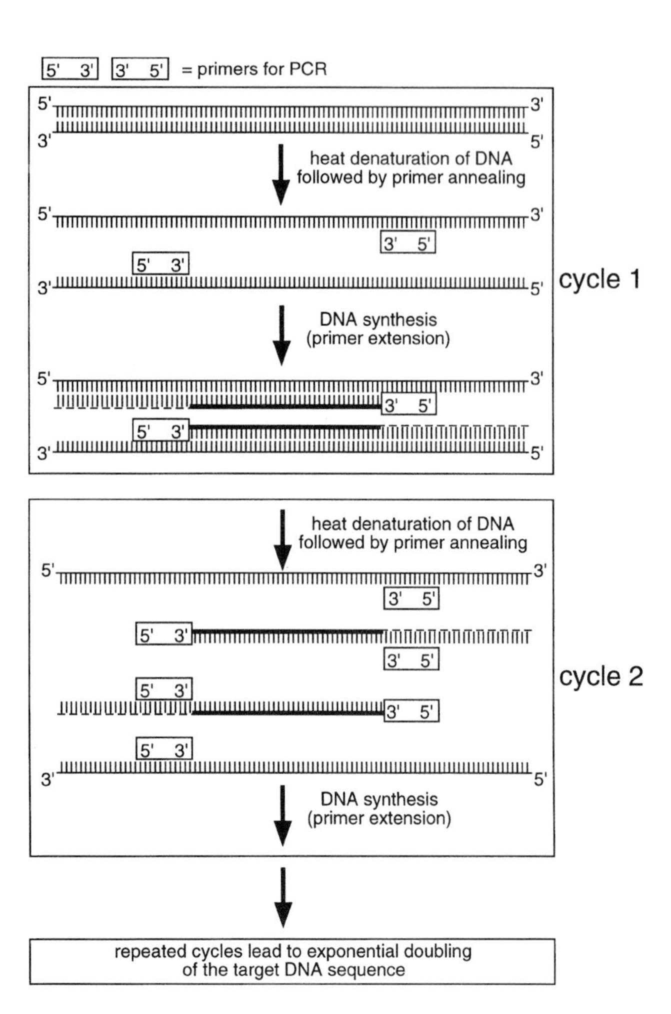

Fig. 19.15. A diagrammatic representation of the polymerase chain reaction (PCR). The introduction of a thermostable DNA polymerase to carry out the PCR (Saiki et al., 1988) was a major advancement in this technology. The actual temperature used to anneal the primers to the DNA template is also crucial in determining the specificity and reproducibility of the final product. This is because in the early stages of the PCR, the reannealing conditions that are optimal for the formation of DNA–DNA hybrids with minimum base-pair mismatch (*see* Chapter 16, Fig. 16.11) vary from sequence to sequence (Don et al., 1991).

19.5 TRADITIONAL STAINS

Traditional stains for chromatin include acetocarmine, aceto-orcein, and eosin. These molecules provide contrast mainly to chromosomes and nuclei, because the planar rings characteristic of many DNA-binding antibiotics are present (Fig. 19.18). These planar rings form stable complexes with the DNA that is present in chromatin. Another traditional staining procedure for chromosomes, Feulgen, uses the dye basic fuchsin and derives its specificity for DNA as a result of a chemical reaction with the deoxyribose sugar of the sugar–phosphate backbone after acid hydrolysis of the DNA. Current developments in fluorescence microscopy have led to the extensive use of DAPI (*see* Chapter 17, Fig. 17.15) to observe

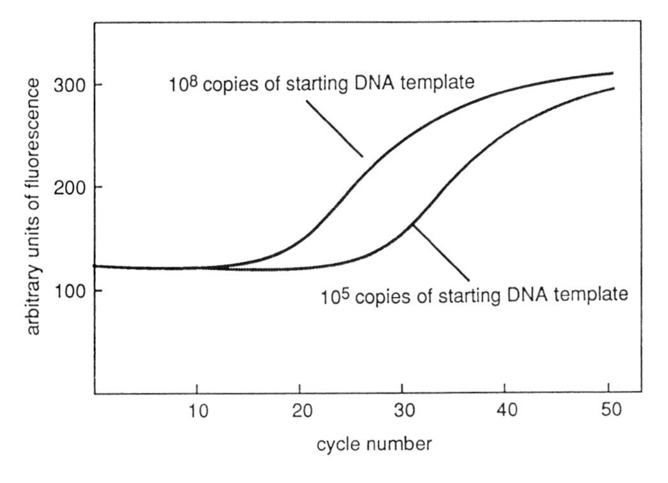

Fig. 19.16. A quantitative comparison of the number of PCR copies produced from differing numbers of initial templates, and the number of cycles needed to produce these final amounts. Continuous monitoring of the polymerase chain reaction by (a) measuring the increase in fluorescence resulting from the binding of ethidium bromide to the newly synthesized DNA, as shown in this figure, or (b) the use of radioactive precursors (Ferré et al., 1994), provides a dramatic demonstration of the principle presented in Fig. 19.14.

chromatin, because the stable complexes that DAPI forms with DNA display a characteristic fluorescence after illumination with light of an appropriate wavelength. Table 19.1 lists examples of molecules that can be applied as stains or as contrast techniques, including some that are used in electron microscopy.

19.6 THE CONCEPT OF REPORTER MOLECULES AND GENES

Although many studies on chromosomes use fixed material, those on living material usually provide critical confirmation of observations from fixed material. Molecules labeled in some specific way can be incorporated into living cells and, after a certain time, they "report" on their positions within cells and chromatin when the cells are fixed for microscopic examination. For example, the incorporation of ^{3}H-thymidine into DNA, in vivo, was important in the analysis of eukaryotic DNA replication and chromosome structure (*see* Chapter 18, Section 18.1.2, and Fig. 18.2). The microinjection of ^{3}H-labeled histones into *Xenopus* oocytes demonstrated the

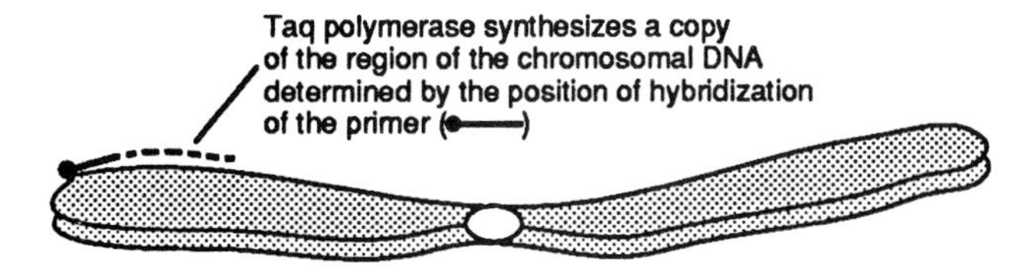

Fig. 19.17. Application of the PCR principle to the *in situ* detection of chromosomal DNA sequences. The technology needed to use PCR for the *in situ* detection of the chromosomal location of genes is still in its infancy. Even so, it is potentially an important technique for the precise detection and location of nonrepetitive DNA sequences (Koch et al., 1991; Hofler, 1993; Long et al., 1993; Gosden and Lawson, 1994; Xie and Troyer, 1996).

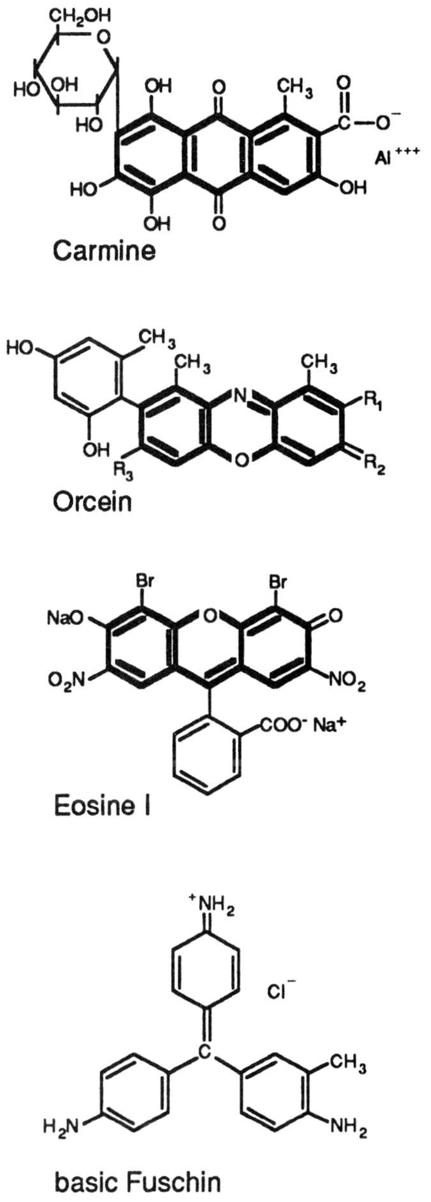

Carmine

Orcein

Eosine I

basic Fuschin

Fig. 19.18. Structures of a range of molecules commonly used for staining chromosomes. With the exception of basic fuchsin, many of the dyes traditionally used to stain chromosomes contain the coplanar-ring structure that is known to intercalate into the DNA double helix. These coplanar ring structures characterize many antibiotics and ethidium bromide, used in the analysis of isolated DNA (*see* Chapter 17, Fig 17.15).

Table 19.1. Commonly Used Chromosome Stains or Contrasting Techniques

Stain or Contrasting Technique	Application
Carmine, orcein, Leishman stain, Giemsa stain, Feulgen stain	Standard detection of DNA in chromosomes and nuclei
Quinacrine, Chromomycin, DAPI, Hoechst 33258	Fluorescent staining of DNA in chromosomes and nuclei
Sulfaflavine, Naphthol yellow	Detection of histones and other nuclear proteins
Cytochrome C (protein) plus shadowing with carbon	Observation of nucleic acids by electron microscopy

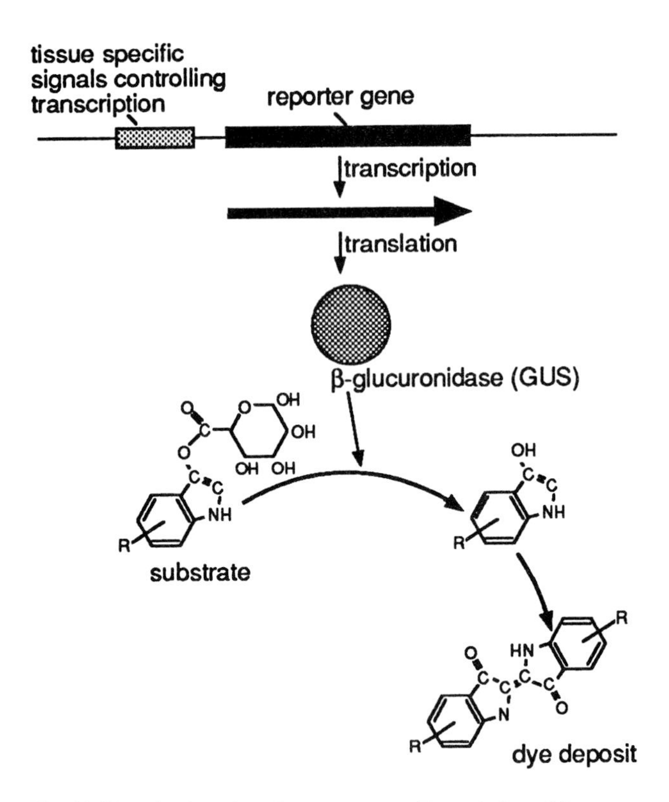

Fig. 19.19. The detection of reporter genes. Transcription of the reporter gene gives an enzyme, GUS, capable of hydrolyzing an indigo-based substrate (the R- indicates a complex part of the indigo molecule), resulting in a localized deposition of dye at the site of activity of the gene product.

rapid migration of these histones into the nucleus, thus suggesting that they could not exist in a free state in the cytoplasm. Fluorescent dyes have been linked to molecules introduced into cells, so that the positions of the molecules in the cells, after fixation, could be monitored by confocal-fluorescence microscopy.

Transformations of plants and animals have provided the means for introducing genes of a specific design (reporter genes) into living cells (Fig. 19.19). Reporter genes are generally used for monitoring the tissue-specific expression of genes, as modulated by control sequences 5′-upstream or 3′-downstream from the gene, but it is clear that they can also report on their environment in the chromosome. If, for example, a reporter gene is located near a nucleoskeleton attachment point, it may not have the same level of activity as when it is located further away.

BIBLIOGRAPHY

Section 19.1

BOEHRINGER MANNHEIM 1994 Biochemicals Catalog.

HARLOW, E., LANE, D. 1988. Antibodies—A Laboratory Manual. Cold Spring Harbor Laboratory Press, Cold Spring Harbor, NY.

HILL, R.J. 1991. Z-DNA: a prodrome for the 1990's. J. Cell Sci. 99: 675–680.

HILL, R.J., STOLLAR, B.D. 1983. Dependence of Z-DNA antibody binding to polytene chromosomes on acid fixation and DNA torsional strain. Nature 305: 338–340.

HOLT, S.J., O'SULLIVAN, D.G. 1958. Studies in cytochemistry I. Principles of cytochemical staining methods. Proc. Roy. Soc. London Ser. B 148: 465–480.

HOLT, S.J., SADLER, P.W. 1958. Studies in cytochemistry II. Synthesis of indigogenic substrates for esterases. Proc. Roy. Soc. London Ser. B 148: 481–494.

MANUELIDIS, L., LANGER-SAFER, P.R., WARD, D.C. 1982. High-resolution mapping of satellite DNA using biotin-labeled DNA probes. J. Cell Biol. 95: 619–625.

RINGERTZ, N.R., CARLSSON, S-A, EGE, T., BOLUND, L. 1971. Detection of human and chick nuclear antigens in nuclei of chick erythrocytes during reactivation in heterokaryons with HeLa cells. Proc. Natl. Acad. Sci. USA 68: 3228–3232.

VÁZQUEZ NIN, G.H., FLORES, E., ECHEVERRÍA, O.M., MERKERT, H., WETTSTEIN, R., BENAVENTE, R. 1993. Immunocytochemical localization of DNA in synaptonemal complexes of rat and mouse spermatocytes and chick oocytes. Chromosoma 102: 457–463.

Section 19.2

APPELS, R. 1989. Three dimensional arrangements of chromatin and chromosomes: old concepts and new techniques. J. Cell Sci. 92: 325–328.

BAUWENS, S. 1994. Procedure for whole mount fluorescence *in situ* hybridization of interphase nuclei in *Arabidopsis thaliana*. Plant J. 6: 123–131.

BOUFFLER, S.D. 1994. Whole-chromosome hybridization. Int. Rev. Cytol. 153: 171–232.

DOMINGUEZ-STEGLICH, M., CARRIER, A., AUFFRAY, C., SCHMID, M. 1992. Assignment of the chicken tyrosine hydroxylase gene to chromosome 6 by FISH. Cytogenet. Cell Genet. 60: 138–139.

ENGLER, J.A., MONTAGU, M.V., ENGLER, G. 1994. Hybridization of whole-mount messenger RNA in plants. Plant Mol. Biol. Rep. 12: 321–331.

FRANZ, P.F., ALONSO-BLANCO, C., LIHARSKA, T.B., PEETERS, A.J.M., ZABEL, P. P., DE JONG, J.H. 1996. High resolution physical mapping in higher plants by fluorescence *in situ* hybridization to extended DNA fibres. Plant J. 9: 421–430.

HEISKANEN, M., PELTONEN, L., PALOTIE, A.. 1996. Visual mapping by high resolution FISH. Trends Genet. 12: 379–382.

HOPMAN, A.H.N., WIEGANT, J., RAAP, A.K., LANDEGENT, J.E., VAN DER PLOEG, M., VAN DUIJN, P. 1986. Bi-color detection of two target DNAs by non-radioactive *in situ* hybridization. Histochemistry 85: 1–4.

KALLIONIEMI, A., KALLIONIEMI, O.-P., WALDMAN, F.M., CHEN, L.-C., YU, L.-C., FUNG, Y.K.T., SMITH, H.S., PINKEL, D., GRAY, J.W. 1992. Detection of retinoblastoma gene copy number in metaphase chromosomes and interphase nuclei by fluorescence *in situ* hybridization. Cytogenet. Cell Genet. 60: 190–193.

LEITCH, I.J., LEITCH, A.R., HESLOP-HARRISON, J.S. 1991. Physical mapping of plant DNA sequences by simultaneous *in situ* hybridization of two differently labeled fluorescent probes. Genome 34: 329–333.

LICHTER, P., TANG, C-J.C., CALL, K., HERMANSON, G., EVANS,

G.A., HOUSMANN, D., WARD, D.C. 1990. High-resolution mapping of human chromosome 11 by *in situ* hybridization with cosmid clones. Science 247: 64–69.

MUKAI, Y., NAKAHARA, Y., YAMAMOTO, M. 1993. Simultaneous discrimination of the three genomes in hexaploid wheat by multicolor fluorescence *in situ* hybridization using total genomic and highly repeated DNA probes. Genome 36: 489–494.

RAAP, A.K., VAN DE CORPUT, M.P.C., VERVENNE, R.A.W., VAN GIJLSWIJK, R.P.M., TANKE, H.J., WIEGANT, J. 1995. Ultra-sensitive FISH using peroxidase-mediated deposition of biotin- or fluorochrome-tyramides. Human Molec. Genet. 4: 529–534.

RAWLINS, D. J., HIGHETT, M.I., SHAW. P.J. 1991. Localization of telomeres in plant interphase nuclei by *in situ* hybridization and 3D confocal microscopy. Chromosoma 100: 424–431.

SCHWARZACHER, T., HESLOP-HARRISON, J.S. 1992. *In situ* hybridization to plant telomeres using synthetic oligomers. Genome 34: 317–323.

SKERRITT, J., APPELS, R. 1995. New Diagnostics in Crop Sciences. CAB International, Biotechnology in Agriculture No. 13, Cambridge University Press, Cambridge.

SMITH, E., ROBSON, L., SCHWANITZ, G. 1992. Fluorescence *in situ* hybridization. Today's Life Science, October, pp. 56–63.

VAN DEKKEN, H., PINKEL, D., MULLIKIN, J., GRAY, J.W. 1988. Enzymatic production of single-stranded DNA as a target for fluorescent *in situ* hybridization. Chromosoma 97: 1–5.

Section 19.3

THIRY, M. 1992. Ultrastructural detection of DNA within the nucleolus by sensitive molecular immunocytochemistry. Exp. Cell Res. 200: 135–144.

Section 19.4

DON, R.H., COX, P.T., WAINWRIGHT, B.J., BAKER, K., MATTICK, J.S. 1991. "Touchdown" PCR to circumvent spurious priming during gene amplification. Nucleic Acids Res. 19: 4008–4015.

FERRÉ, F., MARCHESE, A., PEZZOLI, P., GRIFFIN, S., BUXTON, E., BOYER, V. 1994. Quantitative PCR: an overview. In: The Polymerase Chain Reaction (Mullis, K., Ferré, F., Gibbs, R.A., eds.). Birkhäuser, Boston. pp. 69–81.

GOSDEN, J., LAWSON, D. 1994. Rapid chromosome identification by oligonucleotide-primed in situ DNA synthesis (PRINS). Human Mol. Genet. 3: 931–936.

HOFLER, H. 1993. *In situ* polymerase chain reaction: toy or tool. Histochemistry 99: 103–104.

KOCH, J., HINDKJAER, J., MOGENSEN, J., KOLVRAA, S., BOLUND, L. 1991. An improved method for chromosome-specific labeling of α-satellite *in situ* by using denatured double-stranded DNA probes as primers in a primed *in situ* labeling (PRINS) procedure. Genet. Anal.: Tech. Appl. 8: 171–178.

LONG, A.A., KOMMINOTH, P., LEE, E., WOLFE, H.J. 1993. Comparison of indirect and direct *in situ* polymerase chain reaction in cell preparations and tissue sections. Histochemistry 99: 151–162.

MOORE, G.P.M., RINGERTZ, N.R. 1973. Localization of DNA-dependent RNA polymerase activity in fixed human fibroblasts by autoradiography. Exp. Cell Res. 76: 223–228.

MULLIS, K. 1987. U.S. Patent 4683202: process for amplifying nucleic acid sequences.

SAIKI, R.K., GELFAND, D.H., STOFFEL, S., SCHARF, S.J., HIGUCHI, R., HORN, G.T., MULLIS, K.B., ERLICH, H.A. 1988. Primer-directed enzymatic amplification of DNA with a thermostable DNA polymerase. Science 239: 487-491.

SPEEL, E.J.M., LAWSON, D., HOPMAN, A.H.N. 1995. Multi-PRINS: multiple sequential oligonucleotide primed in situ DNA synthesis reactions label specific chromosomes and produce bands. Human Genet. 95: 29–33.

XIE, H., TROYER, D.L. 1996. Chemiluminescent detection of unique sequences on chromosomes after on-slide PCR. Biotechniques 20: 54–56.

Section 19.5

ALDRICH CHEMICAL COMPANY INC. 1989. Catalogue: handbook of fine chemicals.

DARLINGTON, C.D., LA COUR, L.F. 1975. The Handling of Chromosomes, 6th ed. John Wiley & Sons, New York.

Section 19.6

JEFFERSON, R. A., KAVANAUGH, T.A., BEVAN, M.W. 1987. GUS fusions: β-glucoronidase as a sensitive and versatile gene fusion marker in higher plants. EMBO J. 6: 3901–3907.

20

Molecular Analysis of Chromosomal Landmarks

- Euchromatin contains a relatively high density of actively transcribed genes, which can be observed cytologically in the bands of polytene chromosomes.
- Heterochromatin has a condensed appearance cytologically and contains a low density of transcribed genes in an environment dominated by the presence of long, tandem arrays of DNA sequences.
- The nucleolus-organizer region, or secondary constriction, defines the location of tandem arrays of ribosomal RNA genes.
- The centromere, or primary constriction, is usually a well-defined region of the chromosome, to which microtubules attach during cell division.
- Telomeres define the ends of the chromosomes and often consist of short arrays of tandemly repeated sequences, added postreplicatively to the DNA.

The chromosomes observed using light or electron microscopy are images resulting from the interaction of the biological structure, composed mainly of DNA and protein, with dye molecules or other agents. From observations of mitotic and meiotic chromosomes, cytogeneticists have distinguished a number of features such as centromeres, heterochromatin, nucleolus-organizer regions, and telomeres as "landmarks" for recognizing specific chromosomal segments. Advances in molecular biology have greatly expanded the number and availability of probes that interact with different regions of the chromosomes to reveal new details about their structure and function. Certain features of chromosomes, including the classical "landmarks" mentioned above, are now understood at the molecular level and are discussed in this chapter.

20.1 HETEROCHROMATIN AND EUCHROMATIN

The basic structure of the mitotic chromosome is comprised of two types of chromatin: heterochromatin and euchromatin. Heterochromatin is the term used to describe those portions of chromosomes that at interphase or prophase of the cell cycle are almost always darkly stained because they are highly condensed. Such constitutive heterochromatic regions occur at different locations in different chromosomes depending on the organism under study. They can occur near centromeres, in blocks within chromosome arms, or at the ends of chromosomes. The other kind of heterochromatin, called facultative, is associated with the inactivation of all or parts of the sex chromosomes (as discussed in Chapter 6, Section 6.5). Genetic and molecular studies have shown that heterochromatic regions consist of extremely long, tandem arrays of repetitive DNA sequences, which confer specific physical properties that are sufficiently obvious to be distinguished by the light microscope. Euchromatin, on the other hand, corresponds to the more lightly staining parts of chromosomes. It has been found that euchromatin generally contains shorter arrays of tandemly repeated DNA sequences, a variety of dispersed, repetitive DNA sequences, and, most significantly, a large majority of the genes present in an organism.

20.1.1 The Nature of Euchromatin

One view of chromatin is provided by the larval salivary-gland chromosomes of *Drosophila melanogaster*. These polytene chromosomes are held together at a common chromocenter, which includes the major heterochromatic regions of the *Drosophila* genome. The polytene chromosomes (Fig. 20.1) show characteristic patterns of bands of chromosomes separated by fainter-staining interbands, which are now able to be associated with the actual physical locations of genes. The cloning of more than 80% of the *Drosophila* genome in a variety of cloning vectors (*see* Fig. 20.1) has made large tracts of *Drosophila* DNA available for detailed studies of exact locations of gene sequences and associated control regions, relative to the band/interband structure observed in the polytene chromosomes. For example, the *Notch* locus, which codes for a product that suppresses neurogenesis during early embryogenesis, is located on the X chromosome of *D. melanogaster* at position 3C7. Over 100 kb of DNA containing the *Notch* locus have been cloned and characterized (Fig. 20.2), so that this locus provides a good model to determine the relationship between DNA sequences defining the gene product and the cytologically identified 3C7 band

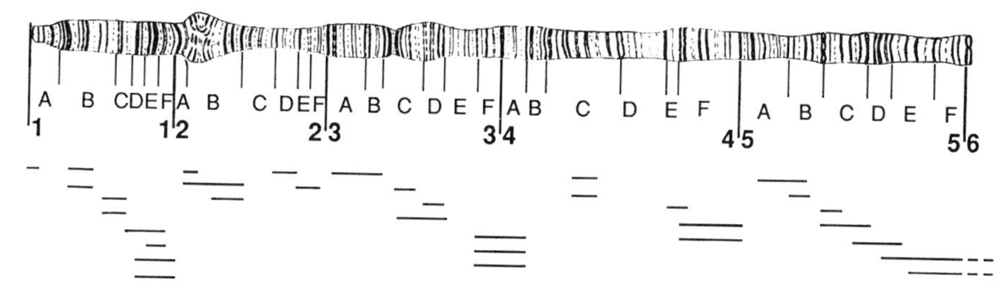

Fig. 20.1. A composite diagram of the end of the X chromosome (sections 1–5) from a salivary gland of *Drosophila melanogaster*, aligned with the original map produced by Bridges (Lindsley and Grell, 1968). The entire X chromosome is divided into 20 sections, each comprised of 6 blocks (A–F). The bands and interbands within each block are then numbered for precise identification. For example, the seventh band in section 3C, 3C7, contains the *Notch* locus (*see* Fig. 20.3). The entire *Drosophila* genome is encompassed within 102 approximately equal sections (Lindsley et al., 1972). The horizontal lines beneath the diagram indicate the approximate locations of those parts of the genome that have been cloned in either yeast artificial chromosomes (Ajioka et al., 1991; Cai et al., 1994), or cosmids (Kafatos et al., 1991; Madueño et. al., 1995). Much of the *Drosophila* genome (90 %) has been cloned, and this collection of DNA segments provides the basis for the complete sequencing of the genome.

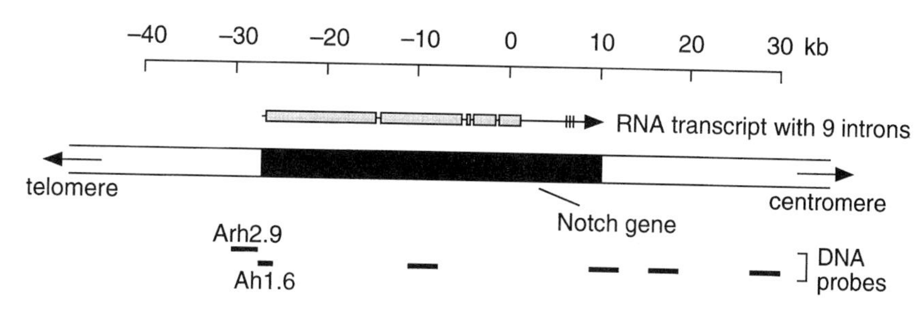

Fig. 20.2. A physical and genetic map of the *Notch* locus of *D. melanogaster*, relating DNA probes to the mRNA transcript of the gene at position 3C7 in polytene chromosomes. The *Notch* gene produces a transcript, about 37 kb in length, which includes nine exons (hatched) ranging in length from 130 to 7250 bp. After the removal of these exons from the RNA transcript, the final mRNA is about 10.5 kb in length and codes for a polypeptide product of 2703 amino acids. The probe Arh2.9 codes for a DNA sequence prior to the start of transcription of the gene, and the probe Ah1.6 codes for the start of transcription and the beginning of the gene sequence. (Modified from Rykowski et al., 1988.)

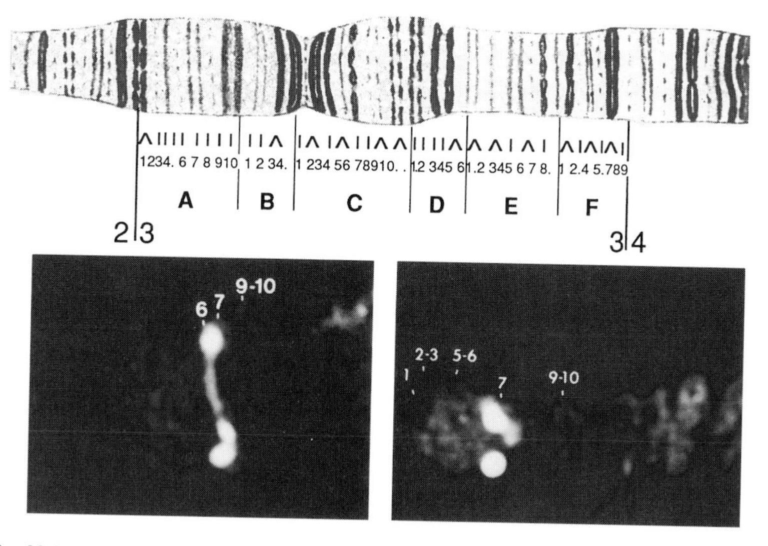

Fig. 20.3. Dissection of the *Notch* locus at position 3C7 of *D. melanogaster* using fluorescence in situ hybridization (FISH) and computer enhancement. *Upper*: The region being analyzed is shown diagrammatically using Bridges' map (*see* Fig. 20.1). Lower: the use of FISH to physically locate the positions on the chromosomes of the fluorescently labeled probes Arh2.9 (left) and Ah1.6 (right) (*see* Fig. 20.2). Arh2.9 is located within the 3C6 band and the interband between 3C6 and 3C7 (left) and Ah1.6 is located within the 3C7 locus (right), allowing the physical DNA map to be related to the cytological map. Multiple colors were originally used in order to observe the hybridization signal and the chromatin bands simultaneously. (From Rykowski et al., 1988.)

associated with the *Notch* locus. Using the high-resolution capabilities of fluorescence microscopy and computer-enhancement technology, DNA probes from this region have been physically located in polytene chromosomes by in situ hybridization. Studies of chromosome regions that were stretched during their preparation allowed the locations of DNA sequences to be precisely established with respect to the band and interband regions. It was found that sequences involved in controlling the time and level of expression of the gene were located in the interband region between 3C6 and 3C7 (Fig. 20.3), whereas the actual gene sequences were entirely within the darkly stained band region of 3C7. Based on the cytological lengths of the band and interband regions, molecular mapping suggests that the level of condensation of DNA is 5 to 20 times greater in the bands than in the interbands. Approximately 100–175 bp of DNA are compacted per nanometer (nm) of a chromosome band, compared to a compaction level of 2.8 bp/nm for naked DNA, 20 bp/nm for a 10-nm nucleosome fiber, and 110–220 bp/nm for a 30-nm chromatin fiber (*See also* section 17.5, Chapter 17).

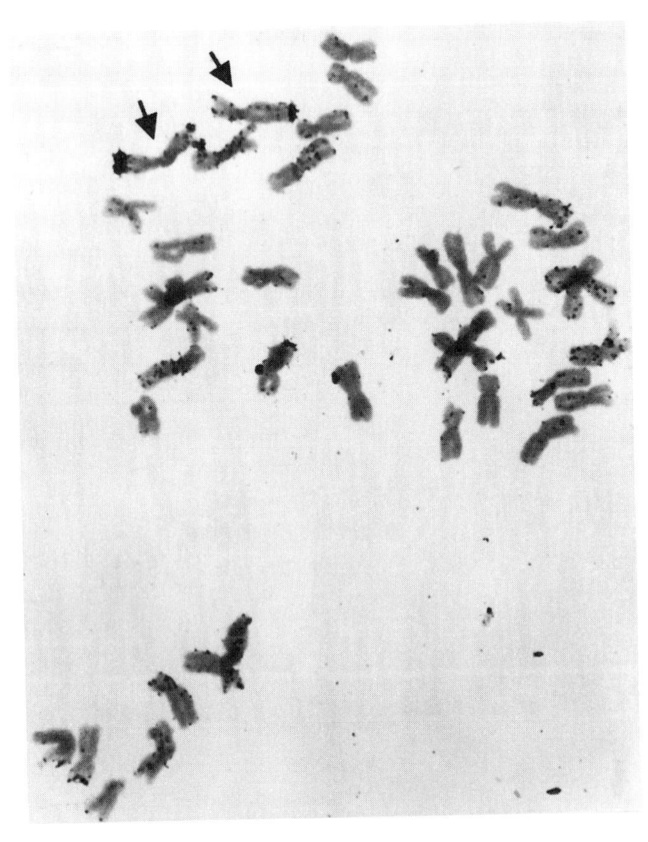

Fig. 20.4. The use of *in situ* hybridization to detect rye chromatin in a wheat background. The radioactively labeled ^{3}H-RNA probe was prepared from heterochromatic DNA sequences isolated from cereal rye (*Secale cereale*) (Appels et al., 1978). This DNA contains well-defined families of tandemly arranged sequences located in rye heterochromatin (Appels 1982). The arrows indicate an entire rye chromosome, 2R (left), in which heterochromatic DNA sequences at the end of both arms are labeled, and a second chromosome (right) in which only the end of the long arm is labeled. The latter chromosome was found to be the result of a centromeric translocation between chromosomes 2R of rye and 2B of wheat, combining the long arm of 2R (2RL) with the short arm of 2B (2BS) (May and Appels, 1980). Whereas most of the heterochromatin is located at the ends of the rye chromosomes, distinctively labeled bands do appear elsewhere, enabling all seven pairs of homologous rye chromosomes to be identified (Appels, 1982).

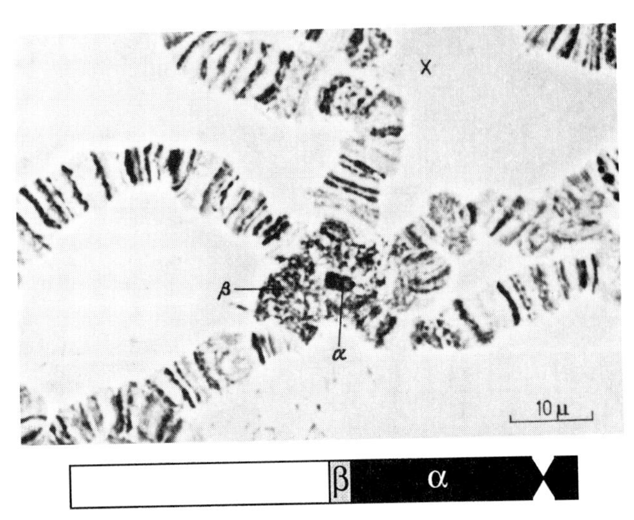

Fig. 20.5. The distribution of α- and β-heterochromatin in *Drosophila* chromosomes. The ratios of the chromatin components differ markedly between salivary-gland polytene chromosomes (top photograph) and normal mitotic chromosomes (lower diagram). The chromocenter (in the photograph), with which all of the centromeres are associated, shows the cytogenetic distribution of α- and β-heterochromatin (From Gall et al., 1971; *see also* Pardue and Hennig, 1990), indicating that heterochromatin often occupies regions of the genome with unusual characteristics such as under-replication in polytene tissue. In other organisms, heterochromatic regions of the genome are completely eliminated from cells undergoing embryogenesis (Beerman, 1977; Muller et al., 1982).

20.1.2 The Highly Condensed Nature of Heterochromatin

The concept that heterochromatin is more condensed at certain stages in the cell cycle suggests a higher density of chromatin per unit length of chromosome. This, in turn, is a direct result of a higher density of DNA compaction within the chromatin, a deduction that is confirmed in organisms such as rye (*Secale*), which shows pronounced heterochromatic blocks in mitotic prophase chromosomes using only phase-contrast microscopy. The use of Feulgen staining, which is specific for DNA, demonstrates that these blocks of heterochromatin contain a greater density of DNA than euchromatin. In fully condensed mitotic chromosomes, the differential condensation of heterochromatin relative to euchromatin is no longer obvious, and special techniques such as C-banding or in situ hybridization, are required to reveal the heterochromatin (Fig. 20.4; *see* also Chapter 6, Section 6.2, and Chapter 19).

In insect salivary glands, the "condensed" nature of heterochromatin results from an absence of the DNA endoreduplication; in the euchromatin, the endoreduplication process leads to the formation of polytene chromosomes. The lack of endoreduplication and the observed late replication of heterochromatin generally appear to result from an unusual DNA-sequence structure within the heterochromatin. This is most clearly seen in the chromocenter, where the centromeres of the polytene chromosomes of *D. melanogaster* are attached. The bulk of the heterochromatin in the chromocenter is normally found in fairly large blocks on either side of the centromeres of mitotic and meiotic chromosome arms (Fig. 20.5). At the junction between heterochromatin of the chromocenter and the endoreduplicated chromosome arms, a further class of heterochromatin, β-heterochromatin, has been defined. The DNA of β-heterochromatin contains expressed genes, some DNA-sequence families in common with the telomeric regions, other sequences found at many sites throughout the euchromatin, and some transposable elements. Molecular studies on the locus suppressor of forked bristles, *su(f)*, suggest that it is located in β-heterochromatin at, or very near, the α/β-heterochromatin boundary and is at least 80% polytenized. The heterochromatin between the centromere and β-heterochromatin is called α-heterochromatin (*see* Fig. 20.5).

20.1.3 The DNA Sequences of Heterochromatin

Heterochromatic DNA is dominated by the presence of long arrays of tandemly repeated sequences. Some chemicals interact with DNA in a sequence-specific manner (*see* Chapter 6), and many of these act as stains that reflect the characteristic DNA-sequence composition of heterochromatin. In prophase chromosomes of *Drosophila*, the entire Y chromosome, almost one-half of the X chromosome, most of chromosome 4, and the regions surrounding the centromeres of chromosomes 2 and 3 are made up largely of heterochromatin. The substructure of the heterochromatic Y chromosome has been investigated in detail using fluorescent dyes such as Hoechst 33258 (Fig. 20.6) and quinacrine (*see* Chapter 17, Fig. 17.15). The bands produced by Hoechst 33258 can be

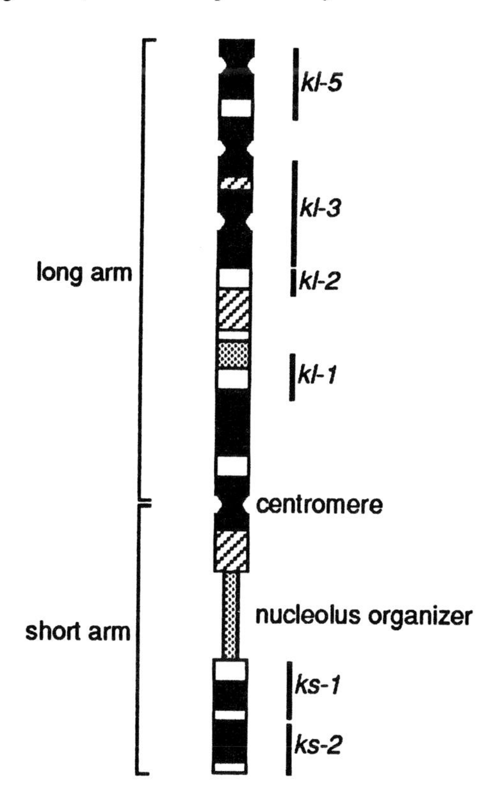

Fig. 20.6. A diagrammatic representation of the Y chromosome of *D. melanogaster* after staining with Hoechst 33258. The chromosome can be divided into defined regions depending on the degree of fluorescence and the presence of constrictions. Filled segments indicate bright fluorescence, hatched segments represent dull fluorescence, and open segments indicate no fluorescence. The lines on the right indicate the maximum physical sizes of the *kl* and *ks* fertility factors located on this chromosome (Bonaccorsi et al., 1988). Although the classical literature (Williamson, 1976) indicates five fertility factors on the long arm of the Y chromosome, *kl-4* has not been mapped in detailed cytogenetic analyses of the type summarized in the figure.

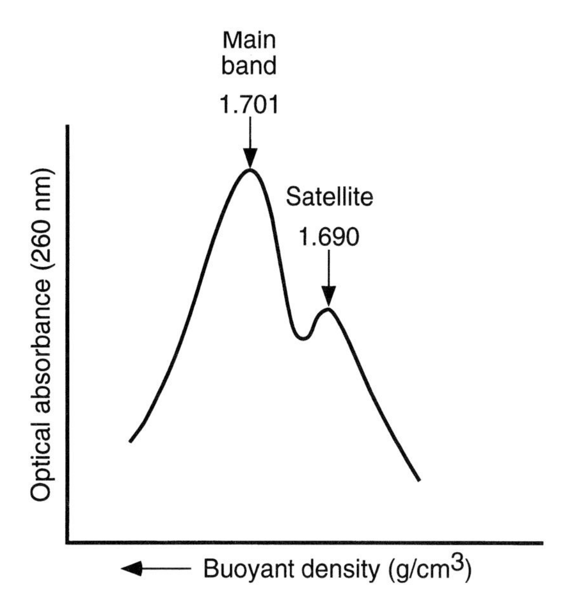

Fig. 20.7. The separation of mouse satellite DNA from genomic DNA by centrifugation in a CsCl density gradient. Differences in the density of satellite DNA can be natural and dependent on contrasting differences in nucleotide content, or they can be artificially induced by the addition of intercalating actinomycin-D or silver (Ag^+) ions. In mouse, the main band comprises about 92% of the total DNA with a G:C content of 42% and a buoyant density of 1.701 g/cm^3. The satellite contains about 8% of the total DNA and, because it has a G:C content of only 30%, it has the distinct buoyant density of 1.690 g/cm^3.

clearly seen in Fig. 20.6 and reflect the distribution of blocks of simple-sequence DNA in the Y chromosome. Other sequences such as long, tandem arrays of AGAAG and AGAGAAG do not bind to the dye, and the regions where these sequences occur appear as nonstaining gaps along the chromosome. The fluorescent dye thereby distinguishes regions within heterochromatin that can be used to provide cytological markers. The presence or absence of these markers in chromosomally engineered genetic stocks of *Drosophila* have been used to locate fertility genes within the Y chromosome (*see* Fig. 20.6).

The discovery of a characteristic DNA-sequence structure in heterochromatin predates the cloning of DNA sequences in bacteria. The technique of cesium-chloride (CsCl) density-gradient centrifugation provided the methodology for the purification of what were originally called satellite-DNA sequences. One of the first sequences isolated in this way was an AT polymer from crab DNA. The nucleotide base-pair composition of this polymer meant that it had a significantly different density from the bulk of the genomic DNA. Similarly, the isolation of mouse satellite DNA using CsCl density centrifugation (Fig. 20.7) was an important step in the study of heterochromatin, because it was one of the first sequences to be used for in situ hybridization experiments. The chromosomal location on mouse satellite DNA was found to correspond to regions of centromeric heterochromatin (Fig. 20.8). Subsequent DNA sequencing studies showed that the heterochromatic DNA of the mouse consisted of a simple repetitive sequence containing tracts of AT base pairs (Fig. 20.9). The repetitive nature of heterochromatic DNA sequences ensures that they renature relatively rapidly, and this characteristic can be used to isolate and purify these sequences (*see also* Chapter 16). Furthermore, if a restriction-endonuclease site is located in one of the units of the repetitive array,

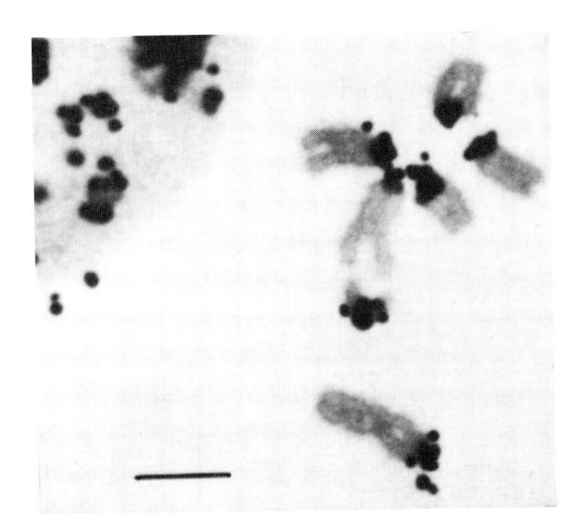

Fig. 20.8. The *in situ* hybridization (ISH) of mouse *(Mus musculus)* mitotic chromosomes with ^{3}H-labeled RNA satellite sequences. The satellite DNA was isolated by ultracentrifugation (*see* Fig. 20.7) and transcribed using *E.coli* RNA polymerase and ^{3}H-labeled ribonucleotide triphosphates. The experiment was one of the first to demonstrate a close physical relationship between the highly repeated satellite DNA sequences and the centromeres of mouse chromosomes. (From Pardue and Gall, 1970.)

it is often located in all the units. This means that the occurrence of the restriction-enzyme sites at identical distances apart results, after digestion, in the characteristic production of large numbers of DNA fragments of identical length. If occasional units in the array do not have the restriction-endonuclease site, some DNA fragments are twice the length of the basic unit. Fragments three, four, and so on, times the size of the basic unit can also occur and thus create a characteristic ladder of DNA fragments.

Not all families of heterochromatic DNA sequences can be isolated by physical methods such as density-gradient centrifugation, and it was not until cloning technology was developed that the full range of heterochromatic sequences could be studied. It has since been found that DNA-sequence families in heterochromatin can comprise units varying in length from 5 bp to 400 bp, arranged in long, tandem arrays. The cloning of these sequences provides the basis for assaying the presence of heterochromatin by in situ hybridization (*see* Chapter 19).

20.1.4 Some Chromosomal Proteins Are More Prominent in Heterochromatin

Although the unique cytological and biological properties of heterochromatin may be totally dependent on the composition of its

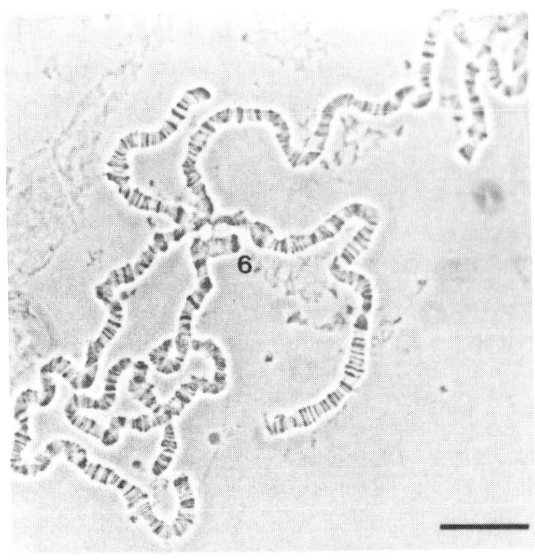

Fig. 20.10. Immunofluorescence staining of the six pairs of *Drosophila virilis* polytene chromosomes using the protein HP1. The chromocenter is heavily labeled as are a few loci on the chromosome arms. Chromosome puffs are unstained. Phase contrast (upper) and fluorescent (lower) images of the same chromosome spread are shown and the sixth chromosome is indicated. (From Tharappel et al., 1989; bar = 40 μm). The HP1 protein contains a domain called the chromobox, which is evolutionally conserved and characteristic of the proteins involved in the repression of gene activity in large chromatin domains within the nucleus (Singh et al., 1991). However, the chromobox is not present in all of the proteins involved in the modulation of chromatin structure (Reuter et al., 1990).

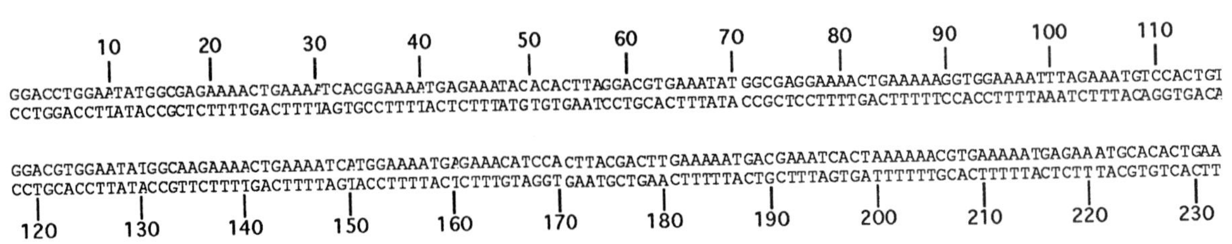

Fig. 20.9. The basic 232-bp repeat sequence of the DNA present in mouse-satellite heterochromatin. This repeat is itself comprised of two subrepeats, each about 116 bp in length. In total, there are 148 A:T base pairs and 84 G:C base pairs, giving a ratio of 0.638:0.362, and this differs from the ratio predicted from buoyant-density analyses (*see* Fig. 20.7) because of the effects of the actual sequence on buoyant density.

unusual DNA sequences, the involvement of certain chromosomal proteins is also a possibility. An investigation of *Drosophila* salivary-gland chromosomes has revealed a high concentration of a single protein, heterochromatin-associated protein No. 1 (HP1) in β-heterochromatin (Fig. 20.10). Immunofluorescent staining has demonstrated that HP1 is also distributed throughout the smallest chromosome and at the ends of chromosomes. Interestingly, the HP1 gene is allelic to the genetic locus *Suppressor of position-effect variegation [Su(var)2-5]*, which modifies the ability of heterochromatin to partially inactivate nearby genes. Position-effect variegation occurs when heterochromatic sequences are transposed to novel sites adjacent to euchromatic genes (*see* Chapter 9). The degree of suppression of these genes by heterochromatin reflects the spread of chromatin condensation into neighboring euchromatin. A second protein, the product of the *Su(var)3-7 locus*, is involved in the modification of position-effect variegation and, therefore, may also contribute to the structure of heterochromatin.

20.1.5 Heterochromatin Contains Genes at a Lower Frequency Than Euchromatin

Detailed genetic and cytogenetic analyses of the chromosomes of *D. melanogaster* have revealed the presence of some gene loci within regions of heterochromatin. These analyses require a distinguishable relationship between the cytological appearance of a chromosome and its genetic map. Chromosomal deletions removing large blocks of heterochromatin from near the centromeres have no noticeable effect on the phenotype and thus, for many years, heterochromatin was thought not to contain any distinct genes; this led to the concept of the genetic inertness of heterochromatin. However, a number of ethyl methanesulfonate (EMS)-induced lethal mutations have since been located in the heterochromatin of chromosome 2 of *Drosophila*, indicating that this heterochromatin must contain unique-sequence genes essential for life (Fig. 20.11). Assuming that EMS-induced mutation analysis is able to detect the majority of the genetic loci in the heterochromatin of chromosome 2, the frequency of genetic loci per unit length of heterochromatin is estimated to be 1% of that found in euchromatin.

To some extent, this analysis of chromosome 2 heterochromatin is similar to the analysis of fertility factors located on the Y chromosome of *D. melanogaster* (*see* Fig. 20.6). The best estimates of the minimum and maximum sizes of chromosome regions encompassing the *kl* and *ks* fertility factors include segments of chromatin known to have a high content of simple repetitive sequences. Therefore, at least some of these fertility factors must be located in regions where the heterochromatic repeat sequences are also highly clustered.

Not all genes in heterochromatin are of the unique-sequence category. The *Responder (Rsp)* element is thought to be composed of tandem arrays of sequences that contribute to a phenotype of *D. melanogaster*, known as the Segregation Distorter, which involves

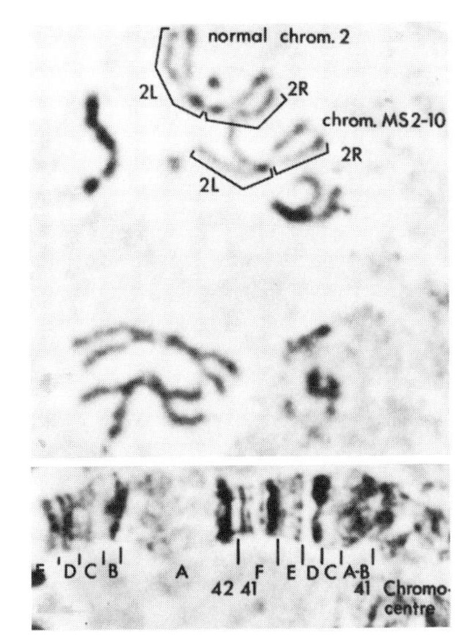

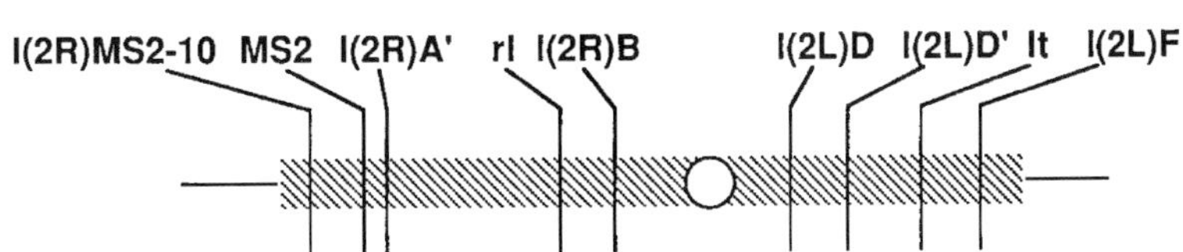

Fig. 20.11. A genetic analysis of the heterochromatin that surrounds the centromere of chromosome 2 of *D. melanogaster*. A mitotic chromosome spread (top photograph) of the translocation line MS2–10 shows the lack of α and β heterochromatin from near the centromere of the 2R arm. The lower photograph shows a section of the right arm of the polytene chromosome 2 pair and the direct attachment of 2R euchromatin to the chromocenter (right) without the presence of 2R α or β heterochromatin. A diagrammatic representation of the genes that are present in the heterochromatin of chromosome 2 and analyzed using deletions such as MS2-10 is also shown (bottom; Hilliker et al., 1980). As an example, the locus *l(2R)A′* indicates a gene located in the heterochromatin of the right arm of chromosome 2.

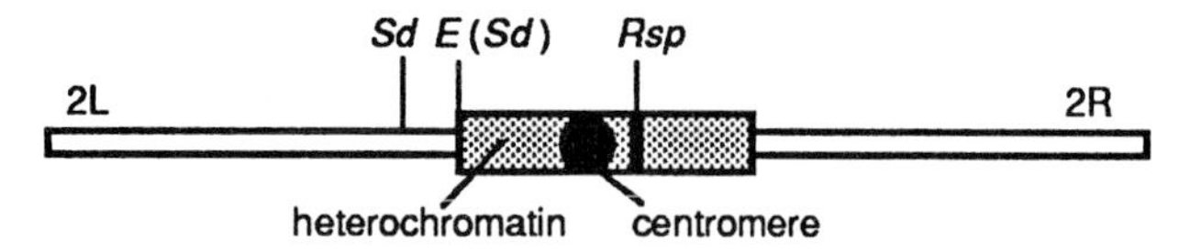

Fig. 20.12. The various genes associated with the Segregation Distorter (SD) phenotype and their locations on chromosome 2 of *Drosophila melanogaster*. (Adapted from Palopoli et al., 1994.) The presence of the SD phenotype leads to the preferential recovery of sperm with chromosome 2 carrying the *Segregation distorter* (*Sd*) locus and the loss of sperm with a homologous chromosome 2 carrying the so-called sensitive allele of *Responder* (*Rsp*). The Enhancer of Sd, *E(Sd)*, is a third locus on chromosome 2 that interacts with *Sd* and *Rsp* (Brittnacher and Ganetsky, 1989; Lyttle, 1989). The *Rsp* locus is composed of a 120-bp sequence tandemly repeated from 25 to 3000 times, depending on the source of the locus analyzed (Wu et al., 1988).

spermiogenesis, with selective survival of particular genotypes of sperm. The controlling genetic complex is comprised of three tightly linked loci surrounding the centromere of chromosome 2 (Fig. 20.12). The *Segregation distorter (Sd)* locus is located in the euchromatin of chromosome arm 2L near the centromere, where it produces a factor that is essential for strong distortion. Sperm carrying chromosome 2 with the *Sd* locus are preferentially recovered in the presence of the so-called sensitive allele of *Rsp*, as described below. The locus *Enhancer of Segregation distorter* [E(Sd)] is also located near the centromere in 2L, but in the β-heterochromatin, and this locus strengthens the degree of distortion. The *Rsp* locus, which is the target for the actions of both *Sd* and *E(Sd)*, is located near the centromere, embedded in the heterochromatin of 2R. It has been shown that *Rsp* is closely associated with a Hoechst 33258 staining band, h39 (Fig. 20.13). Quantitative effects related to the degree of sensitivity of *Rsp* to *Sd* are correlated with the physical size of the h39 band, and with the number of copies of a specific A–T-rich, 240-bp sequence located in the 2R heterochromatin. The Rsp-240 bp sequence has been isolated and cloned from a *Responder-sensitive* genotype *(Rsp^s)*, in which it exists in tandem arrays at least 30 units long with as many as 2000–2800 units per genome; these are not necessarily in a single, tandem array. *Responder-insensitive* genotypes *(Rsp^i)* do not interact with the Segregation Distorter phenotype and have very few Rsp-240 bp sequences, perhaps as low as 25 copies. A working hypothesis, therefore, is that *Sd* (on one chromosome 2 homolog) produces a product that binds to the Rsp-240 sequences (on the other chromosome 2 homolog), resulting in defective chromatin-folding and dysfunction of sperm carrying the chromosome 2 with the *Rsp-sensitive* locus. With fewer Rsp-240 sequences, the *Sd–Rsp* interaction becomes less significant in sperm development.

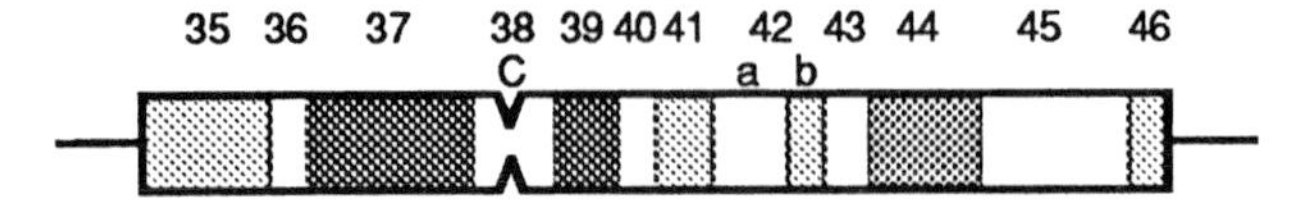

Fig. 20.13. A cytological map of the heterochromatin surrounding the centromere (C) of chromosome 2 of *Drosophila melanogaster* after staining with Hoechst 33258. Fourteen distinct regions can be delineated. The degree of staining of the various regions corresponds to the amount of Hoechst 33258 bound to the DNA. The visual intensity of region h39 (shown as 39) appears to be directly associated with the number of repeat sequences present in the *Rsp* locus (*see* Fig. 20.12), which contributes to of the Segregation Distorter phenotype (Dimitri, 1991).

Ribosomal RNA genes are discussed in more detail later on; they are located in the heterochromatin of *Drosophila* X and Y chromosomes, where they are usually found as tandem repeats of about 200 copies. These loci correspond to the genetic locus first identified as *bobbed (bb)*. Analogous to the existence of various *Rsp* loci, it has been found that variations in the severity of the phenotype expressed by allelic variants of the *bobbed* locus are directly correlated to the number of rRNA genes present, and changes in number can be induced by deletions (as discussed in Section 20.2.1).

20.1.6 Unusual Biological Properties of Heterochromatin

The preceding sections have described a number of unusual properties associated with heterochromatin and these are summarized in Table 20.1. The wide diversity of effects suggest that the presence of long tracts of repetitive DNA sequences can have a variety of pleiotropic effects. Rather than being essential for life, many regions of heterochromatin seem to be involved in quantitative interactions within the genome that lead to adjustments in the overall phenotype of the organism. Consequently, heterochromatin differs from euchromatin in two major ways:

1. The very long tracts of tandemly repeated DNA sequences that are common in heterochromatin are almost lacking in euchromatin. Many of the quantitative-genetic effects associated with heterochromatin are caused by variations in the lengths of tandem arrays of sequences within the heterochromatin.
2. The density of genetic loci/unit length of DNA is much lower in heterochromatin than in euchromatin. This is presumably a result of the presence of long tracts of repetitive-sequence DNA.

20.2 THE NUCLEOLUS AND RIBOSOMAL-RNA GENES

The structure called the nucleolus has been studied since the 1890s. At that time, there were contrasting views on the origin and function of the nucleolus, ranging from its derivation from chromatin by direct transformation of nuclear substance, to the converse view, which suggested that it was of extranuclear origin and not a secretion or excretion of the nucleus. Current views on the structure and function of this chromosomal landmark account for both of the above viewpoints as a result of the combination of microscopic studies and biochemical research.

20.2.1 Identification of the Genes Involved in Forming the Nucleolus

Cytological studies have demonstrated that the nucleolus contains a high content of RNA molecules. In addition, when radioactive precursors of RNA such as ^{3}H-uridine are administered to cells, the nucleolus is always strongly labeled. Ribosomal RNA (rRNA) is also a major constituent of the cytoplasm and provides a crucial component of the protein synthetic processes. These and other observations indicate that the nucleolus is the site of synthesis

Table 20.1. Summary of the Biological Properties Associated with Heterochromatin

Property	Description
1. Condensed, darker staining than euchromatin	Most clearly seen in mitotic-prophase chromosomes of most plants and animals; results from a higher concentration of DNA per unit length of chromosome.
2. Late-replicating	In most eukaryotes, heterochromatin is replicated later in the S phase than the bulk of the genome; in the salivary-gland polytene chromosomes of insects such as *Drosophila*, no endoreplication of heterochromatic DNA occurs.
3. Position-effect variegation	An alteration in the activity of a gene due to a change in its position within the genome relative to the position of heterochromatin (*see* Chapter 9, Section 9.4.1).
4. Pairing sites	Meiotic pairing between nonhomologous chromosomes can occur owing to the presence of common-homologous DNA sequences within the heterochromatin of different chromosomes and can influence segregation.
5. Rescue of abnormal oocyte	A region of *Drosophila* X-chromosome heterochromatin that interacts with the locus *abnormal oocyte* (*abo*) to reduce the severity of its effect.
6. Target for segregation-distorter genes	A region of Drosophila 2R heterochromatin, *Rsp*, that interacts with the *Sd* and *E(Sd)* genes to cause dysfunction of sperm carrying these loci.
7. Low gene density	A more or less universal property of heterochromatin, making large blocks of it inessential for life. Gene low-copy number and repetitive transcribed genes can, however, be present, as well as arrays of repetitive DNA sequences, which provide target sites for the products of euchromatic genes.

and processing of ribosomal RNA in preparation for transmission through nuclear membranes into the cytoplasm. Formal proof of this proposal was obtained when physical studies of the nucleolus were combined with cytogenetic studies.

In the early 1960s, a technique was developed to target cytologically identifiable sites in the cell with a microbeam of ultraviolet light. When this method was used to partially destroy the nucleolus, there was a concomitant loss of RNA synthetic activity and a great reduction in the amount of cytoplasmic RNA, particularly rRNA. This result argued for a central role of the nucleolus in rRNA synthesis. By the mid-1960s, the technique of RNA–DNA hybridization (described in Chapter 19, Section 19.2) could be used to analyze the DNA of *Xenopus* and *Drosophila* mutants that contained differing numbers of nucleolus-organizer regions (NORs). These NORs are cytologically distinct regions of chromosomes, usually seen as constrictions separating a chromosome satellite from the rest of a chromosome arm in mitotic cells, and must be present for a nucleolus to be formed. The *anucleolate* mutation of *Xenopus,* for example, produces tadpoles that can survive up to the tail-bud stage yet synthesize no rRNA, drawing on the large pool of maternal rRNA laid down in the original oocyte. The progeny from crosses between heterozygous individuals carrying one dose of the *anucleolate* mutant segregated in the expected ratio of 1 individual homozygous for *anucleolate* (no NORs), 2 heterozygotes, with 1 homologous chromosome having a NOR and the other homolog without it, and 1 homozygote, with 2 NOR-containing chromosomes. Using RNA–DNA hybridization, it was then demonstrated that the number of rRNA genes in the genomic DNA from the three types of F_2 individuals was directly correlated with the number of NORs that were present in the three genotypes.

The *bobbed (bb)* mutation in *D. melanogaster* results in flies with shorter bristles than wild-type flies, and abnormal coloring of the abdomen (Fig. 20.14). Chromosome-mapping studies suggested that the mutant gene coincided with the NOR at the centromere end of the X chromosome. Hybrids between stocks containing a pair of inversions, $In(1)sc^4$ and $In(1)sc^8$, allowed the number of NORs in an individual fly to be varied (*see* Fig 20.14). RNA–DNA hybridization measurements of the rRNA gene contents in the genomic DNA of progeny carrying recombinant chromosomes indicated that the number of rRNA genes present was proportional to the number of NORs present. Studies on the DNA from flies with different *bobbed* mutations also showed that the degree of severity of the *bobbed* phenotype was a direct result of deletions differing in length and involving rRNA genes in the nucleolus-organizer regions. Direct confirmation of the location of rRNA genes within the *Nor* locus, and within the nucleolus, has been obtained for many organisms using in situ hybridization, as shown for wheat in Fig. 20.15.

In metaphase chromosomes, the NOR appears as a stretched region, or secondary constriction, and in favorable preparations, the rRNA genes can be assayed in the DNA present in the stretched region, as well as in the chromosome region distal to this region. Most models of rRNA gene organization include active and inactive states of the gene complex, and these may well be reflected in the appearance of rRNA genes in the stretched and more condensed regions, respectively, of the metaphase chromosomes.

20.2.2 The Nature of Ribosomal-RNA Genes

When DNA is centrifuged at high speed in salts such as CsCl, not only are major bands of genomic DNA with particular buoyant densities present (*see* Fig. 20.7), but smaller "satellite" bands can also be distinguished. In many cases, the isolation of satellite-DNA sequences can be enhanced by the addition of ions, such as Ag^+, or by antibiotic molecules that bind to specific DNA sequences. When aliquots of *Xenopus* DNA from actinomycin-D/CsCl gradients were assayed for the presence of rRNA genes, a single satellite peak containing all of the rRNA genes was identified. This can only occur if the rRNA genes are clustered together to form large regions of DNA with a characteristic base-pair composition and buoyant density. *Xenopus* genes were the first to be isolated in this way and provided the starting material for the first molecular cloning of eukaryotic DNA in bacteria. It has since been found that all eukaryotes have tandem arrays of rRNA genes, in which a single

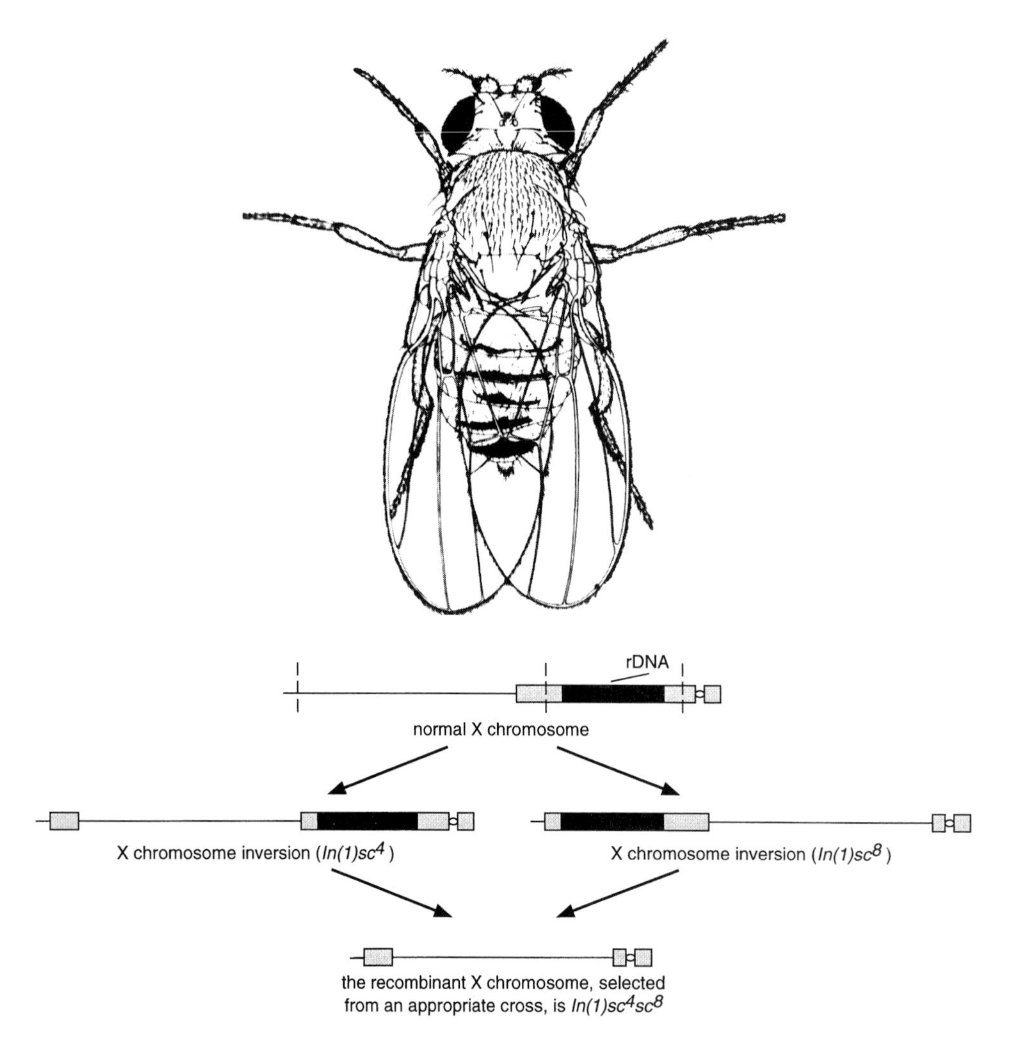

Fig. 20.14. A drawing of an extreme *bobbed (bb)* female of *Drosophila melanogaster* (upper) showing shortened bristles, uneven abdominal color markings, and other abnormalities of the lateral abdomen (Lindsley and Grell, 1968; drawing by E.M. Wallace). This fly can be compared with wild-type illustrated in Chapter 24, Section 24.11, with normal markings on the abdomen. The changes are caused by a severe reduction in the number of rRNA genes present in the nucleolus-organizer region of the X chromosome. The number of these genes can be varied by exploiting the wide range of *bobbed* mutations, or by intercrossing flies with inversion chromosomes (illustrated in lower figure) such as $In(1)sc^4$ and $In(1)sc^8$. If at least partial pairing of the two inverted chromosomes occurs in the F_1 hybrid, a crossover can give the recombinant chromosome (bottom), which lacks the segment with the rDNA genes.

rDNA unit consists of the following rRNA genes (Fig. 20.16) separated by varying lengths of intervening spacer-DNA sequences:

1. *18S or small rRNA gene*: Codes for an RNA molecule varying in size from 1487 bp in the fungus *Anacystis nidulans* to 1869 bp in *Rattus norvegicus*. This molecule is the major RNA component of the small subunit of ribosomes.
2. *5.8S rRNA gene*: Codes for an RNA molecule of about 150–200 bp in length. This molecule is commonly H-bonded to the large rRNA component in the large subunit of ribosomes.
3. *26S or large rRNA gene*: Codes for an RNA molecule ranging in size from 2876 bp in *Anacystis* to 4718 bp in *Rattus*. This molecule is the major RNA component of the large ribosome subunit. In protozoans, dipteran insects, and some bacteria, an extra processing step occurs to cleave this molecule near the middle of the gene sequence.

In addition to these rRNA genes, a further gene sequence codes for a 5S RNA molecule of about 120 bp in length. This gene is part of the rDNA repetitive unit only in some lower eukaryotes and in bacteria. In multicelled eukaryotes, the repetitive arrays of 5S DNA units are found at loci distinct from the nucleolus-organizer regions (Fig. 20.17). The 120-bp 5S RNA molecule is located in the small subunit of the ribosomes.

All of the rRNA gene sequences are separated from one another by varying lengths of intervening-spacer-DNA segments. The major intervening spacer separates the 3′-end of the 26S rRNA gene from the 5′-end of the 18S rRNA gene. Although this sequence is not transcribed, giving it the name nontranscribed spacer (NTS), it is of great importance for the regulation of transcription of the

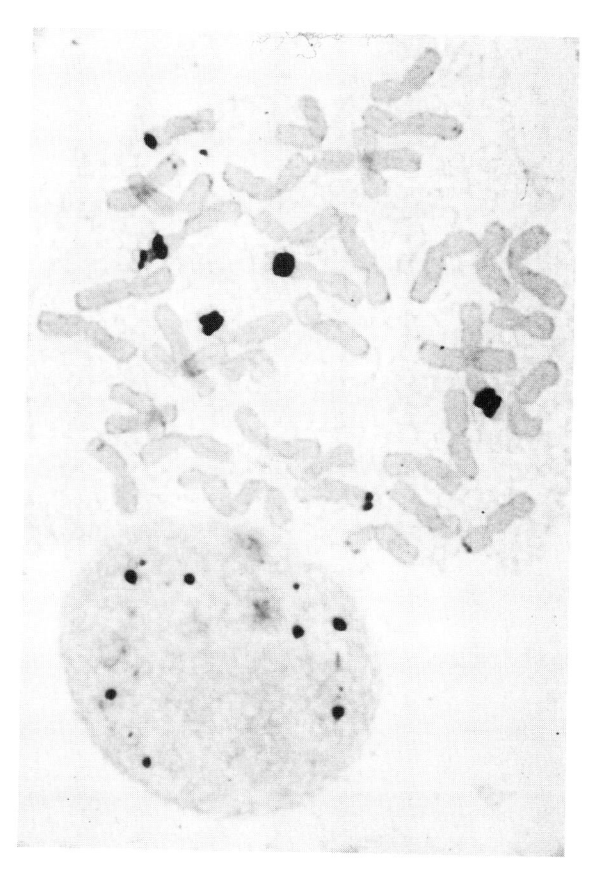

Fig. 20.15. The *in situ* hybridization of ribosomal rDNA spacer probe to the nucleolus-organizer regions of hexaploid wheat, *Triticum aestivum*. In cereals, the nucleolar-organizer regions or highly repeated rRNA gene regions can be designated as either *Nor* or *Rrn*. The two major *Nor* loci of hexaploid wheat, *Nor-B1* and *Nor-B2*, are located on the 1B and 6B pairs of satellited chromosomes, respectively (May and Appels, 1987, 1992). Minor sites are located on chromosomes 5D, 1A, 5A, and 7D (Mukai et al., 1991; Jiang and Gill, 1994).

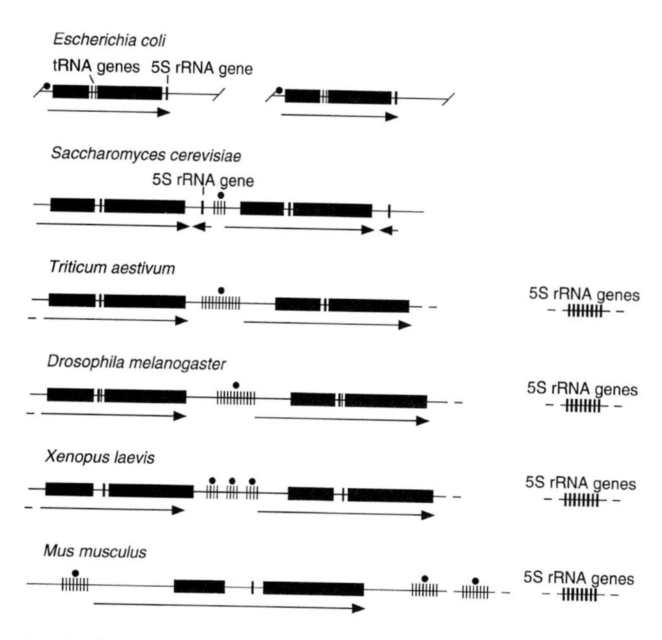

Fig. 20.17. A comparison of rRNA gene sequences from a range of organisms to show the evolutionary similarity of rRNA gene arrangement and transcription. Repetitive-sequence regions are marked by black dots (•) (Appels and Honeycutt, 1985). 5S rRNA genes are found in association with other rRNA genes only in prokaryotes and some unicellular eukaryotes.

actual rRNA genes. The sequence structure of the rDNA repeating unit displays many of the characteristics typical of all eukaryotic genes, as follows:

(a) *The start of transcription*—a DNA sequence closely related to the sequence TATAGTAGGG that has been identified in several plants and animals, 5′-upstream of the 18S gene transcript.

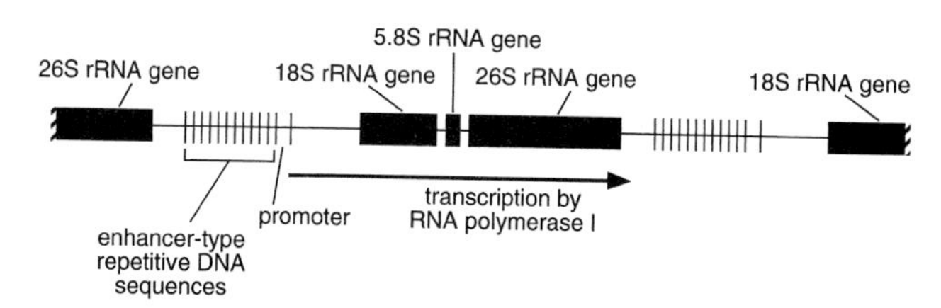

Fig. 20.16. A diagrammatic representation of a typical ribosomal-DNA repeat unit. The repeated sequences contain an 18S, a 5.8S, and a major 26S rRNA gene, and are separated from their neighbors by a stretch of nontranscribed, intervening-spacer DNA, itself composed of varying numbers of short sub-repeat DNA sequences. Promoters, enhancers, and terminators of transcription are present in the nontranscribed, intervening-spacer DNA and are indicated as enhancer-type repetitive-DNA sequences (Appels and Honeycutt, 1985; Firek et al., 1989; Pikaard et al., 1989; Kulkens et al., 1991; Smid et al., 1992).

(b) *A 5′-upstream spacer region*—a region, prior to the start of transcription, characterized by the presence of repetitive sequences that function as promoters/enhancers of transcription by RNA polymerase I. Between organisms, these repetitive sequences range in length from 100 to 300 bp and vary both in the number present and in their nucleotide sequences, which provide an important source of evolutionary variation in the rDNA unit.

(c) *The 3′-downstream spacer region*—contains transcription-termination signals that usually include a TTT motif. They exist in the spacer downstream from the 3′-end of the large, 26S gene sequence. In some situations of high levels of transcriptional activity, transcription can continue through these termination signals, resulting in transcription of some of the spacer region. In such an event, a further termination signal, 5′-upstream from the transcription-promoter region, ensures that termination occurs before further transcription is initiated.

(d) *The gene sequences*—in contrast to the intervening-spacer DNA, the actual rRNA gene sequences are highly conserved during evolution. In particular, the small and large genes contain sections that are recognizable from bacteria and Archaea to eukaryotic plants and animals. As was described in Chapter 1, these sequences have been used to delineate phylogenetic relationships in all of these organisms.

20.2.3 Ribosomal-RNA Synthesis

Electron microscopy has produced images of the ribosomal genes in the process of transcription (*see* Chapter 2, Fig. 2.8),

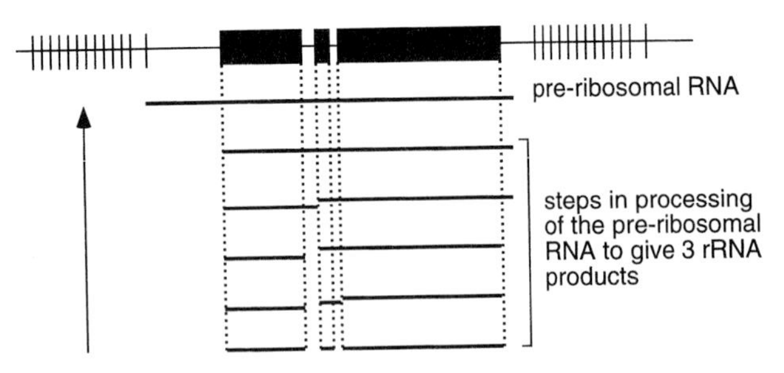

Fig. 20.18. The steps involved in producing individual ribosomal RNAs from the basic mRNA transcript. The elaboration of the processing used pulse-labeling experiments (Gerbi et al., 1990; Savino and Gerbi, 1991) to give an accurate estimation of the different-length fragments found to be present within a defined time frame. The major enzyme involved in this processing is RNase P.

confirming biochemical evidence that one large precursor RNA molecule is synthesized. This molecule is subsequently processed (Fig. 20.18) to generate the three individual rRNA molecules that bind to ribosomal proteins to eventually form ribosomes. Synthesis of rRNA uses an RNA polymerase, together with numerous protein cofactors, and proceeds from the start of transcription to the terminator. The enzyme that produces the precursor RNA is RNA polymerase I, which consists of 8 subunits in the rat and 13 subunits in yeast. The largest subunit binds to zinc in a domain that is common to analogous subunits of RNA polymerases II and III. RNA polymerase II produces messenger RNAs required for the synthesis of proteins, and RNA polymerase III produces small RNA molecules such as 5S RNA and transfer RNAs. Although significant amino-acid-sequence homology exists between the large subunits of RNA polymerases I, II, and III, and the β'-subunit of the RNA polymerase found in *Escherichia coli*, RNA polymerase I alone is insensitive to α-amanitin, a toxic bicyclic octapeptide isolated from the mushroom *Amanita*. The second largest subunits of RNA polymerases I, II, and III share amino-acid-sequence homology with the β-subunit of the *E. coli* enzyme. On the basis of comparison with the *E. coli* RNA polymerase, it is likely that the two largest subunits of RNA polymerases I, II, and III are the major catalytic components of these enzymes. Three-dimensional reconstructions from electron micrographs of crystals of these major catalytic components show a characteristic thumblike projection, defining a channel that could hold a DNA molecule; it is suggested that the conformation of this projection forms the basis for promoter recognition.

A number of other proteins are involved in building the transcription-initiation complex that allows RNA polymerase I to start transcription. The protein factor SL1 and the upstream binding factor (UBF1) from human HeLa cells have been especially well characterized. The two factors act in concert to form a productive transcription-initiation complex, with UBF1 binding to a DNA sequence located 107 bp upstream from the transcription-start site. In cell-free systems, where both mouse and human rDNA are present, the SL1 factor of humans induces human-specific transcription. Even so, SL1 does not interact directly with the human rDNA promoter region, but combines with UBF1 to form human-specific transcription-initiation complexes. The detailed relationships between DNA sequences in the transcription-control region and the proteins binding to them are the subject of extensive research.

Production of the final RNA molecules that are used in ribosome biosynthesis requires processing (*see* Fig. 20.18), and an analysis of this process in the ciliated protozoan *Tetrahymena* led to the elucidation of the biochemistry of self-cleavage of RNA (Fig. 20.19). The RNA component of the rRNA-processing enzyme complex of *Tetrahymena* was the first well-characterized example of an RNA molecule acting as an enzyme or ribozyme, and the details of this reaction (*see* Fig. 20.19) provided a stimulus to the discovery of similar reactions in plant-virus RNAs. Studies of the ribozyme activity of RNA molecules led directly to the design of synthetic RNA molecules that are capable of cleaving other RNA molecules at specific sequences (Fig. 20.20).

20.2.4 The Structure of the Nucleolus

Electron-microscopic studies of the nucleolus have defined three distinguishable regions (Fig. 20.21): a fibrillar center (FC), a dense fibrillar component (DFC), and a granular component (GC). Ribosomal RNA is synthesized near the junction between the FC and the DFC, and as the rRNA matures and binds with ribonuclear proteins, it gradually moves into the GC. The resulting ribonucleoprotein particles (RNPs) transport the rRNA from the nucleolus to the cytoplasm. However, the proteins involved in the production

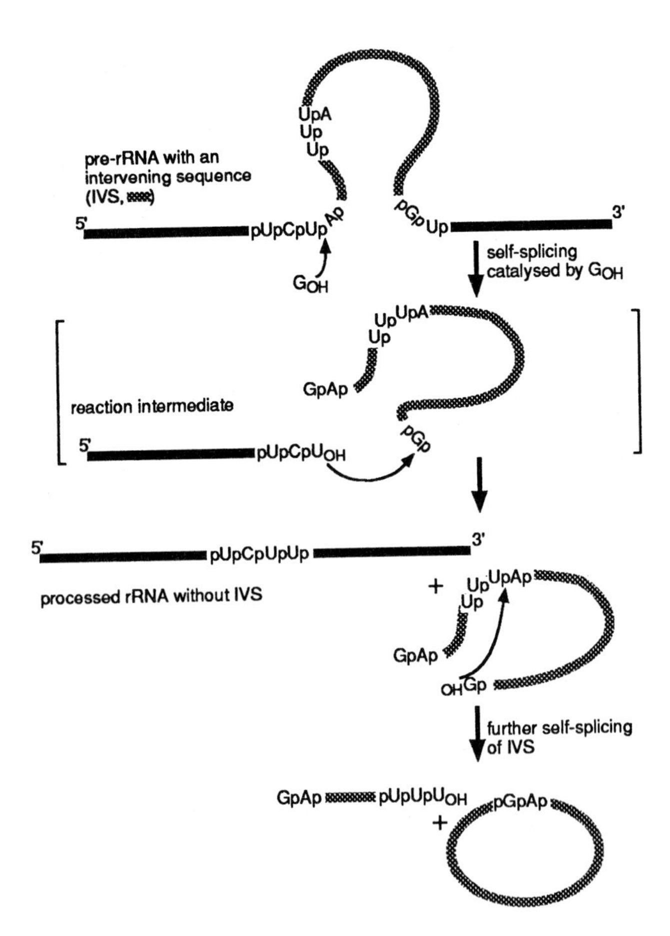

Fig. 20.19. The self-cleavage of RNA in *Tetrahymena* by RNA acting as an enzyme. (Modified from Cech, 1987.) The process is facilitated by the presence of particular base-pair combinations at the 5′- and 3′-ends of the RNA segment destined to be deleted. The enzymatic activity of RNA is also referred to as ribozyme activity and has been observed in a variety of different organisms (Cech, 1986). The 3′OH residues (e.g., G_{OH}) are shown when they are relevant to the chemical reaction. Similarly the 5′-phosphate groups are shown (e.g., pU).

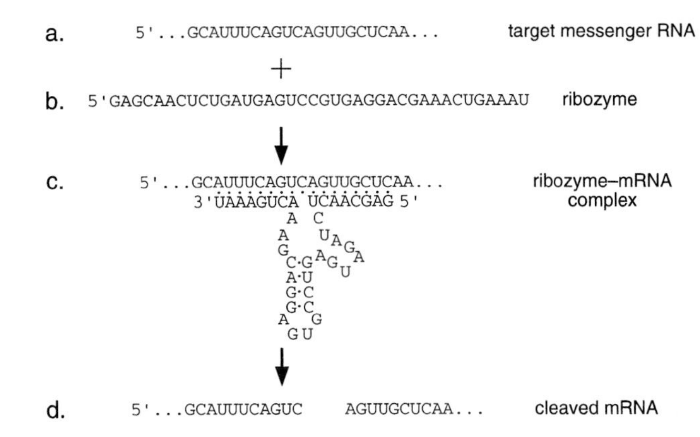

Fig. 20.20. The mode of action of ribozymes and the exploitation of ribozyme activity. By manufacturing ribozyme–mRNA sequences (c) that are specific to a target mRNA molecule (a), the respective mRNA can be destroyed (b) and prevented from providing a template for the production of its normal end product (d). In this way, the expression of specific genes can be blocked (Haseloff and Gerlach, 1988).

of RNPs are themselves synthesized in the cytoplasm, and they need to migrate into the nucleolus to combine with the rRNA molecules.

In addition to RNA polymerase I, other nucleolar-protein genes have also been sequenced and characterized. These include nucleolin, fibrillarin, single-strand-binding protein (SSB1), topoisomerase I, NO38/B23 protein, the protein p67 produced by the *NSR1* gene of yeast, transcription factor UBF, TATA-binding protein and associated factors that comprise transcription factor SL1, and glycine-arginine-rich protein (GAR1). As determined by immunolabeling, most of these proteins are located within the dense fibrillar component of the nucleolus. Nucleolin is particularly well characterized

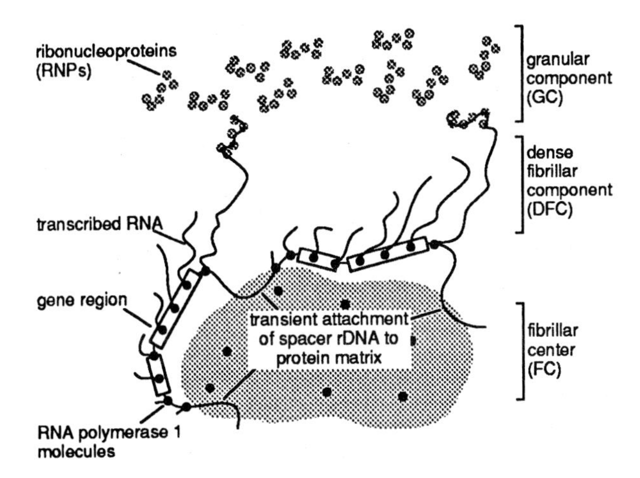

Fig. 20.21. A diagrammatic representation of nucleolar structure at the electron-microscopic level. The nucleolus provides the physical environment in which ribosomal RNA is produced and bound to nucleoproteins for processing and subsequent transportation into the cytoplasm (based on Sommerville, 1985; Nigg, 1988). The proposed organization is supported by *in situ* hybridization studies at the electron-microscopic level (Thiry and Thiry-Blaise, 1991; Wachtler et al., 1992). It is possible that the RNA polymerase 1 molecules (●), shown in the process of transcribing the gene regions, may be attached to a nucleolar skeleton structure (Weipoltshammer et al, 1996) which permeates the entire region illustrated (including the fibrillar center).

as a highly conserved protein and has been identified in humans, *Xenopus,* rat, Chinese hamster, and onions. Another protein in yeast, encoded by the *NOP1* gene, is functionally equivalent to the fibrillarin of vertebrates and is required for many steps in the maturation of ribosomal RNA in the nucleolus, therefore loss of *NOP1* function is lethal. A temperature-sensitive mutation in the NOP1 gene has been used to uncover mutations in other genes that can rescue the cells at the nonpermissive temperature and thus identify new genes coding for proteins that interact with the NOP1 protein.

The precise function of many of the proteins in the nucleolus is still unclear, although they are considered to be important in both the synthesis and processing of rRNA, as well as in providing a structural function in the dense fibrillar component. Many are phosphorylated and most (with the exception of topoisomerase I) share a characteristic glycine/arginine-rich domain, which provides a highly flexible β-type turn in the protein. However, the actual position of the glycine/arginine-rich domain differs in all of the nucleolar proteins, so that its functional significance is unclear.

20.2.5 Nucleolus-Organizer-Region Suppression

One of the classical observations of cytogenetics, the phenomenon of suppression of nucleolus-organizer regions, can now be understood in terms of the synthesis of rRNA. Typically, in the F_1 hybrids of interspecific crosses, the secondary constriction and nucleolus-organizer region of one of the parents is suppressed and unable to be observed. In effect, the NORs of one species are dominant over those of another when the two species are combined in a hybrid. It has been demonstrated that the rDNA from the parent with the suppressed NOR is not transcribed. At the molecular level, the spacer region is an essential component in determining the efficient loading of RNA polymerase I onto the promoter for transcription initiation. It is subsequently possible that in an F_1 hybrid between species, one of the nucleolus-organizer regions has rDNA sequences that are consistently less efficient than the other NOR in competing for available RNA polymerase I or other transcription factors. An alternative suggestion is that the suppression may be a reflection of competition between rDNA regions for a physical site within the nucleolus where rRNA transcription can occur. Failure to compete effectively at such a stage may commit a particular array of rRNA units to inactivity regardless of the level of RNA polymerase I or any other transcription factor. Research continues in an attempt to elucidate this interesting phenomenon.

20.2.6 Ribosomal-RNA Gene Loci and Meiotic-Chromosome Pairing Sites

The analysis of meiotic-chromosome pairing between the heterochromatic regions of the X and Y chromosomes of *Drosophila* indicates that the rDNA regions are the sites of the initial pairing interaction. Information to suggest this possibility was initially obtained by analyzing the behavior of modified X and Y chromosomes, which had been carefully characterized for the presence of different types of rDNA sequences located in the heterochromatic regions of these two chromosomes. Striking confirmation was achieved when a single rDNA unit was introduced into an X chromosome from which rDNA genes had previously been deleted and

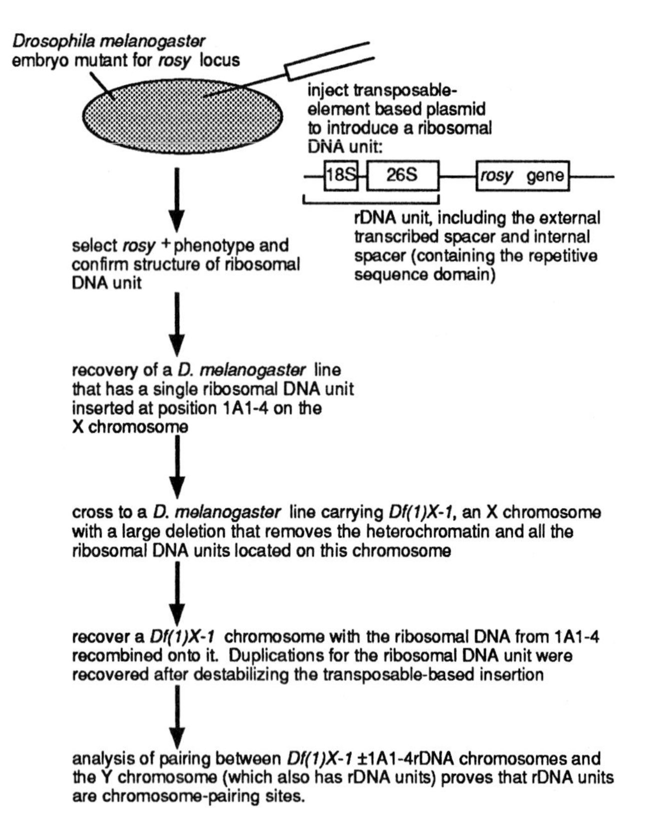

Fig. 20.22. The methodology involved in proving that in *Drosophila melanogaster,* ribosomal RNA functions as a meiotic pairing site between the X and Y chromosomes. The *rosy* gene provides a marker to indicate the successful introduction and transformation of the rDNA unit (McKee and Karpen, 1990; McKee et al., 1992). The introduction of the intervening-spacer unit, with its repetitive-sequence domains, was found to be crucial to the X–Y pairing process.

which had shown no pairing with the Y chromosome. X–Y pairing was recovered, albeit at a reduced efficiency (Fig. 20.22).

20.2.7 Ribosomal-RNA Gene Loci and Fragile Sites

In *Neurospora*, the usual NOR contains from 100 to 200 rDNA units and occupies a terminal site on chromosome V. The analysis of translocations involving the ends of chromosome V has shown that chromosome breakage within the NOR is common; this observation has led to the supposition that in some cases, the NOR can act as a fragile site within a chromosome. Fragile sites not related to the NOR are discussed in Chapter 7, Section 7.4.5. It is also of interest that when NORs break, they are immediately capped with the *Neurospora* telomeric sequences (TTAGGG) in order to maintain stability. Furthermore, in organisms such as *Giardia lamblia* (*see* Chapter 1, Fig. 1.3 for a description of this organism), the rDNA units are located directly adjacent to the telomeres of two chromosomes; in many other organisms, the NORs and rRNA genes are often subtelomeric. The location of rDNA units at or near telomeres has been speculated to reflect breakage of chromosomes, during the course of evolution of the organism, within the rDNA array and capping of the broken ends with telomeric sequences.

20.2.8 Amplification of rDNA Gene Sequences

In *D. melanogaster*, the phenomenon of rDNA amplification was discovered by careful measurement of the rDNA content of *bobbed* males and their progenies. This finding resulted from a breeding experiment involving several generations of flies in which, in succeeding generations, the chromosomes from rDNA-deficient males gradually increased their numbers of rDNA units, so that the original *bobbed* phenotype was eventually corrected by an increase in the number of rRNA units. Although cytogenetic evidence suggests that unequal sister-chromatid exchange could be involved in this magnification phenomenon (Fig 20.23), how a deficiency of rRNA genes can trigger the unequal sister-chromatid exchange leading to magnification is unclear.

Unequal sister-chromatid and intrachromatid exchanges have occurred between tandem arrays of rDNA genes in yeast and *Neurospora,* and they almost certainly contribute to the observed quantitative variations in the numbers of these genes in the nucleolus organizers of other organisms. In allelic variants of the *Nor* loci of wheat and barley, for example, the number of rDNA units at a single locus can vary from a few hundred to several thousand. New alleles that show large quantitative variations in number of rDNA units can, in principle, be formed by unequal crossing over between arrays of same-length rRNA variants or by crossing over between arrays of differing-length variants.

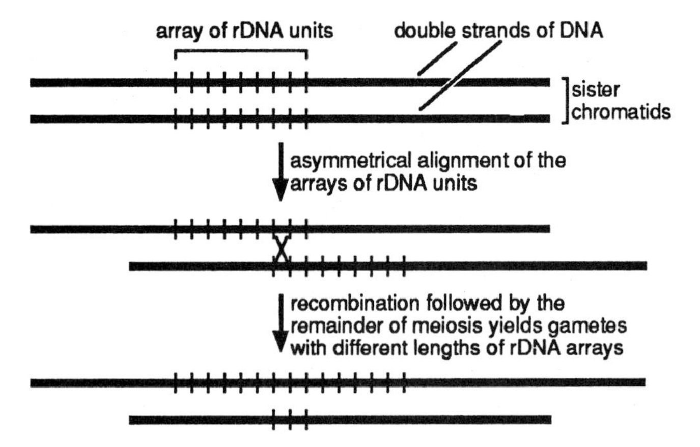

Fig. 20.23. Unequal crossing over as a mechanism for the amplification of rDNA units within a *Nor* locus. In *Drosophila melanogaster,* the phenomenon of amplification occurs when individuals are deficient in rDNA genes due to the presence of *bobbed* mutations (*see* Fig. 20.14). The deficiency in rRNA is thought to physically stress the organism, resulting in increased levels of nicks in replicating genomic DNA and an increase in the frequency of sister-chromatid exchange (SCE). With an increased frequency of SCEs, there is an increased probability of unequal alignment between the *Nor* loci, leading to greater numbers of rDNA units at one of the loci resulting from the exchange (Endow and Atwood, 1988). The reciprocal product of the unequal exchange, with reduced numbers of rDNA units, will tend to be lost because cells with these chromosomes will not be viable. In fungi, plants, and animals, fluctuations in the number of rDNA units at a given *Nor* locus occur readily, mainly as a result of meiotic sister-chromatid exchange (Petes, 1980) and premeiotic, intrachromatid exchange (Butler and Metzenberg, 1990).

20.3 THE CENTROMERE

By the start of the twentieth century, observations on cell division in cells with large nuclei suggested that a fiber network became attached to the constricted regions of chromosomes and pulled the chromosomes to their respective destinations. The primary constriction, to which the fibers attached themselves, became known as the centromere, a significant region of the chromosome with rather poorly defined boundaries.

The centromere is the site at which electron-microscopic studies have differentiated the kinetochore. In this chapter, the kinetochore is treated as a highly specialized structure, composed of an array of proteins and other macromolecules that are intimately associated with DNA sequences within the centromere (*see also* Chapter 4, Section 4.1.2, and Fig. 4.10). Centromeric proteins have been assayed using either autoantibodies isolated from the serum of humans suffering from rheumatic diseases or monoclonal antibodies produced against chromosome-scaffold proteins. The antibodies provide probes for specific centromeric proteins (CENPs) located very close to the kinetochore. These proteins include a 17 kDa histone-like protein, CENP-A, which has proven to be centromere and probably kinetochore-specific; an 80-kDa protein, CENP-B, found within the central domain and spatially associated with tubulin molecules; a 140-kDa protein, CENP-C, of uncharacterized function; a 50-kDa protein, CENP-D, which is specific to the kinetochore itself; and another 140-kDa protein, INCENP or inner-centromere protein (Fig. 20.24). When the respective antibodies are used to stain mitotic chromosomes from the rat kangaroo and Chinese hamster, the CENP-A, B, C, and D proteins are closely associated with the kinetochore plates and adjacent chromatin. The location of INCENP suggests that it may be a chromosome-scaffold protein that is important in holding the replicated centromeric DNA together until mitotic anaphase is initiated.

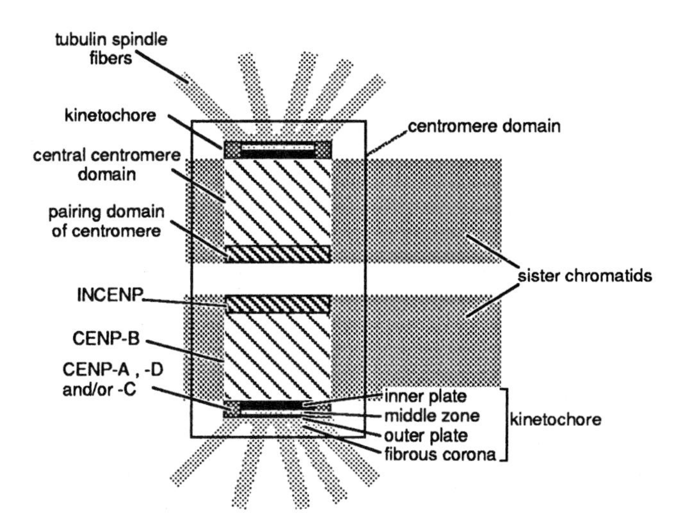

Fig. 20.24. A diagrammatic representation of the mammalian centromere and kinetochore, showing the three major domains (in the upper chromatid) and the diagnostic antigens used to detect proteins within the different domains (in the lower chromatid). The primary source of antibodies with which to assay these protein antigens comes from people suffering from autoimmune diseases. Although the antibody-staining data is consistent in the rat kangaroo and Chinese hamster, caution must be exercised in generalizing the structure to other organisms, because the antibodies show a less-defined activity when used to stain human chromosomes (Earnshaw et al., 1989; Cai and Davis, 1990; Palmer et al., 1990; Pluta et al., 1990; Vallee, 1990; Zinkowski et al., 1991).

20.3.1 The Kinetochore

The kinetochore (*see* Fig. 20.24; *see also* Chapter 4, Section 4.1.2, and Fig. 4.10) is a platelike structure to which the microtubules of the spindle become attached. A central role for the kinetochore in both mitosis and meiosis derives from its interaction with the tubulin-protein subunits, which constitute the spindle microtubules. A variety of studies, including the micromanipulation of chromosomes, have suggested that the kinetochore is the site for the assembly and disassembly of microtubules. It is also the site of attachment for preexisting microtubules, which become stabilized by their attachment. Currently, a model suggesting that the kinetochore is the site of an ATP-driven "motor" actively assembling and disassembling microtubules is being intensively tested. Proteins such as dynein have been localized near the kinetochore and shown to be active in such processes. The model could provide an explanation for the movement of chromosomes during anaphase, because active disassembly of the microtubule at the kinetochore would shorten the tubule and move the kinetochore, and thus the chromosome, toward the pole of the cell. It is also possible that the kinetochore "motor" may simply move along the microtubules in a poleward direction. However, this movement would require an indicator for the direction of movement, a complication that is not required in the disassembly model.

20.3.2 DNA Sequences at or near the Centromeres of Mammalian Chromosomes

Mention has been made of the existence of tandem arrays of repetitive sequences in eukaryotic genomic DNA (*see* Fig 20.9). An intriguing observation is that selected sets of DNA families in this category are located near the centromeres. Research has shown that human centromeric heterochromatin also contains sets of simple, repetitive-sequence DNA. These sequence families, satellites I to IV, comprise 2–5% of the genome and contain tandem arrays of short DNA units ($<$ 100 bp), consisting of variant forms of the sequence GGAAT. These sequences are located at, or very close to, the centromeres of chromosomes 1, 9, and 16, as well as the Y chromosome.

α-Satellite DNA is another sequence family found at the centromeres of humans and mice and is of particular interest because it appears to interact with one of the CENP proteins. This family consists of tandem arrays of a 171-bp unit and can make up as much as 5% of the total DNA. Although the amount of α-satellite DNA varies widely among different chromosomes, as is typical for analogous sequences in plants and other animals, not all of the alphoid family of centromeric sequences are variable. The sequence p82H, for example, a human-derived centromeric sequence, labels the centromeres of all human chromosomes to a similar extent. It also hybridizes to all of the chromosomal centromeres of gorilla, chimpanzee, and orangutan, all primates that are closely related evolutionally to humans. p82H hybridizes to a 170/340-bp multimeric sequence present at the centromeres of these chromosomes.

A subset of α-satellite sequences has been found to contain the nucleotide sequence 5′CTTCGTTGGAAACGGGA3′ in a CENP-B box, which is the binding site for the centromeric protein CENP-

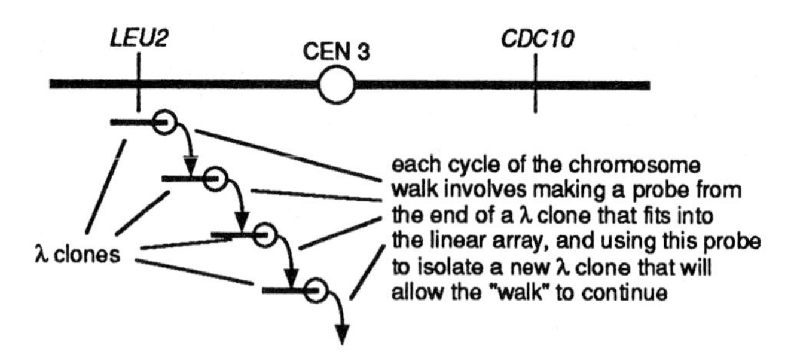

Fig. 20.25. The use of chromosome walking to isolate centromeric DNA (CEN 3) sequences from chromosome 3 of *Saccharomyces cerevisiae,* between the genes *LEU2* and *CDC10.* The chromosome-walk technology relies on the use of a relatively short, single-copy DNA sequence, located at the boundary of one λ clone, as a probe to screen and detect newλ clones with the identical sequence. In the new λ clones, the sequence is located internally rather than at the boundary (by chance), and a new sequence from the new boundary allows the next-neighboring sequence to be isolated. Note that the actual direction of the chromosome walk cannot be controlled, and its determination relies on additional evidence from gene-mapping studies. In the isolation of the CEN3 DNA sequence, cloned sequences that were thought to carry the centromere sequence were tested for their ability to direct the correct segregation of the plasmid in which they were cloned during mitosis (Blackburn and Szostak, 1984).

B, and suggests the potential for formation of a DNA–protein complex in vivo. This is consistent with the finding that both the α-satellite DNA and CENP-B proteins are distributed, in situ, in the same regions of the chromatin complex that are associated with the kinetochore. Unfortunately, the significance of CENP-B for the interactions that lead to the attachment of microtubules and the subsequent movement of chromosomes is unknown.

20.3.3 DNA Sequences from the Centromeres of Yeast Chromosomes

The first centromere to be physically isolated came from an experiment where chromosome-walking was used to clone the DNA spanning the distance between the *LEU2* and *CDC10* genes of *Saccharomyces cerevisiae* (Fig. 20.25). Because these two genes are located on opposite sides of the centromere of chromosome 3, the intervening 25-kb region must contain the centromere. The identification of the precise centromere–DNA sequences was

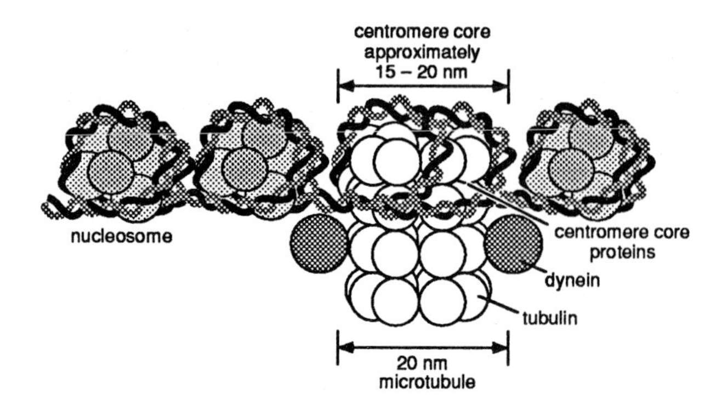

Fig. 20.27. A diagrammatic representation of the attachment of microtubules to centromeric DNA (based on Clarke, 1990; Vallee, 1990; *see* also Chapter 4. Section 4.1.2).

achieved by subcloning DNA segments into plasmids and observing their mitotic stability in actively dividing yeast cells. Plasmids carrying centromere sequences are normally transmitted, along with other chromosomes, in mitotic divisions, whereas plasmids lacking these sequences are rapidly lost due to nondisjunction. Plasmids containing two centromeres are generally unstable and deletion of one centromere often occurs.

After the centromeric DNA sequence of chromosome 3 had been identified, it was found to be characterized by an 88-bp region, of which 82 of the base pairs (93%) were A or T. Other centromeres have since been cloned, using similar chromosome-walking experiments, and some of these sequences, as well as an overall-consensus sequence derived from 10 of the 16 possible yeast-chromosome centromeres, are shown in Fig. 20.26. Within a total length of 120 bp, the A/T-rich core region of the centromeres, centromere DNA element II (CDE II), is consistently flanked by an eight-nucleotide consensus sequence, CDE I, and a 25-bp, bilateral consensus sequence, CDE III. Mutation studies have shown that the highly conserved sequence CDE III is essential for mitotic-centromere function, in contrast to CDE I and CDE II where mutations impair, but do not abolish, centromere function. The 120-bp region is contained within a relatively nuclease-resistant, chromatin-core particle, which may contain CENP-B-type proteins that interact with tubulin protein (Fig. 20.27), as well as other proteins of the centromere region (*see* Fig. 20.24).

In comparison, the centromeres of the fission yeast *Schizosaccharomyces pombe* are strikingly different (Fig. 20.28). This organ-

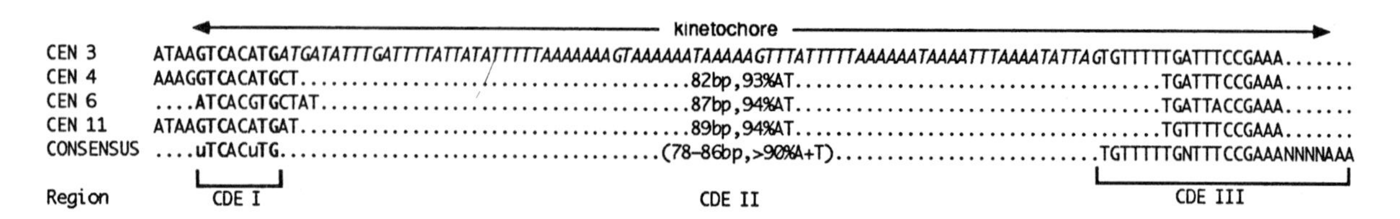

Fig. 20.26. Nucleotide sequences of the kinetochore regions of the centromeres of chromosomes 3, 4, 6, and 11 of yeast (*S. cerevisiae).* The consensus sequence shows essential, conserved nucleotides. Region I (CDE I) consists of 8 essential nucleotides; region II (CDE II) is a very AT-rich region of 78–86 bp with no conserved sequences: the highly conserved sequences of region III (CDE III) show two short arrays of nucleotides with bilateral symmetry (Blackburn and Szostak, 1984; Clarke, 1990). The region III sequences are present in human centromere-associated sequences (Grady et al., 1992), whereas the AT-rich nature of region II is also observed in centromere-associated sequences from *Arabidopsis thaliana* (Richards et al., 1991).

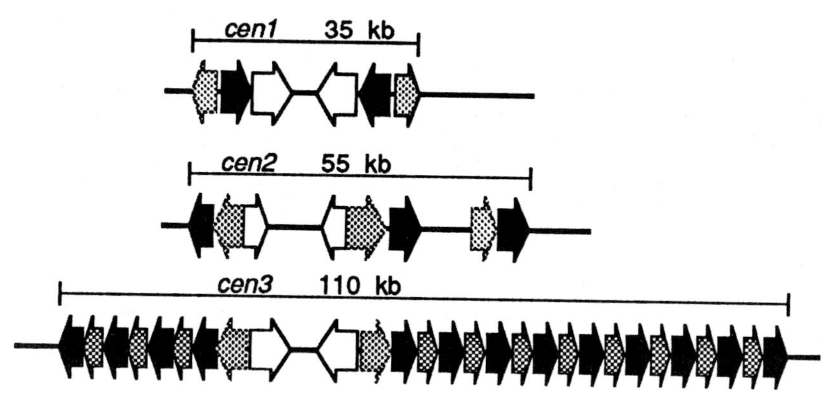

Fig. 20.28. A diagrammatic representation of the DNA-sequence arrays that surround the three centromeres (*cen1*, *cen2*, and *cen3*) of the fission yeast, *Schizosaccharomyces pombe*. The arrows represent DNA sequences, and related DNA sequences are indicated by the same shade of color. A number of tRNA genes are also located within the white arrows (Clarke, 1990; Murakami et al., 1991).

ism has only three chromosomes, and the centromeres are much larger than those of *S. cerevisiae*. The centromere regions contain repetitive DNA sequences, as can be illustrated for *cen3*, which is approximately 110 kb in length and is largely composed of 15 6.6-kb repeat units, each of which is divided into subunits. Figure 20.28 shows 13 pairs of sequences (light- and dark-colored arrows) plus 2 sequences (white arrows) flanking a point of approximate symmetry. Functional assays for centromeric activity, in which the centromeres are subcloned into plasmids, indicate that the entire region is required for centromere function.

Organisms with diffuse centromeres were described in Chapter 4, Section 4.1.1. Although such diffuse centromeres have not been analyzed in sufficient detail to provide DNA-sequence information, it has been speculated that a dispersed-repetitive family of DNA sequences may have evolved that contains the AT-rich sequences characteristic of yeast centromeres. The dispersion of these sequences throughout the chromosomes could result in multiple centromeres.

20.4 THE TELOMERES

Early cytogenetic experiments found that when the ends of chromosomes were broken off, the broken ends were either unstable, resulting in the degradation of the remaining chromatin by nucleases, or they joined with the broken ends of other chromosomes. These results suggested that the ends or telomeres of chromosomes were special entities that were essential for maintaining the stability of eukaryotic chromosomes. The later discovery of the chemical structure of DNA and its mode of replication implied a need for a special structure to be present at the ends of the linear DNA chromosomal molecules to prevent their progressive reduction in length at each replication cycle.

20.4.1 The Requirement of Primers for DNA Synthesis Creates Problems in the Replication of Linear Molecules

Even though prokaryotic DNA is usually circular, whereas eukaryotic DNA is most often linear, the basic molecular features of DNA replication in both are much the same. In particular, all known template-dependent DNA polymerases require a short, double-stranded template or primer for the initiation of 5′–3′ strand synthesis (*see* Chapter 18, section 18.2.1). The initiation of DNA replication from a DNA single-stranded template usually involves a short, single-stranded RNA molecule hybridizing to the origin of replication. In the linear eukaryotic chromosomes, there are many origins of replication, and all the internal primer molecules are replaced by the DNA-repair activity associated with the replicative process. Because the DNA-repair activities also synthesize DNA in the 5′–3′ direction, it is evident that the terminal end of the 3′-ended parental DNA strand will remain unreplicated. This problem does not arise with circular DNA molecules. There are several solutions for linear molecules as follows:

1. The ends of linear molecules can be temporarily joined together so that the replication process from an adjoining molecule can continue into the telomeric region. This process occurs in the replication of the linear DNA molecule of bacteriophage T7.
2. A specific protein can be bound covalently to the end of a linear molecule and can function as the primer for DNA-synthesis initiation. This is observed with mammalian adenovirus, where the 5′-terminal phosphate is covalently linked to a serine-amino acid of a 55-kDa Ad-binding protein. This also occurs in the replication of bacteriophage ϕ29 and in polio-virus RNA.
3. A noncoding DNA can be added post-replicatively, in a reaction that does not use a DNA template. In this way, a primer molecule can initiate DNA synthesis from the added, noncoding DNA to ensure faithful replication of the main DNA molecule. Evidence for this type of modification at the end of a DNA molecule was first obtained from a study of DNA replication in protozoan organisms.

20.4.2 The DNA of Telomeres in the Ciliated Protozoan *Tetrahymena thermophila*

Tetrahymena is a unicellular protozoan with two different types of nuclei, one large macronucleus and a smaller micronucleus,

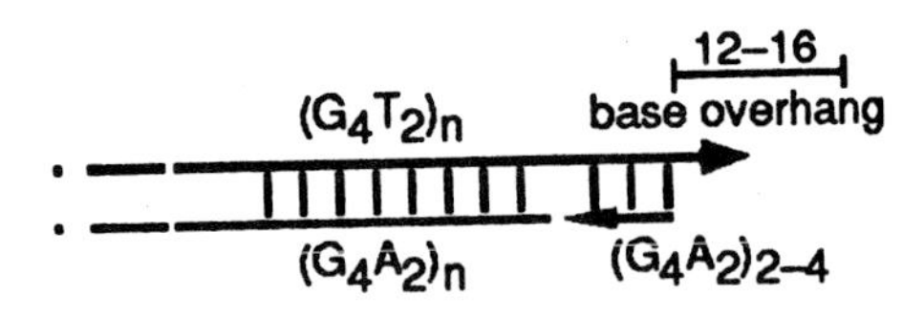

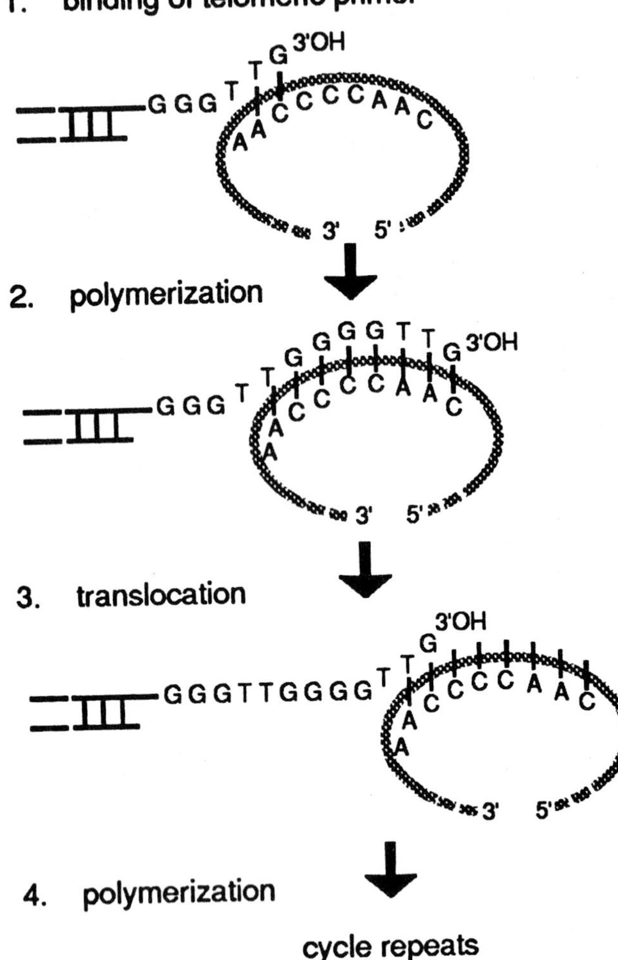

Fig. 20.29. Structure of the telomeric DNA of *Tetrahymena thermophila* (top) and the mechanism of synthesis of the G-rich strand of the telomere by the enzyme telomerase (1–4). Synthesis of the telomere involves (1) pairing between the complementary nucleotides of the telomere and telomerase, (2) elongation of the telomere by simple polymerization, and (3) translocation of the newly elongated 3′-end (Blackburn, 1991b). The macronucleus of *T. thermophila* (*see* Chapter 24, Section 24.9 for a description of the organism) contains the DNA of the germline nucleus in a fragmented and amplified form, with each "minichromosome" capped by tandem arrays of short repetitive sequences, which also are typical of eukaryotic telomeric sequences (*see* Table 20.2). These sequences are a crucial structural feature of genomic DNA to ensure faithful replication of the entire genome. Telomerase, which is responsible for producing these telomeric sequences, is a dimer (Collins et al., 1995), with a mode of action in *Tetrahymena* that is probably typical of other eukaryotes (Blackburn and Szostak, 1984; Cheng et al., 1989; Biessmann et al., 1990; Brown et al., 1990; Blackburn, 1991a; Moysis, 1991; Romero and Blackburn, 1991). The active site of telomerase contains the 5′-CAACCCCAA primer and this is used by the enzyme to carry out steps 1–4 as indicated in the diagram.

```
                          5'
                          |
                          CCCCAAAACCCCAAAACCCCNNNN------
GGGGTTTTGGGGTTTTGGGGTTTTGGGGTTTTGGGGNNNN------
|
3'
```

Fig. 20.30. The DNA sequence of the chromosomal telomeres of the ciliated protozoan *Oxytricha nova.* The ends of the DNA strands, including the 16 exposed nucleotides of the 3′-end, are tightly bound to a protein heterodimer, the telomere-binding protein (Raghuraman et al., 1989). The secondary structure adopted by single-stranded, G.T-rich DNA may be a distinguishing feature recognized by the protein and provides a model for the binding of telomeres to the nuclear matrix of eukaryotes (Raghuraman and Cech, 1989; Moens and Pearlman, 1990; Blackburn, 1991a; de Lange, 1992).

which has the diploid content of DNA and is, in effect, the germline of the organism. The macronucleus is the product of a micronuclear division, and during vegetative development of the cell, its DNA is cleaved at specific sites and the linear molecules are amplified to varying degrees. Most important, the linear molecules of the macronucleus have new ends added to them, to heal the broken ends. Studies on the telomerase enzyme responsible for the healing of broken ends have provided a model for the generation of chromosomal telomeres in eukaryotes (Fig. 20.29). The repetitive DNA sequences, added postreplicatively to the ends of the chromosomes by the telomerase activity, leave a 3′-tail, as shown in Fig. 20.30 for the protozoan *Oxyticha nova.*

20.4.3 Telomeres in Other Organisms

Several observations indicate that the information gained from the study of the telomeres of protozoans may be of broader significance. The nature of DNA sequences that have been located at, or very near, the telomeres of a range of organisms are shown in Table 20.2. All of these sequences have the same general formula and are basically comprised of tandem repeats of the consensus sequence (C or G)$_{1-8}$(T or A)$_{1-4}$. Biochemical and cytogenetic studies are used to argue for a terminal location of nucleotide sequences. Biochemical proof utilizes the degradation of terminal DNA sequences by the exonuclease enzyme Bal 31, which results in the preferential loss of DNA fragments that are located at the telomeres with increased exposure time to the enzyme. Cytogenetic proof includes in situ localization of (C or G)$_{1-8}$(T or A)$_{1-4}$-type sequences to the telomeres (Fig. 20.31).

The telomeric sequences of human chromosomes were identified by their ability to provide stable ends for synthetic yeast chromosomes. The widespread occurrence of telomeric-nucleotide sequences, similar to those of protozoans, suggests that telomerase activity of the type shown in Fig. 20.30 may be widespread. Direct evidence for the presence of telomerase activity in yeast has been established; evidence for humans and *Xenopus* still needs to be obtained. Even so, human-telomeric sequences were cloned by utilizing the fact that they functioned as telomeres in yeast-synthetic chromosomes, as did the *Tetrahymena* sequences indicated in Table 20.2. It is therefore interesting that when *Tetrahymena* telomeres, attached to synthetic chromosomes, were re-examined after propagation in yeast, they had been converted to typical yeast-telomeric sequences. This observation is consistent with the ability of a telomerase enzyme of yeast to add its own characteristic telomeres to suitable DNA templates.

Other more complex repetitive-sequence families are also located near the telomeres and, therefore, in the terminal regions of chromosomes. In the family Triticeae, for example, a DNA sequence with an approximate 350-bp repeat unit is widely distributed in the terminal regions of chromosomes. Other plants such as onion, maize, and tomato have been shown to have repetitive-DNA-sequence families in these terminal positions. In *Saccharomyces,* a 6.7-kb sequence is present in one to four copies in the terminal regions of most chromosomes, and this sequence has been shown to undergo crossing over during meiosis—a natural cause for the difference in the number of copies of the sequence at a given location. The nematode *Ascaris* has at least two types of tandemly repeated-sequence families in the terminal regions of germline chromosomes.

In *Drosophila,* so-called He–T sequences occur at the telomeres. Interestingly, the He–T's are a complex family of repetitive sequences, which appear at the newly acquired telomeres of ring chromosomes when these have spontaneously reconverted to a linear form. The He–T family of sequences contains retrotransposons (*see* Chapter 7, Fig. 7.19) called HeT-A elements, approximately 6 kb in length, and represents another means of adding telomeric sequences to the ends of chromosomes. In this case, transposition of a DNA molecule to the susceptible end of a chromosome replaces the addition of noncoding DNA by telomerase activity.

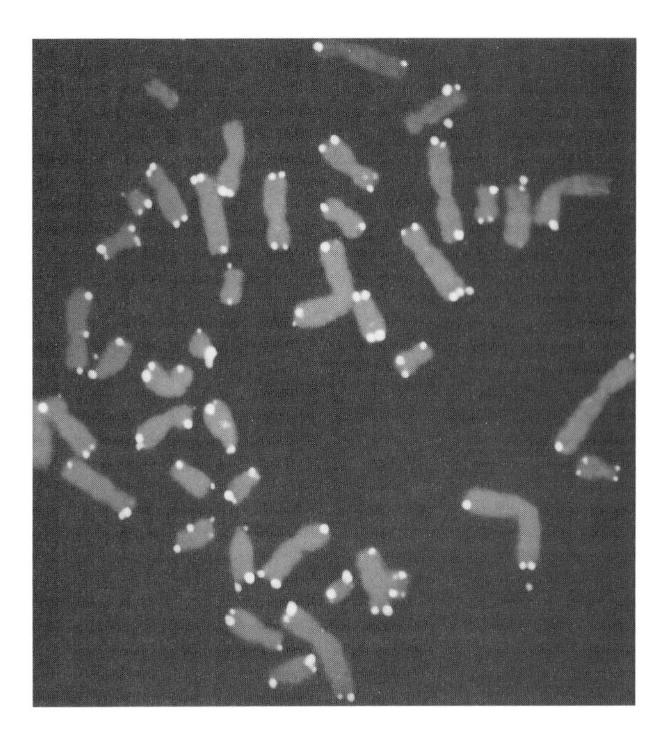

Fig. 20.31. The *in situ* localization of immunofluorescent-labeled human telomeric-DNA sequences on human chromosomes (From Fantes and Cook, 1991); the chromosome spread shown is not a complete karyotype. Hybridization studies of this type—as well as molecular analyses of DNA—have demonstrated that telomeric-DNA sequences can also have interstitial locations within the chromosomes, and it has been suggested that in some organisms, these are, or act as, sites for the integration of foreign DNA (Katinka and Bourgain, 1992).

20.4.5 The Attachment of Telomeres to Nuclear Membranes and Proteins

Based on a number of lines of investigation, it is probable that the ends of chromosomes are usually attached to either the nuclear envelope or the inner nuclear membrane. Classical cytological experiments have demonstrated that the ends of prophase chromosomes commonly occupy a peripheral position at one end of the nucleus in a Rabl configuration, named after the scientist who first observed the phenomenon. Three-dimensional reconstructions of the positions of polytene chromosomes in the salivary-gland nuclei of *Drosophila* have shown that the ends of these chromosomes are positioned at the nuclear membrane. Similarly, meiotic chromosome spreads of a wide range of organisms often show attachment of the chromosome ends to nuclear membranes. However, the nuclear proteins that interact with the telomeric DNA sequences have not been analyzed in any detail. Presumably, one of these proteins is the telomerase enzyme and its interactions are relatively well characterized. Another protein that specifically binds to telomeric-chromosome sequences has been identified in the ciliated protozoan *Oxytricha*. It has been suggested that competition between a telomere-binding protein and the telomerase that binds to the telomeric sequences controls the length of the telomeric sequences added by telomerase. Furthermore, in several organisms including humans, telomeric TTAGGG repeated sequences are attached to the nuclear matrix, and this combination could be important in positioning the complete-chromatin complex within the nucleus.

Table 20.2. Telomeric DNA Sequences Identified from the Ends of Chromosomes; These Repeats from the G-Rich Strand Are Based on a $G_{1-8}(T/A)_{1-4}$ Plan

Organisms	Telomeric DNA Sequences
Ciliated protozoans	
Tetrahymena, Glaucoma	GGGGTT
Paramecium	GGG(G/T)TT
Oxytricha, Stylonychia, Euplotes	GGGGTTTT
Kinetoplastid protozoans	
Trypanosoma, Leptomonas, Leishmannia, Crithidia	AGGGTT
Coccidial protozoans	
Plasmodium	AGGGTTT
Slime molds	
Physarum, Didymium	AGGGTT
Dictyostelium	$G_{1-8}A$
Yeasts	
Saccharomyces	$G_{1-3}T$
Schizosaccharomyces	$G_{2-5}TTAC$
Fungi	
Neurospora	AGGGTT
Algae	
Chlamydomonas	TTTTAGGG
Plant	
Arabidopsis	AGGGTTT
Triticum	AGGGTTT
Mammal	
Homo	AGGGTT

Source: Adapted from Blackburn (1991a).

BIBLIOGRAPHY

General

Appels, R. 1982. The molecular cytology of wheat–rye hybrids. Int. Rev. Cytol. 80: 93–132.

Blackburn, E.H., Szostak, J.W. 1984. The molecular structure of centromeres and telomeres. Annu. Rev. Biochem. 53: 163–194.

Lindsley, D.L., Grell, E.H. 1968. Genetic variations of *Drosophila melanogaster*. Carnegie Institute of Washington, DC.

Miller, O.L., Beatty, B.R. 1969. Visualization of nucleolar genes. Science 164: 955–957.

Section 20.1

Ajioka, J.W., Smoller, D.A., Jones, R.W., Carulli, J.P., Vellek, A.E.C., Garza, D., Link, A.J., Duncan, I.W., Hartl, D.L. 1991. *Drosophila* genome project: one-hit coverage in yeast artificial chromosomes. Chromosoma 100: 495–509.

Appels, R., Dennis, E.S., Smyth, D.R., Peacock, W.J. 1981. Two repeated DNA sequences from the heterochromatic regions of rye (*Secale cereale*) chromosomes. Chromosoma 84: 265–277.

Appels, R., Driscoll, C.J., Peacock, W.J. 1978. Heterochromatin and highly repeated DNA sequences in rye *(Secale cereale)*. Chromosoma 70: 67–89.

Appels, R., Reddy, P., McIntyre, C.L., Moran, L..B., Frankel, O.H., Clarke, B.C. 1989. The molecular cytogenetic analysis of grasses and its application to studying relationships among species of the Triticeae. Genome 31: 122–133.

Beerman, S. 1977. The diminution of heterochromatic chromosomal segments in *Cyclops* (Crustacea, Copepoda). Chromosoma 60: 297–344.

Bonaccorsi, S., Gatti, M., Pisano, C., Lohe, A. 1990. Transcription of a satellite DNA on two Y chromosome loops of *Drosophila melanogaster*. Chromosoma 99: 260–266.

Bonaccorsi, S., Pisano, C., Gatti, M., Puoti, F. 1988. Y chromosome loops in *Drosophila melanogaster*. Genetics 120: 1015–1034.

Brittnacher, J.G., Ganetsky, B. 1989. On the components of segregation distortion in *Drosophila melanogaster*. IV. Construction and analysis of free duplications for the *Responder* locus. Genetics 121: 739–750.

Cai, H., Kiefel, P., Yee, J., Duncan, I. 1994. A yeast artificial chromosome clone map of the *Drosophila* genome. Genetics 136: 1385–1399.

Dernburg, A.F., Sedat, J.W., Hawley, R.S. 1996. Direct evidence of the role for heterochromatin in meiotic chromosome segregation. Cell 86: 135–146.

Dimitri, P. 1991. Cytogenetic analysis of the second chromosome heterochromatin of *Drosophila melanogaster*. Genetics 127: 553–564.

Eissenberg, J.C., Morris, G.D., Reuter, G, Hartnett, T. 1992. The heterochromatin-associated protein HP-1 is an essential protein in *Drosophila* with dosage-dependent effects on position-effect variegation. Genetics 131: 345–352.

Elgin, S.C.R., Amero, S.A., Eissenberg, J.C., Fleischmann, G., Gilmour, D.S., James, T.C. 1988. Distribution patterns of non-histone chromosomal proteins on polytene chromosomes: functional correlations. In: Chromosome Structure and Function: The Impact of New Concepts (Gustafson, J.P., Appels, R., eds.). 18th Stadler Genetics Symp., Plenum Press, New York. pp. 145–156.

Fortini, M.E., Artavanis-Tsakonas, S. 1993. *Notch:* neurogenesis is only part of the picture. Cell 75: 1245– 1247.

Gall, J.G., Cohen, E.H., Polan, M.L. 1971. Repetitive DNA sequences in *Drosophila*. Chromosoma 33: 319–344.

Hilliker, A.J., Appels, R., Schalet, A. 1980. The genetic analysis of *D. melanogaster* heterochromatin. Cell 21: 607–619.

Kafatos, F.C., Louis, C., Savakis, C., Glover, D.M., Ashburner, M., Link, A.J., Siden-Kiamos, I., Saunders, R.D.C. 1991. Integrated maps of the *Drosophila* genome: progress and prospects. Trends Genet. 7: 155–161.

Lefevre, G. 1976. A photographic representation and interpretation of the polytene chromosomes of *Drosophila melanogaster* salivary glands. In: The Genetics and Biology of *Drosophila* (Ashburner, M., Novitsky, E., eds.). Academic Press, London. pp. 31–66.

Lindsley, D.L., Sandler, L., Baker, B.S., Carpenter, A.T.C., Denell, R.E., Hall, J.C., Jacobs, T.A., Miklos, G.L.G., Davis, B.K., Gethmann, R.C., Hardy, R.W., Hessler, A., Miller, S.M., Nozawa, H., Parry, D.M., Gould-Somero, M., 1972. Segmental aneuploidy and the genetic gross structure of the *Drosophila* genome. Genetics 71: 157–184.

Lyttle, T.W. 1989. The effect of novel chromosome position and variable dose on the genetic behavior of the responder (Rsp) element of the Segregation Distorter (SD) system of *Drosophila melanogaster*. Genetics 121: 751–763.

Madueño, E., Rimmington, G., Saunders, R.D.C., Savakis, C., Siden-Kiamos, I., Skavdis, G., Spanos, L., Trenar, J., Adam, P., Ashburner, M., Benos, P., Bolshakov, V.N., Coulson, D., Glover, D.M., Herrmann, S., Kafatos, F.C., Louis, C., Majerus, T., Modolell, J. 1995. A physical map of the X chromosome of *Drosophila melanogaster*: cosmid contigs and sequence tagged sites. Genetics 139: 1631–1647.

May, C.E., Appels, R. 1980. Rye chromosome translocations in hexaploid wheat: a re-evaluation of the loss of heterochromatin from rye chromosomes. Theoret. Appl. Genet. 56: 17–23.

Muller, F., Walker, P., Aeby, P., Neuhaus, H., Felder, H., Back, E., Tobler, H. 1982. Nucleotide sequence of satellite DNA contained in the eliminated genome of *Ascaris lumbricoides*. Nucleic Acids Res. 10: 7493–7510.

Oliver, S.G., et al. 1992. The complete DNA sequence of yeast chromosome III. Nature 357: 38–46.

Palopoli, M.F., Doshi, P., Wu, C.I. 1994. Characterization of two *Segregation distorter* revertants: evidence that the tandem duplication is necessary for Sd activity in *Drosophila melanogaster*. Genetics 136: 209–215.

Pardue, M.L., Gall, J.G. 1970. Chromosomal localization of mouse satellite DNA. Science 168: 1356–1360.

Pardue, M.L., Hennig, W. 1990. Heterochromatin: junk or collector's item? Chromosoma 100: 3–7.

Pimpinelli, S., Dimitri, P. 1989. Cytogenetic analysis of segregation distortion in *Drosophila melanogaster*: The cytological organisation of the Responder (Rsp) locus. Genetics 121: 765–772.

Quiros, C.F. 1976. Meiotic behavior of extra heterochromatin in the tomato: effects on several vital processes. Can. J. Genet. Cytol. 18: 325–337.

Reuter, G., Giarre, M., Farah, J., Gausz, J., Spierer, A., Spierer, P. 1990. Dependence of position-effect variegation in *Drosophila* on dose of a gene encoding an unusual zinc-finger protein. Nature 344: 219–223.

RYKOWSKI, M.C., PARMELEE, S.J., AGARD, D.A., SEDAT, J.W. 1988. Precise determination of the molecular limits of a polytene chromosome band: regulatory sequences for the *Notch* gene are in the interband. Cell 54: 461–472.

SINGH, P.B., MILLER, J.R., PEARCE, J., KOTHARY, R., BURTON, R., PARO, R., THARAPPEL, J.C., GAUNT, S.J. 1991. A sequence motif found in a *Drosphila* heterochromatin protein is conserved in animals and plants. Nucleic Acids Res. 19: 789–794.

SUEOKA, N. 1961. Variation and heterogeneity of base composition of deoxyribonucleic acids: a compilation of old and new data. J. Molec. Biol. 3: 31–40.

THARAPPEL, J.C., EISSENBERG, J.C., CRAIG, C., DIETRICH, V., HOBSON, A., ELGIN, S.C.R. 1989. Distribution patterns of HP1, a heterochromatin-associated nonhistone chromosomal protein of *Drosophila*. Eur. J. Cell Biol. 50: 170–180.

WILLIAMSON, J.H. 1976. The genetics of the Y chromosome. In: The Genetics and Biology of *Drosophila*, Vol. 1b. (Ashburner M., Novitsky, E., eds.). Academic Press, New York. pp. 667–699.

WU, C.I., LYTTLE, T.W., WU, M.L., LIN, G.F. 1988. Association between a satellite DNA sequence and the *Responder* of *Segregation Distorter* in *D. melanogaster*. Cell 54: 179–189.

YAMAMOTO, M-T., MITCHELSON, A., TUDOR, M., O'HARE, K., DAVIES, J.A., MIKLOS, G.L.G. 1990. Molecular and cytogenetic analysis of the heterochromatin–euchromatin junction region of the *Drosophila melanogaster* X chromosome using cloned DNA sequences. Genetics 125: 821–832.

Section 20.2

ADAM, R.D., NASH, T.E., WELLEMS, T.E. 1991. Telomeric location of *Giardia* rDNA genes. Molec. Cell Biol. 11: 3326–3329.

APPELS, R., HONEYCUTT, R.L. 1985. rDNA: evolution over a billion years. In: DNA Systematics, Vol. II: Plants (Dutta, S.K., ed.). CRC Press, Boca Raton, FL. pp. 81–135.

BERGES, T., PETFALSKI, E., TOLLERVEY, D., HURT, E.C. 1994. Synthetic lethality with fibrillarin identifies NOP77p, a nucleolar protein required for pre-rRNA processing and modification. EMBO J. 13: 3136–3148.

BUTLER, D.K. 1992. Ribosomal DNA is a site of chromosome breakage in aneuploid strains of *Neurospora*. Genetics 131: 581–592.

BUTLER, D.K., METZENBERG, R.L. 1990. Expansion and contraction of the nucleolus organizer region of *Neurospora:* changes originate in both proximal and distal segments. Genetics 126: 325–333.

CECH, T.R. 1986. The generality of self-splicing RNA: relationship of nuclear mRNA to splicing. Cell 44: 207–210.

CECH, T.R. 1987. The chemistry of self-splicing RNA and RNA enzymes. Science 236: 1532–1539.

COTTEN, M., BIRNSTIEL, M.L. 1989. Ribozyme mediated destruction of RNA in vivo. EMBO J. 8: 3861–3866.

ENDOW, S.A., ATWOOD, K.C. 1988. Magnification: gene amplification by an inducible system of sister chromatid exchange. Trends Genet. 4: 348–351.

FIREK, S., READ, C., SMITH, D.R., MOSS, T. 1989. Point mutation analysis of the *Xenopus laevis* RNA polymerase I core promoter. Nucleic Acids Res. 18: 105–109.

GAUTIER, T., DAUPHIN-VILLEMANT, C., ANDRE, C., MASSON, C., ARNOULT, J., HERNANDEZ-VERDUN, D. 1992. Identification and characterization of a new set of nucleolar ribonucleoproteins which line the chromosomes during mitosis. Exp. Cell Res. 200: 5–15.

GERBI, S.A., SAVINO, R., STEBBINS-BOAZ, B., JEPPESEN, C., RIVERA-LEON, R. 1990. A role for U3 small nuclear ribonucleoprotein in the nucleolus. In: The Ribosome—Structure, Function and Evolution. (Hill. W.E., Dahlberg, A., Garrett, R.A., Moore, P.B., Schlessinger, D., Warner, J.R., eds.). American Society for Microbiology, Washington, DC. pp. 452–469.

GIRARD, J-P., LEHTONEN, H., CAIZERGUES-FERRER, M., AMALRIC, F., TOLLERVEY, D., LAPEYRE, B. 1992. GAR1 is an essential small nucleolar RNP protein required for pre-rRNA processing in yeast. EMBO J. 11: 673– 682.

HASELOFF, J., GERLACH, W.L. 1988. Simple RNA enzymes with new and highly specific endoribonuclease activities. Nature 334: 585–591.

JANTZEN, H.M., ADMON, A., BELL, S.P., TJIAN, R. 1990. Nucleolar transcription factor hUBF contains a DNA-binding motif with homology to HMG proteins. Nature 344: 830–836.

JIANG, J., GILL, B.S. 1994. New 18S·26S ribosomal RNA gene loci: chromosomal landmarks for the evolution of polyploid wheats. Chromosoma 103: 179–185.

KULKENS, T., RIGGS, D., HECK, J.D., PLANTA, R.J., NOMURA, M. 1991. The yeast RNA polymerase I promoter: ribosomal DNA sequences involved in transcription initiation and complex formation in vitro. Nucleic Acids Res. 19: 5363–5370.

LEE, W.C., XUE, Z., MELESE, T. 1991. The *NSR1* gene encodes a protein that specifically binds nuclear localization sequences and has two RNA recognition motifs. J. Cell Biol. 113: 1–12.

MASSON, C., ANDRE, C., ARNOULT, J., GERAUD, G., HERNANDEZ-VERDUN, D. 1990. A 116000 M_r nucleolar antigen specific for the dense fibrillar component of the nucleoli. J. Cell Sci. 95: 371–381.

MAY, C.E., APPELS, R. 1987. Variability and genetics of spacer DNA sequences between the ribosomal-RNA genes of hexaploid wheat (*Triticum aestivum*). Theoret. Appl. Genet. 74: 617–624.

MAY, C.E., APPELS, R. 1992. The nucleolus organizer regions (*Nor* loci) of hexaploid wheat cultivars. Aust. J. Agric. Res. 43: 889–906.

MCCUSKER, J.H, YAMAGISHI, M., KOLB, J.M., NOMURA, M. 1991. Suppressor analysis of temperature-sensitive RNA polymerase I mutations in *Saccharomyces cerevisiae*: suppression of mutations in a zinc-binding motif by transposed mutant genes. Molec. Cell Biol. 11: 746–753.

MCKEE, B.D., KARPEN, G.H. 1990. *Drosophila* ribosomal RNA genes function as an X–Y pairing site during male meiosis. Cell 61: 61–72.

MCKEE, B.D., HABERA, L., VRANA, J.A. 1992. Evidence that intergenic spacer repeats of *Drosophila melanogaster* rRNA genes function as X–Y pairing sites in male meiosis, and a general model for achiasmatic pairing. Genetics 132: 529–544.

MONTGOMERY, T.H. 1898. Comparative cytological studies with especial regard to the morphology of the nucleolus. J. Morphol. 15: 265–582.

MUKAI, Y., ENDO, T.R., GILL, B.S. 1991. Physical mapping of the

18S·26S rRNA multigene family in common wheat: identification of a new locus. Chromosoma 100: 71–78.

NIGG, E.A. 1988. Nuclear function and organization. Annu. Rev. Biochem. 110: 27–92.

PETES, T.D. 1980. Unequal meiotic recombination within tandem arrays of yeast ribosomal DNA genes. Cell 19: 766–774.

PIKAARD, C.S., MCSTAY, B., SCHULTZ, M.C., BELL, S.P., REEDER, R.H. 1989. The *Xenopus* ribosomal gene enhancers bind an essential polymerase I transcription factor xUBF. Genes Dev. 3: 1779–1788.

POLYAKOV, A., SEVERINOVA, E., DARST, S.A. 1995. Three-dimensional structure of *E. coli* core polymerase: promoter binding and elongation conformations of the enzyme. Cell 83: 365–373.

SAVINO, R., GERBI, S.A. 1991. Preribosomal RNA processing in *Xenopus oocytes* does not include cleavage within the external transcribed spacer as an early step. Biochemistry 73: 805–812.

SCHMIDT-ZACHMANN, M.S., FRANKE, W.W. 1988. DNA cloning and amino acid sequence determination of a major constituent protein of mammalian nucleoli. Chromosoma 96: 417–426.

SEIFARTH, W., PETERSEN, G., KONTERMAN, R., RIVA, M., HUET, J., BAUTZ, E.K.F. 1991. Identification of the genes coding for the second-largest subunits of RNA polymerases I and III of *Drosophila melanogaster*. Molec. Gen. Genet. 228: 424–432.

SMID, A., FINSTERER, M., GRUMMT, I. 1992. Limited proteolysis unmasks specific DNA-binding of the murine RNA polymerase I-specific transcription termination factor TTFI. J. Molec. Biol. 227: 635–647.

SOMMERVILLE, J. 1985. Organizing the nucleolus. Nature 318: 410–411.

THIRY, M., THIRY-BLAISE, L. 1991. Locating transcribed and non-transcribed rDNA spacer sequences within the nucleolus by *in situ* hybridization and immunoelectron microscopy. Nucleic Acids Res. 19: 11–15.

WACHTLER, F., SCHÖFER, C., MOSGÖLLER, W., WEIPOLTSHAMMER, K., SCHWARZACHER, H.G., GUICHAOUA, M., HARTUNG, M., STAHL, A., BERGE-LEFRANC, J.L., GONZALES, I., SYLVESTER, J. 1992. Human ribosomal RNA gene repeats are localized in the dense fibrillar component of nucleoli: light and electron microscopic *in situ* hybridization in human sertoli cells. Exp. Cell Res. 198: 135–143.

WEIPOLTSHAMMER, K., SCHÖFER, C., WACHTLER, F., HOZÁK, P. 1996. The transcription unit of ribosomal genes is attached to the nuclear skeleton. Exp. Cell Res. 227: 374–379.

WOYCHIK, N.A., LIAO, S-M., KOLODZIEJ, P.A., YOUNG, R.A. 1990. Subunits shared by eukaryotic nuclear RNA polymerases. Genes Dev. 4: 313–323.

Section 20.3

BLOOM, K., HILL, A., KENNA, M., SAUNDERS, M. 1989. The structure of a primitive kinetochore. Trends Biochem. Sci. 14: 223.

CAI, M., DAVIS, R.W. 1990. Yeast centromere binding protein CBF1, of the helix-loop-helix protein family, is required for chromosome stability and methionine prototrophy. Cell 61: 437–446.

CLARKE, L. 1990. Centromeres of budding and fission yeast. Trends Genet. 6: 150–154.

EARNSHAW, W.C., RATRIE III, H., STETTEN, G. 1989. Visualization of centromere proteins CENP-B and CENP-C on a stable dicentric chromosome in cytological spreads. Chromosoma 98: 1–12.

GRADY, D.L., RATLIFF, R.L., ROBINSON, D.L., MCCANLIES, E.C., MEYNE, J., MOYZIS, R.K. 1992. Highly conserved repetitive DNA sequences are present at human centromeres. Proc. Natl. Acad. Sci. USA 89: 1695–1699.

JIANG, J., NUSUDA, S., DONG, F., SHERRER, C.W., WOO, S-S., WING, R.A., GILL, B.S., WARD, D.C. 1997. A conserved repetitive DNA element located in the centranese of cereal chromosomes. Proc. Natl. Acad. Sci. USA 93: 14210–14213.

KASZAS, E., BIRCHLER, J.A. 1997. Misdivision analysis of centromere structure in maize. EMBO J. 15: 5246–5255.

MASISON, D.C., BAKER, R.E. 1992. Meiosis in *Saccharomyces cerevisiae* mutants lacking the centromere-binding protein CP1. Genetics 131: 43–53.

MILLER, D.A., SHARMA, V., MITCHELL, A.R. 1988. A human-derived probe, p82H, hybridizes to the centromeres of gorilla, chimpanzee, and orangutan. Chromosoma 96: 270–274.

MURAKAMI, S., MATSUMOTO, T., NIWA, O., YANAGIDA, M. 1991. Structure of the fission yeast centromere *cen3:* direct analysis of the reiterated inverted region. Chromosoma 101: 214–221.

PALMER, D.K., O'DAY, K., MARGOLIS, R.L. 1990. The centromere specific histone CENP-A is selectively retained in discrete foci in mammalian sperm nuclei. Chromosoma 100: 32–36.

PLUTA, A.F., COOKE, C.A., EARNSHAW, W.C. 1990. Structure of the human centromere at metaphase. Trends Biochem. Sci. 15: 181–185.

RATTNER, J.B. 1991. The structure of the mammalian centromere. Bioessays 13: 51–56.

RICHARDS, E.J., GOODMAN, H.M., AUSUBEL, F.M. 1991. The centromere region of *Arabidopsis thaliana* chromosome 1 contains telomere-similar sequences. Nucleic Acids Res. 19: 3351–3357.

VALLEE, R. 1990. Dynein and the kinetochore. Nature 345: 206–207.

ZINKOWSKI, R.P., MEYNE, J., BRINKLEY, B.R. 1991. The centromere-kinetochore complex: a repeat subunit model. J. Cell Biol. 113: 1091–1110.

Section 20.4

BIESSMANN, H., MASON, J.M. 1997. Telomere maintenance without telomerase. Chromosoma 106: 63–69.

BIESSMANN, H., CARTER, S.B., MASON, J.M. 1990. Chromosome ends in *Drosophila* without telomeric DNA sequences. Proc. Natl. Acad. Sci. USA 87: 1758–1761.

BLACKBURN, E.H. 1991a. Telomeres. Trends Biochem. Sci. 16: 378–382.

Blackburn, E.H. 1991b. Structure and function of telomeres. Nature 350: 569–573.

BROWN, W.R.A., DOBSON, M.J., MACKINNON, P. 1990. Telomere cloning and mammalian chromosome analysis. J. Cell Sci. 95: 521–526.

CHENG, J-F., SMITH, C.L., CANTOR, C. 1989. Isolation and characterization of a human telomere. Nucleic Acids Res. 17: 6109–6127.

COHN, M., BLACKBURN, E.H. 1995. Telomerase in yeast. Science 269: 396–400.

Collins, K., Kobayashi, R., Greider, C. 1995. Purification of *Tetrahymena* telomerase and cloning of genes encoding two protein components of the enzyme. Cell 81: 677–686.

Danilevskaya, O., Slot, F., Pavlova, M., Pardue, M-L. 1994. Structure of the *Drosophila* HeT-A transposon: a retrotransposon-like element forming telomeres. Chromosoma 103: 215–224.

de Lange, T. 1992. Human telomeres are attached to the nuclear matrix. EMBO J. 11: 717–724.

Fantes, J., Cook, H. 1991. Visualization of human telomeric sequences by *in situ* hybridization with a human telomeric probe. Nature 350: 6319: front cover photograph.

Gray J.T, Celander D.W., Price C.M., and Cech T.R. 1991. Cloning and expression of genes for the *Oxytricha* telomere-binding protein: specific subunit interactions in the telomeric complex. Cell 67: 807–814.

Katinka, M.D., Bourgain, F.M. 1992. Interstitial telomeres are hotspots for illegitimate recombination with DNA molecules injected into the macronucleus of *Paramecium primaurelia*. EMBO J. 11: 725–732.

Mason, J.M., Biessman, H. 1995. The unusual telomeres of *Drosophila*. Trends Genet. 11: 58–62.

Moens, P.B., Pearlman, R.E. 1990. Telomere and centromere DNA are associated with the cores of meiotic prophase chromosomes. Chromosoma 100: 8–14.

Moysis, R.K. 1991. The human telomere. Sci. Am. 265: 48–55.

Raghuraman, M.K., Cech, T.R. 1989. Assembly and self-association of *Oxytricha* telomeric nucleoprotein complexes. Cell 59: 719–728.

Raghuraman, M.K., Dunn, C.J., Hicke, B.J., Cech, T.R. 1989. *Oxytricha* telomeric nucleoprotein complexes reconstituted with synthetic DNA. Nucleic Acids Res. 17: 4235–4253.

Rawlins, D.J., Highett, M.I., Shaw, P.J. 1991. Localization of telomeres in plant interphase nuclei by *in situ* hybridization and 3D confocal microscopy. Chromosoma 100: 424–431.

Romero, D.P., Blackburn, E.H. 1991. A conserved secondary structure for telomeres RNA. Cell 67: 343–353.

Schechtman, M.G. 1990. Characterization of telomere DNA from *Neurospora crassa*. Gene 88: 159–165.

Traverse, K.L., Pardue, M.L. 1988. A spontaneously opened ring chromosome of *Drosophila melanogaster* has acquired He-T DNA sequences at both new telomeres. Proc. Natl. Acad. Sci. USA 85: 8116–8120.

Tsujimoto, H., Yamada, T., Sasakuma, T. 1997. Molecular structure of a wheat chromosome end healed after gametocidal gene-induced breakage. Proc. Natl. Acad. Sci. USA 94: 3140–3144.

Weber, B., Collins, C., Robbins, C., Magenis, R.E., Delaney, A.D., Gray, J.W., Hayden, M.R. 1990. Characterization and organization of DNA sequences adjacent to the human telomere associated repeat $(TTAGGG)_n$. Nucleic Acids Res. 18: 3353–3361.

Zakian, V.A. 1989. Structure and function of telomeres. Annu. Rev. Genet. 23: 579–604.

21

Genetic and Molecular Mapping of Chromosomes

- Genetic maps include a wide range of markers involving morphological traits, chromosome landmarks, disease-resistance genes, biochemical defects, enzyme and protein characters, and different classes of DNA sequences.
- Following the assignment of markers to chromosomes using chromosome aberrations, a genetic map is constructed by analyzing the progeny from crosses between distinguishable individuals and estimating recombination frequencies between the genetic loci.
- Codominant markers are ideal for genetic mapping because the coupling or repulsion linkage phase of markers with dominant and recessive alleles becomes an additional variable.
- Physical and genetic maps are colinear with respect to gene order, but major distortions occur in apparent distances between loci due to uneven levels of crossing over in different parts of the genome.
- Bulked F_2 segregant analyses and interval mapping can be used to analyze quantitative-trait loci (QTLs).
- Genetic maps provide bases for human genetic counseling, and for plant and animal breeding programs.

Using a combination of morphological, pathological, biochemical, chromosomal, and molecular markers, rapid advances are being made in both plants and animals to determine the precise locations of genes on chromosomes. These advances ensure that the diagnosis of genetic defects, the recognition of resistance and susceptibility to disease, and the presence or absence of any genetic character can be made reliably by assaying for markers closely linked to the gene controlling that character. The more closely a marker is linked to a gene locus, the less likely that the marker and the gene will be separated from each other by recombination. At the molecular level, chromosome walking (*see* Chapter 20, Fig. 20.25) allows a cloned genomic DNA fragment, acting as a marker for a particular gene, to be used to clone neighboring DNA sequences in order to explore that region of the chromosome in more detail. Provided the original gene of interest can be recognized at the DNA level, this type of DNA exploration can lead to its isolation.

21.1 AN OVERVIEW OF MAPPING IN PLANTS, ANIMALS, AND FUNGI

The initial definition of a gene relates to a unit of inheritance that affects a specific feature of the phenotype of an organism. If the unit of inheritance has a distinct or qualitative effect on the phenotype and is simply inherited, it is defined as a single gene, and different forms of this gene are described as alleles. The distinction between quantitative and qualitative effects on phenotype is based primarily on the magnitude of the effect of allelic substitution at the locus controlling the phenotypic trait. If the ratio between the effect of the substitution and the total observed variation is small, the trait is generally regarded as quantitative. Conversely, if the effect of the allelic substitution is large with respect to the total phenotypic variation, the trait is characterized as qualitative. In both cases, it is possible to use genetic markers to further characterize the respective traits.

Different organisms have particular advantages for the analysis of chromosomes at the molecular/genetic marker level. An organism such as the puffer-fish (*Fugu* sp.), for example, may be of particular interest because it has one of the smallest genomes known among vertebrates (400 Mb with 90% unique sequences). A low amount of DNA , however, is only one of many considerations in using a particular organism for studies, and factors such as ease of handling, the available genetic database, and economic significance need to be considered.

21.1.1 Studies in Animals

Since the early 1900s, insects and, in particular, the fruit fly *Drosophila melanogaster*, have been intensively studied. The advantages of *Drosophila* populations for genetic and cytogenetic analyses are that they can be kept in small areas, are simply fed, have a rapid reproduction time, and show easily distinguishable genetic differences. The presence of polytene chromosomes in the salivary glands also is a major advantage in that introduced and deleted chromatin can be related to the presence and absence of genes and their effects. The nematode *Caenorhabditis elegans* has special relevance to developmental genetics, because aspects of cell patterning are well defined.

Among mammals, the mouse genome has been a valuable model for human genetics because so many features of mouse chromosomes such as gene order, structure, and function are common to humans. Agricultural animals such as pigs, and beef and dairy cattle, have been analyzed at the molecular–genetic level primarily because of their economic importance. Mammals vary little in their chromosome numbers and genome sizes, and the genetic map of one organism can often indicate the positions of genes in another organism, an effect described as synteny. The applications of research on related organisms to human molecular/cytogenetic studies are important in medicine and have provided a major stimulus for the analysis of the human genome by the Human Genome Mapping Organization (HUGO).

21.1.2 Studies in Plants

The dicotyledon *Arabidopsis thaliana* is a plant with a small genome (approximately 100 Mb), a short generation time, and an ease of culture, which readily allows transformation and selection experiments. Genetic mapping, cloning, and sequencing of the genome are well advanced, and this organism now provides a source of genes for introduction into other plants. The economically important crop-species tomato (*Lycopersicon esculentum*) and potato (*Solanum tuberosum*) are also widely studied and have extensive genetic databases.

Among monocotyledons, maize, rice, and wheat are targets for genome-mapping projects. Rice (*Oryza sativa*), in particular, because of its small genome and economic importance in the Tropics, is developing into a model organism, with the concept of synteny providing a link between the various genome-mapping studies in grasses. Wheat (*Triticum aestivum*), a more temperate crop, is the model system for polyploid-genetics research.

21.1.3 Studies in Yeast

Efforts to determine the complete DNA sequences of entire genomes provide the basis for the ultimate structural analysis of chromosomes. Yeast (*Saccharomyces cerevisiae*) was the first eukaryotic organism in which not only was the complete sequence of a eukaryotic chromosome determined (in 1993), but the entire genome of 16 chromosomes was completely sequenced by 1996.

21.2 GENETIC MARKERS

A wide range of markers are used to make chromosome maps depicting the locations of genes on chromosomes. These markers include the following:

(a) Morphological markers causing changes in the phenotype of the organism
(b) Physiological markers affecting the growth habit of the plant or animal
(c) Genes conferring resistance or susceptibility to certain diseases
(d) Biochemical–genetic markers resulting from either the modification of the polypeptide protein and enzyme products of gene action, or detectable changes in the products produced by these pathways
(e) Changes in chromosome structure or banding patterns
(f) Molecular markers utilizing specific DNA sequences to recognize identical or closely related DNA sequences within the chromosome

21.2.1 Morphological Markers

The phenotype of an organism describes its usual morphological appearance. In general terms, the external appearance of an organism is determined by the combination of the genotype and the effects of the environment upon that genotype. Although the environment can alter the way in which particular genes are expressed, leading to many of the individual differences between discrete individuals within a population, a normal phenotype is usually readily defined and is referred to as the wild or standard type for that organism. Deviations from wild type can then be recognized and classified as allelic variants caused by mutation.

The recognition of some morphological differences as simply inherited genetic traits is exemplified in Mendel's classic work with the garden pea. One of the crosses used by Mendel in formulating his laws of inheritance was between round- and wrinkled-seeded peas. When plants with these phenotypes were intercrossed, Mendel found that the F_1 plants were all round-seeded, and these seeds produced F_2 plants with three-quarters round and one-quarter wrinkled seeds. Accordingly, round seed is the dominant allele and wrinkled seed the recessive allele. These seed phenotypes have been characterized at a molecular level. The seed wrinkling is the result of a high level of sugar in the developing seed and an associated increase in water accumulation in the embryonic cotyledons. As the seed matures and dries out, the cotyledons shrink and the external seed coat wrinkles. The high sugar levels are the result of a reduction in overall starch synthesis, caused by the absence of one of the starch-branching enzymes (SBE-1). The reduced level of the branched form of starch, amylopectin, resulting from this enzymatic defect causes a decrease in the amount of starch that can accumulate in the seed (Fig. 21.1). Because the absence of SBE-1 leads to the wrinkled-seed phenotype, the round- (normal) seed phenotype is dominant, because a hemizygous gene dosage of SBE-1 can produce sufficient amounts of normal starch to overcome the wrinkled-seed phenotype.

Genes are not always readily categorized as dominant or recessive. They may show incomplete dominance, as in the plant *Antirrhinum* (snapdragon), where particular cultivars with a red-flowered phenotype can be crossed with a white-flowered cultivar to produce F_1 plants that are pink-flowered. The F_2 progenies segregate in the ratio of 1 red 2 pink 1 white. In this case, the hemizygous red-flower gene is unable to produce enough red pigment to compensate for the loss of its partner.

Some morphological characters are not simply or monogenically inherited but require a combination of two or more genes in digenic or polygenic reactions for expression. The importance of knowing the number of genes involved in the production of polygenically controlled characters lies in the fact that genes with a major effect on a morphological character can be identified as a first step in dissecting and understanding that character. The analysis of complex characters and quantitative-trait loci (QTL), within the framework of a genetic map comprised of simply inherited markers, is discussed in Section 21.4.9.

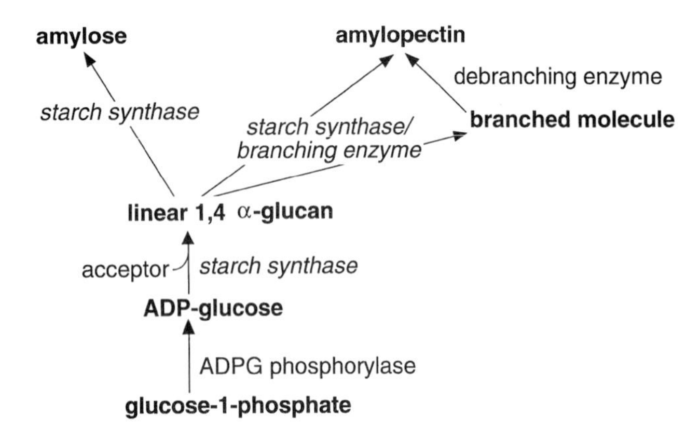

Fig. 21.1. The biosynthetic pathway that produces starch (*see* Morell et al., 1995). In a plant such as Mendel's garden pea, the loss of one of the branching enzymes results in a defect in starch synthesis, leading to reduced starch production in the seed and accumulation of sugar levels. The pathway for starch synthesis starts when glucose sugar is phosphorylated to give glucose-1-phosphate, which then provides a substrate for the enzyme adenosine diphosphate glucose pyrophosphorylase (ADGP phosphorylase). The ADP-glucose moieties are made into linear polymers of 1,4 α-glucan, which form the basis of the starch molecule. The continued formation of linear chains by starch synthase results in amylose. For efficient packaging into the starch granule and more extensive accumulation of starch in the seed, the linear molecule is disrupted by branching enzymes, followed by further elongation of the branches, to form amylopectin.

21.2.2 Chromosomal Landmarks

The discovery of chemical procedures that preferentially stain certain regions of chromosomes to give visible banding patterns was a significant step in the development of cytogenetic maps. Chromosome bands are physical landmarks that distinguish differing chromosome segments, and regions of DNA that have been shown to carry a particular gene can be correlated with the presence or absence of particular bands, and thus placed into the context of a physical map of the chromosome. Furthermore, the banding patterns themselves are often polymorphic in their appearance, so that even within a homologous pair of chromosomes, there can be quantitative differences in the widths and amounts of different bands, and particular bands may be missing completely. Such polymorphism means that bands that stain differentially can also be used as simple genetic markers. C bands, in particular, are often polymorphic and can be used as markers in a genetic cross in the same way as any other marker. The schematic representation in Fig. 21.2 illustrates the appearance of F1 progeny and how crossovers in certain regions could provide the information to estimate genetic linkage between chromosome bands. The relationship between chromosome banding and genetic defects has been pursued to its utmost in humans, where genetic linkage maps are now commonly placed in the context of the physical map of the chromosome.

21.2.3. Disease Resistance and Susceptibility

Organisms are usually resistant to most diseases and susceptible to comparatively few. Conversely, most disease-causing organisms, parasites, or pathogens have a limited range of hosts. Host–pathogen interactions involve a balance between the two organisms to ensure the mutual survival of both. In plants, disease-resistance genes that confer an ability to withstand attack from particular pathogens or their races are extensively used as genetic markers. If these genes are absent, the plants are susceptible. Specific genes conferring resistance to infection by viral, bacterial, fungal, nematode, and insect pathogens can be found by screening plants for their reaction to infection by the particular pathogen or one of its races. In a large number of cases, genes for disease resistance or susceptibility in the host have been shown to have corresponding genes for avirulence or virulence in the pathogen. The interaction between host and pathogen is thus governed by the "gene for gene" hypothesis (Fig. 21.3). Mutations to virulence in the pathogen often result from loss of function, whereas mutations to resistance in the host most likely require a new function and thus do not occur as readily.

Several plant disease-resistance genes have been cloned. In the case of resistance to infection by *Pseudomonas syringae* in tomato, the gene proved to be a protein kinase and likely represents only one step in a longer pathway that leads to resistance. The genes conferring resistance to *Cladosporium fulvum* in tomato, *Pseudomonas syringae* in *Arabidopsis*, and *Melampsora lini* (rust) in flax do not appear to be protein kinases. Although their functions are not clear, it is interesting that they share a characteristic leucine-rich domain often found in proteins that interact with membranes.

21.2.4 Biochemical Defects

Many gene mutations leading to morphological and physiological changes in the individual are related to defects in biochemical

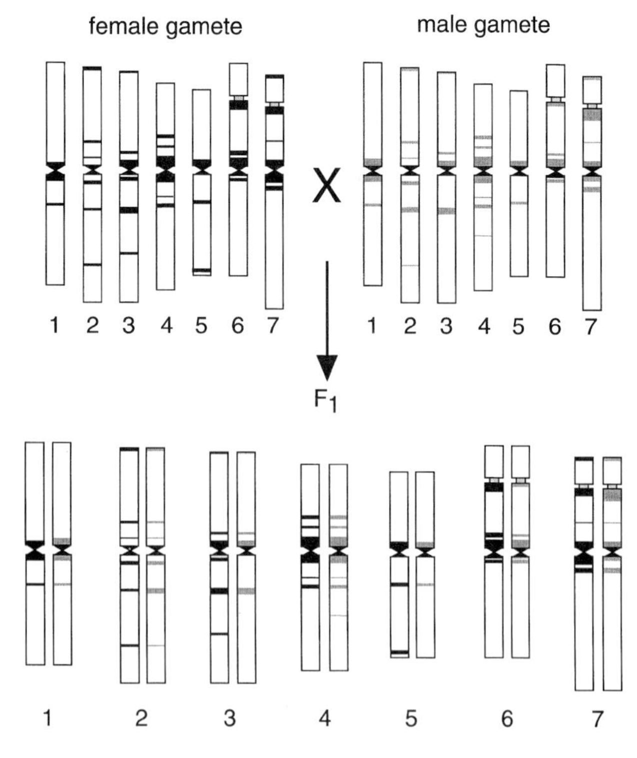

Fig. 21.2. Chromosome C bands can reveal consistent differences between individual, homologous chromosomes within a population and can, consequently, be used as genetic markers. The diagram indicates the variation observed in two different *Hordeum vulgare* (barley) cultivars and in the F_1 hybrid with respect to each chromosome in their mitotic karyotypes, indicated from 1 to 7, with the male chromosomes showing grey bands. In crosses between the two cultivars, the polymorphic bands (for example, on chromosomes 3, 4 and 5) provide markers for recombination (Linde-Laursen, 1978).

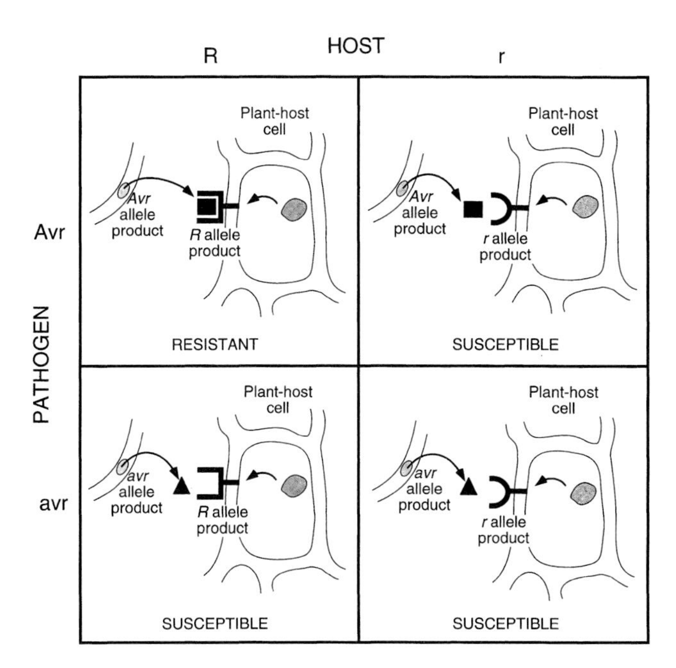

Fig. 21.3. A diagram demonstrating the interactions between genotypes of a pathogen and its host. The balance between virulence and susceptibility can lead to a pseudo-3 susceptible : 1 resistant ratio, where resistance requires both the resistance allele (*R*) of the plant host in combination with the avirulence (*Avr*) allele of the pathogen (top left). All other host/parasite combinations result in susceptibility.

pathways. Alkaptonuria is a good example in humans. This relatively rare disease is inherited in a Mendelian fashion, as an autosomal recessive character, and results from loss of function of a gene. The biochemical lesion is the result of an inability to convert homogentisic acid (or alkapton), a breakdown product of the amino acids phenylalanine and tyrosine, into 4-maleylacetoacetic acid, a further breakdown product (Fig. 21.4). Consequently, homogentisic acid is excreted in the urine, which turns black on exposure to air. There also is a buildup of homogentisic acid in cartilage, causing arthritis. Thus, a distinct, phenotypic character is linked to a biochemical defect resulting from loss of function of the gene that normally produces homogentisic acid oxidase. A number of other physiological defects caused by similar gene mutations have been recognized in this and related pathways (*see* Fig. 21.4).

21.2.5 Enzymes and Other Protein Markers

The genetic code, as read from messenger-RNA molecules, produces a range of polypeptides with either a structural or storage function, or a catalytic function. The polypeptides with enzymic functions are involved in a wide range of metabolic activities, some of which can be readily detected using appropriate substrates to produce colored end products (*see* Chapter 19, Fig. 19.4). The technique makes use of the fact that the activity of many enzymes can be measured after a crude cell extract is placed in a suitable support medium, and electrophoresis is used to separate the components. The support medium commonly consists of a gel of starch, acrylamide, or agarose, which is inert to the electric field. The component proteins in the extract are separated relative to each other by passage through the gel matrix, under the influence of the applied electrical field. In this electrophoretic process, different polypeptides move toward the anode (positive terminal) at rates that are determined by the net negative charge on the molecule and by its size or molecular weight. The final positions of different polypeptides in the gel are determined by color reactions, where a dye is deposited at the site of enzyme action (*see* Chapter 19, Section 19.1), or by the use of biological stains that bind to the proteins themselves. Generally, a band of activity or staining corresponds to a single-polypeptide gene product. Polypeptides with the same or similar enzymatic activities can also occur and are de-

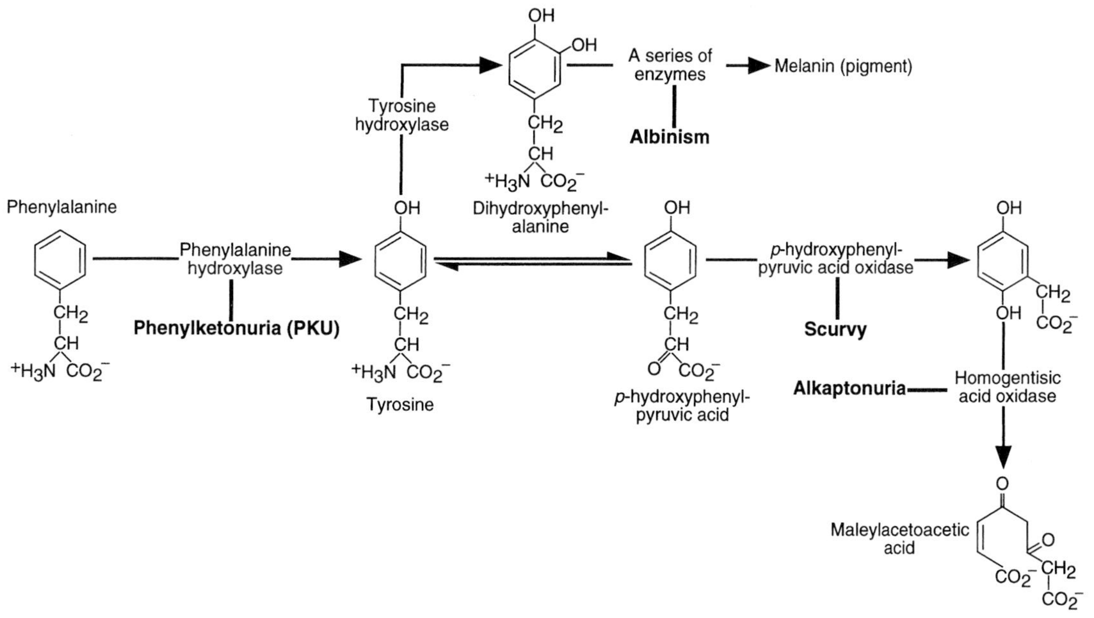

Fig. 21.4. Human metabolic pathways involving the amino acid phenylalanine and some of its derivatives. The absence of or defects in the enzymes controlling the various pathways lead to the disease symptoms and syndromes associated with those specific defects (bold).

scribed as isozymes. Allelic variants of isozymes and proteins are discerned by changes in mobility of the polypeptides, caused by differences in the amino-acid composition, which, in turn, cause a change in the net charge, a change in molecular weight, or a change in the isoelectric point of the molecule. If two parents have allelic protein or isozyme variants, the F_1 heterozygotes commonly show the presence of both bands, and the F_2 progenies show a 1:2:1 segregation. A simple interpretation can sometimes be obscured because enzymes and proteins are often composed of more than one polypeptide chain, so that in heterozygous individuals, new proteins, absent in both parents, are present due to a combination of different subunits (Fig. 21.5).

In certain tissues such as the red blood cells of animals and the seed endosperm of plants, enough of a single protein for a group of proteins is present to allow it or them to be observed with a general stain without the need to assay activity. Variation in the major protein component of human red blood cells, for example, causes the disorder thalassemia. Figure 21.6 shows a range of electrophoretic variants of hemoglobin and the thalassemias associated with them. The defective hemoglobins result from point mutations in the globin genes, as well as reduction in the synthesis of one or more pairs of globin chains. The molecular-biological analysis of globin genes is well advanced (*see* Chapter 16, Fig. 16.25 and Chapter 17, Fig. 17.30), and reduced synthesis has been shown to result from a wide range of alterations such as deletions of upstream-control regions, gene deletions/duplications, gene fusions, inversions, promoter mutations, defective-messenger-RNA splicing, polyA-signal mutations, mutations affecting initiation of translation, introduction of termination codons, and mutations affecting the stability of the final protein product.

A major thalassemia, sickle-cell anemia, is maintained in human populations because heterozygous individuals have increased tolerance to the parasite *Plasmodium falciparum*, which causes malaria. This interaction is a good example of a secondary pleiotropic effect resulting from a single genetic change. In biological interactions of this type, the frequency of the heterozygous state for thalassemia

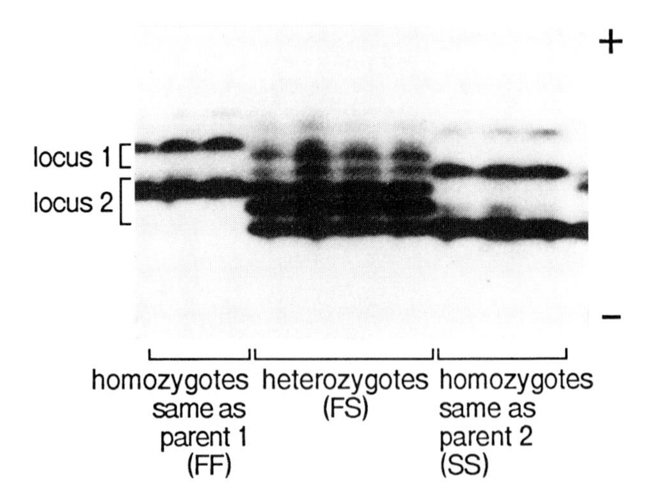

Fig. 21.5. Alcohol dehydrogenase isozyme variation in a segregating progeny from a cross within the grass *Hordeum spontaneum*, with respective homozygotes (FF for faster-moving, or SS for slower-moving) and heterozygotes (FS) indicated. The FS progeny plants show a new band (second from bottom) because the enzymes are usually active as dimers and in the FS individuals, new combinations of heterodimers can form with new, heterologous mobilities. (From Brown, 1980.)

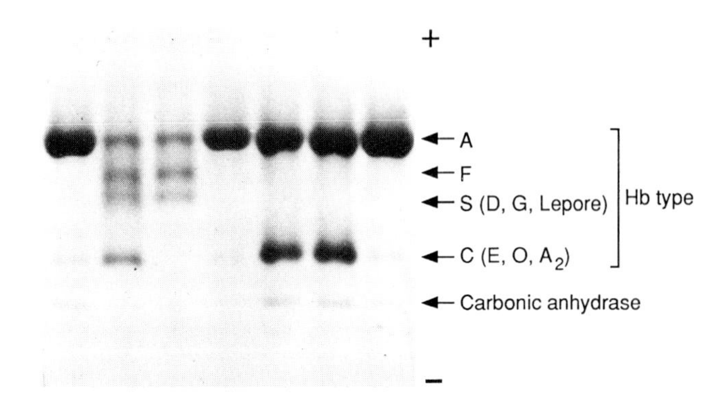

Fig. 21.6. Electrophoretic variants of human-hemoglobin proteins. Colored proteins such as hemoglobin can be observed directly both during and after gel electrophoresis; the band indicated as carbonic anhydrase identifies a protein that is abundant in red blood cells and the letters identify standard hemoglobin electrophoretic variants. Quantitative differences in the amounts of the individual hemoglobins present, and in the electrophoretic variability between different hemoglobin molecules, can be correlated with various thalassemia disorders of the blood. [Photograph kindly supplied by Woden Valley Hospital (Canberra, ACT, Australia).]

increases in human populations exposed to malaria until the incidence of heterozygotes is balanced by the loss of homozygotes from the population. This is known as a balanced polymorphism, but such a balance does not hold true for all defects affecting erythrocytes.

Variations in the proteins that make up wheat endosperm can have significant effects on the bread-making quality of the flour. The nature of the glutenin and gliadin genes that code for a large proportion of the flour proteins has been extensively studied using gel electrophoresis (Fig. 21.7). Environmental as well as genetic factors affect the levels of glutenin and gliadin subunits, and assessing their presence and inheritance is an important part of wheat-breeding programs.

Isozymes and proteins have been used extensively as genetic markers, and they have the following properties: They are codominant; they generally have 2–10 alleles/locus; they provide 5–40 marker loci originating from low-copy regions of the genome; and they can be efficiently applied to a genetic study. The main disadvantage is the relatively low level of polymorphism at loci identified by isozymes and proteins.

21.2.6 DNA Markers

The ability to clone specific DNA sequences from the vast array present in the genome has opened up the possibility of assaying any section of the genome for a genetic map. Furthermore, the cloned sequences can be used in in situ hybridization experiments to relate a sequence to a specific location on a chromosome. The principle underlying the use of DNA markers in genetic analyses is that several different ways are available to discover differences in DNA sequences between individuals. If these differences represent alternatives at a single position in the genome, they define alleles in the same way as the products of wild-type and mutant genes define alleles of these genes. The advantage of studies at the DNA level is that any DNA sequence can be the source of allelic differences between individuals without necessarily knowing the product for which it codes, or even whether it actually codes for a product.

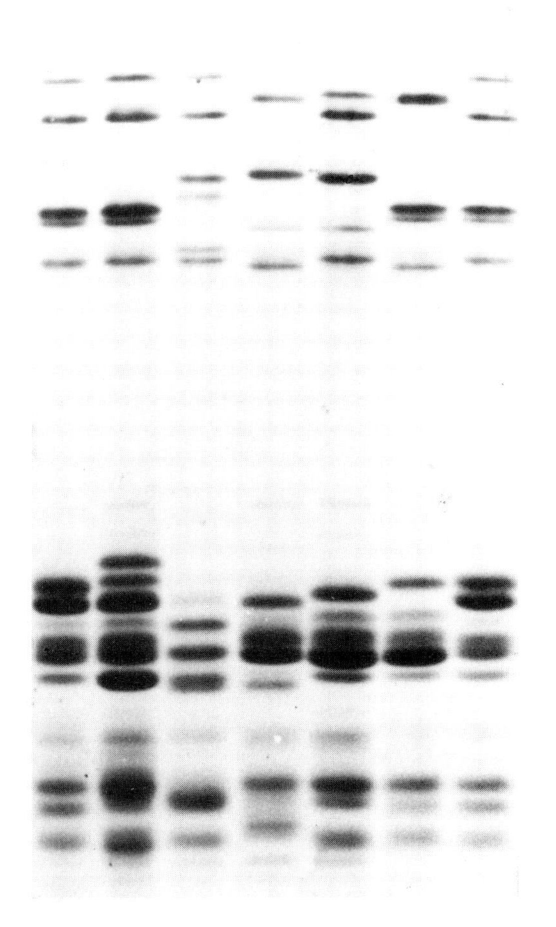

Fig. 21.7. The electrophoretic separation of seed-storage proteins of plants. The gel detects the presence or absence of different gluten proteins in extracts from individual seeds of an F_1 wheat hybrid, with each column derived from a single seed. The separated proteins were stained with Coomassie blue dye. (Photograph kindly supplied by Dr. R. Gupta.)

Restriction fragment-length polymorphisms (RFLPs) are assayed when genomic DNA is digested with a restriction endonuclease, and the resulting fragments are separated by gel electrophoresis, transferred to a nylon-based membrane by blotting techniques, and hybridized to a labeled DNA sequence. The labeling of the DNA sequences is carried out using either radioactive or non-radioactive procedures. The RFLPs revealed by the DNA probe may result either from mutations in one or both of the restriction-endonuclease sites that define the DNA fragments assayed by the probe, or from deletions and insertions of DNA between the sites (Fig. 21.8). In genetic analyses, low-copy RFLP markers are defined as codominant because, in segregating progenies, both parental and heterozygous genotypes are clearly scored (Table 21.1).

The polymerase chain reaction (PCR), described in Chapter 19, Section 19.4, also provides a means to assay DNA variation between individuals by making millions of copies of specific regions of the genome. The PCR-based DNA markers utilize short oligonucleotide sequences, 10–25 bases long, to prime DNA amplification reactions at the specific sites in the genome where DNA sequences complementary to the primer sequences are present. The polymorphisms detected by PCR result from insertions and deletions between the primer-binding sites and from mutations of the actual primer-binding sites. Although it should be possible to observe different-sized fragments in amplification products in different individuals, it is more usual to find that fragments are amplified in one individual but not in another. Amplification of a specific band is prevented by either a mutation in the primer-binding site or a particularly large insertion [100 kilobase pairs (kb) or more] between the primer sites. When individuals are typed for polymorphisms based on PCR products, the presence or absence of a specific DNA fragment is scored after the electrophoretic separation of fragments (Fig. 21.9). In a genetic-mapping study, this means that the PCR-based DNA marker acts as a dominant, because individuals heterozygous for the allele corresponding to the amplified fragment are indistinguishable from homozygous-dominant individuals (*see* also Table 21.1). The following sources of the primers for PCR determine the region of the genome to be assayed for genetic variation:

1. The primers can be based on DNA sequences present at either ends of a gene that is well characterized and has a known function. Different alleles of the gene can be measured by PCR if an adequate DNA sequence database is available for designing the primers (e.g., Fig. 21.10a).
2. Primers can flank a region of DNA containing a simple, tandem array of bases such as CACACACACACACACA. This is an example of a microsatellite sequence formed by repetition of a simple dinucleotide sequence. Microsatellite sequences are prone to changes in the number of repetitive units in the array due to errors in replication of sequences of this type (*see* Chapter 18). The PCR products primed from specific genomic sequences flanking microsatellites fall into the class of probes referred to as sequence-tag sites (STSs), and they have a high probability of showing length polymorphisms useful for genetic studies (e.g.. Fig. 21.10b). Although primers for assaying microsatellites require extensive sequencing work on genomic DNA clones, once developed, they are extremely efficient for a range of applications, including genotype identification, germplasm evaluation, and genetic mapping.
3. Sets of primers, 10 base pairs (bp) in length, with each primer consisting of a random, decamer nucleotide sequence, have been used to assay variation in anonymous regions of the genome. The random-amplified-polymorphic DNA (RAPD) fragments often derive from repetitive families of DNA within the genome and, when used one at a time, commonly give multiple bands that show differences between individuals. RAPDs have been used in genetic

Table 21.1. Comparisons Among DNA Markers Used in Genetic Studies

Marker-Type Difference[a]	Expression	Definition of Alleles	Transfer in Crosses
Low-copy RFLP	Codominant	Unambiguous	Unambiguous
Multicopy RFLP	Dominant	Ambiguous	Ambiguous
RAPD	Dominant	Ambiguous	Ambiguous
STS	Codominant	Unambiguous	Unambiguous
AFLP	Codominant	Unambiguous	Unambiguous

[a] RFLP—restriction fragment-length polymorphism; RAPD—random-amplified-polymorphic DNA; STS—sequence-target sites; AFLP—amplified fragment length polymorphism.

Source: Modified from Hulbert (1993).

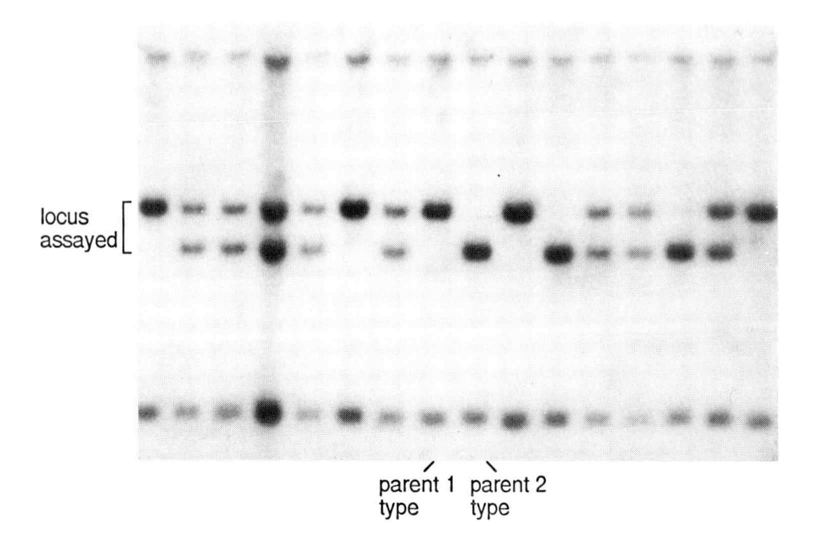

Fig. 21.8. RFLPs in an F_2 wheat progeny. The analysis assayed DNA samples from the leaves of 16 individual F_2 plants. The genomic DNA was digested with the restriction endonuclease EcoRV, and electrophoresed to separate the DNA fragments of different lengths; each lane is the analysis of DNA from a single F_2 individual. The electrophoretically separated DNA fragments were then transferred to a membrane and labeled using hybridization with a radioactively labeled DNA probe complementary to the RFLP sequences (Eastwood et al., 1994). Parental-type and heterozygous genotypes are indicated. As the markers are clearly distinguishable in both the homozygous and heterozygous states, they are codominant. In the individuals shown, the type 1 allele is present in 5 lanes, both alleles are present in 8 lanes, and the type 2 allele is present in 3 lanes; when 50 individuals are assayed in this way, their segregation ratio does not deviate significantly from 1:2:1. (Photograph kindly supplied by Dr. E.S. Lagudah)

studies but create difficulty in defining equivalent regions of the genome in different crosses. This difficulty can be overcome by cloning and sequencing useful polymorphic RAPD products and designing specific primers to assay STSs associated with these sequences.

4. Primers of known sequence can be ligated to genomic DNA cleaved by restriction endonucleases such as Pst (*see* Fig. 16.16, Chapter 16), as part of a procedure for assaying AFLPs (amplified fragment length polymorphisms, Table 21.1). The AFLP procedure is sensitive and reproducible for assaying variation in genomic DNA.

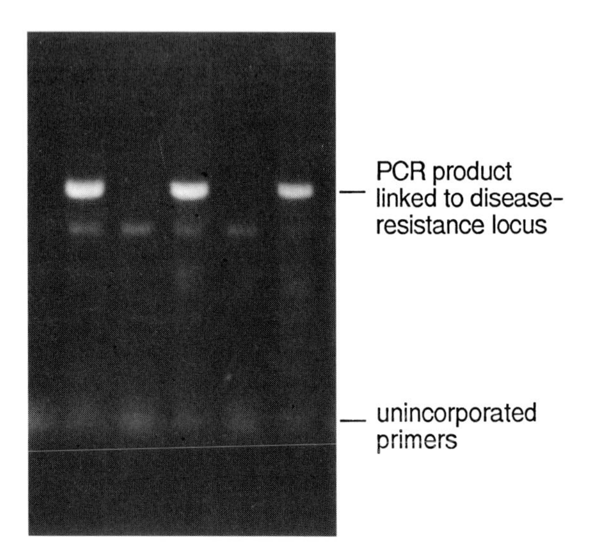

Fig. 21.9. The use of PCR to assay for segregating loci. In the cross illustrated, the PCR assay yields plus (bright bands)/minus (faint bands) reactions in five F_2 plants. Because a plus does not distinguish between a homozygote and a heterozygote, this infers that the loci assayed by PCR are often dominant. (Photograph kindly supplied by Dr. E.S. Lagudah)

21.3 CHROMOSOME ABERRATIONS PROVIDE A FIRST STEP FOR GENE MAPPING

In order to localize genes on chromosomes, it is necessary to be able to identify specific chromosomes and to have a range of chromosome aberrations available. This situation exists in plant species such as maize, tomato, wheat, and barley, and in *Drosophila*, mice, and humans. The steps in gene mapping include the following:

1. The use of sets of trisomics in diploids, or monosomics, usually in polyploids, to associate genes with chromosomes as units. The advantage of using aneuploids at this stage is that the backcross and F_2 segregations of a gene located anywhere on a trisomic or monosomic chromosome deviate from disomic segregations if appropriate genotypes are used in the parental crosses (*see* Chapter 13, Sections 13.1, 13.2, and 13.4; Chapter 14, Section 14.1).
2. The use of telotrisomics in diploids or monotelosomics in polyploids to localize genes to chromosome arms. After the chromosome carrying a gene has been identified, telochromosomes for one or both arms of that chromosome are used to narrow the location of the gene to one arm by means of backcross or F_2 ratios that deviate from a disomic segregation for one arm, but not for the other (*see* Chapter 13, Section 13.3; Chapter 14, Section 14.3).
3. The use of chromosome structural aberrations such as inversions or reciprocal translocations, with breakpoints in the chromosome arms bearing the genes under study. If a long arm is involved, it is desirable to use a paracentric inversion

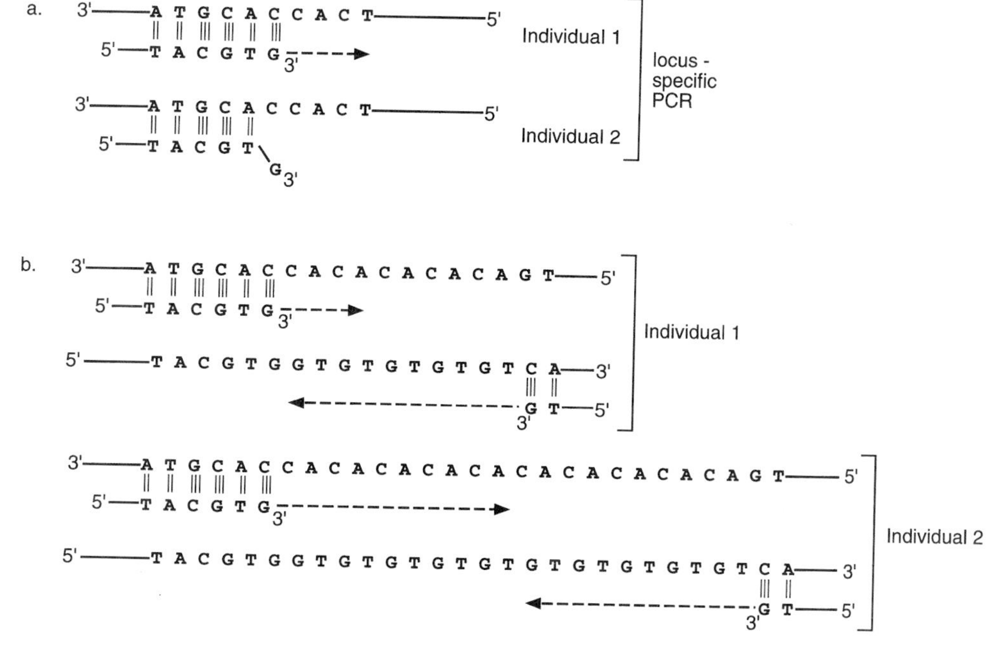

Fig. 21.10. The use of a single primer to identify individuals with single-base differences between alleles or satellite sequences of different lengths. (a) Assaying a single-base difference in a certain gene that is segregating in a population of individuals (homozygotes for a given allele indicated as individual 1 or 2) in a plus (top) and minus (bottom) reaction. The mismatch at the 3′-end of the primer in individual 2 means that the DNA polymerase used in the PCR cannot make a product; (b) Hypervariable region such as a microsatellite, consisting of a tandem array of dinucleotide or trinucleotide sequences. The sequences flanking the microsatellite provide the basis for PCR primers that can differentiate between individuals 1 and 2 based on the length of the segment that is amplified.

with the two breakpoints quite far apart, or two translocations that each have one breakpoint in the pertinent arm but with considerable distance between the breakpoints. Thus, linkage of a gene with one of the breakpoints is more likely to be detected (*see* Chapter 13, Sections 13.5 and 13.6).

21.4 DATA COLLECTION AND LINKAGE ANALYSIS

When genes or DNA sequences are assayed either directly by DNA studies or indirectly using the markers (other than DNA sequences) as described in Section 21.2, a genetic linkage map can be established. Mapping procedures are then used to determine whether the genes segregate independently of each other in the progeny of genetic crosses, family trees, or cell hybrids created in tissue culture. If genes do not segregate independently, they are usually located on the same chromosome and separable from each other only by the process of crossing over at meiosis or by some type of chromosomal aberration. When two genes are very close to each other on the same chromosome, the chance of a crossover event separating them is less than when they are far apart. The frequency of genetic recombination can thus be used as a scale to define genetic distance. However, crossing over is not uniform along the length of a chromosome and can also vary depending on the number of individuals involved in the measurement. Thus, genetic distances are best considered as statistical summaries of the relationships between a set of genes rather than fixed physical distances separating the genes.

21.4.1 The Principles Underlying Linkage Analysis

The principles underlying genetic mapping are best considered in the test-cross situation, where the F_1 individual from a cross is backcrossed to the parent carrying the recessive alleles of the genes under study. The example shown in Fig. 21.11 is the classical mapping of mutant genes for eye color, body color, and wing morphology on the X chromosome of *Drosophila melanogaster*. All recessive loci in the $backcross_1$ progeny can be scored unambiguously and recombinants are clearly identifiable in both male and female progeny because the Y chromosome does not have these loci. It is evident, from the data shown, that recombination is more frequent between *m* and *y* (or *w*) than between *y* and *w*. Furthermore, the recombination between *y* and *m* is slightly larger than that between *w* and *m*, and this allows the following genetic map of the loci to be postulated:

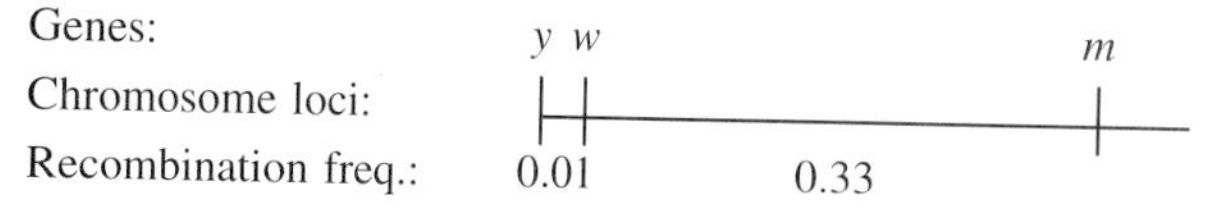

In more abstract terms, the basis of genetic-linkage studies is the recombination fraction (p), where

(i) the probability, or frequency, of an odd number of crossovers between two loci = p.

(ii) If there are r recombinants among N offspring from a cross, then $p = r/N$.

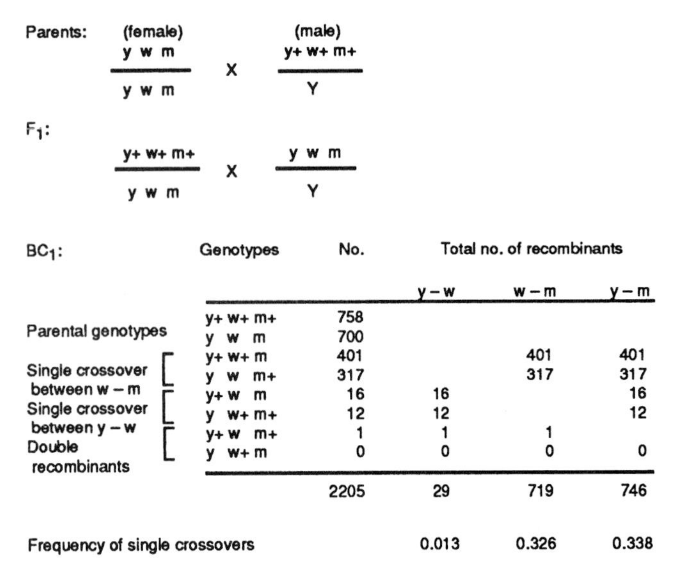

Observed frequency of double crossovers = 1/2205 = 0.00045
Expected frequency of double crossovers = 0.013 x 0.326 = 0.00424
Coefficient of coincidence = 0.00045/0.00424 = 0.11
Interference = 1 - 0.11 = 0.89 or 89% (a high level of interference)

Fig. 21.11. Some of the data used to provide a basis for the linear order of three loci on the X chromosome of *Drosophila melanogaster* (adapted from Sturtevant, 1913). The data demonstrate that double crossovers are fewer than expected based on the observed frequency of single-crossover events. The dominant alleles are designated as +, also used to designate the wild type. The mutant alleles are *y* = yellow, *w* = white eyes, *m* = miniature wings.

(iii) The number of offspring showing no recombination = $1 - p$.

(iv) The standard error can then be calculated as

$$\sqrt{\frac{p(1-p)}{N}}$$

(v) If one considers three closely linked loci, G, H, and J, in that order on the same chromosome, then

$$p_{GJ} = p_{GH} + p_{HJ}$$

(vi) For loci that are unlinked (i.e., on different chromosomes, or far enough apart on the same chromosome to have equal numbers of recombinants and nonrecombinants), $p = 0.5$.

21.4.2 The Analysis of Family and F_2 Progenies

In the analysis of family data, or segregating F_2 progenies, the information available for creating a genetic map is inherently incomplete, because even when all alleles involved are distinguishable or codominant, it is not possible to assess all recombinants unambiguously. For example, two loci (*G* and *H*), located on the same chromosome, may each have two alleles that are homozygous in the two individuals involved in the cross (i.e., *GGHH* and *gghh*). In the segregating F_2 progeny, double heterozygotes with the parental combination of chromosomes (*GH/gh*) cannot be distinguished from double heterozygotes with recombinant chromosomes (*Gh/gH*) (Table 21.2), and this creates uncertainty in the analysis. The problem can be overcome by analyzing backcross progenies, in which a F_1 hybrid is backcrossed to the recessive parent (*gghh*) and the BC_1F_1 progeny assayed for recombinant gametes transmitted through either female or male gametes (*see* Fig. 21.11 where the transmission is via F_1 female gametes). In practice, however, it is not always practical to do this. Furthermore, the analysis of F_2 segregating progenies has the advantage that both female and male gametes of F_1 individuals are considered.

Table 21.2. Expected F_2 Segregations after Selfing F_1 *GH/gh* Hybrids Involving Codominant Loci

	Gametes (♂)			
	Nonrecombinant		Recombinant	
Gametes (♀)	GH	gh	Gh	gH
Nonrecombinant				
GH	GH/GH	GH/gh	GH/Gh	GH/gH
gh	gh/GH	gh/gh	gh/Gh	gh/gH
Recombinant				
Gh	Gh/GH	Gh/gh	Gh/Gh	**Gh/gH**
gH	gH/GH	gH/gh	**gH/Gh**	gH/gH

Note: In using F_2 data to obtain recombination values, the nonrecombinant genotype *GH/gh* (underlined) cannot be distinguished from the recombinant genotype *Gh/gH* (bold).

In calculating the genetic distance between *G* and *H*, the two classes of progeny *GH/gh* and *Gh/gH*, (*see* Table 21.2), do not contribute information to the analysis and are not used. Subsequently,

(i) The frequency of the two types of parental, or nonrecombinant, gametes = $(1 - p)/2$.

(ii) The frequency of the two types of recombinant gametes = $p/2$.

(iii) The frequency of progeny with chromosomes of the parental type = $[(1 - p)/2]^2$.

(iv) The frequency of progeny with any one combination of a parental chromosome and a recombinant chromosome = $[(1 - p)/2](p/2)$.

(v) The frequency of progeny with two recombinant chromosomes = $(p/2)^2$.

Therefore, in an experiment where two codominant markers, *G* (or *g*) and *H* (or *h*) are scored, the numbers of progeny in each of the recombinant classes can be used to calculate *p*, and a final estimate of recombination frequency is derived by averaging the estimates of *p*. If the value of *p* is 0.5 for a given pair of loci, they are said to be unlinked, and if the value is less than 0.3, it is likely that they are linked, although this conclusion depends on the statistical error estimated for the measurement.

Chiasma or chromosome interference is the effect of one crossover on the probability that a second crossover will occur in a nearby segment of the same chromosome. The amount of interference is obtained by dividing the observed frequency of double crossovers by the expected frequency assuming no interference (product of the single crossovers in the two regions; *see* Fig 21.11 for example). The value obtained is called the coefficient of coincidence, which can range from 0 (complete interference) to 1 (no interference) or greater than 1 (negative interference). Values above 0 and below 1 indicate varying amounts of interference. Coincidence and interference are inversely related, and interference = 1 − coincidence.

Although computer programs are available to calculate genetic distances between recombining loci, it is important to understand the underlying principles so that the limitations of the data analyzed are clear. In human genetics, appropriate crosses for measuring p directly are not always available, and the statistical procedure of maximum likelihood provides the basis for an estimate of this parameter and whether the genes for the respective characters are linked. The mathematical treatment, namely maximum likelihood, takes into account the uncertainty in distinguishing double heterozygotes *GH/gh* (parental) and *Gh/gH* (recombinant).

21.4.3 The LOD Score

The likelihood function $L(p)$ is the probability of obtaining a sample of individuals similar to that observed experimentally in a distribution defined by p. The likelihood for a complex distribution is the product of the likelihoods of the component parts, and in this case, the log likelihood is a more convenient measure because the separate log likelihoods can be added together. Furthermore, $L(p)$ is a relative measure and can be multiplied by a constant if necessary. Any constant can also be added to $\log_{10}L(p)$. Originally, the $\log_{10}L(p)$ score for a pair of loci was defined as LOD (= log odds) and was standardized to 0 when $p = 0.5$. It is currently the convention to define the LOD score for a pair of loci as the $\log_{10}$ of the ratio of likelihoods when the loci are taken at their maximum-likelihood recombination, to when the loci are taken to be unlinked. An example of the calculation of LOD values in crosses is illustrated in Fig. 21.12. A value of 3.0 for the LOD score is generally taken as proof of linkage, since this indicates that linkage of the loci is 1000× more likely than nonlinkage. Although in terms of a statistic such as χ^2 (chi-square) this corresponds to a significance level of 0.0002, there remains a chance that two unlinked loci will appear linked. This problem accentuates the need to test many progeny in constructing a genetic map and to consider genetic linkages only as statistical summaries of the relationships between a set of genes.

21.4.4 Bayes' Theorem

Previous observations on the markers involved in a genetic-linkage determination can influence the interpretation of the linkage data. Bayes' theorem formalizes this by defining the probability (P) of a hypothesis (e.g., linkage), based on prior knowledge and additional evidence in the form of new data. Before any new observations are added, an opinion about the value of p can be represented by a prior distribution, $P(p)$. Following new observations in a set of S observations with a likelihood of $L(p)$, the later distribution is then defined by Bayes' theorem:

$$P(p\ S) = kP(p)L(p)$$

The constant k is introduced to normalize the posterior probability to 1 for all possible values of p.

21.4.5 The Influence of Linkage Phase on Gene Mapping

A particular region in the chromosome can be defined by markers that are either dominant, co-dominant, or recessive, and that are present in a coupling or repulsion phase. The coupling phase exists for marker loci when the dominant alleles enter a cross on one homologue and the recessive alleles on the other homologue. In the repulsion phase, there is a mixture of dominant and recessive alleles on each homologue. In a genetic analysis involving F_2 segregants or family data, co-dominant markers are ideal, since heterozygous individuals can be identified and the phase status is not relevant. Often, however, the markers can be either dominant or recessive, and in these situations the coupling or repulsion phase of the markers becomes significant. Figure 21.13 illustrates intervals A–I, defined by dominant and recessive markers in either the coupling or repulsion phase, and separated by 20% recombination units or centiMorgans (cM). Computer simulations show that in small populations of 50 to 100 individuals, genetic distances in the range of 0 to 20 cM can be difficult to distinguish when using dominant markers in repulsion (*see* Fig. 21.13, intervals C, E, F, G, H, I). The definition of these intervals is also affected by their

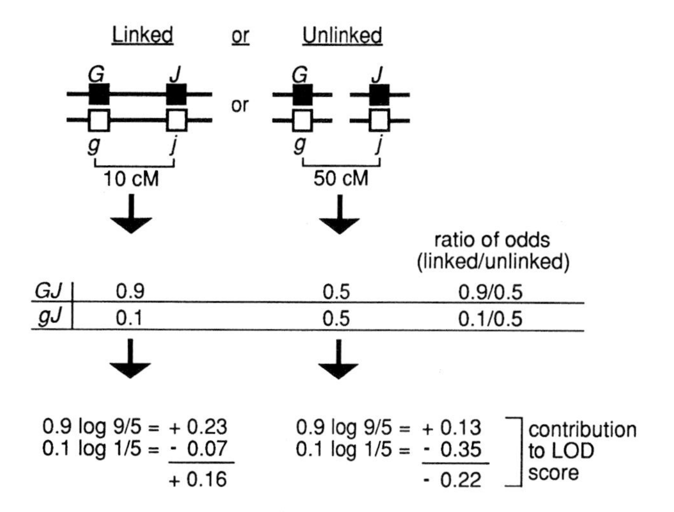

Fig. 21.12. Calculation of a LOD score for the simple situation of two loci linked by a genetic distance of 10 centimorgans (cM). The LOD score is always considered relative to loci that are unlinked and, by definition, separated by at least 50 cM (either on different chromosomes, or far apart on the same chromosome).

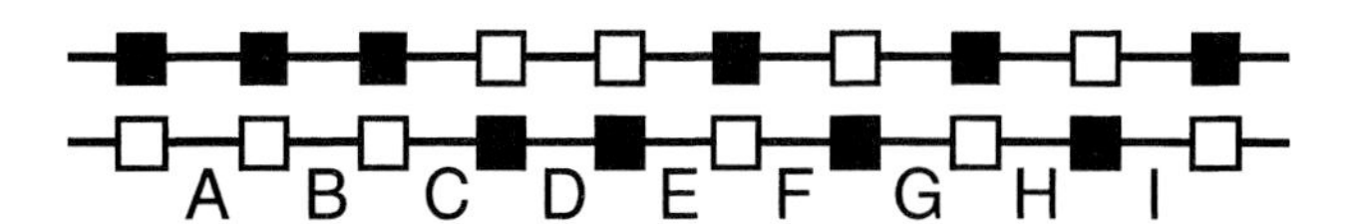

Fig. 21.13. In any specific cross, the actual linkage phase of the loci becomes important when one allele is dominant over the other. The phase can be either in coupling, where adjacent alleles on one strand of the DNA are both dominant or both recessive, or in repulsion, where successive alleles are either dominant or recessive. The black and white squares represent dominant versus recessive alleles of loci, and the letters represent the intervals between the loci.

Table 21.3a. Determination of Linkage in Coupling

	Parental Cross GGHH × gghh F_1 Gametes (♂)			
	Nonrecombinant		Recombinant	
F_1 Gametes (♀)	GH	gh	Gh	gH
Nonrecombinant				
GH	GH/GH	GH/gh	GH/Gh	GH/gH
gh	gh/GH	gh/gh	**gh/Gh**	**gh/gH**
Recombinant				
Gh	Gh/GH	**Gh/gh**	**Gh/Gh**	Gh/gH
gH	gH/GH	**gH/gh**	gH/Gh	**gH/gH**

Note: When the dominant markers are in coupling as indicated, the genotypes that are bold are distinguishable phenotypically from parental types because one locus is homozygous recessive. In this case, they contribute to the recombination frequency between the two genetic loci.

environment, and in an extreme case such as interval C, which is neighbored by sequences of dominant markers in coupling phase, a maximum-likelihood estimate is 0 % recombination, whereas the true distance is 20%. The estimates for intervals E, F, G, H, and I are less ambiguous, although the I interval still creates a problem because of the absence of flanking markers on one side. Best estimates of genetic distances are obtained for intervals A, B, and D.

The cross illustrated in Section 21.4.2 can now be re-examined in the situation where *G* and *H* are dominant to *g* and *h*, respectively, and in a coupling or repulsion configuration (Table 21.3a and b).

21.4.6 Mapping Functions

Although p is related to the physical distance (x) between loci, not all crossovers between two loci are detected, because double crossovers can give what are apparently parental-type chromosomes. Only odd numbers of crossovers result in an exchange of alleles, and as the genes that are being mapped move further apart, the crossover frequency is lower than expected because of the occurrence of undetected double or even higher numbers of crossovers. Mathematically, this underestimation of p can be corrected using a mapping function, and the common equation is the Haldane function:

$$p = 1/2(1 - e^{-2x})$$

where e is the exponential function. The relationship between p and x is further complicated by the lack of uniform crossing over along the length of the chromosome, and by the fact that crossovers do not occur close together because of chiasma interference (*see* Fig. 21.11). Although little is known about interference in plants, in animals the phenomenon was measured using the distribution of cytologically observed chiasmata. These factors complicate the use of mapping functions based on mathematical principles alone, and necessitate functions that are based on empirical observations. These include the Kosambi function, in which

$$4x = \ln(1 + 2p) - \ln(1 - 2p)$$

where ln is a natural log function; and also the Ludwig function:

$$2p = \sin 2x$$

While the Kosambi function is the most widely used mapping function, major distortions in the genetic map, relative to the physical map, are still evident even when this function is used.

21.4.7 Multipoint Mapping

Well before the advent of suitable computer programs designed to implement multipoint procedures for genetic-linkage studies, three-point mapping, using three gene loci, was commonly used to develop genetic-linkage maps. When using the multipoint-mapping procedure to determine the order of five loci such as G, H, I, J, and K, the two loci showing the least recombination (e.g., G and H), based on two- or three-point tests, are taken as a starting point, and a third marker is introduced to determine one of two possible maps (i.e., G–H–J, or H–G–J, ignoring the orientation relative to the centromere). The fourth marker is then introduced and six possible maps are examined. Finally, the fifth marker is studied to determine the most likely gene order from among 30 possible maps. The map that best fits the pattern of recombination for the five loci is accepted as the most likely map.

In multipoint mapping with computers, the pattern of recombination between all possible combinations of the five loci is examined, equal to 5!/2 or 60 maps, and a log-likelihood measure is assigned to each possible map. The absolute value of the log likelihood for a map of five loci is usually very small, because it is a statement defining the likelihood of any given meiosis providing recombination events that exactly fit the respective map. The relative value of log likelihoods between alternate maps is generally a more meaningful measure, and can be used to select the map most likely to account for the data provided by a range of meiotic events.

Table 21.3b. Determination of Linkage in Repulsion

	Parental Cross: GGhh × ggHH F_1 Gametes (♂)			
	Nonrecombinant		Recombinant	
F_1 Gametes (♀)	Gh	gH	GH	gh
Nonrecombinant				
Gh	Gh/Gh	Gh/gH	Gh/GH	Gh/gh
gH	gH/Gh	gH/gH	gH/GH	gH/gh
Recombinant				
GH	GH/Gh	GH/gH	GH/GH	GH/gh
gh	gh/Gh	gh/gH	gh/GH	**gh/gh**

Note: Only the *gh/gh* recombinant progeny are distinguishable phenotypically from the parental types and this is therefore a very inefficient way of measuring recombination. For example, if the loci are one map unit apart, the *gh/gh* class occurs, on average, once in every 40,000 progeny. The statistical error for estimating frequencies for rare events of this type is very high.

21.4.8 Interval Mapping

In three-point crosses, interval mapping (also referred to as marker-bracket mapping) improves the estimation of the probability that a test locus H is linked to marker loci G and J (Fig. 21.14) by considering only offspring in which G and J have not recombined. This allows the chromosome segment defined by G and J to be considered as a single, physically large marker, and establishes the linkage of H based on the chance of a double crossover within the G–J interval. No matter where H falls within the G–J

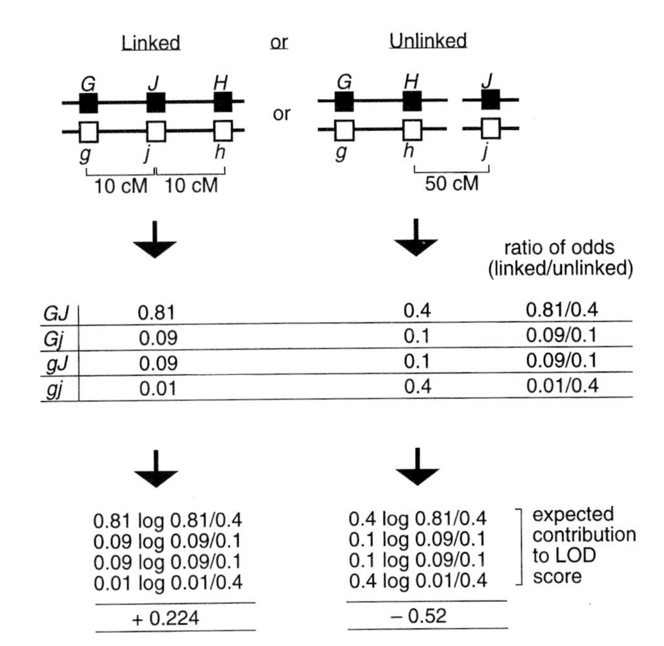

Fig. 21.14. Diagrams summarizing the alternate positions of *H* relative to a pair of markers, *G* and *J*, as used to define an interval for mapping purposes.

interval, linkage will be established, thus the procedure is simplified to looking for evidence of either presence or absence of tight linkage of H to the G–J interval. The rapid advances in producing detailed genetic maps using DNA markers means that a sufficient number of intervals can be defined to allow the genetic dissection of more complex characters. In the field of human genetics, the interval-mapping approach provides a means for analyzing the often limited database that is available.

21.4.9 Quantitative Trait Loci (QTLs)

The combination of extensive genetic maps and interval mapping facilitates the genetic dissection of factors responsible for quantitative changes in a phenotypic trait (*see* also Chapter 14, Section 14.4.3). If environmental influences on the phenotype can be controlled, quantitative differences are usually determined by allelic substitutions at many loci, which are referred to as quantitative trait loci (QTLs). It is important to recognize that the inheritance of quantitative characters is dependent on contributing genes that are subject to the same laws of transmission, and that have the same general properties, as genes for qualitative traits. Genetic markers can be used to recognize the regions of chromosomes that are involved with quantitative characters and to analyze populations that are segregating for these traits. The relative importance of the respective chromosome regions containing the different loci can then be established.

The distribution of a quantitative trait in two populations of individuals (A and B), and in the F_1 progeny from a cross between two A and B individuals, is illustrated in Fig. 21.15. The terms used to describe the observations on the populations include:

μA, μB, μF_1 = means of the variation in the populations;
$\sigma^2 A$, $\sigma^2 B$, $\sigma^2 F_1$ = variances of each population;
D = (μA − μB), the phenotypic difference between populations A and B;
k = number of effective factors (QTLs) in a cross contributing to the quantitative trait.

If QTLs have equal magnitude, are unlinked, and all alleles for the high phenotype contribute to increasing the phenotype, while all alleles for the low phenotype contribute to decreasing the phenotype, then:

D/k = the effect of each QTL;
1/k = variance due to each QTL.

Although the above assumptions do not generally hold in nature, they provide a basic set of definitions for computer-based analyses to determine linkage between genetic markers and QTLs.

The linkage between genetic markers and QTLs can be determined by maximum-likelihood or regression methods. In analyzing QTLs, the individuals in a segregating population showing the extreme phenotypes, i.e., those with alleles of greatest magnitude, are the most informative, and the study can be simplified by determining the genotypes only of these selected individuals, a process of selective genotyping. In a normal, segregating population, progeny more than 1 standard deviation from the mean comprise approximately 33% of the total population, but can contribute approximately 81% of the total linkage information. An initial statistical

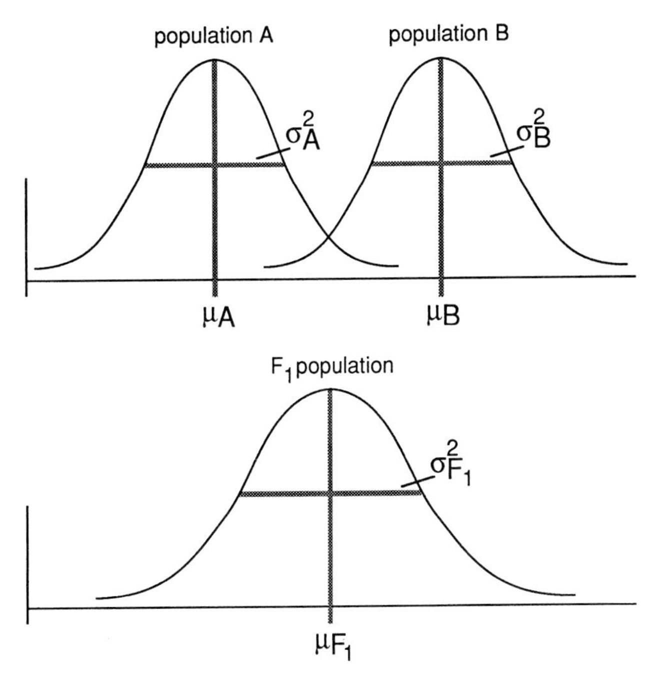

Fig. 21.15. The analysis of quantitative traits. A hybrid population of segregating individuals (bottom) does not give clear parental (top) or heterozygous types, but rather a distribution of values. The diagram summarizes the terms used (*see* text) to describe the distribution of a quantitative trait (Lander and Botstein, 1989). Multiple genes or quantitative trait loci (QTLs) control the output of the trait being measured. The search for marker genes that bracket the QTLs (Thoday, 1961) can provide the basis for genetic mapping, using interval mapping combined with maximum-likehood statistical procedures (Lander and Botstein, 1989) or regression analyses (Haley and Knott, 1992) to establish linkage of flanking markers with a QTL.

test can be applied to the subpopulations to identify markers that are associated with the differentiating phenotypic trait. In a more-detailed analysis, the ideal situation is where a complete genetic map, with markers spaced every 10 cM, is available to dissect the quantitative trait and determine the number and regional locations of the genes that influence the character. The LOD score (*see* Section 21.4.3) for the analysis gives a guide for the probability that the association between a QTL and a genetic-map interval arises by chance. The threshold, often set at LOD = 2–3, is affected by the number of linkage groups, and the extent and length of the genetic map. Complex phenotypes such as blood-pressure regulation, manic depression, and schizophrenia have been analyzed genetically by treating them as QTLs. While studies of this type contribute significantly to understanding a QTL, it is equally important to recognize that an accurate phenotypic definition of a QTL is essential for a genetic understanding of it.

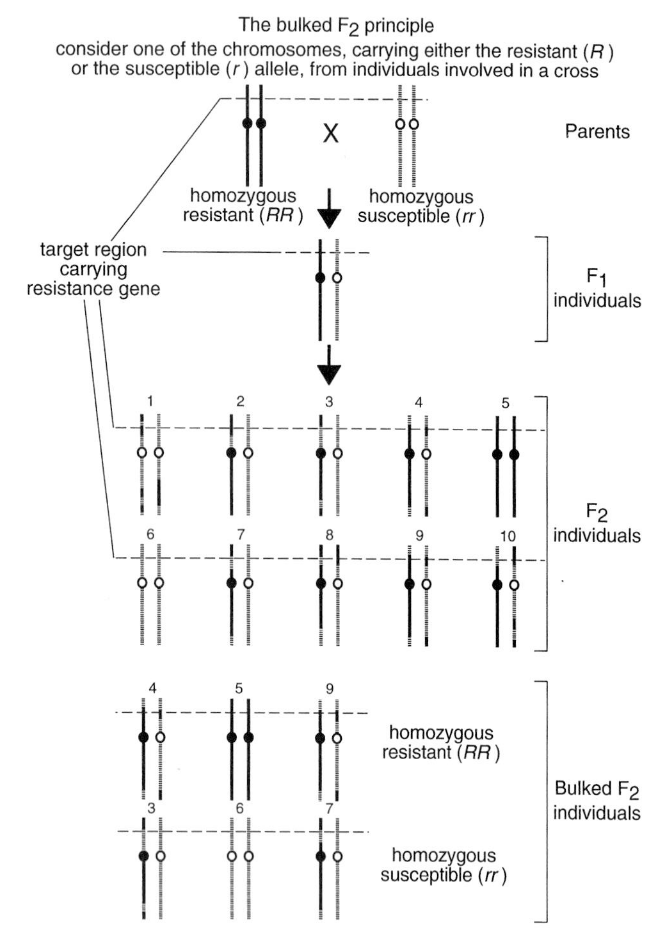

Fig. 21.16. Principle of the bulked F_2 method for uncovering DNA sequences linked to a selected locus. The bulked samples are screened for randomly amplified polymorphic DNA sequences (RAPDs) using PCR primed by random 10-bp primers. Any DNA products that differentiate the bulked DNA samples have a high probability of being linked to the target characteristic for each of the bulked samples. The procedure has proven to be a successful way of obtaining DNA markers that are close to the chromosome region carrying a trait of interest (Arnheim et al., 1985; Giovannoni et al., 1991; Michelmore et al., 1991; Eastwood et al., 1994; Tanksley et al., 1995). A prefractionation of the bulked DNA samples to remove repetitive sequences present in the genome (*see* Chapter 16) can provide a useful enhancement of the method, in that the characteristic sequences are commonly of low copy number that can then be isolated and used as probes in further studies (Eastwood et al., 1994). The prefractionation technique helps to reveal polymorphisms that may otherwise be hidden by the background created by the repetitive DNA sequences. The preparation of new, bulked DNA samples, based on the segregation of DNA markers isolated in the initial screen, provides a mechanism for saturating the genome region of interest with DNA markers. Selective enrichment of DNA linked to the target character, known as representational-difference analysis, has been described by Lisitsyn et al. (1993) and provides a valuable additional approach for analyzing phenotypic traits of specific interest to breeders.

21.4.10 Bulked Segregant Analysis

The procedure of bulked-segregant analysis, in which individuals in a segregating population that share a particular QTL or other genetic trait are bulked together, is a powerful approach for detecting new DNA markers linked to the trait (Fig. 21.16). The principle underlying bulked-segregant analysis was first developed to detect RFLP markers significantly associated with loci controlling human diseases. The procedure is most successful if it is first established that the individuals to be bulked are homozygous for the major genes that determine the QTL or the complex trait being investigated. The main difficulty to be overcome is that the expression of a complex trait can be affected by the environmental conditions under which the organism is maintained. In plants, it is often convenient to separate the genetic and environmental effects on the QTL phenotype by examining F_3 and F_4 progenies in order to establish, retrospectively, the precise genotype of the original F_2 individuals.

21.5 THE PRACTICAL ASPECTS OF GENE MAPPING

Organisms such as *Drosophila*, maize, barley, and tomato have a number of advantages for gene mapping. The primary advantage is that they are diploids, with relatively low chromosome numbers and easily identifiable chromosomes. Each species has a large number of identified spontaneous and induced mutations, which affect different phases of the life cycle; these are commonly maintained at specific germ-plasm centers. In plants, germ-plasm centers are now in operation for almost all agricultural crops. Even if individual mutants are not maintained at all of the locations, the centers are increasingly important for maintaining the genetic diversity of the original populations from which different crops have been developed.

21.5.1 Polymorphism Analysis and Selection of Parents

The most important consideration, when selecting parents in a genetic study, is that they should carry qualitatively different alleles for loci to be mapped. For this reason, hybrids between primitive landraces of plants (or animals) and modern crop cultivars (or highly developed, pedigreed animals) can be expected to produce F_1 and F_2 progenies with considerably higher degrees of heterozygosity than crossbreds between already improved plants and animals. In addition, the parents should be free of obvious chromosomal structural changes, cross easily, and produce fertile progeny. A survey of the gene pool in a species (or related species if required) is generally carried out to select suitable and interesting parents.

21.5.2 Backcross and F_2 Progenies

Genetic mapping using F_2 segregating progenies is efficient from the viewpoint of the numbers of individuals required for study, but, as mentioned earlier, these progenies have the disadvantage that certain recombinant classes cannot be distinguished. Consequently, F_3 progeny testing must be carried out so that all of the F_2 individuals can be classified genotypically. If, on the other hand, the F_1 hybrid is backcrossed to the recessive parent, none of the recombinants are concealed in the progeny. In the development of a backcross (BC_1) population for mapping, it is desirable to obtain approximately 100 progeny for study. The advantage of the BC_1 method is the comparative ease of obtaining a linear map of the markers. The disadvantage is that it is less efficient than a completely classified F_2 progeny, because recombination is sampled in only 50% of the gametes.

21.5.3 Recombinant-Inbred Lines, Single-Seed Descent, and Doubled Haploids

The main advantage of recombinant-inbred lines (Fig. 21.17), in which plants with particular combinations of alleles are continuously inbred to maintain those allelic combinations, is that they allow the development of true-breeding populations, which provide exactly the same genetic material for analyses by many different research groups. Other means for achieving such populations include single-seed-descent procedures, in which single plants with the desired allelic combinations are selected in every generation. Doubled-haploid techniques, where freshly recombined haploid cells are cultured to produce homozygous dihaploid plants, are also important (*see* Chapter 12, Section 12.5). These technologies provide the means for cooperation among different research groups, and allow very extensive genetic maps to be developed quickly. Once these maps are established, any new gene can be precisely located by reason of its proximity to known markers.

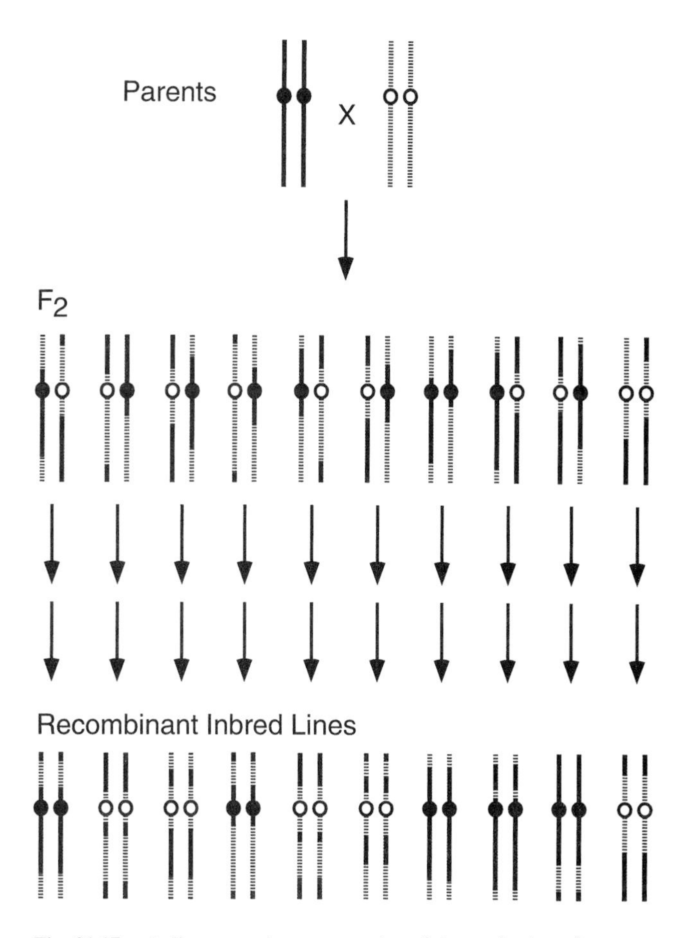

Fig. 21.17. A diagrammatic representation of the production of recombinant-inbred populations for genetic mapping. (From Burr and Burr, 1991.) The final recombinant inbred lines carry chromosomes that have undergone additional recombination in the F_2 and subsequent generations.

21.5.4 Recombinant Disomic Substitution Lines

In polyploid plants, the unique advantage of being able to create maintainable monosomic lines can be exploited to produce mapping populations for individual chromosomes. The procedure (*see* Chapter 14, Sections 14.4 and 14.5) essentially involves cloning individual chromosomes and recombination events, to produce true-breeding plants, which are suitable for the analysis of complex characters.

21.5.5 Genetic Maps for Forest Trees

The long-generation time of many economically important trees makes standard mapping procedures involving controlled crosses difficult. In conifers, haploid megagametophyte tissue is derived from post-meiotic, mitotic divisions of the megaspore that gives rise to the maternal gamete, and sufficient material is available from a single seed for the analysis of both isozyme and DNA variation. Genotyping of megaspores therefore effectively establishes the distribution of alleles of a given gene or DNA sequence in a population of gametes from a single tree. This provides the necessary segregation data for the construction of a genetic map for heterozygous loci in the particular tree. The polymerase-chain reaction (PCR) technology (*see* Section 21.2.6) provides the key to carrying out analyses on the DNA recovered from haploid megagametophyte tissue.

A pseudo-testcross mapping strategy has been developed for trees such as hardwoods, which do not have megagametophytic tissue available for analysis. The strategy is based on the availability of parents and F_1 progeny, and high levels of genetic heterozygosity in the parents. The application of PCR technology, combined with STS primers for PCR (*see* Section 21.2.6) is crucial in assaying for the necessary genetic variation. The procedure analyzes the DNA from parents and F_1 individuals with a wide range of STS primers, to detect the presence or absence of amplified DNA sequences in the F_1 individuals that are characteristic of either parent. The data is then analyzed to define the segregation behavior of the bands of amplified sequences that are present in parent A and absent from parent B and, for those cases, to determine whether the amplified sequence is always present in the F_1 individuals (expected if the locus was homozygous in parent A) or whether the band is segregating 1 : 1 for presence/absence (expected if the locus was heterozygous in parent A). When all of the segregating bands of sequences are analyzed, the F_1 population becomes the equivalent of a test backcross for all of the particular loci these bands represent. All the loci chosen in this way can then be used to generate a linkage map for parent A. A similar procedure can be carried out for the bands of amplified sequences that are present in parent B and absent in parent A, to generate a separate linkage map for parent B. To be effective, the strategy must be based on the availability of large numbers of STS primers to detect heterozygous differences between parents, and these differences can then be used to generate a genetic map. The more similar the parents, the greater the number of STS primers required to detect diversity between the parents.

21.5.6 Somatic-Cell Hybrids

In many organisms, but particularly in the larger mammals, it is not always possible to carry out controlled crosses or to obtain family histories to carry out genetic mapping. Often, however, it is possible to establish somatic-cell cultures and fuse these with established cell lines. The chemical selection of hybrid cells using HAT medium (Fig. 21.18) provides the basis for a different approach to genetic mapping. The random losses of chromosomes and their segments are comparatively frequent during the early stages of tissue culturing or after irradiation of a cell line. C-banding studies on cell lines missing single chromosomes, or carrying an additional chromosome or chromosome segment, can correlate biochemical or DNA markers with the presence or absence of a particular chromosome or segment and yield linkage data for the respective markers. It is also possible to fuse a full nucleus of one mammalian cell with a small part of the nucleus of another, and subsequently characterize the chromosomes and segments present. Using a chemical such as cytochalasin, the subnuclear fractions can be derived by inducing micronucleation and fusing the resulting microcells to normal cells. When a set of markers is consistently lost or gained at the same time as a certain chromosome segment is lost or added, the markers can be assigned to the respective chromosome segment.

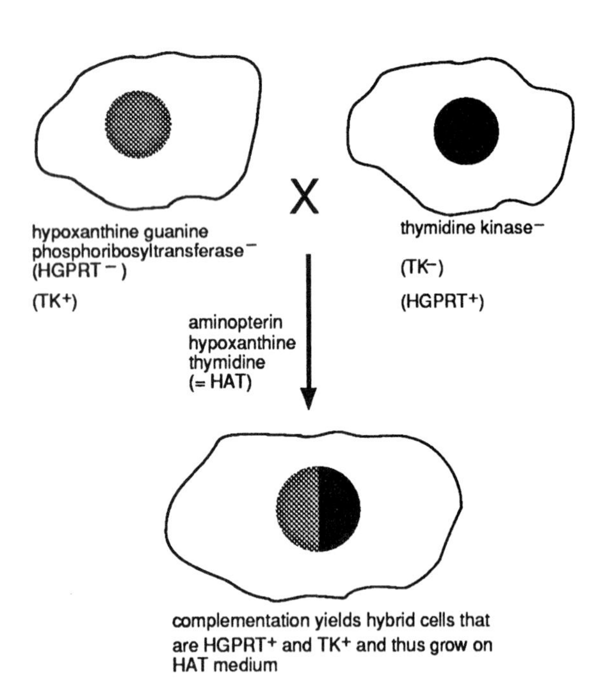

Fig. 21.18. The HAT selection procedure to isolate cell hybrids (Modified from Ringertz and Savage, 1976.) Aminopterin is a drug that blocks the main pathways for the synthesis of purines and pyrimidines, and cells containing this drug can only be rescued by the addition of thymidine and hypoxanthine because they provide substrates for minor, or salvage, biosynthetic pathways in the cell. AS TK- and HGPRT-cells lack the respective salvage pathways, they will not grow on HAT medium unless a hybrid is formed and complementation occurs.

21.5.7 Extended Human Families

A widely used database of families, including their histories and cell lines, is maintained in the Centre European pour le Polymorphism Humain (CEPH) in Paris, France. Information on three generations of individuals from 40 families defines the framework for a detailed, human-genetic map, as the DNA from the individuals in these families is distributed worldwide to research groups studying particular genes.

21.5.8 Homozygosity Mapping

The principle underlying this procedure is that a rare disease, due to a recessive gene(s), or some other recessive phenotype has variants of DNA markers, also rare in the population, that are associated with the respective genes responsible for the phenotype. Comprehensive DNA fingerprinting of inbred F_1 progeny showing the rare disease, or another recessive phenotype, and identifying the markers that are homozygous in the progeny and heterozygous in the parents (not showing the recessive trait) allows the DNA marker to be associated with the respective trait. If the trait of interest, or the DNA marker, is common in the population, the chances that it is homozygous by descent are difficult to separate from the possibility that it is homozygous as a result of random events. With a rare trait, however, a DNA marker that has a low probability of being homozygous in the population can be used to prove significant associations using homozygosity mapping.

21.6 CONSTRUCTION OF CHROMOSOME MAPS

Although large populations minimize errors in genetic mapping, in practical terms, 50–60 segregating F_2 individuals provide a useful map. In the case of test-cross and backcross progenies, recombinant-inbred lines, single-seed descent lines, and doubled-haploid lines, it is usually accepted that 80–100 individuals or lines need to be characterized.

21.6.1 Computer Analysis to Produce a Linkage Map

A broad selection of computer programs are presently available for the analysis of segregation data, and many such as MAPMAKER are based on maximum-likelihood methods. The application of genetic mapping to the analysis of quantitative-trait loci is also an active area of research, which does not necessarily rely on maximum-likelihood methods. In these instances, models using regression methods have been applied and have value because of their relative simplicity.

21.6.2 The Concept of a Consensus Map

A range of genetic maps are usually generated, even for the members of a single species, due to variations in the amount of recombination in different crosses, leading to distinct differences in genetic distances. In addition, the presence of translocations and inversions can lead to possible differences in linkage groups within a given species. The transmission of certain chromosomes can also be adversely affected in crosses between divergent individuals within a species, or in interspecific crosses. This can lead to distortion in the expected segregation ratios in F_2 progenies, but if the data from such crosses are treated with caution, the relative order of markers is generally unaffected. Variation, in fact, is an inherent feature of genetic maps. Because the establishment of a genetic map is usually one step in the process of understanding chromosome structure in more detail, a consensus or average map is often useful in designating a standard map for a species, based on the most common map determined for individuals of the species. Consensus maps can also be used to compare different species within a genus and genera within a family.

21.7 CORRELATION OF LINKAGE MAPS TO PHYSICAL MAPS

The complete nucleotide sequences of chromosomes provide clear end points for relating genetic maps to the physical chromosomes. The technology for large-scale sequencing is progressing rapidly through combining high through-put sequencing of random DNA segments and computer analysis of data; to date 7 bacterial and 1 yeast genome have been completely sequenced. In many organisms, however, it is not feasible to plan such a project, and alternate, less accurate procedures need to be adopted for relating the genetic and physical maps of a chromosome.

21.7.1 Procedures for Placing the Genetic Map into the Context of the Physical Description of a Chromosome

In yeast, many other fungi, and a wide range of other organisms with relatively small chromosomes, the direct linkage of molecular markers to chromosomes is possible. With the use of transverse-pulse-field electrophoresis (TPFE), full-length chromosomes can be separated in agarose gels (*see* Chapter 6, Fig. 6.6). After the DNA molecules are transferred to suitable membranes by blotting, they can be directly assayed with labeled probes to assign RFLP and other DNA markers to particular chromosomes. Using restriction endonucleases that recognize 8-bp or longer nucleotide sequences, which statistically occur much less frequently in the genome, a physical map for the chromosome, based on the distribution of restriction-endonuclease sites, can then be developed. An extension of the use of specific, rare, restriction-endonuclease sites to map long DNA segments is the use of STSs (sequence-target sites). STSs are defined as short, unique sequences that are readily amplified by PCR, using primers that target either ends of the sequence-target site, so that the amplification products provide distinctive-size classes of DNA fragments, which can be used as markers. In combination with in situ hybridization techniques, STSs also provide key landmarks for the process of ordering λ clones and yeast artificial-chromosome clones. In *Drosophila* and humans, extensive regions of the chromosomes have now been mapped in this way.

Other methods for physically mapping chromosomes include banding procedures and in situ hybridization with defined sequences. The genetic mapping of C bands, in the same population of individuals as that used for mapping DNA and other protein markers, provides a means for inserting physical markers into a linkage map. The presence or absence of *Nor* loci can be scored microscopically as well as genetically, using DNA probes for the spacer region (Fig. 21.19). Chromosomes that have been well studied at the level of C bands or by other cytological techniques can also be investigated by creating deletions similar to those seen in Fig. 21.19. The combined analysis of these modified chromosomes and the presence or absence of DNA markers further bridge the relationship between genetic and physical maps.

In situ hybridization is a further means for placing specific DNA markers in a physical location on chromosomes. Whereas few organisms have the advantage of easily accessible polytene chromosomes, double-labeling techniques (discussed in Chapter 19, Section 19.1) are particularly useful in that they allow the ordering of markers with respect to the centromere. Detection of low-copy-number sequences using *in situ* hybridization technology is also attaining satisfactory levels of sensitivity. At present, many chromosome spreads are scored for the locations of the signals; then a statistical interpretation of the chromosomal distribution of the hybridization-signal data is required to tentatively locate the respective sequence.

21.7.2 The Relationship Between Cytogenetic and Genetic Maps

Cytogenetic maps are based on chromosome aberrations that allow genetic markers to be related to the physical description of the chromosome. Genetic-linkage maps are based on recombination ''distances'' between genetic markers. Although the two types of maps are generally collinear, as expected from the fact that every chromosome consists of a single DNA molecule, there is a wide discrepancy in the relative distances between markers in the two maps. Almost always, this is due to a general suppression of crossing over near the centromeres and an enhancement in the more distal regions of chromosomes. In some organisms, it appears that distal regions have a higher density of polymorphic genes and sequences, which are usable in gene-mapping studies. Certain other regions of chromosomes are also known to be hot spots of genetic recombination. The nonrandom distribution of recombination along the length of the chromosome has important implications for chromosome manipulation in breeding programs. If genes of interest are located near the centromeres, much larger populations have to be analyzed to recover the desired recombinants, whereas when the genes are located near the ends of chromosomes, smaller populations can be used. Similarly, strategies for chromosome walking from DNA markers closely linked to genes of significance in breeding will be most successful for linkages in regions of high-recombinational activity, where the genetically linked marker is physically much closer to the gene of interest.

21.7.3 The Length of Genetic Maps

The length of a physical map of a chromosome is fixed and is ultimately defined at the DNA sequence level. The genetic map, on the other hand, is based on a complex biological phenomenon (crossing over) and, as pointed out earlier, represents a statistical summary of the relationships between genes on chromosomes. Historically, chiasma counts in bivalents at meiosis have been used to estimate the maximum genetic length of a pair of homologous chromosomes. Thus, a bivalent with one chiasma per arm has a

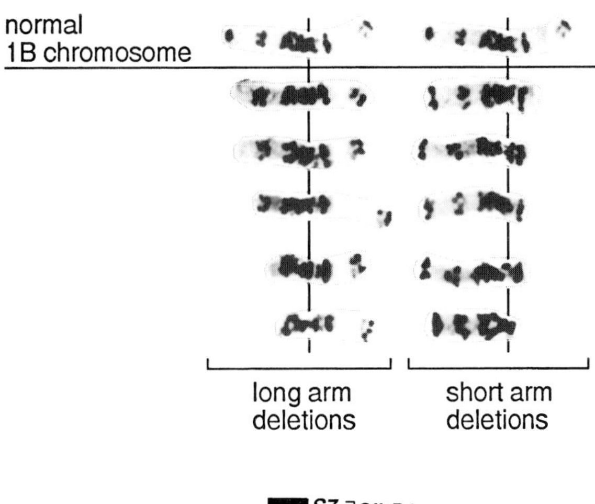

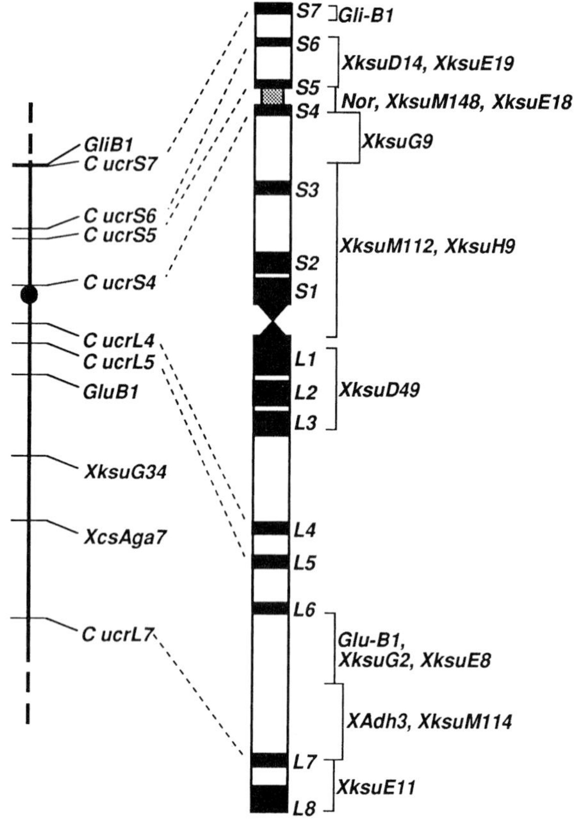

Fig. 21.19. Chromosome deletions characterized by C-banding. An analysis of chromosome 1B of wheat is shown; the top panel demonstrates a series of chromosomes "cloned" in different wheat lines and carrying different-sized deletions (deletions of the short arm and of the long arm as well as the normal 1B chromosome are indicated). The deletions are considered terminal and extending to differing lengths along the chromosome arms. Many deletions of the short arm result in loss of the ribosomal RNA genes, which is readily monitored by DNA probes. A similar scoring of the presence or absence of other DNA and protein markers provides an order of markers that can be directly related to the genetic map of the chromosome, as shown in the lower panel. (From Van Deynze et al., 1995.)

maximum genetic length of 100 cM. However, from intensive gene-mapping studies, two trends have become clear:

1. Genetic maps are considerably longer than expected from chiasma frequencies.
2. In most organisms, genetic maps are expanding as more DNA markers are added.

These trends can be clearly seen in Table 21.4, where the genetic lengths of maize chromosomes are summarized during the course of studies over the past 60 years. The tendency to increasing length as more markers are added to the genetic map is seen in a diverse group of plants including *Brassica* spp., *Hordeum vulgare*, *Lactuca sativa*, *Lycopersicon esculentum*, *Oryza sativa*, *Pisum sativum*, and *Solanum tuberosum*. One implication is that cytologically assayed chiasma counts do not provide a good estimate of recombination in an organism. Furthermore, although errors in scoring can contribute to the expansion of genetic maps, it is clear that recombination hot spots and gene conversion-type events are significant contributors to the tendency for maps to increase in genetic length.

21.8 WIDER APPLICATIONS FOR GENETIC MAPS

A central problem in many genetic situations, including plant and animal breeding, and human genetic counseling, is the assessment of the degree of homozygosity and/or heterozygosity. In plants and insects, lines that are near-isogenic, or homozygous at almost all genetic loci, can be relatively easily developed by either self-pollinating in the case of plants or crossing closely related animals (inbreeding). After several generations of continuous selection for the presence of the wanted traits, near 100% homozygosity is achieved. During this process, the number of heterozygous genes are diminished by half in each generation and the degree of homozygosity is increased by that amount. In simple mathematical terms, this can be expressed as $(1/2)^n$, where n equals the number of selfed generations (denoted by S in plants). Thus, in the S1 generation, $(1/2)^1$ or 50% of the genes are expected to be heterozygous, reflecting the 1:2:1 segregation of Mendel's second law. In subsequent generations, this percentage continues to decrease by half giving 25%, 12.5%, 6.25%, 3.125% heterozygosity, and so on. After six generations, about 98.5% of the genes are expected to be homozygous, and the lines can then be described as near-isogenic. The closer other genes and DNA sequences are to the selected genes, the more likely that they also are isogenic. In practice, as we have seen, recombination does not occur freely in every part of the chromosome, and significant sections remain parental in type. This is seen in breeding programs, where backcrossing and selection for one gene result in the retention of nearby genes in the program, the phenomenon of linkage drag. The use of genetic maps, particularly those with a large number of molecular markers, allows direct selection for homozygous genes. In breeding programs, an extensive genetic map can lead to more objective decisions in the selection of parents and the early selection of desirable individuals in segregating populations, as discussed further in Section 21.8.2.

21.8.1 Genetic Counseling

In human genetics, the construction of genetic maps that include genes for diseases, as well as more complex characters such as

Table 21.4. Historical Development of Linkage Maps for *Zea mays* Chromosomes and Lengths of the Chromosomes (in cM) as Estimated from Chiasma Counts and Linkage Data

	Genetic Estimates Based on Either Chiasma Counts or Linkage Data					
	Chiasma Counts	Linkage Data				
Chromosome	1934[a]	1950[a]	1976[a]	1990[a]	1990[b]	1993[b]
1	187	102	156	161	176	238
2	163	58	128	155	155	229
3	150	92	121	128	167	194
4	148	80	111	143	137	174
5	148	44	72	87	107	235
6	110	52	64	68	78	169
7	123	50	96	112	112	131
8	123	20	28	28	42	173
9	110	52	71	138	140	132
10	98	68	57	99	95	115
Total	1360	618	904	1119	1209	1790

[A] Based on the analysis of gene markers.
[b] Based on the analysis of RFLP markers.
Source: Adapted from Nilsson et al. (1993).

predisposition to heart disease and strokes, will eventually provide much needed data for understanding the bases of these traits. Preventive measures such as appropriate life-style and/or choice of marriage partners can also be considered if a certain predisposition exists in any one individual. In the extreme situation of a prenatal diagnosis indicating, for example, a homozygous state for a severe form of thalassemia, genetic counseling is essential to any decision related to termination of pregnancy made by the parents.

21.8.2 Animal and Plant Breeding

In an ideal situation, a breeder knows the characters that need to be selected, and the following efficient procedures are required to ensure the selection of as many desirable genes as possible:

1. The new technologies help to minimize the number of crosses made by breeders, and to maximize the chances of obtaining new and desirable combinations of genes. The selection of a QTL in a breeding program, for example, is more efficient if a DNA or protein marker is closely linked to the trait. In addition, if two or more genes for a desirable trait are known to be closely linked from mapping studies, they can be reasonably expected to remain together in the breeding program. Conversely, a close linkage of desirable and undesirable genes indicates that special procedures are required to break the linkage.
2. DNA or protein markers can detect the presence of desirable genes when undesirable genes that affect the same or similar characters are present. In this way, it is possible to select for novel combinations of disease-resistance genes, for example.
3. In early-generation material, or as soon as possible in a breeding program, breeders want to remove individuals with deleterious characters and select those with desirable genes and superior characters, thereby minimizing the necessity to grow large populations of progenies.
4. Linkage mapping can be used to locate new mutations, either spontaneous or induced, that may arise in a breeding program and have potential use.
5. Pedigree and variety identifications, using ''fingerprint'' types of DNA-sequence probes, are important in many areas of quality control. The forensic application of this technology has had a major impact in legal challenges. In agricultural animals, these identifications are also known as ''hoofprints.''
6. Germplasm-diversity assessment, arising from the use of DNA probes, is a significant application in maintaining natural populations as genetic resources for future breeding needs.
7. Searches are needed for DNA markers linked to genes controlling the development of diseases that develop slowly and are, therefore, hard to measure and expensive to test. It can be much cheaper and simpler to identify the presence of markers that are closely linked to the wanted resistance genes.
8. Similarly, markers for genes that occur only occasionally in a breeding program and are thus difficult to select by routine testing, can provide a valuable means to maintain the respective genes in the breeding stock.
9. In crosses that are not readily controlled or difficult to achieve, specific DNA markers can provide confirmation of the extent of homozygosity and heterozygosity in the hybrids produced by crossing programs.
10. DNA probes for pathogens can provide the basis for their early detection using PCR technology.

Modern genetic maps are the culmination of a body of knowledge about the molecular biology of genes and other sequences in the genome, the biochemistry of gene action, and powerful computer technology. The maps provide a springboard for the detailed, structural characterization of specific regions of chromosomes and eventually entire chromosomes at the DNA-sequence level. The resulting nucleotide-sequence databases of a variety of plant and animal genomes allow the analysis of breeding populations by studying molecular variation in the genes of individuals of the population. Progress in projects such as Human Genome Sequencing provide a molecular basis to uncover and commercially exploit variation in other organisms that are significant economically but for which genome-sequence data are not available. For example, at the end of 1994, it was estimated that there would be more than 6000 markers on the human genome, and 4000 on the mouse genome. The benefit of these markers is that they can then be used to investigate and produce genetic maps of other mammals. The study of synteny and the development of syntenic maps, in which arrays of genes and DNA sequences are conserved in their order on lengths of chromosomes, provide the basis for interrelating the genetic maps of organisms. The chromosome segments may or may not be rearranged through chromosomal translocations and inversions.

BIBLIOGRAPHY

Section 21.1

BRENNER, S., ELGAR, G., SANDFORD, R., MACRAE, A., VENKATESH, B., APARICIO, S. 1993. Characterization of the pufferfish (*Fugu*) genome as a compact model vertebrate genome. Nature 366: 265.

HEUMANN, K., BÄHR, M., ALBERMANN, K., FRISHMAN, D., GLEIBNER, A., GERSTNER, M., HANI, J., MAIERI, A., PFEIFFER, F., ZOLLNER, A., MEWES, H.W. 1997. The yeast genome. Yeast CD@mips.embnet.org

Section 21.2

AJIOKA, J.W., SMOLLER, D.A., JONES, R.W., CARULLI, J.P., VELLEK, A.E.C., GARZA, D., LINK, A.J., DUNCAN, I.W., HARTL, D.L. 1991. *Drosophila* genome project: one-hit coverage in yeast artificial chromosomes. Chromosoma 100: 495–509.

BROWN, A.A., MUNDAY, J., ORAM, R.N. 1988. Use of isozyme-marked segments from wild barley (*Hordeum spontaneum*) in barley breeding. Plant Breed. 100: 280–288.

BROWN, A.H.D. 1980. Genetic basis of alcohol dehydrogenase polymorphism in *Hordeum spontaneum*. J. Hered. 71: 127–128.

EASTWOOD, R.F., LAGUDAH, E.S., APPELS, R. 1994. A directed search for DNA sequences tightly linked to cereal cyst nematode (CCN) resistance genes in *Triticum tauschii*. Genome 37: 311–319.

HULBERT, S.H. 1993. Molecular markers and the construction of genetic maps. In: Disease Analysis through Genetics and Biotechnology, Interdisciplinary bridges to improved sorghum and millet crops (Leslie, J.F., Frederiksen, R.A., eds.). Iowa State University Press, Ames. pp. 231–254.

LINDE-LAURSEN, I. 1978. Giemsa C-banding of barley chromosomes. I. Banding pattern polymorphism. Hereditas 88: 55–64.

MORELL, M.K., RAHMAN, S., ABRAHAMS, S.L., APPELS, R. 1995. The biochemistry and molecular biology of starch synthesis in cereals. Aust. J. Plant Physiol. 22: 647–660.

PLATT, D.I., PLATT, G.R. 1975. An Introduction to Modern Genetics. Addison-Wesley, Reading, MA.

STUBER, C.W. 1992. Biochemical and molecular markers in plant breeding. Plant Breed. Rev. 9: 37–61.

WEATHERALL, D.J., CLEGG, J.B. 1972. The Thalassaemia Syndromes. 2nd ed. Blackwell Scientific Publications, Oxford.

Section 21.3

LAW, C.N., SNAPE, J.W., WORLAND, A.J. 1981. Intraspecific chromosome manipulation. Phil. Trans. Roy. Soc. London Ser. B 292: 509–518.

Section 21.4

ALLARD, R.W. 1956. Formulas and tables to facilitate the calculation of recombination values in heredity. Hilgardia 24: 235–278.

ARNHEIM, N., STRANGE, C., ERLICH, H. 1985. Use of pooled DNA samples to detect linkage disequilibrium of polymorphic restriction fragments and human disease studies of the *HLA* class II loci. Proc. Natl. Acad. Sci. USA 82: 6970–6974.

DONIS-KELLER, H., ET AL. 1987. A genetic linkage map of the human genome. Cell 51: 319–337.

EDWARDS, M.D., STUBER, C.W., WENDEL, J.F. 1987. Molecular-marker-facilitated investigations of quantitative-trait loci in maize. I. Numbers, genomic distribution and types of gene action. Genetics 116: 113–125.

GIOVANNONI, J.J., WING, R.A., GANAL, M.W., TANKSLEY, S.D. 1991. Isolation of molecular markers from specific chromosomal intervals using DNA pools from existing mapping populations. Nucleic Acids Res. 19: 6553-6558.

HALEY, C.S., KNOTT, S.A. 1992. A simple regression method for mapping quantitative trait loci in crosses using flanking markers. Heredity 69: 315–324.

HANAFEY, M.K., RAFALSKI, A., TINGEY, S.V., WILLIAMS, J.G.K. 1992. Mapping efficiency of dominant and codominant markers in F_2, backcross, and recombinant inbred populations. Maize Genet. Conf. (March).

LANDER, E.S., BOTSTEIN, D. 1986. Mapping complex genetic traits in humans: new methods using a complete RFLP linkage map. Cold Spring Harbor Symp. Quant. Biol. 51: 49–62.

LANDER, E.S., BOTSTEIN, D. 1989. Mapping Mendelian factors underlying quantitative traits using RFLP linkage maps. Genetics 121: 185–199.

LANDER, E.S., GREEN, P., ABRAHAMSON, J., BARLOW, A., DALY, M.J., LINCOLN, S. NEWBURG, L. 1987. MAPMAKER: An interactive computer package for constructing primary genetic linkage maps of experimental and natural populations. Genomics 1: 174–181.

LAW, C.N. 1966. The location of genetic factors affecting a quantitative character in wheat. Genetics 53: 487–498.

LISITSYN, N., LISITSYN, N., WIGLER, M. 1993. Cloning the differences between two complex genomes. Science 259: 946–951.

MICHELMORE, R.W., PARAN, I., KESSELI, R.U. 1991. Identification of markers linked to disease-resistance genes by bulked segregant analysis: a rapid method to detect markers in specific genomic regions by using segregating populations. Proc. Natl. Acad, Sci. USA 88: 9828–9832.

MCINTOSH, R.A. 1987. Gene location and gene mapping in hexaploid wheat. In: Wheat and Wheat Improvement, 2nd ed.

(Heyne, E.G., ed.). American Society Agronomy, Madison, WI. pp. 269—300.

NEUMANN, P.E. 1991. Three-locus analysis using recombinant inbred strains and Bayes' theorem. Genetics 128: 631–638.

SMITH, C.A.B. 1986. The development of human linkage analysis. Ann. Human Genet. 50: 293–311.

SOLLER, M., BECKMANN, J.S. 1983. Genetic polymorphism in varietal identification and genetic improvement. Theoret. Appl. Genet. 67: 25–33.

STURTEVANT, A.H. 1913. The linear arrangement of six sex-linked factors in *Drosophila*, as shown by their mode of association. J. Exp. Zool. 14:43–59.

TANKSLEY, S.D., GANAL, M.W., MARTIN, G.B. 1995. Chromosome landing: a paradigm for map-based cloning in plants with large genomes. Trends Genet. 11: 63–68.

THODAY, J.M. 1961. Location of polygenes. Nature 191: 368–372.

Section 21.5

BURR, B., BURR, F.A. 1991. Recombinant inbreds for mapping in maize: theoretical and practical considerations Trends Genet. 7: 55–60.

GRATTAPAGLIA, D. 1994. Genetic mapping of quantitatively inherited economically important traits in *Eucalyptus*. Ph.D. thesis, North Carolina State University, Raleigh, NC.

GRATTAPAGLIA, D., SEDEROFF, R. 1994. Genetic linkage maps of *Eucalyptus grandis* and *Eucalyptus urophylla* using a pseudo-testcross: mapping strategy and RAPD markers. Genetics 137: 1121–1137.

RINGERTZ, N.R., SAVAGE, R.E. 1976. Cell Hybrids. Academic Press, New York.

Section 21.6

BARKER, D., GREEN, P., KNOWLTON, R., SCHUMM, J., LANDER, E., OLIPHANT, A., WILLARD, H., AKOTS, G., BROWN, V., GRAVIUS, T., HELMS, C., NELSON, C., PARKER, C., REDIKER, K., RISING, M., WATT, D., WEIFFENBACH, B., DONIS-KELLER, H. 1987. Genetic linkage map of human chromosome 7 with 63 DNA markers. Proc. Natl. Acad. Sci. USA 84: 8006–8010.

SUITER, K.A., WENDEL, J.F., CASER, J.S. 1983. LINKAGE-1: a pascal computer program for the detection and analysis of genetic linkage. J. Hered. 74: 203–204.

Section 21.7

VAN DEYNZE, A.E., DUBCOVSKY, J., GILL, K.S., NELSON, J.C., SORRELLS, M.E., DVORAK, J., GILL, B.S., LAGUDAH, E.S., MCCOUCH, S.R., APPELS, R. 1995. Molecular-genetic maps for group 1 chromosomes of Triticeae species and their relation to chromosomes in rice and oats. Genome 38: 45–59.

NILSSON, N-O, SALL, T., BENGTSSON, B.O. 1993. Chiasma and recombination data in plants: are they compatible? Trends Genet. 9: 344–347.

TANG, C.M., HOOD, D.W., MOXON, E.R. 1997. *Haemophilus influenzae*: the impact of whole genome sequencing on microbiology. Trends Genetics 13: 399–404.

VIII
Cytoplasmic DNA, Genetic Engineering, Organisms of Importance

22

Cytoplasmic DNA and Maternally Inherited Traits

- Nucleo-cytoplasmic interactions are an important component in the evolution of organisms.
- Cytoplasmic male sterility is associated with the expression of unusual mitochondrial genes.
- Several maternally inherited human diseases are associated with lesions in mitochondrial DNA.

The phenotype of an organism results from the interaction of the genotype with the environment. The genotype includes genetic information contained in the nucleus, as well as protosomal DNA in mitochondria and chloroplasts, both of which reside in the cytoplasm. Many products of the nuclear genes interact directly or indirectly with the products of mitochondrial and chloroplast genes to drive cell metabolism. Consequently, within a species, nuclear and cytoplasmic genomes co-evolve in a dynamic, yet specific, relationship, so that the genotype is the product of nuclear × cytoplasmic factors. The final phenotype of the organism is modulated by the environment, referred to as the genotype × environment interaction.

The interaction of the nuclear genome with the cytoplasmic genomes of the mitochondria and chloroplasts in relation to male sterility is an important example of how an understanding of the interactions between genomes has commercial implications in plant breeding.

22.1 THE DNA IN MITOCHONDRIA

The presence of DNA in mitochondria (mtDNA) was proven in the mid-1960s. It had previously been recognized that DNA had to be present to explain cases of non-Mendelian inheritance involving metabolic defects that were known to be controlled by the mitochondria. An example is the cytoplasmic *petite* mutants of yeast that are produced when yeast cells are treated with actinomycin. This antibiotic intercalates with the double-stranded DNA of the mitochondria and blocks the biochemical pathways needed for oxidative phosphorylation, so that treated yeast colonies are much smaller than normal. Visible proof of the presence of mtDNA was gained from electron microscopy of mitochondria and from DNA staining with immunofluorescent dyes. Electron microscopy was also used to show that mtDNA is circular with several molecules per mitochondrion, although linear molecules are also believed to occur in the mitochondria of some fungi and plants. The actual similarities of these DNA molecules to bacterial protosomes has been discussed (Chapter 1, Section 1.4) and, as in bacteria, the DNA molecules of mitochondria are self-replicating and are transmitted to new organelles during division. This division is autonomous from normal cell division, so whether new cells are formed by mitosis or egg cells by meiosis, the transmission of defective or normal mitochondria can be a fairly random process. The comparative sizes of the mtDNA molecules of humans, the yeast fungus, and a plant are shown in Fig. 22.1.

22.1.1 Vertebrate mtDNA

The mitochondrial DNAs from several vertebrate animals have now been completely sequenced. The sequences of human (Fig. 22.2), mouse, cow, and frog mtDNA show them to be a model of economy and compactness. In vertebrates, the mtDNA is a circular molecule varying in length from 16.3 to 17.5 kb. This DNA codes for 2 size classes of ribosomal RNA, 22 tRNAs, required for polypeptide synthesis within the mitochondrion, and 11 mRNAs, which specify 13 subunits of the mitochondrial respiratory complexes. These subunits are cytochrome oxidase-I, -II, and -III (CO I, CO II, CO III), two ATPase subunits (A6 and A8), cytochrome-b apoprotein (Cyt b), and seven subunits of the enzyme NADH dehydrogenase (ND1, ND2, ND3, ND4, ND4L, ND5, and ND6). The remainder of the DNA molecule is comprised of a noncoding region called the d-loop (*see* Fig. 22.2), which varies in size from 879 bp in mice to 2134 bp in the amphibian *Xenopus laevis*. In vertebrates, this region contains the replication origin of the so-called heavy or H-strand (O^H), as well as the promoters (P^H and P^L) for transcription from both strands of DNA. Few, if any, nucleotides separate individual gene sequences, and the tRNA genes are more or less evenly distributed around the genome, between the mRNA and ribosomal RNA coding sequences.

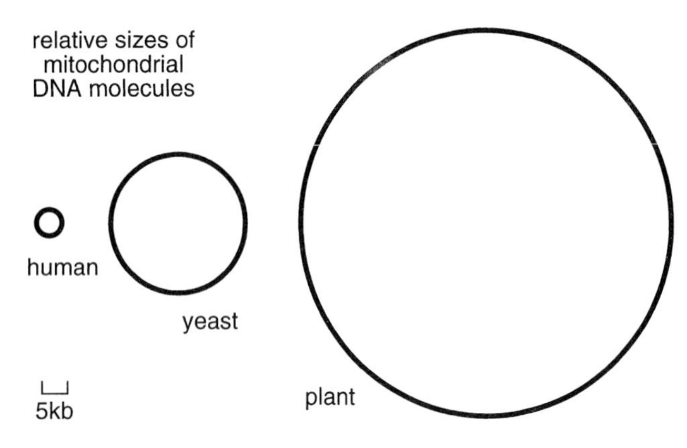

Fig. 22.1. The relative sizes of mitochondrial DNA from humans, fungi, and plants. In evolutionary terms, the major differences in size between the mtDNA of plants and animals can be ascribed to two separate introductions of the bacterial progenitors of mitochondria into what eventually developed into eukaryotic cells (*see* Fig. 1.2).

The order of the mtDNA genes, their nucleotide sequences, and their directions of transcription are virtually identical in all vertebrates. Because of this, the mtDNAs of a wide range of other vertebrates can be mapped by comparison to a master-type mtDNA genome such as that in humans.

22.1.2. Invertebrate mtDNA

The mtDNA of invertebrates is slightly smaller than that of vertebrates, and from the three invertebrate mtDNAs that have been completely sequenced, it is plain that the DNA is almost identical in function to that of vertebrate DNA. Invertebrate mtDNA contains the same genes, the same tRNA genes, and the same rRNA genes as vertebrate mtDNA, but differs significantly in the distribution of genes, the order in which they are arranged, and their nucleotide sequences. In particular, the tRNAs are highly clustered. At the macro level, these changes have been used to plot phylogenetic relationships within the animal kingdom. At the micro level, the changes in nucleotide sequences of the different genes have been, and are being, related to the evolution of different species.

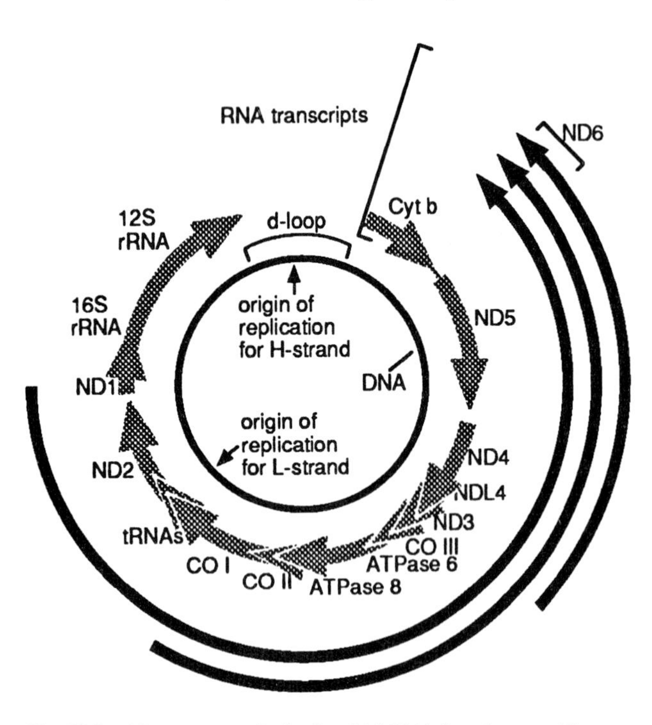

Fig. 22.2. The structure of mitochondrial DNA from humans. The central circle is the DNA molecule. The outer (black) arrows indicate the RNA that is transcribed from the DNA in the direction indicated. The inner (gray) arrows show the positions of the genes in the DNA.

22.1.3. Fungal and Plant mtDNAs

The mtDNA of fungi is generally three to five times the length of animal mtDNAs, and significant changes clearly differentiate the constituent genes and gene sequences from those previously discussed. The mtDNA of plants is even larger (*see* Fig. 22.1), and the complete sequencing of some molecules such as the 373-kb protosome in *Arabidopsis* mitochondria, has been achieved. Plant mtDNA ranges from 200 to 2500 kb in length, much of it probably due to multiple copies of a basic DNA sequence that are combined into a single and larger genome. Within this large genome, DNA sequences can be inverted and, together with the occurrence of deletions, result in a structure that is complex to interpret at a molecular level. In petunia, rice, maize, and beans, the population of mitochondria contains two or three distinct protosomes, which are not simply related to each other in terms of potential recombination products from a larger master molecule. These independent protosomes most likely are located in separate mitochondria, with important implications for understanding the phenomenon of cytoplasmic male sterility.

22.2 THE DNA OF CHLOROPLASTS AND PLASTIDS

Although the evidence for chloroplast DNA, cpDNA, was derived from observations using DNA-binding dyes, chloroplasts are only one of several types of plastids in plants. These different plastids are all derived from a precursor pro-plastid, which is comprised of an outer and an inner membrane with DNA inside. In plant development, the effect of light on the pro-plastids leads to the formation of chlorophyll, as well as the internal arrangement and thylakoid membranes of the mature chloroplast. Etioplasts are commonly present in seedling leaves grown in the absence of light. When the seedlings are transferred into light, the etioplasts become chloroplasts. In other tissues, the types of plastids include amyloplasts, which contain large granules of starch, and chromoplasts, which are filled with colored pigments that give flowers and fruits their color. Usually, plastid types such as chromoplasts are derived directly from chloroplasts in which the chlorophyll and thylakoid membranes have broken down and have been replaced by invaginations containing other pigments such as carotenoids.

In land plants, the cpDNA ranges in size from 120 to 160 kb. Three cpDNAs, from liverwort, tobacco, and rice (Fig. 22.3), have been completely sequenced. Many others have been partially sequenced and/or genetically mapped, including those of petunia, wheat, and the unicellular alga *Chlamydomonas*. All cpDNAs show an essentially similar arrangement of gene sequences. The basic structure is comprised of two inverted repeats, separated by small and large single-copy sequences (*see* Fig. 22.3). Although read in different directions, the inverted repeats are identical in length and always include the four ribosomal RNA genes coding for 5S, 4.5S, 23S, and 16S RNA found in chloroplasts. The lengths of the in-

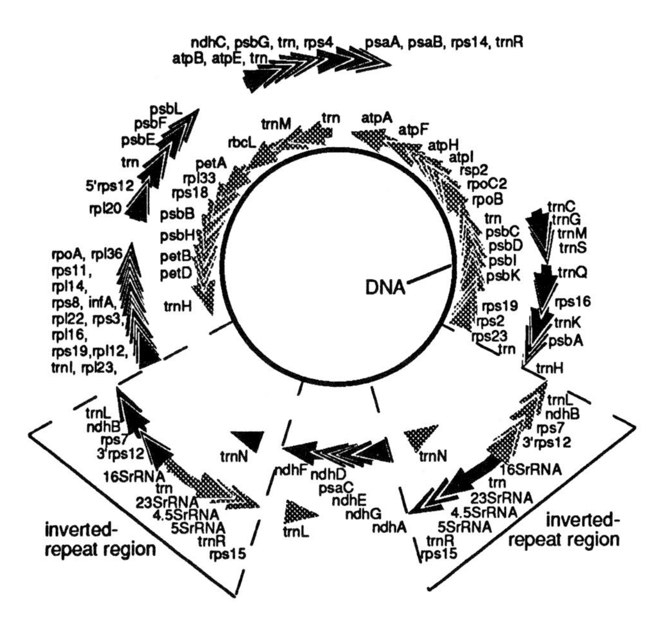

Fig. 22.3. The structure of chloroplast DNA from rice. (From Hiratsuka et al., 1989.) The central circle is the DNA molecule, with outer (black) arrows indicating the RNA, coding for known products, that is transcribed from the DNA in the direction indicated. The inner, grey arrows indicate the coding regions of the RNA. Not all RNA transcripts have been studied in sufficient detail to determine which originate from their own promoters and which are processing products from an RNA precursor. Unidentified, open-reading frames (ORFs or URFs) are not indicated in the figure.

verted-repeat regions differ between species, and in some plants, one of the repeats is missing. Different reading frames for decoding cpDNA into protein products occur throughout the genome. In total, cpDNA commonly contains 30–35 tRNA genes and from 50–70 other genes, not all of which have been identified. The sequences suspected of coding for protein products but not directly demonstrated to do so are designated as open- or unidentified-reading frames (ORFs or URFs).

In general, the overall size of the chloroplast genome is much more stable than that of the mitochondrial genome. Mutations are frequent, however, and often lead to changes in nucleotide sequences, which can be readily assayed by molecular-biology techniques, including restriction fragment-length polymorphisms (*see* Chapter 21, Section 21.2.6). Changes in cpDNA provide valuable markers for comparing different plants.

22.3 THE ANALYSIS OF CYTOPLASMIC TRAITS

Cytoplasmically controlled traits may be distinguished by transferring the nucleus from one species into the cytoplasm of another species by repeated backcrossing (as in plants), by protoplast fusion, or by fusion of an enucleated tissue-culture cell with the nucleus from another cell line (in plants and animals). The direct analyses of mitochondrial and chloroplast DNAs or their gene products can also provide new information on cytoplasmically controlled traits.

22.3.1 Reciprocal Interspecific Hybrids

Nuclear/cytoplasmic specificity is most readily identified during the course of interspecific hybridization. In fact, it was shortly after the rediscovery of Mendel's work that a number of traits were described with uniparental or non-Mendelian inheritance. It has since been realized that, with rare exceptions, chloroplasts and mitochondria are transmitted from one generation to the next exclusively through the egg. Thus, any traits associated with mitochondria and plastids show maternal inheritance. A common way to analyze cytoplasmic traits is to produce reciprocal hybrids, where each parent is used as a male and a female donor. The F_1 hybrids should have an identical complement of nuclear genes, so that any differences in their phenotypes may be ascribed to cytoplasmic inheritance.

If the hybrids are interspecific, it is common for the F_1 progeny to be completely sterile. At least part of this sterility has been linked to cytoplasmic effects, because there is incompatibility between the nuclear genome from one species and the cytoplasm from the other species. For example, F_1 hybrids between the wild relatives of wheat, *Triticum comosum* (female) and *T. uniaristatum* (male), are completely male-sterile. When the F_1 hybrids are backcrossed as females to *T. uniaristatum*, the BC_1 hybrids are male sterile like the F_1 plants. However, when the F_1 hybrids are backcrossed as females to *T. comosum*, the BC_1 hybrids are male-fertile, because they have *comosum* cytoplasm and a nuclear genome that is 75% *comosum*.

Another trait associated with nucleo-cytoplasmic interactions is the phenomenon of hybrid dysgenesis. This was first observed in *Drosophila* when males from certain wild strains were crossed with laboratory-bred females. The F_2 progenies showed a high incidence of sterility, mutations, chromosomal aberrations such as breakage and nondisjunction, and aberrant genetic ratios. The progenies from the reciprocal crosses were normal, indicating a cytoplasmic influence of certain *Drosophila* strains. It is now known that hybrid dysgenesis is due to the insertion and excision of transposable elements (*see* Chapter 7, Section 7.5), with critical events taking place during meiosis in the F_1 hybrids.

22.3.2 Nuclear Substitutions and the Genes Involved in Nucleo-Cytoplasmic Interactions

Following initial crosses between two parents, the F_1 hybrid is backcrossed repeatedly as female to the male parent (nuclear donor), thus retaining the cytoplasm from the female parent. Assuming random genetic assortment and no cytoplasmic transmission through the male gametes, the female parent's nuclear contribution is reduced by half after each backcross while its cytoplasm is exclusively maintained (*see* Fig. 22.4). Thus, after six or more backcrosses to the male parent, the reconstituted nucleus of the resulting alloplasmic lines is, in theory, essentially the same as the male parent, substituted in the cytoplasm of the female parent; in practice, certain blocks of genes from the female parent may be retained as a result of unconscious selection. The alloplasmic individuals often display a number of phenotypic differences affecting seed development, vigor, and fertility. In plants, the most notable trait—and usually the raison d'etre—is the phenomenon of cytoplasmic male sterility (*cms*), which has been exploited to develop crops of immense economic value.

In general, alloplasmic lines segregate for fertile or sterile individuals. The fertility is due to the action of nuclear *fertility-restoration (Rf)* genes, which are transferred from the cytoplasmic donor. The nuclear donor contributes *rf* genes for male sterility, which are used to maintain the CMS (cytoplasmic male-sterile) line (Fig.

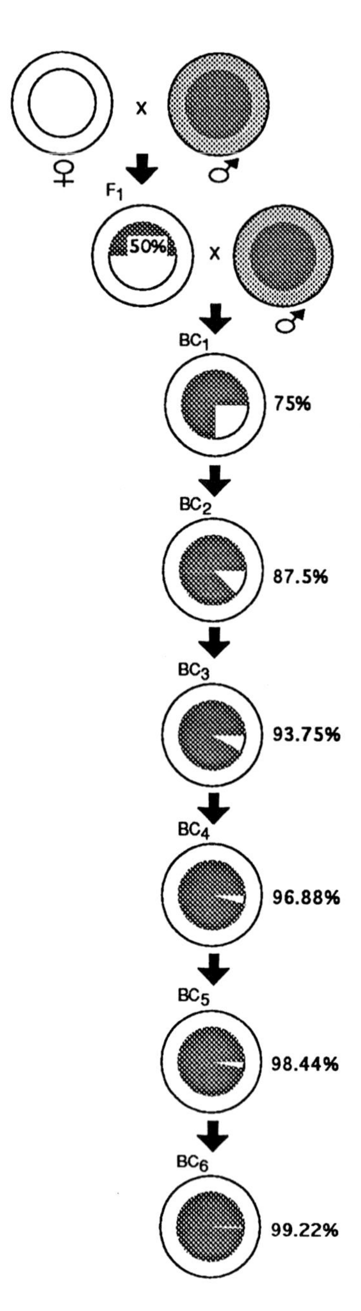

Fig. 22.4. The experimental substitution of the nucleus of one parent into the cytoplasm of another parent using sexual hybridization to produce near-isogenic lines. The mating of the female-parent cytoplasm donor with a male-parent nucleus donor is followed by repeated backcrosses to the male-nucleus donor for six or more generations. BC_1 to BC_6 = backcross 1 to backcross 6. The nuclear percentage figure shows the expected increase in the amount of male-nuclear genes per generation.

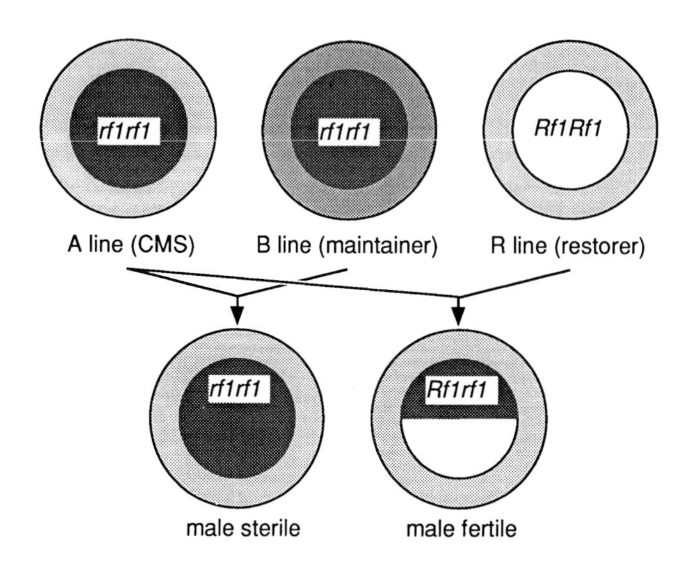

Fig. 22.5. The components of a hybrid-crop breeding system. The A line is male-sterile and is maintained by crossing with the B line (maintainer line) to give progeny that are male-sterile; the B line has a cytoplasm that is compatible with the *rf1 rf1* genotype and thus provides a source of viable pollen. The hybrid seed sold to farmers is produced by crossing the A line with the R line (restorer line) to give fertile progeny (in this case, the dominant *Rf1* allele allows the production of viable pollen). In practice, hybrid-crop breeding is quite complex because the A, B, and R lines must be maintained separately for different objectives.

22.5). By using the *Rf* parent to pollinate the CMS line, hybrid seed and full fertility of the F_1 individuals are restored.

The cytoplasm-specific *Rf* genes of a single species may be used to differentiate cytoplasms. In this method, a single *Rf* gene is substituted into different cytoplasms. Those cytoplasms that respond to a specific *Rf* gene in a similar way—resulting in fertility or sterility—are assumed to be closely related. These analyses are valuable in plant breeding for identifying diverse sources of cytoplasms and *Rf* genes, to guard against genetic uniformity of hybrid-crop plants.

Within a taxonomic group, cytoplasmic relationships can be traced by the nuclear-substitution method (Fig. 22.6). The nucleus

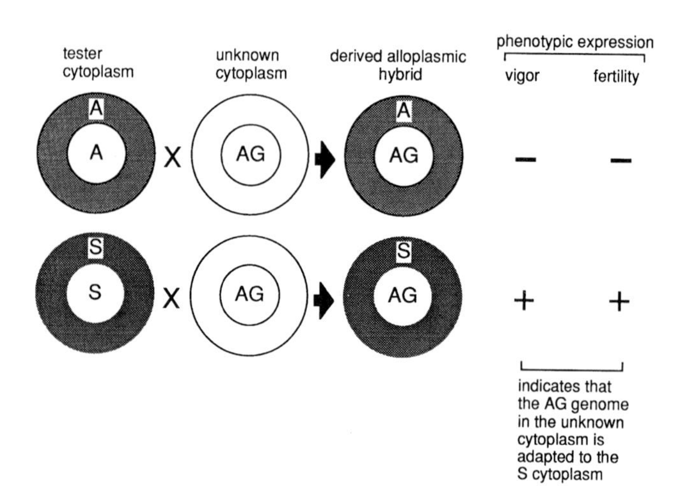

Fig. 22.6. Determination of the cytoplasmic donor of the tetraploid wheat *Triticum timopheevii* (AAGG). The nucleus (AG) of *T. timopheevii* was substituted into the cytoplasm of two possible diploid donor species, *T. monococcum* (A) and *T. speltoides* (S), by the nuclear-substitution method. The results, showing the vigor and fertility of alloplasmic hybrids (right), indicate that the cytoplasm of *T. timopheevii* came from *T. speltoides*.

of the standard type, generally a cultivated plant, is substituted into tester cytoplasms, generally wild species, to produce alloplasmic lines. The observed differences in phenotypic traits, which must result from differential nucleo-cytoplasmic interactions in the various alloplasmic lines, are used to classify cytoplasms. The cytoplasmic donor of the alloplasmic lines most closely resembling the standard type, by inference, must be the cytoplasmic donor and hence also a nuclear donor of the standard type. Cytoplasmic analysis is particularly powerful in polyploid plants that may trace their origins to two or more diploid-progenitor species (*see* Fig. 22.6; Fig. 22.7). The technique has been valuable, for example, in identifying the diploid-progenitor species *Triticum speltoides* (syn. *Aegilops speltoides*) as being most closely related to the species that donated the cytoplasm and B-genome to polyploid wheat.

At a more fundamental level, the nuclear-substitution method has given valuable information on the evolution of cytoplasmic genomes. Cytoplasmic evolution is quite conservative and an autogamous species may contain a single cytoplasmic type, whereas the cytoplasm in allogamous species may be more variable. Each recognized species has its own unique cytoplasm, which differs from other related species. This implies that cytoplasmic differentiation is necessary, along with nuclear genes, to establish reproductive isolation barriers. Nuclear-substitution studies also indicate coevolution of the nuclear and cytoplasmic genomes.

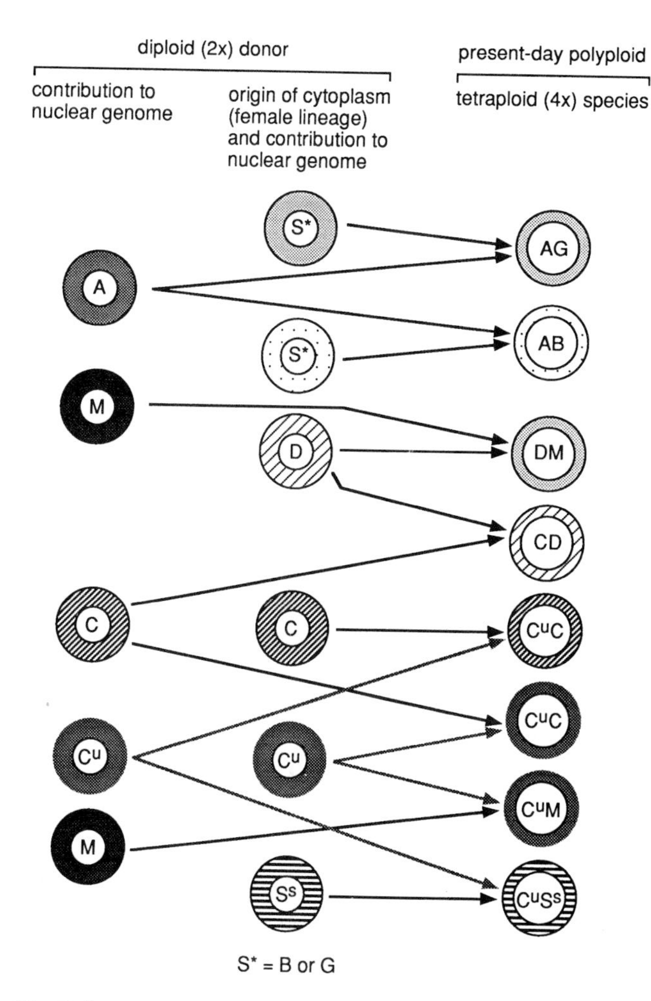

Fig. 22.7. The determination of the cytoplasmic donors of polyploid *Triticum* species as revealed by the nuclear-substitution method described in Fig. 22.6. The diagram illustrates the male (left column) and female (center column) parents of the crosses which must have led to the present-day tetraploid species. (Modified from Tsunewaki, 1991.)

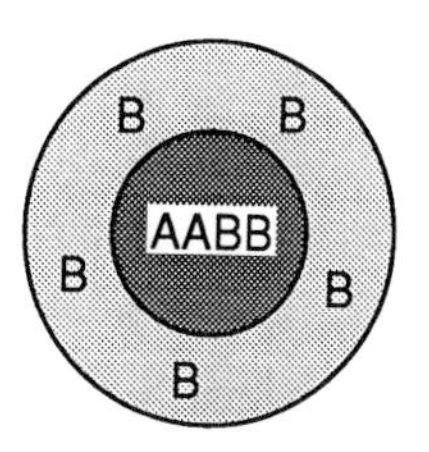

Fig. 22.8. Possible interactions within the nucleus and between the cytoplasm and nucleus, in a polyploid cell. These nuclear × cytoplasmic interactions modulate genome evolution and speciation in polyploid plants.

22.3.3 Nucleo-Cytoplasmic Interactions and Genome Evolution

In polyploid organisms, where two distinct nuclear genomes are brought together in cells with maternal cytoplasm, many types of interactions must occur between the different nuclear genomes and between the cytoplasm and the nucleus (*see* Fig. 22.8). Thus, genes involved in nucleo-cytoplasmic interactions (NCI) must play a major role in genome evolution and speciation in polyploid organisms, and the following postulates are considered particularly important:

1. A new amphiploid must pass through a bottleneck of sterility, resulting from an adverse interaction between the male nuclear genome and both the nuclear and cytoplasmic genomes of the female.
2. Certain mutations or cytogenetic changes (bottleneck-chromosome changes) must occur in the nuclear genome (and probably in the cytoplasm) to restore fertility.
3. Accelerated genic- and chromosomal-mutation rates during introgressive hybridization may provide the needed variation for these processes.

With respect to the first postulate, adverse interactions become apparent when a nucleus of one parent is substituted into the cytoplasm of the other parent. The resulting alloplasmic plants can be semilethal and male-sterile. Therefore, part of the sterility and lack of vigor in the new amphiploid is caused by adverse interactions between the male-nuclear and female-cytoplasmic genomes.

22.4 PROTOPLAST FUSION AND CYTOPLASMIC MANIPULATION

Although the nuclear-substitution method has been widely used for cytoplasmic analysis, its applicability beyond a closely related group of species is limited because of barriers to sexual hybridiza-

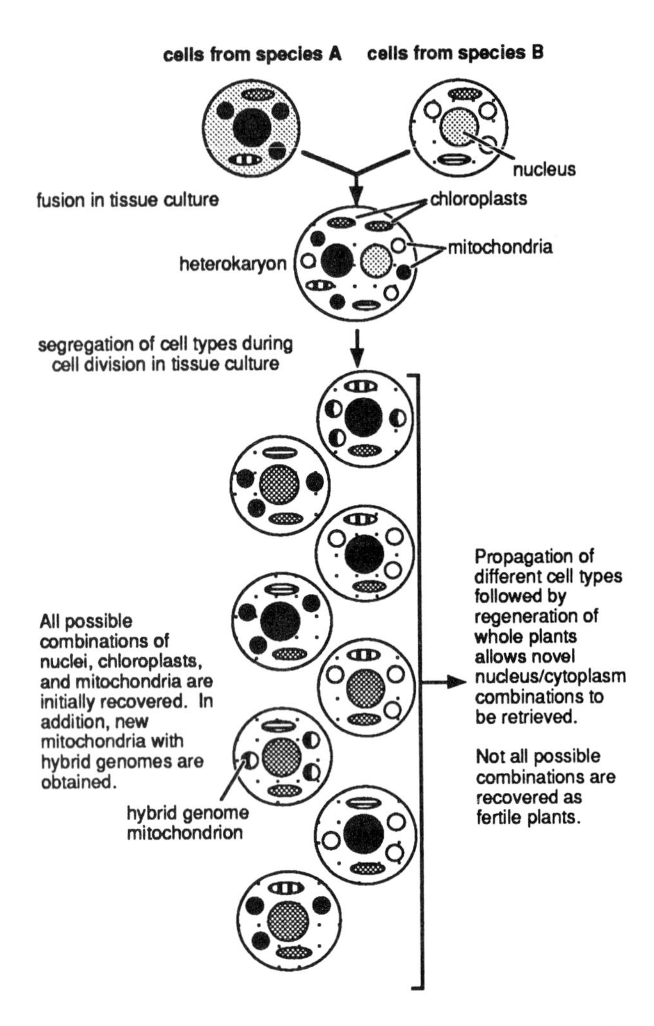

Fig. 22.9. The many types of cybrids with different combinations of cytoplasmic organelles that may arise following protoplast fusion. Although the mitochondria from two species may show recombination, recombination between chloroplasts is rare.

tion. Moreover, it allows unilateral substitution of cytoplasms but not actual mixing of two cytoplasmic types to produce novel combinations of chloroplasts and mitochondria. Therefore, techniques for the fusion of protoplasts by mechanical, chemical, or electrical means, followed by plant regeneration, provide versatile methods for creating new nucleo-cytoplasmic combinations. In most cases, hybrid cytoplasms segregate toward one or the other of the parental types within a few generations. There is evidence that mtDNAs may recombine prior to such segregation. In another technique, uniparental cytoplasm, with the nucleus removed (enucleated) or inactivated by irradiation, may be fused with isolated nuclei from the second parent to produce instant alloplasmic plants. These types of in vitro methods have been used to transfer cytoplasmic-male sterility and other traits within *Nicotiana* and *Petunia*.

22.4.1 Cybrids Versus Somatic Hybrids

Typically, when protoplasts from two species are induced to fuse, the first fusion product is a heterokaryon (Fig. 22.9). If the nuclei fail to fuse, they may segregate in a hybrid cytoplasmic background to form cybrids. A cybrid is defined as a somatic hybrid carrying the nucleus of one species in a hybrid or alien cytoplasm, with mitochondria, chloroplasts, or both, from another species. Cybrids can be experimentally produced by irradiating protoplasts of the cytoplasmic donor (to inactivate the nucleus) and fusing them with protoplasts of the nuclear donor, treated with iodoacetamide (to inactivate its organelle genomes), then using appropriate conditions to regenerate whole plants.

In many cases, cytoplasmic fusion is followed by fusion of the nuclei to form a symmetric-somatic hybrid, which is an amphidiploid ($4x$) containing complete chromosome complements ($2x$) of both parents in the cytoplasm of either parent (after segregation) or a hybrid cytoplasm. Asymmetric–somatic hybrids contain a complete chromosome complement from one parent, but only a few chromosomes from the second parent. Asymmetric hybrids can be experimentally produced by fusing intact-recipient protoplasts with irradiated donor protoplasts.

A large number of cybrids, symmetric– and asymmetric–somatic hybrids, involving sexually compatible and incompatible parental combinations, have been produced. Selected lists are provided in Tables 22.1, 22.2, and 22.3.

22.4.2 Genetic Incompatibility in Somatic Hybrids and Cybrids

Internuclear, intercytoplasmic, and nucleo-cytoplasmic incompatibilities are commonly observed in somatic hybrids, especially

Table 22.1. Some Cybrids and the Composition of Their Cytoplasm

Cybrid Combination		Cytoplasmic Combinations and Plant Phenotype of Cybrids[a]
Recipient (R)	Donor (D)	
Brassica oleracea var. × *botrytis*, cauliflower, Ogura CMS line	*B. campestris* var. *oleifera* cv. 'Candle,' atrazine resistant.	Dct + Dmt, 7 plants all male fertile. Dct + Rmt, 12 plants, 7 male fertile, 5 male sterile.
Brassica napus cv. × 'Regent,' CMS line	*B. campestris*, triazine resistant.	Dct + Dmt, 6 plants. Rct + mt unknown, 7 plants. Dct + Rmt, 3 plants. Dct + Dmt, 1 plant.

[a] ct = chloroplasts; mt = mitochondria; CMS = cytoplasmic male-sterile.

Source: Jourdan et al., 1989.

Table 22.2. Selected List of Somatic Hybrids (Symmetric and Asymmetric) Produced From Protoplast Fusions Between Mostly Sexually Incompatible Species, and Those Developed into Adult Plants

Lycopersicon esculentum	×	*Solanum tuberosum*
Arabidopsis thaliana	×	*Brassica campestris*
Nicotiana tabacum	×	*N. ripenda*
N. tabacum	×	*Solanum melongena*
N. tabacum	×	*Daucus carota*
N. tabacum	×	*Hyoscyamus muticus*
Citrus sinensis	×	*Poncirus trifoliata*
Datura innoxia	×	*Physalis minima*
Solanum tuberosum	×	*S. chacoense*
	×	*S. tuberosum*
	×	*S. brevidens*
	×	*S. berthaultii*
Nicotiana tabacum	×	*N. suaveolens*
	×	*S. tuberosum*
	×	*Atropa belladonna*
	×	*Petunia hybrida*
	×	*Salpiglossus sinuata*

those arising from distantly related species (Tables 22.2 and 22.3). Thus, many symmetric–somatic hybrids, which contain a complete chromosome complement from each parent and in theory should be fertile, are often cytologically unstable and sterile. Part of the sterility in these hybrids may be due to somaclonal variation such as somatic mutations, chromosome breakages, translocations, and duplications, which arise in tissue culture. The symmetric hybrids often change to asymmetric hybrids when chromosomes from one parent are preferentially eliminated (*see* Chapter 15, section 15.2). Some of the asymmetric hybrids may be fertile and are a more attractive alternative for plant-breeding purposes because of the sterility of many symmetric hybrids. The asymmetric hybrids are often produced by irradiation of protoplasts of the parent contributing the cytoplasm. However, in numerous reports of asymmetric hybrids, many donor chromosomes were retained in spite of the irradiation treatment.

Genetic incompatibility is also observed with different cytoplasmic combinations, and between the nucleus and the cytoplasm, in somatic hybrids and cybrids. As a rule, mixed organelles are observed only rarely, and chloroplasts and mitochondria assort independently (*see* Table 22.1; Fig. 22.9). Moreover, although mitochondria from two parents often recombine, recombination between chloroplasts is rare. Although cytoplasmic-organelle segregation is a common phenomenon, a certain combination of chloroplasts and mitochondria may predominate as a result of nucleo-cytoplasmic compatibility.

The sorting of organelles in cybrids has been exploited in plant breeding, where two notable examples of cytoplasmic breeding have been achieved in *Brassica* (*see* Table 22.1) and rice. In cauliflower, broccoli, and canola (rape), the Ogura line has provided a source of cytoplasmic male sterility (CMS) that is highly desirable because of its consistent male sterility and male-fertility restoration. However, at temperatures below 12°C, it exhibits paleness of leaves, presumably due to chloroplast/nucleus incompatibility, which causes a reduction in yield. Several workers were able to introduce different sources of herbicide-resistant chloroplasts, which also overcame the paleness problem. Thus, cybridization not only corrected a genetic defect but allowed herbicide resistance to be combined with a CMS trait. In rice, a valuable CMS trait was transferred from Indica rice to Japonica rice by cybridization (Fig. 22.10).

22.5 MITOCHONDRIAL GENES ASSOCIATED WITH CYTOPLASMIC MALE STERILITY

Recent molecular analyses indicate that novel mitochondrial genes in the cytoplasm, and fertility-restoration genes (*Rf/*) in the nucleus, control expression of cytoplasmic male sterility (CMS) in maize, rice, *Sorghum*, beans, *Brassica*, *Petunia*, radish, and sugar beet,. The CMS phenotype in *rf/rf* plants is associated with the expression of unique mitochondrial genes. The CMS phenotype in *Rf/Rf* plants is suppressed either by transcriptional regulation or posttranscriptional regulation (RNA processing, messenger stability, RNA editing, translation, protein stability) of unique mitochondrial genes, or by elimination of a subpopulation of mtDNA by the nuclear *Rf* gene. The unique mitochondrial CMS genes may arise as chimeras from fusion of two or more normal mitochondrial genes (as in maize, petunia, and sorghum), novel alterations of a duplicated mitochondrial gene (as in rice and bean), or a new open-reading frame (as in sunflower).

22.5.1 RNA Editing

RNA editing is important in the production of proteins by mitochondria, and it is possible that an error in this process can lead to problems in fertility. Extensive DNA-sequence analysis of plant-mitochondrial DNAs has indicated that all the mitochondrial genes undergo posttranscription editing, with the major change being cytosine to uracil (Fig. 22.11). This usually alters the protein for which the gene codes, relative to that expected from "reading" the DNA sequence using the standard genetic code (*see* Chapter 16, Section 16.2 and Fig. 16.3). The various messenger RNAs produced from the mitochondrial DNA are edited to different de-

Table 22.3. Cybrids Obtained by Interspecific- and Intergeneric-Protoplast Fusion of Specified Donor Cytoplasm with *N. tabacum*

Donor of Chloroplasts	Sexual Hybrid	Fertility of Cybrid	Somatic Hybrid Regeneration	Fertility of Somatic Hybrid
N. debneyi	Yes, female	Female fertile	Not attempted	—
N. suaveolens	Yes (rare)	Female fertile	No	—
Solanum tuberosum	No	Yes[a]	Yes	Infertile
Atropa belladonna	No	Yes	Yes	Infertile
Petunia hybrida	No	Yes	Yes	Infertile
Salpiglossis sinuata	No	Yes	Yes	Infertile

[a] Recombinant cpDNA, no cybrids with *S. tuberosum* cpDNA obtained.

Source: Rose et al. (1990).

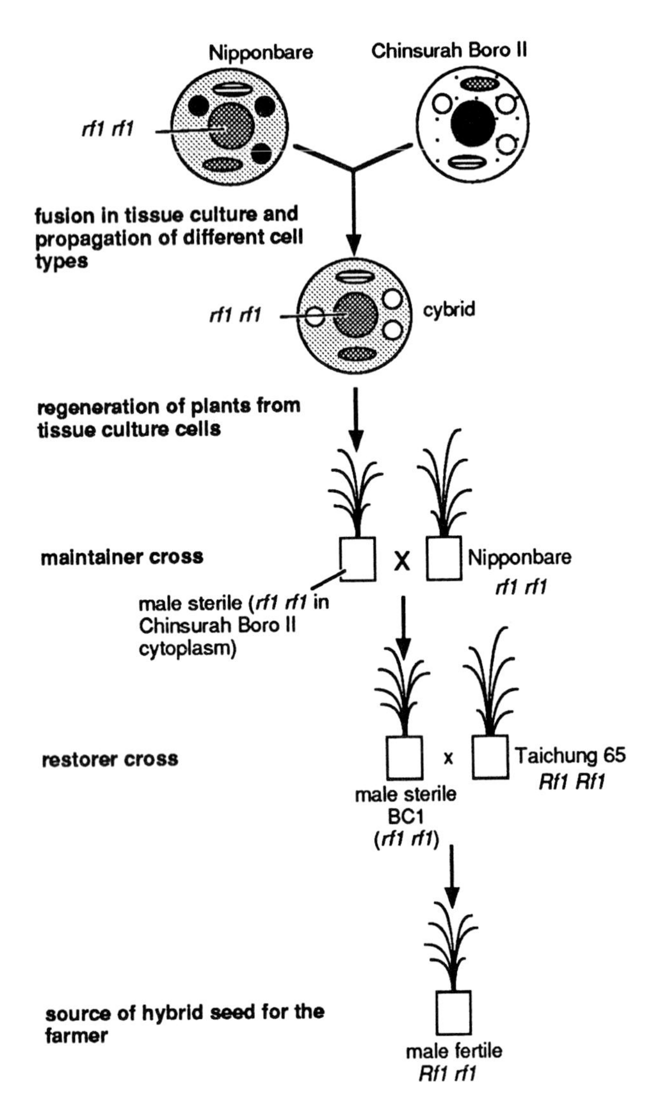

Fig. 22.10. The transfer of cytoplasmic male sterility from Indica rice (cultivar Nipponbare) to Japonica rice (cultivars Chinsurah Boro II and Taichung 65) by cybridization, resulting in a hybrid-rice crop. Nipponbare supplies the *rf1* genotype but is male-fertile because its cytoplasm is compatible with this genotype; when the *rf1* genotype is in the Chinsurah Boro II, cytoplasmic male sterility results as shown in the diagram. The "maintainer" cross provides the male-sterile plants that are crossed to the agronomically desirable Taichung 65 in the "restorer" cross to yield plants that provide the hybrid seed for sale to the farmer.

grees, in some cases resulting in 25% of the final amino acids having a different protein relative to that predicted in the DNA sequence of the gene. The editing process is not well understood, but the two possible mechanisms involve either deamination of cytosine residues (*see* Fig. 22.11) or base substitutions, possibly involving excision/repair enzymes at the RNA level analogous to the processes that operate at the DNA level.

22.5.2 New Proteins in Cytoplasmic-Male-Sterile Plants

The molecular mechanisms underlying CMS expression have been thoroughly investigated in several plant species, and they

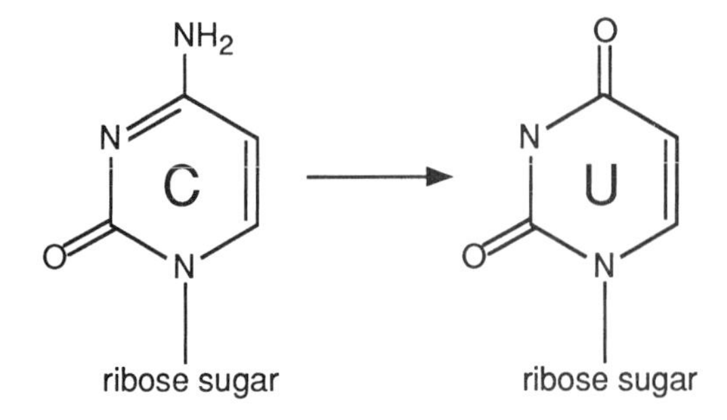

Fig. 22.11. The effect of RNA editing through the conversion of cytosines into uracils by deamination.

indicate that rearrangements associated with mtDNA are responsible for male sterility. The details of the mechanisms for the induction of CMS, and restoration of fertility, are different in the systems that have been studied, as illustrated by three examples.

The T-cytoplasm of maize plants confers male sterility as a result of the presence of a modified *atp6* gene (*T-urf 13*) that produces a 13-kDa protein (Fig. 22.12). The *T-urf 13* gene results from rearrangements in mitochondrial DNA involving the 26S ribosomal RNA genes. Insertions of new DNA into the *T-urf 13* gene, or deletions within this region of the mitochondrial DNA, result in reversion to male fertility. The location of the *T-urf 13* protein in the mitochondrial membrane most likely contributes to a reduction in the metabolic efficiency of mitochondria and results in pollen abortion. Although this may be the basis for the mode of action of the *T-urf 13* protein, it is not clear why the effect should be localized to the reproductive tissue, because most of the plant is not affected by the presence of the *T-urf 13* protein. A pleiotropic effect, associated with the presence of the *T-urf 13* protein, is sensitivity to a toxin produced by race T of the fungal pathogen *Bipolaris maydis*. This unexpected association created a major breeding challenge because of the widespread use of T-cytoplasm in hybrid-maize production in the 1960s. This practice resulted in extensive

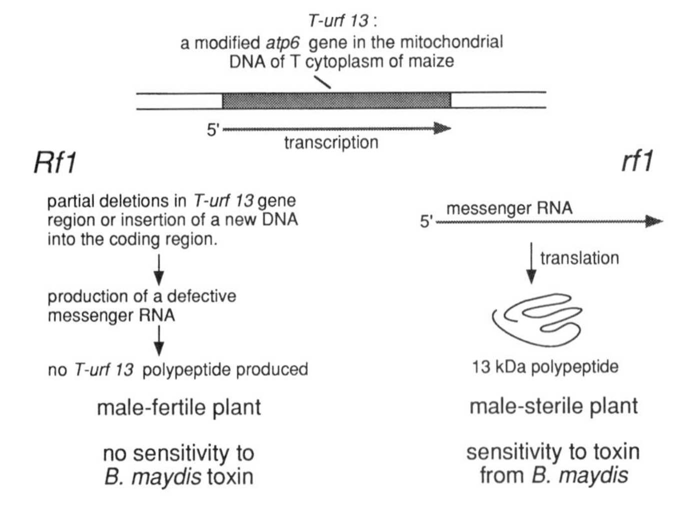

Fig. 22.12. Structure and physiological action of the maize *T-urf 13* locus which is unique to the mitochondria of T-cytoplasm maize. *T-urf 13* is a chimeric gene with 88 codons showing homology to the 3′ flanking region of *rrn26*, 9 codons of unknown origin, and 18 codons with homology to the coding region of *rrn26*. *Rf1* = male-fertile gene; *rf1* = male-sterile gene.

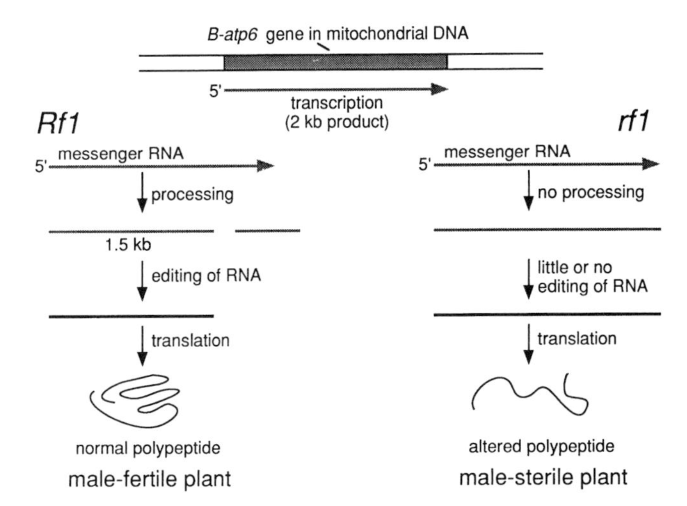

Fig. 22.13. A molecular mechanism of CMS expression in rice. In addition to a normal *atp6* gene, a second gene, *B-atp6*, uniquely present in CMS mitochondria, is first transcribed into a 2.0-kb RNA. In the presence of the *Rf1* gene (left), which may encode an RNA-processing enzyme, the 2.0-kb RNA is processed into 1.5-kb and 0.45-kb RNAs. The processed RNA is efficiently edited and translated into the normal polypeptide. In the absence of the *Rf1* gene (right), the 2.0-kb RNA is neither processed nor efficiently edited and is translated into an altered polypeptide. A large number of altered ATP6 polypeptides may cause CMS by competing with normal ATP6 protein, resulting in a severe reduction of ATPase activity. (From Iwabuchi et al., 1993.)

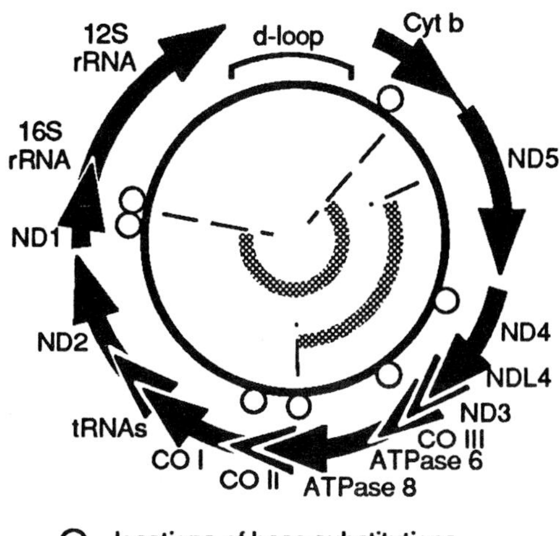

Fig. 22.14. Some of the lesions, caused by both base substitutions and DNA deletions, in human mitochondrial RNA that have been implicated in human diseases. (*See* Fig. 22.2 legend for explanation of the structure of mitochondrial DNA.)

yield losses before the association of *Bipolaris maydis* with the *T-urf 13* protein was identified.

In rice, CMS mitochondria contain two copies of the *atp6* gene. One copy is normal; the other is modified in its 3′-downstream region to produce the *B-atp6* locus. In the presence of the nuclear *Rf1* gene, the product from the *B-atp6* locus is processed normally to form an active *ATP6* protein, whereas in an *rf1* background, the processing is defective and an inactive *ATP6* protein is formed (Fig. 22.13). The male-sterile state is thought to result from the production of a defective *ATP6* protein, which competes with the normal *ATP6* protein and reduces the efficiency of mitochondrial metabolism.

In the common bean (*Phaseolus vulgaris*), CMS is correlated with the presence of a modification of the *atpA* gene, carried on a characteristic 3-kb DNA sequence that is present on a protosome distinct from the other mitochondrial protosomes. Spontaneous, or *Rf*-induced, reversion of the CMS phenotype to male fertility occurs when the protosome carrying the distinctive 3-kb sequence is lost from the plant. The *Rf* nuclear gene therefore acts by either specifically inhibiting replication of the protosome associated with CMS or by modifying the sorting of mitochondria carrying the CMS-associated protosome during cell division.

22.6 MITOCHONDRIAL-DNA MUTATIONS AND HUMAN DISEASES

In recent years, several human diseases have been linked to lesions in mitochondrial DNA (Fig. 22.14). For example, Leber's neuropathy, a disorder characterized by loss of central vision and degeneration of the optic nerve, is most commonly associated with a single nucleotide substitution in the cytochrome-b gene involved in the respiratory pathway. The disease also shows maternal inheritance, which is consistent with a causal relationship between the disease and a defective mitochondrial genome.

Ophthalmoplegia is associated with paralysis of the extraocular muscles and correlates with deletions in a 5-kb region, including the genes coding for ATPase6, cytochrome oxidase III, and NADH dehydrogenases 3, L4, 4, and 5. Other diseases such as Parkinson's disease, Huntington's disease, and cardiomyopathies may also be caused by mutations of mitochondrial genes. In general, the diseases that correlate with defects in mitochondrial DNA are postulated to occur because certain tissues are extremely sensitive to alterations in the levels of oxidative phosphorylation. Some diseases may arise in an individual without any previous family history, because the change in mitochondrial DNA develops de novo. In some cases, it is possible that nuclear-gene mutations interact with mitochondrial genes to cause a disease phenotype. Because each cell usually contains thousands of mitochondrial genomes, there is a chance that a spontaneous mutation may spread throughout the population of molecules in the cell. Such a mutation could be manifested as a disease in the whole organism (if it occurred early in development) or may, during the lifetime of an organism, contribute to cell senescence and aging in particular tissues.

BIBLIOGRAPHY

Section 22.1

Schuster, W., Brennicke, A. 1994. The plant mitochondrial genome: physical structure, information content, RNA editing and gene migration to the nucleus. Annu. Rev. Plant Physiol. Plant Mol. Biol. 45: 61–78.

Wallace, D.C. 1990. The human mitochondrial DNA. In: Genetic Maps, 5th ed. (O'Brien, S.J., ed.). Cold Spring Harbor Laboratory Press, Cold Spring Harbor, NY. pp. 5.246–5.257.

Section 22.2

Hiratsuka, J., Shimada, H., Whittier, R., Ishibashi, T., Sakamoto, M., Mori, M., Kondo, C., Honji, Y., Sun, C-R., Meng, B-Y., Li, Y-Q., Kanno, A., Nishizawa, Y., Hirai, A., Shinozaki, K., Sugiura, M. 1989. The complete sequence of the rice (*Oryza sativa*) chloroplast genome: Intermolecular recombination between distinct tRNA genes accounts for a major plastid DNA inversion during the evolution of cereals. Molec. Gen. Genet. 217: 185–194.

Section 22.3

Gill, B.S., 1991. Nucleocytoplasmic interaction (NCI) hypothesis of genome evolution and speciation in polyploid plants. In: Nuclear and Organellar Genomes of Wheat Species (Sasakuma, T., Kinoshita, T., eds), Kihara Memorial Foundation for the Advancement of Life Science, Yokohama, Japan. pp. 48–53.

Maan, S.S. 1985. Genetic analyses of male-fertility restoration in wheat. II. Isolation, penetrance, and expressivity of *Rf* genes. Crop Sci. 25: 743–748.

Maan, S.S. 1995. The species-cytoplasm-specific gene hypothesis. In: Classical and Molecular Cytogentic Analysis (Raupp, W.J., Gill, B.S. eds). Kansas Agric. Exp. Sta. Dept. Rep. pp. 165–174.

Tsunewaki, K. 1991. A historical review of cytoplasmic studies in wheat. In: Nuclear and Organellar Genomes of Wheat Species (Sasakuma, T., Kinoshita, T., eds.). Kihara Memorial Foundation for the Advancement of Life Science, Yokohama. pp. 6–28.

Section 22.4

Barsby, T.L., Shepard, J.F., Kemble, R.J., Wong, R. 1984. Somatic hybridization in the genus *Solanum*: *S. tuberosum* and *S. brevidens*. Plant Cell Rep. 3: 165–167.

Bates, G.W. 1990. Asymmetric hybridization between *Nicotiana tabacum* and *N. repanda* by donor recipient protoplast fusion: transfer of TMV resistance. Theoret. Appl. Genet. 80: 481–487.

Bonnett, H.T., Glimelius, K. 1990. Cybrids of *Nicotiana tabacum* and *Petunia hybrida* have an intergeneric mixture of chloroplasts from *P. hybrida* and mitochondria identical or similar to *N. tabacum*. Theoret. Appl. Genet. 79: 550–555.

Christey, M.C., Makaroff, C.A., Earle, E.D. 1991. Atrazine-resistant cytoplasmic male-sterile-*nigra* broccoli obtained by protoplast fusion between cytoplasmic male-sterile *Brassica oleracea* and atrazine-resistant *Brassica campestris*. Theoret. Appl. Genet. 83: 201–208.

Dudits, D., Maroy, E., Praznovszky, T., Olah, Z., Gyorgyey, J., Cella, R. 1987. Transfer of resistance traits from carrot into tobacco by asymmetric somatic hybridization: regeneration of fertile plants. Proc. Natl. Acad. Sci. USA 84: 8434–8438.

Hinnisdaels, S., Bariller, L., Mouras, A., Sidorov, V., Del-Favero, J., Veuskens, J., Negrutiu, I., Jacobs, M. 1991. Highly asymmetric intergeneric nuclear hybrids between *Nicotiana* and *Petunia*: evidence for recombinogenic and translocation events in somatic hybrid plants after "gamma"-fusion. Theoret. Appl. Genet. 82: 609–614.

Jourdan, P.S., Earle, E.D., Mutschler, M.A. 1989. Synthesis of male sterile, triazine-resistant *Brassica napus* by somatic hybridization between cytoplasmic male sterile *B. oleracea* and atrazine-resistant *B. campestris*. Theoret. Appl. Genet. 78: 445–455.

Kyozuka, J., Kaneda, T., Shimamoto, K. 1989. Production of cytoplasmic male sterile rice (*Oryza sativa* L.) by cell fusion. Bio/Technology 7: 1171–1174.

Melzer, J.M., O'Connell, M.A. 1990. Molecular analysis of the extent of asymmetry in two asymmetric somatic hybrids of tomato. Theoret. Appl. Genet. 79: 193–200.

Ohgawara, T., Kobayashi, S., Ohgawara, E., Uchimiya, H., Ishii, S. 1985. Somatic hybrid plants obtained by protoplast fusion between *Citrus sinensis* and *Poncirus trifoliata*. Theoret. Appl. Genet. 71: 1–4.

Perl, A., Aviv, D., Galun, E. 1990. Protoplast-fusion-derived *Solanum* cybrids: application and phylogenetic limitations. Theoret. Appl. Genet. 79: 632–640.

Potrykus, L., Jia, J., Lazar, G.B., Saul, M. 1984. *Hyoscyamus muticus* + *Nicotiana tabacum* fusion hybrids selected via auxotroph complementation. Plant Cell Rep. 3: 68–71.

Rose, R.J., Thomas, M.R., Fitter, J.T. 1990. The transfer of cytoplasmic and nuclear genomes by somatic hybridization. Aust. J. Plant Physiol. 17: 303–321.

Schieder, O., Gupta, P.P., Krumbiegel-Schroeren, G., Hein, T., Steffen, A. 1985. Novel techniques in handling and manipulating cells. Hereditas 3(Suppl.): 65–71.

Sproule, A., Donaldson, P., Dijak, M., Bevis, E., Pandeya, R., Keller, W.A., Gleddie, S. 1991. Fertile somatic hybrids between transgenic *Nicotiana tabacum* and transgenic *N. debneyi* selected by dual-antibiotic resistance. Theoret. Appl. Genet. 82: 450–456.

Toki, S., Kameya, T., Abe, T. 1990. Production of a triple mutant, chlorophyll-deficient, streptomycin-, and kanamycin-resistant *Nicotiana tabacum*, and its use in intergeneric somatic hybrid formation with *Solanum melongena*. Theoret. Appl. Genet. 80: 588–592.

Xu, Y.S., Murto, M., Dunckley, R., Jones, M.G.K., Pehu, E. 1993. Production of asymmetric hybrids between *Solanum tuberosum* and irradiated *S. brevidens*. Theoret. Appl. Genet. 85: 729–734.

Yarrow, S.A., Wu, S.C., Barsby, T.L., Kemble, R.J., Shepard, J.F. 1986. The introduction of CMS mitochondria to triazine tolerant *Brassica napus* L., var. "Regent," by micromanipulation of individual heterokaryons. Plant Cell Rep. 5: 415–418.

Section 22.5

Dewey, R.E., Levings III, C.S., Timothy, D.H. 1986. Novel recombinations in maize mitochondrial genome produce a unique transcriptional unit in the Texas male-sterile cytoplasm. Cell 44: 439–449.

Iwabuchi, M., Kyozuka, J., Shimamoto, K. 1993. Processing followed by complete editing of an altered mitochondrial atp6 RNA restores fertility of cytoplasmic male sterile rice. EMBO J. 12: 1437–1446.

Levings III, C.S. 1990. The Texas cytoplasm of maize: cytoplasmic male sterility and disease susceptibility. Science 250: 942–947.

McKenzie, S.A., Chase, C.D. 1990. Fertility restoration is associated with loss of a portion of the mitochondrial genome in cytoplasmic male-sterile common bean. Plant Cell. 2: 905–912.

Section 22.6

Wallace, D.C. 1993. Mitochondrial diseases: genotype versus phenotype. Trends Genet. 9: 128–133.

Zeviani, M., Servidei, S., Gellera, C., Bertini, E., Mauro, S.D., DiDonato, S. 1989. An autosomal dominant disorder with multiple deletion of mitochondrial DNA starting at the D-loop region. Nature 339: 309–311.

23

Engineering the Genome

- DNA transformation has removed most of the barriers that limit the introduction of specific genes from one organism to another.
- The use of Agrobacteria and modified Ti plasmids to transform dicotyledons, and the use of microinjection to transfect the nuclei of animal cells, are being superseded by particle-bombardment biolistic procedures, developed from the need to transform monocotyledonous crops.
- Chimeric-gene constructs combine the gene to be transformed with regulator and promoter sequences suitable for the host, in order to improve the stability or rate of formation of the gene product in the transformed host.
- Through DNA transformation, certain plants and animals can be regarded as factories for the production of novel gene products.
- Transformation by antisense genes can block the formation of product from normal sense genes, if necessary in a specific tissue; the insertion of multiple copies of sense genes can have the same effect.
- Successfully transformed organisms must transmit introduced genes in a Mendelian manner.

In eukaryotic species, individuals located in the same or different populations within a species can interbreed with complete fertility. Different individuals of a species commonly have the same chromosome number, allowing homologous chromosomes to pair during meiosis in intraspecific hybrids. The major differences are whether or not allelic forms of chromosomal genes are present that might result in homozygosity or heterozygosity, or in whether the sex chromosomes are the same or different. Cross-hybridization between species has been studied as a means of introducing novel genetic material from one species into another, and such introductions may even be possible between related genera. However, the farther apart species or genera are in evolutionary terms, the more distant they are in genetic relatedness, chromosomal homology, and the organization of equivalent genes on their chromosomes. As a result, it is more difficult to introduce germplasm from one species or genus into another. The techniques of molecular biology and, in particular, DNA transformation have removed many of these barriers, so that the genes of one species can now be introduced into different genera, different families of plants or animals, or even into completely different kingdoms and domains.

23.1 GENETIC ENGINEERING

The techniques that have been developed for the artificial transformation of DNA from one species into another, are commonly called genetic engineering. When trying to stably incorporate novel DNA sequences into multicellular plants and animals, however, a range of problems arises, including the following requirements:

(a) A satisfactory means of introducing the foreign DNA into the cells of the recipient
(b) Recognition that the DNA has been successfully introduced
(c) Incorporation of these introduced DNA sequences into the genomic DNA of the host
(d) Normal translation of genes on the modified DNA to produce mRNA, with the excision of exons and fusion of introns
(e) Production of the usual polypeptide, protein, or enzyme products of the introduced gene
(f) Manufacture of the gene products in the desired cells and tissues of the host
(g) Minimal disruption to the rest of the genome in the form of unwanted translocations and inversions

All of these requirements have elicited an array of solutions, some more successful than others. In addition, the ability to transform plants is closely tied to the ability to regenerate plants from cells, calli, tissue cultures, or protoplasts. The production of transgenic plants of cereal-crop species is complex because the transformation and regeneration techniques, especially protoplast culture, are still far from routine. There are also major differences between suitable transformation vectors for dicotyledons, such as *Arabidopsis* and tobacco, and those that are suitable for the monocotyledonous crops, such as maize, rice, and wheat. However, given these restrictions, probably over 100 species of plants have now been successfully transformed, with agricultural crops, especially potato, oilseed rape, tobacco, maize, and tomato, leading the way.

In animals, the range of successfully engineered organisms is also ever-increasing. Major developments have entailed the introduction of foreign genes, usually by microinjection, first into tissue cultures, then into egg nuclei, and now, more commonly, into the stem cells of developing embryos. Animals that have been transfected using these procedures range from fish to farm animals, the aims ranging from a simple improvement in growth rate to their use as factories for producing specialty biochemicals.

23.2 *AGROBACTERIUM* TRANSFORMATION OF PLANTS

In 1983, the first report of transgenic plants was a potato, transformed with a gene carried on a bacterial-plasmid vector. Species of the family Solanaceae, including potato (*Solanum tuberosum*), tomato (*Lycopersicon esculentum*), tobacco (*Nicotiana tabacum*), and the garden flower *Petunia hybrida*, have since been much used for research involving genetic engineering because this family is particularly susceptible to infection by the pathogenic gram-negative bacterium *Agrobacterium tumefaciens* (Fig. 23.1), and it is amenable to genetic transformation and subsequent regeneration.

In a normal infection of a susceptible, dicotyledonous host plant by this bacterium, deliberately injured or naturally wounded plant cells are open to infection by *A. tumefaciens*. Various small molecules produced by the wounded plant cells bind to receptors on the surface of the bacterium, allowing the pathogen to colonize the injured cell surfaces. This binding triggers the induction of enzymes that copy the tumor-inducing (Ti) plasmid DNA present inside the bacterium. A small fragment of this DNA, T-DNA or transferred DNA, is then naturally transferred from the Ti plasmid, through the intervening bacterial- and plant-cell walls, into the nuclear DNA

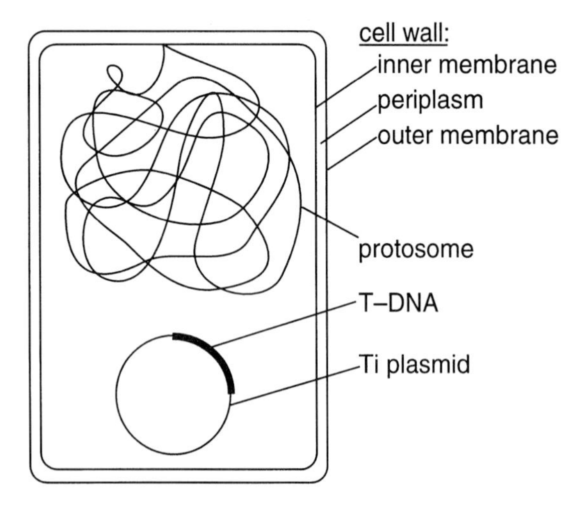

Fig. 23.1. A diagrammatic representation of the DNA present in the bacterium *Agrobacterium tumefaciens*, showing the bacterial protosome and Ti plasmid with T-DNA.

Fig. 23.2. A tomato plant infected with *A. tumefaciens*, resulting in a gall tumor in the injured stem. (Photograph kindly provided by Dr. T.J. Higgins.)

of the plant cells, resulting in the formation of a crown-gall tumor (Fig. 23.2). The gall is formed in direct response to genes on T-DNA that produce excessive amounts of plant hormones, resulting in abnormal or oncogenic growth of the infected cells. In addition, T-DNA carries genes that produce chemicals known as opines, usually nopaline or octopine (Fig. 23.3), although others have also been found, and these chemicals allow the Ti plasmid to spread to all of the agrobacteria in the tumor tissue.

If *Agrobacterium*, the Ti plasmid, and T-DNA are to be used as tools for genetic engineering, it is necessary to determine how the original transformation system works. Different parts of the system can then be modified to allow the introduction of new DNA sequences and new genetic material into the host plant. With this knowledge, modified forms of the Ti plasmid can be genetically engineered to delete the genes causing gall formation, while retaining the opine genes and the ability of T-DNA to transfer foreign DNA into plants. Although many details have still to be exactly determined, the basic transformation system is described in the following subsection.

23.2.1 T-DNA

The two major T-DNAs used for transformation, octopine and nopaline, are both about 22 kilobases (kb) long, about 10% of the total DNA present in the Ti plasmid (Fig. 23.4). A number of oncogenic or tumor-forming genes are present on these DNAs. The basic plan of T-DNA includes a left (L) and a right (R) border at the 5′- and 3′-ends, respectively. The opine-synthase genes *nos*

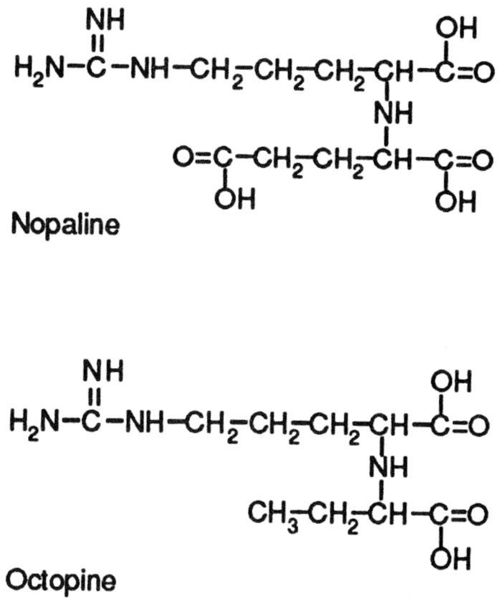

Nopaline

Octopine

Fig. 23.3. The chemical structure of nopaline and octopine, opines produced by *nopaline synthase* (*nos*) and *octopine synthase* (*ocs*) genes, respectively, on the T-DNA of different races of *A. tumefaciens*. Other opines are produced by other races of the pathogen.

(nopaline synthase) or *ocs (octopine synthase)* are adjacent to the R-border and are comprised of a promoter sequence, the gene sequence itself, and a polyadenylation–terminator–signal sequence. At least six or seven other genes are also present in T-DNA, although not all of these have a known function. Those with a recognized function are primarily concerned with hormone production, and the names given to these genes reflect the recognition of their function over time. Thus, genes 1 and 2 were originally called *tms*, indicating *shooty tumor*. The two polypeptide transcripts produced by this complex are now known to encode tryptophan-2-monooxygenase *(iaaM)* and indoleacetamide hydrolase *(iaaH)*, both of which are concerned with the production of the auxin indoleacetic acid or IAA (Fig. 23.5), whose presence primarily affects the stem of the plant. The transcript of gene 4, *tmr* or *rooty tumor,* (*see* Fig.

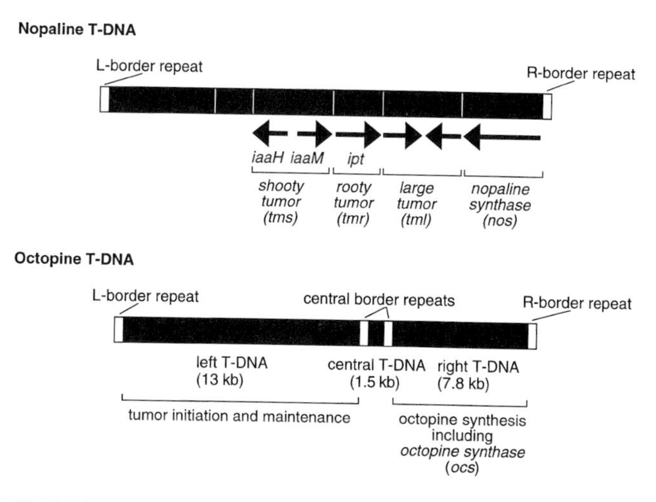

Fig. 23.4. The genetic structure and the direction of transcription of nopaline and octopine T-DNA sequences. The *overdrive* sequence of the octopine Ti plasmid is immediately adjacent to the R-border of the octopine T-DNA.

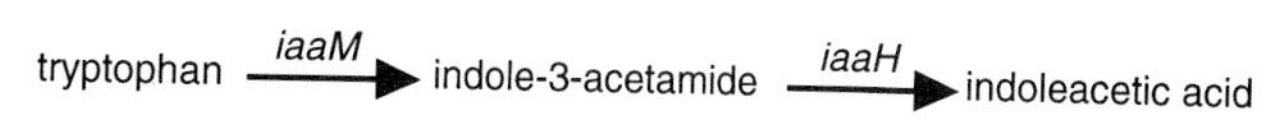

Fig. 23.5. The biosynthetic pathway producing the plant hormone indoleacetic acid, and its control by genes on the T-DNA of the Ti plasmid.

23.4) encodes the enzyme isopentyl transferase (IPT), which is involved in the production of cytokinin, a root-affecting hormone. Finally, the transcript of gene 6b, *tml or large tumor*), is specific to the host plant *Nicotiana*, which is rather apt, considering that most of the above mechanisms have been determined from the tobacco–agrobacterium association.

Octopine T-DNA is divided into three regions, each of which is separated by a border sequence, so that there are four border sequences in all (*see* Fig. 23.4). The first region is a 13-kb fragment of left-T-DNA involved in tumor initiation and maintenance (TL), followed by a border sequence, a 1.5-kb central fragment of unknown function (TC), and another border sequence. The 7.8-kb right-T-DNA (TR) fragment contains several opine synthetic genes and, in particular, the *ocs* gene. Although the basic plan is essentially similar in layout to the nopaline T-DNA, the functional significance of the two interstitial border sequences has yet to be determined.

23.2.2 The Ti Plasmid

The tumor-inducing nopaline Ti plasmid of *A. tumefaciens* is about 200,000 bp (200 kb) in length, although it can vary between 140 and 235 kb (Fig. 23.6). Genes on this plasmid interact with T-DNA during the transformation process, and some, in turn, are affected by products of the T-DNA genes. The most intensively studied of the Ti genes are those affecting the virulence of the plasmid, and many of them are expressed by mediating the transfer of T-DNA. The switching-on of these genes by low-molecular-weight molecules, excreted by wounded cells, activates the transfer of the T-DNA fragments.

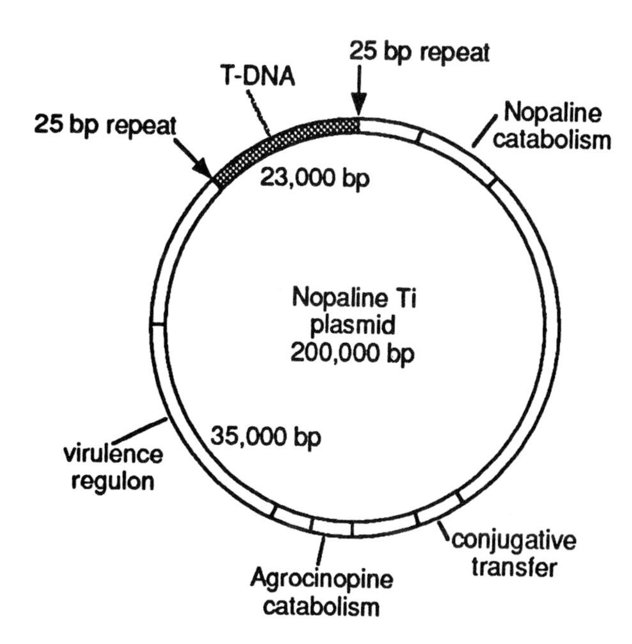

Fig. 23.6. The basic structure of the nopaline Ti plasmid of *A. tumefaciens* showing the positions of identified genes and regulons.

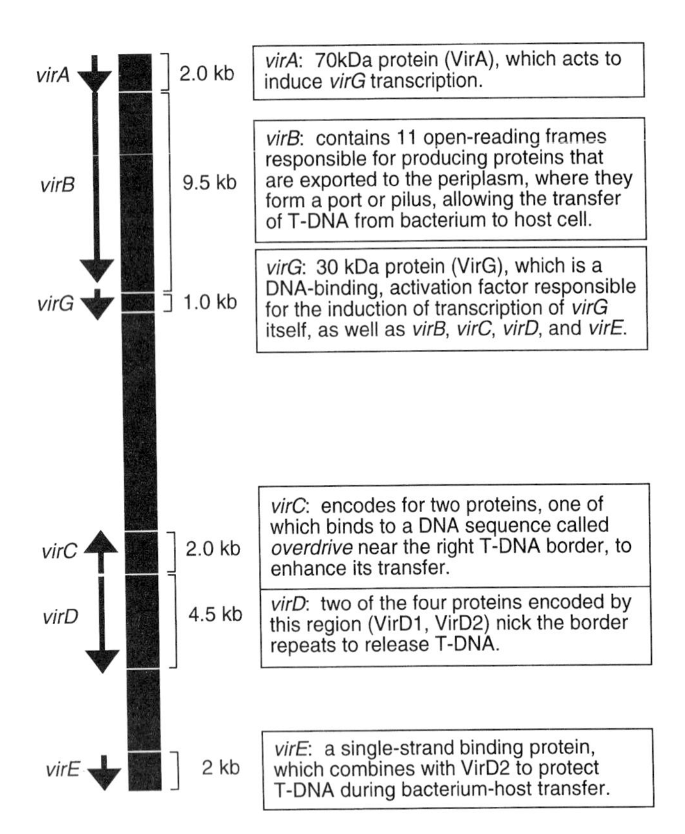

Fig. 23.7. Genetic structure of the virulence regulon of the octopine Ti plasmid. The genes and their lengths and directions of transcription are indicated. The proteins produced by these genes and the functions of these proteins are shown in Table 23.1.

acetosyringone

α-hydroxy-acetosyringone

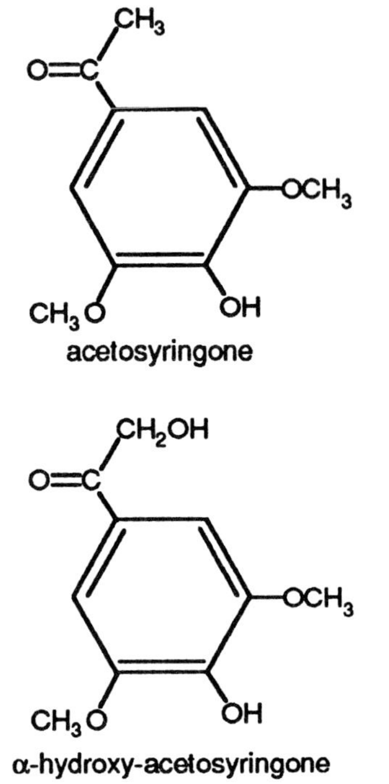

Fig. 23.8. Two of the signal molecules released by wounded plant cells, acetosyringone and α-hydroxy-acetosyringone. These molecules not only induce the *vir* regulon but also attract other cells of *A. tumefaciens* to the wound site, and these cells are, in turn, induced.

The octopine Ti plasmid contains several *virulence* or *vir* genes located in a 35-kb section of the plasmid, as in the nopaline Ti plasmid (*see* Fig. 23.6). This DNA is described as a regulon in that the six major *vir* genes are regulated in a similar fashion. The arrangement of genes within the regulon is shown in Fig. 23.7, and the respective gene functions are summarized in Table 23.1. In tobacco, wounding of the plant cells releases a range of molecules, among which are at least two polyphenolic-signal molecules that interact with the *virA* gene and induce production of the 70-kDa VirA polypeptide. The signal molecules have been identified as acetosyringone (AS) and α-hydroxy-acetosyringone (HO-AS) (Fig. 23.8). Not only do these signal molecules induce the virulence pathway, but they also act as a chemical attractant to other cells of *A. tumefaciens.* Although the modes of action of these *vir* genes are not all known (*see* Table 23.1), it appears that the interaction of the signal molecules with *Agrobacterium* and the Ti plasmid is the extracellular-recognition sequence that induces the intracellular response, beginning with the constitutive production of the VirA polypeptide. It is also possible that the specificity of the VirA protein is highly significant for the noninteraction of *Agrobacterium* with monocotyledonous cells.

The final transfer of T-DNA from bacterium to host involves the production of a single-stranded DNA (ssDNA) of approximately 22 kb in length, which, during the transfer and incorporation process, must be protected from endogenous and exogenous enzymes. Approximately 600 VirE2 protein molecules are required to protect this DNA during the transfer, and the resulting T-complex is large enough to be observed using immunogold-labeled anti-VirE2 antibodies. Other proteins are also induced by acetosyringone and are also under the control of the *vir* regulatory system. These virulence-related proteins (VRPs) include a 45-kDa protein produced by the gene *plant-inducible (pinF),* now more commonly regarded as a seventh complementation group of the vir regulon and redesignated *virH,* and a number of other genes (*see* Table 23.1). However, none of the proteins produced by these genes appear to be essential to the infection process.

The DNA sequences of the R- and L-borders of the Ti plasmids are comprised of identical 25-nucleotide base-pair sequences (Fig. 23.9), as are the interstitial sequences found within octopine T-DNA. However, although changes to the R-border nucleotide array such as deletions of the first 6 bps or of the last 10 bps are always detrimental to the T-DNA insertion procedure, this is seldom the case with modification of the L-border sequences, and the reason for this difference is unknown. Nor is the mechanism known by which changes of the R-border sequence affect the function of the *overdrive* DNA sequence of the octopine plasmid (*see* Fig. 23.4). Finally, *conjugative transfer (tra)* genes, which are induced by the opine product of T-DNA, are also located on the Ti plasmid and allow it to spread to all of the agrobacteria present.

```
                  TG  PuG      AT
                  NC  CT       TC
5'TGGCAGGATATAT     N    TGTAA
3'ACCGTCCTATATA     N    ACATT
                  AC  PyC      TA
                  NG  GA       AG
```

Fig. 23.9. The nucleotide base-pair sequence common to the Ti plasmids on the right and left borders of the T-DNA, and internal in the octopine T-DNA. P_u = a purine; P_y = a pyrimidine; N = any nucleotide. (From Zambryski, 1988.)

Table 23.1. Action of *vir* Regulon Genes, and the Locations and Functions of the Proteins Produced by These Genes

Gene	Action	Protein	Size (kDa)	Location and/or Function in *vir*-Induced *Agrobacterium* Cells
virA	Induced by aceto-syringone or α-hydroxy-acetosyringone.	VirA	70	Induces transcription of *VirG*.
virB	Produces 11 proteins involved in the transfer of the T-complex from the bacterium to the host cell via a pore in the inner membrane, through the periplasm, and through the channel structure.	VirB1	26	Located in the periplasm between the inner and outer membranes.
		VirB2	12	Unknown.
		VirB3	12	Unknown.
		VirB4	87	The largest, and one of the most abundant *VirB* proteins: VirB4 is membrane-bound, and forms part of the outer membrane.
		Vir B5	23	Present in the cytosol and membrane fractions, although most is found in the periplasm; a probable transporter of the T-complex from the cytosol through the membrane.
		VirB6	32	Bound to the inner membrane.
		VirB7	6	The smallest *virB* protein; predicted to be in the outer membrane where it either lyses the bacterial cell wall to release the T-complex, or facilitates the interaction with the plant cell wall.
		VirB8	26	Bound to the inner membrane.
		VirB9	32	Generally membrane.
		VirB10	41	Bound to the inner membrane.
		VirB11	38	Bound to the inner membrane.
virC	Encodes for two proteins.	VirC1	230	Binds to the *overdrive* sequence adjacent to the TR border, through which mechanism it enhances the transfer of the T-complex.
		VirC2	26	Unknown.
virD	Encodes four proteins, two of which have been analyzed.	VirD1	47	This protein nicks the border repeats to release T-DNA.
		VirD2	16	Remains tightly bound to the 5′-end to facilitate migration of T-DNA through cell tissues.
virE	Produces the most abundant protein of any *vir* genes.	VirE2	60.5	A tenacious ssDNA-binding protein; 1 molecule covers 30 nucleotides, so that the 20 kb of T-DNA is fully protected with 600 VirE2 molecules. In combination with VirD2 protein, it forms the T-complex, protecting ssDNA from harm during the transfer process.
virG	Induced by VirA.	VirG	30	Induces self-transcription and transcription of *VirB*, *VirC*, *VirD*, and *VirE*.
virH	Originally *pinF*; encodes two inducible cytochrome P-450-type enzymes.			May protect the cell during the infection process, but is not essential for virulence.

23.2.3 Genes in the *Agrobacterium* Protosome

The transformation process is further dependent upon several genes located on the bacterial protosome. Among these are the genes *chvA* and *chvB*, located in a 15.5-kb fragment of the bacterial protosome and required for the attachment of *A. tumefaciens* to the plant cell. A plant fiber-forming glycoprotein may be the actual surface receptor to which the agrobacteria bind, and a bacterial, calcium-binding surface protein, rhicadhesin, may be involved in this attachment. *chvB* occupies an 8.5-kb DNA fragment that produces a 235-kDa membrane protein involved in the production of cyclic β-1,2-glucan, which binds with a transport protein produced by *chvA*. The combination is transported to the periplasm, where it is involved in the attachment of the bacterial cells to the wound site. Agrobacterial lines that are mutant for *chvB* are deficient in adherence, but the addition of rhicadhesin restores both attachment to the plant surface and virulence to the bacteria. Major neutral and acidic extracellular polysaccharides, synthesized by the 3.0-kb *pscA* gene and the *exoC* gene, also affect the surface composition of the bacterial cells and their attachment to the plant cell.

Overall, the successful transformation of T-DNA into the host genome involves a number of steps, including the following:

(a) The release of exudates by wounded plant cells to attract agrobacteria
(b) Activation of Ti *vir* genes and expression of their products
(c) Production of a single-stranded T-DNA copy from a Ti T-DNA sequence
(d) Transfer of T-DNA copy to the bacterial membrane
(e) Passage of T-DNA from the bacterium into the host cell
(f) Passage of T-DNA through the cytoplasm and nuclear membrane of the host cell
(g) Integration of T-DNA into the plant genome

Analyses of the DNA sequences surrounding inserted T-DNA fragments have not discovered any particular DNA sequence that might be specific to the insertion process or recognized by the T-DNA fragment. It is probable, however, that after T-DNA has been transferred into the plant cell, nuclear-localization signals within the VirD2 and VirE2 proteins physically target the DNA to the host nucleus (*see* Table 23.1). By some means, the T-DNA then becomes stably inserted into the plant genome.

23.2.4 Modifications of the *Agrobacterium* Delivery System

The Ti plasmid is a natural vector able to transfer T-DNA into the genome of suitable dicotyledonous plants. Because T-DNA is

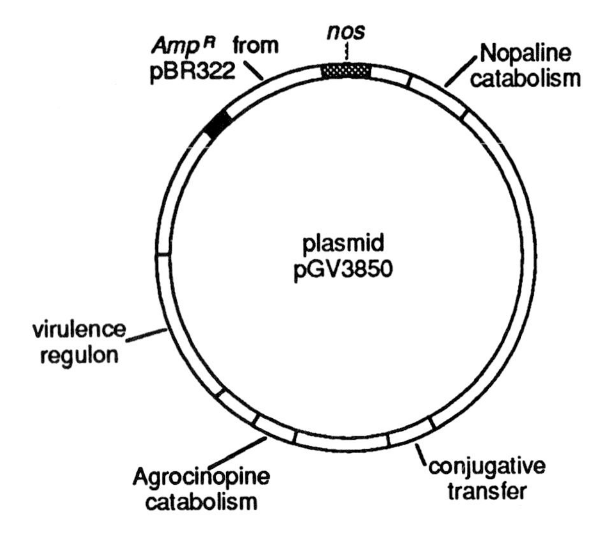

Fig. 23.10. The plasmid pGV3850 produced from the Ti plasmid by engineering the removal of the native oncogenes and replacing them with a large piece of the plasmid pBR322, including the ampicillin-resistance gene (*Amp*^{R}). This DNA substitution gives a disarmed or nononcogenic plasmid incapable of gall formation. (Modified from Zambryski, 1992.)

oncogenic, the natural result of this transfer is that tumors develop in the infected plant tissues. Instead of having the transformation of the host genome by T-DNA evidenced by the development of a gall, and using hormone-independent callus production to select recipient cells, the Ti plasmid can be modified to produce disarmed vectors. These modifications include replacement of the oncogenes by genes with other functions while maintaining the ability of the disarmed T-DNA to transform the host DNA.

One of the first successful modifications of the Ti plasmid involved replacement of the oncogene-containing section by a large piece of the plasmid pBR322, giving the new plasmid pGV3850 (Fig. 23.10). This substitution has the advantage that the pBR322 gene for ampicillin resistance *(Amp*^{R}*)*, which is incorporated into the Ti plasmid, also provides a selectable marker to ensure that only cells transformed with this gene are able to grow in ampicillin-containing media. In addition, because the *vir* genes are still present on the Ti plasmid and the *nos* gene is present on T-DNA, the presence of nopaline in transformed tissues acts as a scorable marker confirming the transformation.

23.2.5 Selectable and Scorable Markers

The incorporation of dominant selectable markers ensures that protoplast, callus, and other tissue transformants can be selected by growing these tissues in the presence of increased concentrations of the respective marker compounds. Scorable markers, on the other hand, are detected by the recognition of novel chemical compounds such as opines or through the enzymatic production of visible end products, which allow the tissues or cell components in which the gene is expressed to be cytologically observed. They are particularly important where genetically modified plants cannot be regenerated from single cells, so that direct selection is neither feasible nor effective.

A range of selectable and scorable markers has been developed to measure transformation efficiency and gene expression. The original scorable markers were the opine products of *nos* and *ocs*. However, nopaline, octopine, and the range of other opines are not usually found in plant cells, and their detection by colorimetric methods can be difficult to quantify. The subsequent development of *lacZ*, a gene producing *β*-galactosidase (GAL), and its introduction from *E. coli* into T-DNA was a step forward, because when transformed gall slices are placed on agar gels containing XGal (5-bromo-4-chloro-3-indolyl-*β*-D-galactopyranoside), the transformed cells turn blue. Again, this assay can be confusing if there exists a relatively high level of endogenous GAL activity in the plant. To overcome this problem, a gene entirely novel to plants has been developed for this purpose, the firefly *luc* gene, which produces the enzyme luciferase (LUC). Tissues expressing LUC are scored when they emit light after incubation with luciferin but, again, the system is rather labile and can be difficult to assay. A gene with fewer of these liabilities is the *E. coli* gene *β-glu*, which produces the enzyme β-glucuronidase (GUS). Genetically modified plant tissues that contain this gene turn blue when incubated with 5-bromo-4-chloro-3-indolyl-1-glucuronide (*See also* Fig. 19-19, Chapter 19). This assay has proven to be excellent for the histochemical localization of transformed tissues, as there is virtually no endogenous plant interference. The utilization of the green fluorescent protein (GFP) gene from the jellyfish *Aequoria victoria* has been a significant advance in monitoring transformation because it does not require the addition of a new substrate and only requires irradiation of tissue with a long wavelength UV light.

Selectable markers, on the other hand, are often the products of genes that confer resistance to bacterial antibiotics or herbicides, so that the presence of these compounds can then be used to screen for the survival of transformed cells under laboratory conditions. Although transformed gall tissues can continue their growth in the presence of relatively high concentrations of specific antibiotics or herbicides, nontransformed tissues usually die. Two bacterial antibiotic transferases were originally used for this purpose: the *bar* gene, which produces neomycin phosphotransferase (NPT, types I and II), and *cat,* producing chloramphenicol acetyltransferase (CAT). NPT inactivates aminoglycoside antibiotics, such as kanamycin and neomycin, by phosphorylation, so that transformed cells survive in the presence of these antibiotics. Even though this system can be difficult to assay, the *bar* gene has already been introduced into more than 36 species of plants. Other enzymes introduced for this purpose include hygromycin phosphotransferase, dihydrofolate reductase, dihydropteroate synthase conferring sulfonamide resistance, bleomycin reductase, gentamycin acetyltransferase, and streptomycin phosphotransferase. Only the last enzyme allows nontransformed tissues to survive, although these cells are recognizable because their subsequent growth is greatly retarded and subject to severe chlorosis. To negate the need for the introduction of either antibiotic or herbicide resistance as selectable markers, regulatory enzymes isolated from amino-acid biosynthetic pathways are being evaluated. Examples of such genes are dihydropicolinate synthase (DHPS) and desensitized aspartate kinase (AK) from bacterial aspartic acid pathways, and a tryptophan decarboxylase gene *(tdc)* from *Catharanthus roseus.*

23.3 THE PRODUCTION OF CHIMERIC GENES

Chimeric genes are artificial constructs in which a defined-promoter DNA sequence is linked to a cDNA gene sequence, which is then attached to a polyadenylated-terminator DNA sequence (Fig.

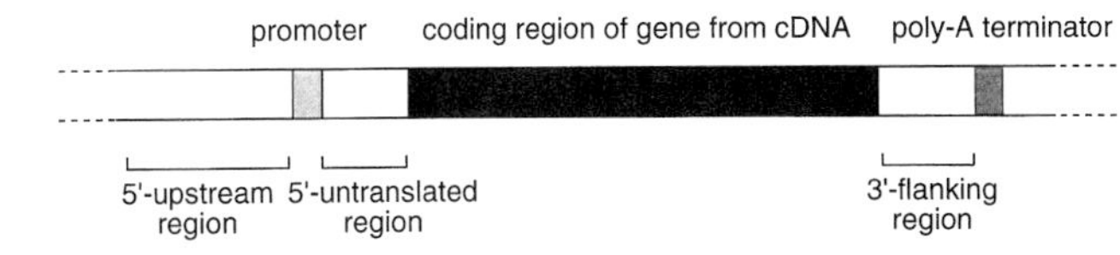

Fig. 23.11. The basic components and DNA regions required for gene action. Any of these regions are capable of being modified or substituted to construct a chimeric gene.

23.11). The combined construct is inserted into a disarmed T-DNA for introduction into an *Agrobacterium*-susceptible host or is introduced by some other means into *Agrobacterium*-recalcitrant hosts. Also, potentially important to gene constructs are leader sequences, 5′- and 3′-untranslated flanking sequences, volunteer plant-regulator sequences, and codon frequency, all of which can affect the secondary structure of the mRNA and the gene product itself. Molecular signals that determine the cellular location of the expressed product have also been utilized.

23.3.1 Regulator Sequences

In the first examples of transformed plants, gene processing after transformation was effected using the natural opine T-DNA promoters and terminators, the most significant being those of the *nos* and *ocs* genes. Another useful promoter, CaMV-35S, was isolated from the 35S gene of cauliflower-mosaic virus. This promoter has been inserted, both singly and in multiple copies, to evaluate the effect of promoter multiplicity on the expression of a particular gene sequence and production of the gene product. The selected promoter must be effective and able to operate in the tissue in which the gene is to be expressed—in many transformations, leaves are involved. The search for 5′-flanking regions of genes strongly expressed in leaves led to the isolation and use of several promoters that are associated with genes encoding abundant leaf proteins such as the chlorophyll a/b-binding protein (*cab*) and the small subunit of ribulose bisphosphate carboxylase (*rbcS*). When the *rbcS* promoter was linked to a bacterial chitinase gene (*chiA*) in an attempt to confer antifungal activity to transformed plants (*see* Section 23.4.4), the amount of ChiA protein produced by the *rbcS/chiA* transformed plants was three times the amount produced by either of two *cab/chiA* fusions. Other promoters used in transformation experiments to regulate the time and tissue location of gene expression are summarized in Table 23.2.

Table 23.2. Some of the Genes That Have Provided Promoters to Enhance the Expression of Transformed Genes in Plants

Promoters	Genes from Which the Promoters Were Isolated
Constitutive promoters	Actin
	Adh-1 intron of maize
	CaMV-35S
	Nos and *Ocs*
	Ubiquitin of maize.
Tissue/environmental	Alcohol dehydrogenase—induced by anaerobic conditions.
	Histone 3—cell cycle active.
	Light-harvesting chlorophyll a/b-binding genes of photosystem II—light inducible, in leaf, stem, and floral organs.
	Bean phytohaemagglutinenin L-gene (*dlec2*)—specific for seed-storage protein.
	Open-reading frame 12 (ORF12) of the Ri plasmid TL-DNA region from *Agrobacterium rhizogenes* (*rol-C*).
	Rubisco, ribulose-1,5-bisphosphate carboxylase/oxygenase small-subunit gene (*rbscS*).
	Potato gene (*Pin2*)—induced by wounding, methyl jasmonate, and abscisic acid.
	Major transcript gene of rice tungro bacilliform virus (RTBV)—for leaf-phloem tissue.
	High molecular weight glutenin promoter from wheat—for mid-endosperm development of the grain.

The promoter can be engineered to increase its activity and stability in different categories of plants. The pEmu promoter, for example, is specifically adapted to monocotyledons (Fig. 23.12). It combines elements of a truncated maize Adh1 promoter, the first intron of the *Adh* gene, six copies of the Anaerobic Response Element (ARE) from the maize *Adh1* gene, and four *ocs* elements. When pEmu was combined with the *β-glu* gene of *E. coli,* production of β-glucuronidase was from 10-fold to 50-fold higher in transformed, monocotyledonous protoplasts than when a CaMV-35S promoter/*β-glu* gene construct was assayed. A related promoter gave a 10-fold increase in expression over the CaMV–35S promoter in tobacco protoplasts. In turn, the maize polyubiquitin promoter and untranslated sequence pAHC27 gave even higher expressions of GUS in maize, wheat, sugarcane, and banana.

The expression of transformed genes may also be enhanced by modifying the 5′-untranslated regions carrying the control of transcription signals. For example, when nucleotide changes were made to the triplet codon immediately prior to the translation-initiation ATG codon of the ChiA-protein DNA sequence, the pre-codon ACC gave rise to transformants with higher levels of ChiA protein than transformants with the pre-codon CAT. The combination of modified DNA sequences in the vicinity of the translation-initiation codon, in conjunction with modifications to the promoter, led to a combined accumulation of ChiA to about 0.25% of the total soluble leaf protein.

Gene expression can be further augmented by altering both the

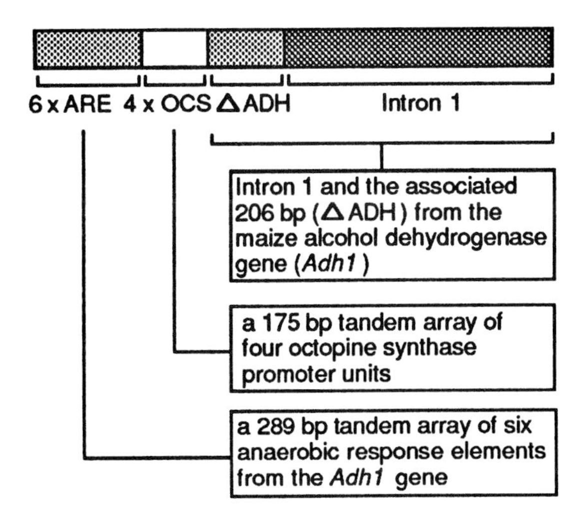

Fig. 23.12. The pEmu promoter/enhancer and the genetic elements used to construct it. pEmu was designed to increase protein production in protoplast cultures of monocotyledons. (Modified from Last et al., 1991.)

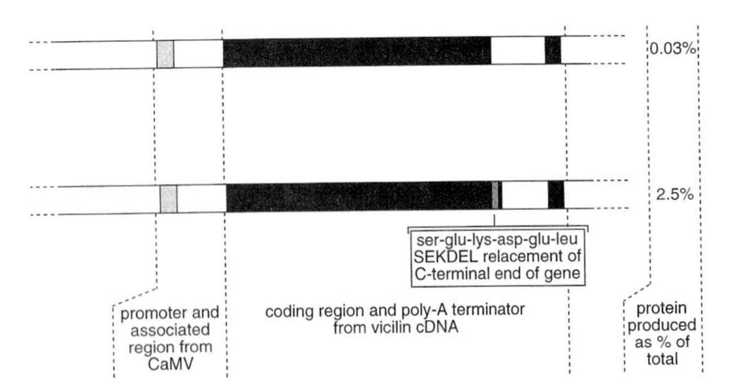

Fig. 23.13. The effect of the SEKDEL nucleotide construct on protein production. The addition of this nucleotide sequence to the C-terminal end of the vicilin gene was designed to ensure that the expressed protein is retained within the endoplasmic reticulum. The benefit of this insertion on the amount of vicilin protein present in transformed tobacco plants is indicated (right-hand side) in the absence (upper) and presence (lower) of the insertion. (Modified from Wandelt et al., 1992.)

5′-upstream and the 3′-downstream regions. For example, in attempts to improve the protein composition and nutritional value of leaves in fodder crops, a novel selection of new proteins has been introduced such as chicken ovalbumin and pea vicilin. In tobacco and alfalfa, the combination of CaMV–35S promoter, ovalbumin cDNA, and ovalbumin 3′-flanking sequence produced one-tenth the amount of protein produced by the combination of CaMV–35S, ovalbumin cDNA, and the 3′-flanking sequence of a pea-albumin gene.

The inclusion of 5′-upstream sequences such as those involved in the control of gene expression in particular regions of chromatin (*see* Chapter 16, Fig. 16.25) can provide independence from the position effect seen in gene introgression. Position-effect variegation is observed when the expression of a gene is modified by the position of the gene within the genome (*see* Chapter 9, Section 9.4.1).

23.3.2 The SEKDEL Modification

One of the most dramatic effects on expression, as measured by accumulation of polypeptide product, is the specific targeting of a protein product to the endoplasmic reticulum by adding nucleotides, coding for the amino-acid sequence SEKDEL, to the cDNA of the introduced protein. The oligopeptide sequence Ser.Glu.-Lys–Asp–Glu–Leu, symbolically abbreviated to SEKDEL, appears to be responsible for localizing proteins within the endoplasmic reticulum (Fig. 23.13). When the DNA sequence coding for SEKDEL was inserted into the carboxy-terminal end of a pea-vicilin gene prior to its introduction into tobacco, the level of vicilin in the transformed tobacco plants increased from 0.03% to 2.5%. In alfalfa, the same construct increased vicilin production 20-fold, from 50 ng/mg to 1020 ng/mg of leaf protein. In contrast to the results discussed in the previous section, changes in the promoter, and in the 5′- and 3′-untranslated regions, had little effect.

23.3.3 Tailoring Gene DNA Sequences

The successful transfer of specific genes by artificial means allows the DNA sequences to be modified prior to transformation, to enhance the expression of the gene. If the gene originates from a nonplant source, codon usage may become an important variable because, owing to the built-in redundancy of the genetic code, any single amino acid in a polypeptide can be coded for by two to six triplet codons (*see* Chapter 16, Fig. 16.3). As the different tRNAs are not equally abundant in plants, animals, and prokaryotes, the modification of an insect gene, for example, to code for a sequence of amino acids utilizing the triplet codons of plants can have major effects on increasing the expression of insect protein in a plant cell. This is thought to be the reason why an artificial combination of two DNA sequences coding for the insect-control protein genes *crylA(b)* and *crylA(c)* of *Bacillus thuringiensis* var. *kurstaki* significantly increased the amount of insect-control protein in cotton leaves from 0.001% to 0.1% of total soluble protein.

Another level of modification relates to ensuring that the introduced gene fulfills the same function and forms the same product as the original gene. There are few problems of this kind in eukaryotes, for the majority share very similar, if not the same, genetic codes. Naturally, this may not be the case when novel genes are inserted into mitochondria, plastids, or bacterial protosomes, for these have their own distinct genetic codes.

23.3.4 Binary Gene Constructs

Binary constructs combine both scorable and selectable marker genes. The best-known and most used example is the BARGUS construct containing the *bar* gene for conferring resistance to the herbicide bialophos, plus *β-gly*, the gene that produces GUS (Fig. 23.14). Binary constructs can be engineered to have the same or differing promoters and can have the same or opposite directions of transcription.

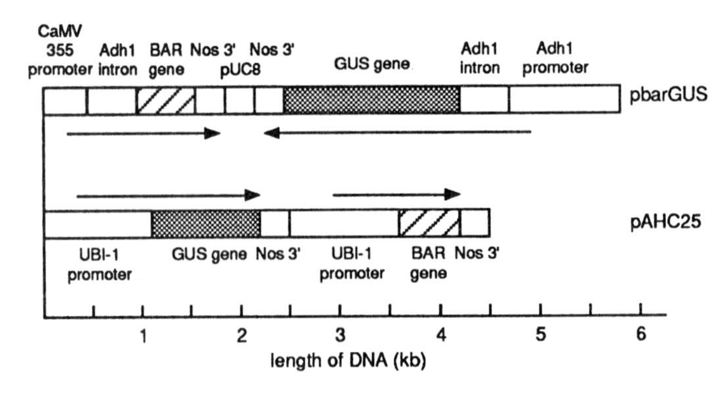

Fig. 23.14. An example of a binary, BARGUS-gene construct, pAHC25. Various modifications of this construct have now been introduced into a broad collection of plant species. (Modified from Vasil et al. 1993.)

23.4 GENETIC TRANSFORMATION IN DICOTYLEDONS

The advances made in modifying the Ti plasmid and the development of a range of procedures for infecting plant cells with DNA have led to the genetic transformation of a wide range of plants (Fig. 23.15). In genetic terms, these are equivalent to extremely wide crosses. As examples, bacterial genes and selectively mutated-plant genes conferring herbicide resistance have been introduced into a wide range of plants. Other genes have been introduced to:

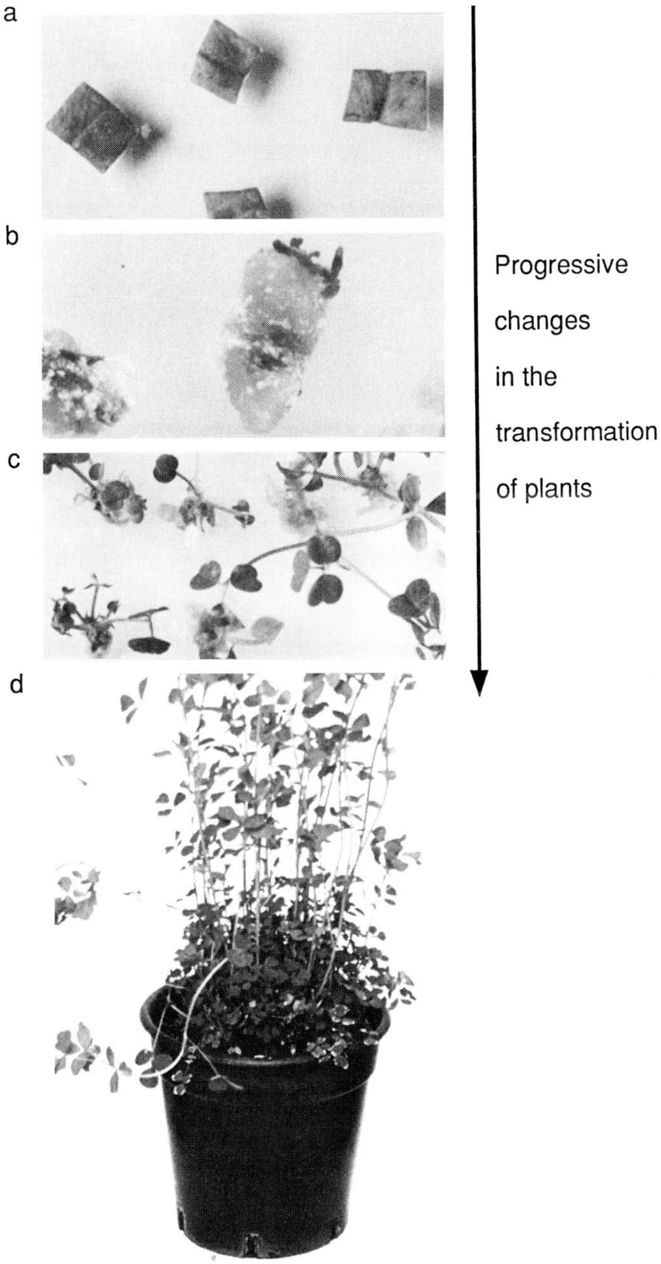

(a) Confer resistance to viruses, fungal pathogens, and insect pests
(b) Improve quality characteristics and marketability
(c) Use the plant as a protein-manufacturing system and biochemical factory

In addition, the introduction of synthetic genes coding for RNA molecules capable of RNAase activity at specific mRNA sites—so-called gene shears (*see* Chapter 20, Fig. 20.20)—provides a mechanism for modifying the levels of specific gene products at different stages of development.

a
b
c
d

Progressive changes in the transformation of plants

Fig. 23.15. Stages in the genetic engineering of alfalfa (*Medicago sativa*) by transformation with *Agrobacterium* containing the herbicide-resistant gene *bar*. (a) Stem sections on nutrient agar after soaking them in agrobacterial solution. (b) Gall formation on the ends of the stem sections where the cells were originally damaged. (c) Regeneration of alfalfa plantlets from the transformed callus. (d) A regenerated mature plant containing the *β-glu* and *bar* genes. (Photographs kindly supplied by T. Wardley-Richardson.)

23.4.1 Introduction of Herbicide Resistance

Novel introductions of genes conferring herbicide resistance on crop plants have the potential to allow very effective weed control, because specific chemicals to which the crop is resistant and the weeds are sensitive can be applied. The majority of resistance genes have been isolated from soil bacteria, selected for their ability to grow in the presence of different herbicides. Other genes have been detected by screening plant-cell cultures for cells able to grow in the presence of increased amounts of the herbicides. With this method, a mutant gene conferring resistance to the herbicide glyphosate was isolated from a cell culture of *Petunia hybrida*, and, by introducing the appropriate cDNA sequence into other plants sensitive to this herbicide, a range of glyphosate-resistant plants have been produced. The modes of action of herbicide resistance can be classified into three gene groups:

1. Genes that overcome the herbicide through overproduction of the herbicide-sensitive target
2. Genes that structurally alter the target of the herbicide, reducing the affinity between herbicide and target
3. Genes that detoxify or otherwise degrade the herbicide before it reaches its target

Examples of bacterial and cDNA genes, and the means by which they inactivate various herbicides, are shown in Table 23.3. The presence of herbicide resistance in transformed cells and tissues provides a new array of selectable markers.

23.4.2 New Resistances to Viruses

Complete viral genomes can be introduced into plants via the agrobacterial system. For example, when a full-length cDNA copy of the genomic RNA of tobacco-mosaic virus (TMV) was introduced into tobacco inside a disarmed Ti plasmid vector, the transformed plants expressed the typical symptoms of TMV infection and produced infectious TMV particles. Even prior to these experiments, it was known that infection of a host plant with a weak or mild viral strain can often provide cross-protection to infection by severe strains of the same virus. Ideas concerning the mechanisms of cross-protection include the following:

Table 23.3. Introduction of Herbicide-Resistant Genes into Crop Plants for More Effective Weed Control; These Genes Are Also Selectable Markers

Gene	Origin	Name and Mode of Action
(a) Incorporating genes from bacteria to inactivate herbicides by detoxification or degradation before they reach their biochemical target.		
bar	*Streptomyces hydroscopicus*	Gives bialophos resistance through the production of phosphinothricin acetyltransferase, which inactivates gluphosinate–ammonium by acetylation.
bxn	*Klebsiella ozaenae*	Bromoxynil nitrilase affects gluphosinate resistance through nitrile hydrolysis.
tfdA	*Allcaligenes eutrophus*	2,4-dichlorophenoxyacetat mono-oxygenase (DPAM) degrades 2,4-D through oxidation.
	Pseudomonas	Produces parathion hydrolase, which degrades parathion by hydroxylation.
(b) Reducing the affinity of the herbicide for the target enzyme by introducing cDNA coding for the production of herbicide-resistant analogs.		
aroA mutant	*Salmonella typhimurium*	Produces a modified form of 5-enolpyruvyl shikimate-3-phosphate synthase (EPSP) that is less sensitive to glyphosate (ROUNDUP®).
cDNA mutant	Bacterial	Produces a modified form of acetolactate synthase (ALS) that is resistant to sulfonylureas and chlorsulfuron herbicides (e.g., GLEAN® and OUST®).
(c) Inducing an overproduction of herbicide-sensitive target.		
EPSP	*Petunia hybrida* cell line	Overproduces EPSP-synthase, giving tolerance to glyphosate.
tms2	*Agrobacterium tumefaciens*	Amidohydrolase effects intolerance to auxin–amide herbicides.
(d) Others with as yet unindentified mechanisms.		
psbA	*Amaranthus hybridus*	Produces Q_B protein conferring resistance to atrazine.
csr1-1	*Arabidopsis thaliana*	An acetolactate synthase conferring resistance to sulfonylureas.

Table 23.4. Introduction of Viral Resistance into Transgenic Plants

Gene Sequences		
From:	To:	Comments:
(a) By inserting viral-coat protein (CP) or nucleocapsid-protein (NCP) gene sequences into plants to give coat protein-mediated protection.		
Tobacco-mosaic virus (TMV)	Tobacco	Field trials of transformed lines have shown good resistance to TMV.
TMV	Tomato	Good field resistance; also protected against strains of tomato-mosaic virus.
Tobacco-streak virus (TSV)	Tobacco	Good resistance.
Alfalfa-mosaic virus (AIMV)	*Medicago sativa*	Also introduced into tobacco and tomato.
Potato-virus X (PVX)	Potato	97% of transformed plants appeared to be normal tetraploids ($2n = 4x = 48$).
Potato-virus Y (PVY)	Potato	Good resistance.
Binary-PVX plus PVY	Potato	Gives resistance to simultaneous infection, and blocks transmission of PVY by viruliferous aphids.
Potato-leafroll virus (PLRV)	Potato	Good resistance.
Soybean-mosaic virus (SMV)	Potato	Also expressed heterologous protection against tobacco etch virus (TEV) and PVY.
Tobacco-rattle virus (TRV)	Tobacco	Also conferred resistance to the early-browning virus (EBV) in pea.
Papaya-ringspot virus (PpRV)	Potato	Delayed-symptom development; also provided resistance to TEV, PVY, and pepper-mottle virus (PeMV).
Cucumber-mosaic virus (CMV)	Tobacco Cucumber	As good as genetically derived resistance.
Tomato-spotted-wilt virus (TSWV)	Tomato	A negative strand RNA virus; also conferred TSWV-resistance on tobacco.
CMV antisense strand	Tobacco	Not as efficient as sense strand.
PVX antisense strand	Potato	Not as efficient as sense strand.
Maize-dwarf-mosaic virus (MDMV)	Maize	Introduction of MDMV, strain-B coat protein also conferred resistance to maize-chlorotic-mottle virus (MCMV).
Tomato-yellow-leaf-curl virus (TYLCV)	Tomato	Transgenic plants expressing the capsid protein of TYLCV are resistant to this virus.
Tomato-spotted-wilt virus N-gene sequences	Tobacco	Both effective. Resistance related to the presence of TSWV sense and antisense N-gene transcripts or high levels of the N-gene protein.
Rice-stripe virus	Rice	Transgenic plants resistant.
(b) Other viral enzymes.		
TMV replicase	Tobacco	Transformed plants completely resistant to TMV strain used.
PVX	Potato	Protection against PVX induced by transformation with mammalian interferon (rat 2′-5′-oligoadenylate synthetase), which activates an endoribonuclease that degrades viral DNA.

(a) Encapsulation of the challenging-strain RNA by the coat protein of the inducing strain
(b) Blockage of the uncoating of the challenged strain
(c) Competition between the two strains for some other factor such as a replicase
(d) Prevention of replication or translation of the severe-strain RNA by genetic interaction with the mild RNA.

Whichever of these or other mechanisms is the cause of this protective reaction, the introduction of viral-coat-protein *(CP)* genes to engineer viral resistance in transgenic plants has been very successful (Table 23.4). Of further benefit, the introduction of a coat-protein gene from one virus often provides protection against other viruses with the same or similar coat proteins.

Other virus-inhibiting genes include viral-RNA replicases, defective-interfering molecules, nonstructural gene sequences, and the antisense insertion of CP RNA genes giving untranslatable or nontranslatable CP-coding sequences. Unfortunately, these are generally not as efficient as CP-sense genes inserted in the normal orientation. The introduction of satellite-gene sequences coding for small, supernumerary, viral-RNA molecules also inhibits virus propagation. For example, satellite RNAs isolated from cucumber-mosaic virus have been introduced into other crop plants with a range of contrasting consequences. Either varying degrees of viral resistance are induced due to a high production of satellite RNA and little viral RNA, or the reverse is true in that satellites that protect nearly every other species induce a spectacular lethal necrosis in certain species, especially tomatoes and other *Lycopersicon* species. This is a pathological change induced by a single base-pair change.

23.4.3 Introduction of Insecticidal Genes

The introduction of *Bt* insect-toxin genes extracted from the entomocidal bacterium *Bacillus thuringiensis* var. *kurstaki* protects a range of crops against insect attacks (Table 23.5). Originally, microbial formulations of this bacterium were sprayed onto plants to produce parasporal crystalline inclusions containing the insecticidal activity. The proteins found in these inclusions are highly toxic to the larvae of Lepidopteran insects, although isolates affecting Dipteran and Coleopteran species have also been found. The toxin proteins disrupt the midgut cells of feeding larvae. The genes that produce Bt proteins have been isolated and introduced into various species of plants. To ensure that the proteins are produced in sufficient quantity, truncated forms of the two genes *cryIA(b)* and *cryIA(c)*, were transformed into cotton. The genetic modifications increased the expression of CryIA(b) and CryIA(c) proteins to 0.05%–0.1% of the total soluble protein. When *cryIA(b)* was subsequently introduced into maize, the toxin protein attained a level of approximately 1.5% of the total soluble protein, an amount easily sufficient to control the European corn borer (*see* Table 23.5). Not surprisingly, attempts are being made to insert these genes into animal genomes for protection against ticks and lice. However, there are indications that transformed *Bt* resistance is naturally breaking down.

Other genes that promote broad-spectrum insecticidal activities include *AalT*, an insect-specific neurotoxin gene from the scorpion *Androctonus australis*, and genes that inhibit insect proteinases and α-amylases. A cholesterol oxidase, isolated from a streptomycetous fungus, is currently being introduced into cotton in an effort to control the boll weevil, which has not been controlled using the Bt protein strategy.

23.4.4 Introduction of Resistance to Bacterial and Fungal Pathogens

In comparison with animal diseases, the effects of bacterial diseases on plants are relatively insignificant compared to the results of fungal diseases. Although attempts have been made to control bacterial infections by the insertion of self-detoxification enzymes such as tabtoxin resistance, control methods in fungi have so far been based on the fact that fungal-cell walls contain a high proportion of chitins (polymers of *N*-acetylglucosamine) and β-1,3-glucans. However, whereas the transformation of plants with chitinase genes commonly increases their resistance to fungal diseases, the introduction of β-1,3-glucanases has shown no similar phenotypic effects. In contrast, when transformed genes containing transcripts of both chitinases and glucanases were combined, the amalgamation gave much greater protection against the tobacco disease "frogeye," caused by the pathogen *Cercospora nicotianae*, than did the chitinase gene alone (Table 23.6).

Other genes that have been assayed for their antifungal activity

Table 23.5. Insecticidal Genes Introduced into Plants by Transformation

Gene	From:	To:	Action
Bt endotoxins *cryIA(b)* and/or *cryIA(c)* and/or modified derivatives of these gene sequences	*Bacillus thuringiensis*	Cotton	Confers resistance to bollworm, budworm, pink bollworm.
		Tomato	Confers resistance to tobacco hornworm, fruitworm, and pinworm.
		Potato	Confers resistance to Colorado potato beetle.
		Cotton	Controls cabbage looper and beet armyworm.
		Corn	Gives excellent resistance to European corn borer.
		Rice	Resistance to striped stem-borer and leaf-folder.
AalT insect-specific neurotoxin	*Androctonus australis*	Corn	Controls *Heliothis* through insertion into the recombinant-baculovirus vector, *Autographa californica* nuclear polyhedrosis virus.
Serine proteinase	Tobacco	Plants	Gives increased resistance to insects.
α-Amylase inhibitor (aAI)		Plants	Gives broad-spectrum insecticidal activity.
	Phaseolus vulgaris	Pea	Seeds contained 1% aAI–Pv protein and were highly resistant to bruchid weevils.
Agglutenin	Wheat germ	Corn	Gives increased resistance to European corn borer.
Cholesterol oxidase	Streptomycetes	Tobacco	Confers lethality to a variety of pests. Being introduced into cotton in attempts to control the major cotton pest, the boll weevil.

Table 23.6. Genes Introduced into Plants to Control Bacterial and Fungal Pathogens

Gene	From:	To:	Action
ttr (tabtoxin resistance)	*Pseudomonas syringae*	Tobacco	Gives protection from "wildfire" caused by infection with *P. syringae*, by a detoxification process effected by an acetlytransferase.
chiA (chitinase)	*Serratia marcescens*	Tobacco	Gives increased tolerance to "brownspot" caused by *Alternaria longipes*.
Chitinase	Bean	Tobacco Potato	With CaMV–35S promoter, conferred increased resistance to *Rhizoctonia solani*.
Chitinase, Glucanase	Rice Alfalfa	Tobacco	Combination of chitinase and glucanase gave greater protection against *Cercospora nicotianae* than either gene alone.
Osmotin	Tobacco	Potato	Induces resistance to potato-late blight caused by *Phytophthora infestans*.
Cecropin	Unknown	Potato	Increases resistance to bacterial infection
RIP (ribosome-inactivating protein)	Barley	Tobacco	Gives increased resistance to *R. solani*.

include ribosome-inactivating proteins (RIPs), which inhibit fungal growth by inactivating fungal ribosomes but have no effect on the protein synthesis of plant ribosomes. Other pathogenesis-related (PR) proteins are being evaluated for their activity, including the tobacco gene *PR5*, which codes for osmotin, an osmosis-regulating protein, whose introduction effectively increases the resistance of potato to potato-late blight. Proteins homologous to osmotin, such as tritin from wheat and ricin from rice, have also been shown to inhibit fungal-protein synthesis. Herbicides can also be toxic to fungal pathogens as exemplified by the control of *Rhizoctonia solani*, the cause of sheath blight of rice, by bialophos application: another reason for the addition of the *bar* gene to rice.

23.4.5 Antisense Genes

Physiological and related responses are being modified by the insertion of antisense genes. These are cDNA gene sequences that are deliberately inserted and read in the reverse direction (Table 23.7). The success of these transformations lies in the fact that antisense sequences are often transcribed at higher rates than the normal-sense genes. Successful examples include melon and tomato genes concerned with fruit maturation and storage. The antisense sequences of these genes have ensured increased resistance to bruising and wounding, improved fruit quality, and delayed ripening through the degradation of ethylene precursors (Fig. 23.16). Other examples include oilseed canola, which has been modified to improve the chemical constitution of the oil extracted from the seed, and alterations to the flower pigmentation of several floricultural-plant species by the insertion of antisense chalcone synthase, an enzyme concerned with flavonoid and anthocyanin synthesis.

23.4.6 Plants as Factories

The development of plant transformation can be viewed in three stages:

1. Establishing the technology of introducing genes into plants and obtaining high levels of expression
2. Utilizing genes that protect plants from external hazards for the improvement of crop yields
3. Using transformed plants as factories for producing new and modified proteins

Some of the genes that have been introduced, their source and recipient species, and the reason for their introduction are listed in Table 23.8, where the major use is to provide new sources of specific commercial products. Potato tubers are especially emphasized, given their ease of harvesting. Improvements in the protein quality

Table 23.7. Production of Transgenic Plants Modified to Express Antisense Genes by Deliberate Insertion of cDNA in the Reverse Direction

Enzyme cDNA	Source	Comment
Polygalacturonase	Tomato	Increases bruising resistance because the antisense gene is transcribed at a higher rate than the sense gene, resulting in diminished cell-wall and pectin degradation. Used in "Flavr Savr®" tomatoes.
Pectinmethylesterase	Tomato Potato	 Increases solids and dry-matter content.
ACC oxidase, deaminase, and synthase	Tomato Melon	Delays ripening by degradation of ethylene precursors (see Fig. 23.16). Used in "Euromelon®" to extend shelf life and ripen on demand.
Prosystemin	Tomato	Antisense DNA decreases systemin expression induced by wounding.
Stearyl ACP desaturase	Canola	Increases stearic acid production.
Thioesterase	Canola	Alters fatty acid content.
Chalcone synthase	Petunia Tobacco *Gerbera*	 Alters flower pigmentation. Dramatically alters flower pigmentation.
Starch synthase	Potato	Eliminates amylose from potato starch.
ADP–glucose pyrophosphorylase	Potato Sweet potato	Sweeter-tasting potatoes as sucrose is not converted to starch.
Fructanase	Chicory	Blocks postharvest conversion of fructans to fructose.

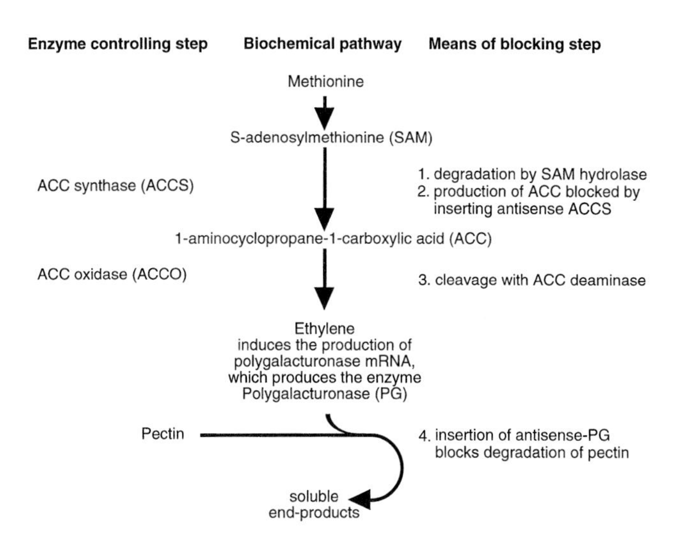

Fig. 23.16. The genes and biosynthetic pathways controlling ethylene biosynthesis and ripening in tomatoes (middle column). Various of the enzymes (left) have been engineered as shown on the right, resulting in the production of the ''Flavr Savr®'' tomato and the ''Euromelon®'' with prolonged shelf life.

COTTON TRANSFORMATION

	PHENOTYPE	GENES REQUIRED	REASON
AGRONOMY	Herbicide resistance	Tolerance to glyphosate, bromoxynil, sulfonyl urea, 2,4-D, bialophos	Increased yield, environmental considerations
	Environmental stress tolerance	Superoxide dismutase and heat, water, cold, and salt stress genes	Improved yield and quality, superior crop adaptation
	Insect resistance	*Bt* toxins, protease inhibitors, venoms	Yield and ecological considerations
SEED QUALITY	Hybrid cotton	Pollen antisense genes, cytotoxic genes	Induction of male steriles for seed production, protection of proprietary seed
	Seed protein	Protein quality genes	Stock feed and food industries
	Oil quality	Oil component genes	Food industry and cosmetics production
FIBER QUALITY	Fiber properties		Textiles
	Fiber modification	Alter genes already present, introduce new plant genes, add bacterial hormones	Improved quality, length, strength, fineness; novel colors
	Novel properties	Genes to increase fiber liquid absorption, fiber heat retention	Biomedical applications, winter clothing, carpets, personal care products

Fig. 23.17. The many and varied applications of genetic engineering to cotton improvement and use. Agronomic and quality parameters are listed on the left, the genes able to modify these characters in the middle, and the benefits that should accrue on the right. It should be noted that Agracetus Co. has been awarded a U.S. patent covering all genetically engineered cotton plants. (Adapted from John and Stewart, 1992.)

of leaves of fodder crops for animal production are also easily accessed, as are attempts to improve the physiological response of crops to enhance their marketability. Some of the possible reasons and explanations for the transformation of cotton are shown in Fig. 23.17 to illustrate the array of novel production and end uses for which the cotton plant may be suitable after appropriate genetic engineering.

23.5 TRANSFORMATION IN CEREALS AND OTHER CROPS

A major constraint on cereal-crop transformation is that only a few cells in monocotyledonous tissues are competent for both transformation and regeneration. In contrast to dicotyledonous tissues, where the wound response is an important basis for regeneration, monocotyledonous tissues display either a rudimentary or no-wound response. Although *Agrobacterium*-mediated transformation of cereals is now possible, the development has been relatively late and alternative methods have been investigated for inserting DNA into the cells of monocotyledons. These procedures generally require the use of electrical or explosive energy.

23.5.1 Electroporation

Electroporation, which renders cell membranes transiently permeable because of an electric shock, has been assayed as a means of transferring novel-gene constructs into cereal protoplasts and calli. Although rice *(Oryza sativa)* protoplasts have been successfully transformed by the bacterial gene *hph* (encoding hygromycin-B resistance) using this technique, the total-transformation frequency showed that only 0.1–0.6% of the clones subjected to selection with hygromycin-B were successfully transformed. The presence of the *hph* gene on a chromosome of the stable transformants was determined by analyzing the DNA of plants regenerated from resistant calli. In these, the copy number of the integrated gene was estimated to be from 2 to 10 per diploid genome. Rice plants have since been successfully regenerated from leaf-mesophyll protoplasts.

23.5.2 High-Velocity Microprojectile Bombardment or Biolistics

An alternative method for transforming plant cells is to blast microprojectiles coated with gene constructs into the cell, and possibly the nucleus, by physically shooting the DNA into the cell. The first successful attempts used a blank cartridge to explosively project a nylon macroprojectile, with tungsten microprojectiles on its tip, toward a stopper plate, which was drilled to allow passage of the DNA-coated tungsten pellets while preventing the passage of the nylon projectile (Fig. 23.18). The cell targets were a few centimeters from the plate, allowing even dispersion of the pellets over a small area of tissue. Other systems use pressurized air or helium as a noncontaminating propulsive force, and even smaller gold pellets to effect the biolistic introduction. Using this technique, the successful transformation and regeneration of maize cells and the production of at least partially fertile, transgenic maize plants have been reported. Cells of embryogenic-maize suspension cultures were bombarded with microprojectiles carrying the bacterial gene *bar*. Stable integration of the phosphoinothricin acetyltransferase (PAT) gene into the genome of transgenic plants was appar-

Table 23.8. Introduction of New and Novel Proteins into Transgenic Plants

Protein	Source	Recipient	Objective/Comment
(a) Commercial production			
α-Amylase	*Bacillus licheniformis*	Tobacco	Industrial production of bulk enzyme in seeds.
Serum albumin (HSA)	Human	Tobacco	
		Potato	Commerical production of HSA in tubers.
Cyclodextrin glycosyltransferase	*Klebsiella*	Potato	Commercial production of cyclodextrins.
Glutamine synthetase	Alfalfa	Tobacco	
β-Conglycinin	Soybean	Petunia	
ADP–Glucose pyrophosphorylase	Bacteria	Potato	Increase starch content of tubers.
Sucrose phosphate synthase	Maize	Tomato	Increase sucrose content and decrease starch content of mature fruit.
Mannitol dehydrogenase	*E. coli*	Tobacco	Increase mannitol production.
Acetoacetyl CoA synthetase and acetoacetyl CoA reductase	*Alcaligenes eutrophus*	*Arabidopsis*	Catalyse steps in the production of poly-β-hydroxybutyrate.
Hyoscyamine-6-β-hydroxylase	*Hyoscyamus niger*	*Atropa belladonna*	Scopolamine production.
Leu-enkephalin peptide	Human	Canola *Arabidopsis*	
Interferon	Human	Turnips	Commercial production.
Antibodies to TMV and nematodes	Mice	Tobacco	Production of "plantibodies" in plants.
(b) Nutritional improvement			
Vicilin	Pea	Tobacco	
		Alfalfa	Nutritional improvement. Adding the KDEL sequence increased vicilin production to 2.5% of the total leaf protein.
Ovalbumin	Chicken	Tobacco	Nutritional improvement.
Zein	Maize	Tobacco	Unstable in recipient.
(c) Physiological improvement			
Glycerol-3-phosphate acyltransferase	Squash	Tobacco *Arabidopsis*	Chloroplast proteins improve chilling resistance by increasing the proportion of cis-unsaturated fatty acids.
ACC deaminase	Bacteria	Tomato	Delays fruit ripening through degradation of the ethylene precursor ACC (Fig. 23.16).
		Broccoli	Controls ethylene production so that the crop stays green longer, improving transportability.
Phytochrome-B	Rice	*Arabidopsis*	Short hypocotyl phenotype and overexpression of phytochrome.
Ribonuclease construct	Fungus	Plants	Destroys tapetal layer in anthers and induces male sterility.
RNase inhibitor	Fungus	Plants	Counteracts ribonuclease construct and restores male fertility.

ently proven by appropriate backcrossing, followed by screening of the progeny for PAT activity. The presence of the *bar* gene was confirmed by DNA analysis.

Similar methods have been used to transform wheat, one of the most recalcitrant of all cereal crops to transformation. Evidence that individual protoplasts and calli were transformed was the presence of the blue-staining product resulting from β-glucuronidase activity associated with resistance to both kanamycin and glyphosate. Although this method provides the necessary conditions for transforming wheat, the ability to produce fertile plants from individual wheat protoplasts lags behind the successes achieved in other plants. Tissue bombardment is also being used to simplify the transformation of dicotyledonous crops, and the embryos of coniferous forest-tree species such as white spruce *(Picea glauca)* have also been transformed with this technology.

With the use of an interesting combination of techniques, immature embryos from rice, wheat and barley, and meristematic tissues of banana *(Musa acuminata),* a member of the monocotyledonous family Zingiberaceae, have now been transformed by *Agrobacterium*. Cells in the tissues to be transformed were initially punctured by biolistic bombardment, left for a few days to develop biochemical wounding responses, and then cocultivated with *Agrobacterium* and acetosyringone.

23.5.3 Other Means of Plant Transformation

A wide range of other procedures have been tried in attempts to produce larger numbers of transformed progenies. Among those that have been tested, with varying degrees of success, are the following:

1. Inserting genes into viruses in the expectation that the viruses will spread the genes throughout susceptible plants. Unfortunately, nonengineered viral genomes do not insert into the host genome.
2. Fusing DNA-containing liposomes with tissue-culture cells, or directly injecting DNA-containing liposomes into tissue cultures. Although DNA can be inserted into cells by these methods, it is effectively degraded by nucleases

in the cytoplasm, and if it escapes this fate, it is still unable to be transmitted through the nuclear membrane.

3. Germinating seeds or soaking plant tissues in the presence of DNA. There is very little evidence that DNA can be simply absorbed through a normal cell wall and then transmitted, unbroken and undamaged, into the nucleus.
4. Coating pollen grains with DNA prior to placing the pollen on the stigma of the flower for fertilization. There is little evidence that this method can be relied upon.
5. Using polyethylene glycol (PEG) to promote DNA uptake into cells grown in tissue culture has had some success. Hygromycin-B resistance encoded by the *hph* gene was used to transform microspore-derived protoplasts of indica-rice using PEG-mediated DNA uptake. Analysis of the DNA from primary-transgenic rice plants and their offspring showed that the plasmid was integrated into DNA of high-molecular weight, presumably chromosomal in origin, and that the *hph* gene was expressed in the leaves. Maize has also been transformed by this procedure.
6. Directly injecting DNA into growing plants by macroinjection. Given the apparent success of the original transformation of rye plants by injecting DNA-containing solution into the region of the stem near the developing head, it is disappointing that this technique has not been particularly reproducible.
7. Directly injecting DNA into cells and nuclei, using microinjection and microscopic imaging, is an effective means of delivery. Even though only one cell can be treated at a time, the advantages are that the host cell and organelle can be deliberately selected, and the quantity of DNA delivered is adjustable. In cereals, there are the usual difficulties of proliferating plants from individual protoplasts, but in animals, in which single egg cells or subordinate tissues can be injected, microinjection is the method of choice (*see* Section 23.6.4).

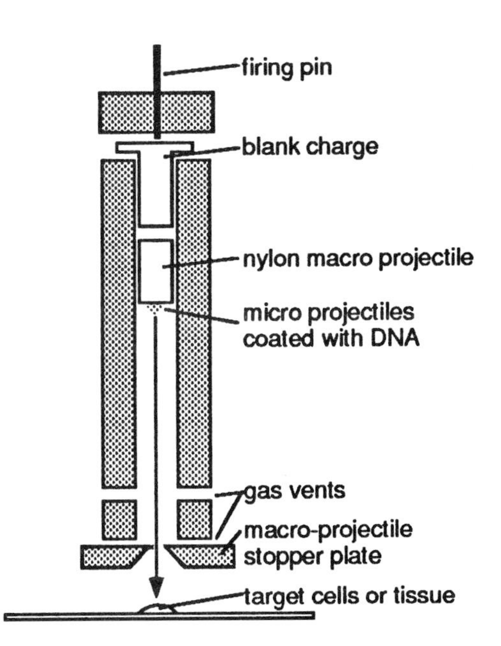

Fig. 23.18. A diagrammatic longitudinal section through a biolistic gun, designed for the high-velocity bombardment of plant and animal cells, with microprojectiles coated with DNA gene constructs. (Adapted from Klein et al., 1987.)

23.6 TRANSGENIC ANIMALS

Among the early examples of transgenic animals were mice that had been transformed with a rat growth-hormone gene. A cover photograph in the journal *Nature* (December 1982) showed a transformed mouse that was almost twice the size of its nontransformed sibling. Animals that have since been successfully transformed include nematodes, *Drosophila,* sea urchins, frogs, and fish. On a larger scale, farm animals, including pigs, sheep, and cows, have also been transformed.

23.6.1 Transformation in Animal-Cell Cultures

Beginning in the late 1970s, animal-transformation research was principally concerned with problems related to gene expression and regulation in cells maintained in tissue culture. Electroporation was used to enhance transformation frequencies, a methodology closely allied to bacterial transformation, as were the genes involved. Usually, the transgenes were introduced into cultures of embryo-derived stem cells extracted from mammalian blastocysts, but yeast and insect cell-culture systems also were favored for production of the complex proteins of higher eukaryotes, as these systems also produce the appropriate posttranslational modifications to the final product. Consequently, even though these cultured cell lines are incapable of development into organisms, they have formed the basis for the factory production of specific biochemicals. Some examples of the range of human genes currently being investigated for their suitability for cell-culture production are listed in Table 23.9. As might be expected, protein production in cell-culture systems can occasionally be inhibited through the accumulation of polypeptides in inclusion bodies or, as in plants, can completely lack activity due to deficiencies in posttranslational modifications and protein conformation.

23.6.2 Mammalian Transfection

Changes introduced by genetic engineering are little different from, and certainly complementary to, other genetic procedures for plant and animal improvement. The prominent gene for the production of metallothionein was used in many early animal-transformation studies, and since then, the many derivatives of this gene utilizing the metallothionein promoter alone (Fig. 23.19). Gene constructs under the control of this promoter are induced by surges of heavy metals. Other genes and other genetic constructs that have been used to transform animals are described in Table 23.10.

Apart from the biotechnological aspects of animal transformation, mostly to produce compounds of therapeutic interest, animals have also been transfected to introduce physiological changes such as improvements in growth rates and the efficiency of food conversion, to improve defective characters, or to generate animals with improved characteristics. The major difficulty with mammalian transformation lies in transformed cells that are subsequently capable of generation or regeneration into a complete animal. A number of methods have been tried and tested for animal transfection with the aim of introducing a new gene either directly into the fertilized

Table 23.9. Biotechnological Aspect of Recombinant Transgenesis with Some of the Organisms and Cell Lines Used to Produce Human Proteins

Proteins	Organisms/Cell Lines Being Used to Express Protein			
	E. coli	Fungi[a]	Cell Lines[a]	Other Organisms
Immunomodulators				
Interferons, α, β, χ	+	Y	Mu, Mn	
Interleukins, -1, -2, -3, -6	+	Y	SAC, CHO	
Hormones				
Human-growth hormone (hGH)	+	Y	CHO, SAC	Mice
Somatostatin	+			
Calcitonin	+	Y		
Chorionic gonadotrophin		Y	Mu	
Luteinizing hormone			Mu	
Relaxin	+			
Insulin	+	Y		
Proinsulin	+	Y		
β-Endorphin	+	Y		
Insulinlike growth factors	+	Y,P		
Growth factors (GF)				
G- and M-colony stimulating factor	+	Y	CHO	
Fibroblast GF	+	Y		
Epidermal GF	+	Y		
Platelet-derived GF	+	Y	SAC	
Connective-tissue activator	+			
Plasminogen activator		Y		
Angiogenin-inducing factor			SAC	
Fibronectin		Y		
TGF α and β	+	Y		
Lactoferrin		A		
Blood proteins				
Human-serum albumin	+	Y, P		Potato, tomato
Hemoglobin	+	Y		Mice, pigs
Antithrombin III			CHO	
Factor VII		Y		
Factor VIII		S	SAC, CHO	
Factor IX				Sheep
Factor XIII	+	Y		
Urokinase			SAC	
Prourokinase	+		SAC	
Streptokinase		P		
Hirudin		Y		
Protein C			SAC	
Thrombomodulin			Mn	
Alpha-1-antitrypsin	+	Y		Sheep
Apolipoproteins	+			
Platelet factor 4	+			
Inhibitors				
Mullerian-inhibiting substance			CHO	
Elastase inhibitor	+			
Lipocortin 1	+	S		
Leucocyte-protease inhibitor		Y		
Enzymes				
Lysozyme	+	Y		
SOD	+	Y, P	SAC	
Rennin	+		CHO	
Gastric Lipase		Y		
Vaccines				
Hepatitis B		Y, P	CHO	
Whooping cough	+			
Malaria		Y		

Note: Those genes used to transfect other animals are also indicated.

[a] Y = *Saccharomyces cereviseae;* S = *Schizosaccharomyces pombe;* P = *Pichia pastoris;* A = *Aspergillus oryzae;* SAC = surface-adherent cells; CHO = Chinese-hamster ovary cells; Mu = Murine cells; Mn = mammalian cells.

Source: Adapted from Hodgson (1993).

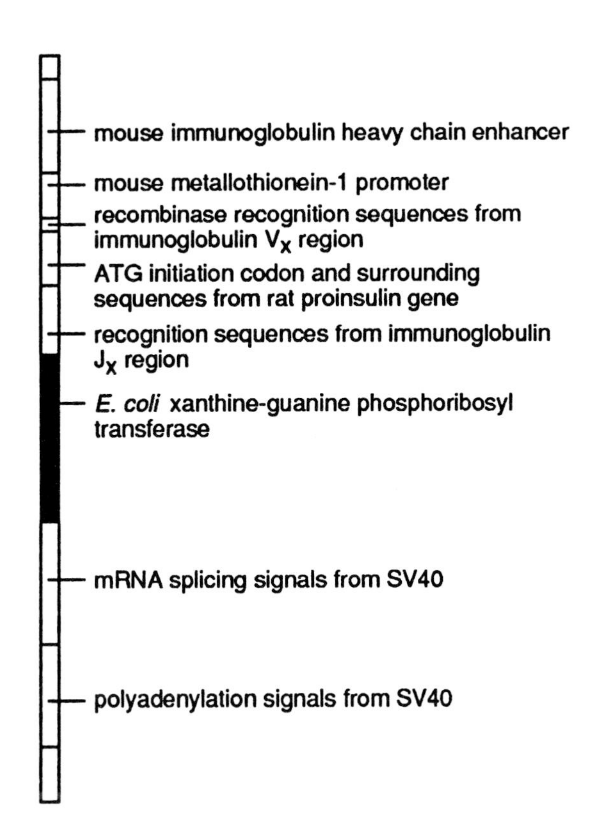

Fig. 23.19. The pHRD transgene for mammalian transgenesis, using a combination of mouse, rat, bacterial, and SV40 DNA segments. The addition of a heavy metal to the transformed animal switches on the metallothionein promoter, inducing production of the transferase. (Modified from Engler et al., 1991.)

ovum or, more indirectly, into the sperm that will fertilize the ovum. By whatever means, the gene has to be transmitted through the relatively impermeable cell membrane and into the cell nucleus before it can be incorporated into the host genome.

23.6.3 Retroviral Vectors

Among the biological techniques, the use of retroviral vectors was for some time the only such method. However, although retroviruses are very efficient at inserting foreign DNA into cells and expressing that DNA with high efficiency, they have so far been fairly ineffective in the production of transgenic animals. Some of the reasons limiting the effectiveness of retroviruses for this purpose include the following:

1. Special containment facilities are required to work with retroviruses.
2. Retroviruses are incapable of self-replication, and complicated host–vector constructs are required for their use.
3. When integration does occur, the inserted gene is commonly flanked by long, terminal-repeat sequences, which often interfere with the expression of the gene itself.
4. Integration occurs in multicelled embryos, so that not all founder cells are transfected, reducing the efficiency of future germline transfer.
5. The presence of retroviruses in transgenic animals is a severe restraint to the use of that animal for human consumption.

To overcome these defects, the physical introduction of DNA into the nucleus using direct microinjection was developed.

23.6.4 Microinjection in Animals

Most transgenic animals have been produced by the direct injection of DNA into the male pronucleus of fertilized, one-cell eggs (*See also* Fig. 20.22, Chapter 20). Whereas this methodology is somewhat limited due to the difficulties in being able to observe and directly inject DNA into the nucleus, microinjection is presently the method of choice. The technical steps involved are as follows:

1. Several hundred copies of linearized, cloned genes are suspended in a small volume of buffer and injected directly into the nuclei of one-celled embryos, using a glass micropipette. Commercial equipment and micromanipulators are available for this purpose.
2. Viable embryos that have withstood the injection and have continued cell division are reimplanted into the uterus of pseudopregnant females to develop into progeny.
3. DNA is extracted from tissue samples of the presumptive "transgenics" and analyzed to confirm the presence of the heterologous gene.
4. Confirmed transgenics are raised for further study.

Although this sounds relatively simple and straightforward, an example of the actual success rate of these and related procedures is listed in Table 23.11. In large animals, the percent success rate, from microinjected ovules to transgenic organisms, is of the order of 0.08%. It is also technically difficult to genetically transform large animals for the following reasons:

(a) The ova are more opaque as a result of a high-lipid content.
(b) Major surgery is required to obtain small numbers of recently fertilized ova.
(c) After microinjection, the ova need to be surgically implanted into synchronous recipient animals.

Furthermore, even though confirmed transgenics may correctly express the introduced gene in the correct and/or desired tissues of the new host, there may be novel and unexpected effects elsewhere in the organism. For example, when bovine-growth hormone was transformed into pigs, there was a significant improvement in daily weight gain, feed efficiency, and changes in carcass composition, including a reduction in subcutaneous fat. At the same time, unfortunately, the transformation induced a number of detrimental effects, including gastric ulcers, arthritis, cardiomegaly, dermatitis, and renal defects. This study included two successive generations of two lines of transgenic pigs, which were selected for homozygosity of the introduced gene. Further studies on these animals suggested that it might be possible to minimize the deleterious effects by using pigs with different pedigrees or changing the pig-husbandry methods.

The transformation of farm animals raises the possibility that it may be efficient to produce novel proteins in large animals, using the obvious lactation systems of bovines and ovines, which are capable of producing several liters of milk per day. This requires

Table 23.10. Transgenesis in Animals for Genetic Change

Gene	Source	Recipient	Comments
(a) Entire gene sequence			
AαIT	Scorpion	Mouse fibroblast cells	
Antifreeze protein	Wolffish	*Drosophila*	Retained full biological activity.
β-globin	Rabbit	*Xenopus* oocyte	
β-gal	*E. coli*	Fish	
Growth hormone	Rat	Mice	The first transgenic animals.
Growth hormone	Rat	Fish	Use of promoter sequences from fish improves activity.
Growth hormone	Human	Fish	Now successful in many species.
Egg-white lysozyme	Hen	Mice	
α- and β-globin	Human	Mice	Produced HbS, human sickle cell hemoglobin, in the mice.
$tRNA^{leu}$	Yeast	Rabbit	
Erythropoietin	Human	Mice	Function and tissue expression of trangene.
Lysozyme	Chicken	Mice	Lysozyme expressed specifically in macrophages, independently of its chromosomal position.
(b) Hybrid gene constructs			
Ovine β -lactoglobulin promoter/human α-1-antitrypsin (*hα1At*) gene		Mice	Mice yielded up to 7 g/l of active *hα1AT*.
hα1AT gene		Sheep	Four (out of six) trangenic ewes produced up to 35 g/l of *hα1At* in milk. Seven (G1) daughters of transgenic ram secreted at least 15 g/l. Two G2 ewes produced similar amounts.
Bovine α-S1-casein promoter/human urokinase gene/casein poly (A) terminator		Mice	Secreted 1–2 g/l of human urokinase in milk.
Murine metallothionein promoter/ herpes simplex virus (HSV)-thymidine kinase gene and downstream sequence (pMK)		Mice	Thymidine kinase gene transmitted only by female mice.
Carp β-actin promoter/CAT/salmon growth hormone poly(A) terminator		Zebrafish	High level of gene expression using piscine regulator.
pHRD transgene containing *E. coli* xanthine-guanine phosphoribosyl transferase		Mice	Methylation of introduced genes

the introduction of genetic constructs specifically devised to be activated in the milk-producing glands of mammals, so that the new product can be simply extracted from the milk. Consequently, genes constructed for these systems generally include the activators and promoters of genes that are already producing milk proteins.

Physical microinjection has been used to insert human metaphase-chromosome fragments in mouse-egg cells, without cloning specific gene sequences. The success of this procedure was evaluated using biotin-labeled satellite DNA sequences that specifically assay human-chromosome centromeres. Increasingly, biolistic-bombardment methods are being used to transform mammalian cells and tissues.

Table 23.11. Success Rate of Bovine Transgenesis

Stage in Transgenesis Process	Number
Oocytes extracted	2470
Matured oocytes	2297
Fertilized egg cells	1358
Successfully injected	1154
Survived	981
Successful cleavage	687
Transferred to foster-mothers	129
Successful pregnancies	21
Gene integration	2

Source: Modified from Krimpenfort et al. (1991).

23.6.5 Homologous Recombination for Gene Introgression

The biological transfer and integration of genes by homologous recombination is becoming increasingly important. Cloned genes that have been genetically modified are permanently introduced into chromosomal DNA by recombination between the original chromosomal DNA and the corresponding cloned, homologous DNA sequences. With such precise gene targeting, any modification of a cloned gene should be able to be directly introduced into the genome of the host organism at the original locus for that gene. This procedure removes variables such as the integration of the cloned DNA fragments into random sites throughout the genome and the formation of the novel DNA product at sites and in tissues that are inappropriate.

In animals such as mice, the technique normally involves the introduction of the cloned, modified gene into embryo-derived stem (ES) cells (Fig. 23.20). In most of these cells, the vector inserts randomly. However, in some cells, the target vector manages to pair with its homologous, chromosomal DNA sequence, so that the

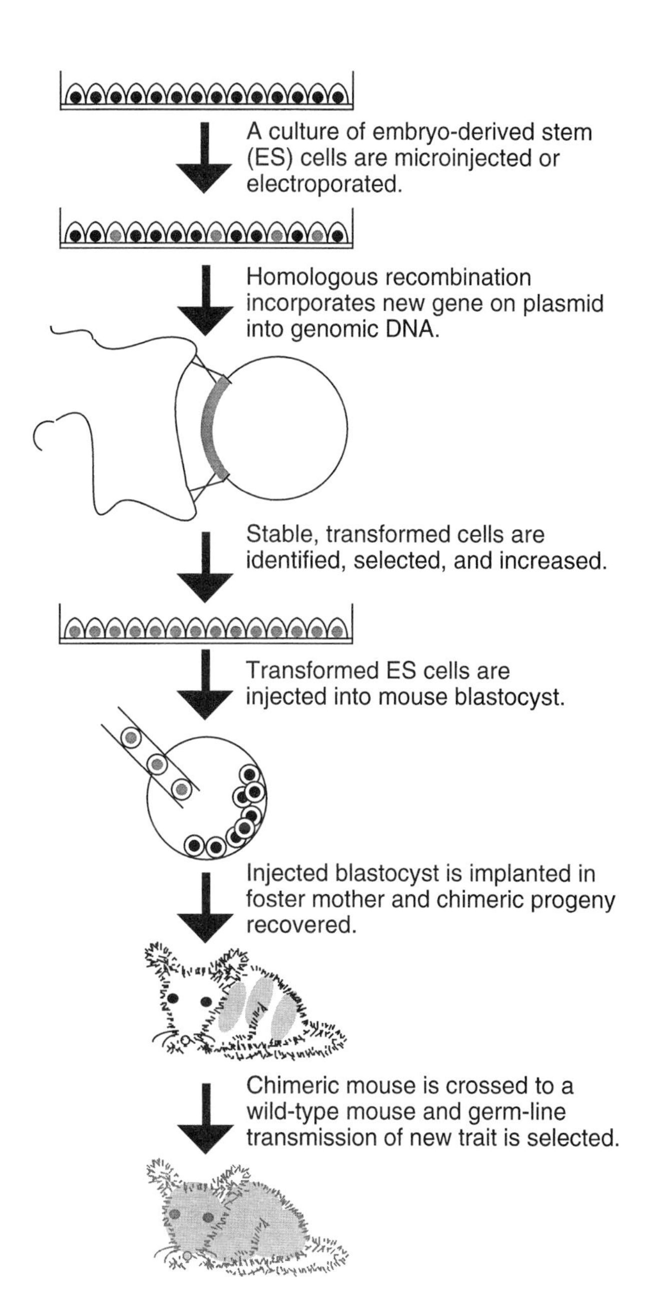

Fig. 23.20. Homologous recombination in mice, whereby a gene introduced via a plasmid is transformed into genomic DNA through common DNA sequences at either ends of both sequences.

modified gene is incorporated into the original genome. The cells that have undergone homologous recombination are then selected, multiplied by direct screening, injected into blastocysts, and implanted in foster mothers. In turn, the chimeric progeny of these females are reared, intercrossed with normal mice, and their progeny selected for homozygotes carrying the modified gene (*see* Fig. 23.20). The essential skill is in being able to select for the specific stem cells in which homologous recombination has occurred, although there are a number of ways in which this can be done, one of which is by insertional mutagenesis; that is, when a modified transgene is inserted into a functioning gene, the lack of product of the normal gene or the defective product of the mutated gene allows the function of the normal gene to be analyzed and described. Alternatively, sequence-replacement vectors can be used that insert the whole target vector into the chromosomal locus, replacing the normal DNA sequence with the novel sequence. Such specific-gene targeting may be used to create dominant, negative mutations, which block the production of protein products from mutant genes by interfering with the normal function of other genes. The possibility that normal genes can be introduced via pluripotent embryo-stem cells into the somatic tissues of a developing organism, which has an inherited mutant version of the original gene, is currently an active area of research.

23.6.6 Antisense Oligonucleotides

These are short pieces of nucleic acid that recognize and bind to specific mRNAs, thereby blocking the translation of the messenger into polypeptide product. In animals, it has frequently proven difficult to keep the antisense molecule tightly bound to the RNA, and protein synthesis is often incompletely blocked. Even so, antisense oligonucleotides have been used to inhibit the growth of the papillomavirus that causes genital warts. It is also possible for antisense drugs to be mixed with the leukemic cells of bone marrow and, after the remaining marrow has been destroyed with chemotherapy, transplanted back into the organism to restore the immune system. Experiments of this type have been carried out in a range of mammals, including humans. The alternative to antisense suppression of a gene product is sense suppression, where the production of gene product from an additional copy of the respective gene causes interference, and an overall reduction in the amount of gene product.

The ability to inhibit the expression of defective-gene products may also be carried out using ribozymes, which are catalytic RNA sequences designed to cut specific pieces of mRNA (*see* Chapter 20, Section 20.2.3) and which are being designed to attack the RNA of viruses that infect humans. In addition, it may be possible to design oligonucleotides that bind to specific mRNA sequences, allowing the natural enzyme RNAase-P to cleave the target double-stranded RNA construct.

23.6.7 Transfer of Mice Technology to Larger Animals and Humans

The use of mice in transformation experiments allows basic molecular–genetic studies to be carried out in the laboratory. The database created in this way can then be used to benefit larger animals such as pigs, cows, goats, and sheep, providing an alternate use for these animals in addition to their traditional uses as sources of meat, milk, or fiber. The technology is also available for application to humans.

As an example, the ability to treat mice successfully for a disease such as sickle cell anemia, could well lead to improved techniques for the control of this genetic disease. For this purpose, two vectors carrying the human genes for α- and β^S-globins were introduced into mice. The transgenic mice not only synthesized human sickle cell hemoglobin but exhibited many other symptoms of the disease, including sickling of the red blood cells when deoxygenated and general spleen defects. The development of a successful treatment for this genetic disease through the use of transformed mice would be of great significance. For similar reasons, the introduction into mice of oncogenes that are directly or indirectly involved in the development of human cancers is also being studied.

These and many other genes that can interfere with the perpetua-

a. transformed mice (Line 1) containing gene construct with a defective target gene that is not expressed because of the inactive promoter

inactive promoter
yeast activator gene, *GAL4*, transacting
defective gene

b. transformed mice (Line 2) containing a second construct with the activator gene sequences

active promoter
GAL4

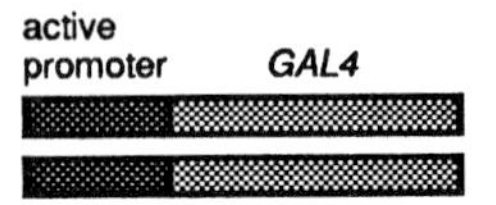

c. genotype of hybrid mice produced by crossing Line 1 and Line 2 parents. The active promoter switches on the transacting *GAL4* resulting in the production of defective gene product

active promoter
GAL4

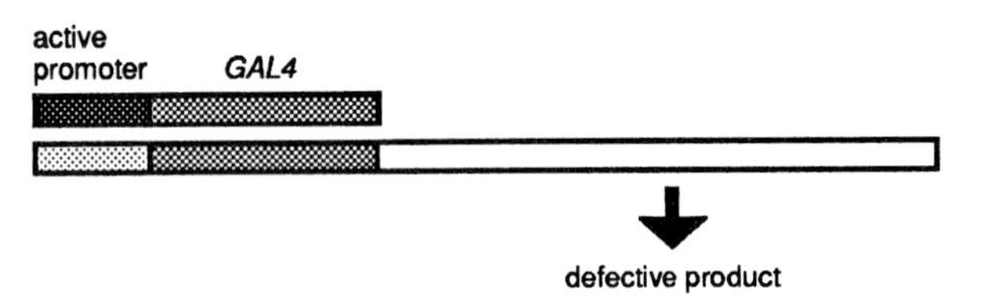

Fig. 23.21. A binary mouse system designed to perpetuate what would otherwise be defective or lethal phenotypes. The conformation must be able to interact in transposition.

tion of genetic traits by causing the development of sublethal or lethal phenotypes, might also be maintained synthetically. The solution is to develop transformable gene constructs, which by themselves are ineffective and correspondingly simple to maintain, but when added together through normal crossing, cause the defective phenotype. These constructs would be used to produce two genetically engineered parental lines, which are in themselves viable but express the defective or lethal phenotypes in their hybrid progeny. An example of such a system is shown in Fig. 23.21.

23.7 DELETION OF UNNECESSARY GENES

Throughout this chapter, we have seen how novel genes are transformed into plants and animals. However, because of the necessity to transform both scorable and selectable genes to ensure that the transformation has been successful, unwanted genes such as those conferring resistance to antibiotics may also be transferred. If the products of these genes are unacceptable, as is possible, for example, with food products, it may be necessary to remove the selectable genes before the transformed product can be used. The presence of any one selectable marker also precludes the use of that marker in subsequent transformations. In most situations, the use of genetic recombination as a means of breaking the linkage between a wanted and an unwanted gene is highly unlikely, because the genes present in single constructs are so intimately linked. Moreover, several copies of the construct may have been inserted. Assorted methods have been developed to overcome this problem.

23.7.1 Removal of Genes Using CRE/LOX Recombination

A system known as CRE/LOX has been devised and tested for the removal of specific marker genes. The system involves the introduction of enzymic scissors, CRE, produced by the gene *control of recombination* (*Cre*), and originally obtained from bacteriophage P1. The CRE enzyme excises DNA sequences located between a pair of identical 34-base-pair DNA sequences called *locus of crossing over* (*Lox*), by catalyzing the exchange of sequences flanking the LOX sites. Thus, if the introduced marker gene is first positioned between two *Lox* sequences, organisms that have been successfully transformed with the marker gene can be crossed to sister lines transformed with *Cre*. In these hybrids, the CRE enzyme excises the marker gene, leaving only the desired gene behind. If necessary, the *Cre* gene can itself be deleted in later generations, leaving only the wanted gene in the transformed organism. Although this system clearly has potential in that early results reported the loss of marker in about one-quarter of an F_2 progeny, this turned out to be a quasi-genetic rather than an actual F_2 segregation, because the two genes and the CRE/LOX reaction were present in the F_1 plant, and not all of the marker genes were, in fact, deleted. It has since been found that this reaction was related to the fact that different CRE transformants give differing levels of chimerism in F_1individuals.

23.7.2 Removal of Genes Using Maize *Ac/Ds* Transposable Elements

Optionally, if the selectable marker cotransformed along with a desired gene can be shifted elsewhere in the genome, it can be removed by simple crossing over. Such genomic shifts are possible using suitably modified elements of the maize *Ac/Ds* transposon system (*see* Chapter 7, Section 7.5). Two types of vector have been designed to utilize this system, both inserted between right- and left-handed T-DNA borders. One of the vectors encloses the gene of interest between inverted-repeat *Ds* elements, and the second vector encloses the selectable marker gene between inverted *Ds* elements. In the presence of an *Ac* element comprised of a transposase gene and the same two inverted repeats, so that, in effect, the *Ds* elements are *Ac* elements without the transposase gene, genes within the *Ds* elements are transposed to other locations throughout the genome and are thereby genetically separated. The individual components can then be selected or deleted by chromosomal crossing over. The Ac transposase can itself be included within the transformed T-DNA.

23.8 POST-TRANSFORMATION CONTROL OF TRANSGENES AND GENETIC SILENCING

Even though specific transgene sequences are proven to be present in transformed plants, the presence of multiple copies of homologous transgenes can lead to the genetic silencing of some or all of the copies, with two effects. First, it creates major problems through the unwanted silencing of transgenes in plants known to be successfully transformed. Second, it raises the question of how plants regulate gene expression.

The evidence for genetic silencing has gradually accumulated from the results of what would otherwise be unrelated experiments (*see* Chapter 16, Section 16.9). Among these were experiments in which attempts to increase the number of copies of a particular gene sequence often resulted in a reduction in the expression of that gene. This occurred not only when multiple, linked copies of a gene construct were inserted as a single unit but also when the same gene sequence was inserted on different occasions, using different antibiotic-selector genes with the same promoter. Genes on the first construct were then found to be inactivated through methylation (*see* Chapter 17, Section 17.2). Again, when sense copies of genes were introduced as controls for the insertion of antisense DNA sequences, it was often found that both constructs were just as effective in silencing the endogenous genes. Finally, when sense genes, under the control of strong promoters, were introduced in attempts to overexpress endogenous genes, the expressions of both the transgene and the endogenous genes were commonly completely abolished. Further work has shown that the phenomenon of homology-dependent genetic silencing can be caused by regions of homology within promoter DNA sequences, gene-coding regions, or both (discussed in Chapter 17, Section 17.2).

23.9 MENDELIAN INHERITANCE OF TRANSFORMED DNA

Finally, whether in plants or animals, the definitive expression of genetic transformation must lie in the knowledge that the transformed gene has the ability to be transferred from one generation to another. The supposition is that the introduced genes are fixed in position within a single chromosome, allowing them to be genetically mapped with respect to preexisting genes on that chromosome. Although scorable, selectable, and DNA markers have been used to determine the presence of new genes in transformed plants and animals, and their progenies, few transformed gene been definitively mapped. Even the number of generations through which transformed genes are inherited remains in a state of flux. In addition, although some researchers have reported that the new DNA sequences are transmitted in a Mendelian fashion, the actual genes may not be expressed. In even fewer instances is there data showing that Mendelian inheritance has occurred.

BIBLIOGRAPHY

Section 23.1

Old, R.W., Primrose, S.B. 1989. Principles of Gene Manipulation: an Introduction to Genetic Engineering, 4th ed. Blackwell Scientific, Oxford.

Section 23.2

Boxer, S.G. 1996. Green fluorescent protein: another green revolution. Nature 383: 484–485.

Cocking, E.C., Davey, M.R., Pental, D., Power, J.B. 1981. Aspects of plant genetic manipulation. Nature 293: 265–270.

De Block, M., Herrera-Estrella, L., Van Montagu, M., Schell, J., Zambryski, P. 1984. Expression of foreign genes in regenerated plants and in their progeny. EMBO J. 3: 1681–1689.

Flavell, R.B., Dart, E., Fuchs, R.L., Fraley, R.T. 1992. Selectable marker genes: safe for plants? Bio/Technology 10: 141–144.

Fraley, R.T., Rogers, S.G., Horsch, R.B., Sanders, P.R., Flick, J.S., Adams, S.P., Bittner, M.L., Brand, L.A., Fink, C.L., Fry, J.S., Galluppi, G.R., Goldberg, S.B., Hoffmann, N.L., Woo, S.C. 1983. Expression of bacterial genes in plant cells. Proc. Natl. Acad. Sci. USA 80: 4803–4807.

Herrera-Estrella, L., DeBlock, M., Messens, E., Hernalsteens, J-P., Van Montagu, M., Schell, J. 1983. Chimeric genes as dominant selectable markers in plant cells. EMBO J. 2: 987–995.

Hooykaas, P.J.J. 1989. Transformation of plant cells via *Agrobacterium*. Plant Molec. Biol. 13: 327–336.

Jefferson, R.A., Kavanaugh, T.A., Bevan, M.W. 1987. GUS fusions: β-glucuronidase as a sensitive and versatile gene fusion marker in higher plants. EMBO J. 6: 3901–3907.

Jones, J.D.G., Svab, Z., Harper, E.C., Hurwitz, C.D., Maliga, P. 1987. A dominant nuclear streptomycin resistance marker for plant cell transformation. Molec. Gen. Genet. 210: 86–91.

Marx, J.L. 1982. Ti plasmids as gene carriers. Science 216: 1305–1307.

Matsumoto, S., Takebe, I., Machida, Y. 1988. *Escherichia coli lacZ* gene as a biochemical and histochemical marker in plant cells. Gene 66: 19–29.

Perl, A., Galili, S., Shaul, O., Ben-Tzvi, I., Galili, G. 1993. Bacterial dihydropicolinate synthase and desensitized aspartate kinase: two novel selectable markers for plant transformation. Bio/Technology 11: 715-718.

Prasher, D.C. 1995. Using GFP to see the light. Trends Genet. 11: 320–323.

Ratner, M. 1989. Crop biotech '89: research efforts are market driven. Bio/Technology 7: 337–347.

Thomashow, M.F., Nutter, R., Postle, K., Chilton, M.D., Blattner, F.R., Powell, A., Gordon, M.P., Nester, E.W. 1980. Recombination between higher plant DNA and the Ti plasmid of *Agrobacterium tumefaciens*. Proc. Natl. Acad. Sci. USA 77: 6448–6452.

Vasil, I.K. 1990. The realities and challenges of plant biotechnology. Bio/Technology 8: 296–301.

Waldron, C., Murphy, E.B., Roberts, J.L., Gustafson, G.D., Armour, S.L., Malcolm, S.K. 1985. Resistance to hygromycin B. Plant Molec. Biol. 5: 103–108.

Zambryski, P. 1988. Basic processes underlying *Agrobacterium*-mediated DNA transfer to plant cells. Annu. Rev. Genet. 22: 1–30.

Zambryski, P.C. 1992. Chronicles from the *Agrobacterium*-plant cell DNA transfer story. Annu. Rev. Plant Physiol. Plant Mol. Biol. 43: 465–490.

Section 23.3

Dong, J-Z., Yang, M-Z., Jia, S-R., Chua, N-H. 1991. Transformation of melon (*Cucumis melo* L.) and expression from the cauliflower mosaic virus 35S promoter in transgenic melon plants. Bio/Technology 9: 858–863.

JEFFERSON, R.A. 1987. Assaying chimeric genes in plants: the GUS gene fusion system. Plant Molec. Biol. Rep. 5: 387–405.

JONES, J.D.G., DEAN, C., GIDONI, D., GILBERT, D., BOND-NUTTER, D., LEE, R., BEDBROOK, J., DUNSMUIR, P. 1988. Expression of bacterial chitinase protein in tobacco leaves using two photosynthetic gene promoters. Molec. Gen. Genet. 212: 536–542.

KOTULA, L., CURTIS, P.J. 1991. Evaluation of foreign gene codon optimization in yeast: expression of a mouse IG kappa chain. Bio/Technology 9: 1386–1389.

LAST, D.I., BRETTELL, R.I.S., CHAMBERLAIN, D.A., CHAUDHURY, A.M., LARKIN, P.J., MARSH, E.L., PEACOCK, W.J., DENNIS, E.S. 1991. pEmu: an improved promoter for gene expression in cereal cells. Theoret. Appl. Genet. 81: 581–588.

LU, C-Y., NUGENT, G., WARDLEY-RICHARDSON, T., CHANDLER, S.F., YOUNG, R., DALLING, M.J. 1991. *Agrobacterium*-mediated transformation of carnation (*Dianthus caryophyllus* L.). Bio/Technology 9: 864–868.

MANTE, S., MORGENS, P.H., SCORZA, R., CORDTS, J.M., CALLAHAN, A.M. 1991. *Agrobacterium*-mediated transformation of plum (*Prunus domestica* L.) hypocotyl slices and regeneration of transgenic plants. Bio/Technology 9: 853–857.

MULLINS, M.G., ARCHIE TANG, F.C., FACCIOTTI, D. 1990. *Agrobacterium*-mediated genetic transformation of grapevines: transgenic plants of *Vitis rupestris* Scheele and buds of *Vitis vinifera* L. Bio/Technology 8: 1041-1045.

MUNRO, S., PELHAM, H.R.B. 1987. A C-terminal signal prevents secretion of luminal ER proteins. Cell 48: 899-907.

RAINERI, D.M., BOTTINO, P., GORDON, M.P., NESTER, E.W. 1990. *Agrobacterium*-mediated transformation of rice (*Oryza sativa* L.). Bio/Technology 8: 33–38.

SÁGI, L., PANIS, B., REMY, S., SCHOOFS, H., DE SMET, K., SWENNEN, R., CAMMUE, B.P.A. 1995. Genetic transformation of banana and plantain (*Musa* spp.) via particle bombardment. Bio/Technology 13: 481–485.

TAYLOR, M., VASIL, V., VASIL, I.K. 1993. Enhanced GUS gene expression in cereal/grass cell suspensions and immature embryos using the maize ubiquitin-based plasmid pAHC25. Plant Cell Rep. 12: 491–495.

TERADA, R., SHIMAMOTO, K. 1990. Expression of CaMV35S-GUS gene in transgenic plants. Molec. Gen. Genet. 220: 389–392.

VASIL, V., SRIVASTAVA, V., CASTILLO, A.M., FROMM, M.E., VASIL, I.K. 1993. Rapid production of transgenic wheat plants by direct bombardment of cultured immature embryos. Bio/Technology 11: 1553–1558.

WANDELT, C.I., KHAN, M.R.I., CRAIG, S., SCHROEDER, H.E., SPENCER, D., HIGGINS, T.J.V. 1992. Vicilin with carboxy-terminal KDEL is retained in the endoplasmic reticulum and accumulates to high levels in the leaves of transgenic plants. Plant J. 2: 181–192.

WANDELT, C., KNIBB, W., SCHROEDER, H.E., KHAN, M.R.I., SPENCER, D., CRAIG, S., HIGGINS, T.J.V. 1991. The expression of an ovalbumin and a seed protein gene in the leaves of transgenic plants. In: Plant Molecular Biology 2 (Herrman, R.G., Larkins, B., eds). Plenum Press, New York. pp. 471–478.

Section 23.4.1

COMAI, L., FACCIOTTI, D., HIATT, W.R., THOMPSON, G., ROSE, R.E., STALKER, D.M. 1985. Expression in plants of a mutant aroA gene from *Salmonella typhimurium* confers resistance to glyphosate. Nature 317: 741–744.

DE BLOCK, M., BOTTERMAN, J., VANDEWIELE, M., DOCKX, J., THOEN, C., GOSSELÉ, V., MOVVA, N.R., THOMPSON, C., VAN MONTAGU, M., LEEMANS, J.J. 1987. Engineering herbicide resistance in plants by expression of a detoxifying enzyme. EMBO J. 6: 2513–2518.

D'HALLUIN, K., BOSSUT, M., BONNE, E., MAZUR, B., LEEMANS, J., BOTTERMAN, J. 1992. Transformation of sugarbeet (*Beta vulgaris* L.) and evaluation of herbicide resistance in transgenic plants. Bio/Technology 10: 309–314.

SHAH, D.M., HORSCH, R.B., KLEE, H.J., KISHORE, G.M., WINTER, J.A., TUMER, N.E., HIRONAKA, K.M., SANDERS, P.R., GASSER, C.S., AYKENT, S., SIEGEL, N.R., ROGERS, S.G., FRALEY, R.T. 1986. Engineering herbicide tolerance in transgenic plants. Science 233: 478–481.

STALKER, D.M., MCBRIDE, K.E., MALYJ, L.D. 1988. Herbicide resistance in transgenic plants expressing a bacterial detoxification gene. Science 242: 419–423.

STREBER, W.R., WILLMITZER, L. 1989. Transgenic tobacco plants expressing a bacterial detoxifying enzyme are resistant to 2,4-D. Bio/Technology 7: 811–816.

THOMPSON, C.J., MOVVA, N.R., TIZARD, R., CRAMERI, R., DAVIES, J.E., LAUWEREYS, M., BOTTERMAN, J. 1987. Characterization of the herbicide-resistance gene bar from *Streptomyces hygroscopicus*. EMBO J. 6: 2519–2523.

Section 23.4.2

VAN DEN ELZEN, P.J.M., HUISMAN, M.J., WILLINK, D.P-L., JONGEDIJK, E., HOEKEMA, A., CORNELISSEN, B.J.C. 1989. Engineering virus resistance in agricultural crops. Plant Molec. Biol. 13: 337–346.

GADANI, F., MANSKY, L.M., MEDICI, R., MILLER, W.A., HILL, J.H. 1990. Genetic engineering of plants for virus resistance. Arch. Virol. 115: 1–21.

GIELEN, J.J.L., DE HAAN, P., KOOL, A.J., PETERS, D., VAN GRINSVEN, M.Q.J.M., GOLDBACH, R. 1991. Engineered resistance to tomato spotted wilt virus, a negative-strand RNA virus. Bio/Technology 9: 1363–1367.

GOLEMBOSKI, D.B., LOMONOSSOFF, G.P., ZAITLIN, M. 1990. Plants transformed with a tobacco mosaic virus nonstructural gene sequence are resistant to the virus. Proc. Natl. Acad. Sci. USA 87: 6311–6315.

GONSALVES, D., CHEE, P., PROVVIDENTI, R., SEEM, R., SLIGHTOM, J.L. 1992. Comparison of coat protein-mediated and genetically-derived resistance in cucumbers to infection by cucumber mosaic virus under field conditions with natural challenge inoculations by vectors. Bio/Technology 10: 1562–1570.

HILL, K.K., JARVIS-EAGAN, N., HALK, E.L., KRAHN, K.J., LIAO, L.W., MATHEWSON, R.S., MERLO, D.J., NELSON, S.E., RASHKA, K.E., LOESCH-FRIES, L.S. 1991. The development of virus-resistant alfalfa, *Medicago sativa* L. Bio/Technology 9: 373–377.

HOEKEMA, A., HUISMAN, M.J., MOLENDIJK, L., VAN DEN ELZEN, P.J.M., CORNELISSEN, B.J.C. 1989. The genetic engineering of two commercial potato cultivars for resistance to potato virus X. Bio/Technology 7: 273–278.

KANIEWSKI, W., LAWSON, C., SAMMONS, B., HALEY, L., HART, J., DELANNAY, X., TUMER, N.E. 1990. Field resistance of transgenic Russet Burbank potato to effects of infection by potato virus X and potato virus Y. Bio/Technology 8: 750–754.

KUNICK, T., SALOMON, R., ZAMIR, D., NAVOT, N., ZEIDAN, M., MICHELSON, I., GAFNI, Y., CZOSNEK, H. 1994. Transgenic tomato plants expressing the tomato yellow leaf curl virus capsid protein are resistant to the virus. Bio/Technology 12: 500–504.

LAWSON, C., KANIEWSKI, W., HALEY, L., ROZMAN, R., NEWELL, C., SANDERS, P., TUMER, N.E. 1990. Engineering resistance to mixed virus infection in a commercial potato cultivar: resistance to potato virus X and potato virus Y in transgenic Russet Burbank. Bio/Technology 8: 127–134.

LING, K., NAMBA, S., GONSALVES, C., SLIGHTOM, J.L., GONSALVES, D. 1991. Protection against detrimental effects of potyvirus infection in transgenic tobacco plants expressing the papaya ringspot virus coat protein gene. Bio/Technology 9: 752–758.

MURRY, L.E., ELLIOTT, L.G., CAPITANT, S.A., WEST, J.A., HANSON, K.K., SCARAFIA, L., JOHNSTON, S., DELUCA-FLAHERTY, C., NICHOLS, S., CUNANAN, D., DIETRICH, P.S., METTLER, I.J., DEWALD, S., WARNICK, D.A., RHODES, C., SINIBALDI, R.M., BRUNKE, K.J. 1993. Transgenic corn plants expressing MDMV strain B coat protein are resistant to mixed infections of maize dwarf mosaic virus and maize chlorotic mottle virus. Bio/Technology 11: 1559–1564.

PANG, S-Z., SLIGHTOM, J.L., GONSALVES, D. 1993. Different mechanisms protect transgenic tobacco against tomato spotted wilt and Impatiens necrotic spot Tospoviruses. Bio/Technology 11: 819–824.

STARK, D.M., BEACHY, R.N. 1989. Protection against potyvirus infection in transgenic plants: evidence for broad spectrum resistance. Bio/Technology 7: 1257–1262.

TEPFER, M. 1993. Viral genes and transgenic plants. Bio/Technology 11: 1125–1132.

TRUVE, E., AASPÔLLU, A., HONKANEN, J., PUSKA, R., MEHTO, M., HASSI, A., TEERI, T.H., KELVE, M., SEPPÄNEN, P., SAARMA, M. 1993. Transgenic potato plants expressing mammalian 2′-5′-oligoadenylate synthetase are protected from potato virus X infection under field conditions. Bio/Technology 11: 1048–1052.

YAMAYA, J., YOSHIOKA, M., MESHI, T., OKADA, Y., OHNO, T. 1988. Expression of tobacco mosaic virus RNA in transgenic plants. Molec. Gen. Genet. 211: 520–525.

Section 23.4.3

DELANNAY, X., LAVALLEE, B.J., PROKSCH, R.K., FUCHS, R.L., SIMS, S.R., GREENPLATE, J.T., MARRONE, P.G., DODSON, R.B., AUGUSTINE, J.J., LAYTON, J.G., FISCHHOFF, D.A. 1989. Field performance of transgenic tomato plants expressing the *Bacillus thuringiensis* var. *kurstaki* insect control protein. Bio/Technology 7: 1265–1269.

FEITELSON, J.S., PAYNE, J., KIM, L. 1992. *Bacillus thuringiensis*: Insects and beyond. Bio/Technology 10: 271–275.

FUJIMOTO, H., ITOH, K., YAMAMOTO, M., KYOZUKA, J., SHIMAMOTO, K. 1993. Insect resistant rice generated by introduction of a modified-endotoxin gene of *Bacillus thuringiensis*. Bio/Technology 11: 1151–1155.

KOZIEL, M.G., BELAND, G.L., BOWMAN, C., CAROZZI, N.B., CRENSHAW, R., CROSSLAND, L., DAWSON, J., DESAI, N., HILL, M., KADWELL, S., LAUNIS, K., LEWIS, K., MADDOX, D., MCPHERSON, K., MEGHJI, M.R., MERLIN, E., RHODES, R., WARREN, G.W., WRIGHT, M., EVOLA, S.V. 1993. Field performance of elite transgenic maize plants expressing an insecticidal protein derived from *Bacillus thuringiensis*. Bio/Technology 11: 194–200.

MCCUTCHEN, B.F., CHOUDARY, P.V., CRENSHAW, R., MADDOX, D., KAMITA, S.G., PALEKAR, N., VOLRATH, S., FOWLER, E., HAMMOCK, B.D., MAEDA, S. 1991. Development of a recombinant baculovirus expressing an insect-selective neurotoxin: potential for pest control. Bio/Technology 9: 848–852.

MCGAUGHEY, W.H., WHALON, M.E. 1992. Managing insect resistance to *Bacillus thuringiensis* toxins. Science 258: 1451–1455.

MAY, R.M. 1993. Resisting resistance. Nature 361: 593–594.

PERLAK, F.J., DEATON, R.W., ARMSTRONG, T.A., FUCHS, R.L., SIMS, S.R., GREENPLATE, J.T., FISCHHOFF, D.A. 1990. Insect resistant cotton plants. Bio/Technology 8: 939–943.

SHADE, R.E., SCHROEDER, H.E., PUEYO, J.J., TABE, L.M., MURDOCK, L.L., HIGGINS, T.J.V., CHRISPEELS, M.J. 1994. Transgenic pea seeds expressing the α-amylase inhibitor of the common bean are resistant to bruchid beetles. Bio/Technology 12: 793–796.

Section 23.4.4

ANZAI, H., YONEYAMA, K., YAMAGUCHI, I. 1989. Transgenic tobacco resistant to a bacterial disease by the detoxification of a pathogenic toxin. Molec. Gen. Genet. 219: 492–494.

LOGEMANN, J., JACH, G., TOMMERUP, H., MUNDY, J., SCHELL, J. 1992. Expression of a barley ribosome-inactivating protein leads to increased fungal protection in transgenic tobacco plants. Bio/Technology 10: 305–308.

MESTEL, R. 1995. Fungus hits bugs in the gut. New Sci. 1967: 15.

UCHIMIYA, H., IWATA, M., NOJIRI, C., SAMARAJEEWA, P.K., TAKAMATSU, S., OOBA, S., ANZAI, H., CHRISTENSEN, A.H., QUAIL, P.H,. TOKI, S. 1993. Bialophos treatment of transgenic rice plants expressing a bar gene prevents infection by the sheath blight pathogen (*Rhizoctonia solani*). Bio/Technology 11: 835–836.

ZHU, Q., MAHER, E.A., MASOUD, S., DIXON, R.A., LAMB, C.J. 1994. Enhanced protection against fungal attack by constitutive co-expression of chitinase and glucanase genes in transgenic tobacco. Bio/Technology 12: 807–812.

Section 23.4.5

ELOMAA, P., HONKANEN, J., PUSKA, R., SEPPÄNEN, P., HELARIUTTA, Y., MEHTO, M., KOTILAINEN, M., NEVALAINEN, L., TEERI, T.H. 1993. *Agrobacterium*-mediated transfer of antisense chalcone synthase cDNA to *Gerbera hybrida* inhibits flower pigmentation. Bio/Technology 11: 508–511.

SHEEHY, R.E., KRAMER, M., HIATT, W.R. 1988. Reduction of polygalacturonase activity in tomato fruit by antisense RNA. Proc. Natl. Acad. Sci. USA 85: 8805–8809.

VAN BRUNT, J. 1992. The battle of engineered tomatoes. Bio/Technology 10: 748.

VISSER, R.G.F., SOMHORST, I., KUIPERS, G.J., RUYS, N.J., FEENSTRA, W.J., JACOBSEN, E. 1991. Inhibition of the expression of the gene for granule-bound starch synthase in potato by antisense constructs. Molec. Gen. Genet. 225: 289–296.

Section 23.4.6

BECK, C.I., ULRICH, T. 1993. Biotechnology in the food industry. Bio/Technology 11: 895–900.

FRALEY, R. 1992. Sustaining the food supply. Bio/Technology 10: 40–43.

HIATT, A., CAFFERKEY, R., BOWDISH, K. 1989. Production of antibodies in transgenic plants. Nature 342: 76-78.

JOHN, M.E., STEWART, J.M. 1992. Genes for jeans: biotechnological advances in cotton. Trends Biotech. 10: 165–170.

MURATA, N., ISHIZAKI-NISHIZAWA, O., HIGASHI, S., HAYASHI, H., TASAKA, Y., NISHIDA, I. 1992. Genetically engineered alteration in the chilling sensitivity of plants. Nature 356: 710–713.

OAKES, J.V., SHEWMAKER, C.K., STALKER, D.M. 1991. Production of cyclodextrins, a novel carbohydrate, in the tubers of transgenic potato plants. Bio/Technology 9: 982–986.

PEN, J., MOLENDIJK, L., QUAX, W.J., SIJMONS, P.C., VAN OOYEN, A.J.J., VAN DEN ELZEN, P.J.M., RIETVELD, K., HOEKEMA, A. 1992. Production of active *Bacillus licheniformis* alpha-amylase in tobacco and its application in starch liquefaction. Bio/Technology 10: 292–296.

ROCHAIX, J-D. 1992. Plant molecular biology—moving towards application. Trends Biotech. 10: 78–79.

SIJMONS, P.C., DEKKER, B.M.M., SCHRAMMEIJER, B., VERWOERD, T.C., VAN DEN ELZEN, P.J.M., HOEKEMA, A. 1990. Production of correctly processed human serum albumin in transgenic plants. Bio/Technology 8: 217–221.

STARK, D.M., TIMMERMAN, K.P., BARRY, G.F., PREISS, J., KISHORE, G.M. 1992. Regulation of the amount of starch in plant tissues by ADP glucose pyrophosphorylase. Science 258: 287–292.

VANDEKERCKHOVE, J., VAN DAMME, J., LIJSEBETTENS, M., BOTTERMAN, J., DE BLOCK, M., VANDEWIELE, M., KREBBERS, E. 1989. Enkephalins produced in transgenic plants using modified 2S seed storage proteins. Bio/Technology 7: 929–932.

Section 23.5

CHRISTOU, P., FORD, T.L., KOFRON, M. 1991. Production of transgenic rice (*Oryza sativa* L.) plants from agronomically important indica and japonica varieties via electric discharge particle acceleration of exogenous DNA into immature zygotic embryos. Bio/Technology 9: 957–962.

ELLIS, D.D., MCCABE, D.E., MCINNIS, S., RAMACHANDRAN, R., RUSSELL, D.R., WALLACE, K.M., MARTINELL, B.J., ROBERTS, D.R., RAFFA, K.F., MCCOWN, B.H. 1993. Stable transformation of *Picea glauca* by particle acceleration. Bio/Technology 11: 84–89.

FROMM, M.E., MORRISH, F., ARMSTRONG, C., WILLIAMS, R., THOMAS, J., KLEIN, T.M. 1990. Inheritance and expression of chimeric genes in the progeny of transgenic maize plants. Bio/ Technology 8: 833–839.

FROMM, M., TAYLOR, L.P., WALBOT, V. 1985. Expression of genes transferred into monocot and dicot plant cells by electroporation. Proc. Natl. Acad. Sci. USA 82: 5824–5828.

GORDON-KAMM, W.J., SPENCER, T.M., MANAGANA, M.L., ADAMS, T.R., DAINES, R.J., START, W.G., O'BRIEN, J.V., CHAMBERS, S.A., ADAMS, JR., W.R., WILLETS, N.G., RICE, T.B., MACKEY, C.J., KRUEGER, R.W., KAUSCH, A.P., LEMAUX, P.G. 1990. Transformation of maize cells and regeneration of fertile transgenic plants. Plant Cell 2: 603–618.

GUPTA, H.S., PATTANAYAK, A. 1993. Plant regeneration from mesophyll protoplasts of rice (*Oryza sativa* L.). Bio/Technology 11: 90–94.

KLEIN, T.M., ARENTZEN, R., LEWIS, P.A., FITZPATRICK-MCELLIGOTT, S. 1992. Transformation of microbes, plants and animals by particle bombardment. Bio/Technology 10: 286–291.

KLEIN, T.M., WOLF, E.D., WU, R., SANFORD, J.C. 1987. High-velocity microprojectiles for delivering nucleic acids into living cells. Nature 327: 70–73.

MAY, G.D., AFZA, R., MASON, H.S., WIECKO, A., NOVAK, F.J., ARNTZEN, C.J. 1995. Generation of transgenic banana (*Musa acuminata*) plants via *Agrobacterium*-mediated transformation. Bio/Technology 13: 486–492.

MCCABE, E., MARTINELL, B.J. 1993. Transformation of elite cotton cultivars via particle bombardment of meristems. Bio/Technology 11: 596–598.

MOFFATT, A.S. 1991. Excess genetic baggage dumped. Science 254: 1452.

MOFFATT, A.S. 1992. High-tech plants promise a bumper crop of new products. Science 256: 770–771.

ODELL, J.T., CAIMI, P.G., SAUER, B., RUSSELL, S.H. 1990. Site-directed recombination in the genome of transgenic tobacco. Molec. Gen. Genet. 223: 369–378.

POTRYKUS, I. 1990. Gene transfer to cereals: an assessment. Bio/ Technology 8: 535–542.

POTRYKUS, I. 1992. Micro-targeting of microprojectiles to target areas in the micrometre range. Nature 355: 568–569.

VASIL, V., BROWN, S.M., RE, D., FROMM, M.E., VASIL, I.K. 1991. Stably transformed callus lines from microprojectile bombardment of cell suspension cultures of wheat. Bio/Technology 9: 743–747.

Section 23.6

BERKOWITZ, D.B. 1990. The food safety of transgenic animals. Bio/ Technology 8: 819–825.

BISHOP, J.O., SMITH, P. 1989. Mechanism of chromosomal integration of microinjected DNA. Molec. Biol. Med. 6: 283–298.

BONIFER, C., VIDAL, M., GROSVELD, F., SIPPEL, A.E. 1990. Tissue specific and position independent expression of the complete gene domain for chicken lysozyme in transgenic mice. EMBO J. 9: 2843–2848.

BRINSTER, R.L., CHEN, H.Y., TRUMBAUER, M.E., YAGLE, M.K., PALMITER, R.D. 1985. Factors affecting the efficiency of introducing foreign DNA into mice by microinjecting eggs. Proc. Natl. Acad. Sci. USA 82: 4438–4442.

CARVER, A.S., DALRYMPLE, M.A., WRIGHT, G., COTTOM, D.S., REEVES, D.B., GIBSON, Y.H., KEENAN, J.L., BARRASS, J.D., SCOTT, A.R., COLMAN, A., GARNER, I. 1993. Transgenic livestock as bioreactors: stable expression of human alpha 1-antitrypsin by a flock of sheep. Bio/Technology 11: 1263–1270.

CHEN, T.T., POWERS, D.A. 1990. Transgenic fish. Trends Biotech. 8: 209–215.

CREGG, J.M., VEDVICK, T.S., RASCHKE, W.C. 1993. Recent advances in the expression of foreign genes in *Pichia pastoris*. Bio/Technology 11: 905–910.

DEE, A., BELAGAJE, R.M., WARD, K., CHIO, E., LAI, M-H.T. 1990. Expression and secretion of a functional scorpion insecticidal toxin in cultured mouse cells. Bio/Technology 8: 339–342.

ENGLER, P., HAASCH, D., PINKERT, C.A., DOGLIO, L., GYLMOUR, M., BRINSTER, R., STORB, U. 1991. A strain-specific modifier on mouse chromosome 4 controls the methylation of independent transgene loci. Cell 65: 939–947.

GANNON, F., POWELL, R., BARRY, T., MCEVOY, T.G., SREENAN, J.M. 1990. Transgenic farm animals. J. Biotech. 16: 155–170.

GIGA-HAMA, Y., TOHDA, H., OKADA, H., OWADA, M.K., OKAYAMA, H., KUMAGAI, H. 1994. High-level expression of human lipocortin I in the fission yeast *Schizosaccharomyces pombe* using a novel expression vector. Bio/Technology 12: 400–404.

HODGSON, J. 1992. Whole animals for wholesale protein production. Bio/Technology 10: 863–866.

HODGSON, J. 1993. Expression systems: a user's guide. Bio/Technology 11: 887–893.

KRIMPENFORT, P., RADEMAKERS, A., EYESTONE, W., VAN DER SCHANS, A., VAN DEN BROEK, S., KOOIMAN, P., KOOTWIJK, E., PLATENBURG, G., PIEPER, F., STRIJKER, R., DE BOER, H. 1991. Generation of transgenic dairy cattle using "in vitro" embryo production. Bio/Technology 9: 844–847.

LIU, Z., MOAV, B., FARAS, A.J., GUISE, K.S., KAPUSCINSKI, A.R., HACKETT, P.B. 1990. Development of expression vectors for transgenic fish. Bio/Technology 8: 1268–1272.

MEADE, H., GATES, L., LACY, E., LONBERG, N. 1990. Bovine alpha s1-casein gene sequences direct high level expression of active human urokinase in mouse milk. Bio/Technology 8: 443–446.

PALMITER, R.D., BRINSTER, R.L., HAMMER, R.E., TRUMBAUER, M.E., ROSENFELD, M.G., BIRNBERG, N.C., EVANS, R.M. 1982. Dramatic growth of mice that develop from eggs microinjected with metallothionein-growth hormone fusion genes. Nature 300: 611–615.

PURSEL, V.G., PINKERT, C.A., MILLER, K.F., BOLT, D.J., CAMPBELL, R.G., PALMITER, R.D., BRINSTER, R.I., HAMMER, R.E. 1989. Genetic engineering of livestock. Science 244: 1281–1288.

RANCOURT, D.E., PETERS, I.D., WALKER, V.K., DAVIES, P.L. 1990. Wolffish antifreeze protein from transgenic *Drosophila*. Bio/Technology 8: 453–457.

RICHA, J., LO, C.W. 1989. Introduction of human DNA into mouse eggs by injection of dissected chromosome fragments. Science 245: 175–177.

RYAN, T.M., TOWNES, T.M., REILLY, M.P., ASAKURA, T., PALMITER, R.D., BRINSTER, R.L., BEHRINGER, R.R. 1990. Human sickle hemoglobin in transgenic mice. Science 247: 566–568.

SEMENZA, G.L., DUREZA, R.C., TRAYSTMAN, M.D., GEARHART, J.D., ANTONARAKIS, S.E. 1990. Human erythropoieten gene expression in transgenic mice: multiple transcription initiation sites and cis-acting regulatory elements. Molec. Cell. Biol. 10: 930–938.

TEMIN, H.M. 1989. Retrovirus vectors: promise and reality. Science 246: 983.

VAN BRUNT, J. 1990. Transgenics primed for research. Bio/Technology 8: 725–728.

WAGENBACH, M., O'ROURKE, K., VITEZ, L., WIECZOREK, A., HOFFMAN, S., DURFEE, S., TEDESCO, J., STETLER, G. 1991. Synthesis of wild type and mutant human hemoglobins in *Saccharomyces cerevisiae*. Bio/Technology 9: 57–61.

WARD, P.P., LO, J-Y., DUKE, M., MAY, G.S., HEADON, D.R., CONNEELY, O.M. 1992. Production of biologically active recombinant human lactoferrin in *Aspergillus oryzae*. Bio/Technology 10: 784–789.

WILKIE, T.M., BRAUN, R.E., EHRMAN, W.J., PALMITER, R.D., HAMMER, R.E. 1991. Germ-line intrachromosomal recombination restores fertility in transgenic MyK-103 male mice. Genes Dev. 5: 38–48.

WRIGHT, G., CARVER, A., COTTOM, D., REEVES, D., SCOTT, A., SIMONS, P., WILMUT, I., GARNER, I., COLMAN, A. 1991. High level expression of active human alpha-1-antitrypsin in the milk of transgenic sheep. Bio/Technology 9: 830–834.

Section 23.7

DALE, E.C., OW, D.W. 1991. Gene transfer with subsequent removal of the selection gene from the host genome. Proc. Natl. Acad. Sci. USA 88: 10558–10562.

GOLDSBROUGH, A.P., LASTRELLA, C.N., YODER, J.I. 1994. Transposition mediated re-positioning and subsequent elimination of marker genes from transgenic tomato. Bio/Technology 11: 1286–1292.

MATZKE, M.A., MATZKE, A.J.M. 1995. Homology-dependent gene silencing in transgenic plants: what does it really tell us? Trends Genet. 11: 1–3.

YODER, J.I., GOLDSBROUGH, A.P. 1994. Transformation systems for generating marker-free transgenic plants. Bio/Technology 12: 263–268.

24

Organisms of Importance to Genetics and Cytogenetics

- Organisms that are of major significance to the study of genetics and cytogenetics are characterized.
- The reasons for the genetic importance of these species are summarized.

In classical and molecular cytogenetics, and in all the subdivisions of genetics generally, scientific progress has often depended on the unique properties of particular organisms that have allowed a particular process or structure to be clearly observed or visualized. In this final chapter, we describe some of the organisms that have been intensively studied by geneticists and cytogeneticists, emphasizing the diversity and range of features for which particular organisms have been investigated. Whereas many of these organisms are of economic importance, others are examples of a certain type of organism that is useful in genetic research and the study of chromosomes generally.

Some approximations are relevant to this chapter. In general terms, 1 picogram (pg) of DNA contains about 1.17×10^9 nucleotide base pairs, and reciprocally, 1×10^9 bp = 0.85 pg. In addition, a mass of 1 pg of DNA has a molecular weight of 6.1×10^{11} D, and, freely extended, a length of about 30 cm. At the gene level, a triplet codon codes for one amino acid and an average enzyme or protein polypeptide chain is encoded by 1000–2000 nucleotide base pairs (bp). In the protosome of the bacterium *Haemophilus influenzae*, a total of 1749 genes are contained within a genome of 1,830,121 bp. In eukaryotes, the presence of introns between coding exons can add several thousand nucleotide base pairs, although the introns are excised before transcription. Some genes may also be present in multiple copies, up to several thousands in some cases.

24.1 BACTERIOPHAGE λ (A BACTERIAL VIRUS)

24.1.1 Description

Bacteriophages such as phage λ (Fig. 24.1) are viruses that code for bacteriocins which are toxic to specific bacteria in order to enhance their pathogenicity. Their presence or absence is commonly equated with the presence or absence of specific characteristics conferred upon the bacterial host. These include resistance to antibiotics and differing nutritional requirements.

24.1.2 Genome Size

A circular, double-stranded DNA molecule, 48,502 bp in length. Gene number: 62.

24.1.3 Features of Interest to Genetics and Molecular Biology

Bacteriophage λ was originally discovered in the K12 strain of *E. coli,* in which it was first recognized and described as the F factor. The phage was completely sequenced in 1982. The strain that was sequenced was λc1*ind*1ts857*S*7, which differs from the wild-type by mutations in three genes, *ind*1 to ind^+, *clts857* to cl^+, and *Sam7* to S^+. In general, bacteriophage λ has two sets of genes, one for lytic growth, the other for lysogenic growth. These genes are grouped into operons, with the lysogenic operons involved in the production of head and tail proteins, recombination, immunity, and DNA replication. The regulatory genes *N* and *Q* are also required. The amino groups of adenine and cytosine are usually methylated, presumably to reduce bacterial nuclease attack.

After insertion of the linearized bacteriophage DNA into the host cell, a host DNA ligase seals the ends of the DNA to re-form a circular DNA molecule. This DNA may then replicate independently, using bacteriophage λ proteins and host DNA replication enzymes, or become inserted into the bacterial protosome as a prophage. This insertion is effected between a host attachment site, *attB*, located between the *galE* and *bioA* operons of the *E. coli* protosome, and the bacteriophage λ-specific attachment site, *attP*. It can then be replicated for many generations in time with replication of the protosome. Excision of the prophage requires two enzymes, the original bacterial host integrase in combination with bacteriophage λ excisionase, an enzyme that is normally suppressed by a bacteriophage λ repressor molecule. For the lytic life-style, the DNA between the genes *J* and *attP* can be replaced or modified by the addition of new genes. Modifications to this region allow bacteriophage λ to be used as a cloning vehicle.

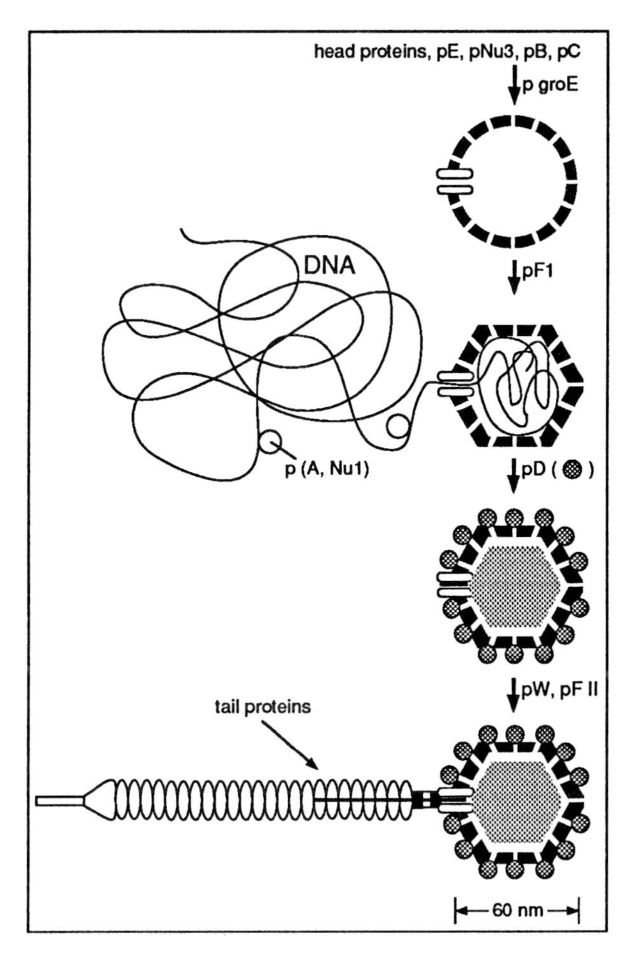

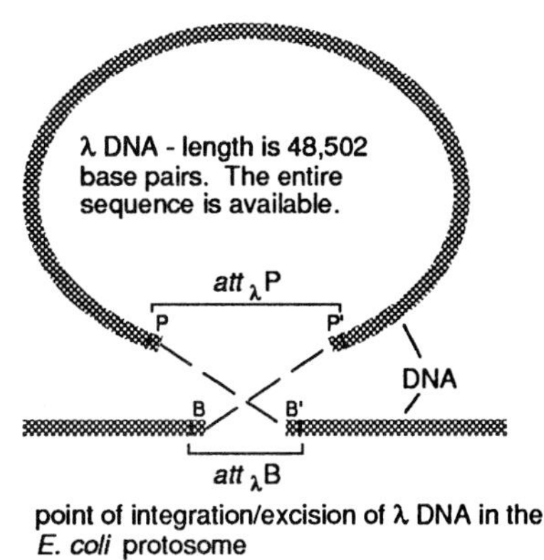

Fig. 24.1. Phage λ and the insertion and excision of prophage λ into and from a bacterial protosome.

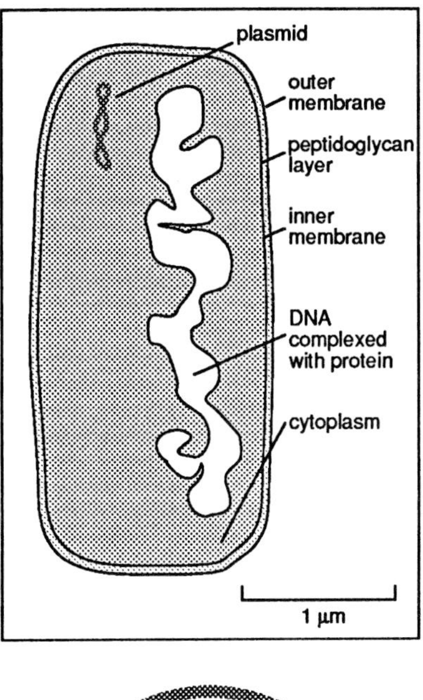

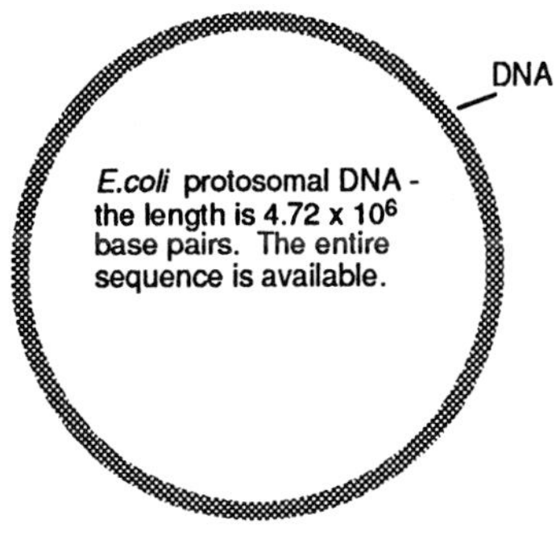

Fig. 24.2. A cell of *E. coli* and a simplified version of its protosome.

24.2 *ESCHERICHIA COLI* (COLIFORM BACTERIUM)

24.2.1 Description

Escherichia coli is classified as a gram-negative rod bacterium of the family Enterobacteriaceae (Fig. 24.2). Gram-negative bacteria are characterized by the presence of multilayered cell walls, the outer layers of which are composed of a roughly textured lipopolysaccharide membrane that is relatively permeable to the Gram stain. *E. coli* was originally isolated from the hind-gut of the oriental cockroach and is a relatively harmless member of the intestinal flora of most mammals. No records are available to elucidate the evolutionary history of enteric bacteria such as *E. coli,* so that most of the recent advances in establishing relationships between bacteria have been obtained from comparative amino-acid-sequence data of related proteins and enzymes, and nucleotide-base-pair sequence data.

24.2.2 Genome Size

DNA content: A circular molecule containing 0.0044 pg of DNA.

Length of protosome: 4.2×10^6 bp = 1.36 mm.

Gene number: ~ 4000.

24.2.3 Features of Interest to Genetics and Molecular Biology

Although both vertical and horizontal transfers of genetic information can occur (i.e., from one generation to the next or between sibling cells, respectively), the patterns of bacterial evolution are predominantly vertical. This is because horizontal gene transfer and recombination are probably rare—though not unique—in natural populations. Although natural populations of *E. coli* may be evolutionarily stable, DNA sequences can be introduced in three ways:

1. By transformation, in which fragments of free DNA insert directly into competent recipient cells.
2. By transduction, in which DNA is transferred from one cell to another by viral phage particles, such as bacteriophage λ.
3. By conjugation, whereby DNA is transferred by direct contact between the host and the recipient cells, and both protosomal and plasmid DNA can be transferred in this way. The original genetic maps of the *E. coli* protosome were developed by investigating the length of time it took for genes to be transferred from one bacterium to another, starting at the point of origin *(ori)* on the bacterial protosome.

The overall details of DNA replication and control of gene expression were revealed in *E. coli* using a combination of mutational and genetic analyses. In addition, most of the routine gene manipulations that form the basis for molecular biology are carried out in *E. coli* because of the enormous microbiological background known about this organism. The first cloning of eukaryote DNA utilized *E. coli* and a plasmid, pSC101, that was found in certain strains of the bacterium. More recently, large segments of eukaryotic genomic DNA, approximately 100,000 bp in length, have been introduced into *E. coli* as so-called bacterial artificial chromosomes (BACs).

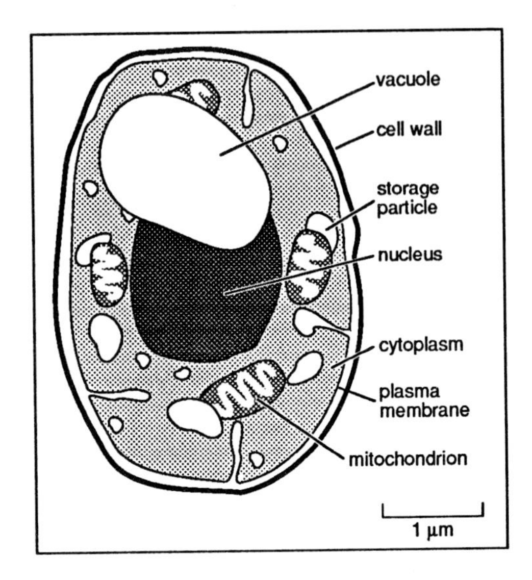

chromosome	I	II	III	IV	V	VI	VII	VIII
size in kb	240	820	320	1200	280	280	1100	570

chromosome	IX	X	XI	XII	XIII	XIV	XV	XVI
size in kb	440	750	670	1080	930	790	1100	950

Fig. 24.3. A yeast cell with a numerical karyotype of its haploid set of 16 chromosomes. (Abstracted from http://genome-www.stanford.edu/cgi-bin/SGD/pgMAP/MAP).

24.3 *SACCHAROMYCES CEREVISIAE* (BREWER'S YEAST)

24.3.1 Description

Yeast is an ascomycete fungus (Fig. 24.3). *S. cerevisiae* is utilized in producing food products such as bread and fermented drinks. This is because yeasts are capable of both respiration in the presence of oxygen and anaerobic fermentation in its absence. The former is used to manufacture new yeast cells and produce CO_2 and is of great importance to the bread-making process. The latter is of great importance to the manufacture of beer, wine, and spirituous liquors through the conversion of sugars to alcohol. The former relies on the ability of mitochondria to produce ATP for growth, whereas in the latter, ATP is obtained from the conversion process. For this reason, the mitochondria of *S. cerevisiae* are also well studied, the best known mutants being those that result in the formation of "petite" morphological colonies which completely lack mitochondria. It is of interest that most yeasts also contain multiple copies of an autonomously replicating 2 μm circular DNA, of unknown function, which has been used to introduce new genes into yeast. *S. cerevisiae* is only distantly related to the fission yeast *Schizosaccharomyces pombe,* which has three pairs of chromosomes.

24.3.2 Genome Size and Karyotype

Chromosome number: $2n = 32$. Individual chromosomes are designated from I to XVI, with left (L) and right (R) arms.

2C DNA content: 0.3 pg.

1C DNA content: 12.06×10^6 bp.

Gene number: ~ 6,000.

24.3.3 Features of Interest to Genetics and Molecular Biology

Saccharomyces cerevisiae is of continuing benefit to the study of genetics. Individual cell cultures, whether haploid or diploid, normally reproduce through asexual budding. Under certain conditions, haploid cells of different mating strains fuse to form diploid budding cells that can, in turn, be induced to undergo meiosis, resulting in a tetrad containing four haploid ascospores, two of which are of mating type a, the other two of mating type α. In cell fusion and the re-formation of diploid cells, a-type ascospores can only pair with α-types. The individual cells of the tetrad can be isolated by micromanipulation and grown separately. Because the wild-type fungus can be cultured and grown on a relatively simple medium, a large collection of biochemical, auxotrophic, temperature-sensitive, mating-type, and morphological mutants have been isolated. Genetic mapping is commonly performed by tetrad analysis, during which the four ascospores are separated into parental and nonparental tetrads, or into tetratypes, which indicate crossing over and linkage. Mitotic crossing over also occurs naturally in this fungus but is increased with ultraviolet (UV) light.

The majority of yeast genes and DNA sequences can now be located without genetic mapping. This is because gene probes can either be hybridized to Southern blots of whole chromosomes separated by pulse-field gel electrophoresis—so-called chromoblots—or, as most if not all of the yeast genome has now been cloned into bacteriophage λ vectors, to the array of DNA clones immobilized on filters. Furthermore, while the entire nucleotide sequence of chromosome III was first determined, the entire genome, all 16 chromosomes, has now been completely sequenced, providing unambiguous locations of all genes and ORFs, although not all the functions of these have been identifiable.

Molecular–genetic studies of yeast have been of great use in elucidating the cell cycle, the structure of centromeres and telomeres, and phenomena such as telomeric silencing, by which genes inserted adjacent to poly-($G_{1-3}T$) tracts are transcriptionally repressed. With the use of yeast artificial chromosomes (YACs), large segments of eukaryotic chromatin from other organisms are being introduced into yeast cells for cloning purposes. Because homologous YACs undergo relatively normal meiosis and recombination, YAC libraries are now being used to clone the entire genomes of other eukaryotes.

24.4 *ARABIDOPSIS THALIANA* (THALE WEED OR WALL CRESS)

24.4.1 Description

Arabidopsis thaliana is a small flowering plant native to Europe, Asia, and northern Africa (Fig. 24.4). The species is a member of the family Cruciferae, which contains a number of interesting crops such as *Eutrema wasabi* (wasabi), *Nasturtium officinale* (water cress) and *Raphanus* spp. (Radish), as well as the significant oilseed crops *B. nigra* (black mustard), *B. campestris* and *B. napus* (rapeseed and canola), and *B. juncea* (mustard). A variety of horticultural crops such as cabbage, turnip, swede, and rutabaga also occur in this family. These, most likely, share a high degree of genomic collinearity with the *Arabidopsis* genome, and so are particularly well suited to genetic engineering through the introduction of gene sequences recognized in their weedy relative, *A. thaliana*.

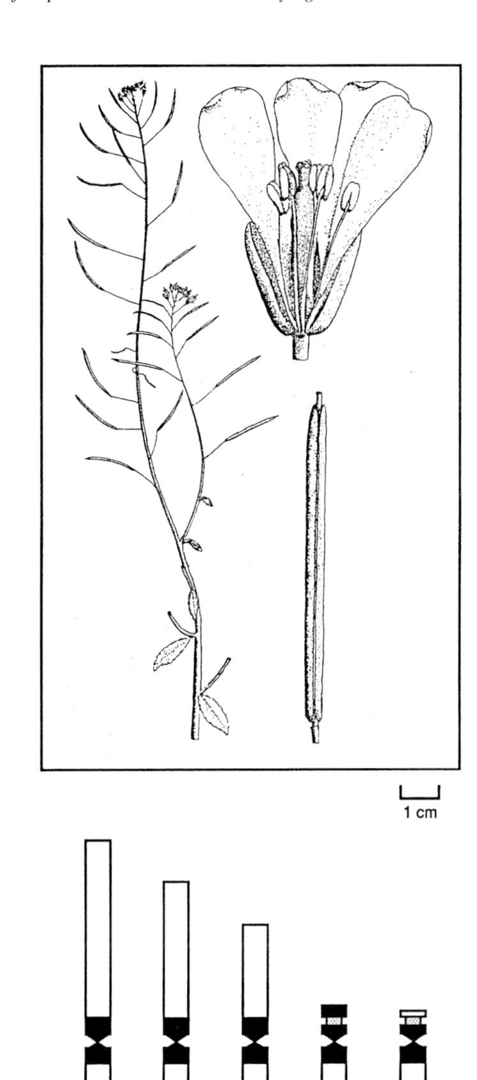

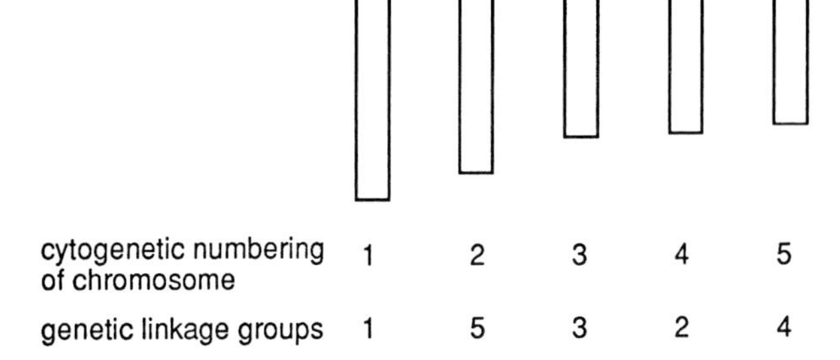

Fig. 24.4. *Arabadopsis thaliana* and its chromosomes. The scale in the drawing of the plant refers to the complete plant on the left.

24.4.2 Genome Size and Karyotype

Arabidopsis has one of the smaller genomes of the plant kingdom and some of the smallest chromosomes, averaging only 14 Mb in length.

Chromosome number: $2n = 10$.

2C DNA content: 0.45 pg.

1C DNA content: 70×10^6 bp.

Map length: ~ 600 cM.

24.4.3 Features of Interest to Genetics, Cytogenetics, and Molecular Biology

Originally, *A. thaliana* was investigated genetically to discover nutritional mutants similar to those that had previously been found in biosynthetic studies of yeast and bacteria; these mutants required the addition of specific amino acids, vitamins, and other nutrients for growth. *A. thaliana* has now become widely recognized as the model for the study of plant development and genome organization because of its small size, ability to thrive at high sowing densities, the presence of self-pollinating hermaphrodite flowers, the ease with which seed can be obtained by cross-pollination, a relatively prolific seed set, and a short life cycle (5 weeks in suitable conditions). The small genome has also helped to make *Arabidopsis* a valuable experimental organism in that only small quantities of repetitive DNA sequences are present and the distance between genetic markers is correspondingly decreased, with 1 cM being the equivalent of only 160,000 bp.

The impact of *Arabidopsis* on plant molecular biology is of great significance because of the following attributes:

1. The five chromosomes are now well mapped with a combination of morphological, developmental, and metabolic genes, and an array of DNA markers. The chromosomes

and their arms are being sequenced in a coordinated approach by laboratories throughout the world. A 1.9 mb contiguous sequence of chromsome 4 was analysed by 1998.

2. A statistically complete coverage of the entire genome in the form of DNA sequences cloned into bacteriophage l, cosmid, YAC, and BAC libraries is now available.
3. Transformed plants of *Arabidopsis* are readily produced.
4. Transposable-element mutagenesis is possible to determine gene function.
5. *Arabidopsis* contains representatives of all of the genes that are necessary for normal plant growth and development, as well as genes that are common to varied life-forms such as the ribosomal RNA genes and chromosomal histone proteins. Consequently, DNA sequences specific to *Arabidopsis* genes can be used to distinguish genes with similar functions in many other plant species.

24.5 *LYCOPERSICON ESCULENTUM* (TOMATO)

24.5.1 Description

Tomato is a member of the cosmopolitan plant family, the Solanaceae (Fig. 24.5). This family includes many economically important plant species such as potato (*Solanum tuberosum*) and tobacco (*Nicotiana tabacum*), as well as Jimson weed (*Datura stramonium*), an earlier model in genetic history. The genus *Lycopersicon* contains 8–10 species and is native to Central America where it was first domesticated. Tomatoes and potatoes were first introduced into Europe in the sixteenth century and they have since been spread around the world.

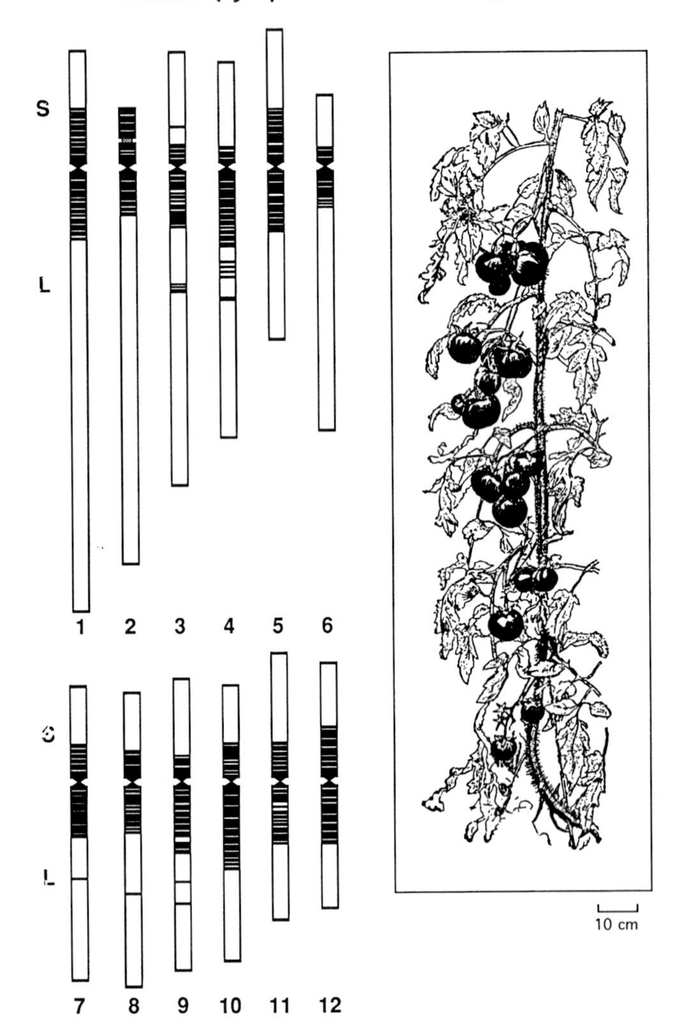

Fig. 24.5. Cultivated tomato and its haploid set of chromosomes.

24.5.2 Genome Size and Karyotype

Chromosome number: $2n = 24$.

2C DNA content: 2.0 pg.

1C DNA content: 1.2×10^9 bp.

Map length: 1276 cM.

24.5.3 Features of Interest to Genetics, Cytogenetics, and Molecular Biology

The tomato plant has long been a favored organism for genetic studies. In large part, this is due to desirable reproductive features, namely its self-pollinated nature with each controlled pollination providing up to 300 seeds per fruit, and single plant being able to produce up to 25,000 seeds. In addition, the tomato phenotype, especially the foliage, is diverse with respect to pattern, color, and texture, so that several hundred morphological mutants have been isolated. As in most of the Solanaceae, tomato is amenable to tissue culture, allowing protoplast fusion to be used to produce hybrids with other members of the genus and family. In addition, it is also very responsive to agrobacterial transformation. The species has a comparatively small genome and a large inventory of aneuploid chromosome stocks has been established. Additional features of interest include the following:

1. Several hundred morphological, physiological, biochemical, and developmental genetic mutants have been mapped.
2. A saturated RFLP linkage map with over 1000 markers has been established.
3. A functional *Ac* mutation system is available for directed, transposon mutagenesis studies.
4. The comparatively small genome and the associated low level of repetitive DNA sequences (~ 10–15%) make the plant relatively amenable to chromosome walking.

24.6 *ZEA MAYS* (MAIZE OR CORN)

24.6.1 Description

Maize is a major monocotyledonous crop plant, which has been grown and selectively improved over the last 7000 years (Fig. 24.6). The species is considered to be meso-American in origin although no wild forms of the species have been described. Genetically, maize appears to be more closely related to the wild, relatively nonedible teosinte (*Euchlaena mexicana*) than to the more edible but genetically unrelated *Tripsacum*. However, there are major morphological differences between maize and teosinte that argue against a direct origin of maize from teosinte.

24.6.2 Genome Size and Karyotype

Chromosome number: $2n = 20$.

2C DNA content: 7.5 pg.

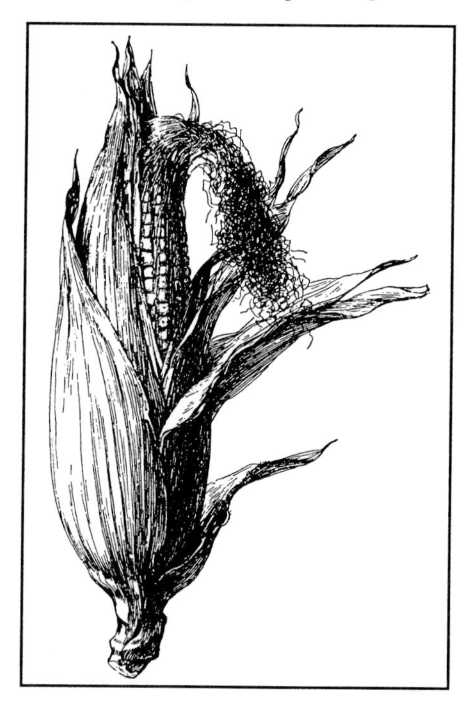

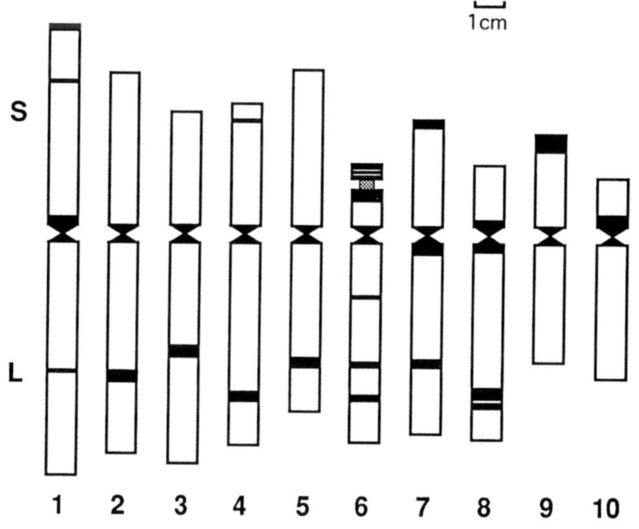

Fig. 24.6. Corn and its karyotype.

1C DNA content: 3×10^9 bp.

Map length: 2200 cM.

24.6.3 Features of Interest to Genetics, Cytogenetics, and Molecular Biology

Zea is well known for the morphology of its pachytene chromosomes and the early work that led to the discovery of transposable elements. The major genetic advantage of the species, for a monocotyledon at least, lies in the fact that it is essentially a diploid organism. The physical separation of male anthers from female organs enhances the ease and reliability of hybridization, as does the copious volume of pollen produced by the male florets. The large numbers of seed capable of being produced by the female ears makes it comparatively simple to isolate and maintain mutants of this species. The mutants characterized in maize cover a broad range of phenotypes such as seed and leaf color, shape and size, as well as many other morphological and biochemical mutants.

Recent comparative genetic-mapping studies analyzing syntenic relationships between the chromosomes of different monocotyledons indicate that the maize genome is possibly a relic tetraploid, with duplicate copies of related genes in different orders on nonrelated chromosomes. The evidence suggests alloploidy and a hybrid origin for the genome.

24.7 *TRITICUM AESTIVUM* (BREAD WHEAT)

24.7.1 Description

Wheats and plants in the genus *Triticum* are primarily of Middle Eastern origin (Fig. 24.7). Diploid einkorn wheats were used as a very early food resource, tetraploid emmer or durum wheats became cultivated about 8000 years ago, and hexaploid bread wheat appeared about 6000 years before the present. The common or bread wheat *T. aestivum* is derived from the natural hybridization of the diploid species *T. tauschii* (D genome) with a tetraploid wheat (A and B genomes). The tetraploid wheat, in turn, arose from the progeny of a cross between *T. monococcum* (A genome) and a further *Triticum* sp. or combination of species (B genome).

24.7.2 Genome Size and Karyotype

Chromosome number: $2n = 6x = 42$.

Genome: AABBDD, triplicated sets of chromosomes belonging to seven homoeologous groups of genetically related A, B, and D genome chromosomes.

2C DNA content: 34.6 pg.

1C DNA content: 16×10^9 bp or ~ 11 m of DNA.

24.7.3 Features of Interest to Genetics, Cytogenetics, and Molecular Biology

The red-seeded genotype of the wheats provided an early example of a 63 : 1 genetic F_2 ratio, indicating, retrospectively, the hexaploid nature of the wheat chromosome complement. Wheat is an allohexaploid, and genetic proof for the relatedness of potential diploid progenitors to one-third of the wheat chromosomes was obtained by hybridizing the hexaploid plant to various diploid and tetraploid relatives of wheat. The diploid progenitors are now generally included in the genus *Triticum* (syn. *Aegilops*). Both *T. monococcum* and *T. tauschii* chromosomes show almost complete pairing with their respective partners in the A and D genomes, respectively. Although *T. speltoides* showed some structural homologies with the B-genome chromosomes, this relationship has since been revised and the B genome is now generally regarded as being a composite of chromosomes from a combination of potential donors.

Homoeology at the chromosomal level was proven cytogenetically by selecting wheats that had lost a pair of homologous chromosomes and replacing them with an additional pair of related chromosomes. This work showed that the addition of either of two related homoeologous pairs of chromosomes generally compensated for the loss of the third pair of related chromosomes. These nullisomic–tetrasomic lines and other aneuploids that were devel-

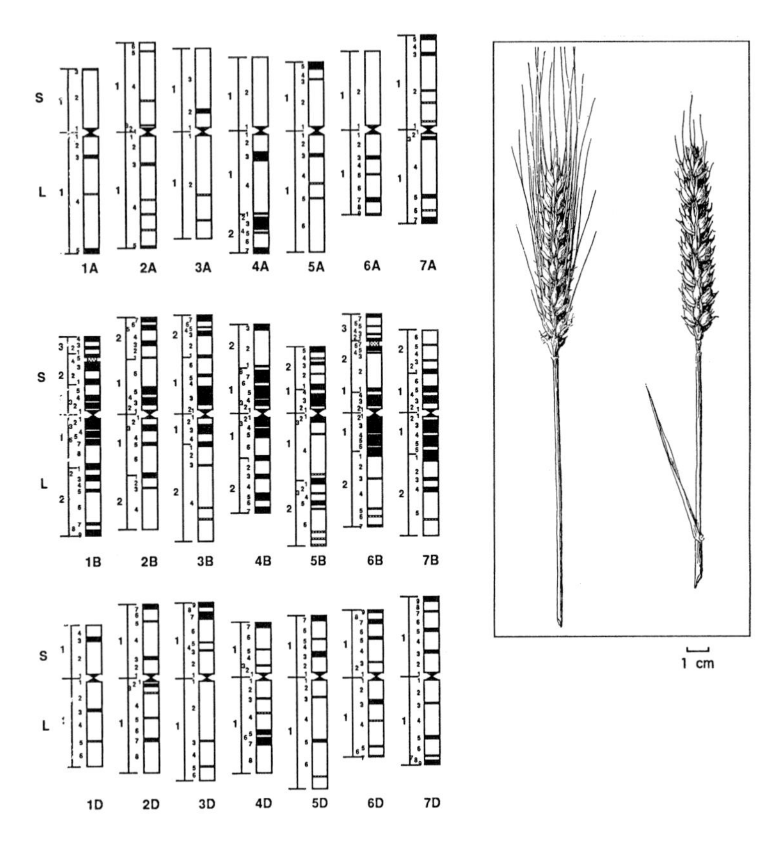

Fig. 24.7. Bread wheat and its three sets of homoeologous G-banded chromosomes.

oped, including ditelosomic lines lacking a single pair of chromosome arms, and addition lines containing an extra pair of chromosomes from various species of *Triticum*, *Secale*, and *Hordeum* rapidly provided a basis for the mapping of morphogenetic characters and for the biochemical genetic analyses of proteins and isozymes. These cytogenetic stocks are now used widely to find the locations of DNA markers. Good genetic maps for each of the homoeologous chromosomes have been published, and in many cases, cytogenetically characterized deletions have been used to relate the genetic map to the physical map of the chromosomes. Although the hexaploid nature of wheat makes genetic mapping more difficult, the mapping of diploid progenitors including *Triticum tauschii* and *T. monococcum*, as well as the more distantly related *Secale cereale* (cereal rye) and *Hordeum vulgare* (barley), have helped resolve ambiguities.

24.8 *ORYZA SATIVA* (RICE)

24.8.1 Description

Rice is an important cereal crop ideally suited to growth in the tropics where the combination of water supply, soil fertility, and breeding results in abundant yields (Fig. 24.8). Rice is basically an aquatic plant and was almost certainly domesticated independently in Asia and Africa. The genus *Oryza* includes from 25–30 species, most of which are tropical or semitropical in origin. About 20 of the species are diploids and capable of forming interspecific hybrids, in part because about two-thirds of the diploids share the A genome of *O. sativa.* However, no known tetraploid species includes the A genome: Asian and African accessions are BBCC and American tetraploids are CCDD. Although B- and C-genome progenitor species have been identified, no diploid species with the D-genome complement have been identified. Other genomes (EE and FF) have been identified further from the centers of origin of the genus. Cultivated rice can be subdivided into three agronomic types, vars. japonica, indica, and javanica, all with an array of differing agronomic characters.

24.8.2 Genome Size and Karyotype

Chromosome number: $2n = 24$.

2C DNA: 0.9 pg.

1C DNA content: 40×10^7 bp.

Map length: 1491 cM.

24.8.3 Features of Interest to Genetics, Cytogenetics, and Molecular Biology

Recent advances in microscopic imaging and molecular technologies have simplified chromosomal identification in *Oryza sativa.* For example, fluorescence in situ hybridization (FISH) has shown that most japonica types contain only a single pair of NORs, whereas indica and javanica types commonly have two pairs. In retrospect, although the actual number of chromosomes in *O. sativa* was first determined in 1910, it was not until 1960 that 12 linkage

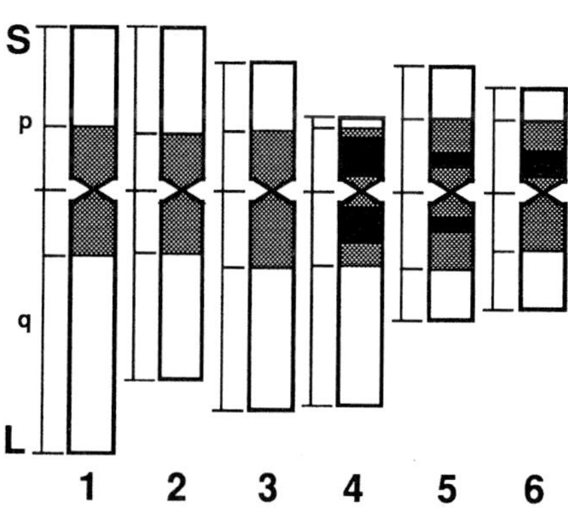

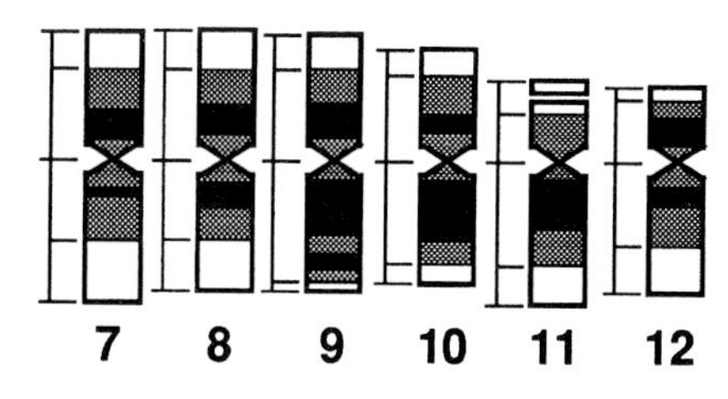

Fig. 24.8. Rice and its karyotype.

groups were finally established. This delay was In large part due to the small size of the chromosomes, each having an average DNA content of about 33 Mb. However, it is this small size that is now proving to be an advantage in mapping the rice genome, and a detailed molecular linkage map has been constructed using a backcross population between *O. sativa* and one of its wild African relatives, *O. longistaminata*. Several of the diploid *Oryza* species that share the A genome with *O. sativa* have also been employed in an array of wide crosses to introduce new genes for quality and agronomic traits into this species. With the success in obtaining transformed rice plants with specific gene constructs, and the continuing genome-mapping projects combined with cytogenetic analyses, our knowledge of rice is quickly outstripping the information available for any other plant and certainly results in the use of rice as a model genome for monocotyledonous crops.

24.9 *TETRAHYMENA THERMOPHILA* (CILIATED PROTOZOAN)

24.9.1 Description

Tetrahymena thermophila is an organism of freshwater ponds and streams (Fig. 24.9). It is a member of the protozoa, a group of motile, unicellular, almost colorless eukaryotes, which lack defined cell walls and obtain food by ingesting it via a funnel-shaped gullet and enveloping the food particles with cell membrane. Related

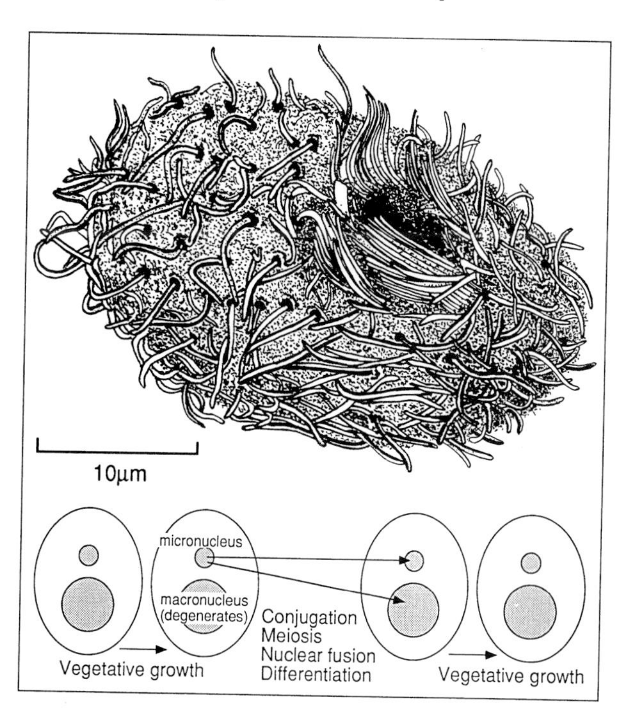

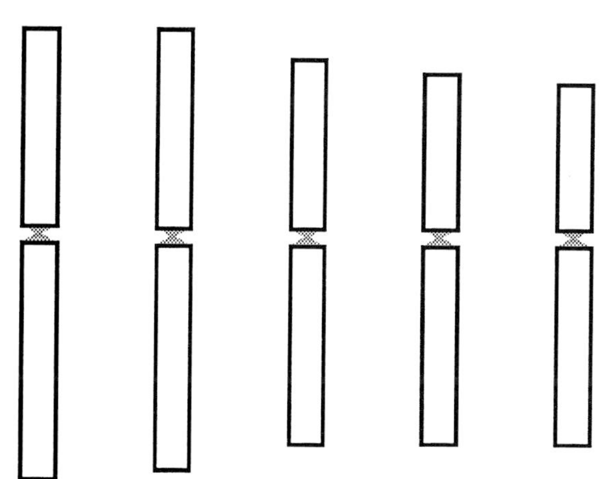

Fig. 24.9. *Tetrahymena* and its five haploid chromosomes.

protozoans include species of *Paramecium,* which have been examined intensively to investigate nucleo-cytoplasmic relationships. Ciliated protozoans move by means of linear arrays of cilia, which almost completely cover the surface of the cell. In *T. thermophila* and in protozoans in general, it is not possible to differentiate between sexes at the morphological level as there are no egg or sperm cells. Rather, the sexual process involves a transfer of nuclei and is distinguished by the presence of inheritable physiological mating types.

24.9.2 Genome Size and Karyotype

Chromosome number: $2n = 10$.

2C DNA content: ~ 0.5 pg.

Gene number: ~ 12,000.

24.9.3 Features of Interest to Genetics, Cytogenetics, and Molecular Biology

Much of the genetic interest in protozoans such as *T. thermophila* lies in the fact that individual cells contain two types of nuclei, a diploid, germinal micronucleus and a highly polyploid macronucleus. Genes must be present in the micronucleus to be inherited and in the macronucleus to be expressed. The micronucleus is transcriptionally silent and consists of five pairs of chromosomes. The transcriptionally active macronucleus arises by the fragmentation of the original five chromosomes into several hundred chromosomes with an average size of 600 kb and the replication of most of this chromatin to about 45-ploid. During sexual reproduction, cells of differing mating types conjugate, their micronuclei undergo meiosis to give four products, three of which are resorbed by the cell, with the fourth undergoing mitosis to give two haploid nuclei. One of these is exchanged with a nucleus of the alternative mating type, and the two haploid nuclei then fuse to form a zygotic, diploid nucleus. The original macronuclei then break down and their contents are, in turn, resorbed by the cells. A new macronucleus then forms from one of the mitotic products of the germline nucleus, and the cells then separate and undergo fission to give four daughter cells or karyonides, each with a reconstituted germline micronucleus and a somatic macronucleus.

A process similar to self-fertilization can occur or be induced in cultures of *Tetrahymena* to give haploid lines, and genomic exclusion by heat-induced cytogamy also allows haploid lines to be produced and homozygous recessive genes to be expressed. With suitable crosses, haploid strains can be used to construct monosomic and nullisomics. Genomic exclusion also occurs when ''star'' strains, designated with an asterisk–as n C*III–with defective micronuclei are hybridized with normal diploids to give haploid strains. The ''star'' effect is usually caused by severe aneuploidy, especially hypodiploidy, in the micronucleus and is commonly one of the end results of trying to maintain pure cultures with no sexual recombination.

At the molecular level, species of *Tetrahymena* are the only eukaryotes known to possess only one copy of the rRNA gene, located on the left arm of chromosome 2. The multiple replications in the macronucleus increase the number of rRNA genes present within the organism and this quantitative increase allows the structure and function of these genes to be closely studied. The analysis of the intervening sequences of *Tetrahymena* rRNA, for example, led to the discovery that the RNA transcribed from these genes was capable of self-splicing. This was the first demonstration that RNA was capable of an enzymatic reaction. The fragmentation and replication process to form the multiplicity of chromosomes in the macronuclei also led to characterization of the telomeres at the ends of these chromosomes., The telomeres are essential for replication and the stability of the linear chromosomes, and they have the sequence structure $d(G_4T_2)(A_2C_4)$.

24.10 *CAENORHABDITIS ELEGANS* (NEMATODE EELWORM)

24.10.1 Description

Caenorhabditis elegans has been simply described as a 1-mm-long, bacteria-eating, soil-dwelling worm (Fig. 24.10). It is a nema-

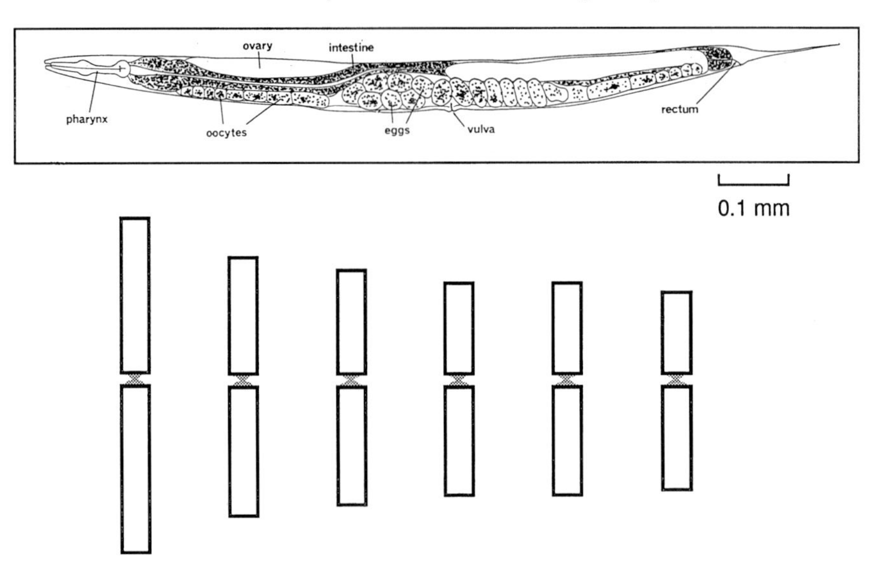

Fig. 24.10. *C. elegans* and its six haploid chromosomes.

Table 24.1. The Four Stages of Embryogenesis in *Caenorhabditis elegans,* the Common Eelworm

	Stage 1		Stage 2		Stage 3		Stage 4	
Zygote	→	28 cells	→	~550 cells	→	959 cells	→	Fertile adult
	140 min Embryonic axes, germline, and founder cells established.		140–360 min. Gastrulation and cell proliferation.		Differentiation, morphogenesis and organogenesis.		Growth	

tode, a metazoan organism, and a member of the class Platyhelminthes. This class features a great variety of flat and round worms, many of which are parasitic in nature.

24.10.2 Genome Size and Karyotype

There are six linkage groups comprised of five pairs of autosomes (I to V) and one pair of XX chromosomes. The eelworm has two sexes, XO males and XX hermaphrodites. The latter are basically female but in the absence of males, they are self-fertile and able to make their own sperm, producing XX hermaphrodite progenies.

Chromosome number: $2n = 12$ (female); $2n = 11$ (male).
DNA content: 8×10^7 bp.
Gene number: 17,800.

24.10.3 Features of Interest to Genetics, Developmental Genetics, and Molecular Biology

A normal adult worm has exactly 959 cells produced in 3 days from ovum to adult. Four stages are involved in embryogenesis, which takes about 14 h at 22°C (Table 24.1). These stages illustrate the importance of the nematode to studies of developmental genetics in that the precursors of every adult cell have been identified from the ovum onward. Neurogenetic studies are continuing to show the development of synapses and the control of bodily functions by the cells of the nervous system.

The chromosomes are holocentric with diffuse kinetochores during mitosis. During meiosis, however, each of the six chromosomes has a single primary meiotic pairing center near one end. Over 80% of the total genome has been deleted in 280 separate deficiency mutants, enabling efficient mapping of the genes involved in embryogenesis. Five families of transposons (*Tc1* to *Tc5*) have been identified, although the activity of each family is restricted to particular strains of the nematode. The entire mitochondrial genome of 13,794 bp has been sequenced. The *Caenorhabditis elegans* genome-sequencing project has started with the sequencing of 121,298 bp of chromosome III cloned in three cosmids. Significant numbers of genes of known function that were not previously identified were discovered in these analyses. The development of automatic DNA sequencing methods and robotic systems for large-scale, shotgun sequencing of nematode DNA has been a source of innovation for the analysis of other organisms.

24.11 *DROSOPHILA MELANOGASTER* (FRUIT FLY)

24.11.1 Description

The fruit fly, *D. melanogaster*, is a member of the cosmopolitan insect order Diptera (Fig. 24.11). Over 100 species of *Drosophila*

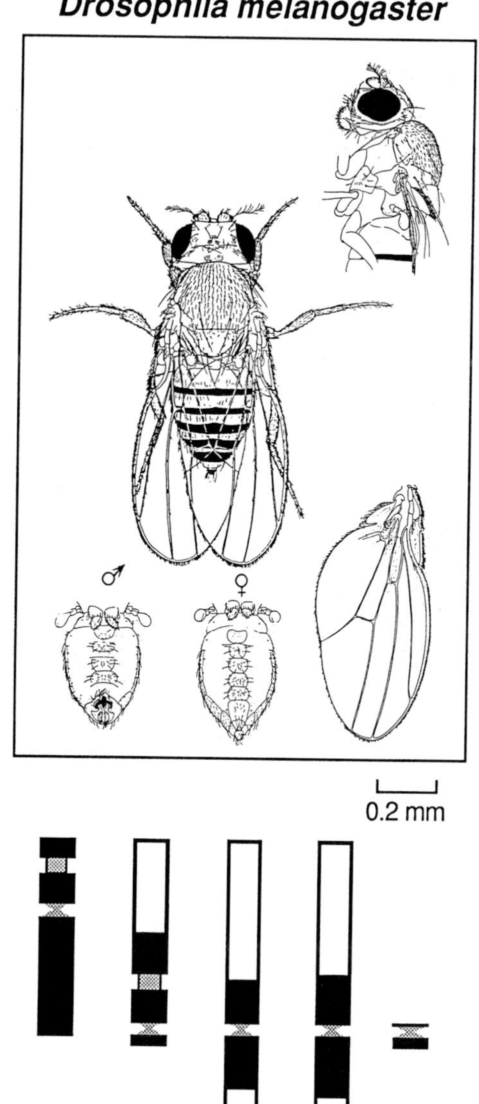

Fig. 24.11. The fruit fly, *D. melanogaster,* and its N-banded karyotype showing heterochromatin.

are found in the United States alone, and there are presently 3318 species within the family. Morphological, chromosomal, genic, and DNA comparisons of these species have been, and are being, used to link the species into groups and subgenera and to study the genetic basis of evolution. Individual species are often strictly environmentally limited as related to humidity, temperature, light intensity, and oviposition sites.

24.11.2 Genome Size and Karyotype

There are two large (2, 3) and one small (4) pair of autosomes plus the sex chromosomes (X or 1, and Y). The chromosomes have left (L) and right (R) arms; the telomeres of these arms are designated as t (e.g. 2Rt is the telomere on the right arm of chromosome 2), and the centromeres as cen (e.g., 2cen).

Chromosome number: $2n = 8$ (male, $3'' +$ XY and female, $3'' +$ XX).

2C DNA content: 7 pg.

1C DNA content: 1.4×10^8 bp.

Gene number: ~ 15,000.

24.11.3 Features of Interest to Genetics, Cytogenetics, and Molecular Biology

Since the early 1900s, *D. melanogaster* has been one of the most intensively studied genetic organisms. It has a short generation time, an ability to produce large numbers of progenies, is well adapted to laboratory life, and has been spread to laboratories throughout the world. The accessibility to *Drosophila* germplasm is one of the major reasons for the genetic success of this fly. Over 6000 genes have been physically identified in *D. melanogaster*, and transposon mutagenesis and enhancer-trapping techniques are being used to detect the DNA sequence elements that control the tissue and developmental specificity of different genes. In turn, this knowledge has led to the use of *Drosophila* in developmental biology, evolution, and embryology. For example, *Drosophila* embryogenesis results in an embryo with about 60,000 cells and these cells, as in *Caenorhabditis*, can now be developmentally studied.

An array of chromosomal aberrations has been developed and maintained in *Drosophila*, including ring chromosomes, translocations between different chromosomes, inversions and transpositions within chromosomes, and numerous deficiencies and duplications. The X chromosomes have been instrumental in the genetic success of this organism and the euchromatic X chromosome has now been physically mapped using bacteriophage I, BAC, and YAC technologies. Sequence tagged sites (STSs) have provided valuable molecular markers. *Drosophila* chromosomes are somewhat unusual in that their telomeres are comprised of retrotransposable elements rather than the simple telomeric nucleotide repeats found in other organisms. Between species, the Y chromosomes vary greatly in size and are completely heterochromatic. Genetically, they appear to carry only male-fertility genes, of which *D. melanogaster* carries two on the short arm of the Y chromosome and four on the long arm. These genes are extremely large, about 1×10^6 bp in length, and the remainder of the Y chromosome, about 40 Mb, comprises mainly repetitive DNA families.

24.12 *STRONGYLOCENTROTUS PURPURATUS* (SEA URCHIN)

24.12.1 Description

Strongylocentrotus purpuratus is but one of a range of echinoderms that have been studied in genetics. Other important sea urchin species include *Paracentrotus lividus*, *Psammechinus miliaris*, *Clypeaster japonicus*, and *Lytechinus pictus*. The benefit of sea urchins as experimental organisms lies in the fact that females produce massive numbers of eggs and males produce millions of sperm. Fertilization and embryo development can be synchronized, enabling large numbers of organisms at similar stages of development to be analyzed, both microscopically and biochemically.

24.12.2 Genome Size and Karyotype

Chromosome number: $2n = 36$.

1C DNA content: 8.6×10^8 bp.

Gene number: ~ 25,000

24.12.3 Features of Interest to Genetics, Cytogenetics, and Molecular Biology

Many of the early depictions of mitotic division were gained from observations of the dividing cells of sea-urchin embryos. Early studies of embryology were also based on this material, owing to the hardiness of the cells and their ease of maintenance. Thus, the effect of enucleation of the egg cell before and after fertilization could be studied; the cells resulting from mitotic divisions could be easily separated to measure their degrees of recovery with respect to totipotency; pigmented markers helped to orient the developing embryos. The biochemistry of mitosis was subsequently studied by physically isolating the complete mitotic apparatus from the cytoplasm of dividing cells. The ready availability of this material enabled bulked quantities of the mitotic apparatuses of dividing cells to be analyzed chemically and biochemically.

Strongylocentrotus has been analyzed at the RNA and DNA levels to determine the number of genes active at differing stages of the life cycle. In embryological order, the oocyte manifests about 18,000 genes, the blastula 13,000, the gastrula 8500, the pluteus 7000, and individual adult tissues from 2500 to 3000. From these figures, it can be estimated that *S. purpuratus* contains somewhat less than 25,000 genes, comprising about 3% of the total DNA, the rest being a mixture of highly repeated and moderately repeated DNA sequences. Some gene sequences are also moderately repetitive in nature, including arrays of histone genes (300–600 copies) and rRNA genes. These arrays can be differentiated by variation in the nontranscribed spacer DNA sequences separating the genes.

24.13 *XENOPUS LAEVIS* (CLAWED TOAD)

24.13.1 Description

Frogs and toads are amphibians of the order Anura (Fig. 24.12). Anurans, generally, and *X. laevis*, in particular, are further examples of organisms renowned for their production of prolific quantities of eggs and sperm such that the abdomen of the females are

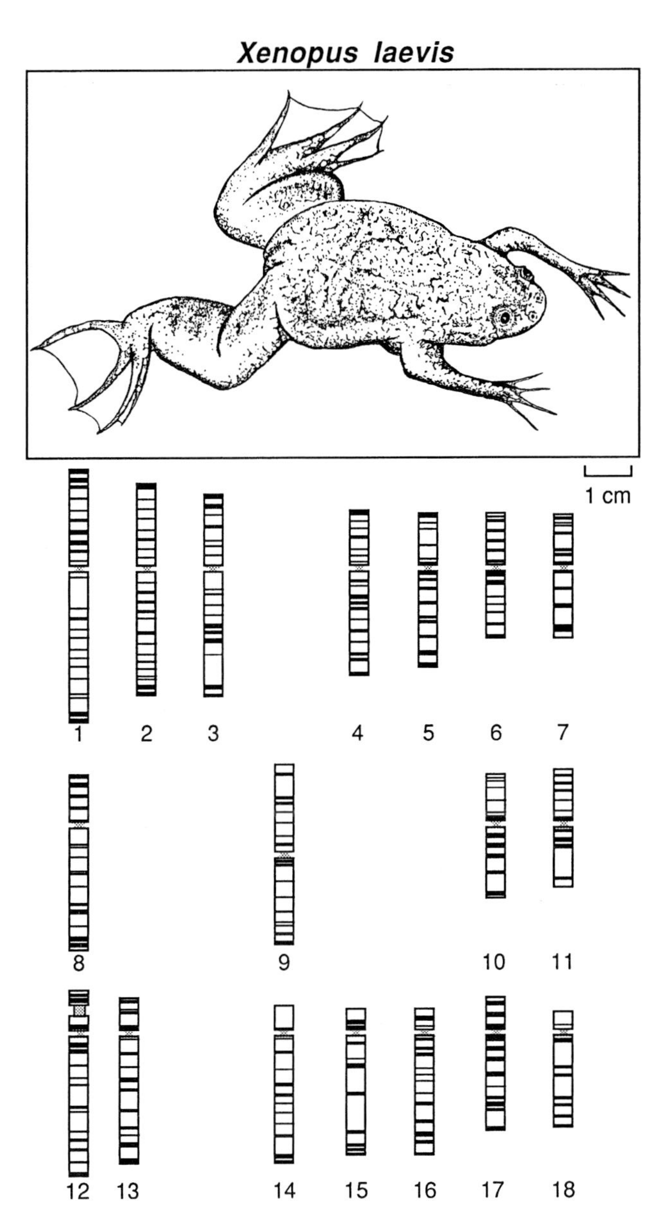

Fig. 24.12. The clawed toad and its BrdU-labeled karyotype.

commonly almost entirely occupied by eggs in the two ovaries. These eggs are large enough to support the cell division that takes place after fertilization, when up to 4000 cells are produced within each embryo in a period of 8 h. Each egg, for example, contains enough histone to construct 20,000 nuclei. These quantities almost certainly explain the presence of diplotene lampbrush chromosomes because these chromosomes are considered to reflect the high rates of gene transcription that occur in these cells in order to ensure the necessary building blocks for the growth of the embryo. The resumption of meiosis from diplotene onward is fertilization dependent. In vitro, this resumption is triggered by a number of biochemical agents including the male hormone progesterone. In vivo, the reaction may not be that simple.

24.13.2 Genome Size and Karyotype

Chromosome number: $2n = 36$.

In amphibians generally, genome sizes range from about 10^9 to 10^{11} bp.

24.13.3 Features of Interest to Genetics, Cytogenetics, and Molecular Biology

As an amphibian experimental organism, *Xenopus* has three advantages:

1. By injection with gonadotrophic hormone, eggs are available throughout the year.
2. It is easily maintained in laboratory aquaria.
3. Its life cycle is complete within 12 months.

This life cycle does take a comparatively long period of time and, in association with the difficulties associated with the production of inbred strains of amphibians, has meant that genetic maps of anurans are few. Most of the markers to date have been determined using biochemical–genetic methods to separate proteins and enzymes, although molecular markers and in situ hybridization of cloned genes are becoming increasingly important. Among such genes, the 5S rRNA genes (24,000, several of which have been completely sequenced) occupy terminal positions on the ends of all of the chromosomes. The toad genome also contains about a single tandem cluster of 450 rRNA genes separated by two nontranscribed spacers of different lengths. These ribosomal RNA genes can be isolated from genomic DNA preparations by buoyant-density centrifugation, and their availability meant they were the first eukaryotic genes to be transformed into bacteria. These gene arrays were also involved in early demonstrations to relate a cytologically identifiable chromosomal landmark (nucleolus organizer) to the presence of the ribosomal RNA genes.

Xenopus has been used to determine the duration of meiotic DNA replication. In cultured cell lines of *X. laevis*, replicons were found to occur at intervals of 57.5 μm and replicate at rates of 9 μm/h. The chromosomes also have been investigated with BrdU labeling and the mtDNA has been completely sequenced.

24.14 *GALLUS DOMESTICUS* (CHICKEN)

24.14.1. Description

The common chicken, *Gallus domesticus* or *Gallus gallus*, is of Asian origin. Although hybrids between *Gallus* spp. are generally viable, those between fowl and turkey, for example, are not. Even interspecific fowl where F_1 hybrids are viable often yield F_2 progeny that are not viable, because they are incapable of producing normal gametes due to a premeiotic defect in the hybrids that causes genotypically controlled hybrid sterility.

24.14.2 Genome Size and Karyotype

Macro-sized Z and W sex chromosomes with females being the heterogametic sex, plus 4 pairs macrochromosomes, 5 pairs minichromosomes, and 29 pairs microchromosomes.

Chromosome number: $2n = 78$ (males, $38'' + ZZ$; females, $38'' + ZW$).

2C DNA content: 2.50 pg.

1C DNA content: 1.2×10^9 bp.

Map length: ~ 3000 cM.

24.14.3 Features of Interest to Genetics, Cytogenetics, and Molecular Biology

In contrast to mammals, birds contain nucleated red blood cells, a feature that greatly simplifies DNA extraction procedures. The nucleated red blood cells have also provided the standard reference point of cytophotometric Feulgen DNA measurements. The total map length of *G. domesticus* is very long compared to its genomic DNA content, because chromosomal recombination rates in chickens are three times higher than they are in mammals. This refers mainly to the macrochromosomes. The minichromosomes and microchromosomes have only a single chiasma located in their telomeric regions and thus, these chromosomes are commonly detected as single-point linkage groups. The other significant feature dividing the macrochromosomes from the microchromosomes from the minichromosomes is the ability to identify the two former classes by their morphology and banding patterns: The minichromosomes are indistinguishable using these techniques. Molecular maps have been derived using the parents and hybrids obtained from a cross between a red jungle fowl male and a highly inbred white leghorn female.

Although chickens can be induced to produce tetraploid oocytes, most aneuploids such as trisomics and mixoploids cause embryo death.

24.15 *MUS MUSCULUS* (COMMON MOUSE)

24.15.1 Description

The species *Mus musculus* is divided into two subspecies to differentiate the wild type *M. m. musculus* from the house mouse, *M. m. domesticus* (Fig. 24.13). The two subspecies, in turn, have a number of variable and not always stable races. This population instability is commonly man-made in that it is caused by abrupt local increases and decreases of stored food, leading to explosions and crashes, respectively, of mouse numbers. There is then a significant potential for founder effects, whereby the genes present in a breeding pair of mice or even a single pregnant female are the only genes present in the mouse population derived from that pair or female, which can lead to rapid divergences between mouse populations. Conversely, this characteristic also enables the production of inbred lines of laboratory mice defined as being produced by brother × sister matings for at least 20 generations or more from a single ancestral pair. A second species, *M. spretus*, is regarded as a wild relative for genetic-mapping purposes.

24.15.2 Genome Size and Karyotype

The symbols p and q are used to denote the short and long arms of chromosomes, respectively.

Chromosome number: $2n = 40$ (males, $19'' + XY$; females, $19'' + XX$).

2C DNA content: 6.3 pg.

1C DNA content: 2.7×10^9 bp.

Gene number: 80,000.

Map length: 1600 cM.

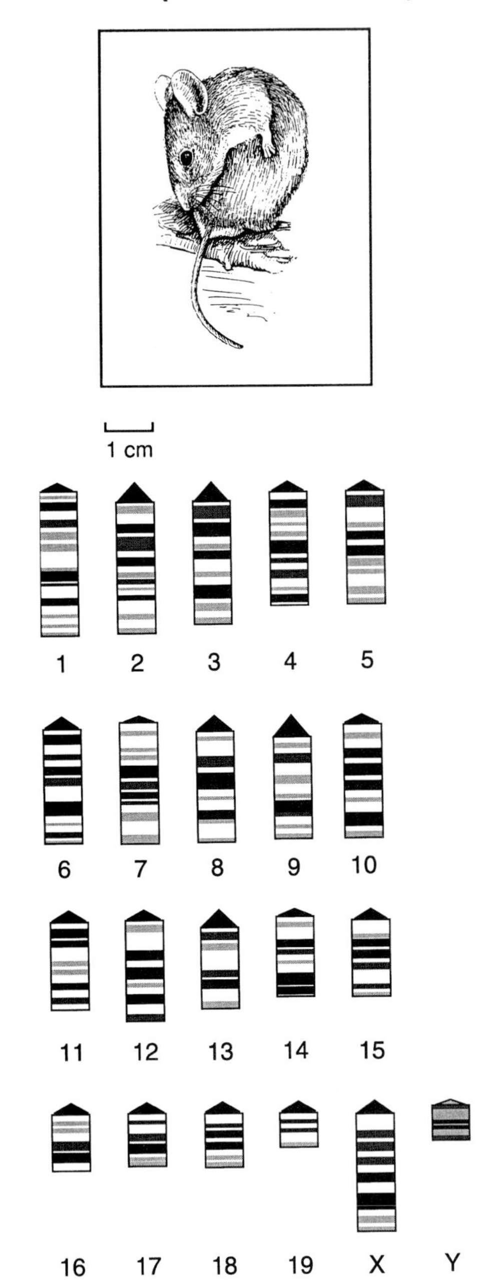

Fig. 24.13. The acrocentric chromosomes of *M. musculus*.

24.15.3 Features of Interest to Genetics, Cytogenetics, and Molecular Biology

Mice are an easily maintained, laboratory-size mammal, originally analyzed for coat color variants. They yield excellent diplotene/diakinesis meiotic chromosome spreads and the bivalents are easily C-banded. In fact, chromosome banding is required to distinguish between the 20 pairs of similarly sized acrocentric chromosomes. The chiasmata show no signs of terminalization with the three longest chromosomes displaying a tendency for subcentromeric and subtelomeric chiasmata; the remaining chromosomes have predominantly subtelomeric chiasmata, and there are few interstitial chiasmata. Among different races of mice, there are varying degrees of chromosomal divergence associated with the accumulation and fixation of metacentric Robertsonian (Rb) translocation chromosomes through the centric fusion of different acrocentrics. The two major subspecies are separated by a hybrid zone running across Europe from Denmark to Bulgaria and from Belgium to northern Italy. The southern subspecies, *M. m. domesticus*, is distinguished from the northern *M. m. musculus*, by the presence of a Robertsonian (Rb) 4.12 translocation chromosome. The actual hybrid zone between the two subspecies, in which Rb 4.12 heterozygotes can be found, is approximately 40 km wide. The presence of the translocation appears to have no effect on fertility. Other races are found elsewhere. For example, in northeastern Scotland, a race with $2n = 32$ containing four Rb metacentrics (4.10, 9.12, 6.13, and 11.14) is separated by a hybrid zone dominated by individuals with $2n = 36$ (4.10 and 9.12). A Tunisian race of *M. m. domesticus* has $2n = 22$ with nine pairs of fusion metacentrics. Consequently, a large selection of different inbred strains has now been produced featuring an array of isogenic or near-isogenic inversions and translocations between the wild-type acrocentric chromosomes. These defined inbred strains, in common with several hundred known mutants, a well-developed transgenic technology, and a direct application and synteny to hundreds of human genes, provide a powerful model for human studies.

24.16 *HOMO SAPIENS* (HUMANS)

24.16.1 Description

At the genetic and chromosomal levels, the genus *Homo*, of which humans are the sole representative species, is closely related to the family Pongidae, which contains the great apes (Fig. 24.14). These include *Pan troglodytes* (chimpanzee), *Gorilla gorilla* (gorilla), and *Pongo pygmaeus* (orangutan). All of these species have karyotypes containing 48 chromosomes, which, with two exceptions, are almost identical and certainly homoeologous if not homologous with the chromosomes of *H. sapiens*. The exceptions involve what is clearly the result of either a centromeric cleavage of the two arms of chromosome 2 of humans to give two large acrocentric chromosomes in the apes, or the fusion in human evolution of these acrocentrics to give the large metacentric chromosome 2 of the human karyotype. An array of genetic markers have been used to show these homologies. Chromosome banding has also revealed that the genus *Pan* differs from the other three genera by the presence of a pericentric inversion in chromosome 3. All of the Pongidae show a high percentage of DNA sequence homology with the human genome.

24.16.2 Genome Size and Karyotype

The normal male karyotype is given as 46, XY, the normal female as 46, XX. The symbols p and q are used to denote the short and long arms, respectively.

Chromosome number: $2n = 46$ (male, 22″ + XY; female, 22″ + XX).

2C DNA content: generally standardized at 7.0 pg.

1C DNA content: 3×10^9 bp. The smallest human chromosome contains about 4.6×10^7 bp and is about 1.4 cm in length.

Gene number: 80,000.

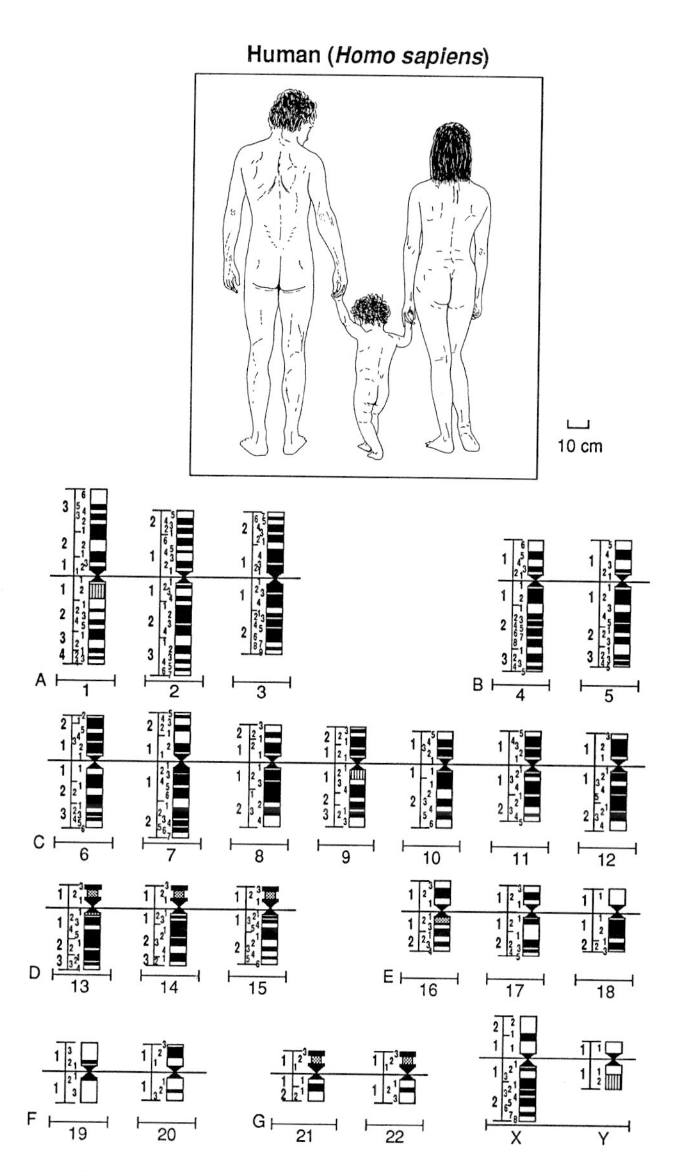

Fig. 24.14. The G-banded karyotype of human chromosomes.

24.16.3 Features of Interest to Genetics, Cytogenetics, and Molecular Biology

Research into the genetic aspects of human biology is largely driven by concerns for human health and intelligence. At least 10% of the total number of human genes have been cloned and sequenced, including many genes associated with diseases or other abnormalities. A complete sequencing of the human genome is underway, organized by the Human Genome Organization (HUGO), an international collaborative group. Aspects of genetics that have been improved by research into human genetics include the production of man–mouse somatic cell hybrids, the maintenance of somatic cell lines with aberrant karyotypes, the development of specific chromosome and genome DNA libraries, and the development of extended human family-banks for associative and correlative mapping. The male Y chromosome is of particular interest in that with the exception of a small pseudoautosomal region, no crossing over occurs within the 60 Mb linear DNA molecule. There are relatively few polymorphic DNA sequences. In an evolutionary sense, it should eventually be possible to relate all Y chromosome variants back to a single ancestor. The inactivation of one of the X chromosomes of the normal (mammalian) female also requires the presence of a pseudoautosomal region.

BIBLIOGRAPHY

ARUMUGANATHAN, K., EARLE, E.D. 1991. Nuclear DNA content of some important plant species. Plant Molec. Biol. Rep. 9: 208–218.

BENNETT, M.D., SMITH, J.B. 1976. Nuclear DNA amounts in angiosperms. Proc. Roy. Soc. London Ser. B. 274: 227–274.

BIRD, A.P. 1995. Gene number, noise reduction and biological complexity. Trends Genet. 11: 94–100.

BOURSOT, P., BONHOMME, F., BRITTON-DAVIDIAN, J., CATALAN, J., YONEKAWA, H., ORSINI, P., GERASIMOV, S., THALER, L. 1984. Introgression différentelle des génomes nucléaires et mitochondriaux chez deux semi-espèces de souris. C.R. Acad. Sci. Paris 299: 365–370.

BURNETT, J.H. 1975. Mycogenetics. John Wiley & Sons, London.

BURT, D.W., BUMSTEAD, N., BITGOOD, J.J., PONCE DE LION, F.A., CRITTENDEN, L.B. 1995. Chicken genome mapping: a new era in avian genetics. Trends Genet. 11: 190–194.

CAPANNA, E. 1982. Robertsonian numerical variation in animal speciation: *Mus musculus*, an emblematic model. Prog. Clin. Biol. Res. 96: 155–177.

CAUSSE, M.A., FULTON, T.M., CHO, Y.G., AHN, S.N., CHUNWONGSE J., WU, K., XIAO, J., YU, Z., RONALD, P.C., HARRINGTON S.E., SECOND, G., MCCOUCH, S.R., TANKSLEY, S.D. 1994. Saturated molecular map of the rice genome based on an interspecific backcross population. Genetics 138: 1251–1274.

COPELAND, N.G., JENKINS, N.A., GILBERT, D.J., EPPIG, J.T., MALTAIS, L.J., MILLER, J.C., DIETRICH, W.F., WEAVER, A., LINCOLN, S.E., STEEN, R.G., STEIN L.D., NADEAU, J.H., LANDER, E.S. 1993. A genetic linkage map of the mouse: current applications and future prospects. Science 262: 57–66.

DARLINGTON, C.D., WYLIE, A.P. 1955. Chromosome Atlas of Flowering Plants. George Allen and Unwin Ltd, London.

European Union Arabidopsis Genome Project. 1998. Analysis of 1.9 Mb of contiguous sequence from chromosome 4 of Arabidopsis thaliana. Nature 391: 485–488.

FUKUI, K., IIJIMA, K. 1991. Somatic chromosome map of rice by imaging methods. Theoret. Appl. Genet. 81: 589–596.

HAYASHIZAKE, Y., HIROTSUNE, S., OKAZAKI, Y., SHIBATA, H., AKASAKO A., MURAMATSU, M., KAWAI, J., HIRASAWA, T., WATANABE, S., SHIROISHI, T., MORIWAKI, K., TAYLOR, B.A., MATSUDA, Y., ELLIOTT, R.W., MANLY K.F., CHAPMAN, V.M. 1994. A genetic linkage map of the mouse using restriction landmark genomic scanning (RLGS). Genetics 138: 1207–1238.

JOBLING, M.A., TYLER-SMITH, C. 1995. Fathers and sons: the Y chromosome and human evolution. Trends Genet. 11: 449–456.

KAKEDA, K., YAMAGATA, H., FUKUI, K., OHNO, M., WEI, Z.Z., ZHU, F.S. 1990. High resolution bands in maize chromosomes by G-banding methods. Theoret. Appl. Genet. 80: 265–272.

KOWALSKI, S.P., LAN, T-H., FELDMANN, K.A., PATERSON, A.H. 1994. Comparative mapping *of Arabidopsis thaliana* and *Brassica oleracea* chromosomes reveals islands of conserved organization. Genetics 138: 499-510.

MADUEÑO, E., PAPAGIANNAKIS, G., RIMMINGTON, G., SAUNDERS, R.D.C., SAVAKIS, C., SIDÈN-KIAMOS, I., SKAVDIS, G., SPANOS, L., TRENEAR, J., ADAM, P., ASHBURNER, M., BENOS, P., BOLSHAKOV, V.N., COULSON, D., GLOVER, D.M., HERRMANN, S., KAFATOS, F.C., LOUIS, C., MAJERUS, T., MODOLELL, J. 1995. A physical map of the X chromosome of *Drosophila melanogaster:* cosmid contigs and sequence tagged sites. Genetics 139: 1631–1647.

MAZIA, D. 1961. Mitosis and the physiology of cell division. In: The Cell: Biochemistry, Physiology, Morphology (Brachet, J., Mirsky, A.E., eds.). Academic Press, New York. pp. 77–412.

MOORE, G., FOOTE, T., HELENTJARIS, T., DEVOS, K., KURATA, N., GALE, M. 1995. Was there a single ancestral cereal chromosome? Trends Genet. 11: 81–82.

MORISHIMA, H., SANO, Y., OKA, H-I. 1992. Evolutionary studies in cultivated rice and its wild relatives. In: Oxford Surveys in Evolutionary Biology 8: 135–146.

NEIDHART, F.C. 1987. *Escherichia coli* and *Salmonella typhimurium*: Cellular and Molecular Biology. American Society for Microbiology, Washington, DC.

O'BRIEN, S.J. (ED). 1990. Genetic Maps: Locus Maps of Complex Genomes, 5th ed. Cold Spring Harbor Laboratory Press, Cold Spring Harbor, NY.

PETERSON, D.G., STACK, S.M., PRICE, H.J., JOHNSTON, J.S. 1995. Distribution of DNA in heterochromatin and euchromatin of *Lycopersicon esculentum* pachytene chromosomes. TGC Rep. 45: 35.

ROGERS, J. 1993. The phylogenetic relationships among *Homo*, *Pan* and *Gorilla*: a population genetics perspective. J. Human Evol. 25: 201–215.

SHERMAN, J.D., STACK, S.M. 1995. Two-dimensional spreads of synaptonemal complexes from solanaceous plants. VI. High resolution recombination nodule map for tomato (*Lycopersicon esculentum*). Genetics 141: 683–708.

Index